Lecture Notes in Mechanical Engineering

Lecture Notes in Mechanical Engineering (LNME) publishes the latest developments in Mechanical Engineering—quickly, informally and with high quality. Original research reported in proceedings and post-proceedings represents the core of LNME. Volumes published in LNME embrace all aspects, subfields and new challenges of mechanical engineering.

To submit a proposal or request further information, please contact the Springer Editor of your location:

Europe, USA, Africa: Leontina Di Cecco at Leontina.dicecco@springer.com
China: Ella Zhang at ella.zhang@springer.com
India: Priya Vyas at priya.vyas@springer.com
Rest of Asia, Australia, New Zealand: Swati Meherishi at swati.meherishi@springer.com

Topics in the series include:

- Engineering Design
- Machinery and Machine Elements
- Mechanical Structures and Stress Analysis
- Automotive Engineering
- Engine Technology
- Aerospace Technology and Astronautics
- Nanotechnology and Microengineering
- Control, Robotics, Mechatronics
- MEMS
- Theoretical and Applied Mechanics
- Dynamical Systems, Control
- Fluid Mechanics
- Engineering Thermodynamics, Heat and Mass Transfer
- Manufacturing
- Precision Engineering, Instrumentation, Measurement
- Materials Engineering
- Tribology and Surface Technology

Indexed by SCOPUS, EI Compendex, and INSPEC

All books published in the series are evaluated by Web of Science for the Conference Proceedings Citation Index (CPCI).

To submit a proposal for a monograph, please check our Springer Tracts in Mechanical Engineering at https://link.springer.com/bookseries/11693.

Volodymyr Tonkonogyi · Vitalii Ivanov ·
Justyna Trojanowska · Gennadii Oborskyi ·
Ivan Pavlenko
Editors

Advanced Manufacturing Processes V

Selected Papers from the 5th Grabchenko's International Conference on Advanced Manufacturing Processes (InterPartner-2023), September 5–8, 2023, Odessa, Ukraine

Springer

Editors
Volodymyr Tonkonogyi
Odessa Polytechnic National University
Odessa, Ukraine

Vitalii Ivanov
Sumy State University
Sumy, Ukraine

Justyna Trojanowska
Poznań University of Technology
Poznan, Poland

Gennadii Oborskyi
Odessa Polytechnic National University
Odessa, Ukraine

Ivan Pavlenko
Sumy State University
Sumy, Ukraine

ISSN 2195-4356 ISSN 2195-4364 (electronic)
Lecture Notes in Mechanical Engineering
ISBN 978-3-031-42777-0 ISBN 978-3-031-42778-7 (eBook)
https://doi.org/10.1007/978-3-031-42778-7

This Springer imprint is published by the registered company Springer Nature Switzerland AG
The registered company address is: Gewerbestrasse 11, 6330 Cham, Switzerland

Paper in this product is recyclable.

Preface

This volume of Lecture Notes in Mechanical Engineering contains selected papers presented at the 5th Grabchenko's International Conference on Advanced Manufacturing Processes (InterPartner-2023), held in Odessa, Ukraine, on September 5–8, 2023. The conference was organized by Odessa Polytechnic National University, National Technical University "Kharkiv Polytechnic Institute", Sumy State University, and International Association for Technological Development and Innovations (IATDI) in partnership with Poznan University of Technology (Poland).

InterPartner Conference Series promotes research and developmental activities, intensifying scientific information interchange between researchers, developers, and engineers.

InterPartner-2023 received 87 contributions. After a thorough peer-review process, the Program Committee accepted 53 papers written by authors from 13 countries. Thank you very much to the authors for their contribution. These papers are published in the present book, achieving an acceptance rate of 60%.

We thank members of the Program Committee and invited external reviewers for their efforts and expertise in contributing to reviewing, without which it would be impossible to maintain the high standards of peer-reviewed papers.

Thank you very much to the keynote speakers Dr. Slawomir Luscinski (Kielce University of Technology, Poland) and Prof. Vasily Larshin (Odessa Polytechnic National University, Ukraine).

The book "Advanced Manufacturing Processes V" was organized into six parts according to the main conference topics: Part 1—Design Engineering and Production Planning; Part 2—Manufacturing Technology and Machining Processes; Part 3—Advanced Materials; Part 4—Quality Assurance; Part 5—Mechanical Engineering; and Part 6—Process Engineering.

The first part "Design Engineering and Production Planning" includes recent advancements in designing automatic control systems and mechatronic actuators, parametric synthesis of electrohydraulic control systems, computer modeling of centrifugal pump parts, and shaping gear wheels. It also presents studies in designing installations for ion-plasma deposition, thermomechanical phenomena during machining, and designing control configured mechatronic mechanisms. Moreover, this part includes studies in determining an effective supply chain and optimizing lifting mechanisms. Recent developments in the 3D reconstruction of a virtual building environment are also presented in this part.

The second part "Manufacturing Technology and Machining Processes" includes modeling workpiece movements in advanced manufacturing and tool production, multifractal analysis of periodic surface relief of parts after face milling, and regularities of oscillations during turning and end-milling. It also presents recent developments in optimizing cutting modes during sustainable machining and ways to ensure the machining accuracy of difficult-to-machine steel. Additionally, this part contains studies on contact processes while drilling stainless steel and gear machining by power skiving. Combined technology for processing parts based on applying an antifriction coating and deforming broaching, caliber wear problems, and modeling vibrational–centrifugal strengthening of functional surfaces are also considered. Finally, this part includes an approach for integrated modeling and manufacturing customized orthopedic implants.

The third part "Advanced Materials" is devoted to forming 2D carbon nanosheets, carbon-shelled nanoparticles, and copper oxide nanostructures. The wear resistance of steel with nanocrystalline layers, the thermomechanical properties of microalloyed steel, and the structure-phase state of medium-carbon steel are also presented in this part. Additionally, the third part presents technological regulations of furnace equipment for carbon graphite electrode production, structure and thermal stability of vacuum condensates, and highly efficient titanium porous materials. Moreover, features of magnesium alloy protection technologies in die casting, structure, and mechanical properties of trip-assisted steel after treatment, as well as composite construction materials, are also discussed in this part.

The fourth part "Quality Assurance" presents standardization in additive manufacturing, experimental studies of strength quality parameters for interference fit joints, and modeling of thermal and dynamic conditions during intermittent grinding. A study of the scaling effect of cutting elements of the abrasive wheels' discretized working surface on the surface layer quality of ground parts is also included in this part. The fourth part also describes problems regarding evaluating fractal dimensions using the infrared thermal inspection method for non-metallic heterogeneous materials. Moreover, ways to improve professional competence in engineering education are also presented in this part.

The fifth part "Mechanical Engineering" is based on recent developments in the energy efficiency of an elevator winch, experimental studies of longevity in the metallic structure of a portal crane, and the diagnosis of an autotransformer with defects. It also includes studies in mass absorption and technological damage of concrete structures and the optimality of mechanical characteristics of foams. Moreover, the fifth part is based on advancements in ensuring geometric and functional parameters of the rotors for planetary hydraulic motors and assessing the lifecycle cost of torque-flow pumps. Numerical simulation of turbine blade vibrations with single-crystal anisotropy and designing of spiral springs are also presented in this part.

The sixth part "Process Engineering" presents research studies on the energy efficiency of combined heating systems based on heat pumps. Research works on the numerical evaluation of dust concentration, operating parameters of modified nozzle for cold spraying, and energy characteristics of the oil vortex chamber supercharger are also presented in this part. Moreover, the sixth part includes applications of the similarity theory in designing vibroconveyor dryers for grain.

The editors appreciate the outstanding contribution of all the authors. We are deeply convinced that the research papers presented in the book will be helpful to scientists, industrial engineers, and highly qualified practitioners worldwide.

We appreciate a reliable partnership with Springer Nature and iThenticate for their support during the preparation of InterPartner-2023.

Thank you very much to InterPartner Team. Their involvement, devotion, and hard work were crucial to the success of the conference.

InterPartner's motto is "Science unites people together".

September 2023

Volodymyr Tonkonogyi
Vitalii Ivanov
Justyna Trojanowska
Gennadii Oborskyi
Ivan Pavlenko

Organization

General Chair

Volodymyr Tonkonogyi	Odessa Polytechnic National University, Ukraine

Co-chair

Vitalii Ivanov	Sumy State University, Ukraine

Steering Committee

Vitalii Ivanov	Sumy State University, Ukraine
Gennadii Oborskyi	Odessa Polytechnic National University, Ukraine
Ievhen Ostroverkh	National Technical University "Kharkiv Polytechnic Institute", Ukraine
Ivan Pavlenko	Sumy State University, Ukraine
Volodymyr Tonkonogyi	Odessa Polytechnic National University, Ukraine
Justyna Trojanowska	Poznan University of Technology, Poland

Program Committee

Katarzyna Antosz	Rzeszow University of Technology, Poland
Yevheniia Basova	National Technical University "Kharkiv Polytechnic Institute", Ukraine
Khrystyna Berladir	Sumy State University, Ukraine
Jozef Bocko	Technical University of Kosice, Slovak Republic
Dagmar Caganova	Slovak University of Technology, Slovak Republic
Vasile George Cioata	Polytechnic University of Timisoara, Romania
Olaf Ciszak	Poznan University of Technology, Poland
Radu Cotetiu	Technical University of Cluj-Napoca, Romania
Milan Edl	University of West Bohemia, Czech Republic
Volodymyr Gurey	Lviv Polytechnic National University
Ihor Hurey	Lviv Polytechnic National University, Ukraine
Vitalii Ivanov	Sumy State University, Ukraine
Jerzy Jozwik	Lublin University of Technology, Poland
Maryna Ivanova	National Technical University "Kharkiv Polytechnic Institute", Ukraine

Gennadii Khavin	National Technical University "Kharkiv Polytechnic Institute", Ukraine
Kateryna Kostyk	National Technical University "Kharkiv Polytechnic Institute", Ukraine
Agnieszka Kujawinska	Poznan University of Technology, Poland
Jaroslav Kusyj	Lviv Polytechnic National University, Ukraine
Dmytro Lesyk	National Technical University of Ukraine "Igor Sikorsky Kyiv Polytechnic Institute", Ukraine
Dmytro Levchenko	Lodz University of Technology, Poland
Oleksandr Liaposhchenko	Sumy State University, Ukraine
Natalia Lishchenko	Trinity College Dublin, Ireland
Slawomir Luscinski	Kielce University of Technology, Poland
Jose Machado	University of Minho, Portugal
Angelos Markopoulos	National Technical University of Athens, Greece
Mykola Melnychuk	Lutsk National Technical University, Ukraine
Arun Nagarajah	University of Duisburg-Essen, Germany
Marek Ochowiak	Poznan University of Technology, Poland
Gennadii Oborskyi	Odessa Polytechnic National University, Ukraine
Oleh Onysko	Ivano-Frankivsk National Technical University of Oil and Gas, Ukraine
Vitalii Pasichnyk	National Technical University of Ukraine "Igor Sikorsky Kyiv Polytechnic Institute", Ukraine
Ivan Pavlenko	Sumy State University, Ukraine
Oleksandr Povstyanoi	Lutsk National Technical University, Ukraine
Andrii Rogovyi	National Technical University "Kharkiv Polytechnic Institute", Ukraine
Vira Shendryk	Sumy State University, Ukraine
Robert Sika	Poznan University of Technology, Poland
Marcin Sosnowski	Jan Dlugosz University in Czestochowa, Poland
Vadym Stupnytskyy	Lviv Polytechnic National University, Ukraine
Ihor Sydorenko	Odessa Polytechnic National University, Ukraine
Marek Szostak	Poznan University of Technology, Poland
Valentyn Tikhenko	Odessa Polytechnic National University, Ukraine
Volodymyr Tonkonogyi	Odessa Polytechnic National University, Ukraine
Justyna Trojanowska	Poznan University of Technology, Poland
Nicolae Ungureanu	Technical University of Cluj-Napoca, Romania
Alper Uysal	Yildiz Technical University, Turkey
Djordje Vukelic	University of Novi Sad, Serbia
Jerzy Winczek	Czestochowa University of Technology, Poland
Szymon Wojciechowski	Poznan University of Technology, Poland
Oleg Zabolotnyi	Lutsk National Technical University, Ukraine

Invited External Reviewers

Ivan Dehtiarov	Sumy State University, Ukraine
Yuliia Denysenko	Sumy State University, Ukraine
Kostiantyn Dyadyura	Odessa Polytechnic National University, Ukraine
Oksana Haponova	Sumy State University, Ukraine
Maryna Holofieieva	Odessa Polytechnic National University, Ukraine
Vlodymyr Klitnoi	National Technical University "Kharkiv Polytechnic Institute", Ukraine
Vladyslav Kondus	Sumy State University, Ukraine
Olha Kostiuk	Lviv Polytechnic National University
Andrii Kuk	Lviv Polytechnic National University
Vadym Makarov	National Technical University "Kharkiv Polytechnic Institute", Ukraine
Ruslan Ostroha	Sumy State University, Ukraine
Kseniya Rezvaya	National Technical University "Kharkiv Polytechnic Institute", Ukraine
Serhii Sharapov	Sumy State University, Ukraine
Volodymyr Topilnytskyy	Lviv Polytechnic National University
Alona Tulska	National Technical University "Kharkiv Polytechnic Institute", Ukraine
Olha Turchyn	National Technical University "Kharkiv Polytechnic Institute", Ukraine

InterPartner Team

Anna Balaniuk	Odessa Polytechnic National University, Ukraine
Kristina Berladir	Sumy State University, Ukraine
Vitalii Ivanov	Sumy State University, Ukraine
Vadym Khamrai	Odessa Polytechnic National University, Ukraine
Kateryna Kirkopulo	Odessa Polytechnic National University, Ukraine
Gennadii Oborskyi	Odessa Polytechnic National University, Ukraine
Ievhen Ostroverkh	National Technical University "Kharkiv Polytechnic Institute", Ukraine
Ivan Pavlenko	Sumy State University, Ukraine
Andrey Pavlyshko	Odessa Polytechnic National University, Ukraine
Volodymyr Tonkonogyi	Odessa Polytechnic National University, Ukraine
Alla Toropenko	Odessa Polytechnic National University, Ukraine
Justyna Trojanowska	Poznan University of Technology, Poland

Contents

Manufacturing Technology and Machining Processes

Advanced Materials

Design Engineering and Production Planning

Computer Modeling of Casting Processes for Centrifugal Pump Parts

Khrystyna Berladir[1,2(✉)], Tetiana Hovorun[1], and Jozef Zajac[2]

[1] Sumy State University, 2, Rymskogo-Korsakova Street, Sumy 40007, Ukraine
kr.berladir@pmtkm.sumdu.edu.ua
[2] Technical University of Kosice, 1, Bayerova Street, Presov 08001, Slovak Republic

Abstract. The paper describes the application of computer programs for modeling foundry processes – casting metal blanks. Several casting objects (samples) of varying complexity were used in the simulation. A 3D model of the part and casting of a higher-quality multifunctional mold for casting in the SolidWorks program was developed. Defects in drawings were also automatically eliminated. The NovaFlow&Solid and MAGMASOFT programs were used to calculate the formation of shrinkage shells. The NovaFlow&Solid program is best used for checking simple and small models. It performs calculations quickly enough, but the accuracy of this program is low; therefore, when calculating complex models, it may give incorrect results. When using the MAGMASOFT program, more time is needed for the calculation process, but more correct results can be obtained. Also, the interface and functions of this program are more advanced and more convenient. After comparing the results of the formation of shrinkage shells during casting simulation using both programs, it was found that they practically do not differ. Therefore, this model can be considered successful. Based on the obtained results, it was shown that the stage of virtual design of casting technology (before the production of castings) allows for minimizing possible miscalculations and errors that inevitably occur in the development process, reducing financial and time costs, increasing efficiency, competitiveness, quality and reliability of products.

Keywords: Process Innovation · Foundry · Casting · Centrifugal Pump · Guide Vanes · Computer Modeling · Novaflow&Solid · MAGMASOFT

1 Introduction

Centrifugal pumps are very responsible units, so their operation should be as reliable and durable as possible [1]. Previously, this problem was solved by designing centrifugal pumps with a large safety margin, increasing their dimensions, weight, and metal consumption [2]. However, with the development of computer technology, it became possible to pre-model various components of centrifugal pumps to reduce the use of materials without losing quality indicators in work [3].

Modern mechanical engineering requires rapidly updating the model range and design of centrifugal pumps with characteristics superior to previously created samples. It is possible to reduce the development time of pumps with improved features due

V. Tonkonogyi et al. (Eds.): InterPartner 2023, LNME, pp. 3–15, 2024.
https://doi.org/10.1007/978-3-031-42778-7_1

to computer simulation [4]. This requires a system of automated design and engineering analysis using automated decision-making algorithms [5, 6] and information management systems [7]. This system should allow the designer to quickly obtain a comparative study of various options and conduct a multi-criteria design development [8].

The application of computer modeling in foundry processes is essential since many parts of centrifugal pumps can be manufactured by casting methods. The casting method is universal and straightforward for manufacturing such essential components of centrifugal pumps as impellers and guide vanes. Due to their complex shape, these parts are quite challenging to process mechanically [9]. It allows getting a casting as close as possible to the part's final shape. This helps to avoid complex and long-term mechanical processing, which is economically feasible in serial or multi-series production.

The purpose of the work is a comparative analysis of the use of computer modeling in foundry processes for parts of centrifugal pumps. To solve the goal, a modern computer base was used to model foundry processes for parts of centrifugal pumps obtained by casting. It allows creating 3D models in the SolidWorks program and using them for further modeling in the NovaFlow&Solid and MAGMASOFT programs. This made it possible to scientifically substantiate the casting manufacturing technology, design a 3D model, select technological casting parameters, and conduct experimental studies in laboratory and industrial conditions.

2 Literature Review

The analysis of the main spheres of the Industry 4.0 concept made in [10] showed that foundries are at the initial stage of its implementation. The authors attribute this to the laboriousness of foundry processes, the number of input and output parameters (final product properties), the number and type of foundry defects (quality parameters), communication problems, and the mathematical description of several process stages.

Improving the quality of products and reducing metal and energy intensity is an urgent task in the market economy of Ukraine. For this purpose, to reduce time and financial costs for production preparation, foundries began to widely implement computer modeling of forming castings in the design of technology [11] and the use of CAD systems in manufacturing equipment [12, 13].

Visualization of the physical processes of foundry technology, such as filling the mold cavity with melt [14], cooling and solidification of the metal [15], and its grooving under the action of thermal stresses, allows for a better understanding of the features of these processes, and, therefore, to more effectively manage them to reduce casting defects [16] and to increase the yield of workable parts [17, 18].

Modern design in foundry production includes the development of a 3D model of a part, foundry system, and equipment. Creating a 3D model of components and casting is most often implemented in SolidWorks and AutoCAD programs [19]. SolidWorks is a software complex for automating the work of an industrial enterprise at the stages of design and technological preparation of production, which ensures the development of products of any degree of complexity and purpose [19].

Modern computer modeling programs, based on physical theories of thermal, diffusion, hydrodynamic, and deformation phenomena, can adequately reflect the picture of physicochemical processes that occur when filling a mold with liquid metal, crystallization of a multicomponent alloy, and annealing. The leading software for numerical modeling of filling [20] and solidification of castings include [21]: MAGMASoft (Germany) [22], Novaflow&Solid (Sweden) [23], Procast (USA) [24], Flow3D (USA) [25], and CastCAE (Finland).

In particular, the MAGMASOFT program simulates the filling of the pouring system and the mold cavity with melt-in conditions as close as possible to reality [22]. It is also possible to solve three-dimensional temperature problems during filling and casting crystallization and simultaneously consider phenomena in the two-phase zone [26].

The application of NovaFlow&Solid allows to optimize of the modes of alloy pouring and casting hardening [23]; optimization of the gating system; analysis of casting processes when using different materials; improvement of the quality of castings and, as a result, finished products, increase the competitiveness of products; to reduce the total time spent on the production of products; reduce the cost of finished products [27, 28]; minimize foundry waste; reduce material capacity; significantly reduce, and in some cases eliminate the defect [29].

Thus, special software allows visualizing the casting process and understanding its nature, improving the process of designing and producing castings significantly, reducing the consumption of resources – metal, working time, and electricity.

3 Research Methodology

The paper examines the process of obtaining and computer modeling a casting part "guide vanes" of the centrifugal sectional pump type with a maximum outer diameter of 410 mm. The weight of the finished part is 19.5 kg, and the weight of the casting is 95 kg. The mass calculation was conducted in the SolidWorks program.

A corrosion-resistant cast alloy steel 08Cr14Ni7MoL of the austenitic-martensitic class was chosen to manufacture this part.

Obtaining the part «guide vanes» refers to serial production, so it would be appropriate to use the casting method. The development of the technological process of manufacturing a part by casting consists of the following operations: obtaining cast iron; production of steel by the duplex process; obtaining a casting in a sand-clay mold; mechanical and heat treatment. Obtaining a casting in a sand-clay mold includes a few operations and is presented in Fig. 1.

Figure 2 shows a simplified sketch of the assembly in sand-clay molds for casting for the guide vanes. Formation in sand-clay forms included the following stages. The model was covered with a special separation mixture; the lower model was placed in the lower casting box 2 and fixed with false stands 4; the molding mixture was poured into the lower casting box, and the excess quantity was removed. Gating system 6 was obtained, and the upper casting box 1 was installed on the lower casting box. The process was repeated for the upper casting box 1; the profit model 3 and a model of the guide vanes 5 were removed. Then the created imprint was covered with a thin layer of a special heat-resistant and non-stick mixture; in the place of the connector, special separating sand was applied, and the lower 2 and upper 1 casting boxes were fastened together. The model using a false stand is shown in Fig. 3.

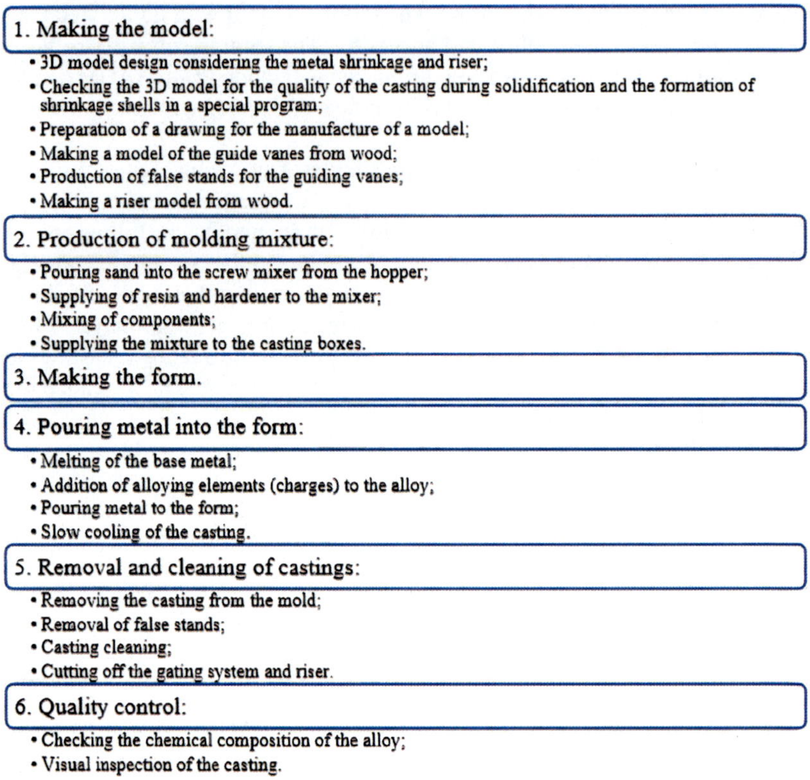

Fig. 1. The sequence of the obtaining technology a casting in a sand-clay mold.

After the formation is completed, the process of remelting the metal begins. Then the remelted steel is poured into a mold and cooled in air. After cooling, the casting is removed from the casting boxes and cleaned; the riser and the gating system are cut off. After cleaning the casting, it is inspected for cracks and other defects, and in the case of alloying during melting.

Since this steel was alloyed during remelting for casting, it is essential to check its actual chemical composition after the solidification of the casting. A portable X-ray fluorescence analyzer S1 TITAN SMX 524 and a stationary optical emission spectrometer PolySpek Junior were used to determine the chemical composition of steel after remelting. Table 1 shows the chemical composition of steel 08Cr14Ni7MoL, which fully complies with the standard DSTU 8781:2018.

The finished form of the obtained casting before mechanical processing is shown in Fig. 4.

The resulting casting is heading to the machine shop, where it is checked for cracks and other defects. After the inspection, the stage of rough mechanical processing of the casting begins. After mechanical processing, quality control of the part is carried out.

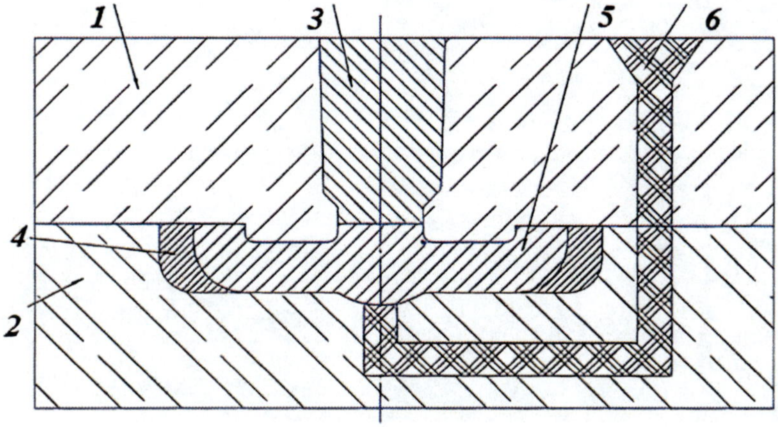

Fig. 2. Simplified sketch of the assembly of the casting box for casting for the guide vanes: 1 – the upper casting box; 2 – the lower casting box; 3 – the riser model; 4 – false stands; 5 – a model of the guide vanes; 6 – the gating system.

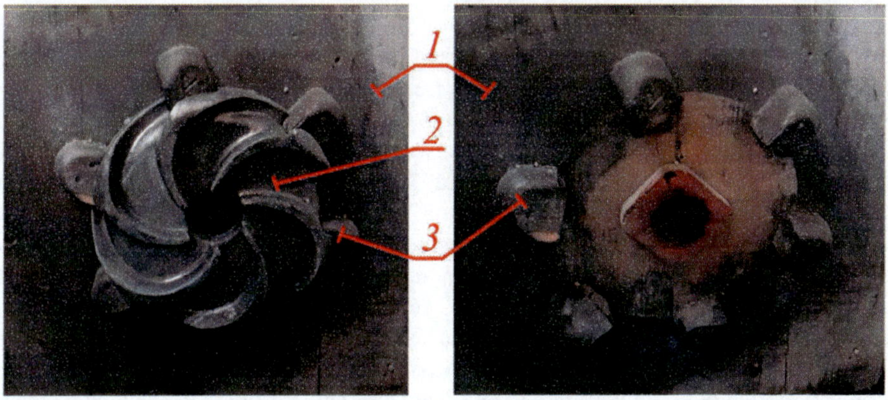

Fig. 3. Model using false stand: 1 – sub-model plate; 2 – guiding vanes model; 3 – false stand.

Table 1. The composition of steel 08Cr14Ni7MoL.

	C	Si	Mn	Cr	Ni	Mo	S	P
	No more						No more	
After remelting	0.080	0.75	0.90	13.0–15.0	6.0–8.5	0.5–1.0	0.025	0.025
DSTU 8781:2018	0.078	0.72	0.77	14.34	6.46	0.87	0.024	0.027

Fig. 4. The appearance of the casting after receiving it before mechanical processing.

Since this part was made by the casting method, it is subjected to capillary defectoscopy to detect small cracks on the part's surface.

After a successful inspection and control, the guiding vanes arrive at the thermal section. During operation, the guiding vanes of the centrifugal sectional pump type are subjected to a constant load; therefore, heat treatment is carried out to give the part the necessary properties, including quenching and cold treatment, and tempering.

Quenching of steel 08Cr14Ni7MoL was carried out at a temperature of 1100–1120 °C with an exposure time of 75–90 min and cooling in air. Since 08Cr14Ni7MoL steel contains 0.08% carbon, the alloy structure consists of acicular martensite after quenching. The casting cooled to shop temperature no later than 2 h after quenching. It was subjected to cold treatment in a cryogenic chamber using liquid nitrogen evaporation with cooling to −70 °C and exposure for 120–150 min. After this, the casting was removed from the freezer and heated to shop temperature by self-heating in the air. This operation reduces residual stresses in steel. The last heat treatment operation was carried out – tempering – to eliminate internal stresses in the steel and create a more uniform structure. Tempering was carried out at a temperature of 250–350 °C with a holding time of 120–150 min. The structure after tempering consists of acicular martensite and austenite.

In contrast to the cast steel structure, heat-treated steel 08Cr14Ni7MoL has a finer grain, which gives the steel higher hardness and wear resistance and, in turn, increases the service life of the guide vanes. The grain size of steel after casting is 3–4 points, and after heat treatment of steel, the grain size is 6–7 points. After heat treatment, the hardness of the steel is 36 HRC.

After heat treatment, part quality control occurs hardness check, structural analysis, and corrosion resistance check. If all the results of the mechanical tests of the part meet the requirements, then it is subjected to finishing machining, in which the part is threaded, ground, and polished to give the part a shiny appearance. After that comes the final quality control, in which the size of the part is checked and, if necessary, additional analyses are carried out, such as ultrasonic flaw detection.

4 Results and Discussion

The casting modeling process can be divided into six stages (Fig. 5).

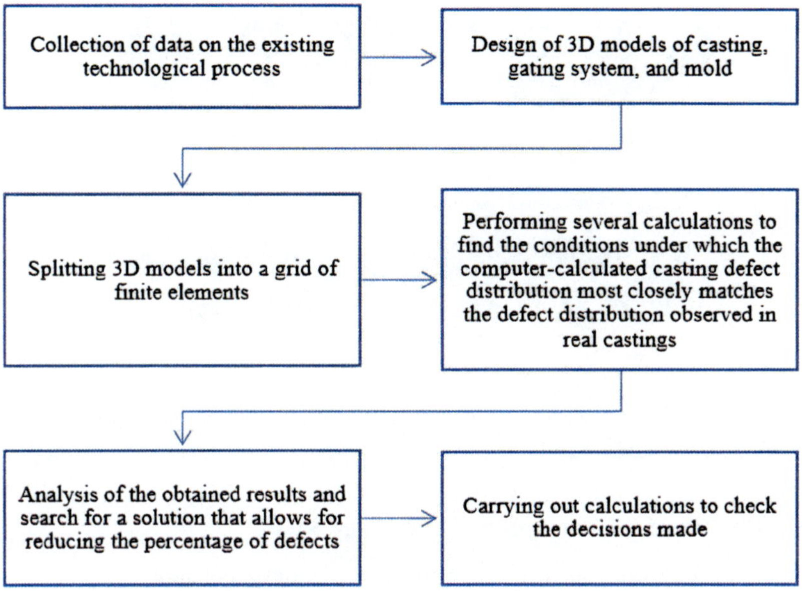

Fig. 5. Stages of the casting modeling process.

4.1 Design of a 3D Model of the "Guide Vanes" Part

The SolidWorks program was used to create a 3D model of the "guide apparatus" part (Fig. 6). This part model will be used to design the 3D model of the casting further and to further process the casting on CNC milling machines.

For the 3D development of a steel model, the number, size, and location of profits must be considered in addition to allowances. Also, when creating such a model, it is necessary to consider the possibility of its formation for casting. Since the guiding vanes have a complex shape, it is impossible to shape them in the sand without breaking the imprint without additional means. When creating a model, the foundry should use rods or false stands (Fig. 2).

Also, when developing a 3D model of a casting, it is necessary to consider economic feasibility. Figure 7 shows an example of two different models of the same part. It will not be economically feasible to use the Model 1 since much more metal will be needed for such a casting, which will increase the cost of the part.

Considering the requirements listed above, when creating a 3D model of the casting of the guide vanes, it is advisable not to increase the volume of the model itself relative to the part and to use only one central wooden riser. Figure 8 shows the casting and gating system (in gray) and the riser (in orange) of the 3D model of the guide vanes.

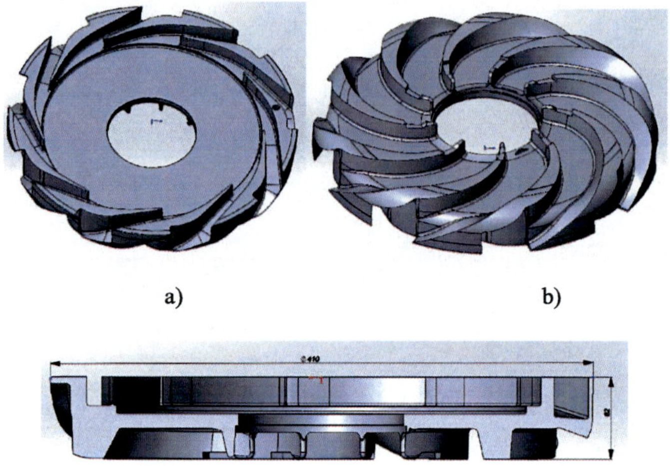

c)

Fig. 6. 3D model of the “guide apparatus” part: a) isometric view from above; b) isometric view from below; c) sectional view.

4.2 Calculation of the Formation of Shrinkage Shells

After creating a 3D model of the casting of the guide apparatus, it is necessary to calculate the formation of shrinkage shells, which is performed in special programs NovaFlow&Solid and MAGMASOFT.

For NovaFlow&Solid and MAGMASOFT software to work with 3D casting models made in SolidWorks, the model file must be saved in a special STEP AR214 format.

The NovaFlow&Solid program, unlike MAGMASOFT, is not able to work with models that are divided into several solid elements, so it is necessary to create a separate composite model for it, which is a significant, but not critical, drawback of this program since the creation of such a model takes more time.

However, at the same time, such a system can be more convenient for engineers because it is more modular. It helps to change or add different model elements more quickly if necessary, especially if they have been saved separately.

The results of shrinkage modeling in the NovaFlow&Solid program are shown in Fig. 9. The main shrinkage is in the central riser, and there are defects in the area of the liquid drainage channels. The NovaFlow&Solid program shows a small number of defects in the X-Ray mode (Fig. 9, a).

Further, calculations were made on the formation of shrinkage shells of the 3D casting model of the guide vanes in the MAGMASOFT program (Fig. 10). Almost all the shrinkage is in the central open riser and reaches almost its beginning. This means that the volume of metal in the riser is enough to pull most of the defects on itself and, simultaneously, not to use excess metal to fill the casting.

It also shows defects in the area of the liquid drainage channels. Their presence in this location is predicted because the casting model was designed with a clamp in

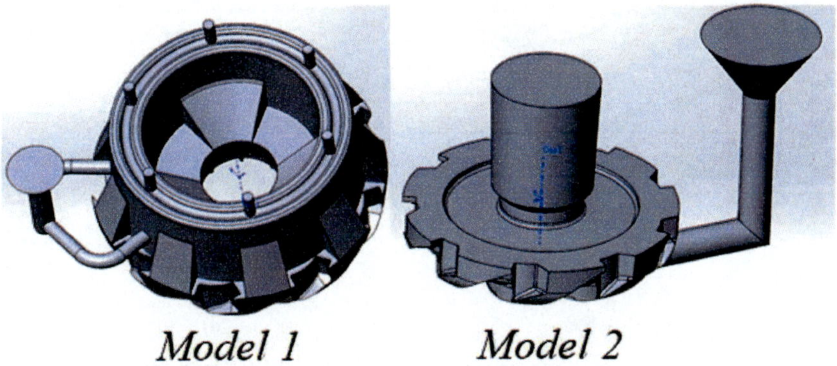

Fig. 7. An example of casting models of the guiding vanes.

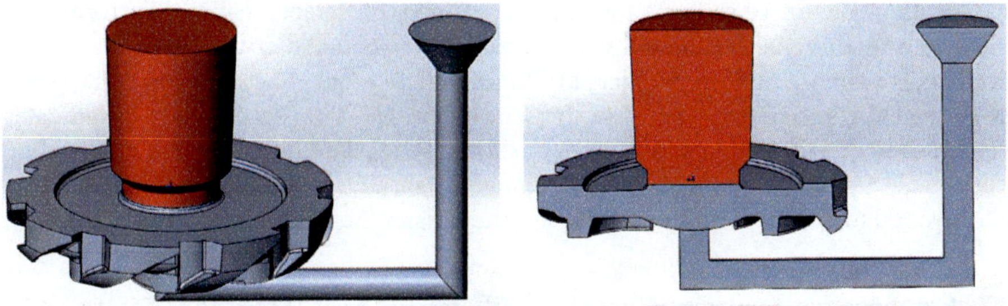

Fig. 8. 3D model of the casting of the guide vanes in the assembly: a) isometric view; b) sectional view.

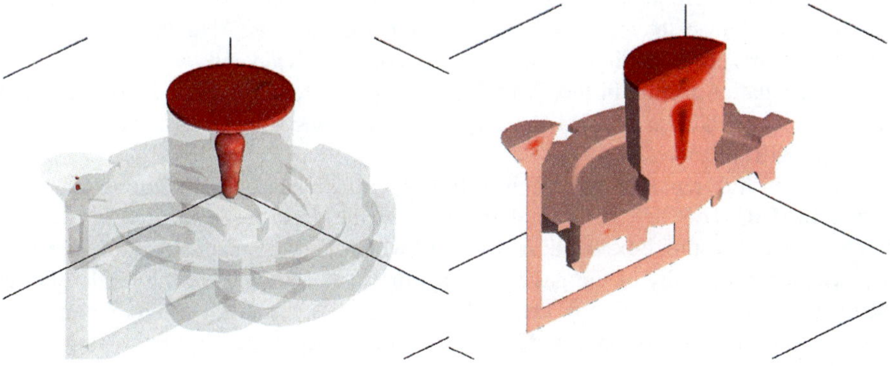

a) b)

Fig. 9. Calculation results of shrinkage shells in the NovaFlow&Solid program: a) view in X-Ray mode; b) sectional view.

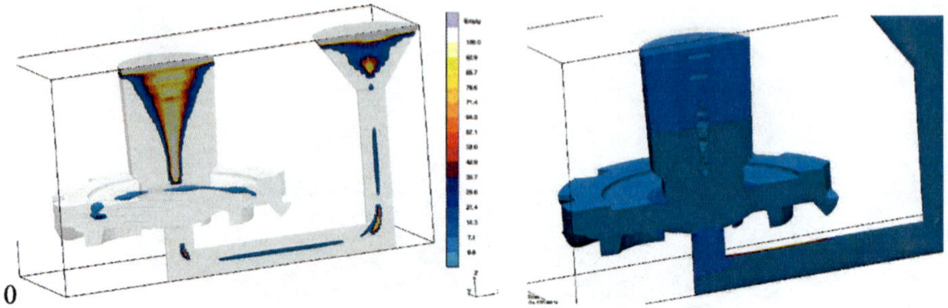

a) b)

Fig. 10. Calculation results of shrinkage shells in the MAGMASOFT program: a) view in X-Ray mode; b) sectional view.

that location. Figure 10 shows a much larger number of defects that are not formed in practice, or their value is so small that they are not considered critical defects, and the products are not missing. In practice, after mechanical processing, defects in that place are formed very rarely, and otherwise, they can be welded. The scale in Fig. 10a indicates the percentage of formation and the size of shrinkage shells. The area without shrinkage is indicated by white, and then the percentage of shrinkage formation is indicated by color from 0 to 100. The brighter the color, the greater the chance of defects.

4.3 Comparative Analysis of NovaFlow&Solid and MAGMASOFT Programs

The NovaFlow&Solid program is best used for checking simple and small models. It performs calculations quickly enough, but the accuracy of this program is low, so when calculating complex models, incorrect results may appear. Although the MAGMASOFT program takes longer to calculate, it gives more correct results. Also, the interface and functions of this program are more advanced and more convenient.

The simulation results in the NovaFlow&Solid and MAGMASOFT programs are almost the same. Unlike MAGMASOFT, the NovaFlow&Solid program shows fewer defects in X-Ray mode.

The main difference between this software is their gas output settings from the casting. In MAGMASOFT, this is implemented with the help of a special area that mimics the shape of the casting box, the dimensions of which must be manually configured. In this case, the peculiarity of the gas output setting is that the part of the model that is outside the border of this selected area is the casting box because the program considers that part of the model to be open and has access to air. There are no virtual casting boxes in the NovaFlow&Solid program, so it needs to specify where the gas will exit using a special function.

After comparing the results of the formation of shrinkage shells during casting modeling using both programs (Figs. 9 and 10), it can be concluded that they are practically not different, and therefore this model of the casting of the guide vanes can be considered successful. The selected casting technological parameters meet the requirements. The primary defects are in the gating system, riser, and the middle of the model. Defects were not included in the machining allowance. The results' reliability can be estimated with a probability of 90–95%, which was confirmed by the experience of practical use of programs for computer modeling of casting processes.

The analysis and calculations must be performed to eliminate shrinkage defects while developing the virtual model. This makes it possible to significantly shorten the production time of castings by making timely changes to the design of the gating and feeding systems, which ensures an increase in the quality of castings and a reduction in the cost of the part. Therefore, when solving issues related to the foundry production of parts of centrifugal pumps, modeling of parts and casting is an indispensable tool for an engineer and foundry technologist.

5 Conclusions

The technological process of obtaining the «guide vanes» part of the centrifugal sectional pump type was developed by casting it into sand-clay molds. The material for its manufacture was steel for casting 08Cr14iN7MoL. It is highly alloyed, corrosion-resistant steel of the austenitic-martensitic class, which has relatively high strength and hardness, high resistance to plastic deformation, and good cold-resistant properties.

SolidWorks software was used to create a 3D model of the part and casting, which is a software complex for automating the work of an industrial enterprise at the stages of design and technological preparation of production and ensures the development of products of any degree of purpose and complexity.

Shrinkage shells were checked in the NovaFlow&Solid and MAGMASOFT programs. After comparing the results of the formation of shrinkage shells during casting modeling using both programs, it can be concluded that they are practically not different, and therefore this model of the casting of the guide vanes can be considered successful. The main defects are in the gating system, riser, and the middle of the model. The results' reliability can be estimated with a probability of 90 – 95%, which was confirmed by the experience of practical use of programs for computer modeling of casting processes.

Computer modeling allows the creation of an optimal structure for the gating system at the design stage. It avoids expensive operations of proving the technology and changing equipment directly in organizing the production of the part.

Acknowledgment. This research work has been supported by the projects VEGA1/0080/20 and KEGA 028TUKE-4/2021 were granted by the Ministry of Education, Science, Research and Sport of the Slovak Republic. This publication is the result of the Project implementation: "Development of excellent research capacities in the field of additive technologies for the Industry of the 21st century", ITMS: 313011BWN5, supported by the Operational Program Integrated Infrastructure funded by the ERDF.

The results have been partially obtained within the research project "Fulfillment of tasks of the perspective plan of development of a scientific direction "Technical sciences" Sumy State University" funded by the Ministry of Education and Science of Ukraine (State reg. no. 0121U112684). The authors appreciate the support of the Research and Educational Center for Industrial Engineering (Sumy State University) and the International Association for Technological Development and Innovations.

References

1. Pavlenko, I., Simonovskiy, V., Verbovyi, A., Ivchenko, O., Ivanov, V.: Rotor dynamics and stability of the centrifugal pump CPN 600-35 for nuclear power plants. In: Ivanov, V., Pavlenko, I., Liaposhchenko, O., Machado, J., Edl, M. (eds.) Advances in Design, Simulation and Manufacturing V: Proceedings of the 5th International Conference on Design, Simulation, Manufacturing: The Innovation Exchange, DSMIE-2022, June 7–10, 2022, Poznan, Poland – Volume 2: Mechanical and Chemical Engineering, pp. 33–46. Springer International Publishing, Cham (2022). https://doi.org/10.1007/978-3-031-06044-1_4
2. Pavlenko, I., Trojanowska, J., Gusak, O., Ivanov, V., Pitel, J., Pavlenko, V.: Estimation of the reliability of automatic axial-balancing devices for multistage centrifugal pumps. Periodica Polytechnica Mech. Eng. **63**(1), 52–56 (2019). https://doi.org/10.3311/PPme.12801
3. Pavlenko, I., Kulikov, O., Ratushnyi, O., Ivanov, V., Pitel, J., Kondus, V.: Effect of impeller trimming on the energy efficiency of the counter-rotating pumping stage. Appl. Sci. **13**(2), 761 (2023). https://doi.org/10.3390/app13020761
4. Pavlenko, I., Ivanov, V., Kuric, I., Gusak, O., Liaposhchenko, Ol.: Ensuring vibration reliability of turbopump units using artificial neural networks. In: Trojanowska, J., Ciszak, O., Machado, J.M., Pavlenko, I. (eds.) Advances in Manufacturing II: Volume 1 – Solutions for Industry 4.0, pp. 165–175. Springer International Publishing, Cham (2019). https://doi.org/10.1007/978-3-030-18715-6_14
5. Javanbakht, T.: Optimization of graphene oxide's characteristics with TOPSIS using an automated decision-making process. J. Eng. Sci. **10**(1), E1–E7 (2023). https://doi.org/10.21272/jes.2023.10(1).e1
6. Kotliar, A., et al.: Ensuring the economic efficiency of enterprises by multi-criteria selection of the optimal manufacturing process. Manag. Product. Eng. Rev. **11**(1), 52–61 (2020). https://doi.org/10.24425/mper.2020.132943
7. Haidabrus, B., Grabis, J., Protsenko, S.: Agile project management based on data analysis for information management systems. In: Ivanov, V., Trojanowska, J., Pavlenko, I., Zajac, J., Peraković, D. (eds.) DSMIE 2021. LNME, pp. 174–182. Springer, Cham (2021). https://doi.org/10.1007/978-3-030-77719-7_18
8. Monkova, K., et al.: Condition monitoring of Kaplan turbine bearings using vibro-diagnostics. Int. J. Mech. Eng. Robot. Res. **9**(8), 1182–1188 (2020). https://doi.org/10.18178/ijmerr.9.8.1182-1188
9. Haldar, P., Sutradhar, G.: Simulation and validation of castings in shop floor. In: Abdallah, Z., Aldoumani, N. (eds.) Casting Processes and Modelling of Metallic Materials. IntechOpen (2021). https://doi.org/10.5772/intechopen.94596
10. Kozłowski, J., Sika, R., Górski, F., Ciszak, O.: Modeling of foundry processes in the era of Industry 4.0. In: Ivanov, V., et al. (eds.) DSMIE 2018. LNME, pp. 62–71. Springer, Cham (2019). https://doi.org/10.1007/978-3-319-93587-4_7
11. Halchuk, T.N., Povstyanoy, O., Bembenek, M., Redko, R.G., Chetverzhuk, T.I., Polinkevych, R.M.: Impact of technological system's characteristics on the machining accuracy of bearing rings. J. Eng. Sci. (Ukraine) **10**(1), A22–A30 (2023). https://doi.org/10.21272/jes.2023.10(1).a4
12. Małysza, M., Puzio, S., Major-Gabryś, K., Głowacki, M., Wilk-Kołodziejczyk, D., Kamińska, J.: Computer modelling of the ablation casting process and prediction of the strength properties of AC-42000 castings. Comput. Methods Mater. Sci. **22**(2), 79–88 (2022). https://doi.org/10.7494/cmms.2022.2.0774
13. Zaleta, O.M., Povstyanoy, O.Y., Ribeiro, L.F., Redko, R.G., Bozhko, T.Y., Chetverzhuk, T.I.: Automation of optimization synthesis for modular technological equipment. J. Eng. Sci. **10**(1), A6–A14 (2023). https://doi.org/10.21272/jes.2023.10(1).a2

14. Šabík, V., Futáš, P., Pribulová, A.: Failure analysis of a clutch wheel for wind turbines with the use of casting process simulation. Eng. Failure Anal. **135**, 106159 (2022). https://doi.org/10.1016/j.engfailanal.2022.106159
15. Dou, K., Lordan, E., Zhang, Y.J., Jacot, A., Fan, Z.Y.: A complete computer aided engineering (CAE) modelling and optimization of high pressure die casting (HPDC) process. J. Manuf. Process. **60**, 435–446 (2020). https://doi.org/10.1016/j.jmapro.2020.10.062
16. Zheng, J., et al.: Method for evaluating the resource, energy, and environmental impact of the casting fault rectification process in patternless sand castings. Sustain. Mater. Technol. **35**, e00565 (2023). https://doi.org/10.1016/j.susmat.2022.e00565
17. Patil, M.A., Patil, S.D., Yadav, P.H., Desai, A.A.: Methoding and defect minimization of center plate casting by auto-CASTX1 software. Mater. Today: Proc. **77**, 662–672 (2023). https://doi.org/10.1016/j.matpr.2022.11.286
18. Kusyi, Y., Stupnytskyy, V.: Optimization of the technological process based on analysis of technological damageability of castings. In: Ivanov, V., Trojanowska, J., Pavlenko, I., Zajac, J., Peraković, D. (eds.) DSMIE 2020. LNME, pp. 276–284. Springer, Cham (2020). https://doi.org/10.1007/978-3-030-50794-7_27
19. Hernández, F., Fragoso, A.: Fabrication of a stainless-steel pump impeller by integrated 3D sand printing and casting: mechanical characterization and performance study in a chemical plant. Appl. Sci. **12**, 3539 (2022). https://doi.org/10.3390/app12073539
20. Chen, Z., Li, Y., Zhao, F., Li, S., Zhang, J.: Progress in numerical simulation of casting process. Measur. Control **55**(5–6), 257–264 (2022). https://doi.org/10.1177/00202940221102656
21. Rajkumar, I., Rajini, N.: Metal casting modeling software for small scale enterprises to improve efficacy and accuracy. Mater. Today: Proc. **46**, 7866–8787 (2021). https://doi.org/10.1016/j.matpr.2021.02.542
22. Aravind, S., Ragupathi, P., Vignesh, G.: Numerical and experimental approach to eliminate defects in al alloy pump-crank case processed through gravity die casting route. Mater. Today: Proc. **37**, 1772–2177 (2021). https://doi.org/10.1016/j.matpr.2020.07.365
23. Jezierski, J., Dojka, R., Janerka, K.: Optimizing the gating system for steel castings. Metals **8**, 266 (2018). https://doi.org/10.3390/met8040266
24. Liao, Q., et al.: Simulation study on the investment casting process of a low-cost titanium alloy gearbox based on ProCAST. Adv. Mater. Sci. Eng. **2022**, 4484762 (2022). https://doi.org/10.1155/2022/4484762
25. Małysza, M., et al.: Technological optimization of the stirrup casting process with the use of computer simulations. Materials **15**(19), 6781 (2022). https://doi.org/10.3390/ma15196781
26. Dojka, R., Jezierski, J., Szucki, M.: The importance of the geometry of the down sprue in the gravity casting process. Materials **15**, 4937 (2022). https://doi.org/10.3390/ma15144937
27. Lesyk, D., Martinez, S., Mordyuk, B., Dzhemelinskyi, V., Lamikiz, A.: Surface finishing of complexly shaped parts fabricated by selective laser melting. In: Tonkonogyi, V., et al. (eds.) Advanced Manufacturing Processes: Selected Papers from the Grabchenko's International Conference on Advanced Manufacturing Processes (InterPartner-2019), September 10–13, 2019, Odessa, Ukraine, pp. 186–195. Springer International Publishing, Cham (2020). https://doi.org/10.1007/978-3-030-40724-7_19
28. Dzhemelinskyi, V., Lesyk, D., Goncharuk, O., Danyleika, O.: Surface hardening and finishing of metallic products by hybrid laser-ultrasonic treatment. Eastern-European J. Enterp. Technol. **1**(12–91), 35–42 (2018). https://doi.org/10.15587/1729-4061.2018.124031
29. Sertucha, J., Lacaze, J.: Casting defects in sand-mold cast irons—an illustrated review with emphasis on spheroidal graphite cast irons. Metals **12**, 504 (2022). https://doi.org/10.3390/met12030504

Automatic Control "By Disturbance" Based on a Mechatronic Actuator

Anatoly Gushchin[1], Vasily Larshin[1(✉)], Oleksandr Lysyi[2], Alina Tselikova[3], and Oleksandr Lymarenko[1]

[1] Odessa Polytechnic National University, 1, Shevchenko Avenue, Odessa 65044, Ukraine
vasilylarshin@gmail.com

[2] Odessa Military Academy, 10, Fontanskaya Doroga Street, Odessa 65009, Ukraine

[3] Odessa State Academy of Civil Engineering and Architecture, 4, Didrihson Street, Odessa 65029, Ukraine

Abstract. The evolution of a change in the design of a mechatronic mechanism with a sensitive element of an automatic control system "by disturbance" was shown based on an electrodynamic coupling mechanism. The sensing element converts the difference in torque on the input and output shafts of this coupling into an axial linear mechanical displacement of the driven coupling half relative to the driving one. This displacement was proportional to the difference between the above torques and can be used either as an information signal or as a control action in the automatic control system "by disturbance". The advantage of automatic control "by disturbance" and constructive ways to increase the ratio of this displacement to the difference in the torques of the electrodynamic coupling were shown. For this purpose, the mechatronic mechanism designs were based on ball and cam couplings, as well as based on a ball-bearing screw converter with a range of regulation of the above axial displacement up to 2 mm, up to 8 mm, and up to 40 mm, respectively, were developed. Experimental data were presented to confirm the effectiveness of a mechatronic drilling system for small diameter (up to 5 mm) holes in aluminum alloy panels. Automatic control "by disturbance" for the torque value on the drill due to a change in the axial drilling force made it possible to eliminate burrs when drilling these holes.

Keywords: Technological Machine · External Environment · Cutting Zone · Linear Actuator · Energy Parameters · Axial Force · Cutting Torque · Automatic Control · Pressure Element · Transfer Coefficient · R&D Investment

1 Introduction

The product quality obtained with the help of a technological machine is determined by the force and temperature factors in the zone of the technological impact of the working body on the object to be machined. In turn, the above energy parameters (force and temperature) result from the machining kinematic parameters, the type of tool used, and other machining conditions.

V. Tonkonogyi et al. (Eds.): InterPartner 2023, LNME, pp. 16–25, 2024.
https://doi.org/10.1007/978-3-031-42778-7_2

For example, the technological impact zone is the cutting zone for a CNC machine. Axial force and cutting torque are the most common force parameters of machining, which are the result of the programmed operating modes of the CNC machine. It is necessary to ensure the values of the above force and temperature parameters to control the product quality. For this purpose, it is necessary to apply appropriate control, for example, control using information from a monitoring or automatic control.

The mechatronic mechanism (MM) described in the paper is a kind of mechatronic linear actuator based on the design of an electrodynamic coupling, a sensitive element of the automatic control system "by disturbance". This means that any torque deviation from the threshold value (required by technology in the technological machine) can be converted by a linear actuator into a change in the position of the working body of the technological machine. Consequently, the external environment, e.g., the cutting zone, having reacted, will change its response force reaction to a value equal to the setpoint (task). At the same time, in the contact interaction zone "working body – environment", only power parameters regulated in magnitude (axial force and torque) arise, which ensures the dynamic balance of the MM working body in the technological machine at each moment. This dynamic balance is due to the interaction of oppositely directed forces of action and reaction applied to the working body of the technological machine.

2 Literature Review

Force parameters are widely used, e.g., in tapping monitoring of blind or through holes [1]. Monitoring the cutting forces and torque helps both anticipating the tool's fracture and controlling, e.g., the crest edge roundness progression [2], including both detection of the tool wear types [3] and thread quality control [4]. For each tap, a multivariate statistical process control chart is presented based on the principal components of the torque signal directly measured from the spindle motor drive to diagnose the thread profile quality [5]. However, the possibility of automatic torque control to reduce (e.g., the tap wear) was not considered.

An integrated technological system is proposed with the following three levels of control: intelligent (upper), adaptive (middle), and robust (lower) [6]. However, nothing is said about the automatic control mechanism "by disturbance".

The design of a mechatronic module (based on a linear actuator idea) is proposed, which contains two design variants of this module: a mechatronic transducer and a mechatronic power converter [7].

The paper [8] proposes a method for controlling contact loads along the generatrix of the workpiece when using various compositions of cutting fluids and assessing their effect on the roughness of the resulting surface. Analogically, the paper [9] focuses on studying the influence of solid lubrication on the grinding characteristic of difficult-to-machine materials. Nevertheless, the "mechanism" of automatic control "by disturbance" for this aim was not offered.

The well-known control principles (by deviation and by disturbance) in modern control theory are accompanied by a new direction: distributed and hierarchical control systems with four system levels (from bottom to top): a component level, an information preprocessor level, an intelligent preprocessor level, and a top-level [10]. Essential control theory, transfer functions, and state space approaches in this theory are the base for developing the related systems mentioned above [11]. Besides, machine tools are mechatronic systems themselves [12]. Hence, they are open to including mechatronic mechanisms, but there are no hints of it.

There is a design of a mechatronic electro-spindle containing a mechatronic actuator [13]. However, the hierarchy of the mechatronic servo system containing the machine CNC at the upper control level was not disclosed.

Measuring thrust force and torque during tapping helps overcome the tendency to adhere to the tool surface since the machinability of the hypereutectic Al-Si alloys is challenging [14]. The same applies to intelligent automated drilling in laminate composites and hybrid materials [15].

At the end of the review, it is necessary to mention works on a study of the quality of machined parts made of CFRP [16], composites in drilling [17] and in general [18], as well as works on the physics of electromagnetism [19] and grinding technology feature [20].

The paper aims to create a MM design based on the electrodynamic coupling mechanism, which can be both a power and a sensitive element for automatic control "by disturbance" and "by deviation", respectively.

3 Research Methodology

3.1 Coupling Mechanism Conversion into a Power and Sensing Element

The literature review shows that an urgent task in machining hard-to-cut materials is the creation of functionally complete mechatronic systems for the implementation of adaptive and intelligent control of technological machines for a wide range of applications, including CNC machines.

MM shown below contains the coupling mechanism in which the armature of the linear electric actuator was used as a pressure element. In turn, the axial displacement ΔS was used as an information source for the cutting tool torque regulating through the feedback system. For this purpose, the axial displacement ΔS through the inductive sensor was converted into a control electrical signal for the automatic control system with negative feedback by deviation.

In a system configuration with negative feedback, the influence of the system parameter variations becomes less as the gain coefficient increases. However, an increase in the gain coefficient may result in an unstable system. Thus, one of the possible methods of solving the problem reduces to the design of structures that will allow the required loop gain coefficient to be increased without affecting the system stability. The possible ways to do this are both to increase the ratio $\Delta S/\Delta\varphi$ by increasing ΔS and to use the open loop automatic control "by disturbance".

This MM contained a sensitive element (SE). The purpose of the SE, for example, during machining is to convert the torque increment ΔM on the output shaft into the axial linear movement ΔS of the SE of the movable pressure element. The linear displacement ΔS was converted into a control signal to compensate for the specified increment ΔM. SE (Fig. 1) of such a transformation is made in the form of a mechanical part of the above coupling, containing a driven (movable in the axial direction) 1 and a leading (fixed in the axial direction) 2 half-couplings, between which the rolling elements balls 3 are located. Torque increment ΔM leads to an increase in the angular displacement $\Delta\varphi$ of the movable coupling half relative to the fixed one and, as a result, to an increase in the axial displacement ΔS (Fig. 1) of the pressure element – the armature of the linear power actuator.

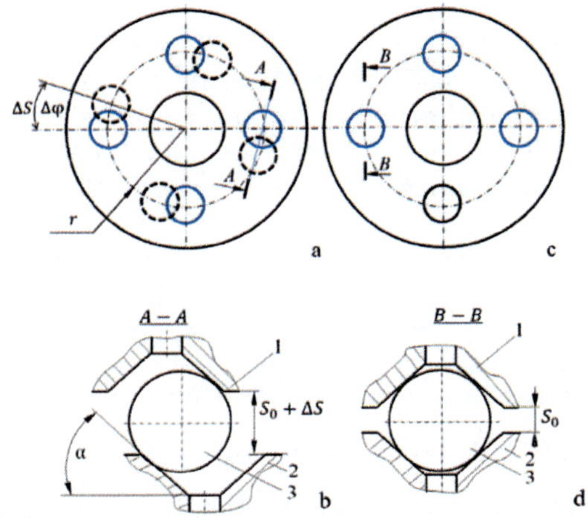

Fig. 1. Angular displacement between coupling halves when the transmitted torque exceeds the preset level (a, b) and does not exceed (c, d).

Figure 1 introduced the following notations: r – radius of the rolling circle of the balls, m; $\Delta\varphi$ is the shift in the angle of rotation of one coupling half relative to the other when the resistance torque exceeds its predetermined value, deg; Δl is the length of the arc of displacement of the centers of the balls along a circle of radius r due to the angular displacement $\Delta\varphi$, m.

The SE considered in the paper is part of the MM. It has been perfected for a long time as a source of information (i.e., transducer) about the torque to ensure the linearity of the relationship between the pressure force and the displacement ΔS, on the one hand, and to increase the displacement ΔS. This made it possible to use SE to control the technological process according to its power parameters (Fig. 2, a) by comparing the torques on its input and output shafts.

The transformation of the difference in torques on the input and output shafts of the coupling into the axial movement of the movable coupling half is an integral property of all designs of the developed couplings: ball coupling, cam one, and that based on a ball-bearing screw converter.

The angular displacement of the movable coupling half $\Delta\varphi$ relative to the fixed one leads to axial displacement ΔS of the movable coupling half. This axial displacement (ΔS) can be used in two ways. Firstly, as a feedback signal for the automatic deviation control system. To do this, the displacement ΔS must be converted into an electrical signal. In this case, the variable voltage in volts Δu will be proportional to ΔS, i.e., Δu

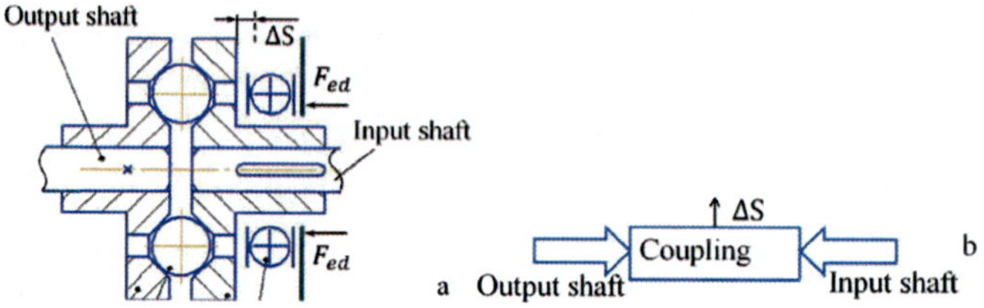

Fig. 2. The SE, based on a safety ball clutch (a) and its replacement scheme (b).

$= k\Delta S$, where k is the proportionality factor. Secondly, the displacement ΔS can be used as a force action (discussed below) in a linear actuator in order to stabilize, for example, the cutting torque [9] with automatic control "by disturbance". The automatic control range ΔS was successively increased from the minimum (1 mm $\leq \Delta S_{\min} \leq$ 2 mm) in the ball coupling to the average (5 mm $\leq \Delta S_{\text{ave}} \leq$ 8 mm) in the cam coupling and further to the maximum (20 mm $\leq \Delta S_{\max} \leq$ 40 mm) in a ball-bearing screw converter.

Thus, the created "line of actuators" allows selecting the desired actuator (from the specified range of linear displacements (1 mm $\leq \Delta S_{\max} \leq$ 40 mm) in the corresponding MM, considering the requirements of a particular technological machine.

3.2 Sensing Element Static Transfer Function

To derive the dependence $\Delta S(r, \Delta\varphi, \alpha)$ based on Fig. 1, a calculation scheme was constructed (Fig. 3).

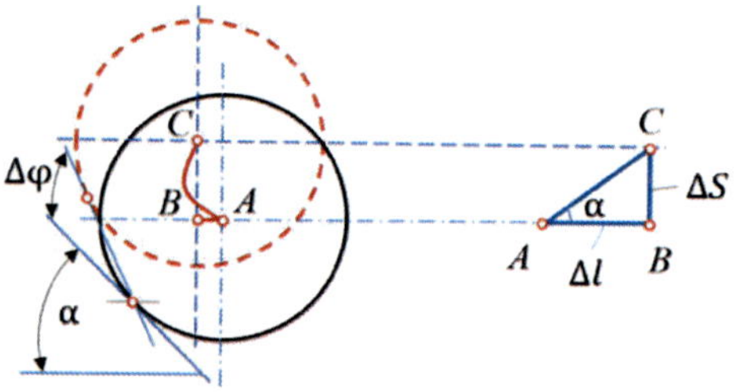

Fig. 3. A calculation scheme to derive the SE static transfer function.

This scheme shows the trajectory AC of the movement of the ball center, which consists of two elementary movements, respectively, in the horizontal (AB) and vertical (BC) planes. The arc length Δl is determined by the formula: $\Delta l = r\Delta\varphi$, where the angle $\Delta\varphi$ is measured in radians. When measuring $\Delta\varphi$ in degrees, we get

$$\Delta l = \frac{\pi r \Delta\varphi}{180}, \tag{1}$$

From the triangle ABC (Fig. 2), we get: $BC = AB \cdot \tan\alpha$ or $\Delta S = \Delta l \cdot \tan\alpha$. Therefore,

$$\Delta S = \frac{\pi r \Delta\varphi}{180} \tan\alpha \tag{2}$$

Equation (2) gives the SE transfer coefficient (gain factor)

$$\frac{\Delta S}{\Delta \varphi} = \frac{\pi r}{180} \tan \alpha \tag{3}$$

3.3 Increasing the Disturbance Control Range

The transition from a ball coupling to a cam one with an increased range of linear axial displacement ΔS made it possible to increase the SE transfer coefficient by order of magnitude in the equation $\Delta S = k\,\Delta \varphi$, where k is a proportionality factor that depends on the design parameters of the coupling halves.

An example of a cam coupling with a larger range of axial movement ΔS (compared to a ball coupling) is the design described in [8]. In this design, the range of possible axial displacements of the driven coupling half is $\Delta S_{\min} \le \Delta S \le \Delta S_{\max}$ (Fig. 4).

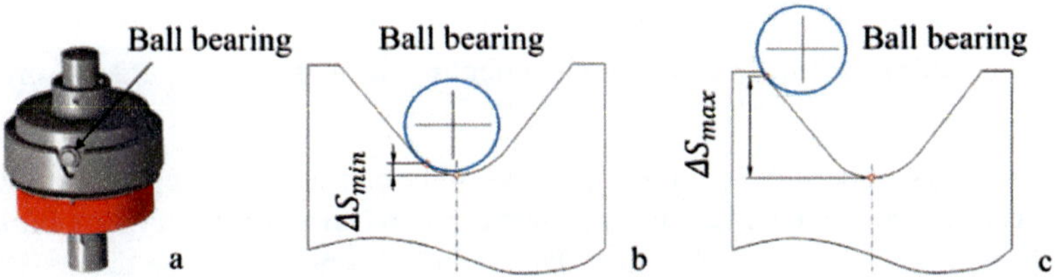

Fig. 4. 3D model of the cam cOupling (a) and two extreme axial positions of its driven half-coupling: lower (b) and upper (c).

3.4 Ball-Bearing Screw Converter

In the design of a ball-bearing screw converter (Fig. 5, a), the driving half-coupling is a "nut", which is a hollow cylindrical body 1, on the side surface of which slotted grooves are made at an angle α (two-start thread turns with pitch t). Such a "nut" is made of non-magnetic material (brass, bronze), and its two ends are equipped with PTFE guide bushings, which provide the spindle ("screw") with the possibility of free (without backlash and jamming) reciprocating movements along the longitudinal axis.

The "screw" (aka the driven coupling half, made in the form of a spindle) has the shape of a cylindrical rod 1 (Fig. 5, b) with two support ball bearings 2, fixed on it (threads on the "screw") with the help of centering elements 3 and 4. The symmetry of the installation of ball bearings 2 on the spindle is adjusted by compensators 5.

Thus, the evolution of SE designs made it possible to switch from automatic control "by deviation" to that "by disturbance" with a control range ΔS, which is determined by the length of the two-start thread of the ball screw nut (Fig. 5) turning ΔS into a disturbance control range, i.e., automatic control range "by disturbance".

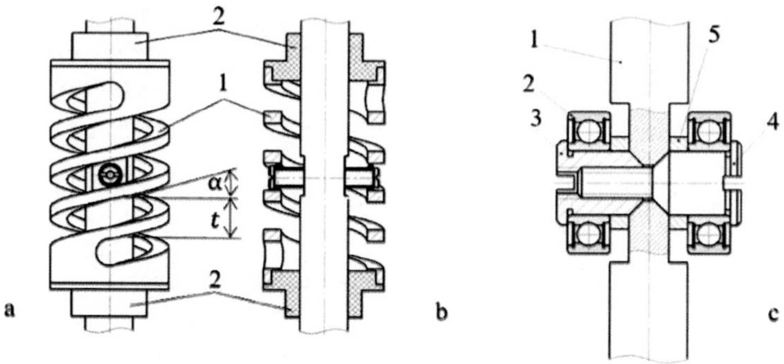

Fig. 5. The driving (a, b) and driven (c) parts of the ball-bearing screw converter.

4 Results and Discussion

4.1 Patterns of the SE Design

The general patterns of the SE design for all coupling designs (as parts of MM) are listed below.

1. In all designs of ball and cam movable coupling halves, including that based on a ball-bearing screw converter, linear movements are possible within the normalized (relative) interval $-1 \leq \Delta S \leq \pm 1$. This interval's absolute value depends on the coupling's functional purpose, design, and geometric parameters. The required range of axial linear movement ΔS of the movable coupling half for performing a technological operation is a condition for choosing one of its three constructs as part of the MM: a ball coupling, a cam coupling, or a coupling based on a ball-bearing screw converter.
2. In all coupling designs to transfer the regulated torque, the effect of the dynamic balance of rolling elements on an inclined surface is used. For example, the balance of a ball in a conical hole (Fig. 1), a ball bearing in an angular groove (Fig. 4), or a ball bearing on an inclined threaded surface (Fig. 5).
3. As a pressure element, first of all, the armature of a linear actuator is used, the electric winding of which is switched on in the computer control circuit [7]. Changing the magnitude of the electric current in the armature winding makes it possible to adjust the closing force of the coupling halves. The force of closing (compression) of the coupling halves through the rolling elements is a reference torque (torque setpoint) required by the technology, for example, in the cutting zone.
4. Computer control of the pressing force is performed by changing the electric current strength in the winding of the linear actuator armature or the linear actuator field winding. Combined control is also possible, i.e., simultaneous currents change in the armature and field windings.
5. In all coupling designs, the possibility of summing circular and reciprocating linear displacements has been implemented to convert them into a helical movement of a movable coupling half.

6. In all coupling designs, elements of a non-self-braking pair "screw – nut" are implemented, in which the "nut" is the driving half-coupling, and the "screw" is the driven one that is moving in the axial direction. This is also fully implemented in the SE design with a moving element working stroke of 40 mm (Fig. 6).

Thus, in all the considered coupling designs, the SE is an element of an equivalent non-self-braking ball screw transmission in which the movable element can reciprocate from both the axial force and the torque applied to it (Fig. 6). As a result of the SE evolutionary development, a ball-bearing screw converter with a control range of $\Delta S =$ 40 mm was created as part of the MM with automatic control "by disturbance" (Fig. 6).

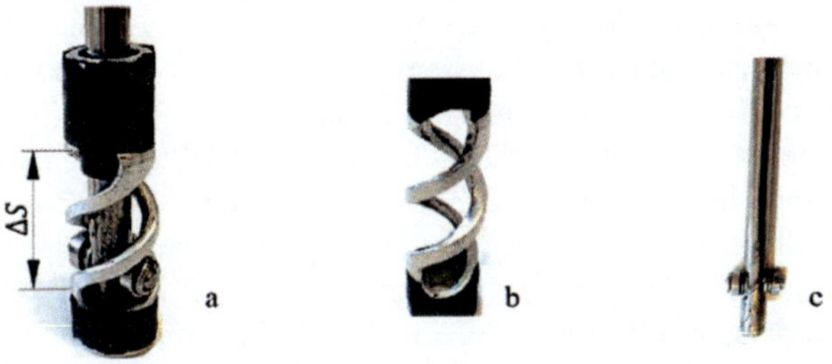

Fig. 6. Ball-bearing screw converter (a) and its elements: "nut" (b) and "screw" (c).

As a MM block, the created design of the ball screw transmission made it possible to implement technologies for defect-free shaping machine parts from hard-to-machine materials: viscous, high-strength, and anisotropic, for example, for drilling viscous materials such as aluminum alloys, as shown below.

4.2 Testing a Drill Head with MM Built-In

A drill head with built-in MM was tested in cooperation with HUGO RECKERTH GmbH when drilling holes in an aluminum alloy plate (Fig. 7).

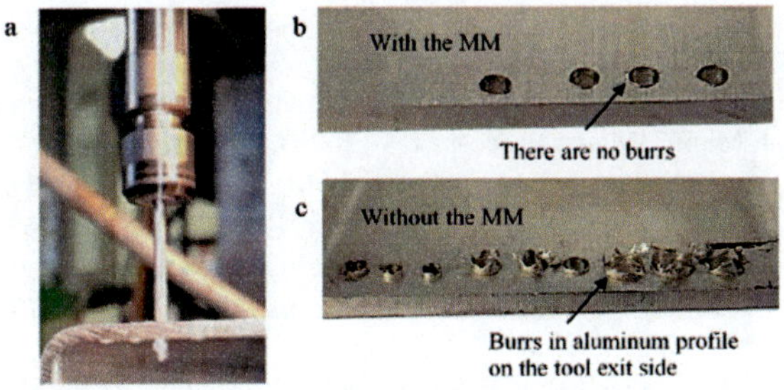

Fig. 7. Motorspindle based on the MM (a) and drilling result with (b) and without MM (c).

It can be seen (Fig. 7, c) that when drilling holes with torque stabilization on the spindle, due to the regulation of the axial cutting force, there are no burrs on the output side of the drilled holes.

5 Conclusions

The evolution of a change in the design of a mechatronic mechanism containing a sensitive element of an automatic control system "by disturbance", made based on an electrodynamic coupling mechanism, is shown. The sensing element converts the difference in torques on the input and output shafts of this coupling mechanism into axial linear displacement of the driven half-coupling relative to the driving one and is a source of force in the automatic control system "by disturbance", and can also be a source of information for the automatic control system "by deviation". Such a feature underlines this paper's added value compared with similar ones.

The advantage of automatic control "by disturbance" and constructive ways of increasing the range of variation of the control parameter from 2 mm to 8 mm for ball and cam couplings, respectively, and up to 40 mm for a ball-bearing screw converter are shown.

Experimental data confirming the effectiveness of the mechatronic drilling system are presented in the example of drilling through holes of small diameter (up to 5 mm) in aluminum alloy panels. Automatic control "by disturbance" of the torque value on the drill due to a change in the axial cutting force made it possible to eliminate burrs when drilling these holes.

References

1. Landeta, J.F., Valdivielso, A.F., López de Lacalle, L.N., Franck Girot, J.M., Pérez, Pérez.: Wear of form taps in threading of steel cold forged parts. J. Manuf. Sci. Eng. **137**(3), 031002 (2015). https://doi.org/10.1115/1.4029652
2. Polvorosa, R., de Lacalle, L.L., Egea, A.J.S., et al.: Cutting edge control by monitoring the tapping torque of new and resharpened tapping tools in Inconel 718. Int. J. Adv. Manuf. Technol. **106**, 3799–3808 (2020). https://doi.org/10.1007/s00170-019-04914-5
3. Monka, P., Monkova, K., Modrak, V., Hric, S., Pastucha, P.: Study of a tap failure at the internal threads machining. Eng. Fail Anal. **100**, 25–36 (2019). https://doi.org/10.1016/j.engfailanal.2019.02.035
4. Gil Del Val, A., Veiga, F., Suárez, A., Arizmendi, M.: Thread quality control in high-speed tapping cycles. J. Manuf. Mater. Process. **4**, 9 (2020). https://doi.org/10.3390/jmmp4010009
5. Gil Del Val, A., Veiga, F., Penalva, M., Arizmendi, M.: Oversizing thread diagnosis in tapping operation. Metals **11**(4), 537 (2021). https://doi.org/10.3390/met11040537
6. Larshin, V.P., Gushchin, A.M.: Mechatronic technological system information support. Appl. Aspects Inform. Technol. **4**(2), 153–167 (2021). https://doi.org/10.15276/aait.02.2021.3
7. Gushchin, A., Larshin, V., Marchuk, V.: CNC machine adaptive control mechatronic module. In: Radionov, Andrey A., Gasiyarov, Vadim R. (eds.) Proceedings of the 8th International Conference on Industrial Engineering: ICIE 2022, pp. 154–163. Springer International Publishing, Cham (2023). https://doi.org/10.1007/978-3-031-14125-6_16

8. Dzhemilov, E., Bekirov, E., Uysal, A., Dzhemalyadinov, R.: An impact of the cutting fluid supply on contact processes during drilling. In: Ivanov, V., Trojanowska, J., Pavlenko, I., Rauch, E., Piteľ, J. (eds.) Advances in Design, Simulation and Manufacturing VI. DSMIE 2023. Lecture Notes in Mechanical Engineering, pp. 195–204. Springer Nature Switzerland, Cham (2023). https://doi.org/10.1007/978-3-031-32767-4_19
9. Sevidova, E., Rudnev, A., Gasanov, M., Kotliar, A., Titarenko, O.: An impact of solid lubrication on the diamond grinding characteristics of difficult-to-machine materials. In: Ivanov, V., Trojanowska, J., Pavlenko, I., Rauch, E., Piteľ, J. (eds.) Advances in Design, Simulation and Manufacturing VI. DSMIE 2023. Lecture Notes in Mechanical Engineering, pp. 337–346. Springer, Cham (2023). https://doi.org/10.1007/978-3-031-32767-4_32
10. De Silva, C.W.: Mechatronics: a Foundation Course. CRC Press/Taylor & Francis, Boca Raton (2010)
11. Billingsle, J.: Essential of mechatronics. John Wiley & Sons, Inc., Hoboken, New Jersey (2006)
12. de Silva, C.W. (ed.): Mechatronic Systems: Devices, Design, Control, Operation and Monitoring. CRC Press (2007). https://doi.org/10.1201/9780849307768
13. Dobrinski, A., Möhring, H.-C., Stehle, T.: Development of an adaptronic spindle for a faultless machining of homogeneous and inhomogeneous materials. J. Mach. Eng. **21**(2), 5–23 (2021). https://doi.org/10.36897/jme/136277
14. Steininger, A., Siller, A., Bleicher, F.: Investigations regarding process stability aspects in thread tapping AL-SI alloys. Procedia Eng. **100**, 1124–1132 (2015). https://doi.org/10.1016/j.proeng.2015.01.475
15. Dobrinski, A., Dudarev, A.: Intelligent automated drilling in the laminate composites and hybrid materials. Mater. Today: Proc. **38**(4), 1980–1983 (2021). https://doi.org/10.1016/j.matpr.2020.09.723
16. Shimana, K., et al.: Surface integrity of machined surface in simultaneous cutting of CFRP. J. Mach. Eng. **20**(1), 98–106 (2020). https://doi.org/10.36897/jme/117781
17. Sultana, I., Shi, Z., Attia, M.H., Thomson, V.: Surface integrity of holes machined by orbital drilling of composites with single layer diamond tools. Procedia CIRP **45**, 23–26 (2016)
18. Fleischer, J., Teti, R., Lanza, G., Mativenga, P., Möhring, H.-C., Caggiano, A.: Composite materials parts manufacturing. CIRP Annals **67**(2), 603–626 (2018). https://doi.org/10.1016/j.cirp.2018.05.005
19. de Silva, C.W.: Sensors and Actuators: Engineering System Instrumentation, 2nd edn. CRC Press, Boca Raton (2015). https://doi.org/10.1201/b18739
20. Tonkonogyi, V., Yakimov, A., Bovnegra, L., Sidelnykova, T., Dašić, P.: The use of intermittent wheels, impregnated by the contact method to reduce the thermal stress of the grinding process. IOP Conf. Ser.: Mater. Sci. Eng. **708**(1), 012034 (2019). https://doi.org/10.1088/1757-899X/708/1/012034

Influence of the Shape of Bevel Gear Wheel Bodies on Their Deformability

Viktor Ivanov[1(✉)], Lubomir Dimitrov[2], Svitlana Ivanova[3], and Mariia Volkova[4]

[1] Odesa Polytechnic National University, 1, Shevchenko Avenue, Odessa 65044, Ukraine
ivv@op.edu.ua

[2] Technical University of Sofia, 8, Kliment Ohridski Blvd., Sofia 1000, Bulgaria

[3] South Ukrainian National Pedagogical University Named After K. D. Ushynsky, 26, Staroportofrankivs'ka Street, Odessa 65020, Ukraine

[4] State University of Intellectual Technologies and Communications, 1, Kuznechnaya Street, Odessa 65029, Ukraine

Abstract. The load distribution in bevel gears depends on many factors, one of which is the deformation of the gear wheel body, which most researchers have neglected. The effect of rim deformation is not reflected in the ISO and AGMA standards. Many CAD Software packages include a finite element analysis module. Building a 3-D model of a gear bevel wheel is a time-consuming task. Replacing the ring gear with a smooth disk makes it possible to analyze the deformation of the rim as one of the stages of gear design. Calculations were carried out using the finite element method to check the adequacy of the simplified model. The calculation data were compared with experiments on disks that repeated the shape of the bevel wheel body. Also, full-scale tests of models of spur and spiral bevel gears made of Plexiglas were carried out. The applicability of the simplified model for most gear designs was proven. Particular attention was paid to wheels with a conical shape body of the wheel. The following body shapes of the wheel were considered: flat, conical surface (with the vertex located on the same side as the pith cone vertex), and conical surface (with the vertex located on the side opposite from the pith cone vertex).

Keywords: Bevel Wheel Bodies · Plexiglas Gear Models · Rim Bending Deformation · Product Innovation

1 Introduction

In bevel gears, particularly spiral bevel gears, the focus is on the influence of tooth shape on contact and bending stresses. Other factors affecting contact and bending stresses remain without proper attention. The effect of mounting errors, bearing, and shaft deformation on the mating tooth surfaces has not been sufficiently studied. Very few publications are devoted to the gear wheel body deformation. With large gear ratios, the gear wheel body has the shape of a thin disk and is subject to deformation due to

V. Tonkonogyi et al. (Eds.): InterPartner 2023, LNME, pp. 26–37, 2024.
https://doi.org/10.1007/978-3-031-42778-7_3

factors of various natures [1]. These factors include centrifugal forces, wheel heating, and axial force acting on the wheel. Centrifugal forces lead to an increase in the diameter of the wheel [2]. These displacements are insignificant and can affect the position of the teeth in contact at high speeds $n > 3000$ rpm [3]. Moreover, as with the action of centrifugal forces, the displacement of the outer end of the teeth will be greater than the displacement of the inner end. When the width of the gear rim is greater than the width of the body wheel, the effect of uneven heating can be significant. The areas of the inner surface of the gear rim, washed by the air-oil mist, have a lower heating temperature than the areas in contact with the body wheel [4]. In cylindrical gear wheels, this deformation results in a teeth barrel shape, as with barrel contact. In bevel gears, the character of the deformation is more complex. Of the three factors mentioned, the deformation of the body wheel due to the action of the axial force has the most significant effect on the stresses in the contact. At high gear ratios, the deformation of the body wheel is comparable to the contact errors due to the deformation of the bearing unit and much more critical than the deformation of the shaft, which is given incomparably more attention. One interesting feature of the deformation of the bevel body wheel must be noted. All three factors – centrifugal forces, thermal expansion, and body wheel bending – lead to a displacement of the contact spot towards the inner tooth end. Typical for a bevel gear is the problem of shifting the contact spot to the outer end due to mounting errors, deformation of the support units, and shaft bending [5]. Thus, body wheel deformation is not always a negative factor and, in some cases, can compensate for the influence of traditionally considered factors.

2 Literature Review

Thermal phenomena in the contact of the spiral teeth of a bevel gear are comprehensively investigated. The temperature of the tooth surface was determined from the combined consideration of thermo-elasto-hydrodynamic phenomena, taking into account shear load, characteristics of the mixed lubricant, and heat generation in the contact zone. Based on the analysis of contact stresses and the thermo-hydrodynamic lubrication process, the change in temperatures during teeth reconjugation has been studied. The reliability of the obtained results was experimentally confirmed [6].

The work [7] is also devoted to thermal phenomena in a spiral bevel gear. The contact temperature and the teeth' temperature distribution were determined [8]. The shape of the spiroid body wheel is close to the bevel body wheel shape at large gear ratios. The heating temperature may exceed the permissible values at high rotation speed and load. In this regard, a method for measuring the body temperature of a spiroid wheel is proposed instead of measuring the oil temperature [8].

The ISO 10300 standard for calculating spur bevel gears considers mounting errors and bearing unit deformation. Factors leading to the body wheel's deformation are not considered [9].

When calculating the parameters of the tooth modifications contact of the bevel gear teeth, the random nature of mounting errors is considered first considered [10]. This problem has also been solved for such rare bevel gears as skew gears. A skew bevel gear is more sensitive to mounting errors than a spur gear and even more so with spiral teeth. Therefore, using a crowning contact face width profile for such gears is relevant [11].

Studies of the effect of deformation of the gear body on the tooth contact are more widely presented for cylindrical gears. Wheels of large diameter are made with a thin wheel body or spokes for mass reduction [12]. This is especially true for ship transmissions. The data obtained for cylindrical gears can be used for large bevel gears with small gear ratios [13].

FEM is widely used to assess the deformability of transmission parts [14]. With the help of FEM, the effect of mounting errors and deformation of bearing units on the position of the contact spot in the gearing is comprehensively considered. Most often, these two factors are considered. Along with these factors, the pinion shaft's bending deformation is considered [15]. Finite element modeling and analysis of a straight bevel gear tooth that is used in pericyclic mechanical transmission was performed in the article [16]. The distribution of stress in the root of the tooth is found considering the parameters of gearing, shaft misalignment, and rim parameters. The obtained results are confirmed experimentally [16].

For spiral bevel gears, a comprehensive analysis was performed, considering bending and shear deflections of gear teeth, local contact deformations of mating surfaces, manufacturing and installation errors, and deflections of support shafts. Moreover, the bending and torsion of the gear pinion are considered in [17]. It should be noted that a detailed analysis of bending stresses requires consideration of the rim area adjacent to the tooth roots. In this case, the design scheme for bevel gears is chosen by analogy with cylindrical gears.

The most relevant is the study of body wheel deformation for bevel gears used in aviation. For such gears, the task is always to reduce the mass, making them thin-walled. Deformation analysis taking into account contact and bending deformation of the teeth, bending and torsion of the shaft, and deformation of the body wheel, are performed [18]. Moreover, the part of the body wheel adjacent to the toothed rim, the spokes, and the hub are considered [18]. The analysis of the deformation of the rim was carried out using the FEM.

Along with the influence of body wheel deformation on tooth damage, it is necessary to consider the possibility of damage to the body wheel itself. In a number of cases, at significant speeds, cases of body wheel destruction were recorded. This is especially true for welded body wheels with spokes [19].

In bevel gears with gear ratios $u > 5$, the wheel rim's deformation under axial force is one of the main factors that should be considered. It should be borne in mind that bending the body wheel can reduce the load factor by compensating for mounting errors and deformation of the bearing unit. The analysis carried out indicates the importance of taking this factor into account.

The body of a gear wheel can be a conical shape, in which case centrifugal deformations lead to an increase in the diameter and a change in the wheel's shape. The deformation of such cylindrical helical gears under centrifugal and axial forces is considered [20]. It is taken into account that the deformation of the wheel depends on the direction of the axial force relative to the generatrix of the wheel body.

The impact of centrifugal and bending deformations of the body of the spiral bevel gear wheel on dynamic performance was studied [21]. The angle of inclination of the bevel gear generatrix plays the main role. It is indicated that the improved method of finite elements made it possible to reduce the calculation time of one transmission to eight hours.

Summing up the review of literary sources, we note that the thickness of the gear body is considered only from the point of view of the influence of deformation on the gear's meshing performance. However, insufficient thickness can lead to the destruction of the gear wheel body. The interdependence of the location of the generatrix of the wheel body and the axial force on the amount of deformation for bevel wheels is not considered. The complexity of constructing a calculation model that reflects the geometry of the gearing and the calculation time for one gear makes it impossible for design engineers to consider the wheel body deformation.

3 Research Methodology

There is a wide variety of bevel wheel body shapes. This is because the shape of the bevel wheel body is highly dependent on the gear ratio. With gear ratios of 1.5… 2.0, the bevel wheel has the shape of a cone. With gear ratios of 3.15… 6.3, the bevel wheel is a disk with a ring gear at the end of the wheel. With large gear ratios, a large axial force acts on the wheel rim, which bends the disc-shaped wheel body. The study of bevel wheels' body stiffness follows similar studies for cylindrical wheels. That is, the location of the ring gear concerning the wheel body is considered: in the middle or at the outer end. This takes into account the thickness of the ring gear and the thickness of the wheel body. For bevel wheels, fundamentally different body shapes than cylindrical gear wheels are used. The body wheel may have a conical shape. The vertex of the conical surface of the body wheel can be located on the side of the inner end, the vertex of the gear rim's pith cone, and the outer end.

This shape of the body wheel is often due to the layout requirements of the bevel gear unit concerning other mechanism units. However, sometimes, such a shape is chosen intentionally, taking into account the effect of deformation on the load distribution in the gearing.

The bending deformation of the body wheel leads to a shift of the resultant load distribution to the inner end of the gear rim. To the greatest extent, this phenomenon is manifested at large gear ratios. The displacement of the resultant load distribution to the inner end of the ring gear may not be acceptable and, in some cases, maybe a positive factor. The deformation will depend not only on the thickness of the body wheel and gear rim but also on the shape of the body wheel. By changing the shape of the body wheel, without increasing the mass of the gear wheel, we can significantly increase its rigidity.

The finite element method has become a tool not only for scientific but also for engineering analysis. It is included in many engineering design packages: SolidWorks, Inventor, and Mechanical Desktop. The study of a particular body wheel design can be performed at the preliminary design stage. Building a solid body wheel model is a simple task. Moreover, building a solid model of a bevel gear, especially with spiral teeth that reproduce the geometry of the tooth surface according to the technologies of Gleason, Oerlikon, and Klingelnberg, is not at all an easy task.

In the first stage of this study, we will investigate how the shape of the gear wheel body affects the amount of deformation. In the second stage, we will check the admissibility of such a simplified engineering approach to solving this complex problem.

The calculation technique was initially worked out for the case with a flat rim. A disk of a complex shape replaces the bevel wheel. An assumption is made that the rigidity of the gear rim can be modeled by increasing the thickness of the body wheel. Namely, it is assumed that the surface of the bodywheel's end face coincides with the gear rim's pith cone. The 3D solid model was created using the Mechanical Desktop PowerPack. This package allows solving plane and spatial problems by the finite element method.

One more assumption is made, we believe that this body wheel is loaded only by axial force. The axial force in gearing is perceived by the end face of the bevel gear rim interacting with the shoulder on the shaft. We select as a support a number of pivotally movable supports placed along the circumference (Fig. 1). In the radial direction, the shaft takes the load. As the experience of calculations and further experimental studies has shown, the absence of circumferential force and bearings that perceive torque does not reduce the accuracy of calculations and, in some cases, allows getting a more accurate result. That is, the selected type and arrangement of supports in the form of two rows of hinged, movable supports in the hub's left and right end sections, to the greatest extent, meet the conditions of the problem being solved. The point of application of the concentrated axial force on the generatrix of the conical surface replacing the gear rim in the middle of the tooth length is chosen.

Initially, steel S235J2G4 was specified as the wheel material (Poisson's ratio – $\nu =$ 0.3, modulus of elasticity – $E = 210000$ N/mm^2, allowable stresses – $[\sigma] = 235$ N/mm^2).

We use automatic splitting of the part into finite elements. The splitting step was reduced in the most essential part of the study. The result of the calculation is presented in the form of a deformed finite element frame (Fig. 1).

The maximum stress from the wheel tea bending is 72 MPa, which is much less than the allowable ones. Thus, the strength of the wheel body is ensured. The thickness of the wheel body is reduced to 15 mm. This is the minimum value for strength (Fig. 2).

A comparative analysis of wheel bodies of various shapes has been carried out. Solid models were developed for schemes of wheel bodies a, b, and c, and deformations were found. As expected, scheme *b* has the most significant rigidity of 57 μm, scheme *c* has the least rigidity of 90 μm, and scheme *a* – has the maximum deformation of 66.5 μm.

The calculation model contained a number of assumptions. Approvals are made not only due to replacing the gear wheel with a smooth rim but also when considering the stiffness of the rim. Assumptions are related to the choice of the method of fastening, the method of application, and the type of load. We also used automatic partitioning and a rather significant step. Plexiglas rim models were made to check the adequacy of the data obtained for a smooth rim.

The mechanical characteristics of Plexiglas grade SO-120-A are as follows: modulus of elasticity $E = 3000$ MPa, allowable tensile stress 77.5 MPa. The deformation of Plexiglas models is much greater than that of steel elements. This reduces the deformation measurement error many times over.

The results obtained for Plexiglas models can be applied to metal wheels using similarity theory [22].

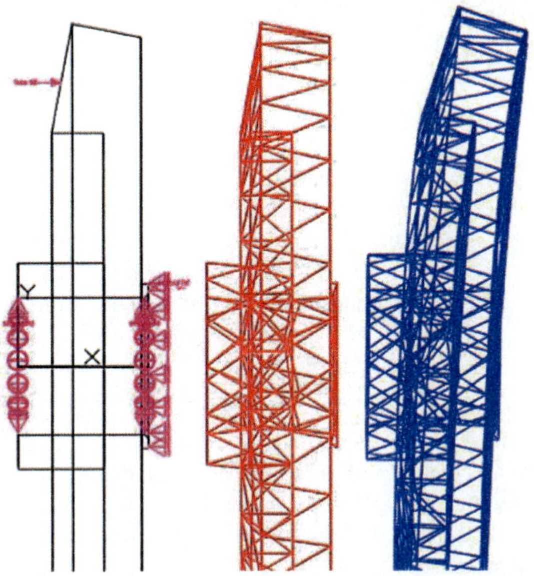

Fig. 1. Calculation model and initial and deformed frames of the wheel body.

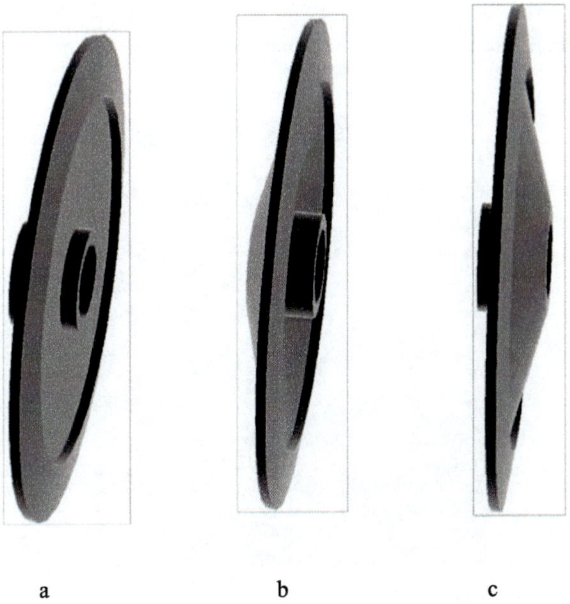

Fig. 2. Various shapes of wheel bodies.

Displacements on the model Δl_M и and steel wheel Δl_0, are related by the relationship:

$$\Delta l_0 = \frac{E_M \beta}{E_0 \alpha} \Delta l_M,$$

where E_0 и E_M – are the elasticity moduli of the object and the model, respectively, $\beta = F_0/F_M$ is the ratio of the force F_0, acting on the tooth of the object to the force F_M acting on the model; α is the ratio of the linear dimensions of the steel rim l_0 and model l_M.

It is also necessary that the dimensionless similarity conditions be satisfied [23].

$$\frac{\delta_M}{[\delta_M]} = \frac{\delta_0}{[\delta_0]} \text{ and } \frac{\delta_0}{\delta_M} = \frac{\beta}{\alpha},$$

where δ_M, $[\delta_M]$, δ_0, $[\delta_O]$ – stresses in the model and object, and allowable stresses for Plexiglas (105 MPa) and metal (320 MPa), respectively; β and α are the scales of linear displacements of steel and Plexiglas rims, respectively. The ratio of the elastic moduli of steel and Plexiglas $E_M/E_0 = 100$. The force on the tooth of the model was calculated from the condition $\beta = 18°$.

4 Results and Discussion

The end of the Plexiglas model's rim hub was located on the plate (Fig. 3).

Fig. 3. Model of a Plexiglas conical disk.

The model was fastened to the plate with a bolt, the axis of which coincided with the axis of the model. The bolt head was placed in the groove of the plate. A two-arm lever was placed above the model. A puncheon was attached to one end of the lever, and a dynamometer DOSM – 0.2 was attached to the other. The force was created with the help of a jack. The displacement was measured with a micrometer at a point lying on the maximum diameter of the same generatrix on which the force application point lies.

The results of the experiments showed good agreement between the deformation values for a flat rim of 6% and a rim with a low stiffness of 5%. The most significant error was obtained for a rim with a higher rigidity of 8%.

In addition to Plexiglas rims, models of Plexiglas bevel wheels with straight and spiral teeth were used (Fig. 4).

a b

Fig. 4. Gear wheel model: a) with straight teeth with parameters: $m_e = 10$, $z = 30$, $R_e = 172.41$ mm, $b = 50$ mm; b) with spiral equally high teeth with parameters: $m_{te} = 10$, $z = 45$, $\beta_m = 40°$, $R_e = 250{,}45$ mm, $b = 50$ mm.

A stand was used to test the gears (Fig. 5). All parts of the stand were made of steel, so their deformation could be neglected [24]. The wheel was fixed from rotation. The gear shaft rotated with the help of a lever on the end of which the jack pressed. The force was measured with a dynamometer installed between the end of the lever and the jack. Thus, the loading conditions of the body wheel corresponded to the real ones. The displacements were measured along three generatrices coinciding with the axes of three adjacent teeth for spur gears and passing through the middle of the tooth for spiral gears. For spur gears, a tooth engaging in the pole was chosen as the middle tooth. Thus, two neighbors did not participate in the transfer of the load.

The overlap ratio was more than two in gears with spiral equally high teeth. This type of bevel gear also has line contact. The middle tooth was in contact along its entire length. Neighboring teeth engaging and disengaging had the same contact line length. Measurements were taken with a micrometer at three points on each of the generatrices: on the inner, outer, and middle of the tooth (Table 1).

The tests were carried out for four wheels, two spurs, and two with spiral teeth. The wheels differed in body thickness. The ratio of the body thickness to the outer end diameter ψ_{bd} for spur wheels was 0.14 and 0.18 and for wheels with spiral teeth – 0.12 and 0.15.

Fig. 5. Stand with mounted models of gear pinion and wheel.

Table 1. The discrepancy between calculated values and experimental data in percent.

Gear type Measu-rement point	Spur bevel wheel $\psi_{bd} = 0{,}14$			Spur bevel wheel $\psi_{bd} = 0{,}18$			Spiral bevel wheel $\psi_{bd} = 0{,}15$			Spiral bevel wheel $\psi_{bd} = 0{,}12$		
№	1	2	3	1	2	3	1	2	3	1	2	3
Inner end	13	11	12	10	9	9	12	8	14	22	20	20
Middle section	11	9	10	8	9	8	11	7	12	17	16	17
Outer end	10	9	9	6	7	8	11	8	10	14	13	16

The results of the experiments showed that the analysis of deformations with a simplified gears model can be used in engineering practice. The error in the vast majority of cases is within 15%. A significant error was obtained at the inner end. A discrepancy between the results of the calculation and experiments was obtained only for a wheel with spiral teeth with a smaller body wheel thickness. The error is more than 20%. In this case, the results cannot be used to determine the load distribution in the gearing.

Nevertheless, they allow the engineer to indicate that this factor is significant and requires attention and further research. In our opinion, the error in the latter case is since the bending of the teeth in gearing leads to the deformation of the adjacent section of the body. The body wheel bends not only along the generatrix but also in the section perpendicular to the generatrix passing through the middle section of central tooth in contact.

5 Conclusions

The analysis showed that for bevel gears, it is essential to consider the body wheel's deformation. With gear ratios greater than $u \geq 5$ and small values of the coefficient ψ_{bd}, the body wheel deformation must be considered. The values of the body wheel deformation are comparable to the deformation of the support unit. In bevel gears, body wheel deformation leads to a displacement of the contact spot towards the inner end of the tooth. Thus, the deformation of the body wheel can compensate for mounting errors and deformation of the bearing unit. This phenomenon can be used to reduce the load factor.

The proposed simplified method for determining the deformation of the body of bevel gears has shown its adequacy and applicability. Replacing the gear with a smooth disk allows the engineer to calculate the body wheel deformation using the FEM module included in the CAD package when designing the transmission. In the calculation scheme, the following assumptions can be used: only a radial force is applied, the body wheel's end surface corresponds to the gear rim's pith cone, and pivotally movable supports are placed along the outer diameter of the hub parallel to the shaft axis.

The adequacy of the simplified model in the form of a smooth disk was checked on Plexiglas disk models. It has been established that a body wheel with a vertex of the conical surface located on the opposite side of the element, relative to the point of application of the axial force, has greater rigidity than a flat body wheel. A body wheel with a vertex of the conical surface located on one side of the element, relative to the point of application of the axial force, has greater deformability than a flat body wheel.

Acknowledgment. This work has been accomplished with financial support from Grant No. BG05M2OP001-1.002-0011 "MIRACle (Mechatronics, Innovation, Robotics, Automation, Clean technologies)".

References

1. Li, Z., Wang, H., Zhu, R., Ye, W.: Solutions of active vibration suppression associated with web structures on face gear drives. J. Vibroeng. **14**, 146–150 (2017). https://doi.org/10.21595/vp.2017.19009
2. Guilbert, B., Velex, P., Cutuli, P.: Quasi-static and dynamic analyses of thin-webbed high-speed gears: centrifugal effect influence. Proc. Inst. Mech. Eng., Part C: J. Mech. Eng. Sci. **233**(21–22), 7282–7291 (2019). https://doi.org/10.1177/0954406219855411
3. Hou, L., Lei, Y., Fu, Y., Hu, J.: Effects of lightweight gear blank on noise, vibration and harshness for electric drive system in electric vehicles. Proc. Inst. Mech. Eng., Part K: J. Multi-Body Dyn. **234**(3), 447–464 (2020). https://doi.org/10.1177/1464419320915006
4. Gan, L., Xiao, K., Wang, J., Pu, W., Cao, W.: A numerical method to investigate the temperature behavior of spiral bevel gears under mixed lubrication condition. Appl. Therm. Eng. **147**, 866–875 (2019). https://doi.org/10.1016/j.applthermaleng.2018.10.125
5. Han, X., Hua, L., Deng, S., Luo, Q.: Influence of alignment errors on contact pressure during straight bevel gear meshing process. Chin. J. Mech. Eng. **28**(6), 1089–1099 (2015). https://doi.org/10.3901/CJME.2015.0413.041

6. Zhang, J.G., Liu, S.J., Fang, T.: Determination of surface temperature rise with the coupled thermo-elasto-hydrodynamic analysis of spiral bevel gears. Appl. Therm. Eng. **124**, 494–503 (2017). https://doi.org/10.1016/j.applthermaleng.2017.06.015
7. Klingelnberg, J.: Load capacity and efficiency. In: Klingelnberg, J. (ed.) Bevel Gear: Fundamentals and Applications, pp. 101–195. Springer Berlin Heidelberg, Berlin, Heidelberg (2016). https://doi.org/10.1007/978-3-662-43893-0_4
8. Chodoła, Ł, Mazurkow, A., Surowaniec, M., Markowski, T., Homik, W.: Measurement method of temperature of the face gear rim of a spiroid gear. Sensors **22**(22), 8860 (2022). https://doi.org/10.3390/s22228860
9. Osakue, E.E., Anetor, L.: Comparing contact stress estimates of some straight bevel gears with ISO 10300 standards. In: ASME International Mechanical Engineering Congress and Exposition (2018), vol. 52187, p. V013T05A025. American Society of Mechanical Engineers (2018). https://doi.org/10.1115/IMECE2018-86572
10. Simon, V.V.: Influence of tooth modifications on tooth contact in face-hobbed spiral bevel gears. Mech. Mach. Theory **46**(12), 1980–1998 (2011). https://doi.org/10.1016/j.mechmachtheory.2011.05.002
11. Fuentes, A., Iserte, J.L., Gonzalez-Perez, I., Sanchez-Marin, F.: Computerized design of advanced straight and skew bevel gears produced by precision forging. Comput. Methods Appl. Mech. Eng. **200**(29–32), 2363–2377 (2011). https://doi.org/10.1016/j.cma.2011.04.006
12. Medvecká-Beňová, S.: Analysis of gear wheel body influence on gearing stiffness. Acta Mechanica Slov. **21**(3), 34–39 (2017)
13. Maláková, S., Puškár, M., Frankovský, P., Sivák, S., Harachová, D.: Influence of the shape of gear wheel bodies in marine engines on the gearing deformation and meshing stiffness. J. Marine Sci. Eng. **9**(10), 1060 (2021). https://doi.org/10.3390/jmse9101060
14. Mieth, F., Ulrich, C., Schlecht, B.: Stress calculation on bevel gears with FEM influence vectors. Forsch Ingenieurwes **86**, 491–501 (2022). https://doi.org/10.1007/s10010-021-00550-2
15. Peng, S., Ding, H., Zhang, G., Tang, J., Tang, Y.: New determination to loaded transmission error of the spiral bevel gear considering multiple elastic deformation evaluations under different bearing supports. Mech. Mach. Theory **137**, 37–52 (2019). https://doi.org/10.1016/j.mechmachtheory.2019.03.013
16. Mathur, T.D., Smith, E.C., DeSmidt, H., Bill, R.C.: Load distribution and mesh stiffness analysis of an internal-external bevel gear pair in a pericyclic drive. In: 72nd American Helicopter Society International Annual Forum 2016: Leveraging Emerging Technologies for Future Capabilities, vol. 1, p. 2646–2657. American Helicopter Society (2016)
17. Simon, V.V.: Load distribution in spiral bevel gears. ASME. J. Mech. Des. **129**(2), 201–209 (2006). https://doi.org/10.1115/1.2406090
18. Lu, S., Ding, H., Rong, K., Rong, S., Tang, J., Xing, B.: Composite mechanical deformation based semi-analytical prediction model for dynamic loaded contact pressure of thin-walled aerospace spiral bevel gears. Thin-Walled Struct. **171**, 108794 (2022). https://doi.org/10.1016/j.tws.2021.108794
19. Ivanov, V., Dimitrov, L., Ivanova, S., Volkova, M.: Reverse engineering in the remanufacturing: metrology, project management, redesign. In: Karabegović, I. (ed.) NT 2021. LNNS, vol. 233, pp. 169–176. Springer, Cham (2021). https://doi.org/10.1007/978-3-030-75275-0_20
20. Li, S.: Effects of centrifugal load on tooth contact stresses and bending stresses of thin-rimmed spur gears with inclined webs. Mech. Mach. Theory **59**, 34–47 (2013). https://doi.org/10.1016/j.mechmachtheory.2012.08.011
21. Hou, X., Qiu, L., Zhang, Y., Li, Z., Zhu, R., Lyu, S.K.: High-speed spiral bevel gear dynamic rules considering the impact of web thicknesses and angles. Appl. Sci. **12**(6), 3084 (2022). https://doi.org/10.3390/app12063084

22. Nam, J.S., Park, Y.J., Kim, J.K., Han, J.W., Nam, Y.Y., Lee, G.H.: Application of similarity theory to load capacity of gearboxes. J. Mech. Sci. Technol. **28**, 3033–3040 (2014). https://doi.org/10.1007/s12206-014-0710-5
23. Wang, A.L., Zhu, C.: Research for scale effect based on similarity theory. Appl. Mech. Mater. **364**, 833–837 (2013). https://doi.org/10.4028/www.scientific.net/AMM.364.833
24. Ivanov, V., Ivanova, S., Urum, G., Purich, D.: An experimental study of the influence of mounting errors on the load distribution along the face width in a spur bevel gear. In: Karabegović, I. (ed.) NT 2021. LNNS, vol. 233, pp. 38–45. Springer, Cham (2021). https://doi.org/10.1007/978-3-030-75275-0_4

Design of an Operator Interface for Controlling the Installation of Ion-Plasma Deposition

Kateryna Kirkopulo[1(✉)], Volodymyr Tonkonogyi[1], Vladimir Litvinov[1], Alla Toropenko[1], and Predrag Dasic[2]

[1] Odessa Polytechnic National University, 1, Shevchenko Avenue, Odesa 65044, Ukraine
kateryna.grygoryvna@gmail.com

[2] High Technical Mechanical School of Professional Studies, 19, Radoja Krstića Street, 37240 Trstenik, Serbia

Abstract. The technology of ion-plasma coating is a continuous process that can be divided into three strictly consecutive stages. Compliance with the strict regulations of the technological process directly affects the quality of the finished product and the absence of defective products. Installation of ion-plasma spraying can operate almost automatically. However, some processes require an operator: loading and unloading a tool, selecting a recipe to work with, and setting up a plant. The automation system must implement a number of control algorithms that are simple from the point of view of software implementation on modern microprocessors. More significant is the possibility of modifications and adaptation of the developed software. In this case, it is necessary to have a computer and develop a human-machine interface to interact with the operator. Previously developed interfaces do not meet the requirements. This is explained by the fact that the task of developing an interface was considered separately from the task of developing an automatic control system. If automated control is implemented in such interfaces, then, as a rule, only with the help of algorithms for implementing sequential execution of operations according to specified conditions.

Keywords: Ion-Plasma Deposition Technology · Human-Machine Interface · R&D Investment

1 Introduction

Ion-plasma deposition technology is a world-renowned method that allows tools to acquire necessary qualities such as wear resistance, reliability, and strength. Additionally, ion-plasma deposition enables the production of highly sought-after tools and other objects, not only from a technical standpoint but also from a decorative perspective. For example, there is a production of watches whose cases are coated with a special composition obtained through ion-plasma coating. Kitchen sinks, whose surfaces are covered with a protective layer, have a longer service life and stand out favorably from competitors in terms of design and external attractiveness.

V. Tonkonogyi et al. (Eds.): InterPartner 2023, LNME, pp. 38–47, 2024.
https://doi.org/10.1007/978-3-031-42778-7_4

It can also be said that improving the quality of tools by applying thin coating layers by ion-plasma deposition significantly saves time. For example, increasing the service life will reduce the time for technological downtime during tool replacement.

This technology is used in mass production installations such as "Bulat" (Ukraine), MR-333 and MR-383 (USA), and others. However, many of these advantages can be lost. On the one hand, due to the lack of a highly organized and modern system for automated control of the coating application process. On the other hand, it should be noted that the operator interface also requires modification. This is due to the emergence of new standards and requirements for interfaces. Adapting the interface to new realities will also significantly reduce the number of defective products.

2 Literature Review

The interface must comply with the international standard ANSI/ISA-101.01-2015 requirements and support the high level of automation embedded in the developed plant control system.

Previously developed interfaces do not meet the requirements specified. This is because the interface development task was considered separately from developing an automatic control system. If automated control is implemented in such interfaces, it is usually achieved only through algorithms for sequentially executing operations under certain conditions.

The work [1] describes the developed automation system for the ANGA-1 installation for coating small parts. The automation system is implemented according to modern requirements, but the details of implementing the human-machine interface and the models used are not provided.

Of the existing interfaces designed for coating application processes, the most advanced is interface [2], which is oriented towards a touch screen and includes recipe setting and reporting systems.

At the same time, interfaces for the most in-demand installations [3] are overloaded with details, shadows, and colors, which makes it difficult for operators to perceive information.

The work [4] considers a computer automation system for the installation. The disadvantage of the system is the use of sensors with RS-232C and USB interfaces with converters between them. These interfaces only provide signal transmission over a short distance, and the presence of intermediate converters reduces the system's reliability.

Interface [5] does not meet modern computer ergonomics requirements: the screen form's video frame does not display the dynamics of the process, and the recipe system shown corresponds only to logical process control.

Thus, the task of developing an efficient interface for ion-plasma installations remains relevant at present.

3 Research Methodology

The work [6] contains the development of a complex of mathematical models of the dynamics of all stages of the technological process of ion-plasma deposition of coatings on a metal-cutting tool.

This complex of mathematical models was used to develop nonlinear control systems for all stages of the technological process [7].

A comprehensive approach is proposed to solve the task of developing an effective human-machine interface. Such an approach includes an analysis of the technological process of ion-plasma spraying installation operation, determination of the requirements and features of the technical implementation of the control system, and formulation of requirements for the human-machine interface for interaction between the operator and the ion-plasma spraying installation.

3.1 The Main Stages of the Operation of the Ion-Plasma Facility

The «Bulat» installation involves three stages of operation: ion cleaning, coating application, and cooling. Each of these stages requires the operation of a corresponding automatic control system. The process can be schematically represented in Fig. 1.

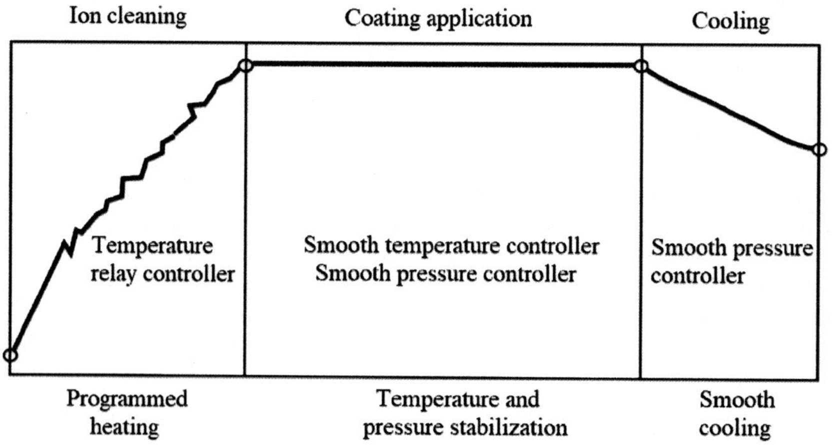

Fig. 1. Stages of operation of the "Bulat" setup and required control systems to ensure the quality of their implementation.

Let us look at the equipment's operation principle for implementing the vacuum arc deposition technology. The technology is based on using a vacuum arc discharge to create a stream of ionized atoms of the material that is evaporated. The cathode of the evaporator is supplied with a negative potential from the power source of the arc discharge. In this case, the anode is the body of the vacuum chamber. The material that evaporates in the cathode area is in a plasma state and ionizes the atoms that are evaporated from the surface of the cathode. The degree of ionization of the substance being evaporated reaches 70–80%. The ion flux can be controlled by changing the energy of the ions, which, in turn, is determined by the potential on the substrate.

The technology of ion-plasma coating deposition is a continuous process, which can be clearly divided into three strictly sequential stages, the quality of which directly affects the result – a finished product with a coating without defects.

The "Bulat" installation is equipped with three evaporators. The installation includes a vacuum-arc plasma source, a voltage source, a temperature sensor, and a reagent gas injector with fine adjustment of the amount of supplied gas.

The main problem of technical measurements in the technological process of ion-plasma sputtering is the problem of temperature measurement. If the choice of technical means of automation for measuring pressure, flow, and rotation speed is quite simple, then the problem of measuring the temperature of a tool that rotates in a vacuum has remained relevant for more than a decade. Temperature control is critically important because a violation of the temperature regime affects the adhesion between the coating and the tool, which leads to a critical decrease in the microhardness and stability of the tool.

Traditionally used pyrometers do not meet modern requirements on several indicators, such as measurement range, significant measurement error, and lack of a modern network interface.

In general, installation models are vulnerable from a management point of view. The main parameters of the technological process are controlled visually by the operator, so the efficiency of the process depends on the operator's qualifications and attention. Almost all existing plant models are manually controlled, and existing software control implementations are built without the ability to adapt the control system to a specific tool.

3.2 Requirements and Features of the Technical Implementation of the Automated Process Control System for the Facility of Ion-Plasma Spraying

The level of requirements and the features of technical implementation depends on many factors: application areas, industries, product quality requirements, the complexity of the technological process, and explosion safety.

The automated process control system for ion-plasma spraying corresponds to the following classification:

- the system belongs to the lower level of the hierarchy of possible management;
- management is carried out by a continuous-discrete technological process;
- of medium reliability, where the automated process control system stops in case of failure;
- of small information capacity, where the number of information channels does not exceed 40 units.

Based on the provided classification, it is possible to analyze the main components of the software and the technical structure of the automated process control system.

The level of automated computer control of the automation system includes the following software:

1. OPC server – used to represent the technical equipment of the automation system as a set of structural variables (tags) for ease of interaction with the software.
2. Management service – used for background technological process management using automatic controllers that perform the specified program.
3. Service Management – is used for background control of the technological process using automatic regulators that perform a specified program.

4. The database server stores a table of available recipes, the history of previous SCADA launches, and the system control activity history. A relational database server with support for standard software interaction interfaces is convenient for a database server.
5. The recipe editor allows editing and creating new recipes. The editor implements algorithmic support for recipe configuration.
6. SCADA (Supervisory Control And Data Acquisition) [8] – automatically implements a recipe under human control using a human-machine interface, process control tools, and post-completion process analysis. Third-generation SCADA [9], built using client-server architecture and oriented towards the OPC protocol, is necessary. The server and client sides implement algorithmic support for operation according to the specified schedule.
7. Simulation software is also required, which includes all the means for debugging control system elements and forming a recipe (processing schedule) using a simulation model of the technological process [7].
8. Connecting sensors and actuators to the control device can now be done with a separate wire or an industrial sensor network. An industrial network provides reliable connectivity and high data transmission accuracy between various devices. Using sensor networks in automated process control systems allows for increased flexibility in construction, simplifies integration and maintenance, and meets the recommendations of modern industrial standards.

Currently, there are no fewer than 50 sensor network protocols, such as Profibus [10], HART [11], and CAN [12]. However, most manufacturers' most common network embedded in sensors, actuators, and industrial computers worldwide is the RS 485 Modbus RTU [13] network. The disadvantage of such a network is its limitation to 32 devices in one network segment and relatively low data transmission speed. However, this is insignificant for the automation of ion-plasma spraying.

Sensors for position, weight, speed, temperature, pressure, and actuators for nitrogen supply are selected with an RS-485 interface, simplifying their integration. Control of the relay and substrate voltage is implemented using special regulators, which will also be connected to the sensor network.

3.3 Formulation of Requirements for a Human-Machine Interface for Interaction Between the Operator and the Facility

Installation of ion-plasma spraying can operate practically in automatic mode. However, the loading and unloading processes of the tool, recipe selection for work, and setup of the installation are performed by the operator.

The automation system should implement a series of straightforward control algorithms in terms of software implementation on modern microprocessors.

The question of the possibility of modifications and adaptation of the developed software is more significant. In this case, it is necessary to have a computer and develop a human-machine interface as a means of interaction with the operator.

Since the task is to control a single industrial unit with constant operator intervention and the possibility of detecting the need for modification of the control system software algorithms, it is appropriate to use an industrial computer with a touch screen and a touch network interface as a control device.

It is necessary to use a hard real-time operating system to ensure reliable computer control of the technological process.

Some systems (e.g., QNX, VxWorks [14]) have high technical characteristics, but automation software developers rarely develop software products for such systems. This is due to the relatively small number of users. The small number of existing software products for solving typical automation tasks has relatively low functionality and a low degree of debugging in various situations, leading to software failures due to the low quality of the algorithms used.

A preferred option for automation systems is a real-time embedded operating system, such as Windows Embedded. Although this system is less reactive, it allows using almost all software loads developed for a standard Windows operating system in the natural state or with some modification.

Modern computer-integrated automation systems use information parameter servers. The use of information parameter servers simplifies the interaction of standard software tools with sensors and mechanisms of the automation system.

Industrial automation systems' information parameter servers use OPC, OPC UA, and MQTT protocols [15]. The advantage of the latter two is their independence from the operating system, but OPC is a more standard and developed mechanism, so in our case, it is advisable to use OPC as the protocol for interacting with the information parameter server.

Many information parameter servers are associated with interfaces of programmable logic controller interfaces from different manufacturers. Among OPC servers that can be directly connected to the sensor network via the Modbus protocol [16], the most developed is the Modbus Universal MasterOPC server [17].

The OPC server supports a simple scripting language that allows for preliminary data processing after reading from external devices and before writing to them. Additionally, it is possible to simulate changes in information parameters for debugging automation system components using the scripting language.

The primary software element of computer automation systems is the SCADA.

Despite the variety of functions performed by SCADA, its main feature is the presence of a user interface.

The convenience of the workplace, interface clarity, presence of prompts, and blocking of operator errors are essential features of SCADA, and their further development is primarily focused on improving ergonomics.

SCADA uses the concepts of messages and events.

Examples of such messages could be the critical temperature of a tool during processing or reaching a critical pressure value in a chamber, and so on. Since notifications require decision-making, they are divided into confirmed and unconfirmed messages. When confirming a message, the system must record the confirmation time.

One of the main functions of SCADA is the development of a human-machine interface (HMI), which means that SCADA is simultaneously an HMI and a tool for creating it. The development speed is the main indicator of SCADA's property from the system integrator's point of view. The development process includes the following operations:

- creation of a graphical interface (schematics, graphs, tables, and elements for entering operator commands);
- programming and debugging of automation system operation algorithms;
- setting up the communication system with automation system devices;
- creating databases and connecting SCADA to them.

4 Results and Discussion

The presented technological procedure characterizes the process as strictly staged with rigid time regulations. It can be violated for the following three reasons: inaccurate regulations, incorrect controller settings, and equipment failure.

An effective computer interface should help easily identify the violation and accurately localize the source of the error.

A review of modern standards in the field of ergonomics of computer interfaces is presented in the work [18].

The following problems relating to common computer interfaces for ion-plasma devices can be highlighted [19]:

1. The main screen displays a diagram with a multitude of digital parameters that do not allow for assessing the dynamics and predicting the course of the process (since trends are displayed on another screen or not displayed at all).
2. A multitude of distracting factors is used, such as excessive color design, 3D shadows, and animation.
3. The organization of screens is oriented towards presenting information of different types and is not a result of analyzing typical operator actions and tasks they must constantly solve.
4. The interfaces' component base and digital alphabet are not sufficiently justified, which does not allow for effectively directing the operator's attention.

Let us summarize the rational requirements and methods of their implementation in Table 1.

A recipe editor was considered as an example of an interface implemented according to the specified requirements. This screen allows creating and editing the current recipe for a particular type of tool and batch of that tool. The goal of forming the recipe is to achieve specified quality indicators of the product, primarily its durability, microhardness, and coating thickness for metal cutting tools. Accordingly, the panel contains interface elements that can be used for editing, deleting, adding, and storing information related to the recipe. Additionally, the screen has fields that can be filled with data according to the coating requirements.

The operator uses the recipe parameter correction screen for prompt editing of recipe parameters before starting the process and for correction during operation. All operational recipe configurations are recorded.

Simple SCADA is used as the SCADA system. It is a software system that belongs to the third generation of SCADA. It has a client-server architecture and is oriented towards interacting with the server of information parameters. Access to the server is implemented using an industrial protocol, such as OPC.

Table 1. Main interface development requirements and implementation methods.

Requirement	Requirement Implementation method
Using visual representation of screens according to the recommendations of the ANSI/ISA 5.1 standard	1. A light gray background is used 2. A simplified schematic diagram of the installation is used with the use of dark gray lines 3. Yellow is assigned to pre-alarm signaling, while red is assigned to alarm signaling 4. Operators' actions should be confirmed
Achieving situational awareness	1. The main parameters are displayed in trends with dashed lines of dark green color indicating the permissible boundaries of the parameters, allowing the operator to forecast the process flow 2. The trend is displayed for a rational interval of the process, in this case, 10 min, rather than for the entire process, enabling the operator to assess the situation 3. The progress of the current stage of the process can be assessed using a bar indicator
Logically justified organization of screens with a focus on tasks that the operator must solve	1. Five screens are allocated: process, reports, recipe, diagnostics, and settings 2. All screens have three mandatory zones: screen switching menu, main zone, and message display zone 3. The "process" screen is intended for viewing the state of the process flow. Its content dynamically changes depending on the current stage of the technological process
Using an intuitive and understandable element base of the interface that corresponds to the color alphabet	1. The color of the simplified process diagram lines and the text of its elements is dark gray 2. Dashed black lines are used to display the connection between control circuit elements 3. Blue lines with white fill are used to display active interface elements (sensors, actuators, and progress slider) 4. The text color of active technological parameters is blue 5. The text color of parameters that require input is light green 6. The text color of clickable elements is purple

The SCADA system implements a client-server architecture, oriented towards interaction only via the OPC protocol, and is easily integrated with the MySQL and MS SQL Server DBMS. The developed interfaces scale well to any touchscreen expansion.

5 Conclusions

Analyzing regulations and up-to-date practices in developing efficient interfaces allowed for formulating requirements for an operator interface that meets modern ergonomics and situational awareness requirements.

The computer interface is implemented in a modern SCADA system. The developed interface improves the quality of processing tools and other objects by reducing the number of defective products caused by operator errors in the installation.

The results can be applied to developing new computer systems to automate ion-plasma installations.

References

1. Dyadyun, K.V., Chebukina, V.F.: The process of applying ion-plasma coatings and a systematic approach to process control. New Mater. Technol. Mach.-Build. **1**, 7–10 (2016). [in Ukrainian]
2. Smirnov, I.V., Dolgov, M.A., Ivanchenko, O.V., et al.: Multichannel system of automation of the robotic installation of vacuum-arc sawing. Eastern-Eur. J. Enterp. Technol. **61**(1/5), 60–64 (2013)
3. Mazurkiewicz, A., Smolik, J.: Development of novel nano-structure functional coatings with the use of the original hybrid device. Mater. Sci. Forum **674**, 1–9 (2011)
4. CatArc 2500 PVD Coating System. Brochure. Vergason Technology Inc, Van Etten, NY, USA (2018)
5. Marszałek, K., Małek, A., Winkowski, P., et al.: LabVIEW controller for storage results and control parameters of low thickness antireflection coatings deposition processes. Elektronika **2**, 31–34 (2016)
6. Kirkopulo, K., Tonkonogyi, V., Stopakevych, O., Stopakevych, A.: Design of a set of nonlinear control systems of the arc PVD ion plasma installation. Eastern-Eur. J. Enterp. Technol. **2**(292), 65–74 (2018). https://doi.org/10.15587/1729-4061.2018.127708
7. Kirkopulo, K., Tonkonogyi, V.: Automated control system of technological process of applying wear-resistant coating by ion-plasma method). J. Autom. Inf. Sci. **52**(10), 52–61 (2020). https://doi.org/10.1615/JAutomatInfScien.v52.i10.40
8. Blokdyk, G.: SCADA: a complete guide. Emereo Pty Limited, Brisbane, Australia (2018)
9. Radvanovsky, R., Brodksy, A.: Handbook of SCADA Control Systems. CRC Press, Boca Raton, FL, USA (2016)
10. Cuayo, L.D., Kerbee Culla, J., Gualvez, J., Padua, S.E., John Gallano, R.: Development of a wireless microcontroller-based Scada RTU. In: TENCON 2018-2018 IEEE Region 10 Conference, pp. 2566–2570. IEEE (2018). https://doi.org/10.1109/TENCON.2018.8650114
11. Pfeiffer, O., Ayre, A., Keydel, C.: Embedded networking with CAN and CANopen. RTC Books, San Clemente, CA (2008)
12. Xiaocong, F.: Real Time Embedded Systems: Design Principles and Engineering Practices. Elseiver, Walthman, MA, USA (2015)
13. Cho, C., Seong, Y., Won, Y.: Mandatory access control method for windows embedded OS security. Electronics **10**(20), 2478 (2021)
14. Zhang, Z., Jiansong Mo, L., Niu, S.: Vulnerabilities analysis and solution of VxWorks. In: 2nd International Conference on Teaching and Computational Science (ICTCS 2014), pp. 94–97 (2014)
15. Silveira Rocha, M., Serpa Sestito, G., Luis Dias, A., Celso Turcato, A., Brandão, D.: Performance comparison between OPC UA and MQTT for data exchange. In: Conference: 2018 Workshop on Metrology for Industry 4.0 and IoT, pp. 175–179. IEEE (2018). https://doi.org/10.1109/METROI4.2018.8428342
16. Herath, H.M.K.K.M.B., Ariyathunge, S.V.A.S.H., Priyankara, H.D.N.S.: Development of a data acquisition and monitoring system based on MODBUS RTU communication protocol. Int. J. Innov. Sci. Res. Technol. **5**(6), 433–440 (2020). https://doi.org/10.38124/IJISRT20JUN479

17. Ádámkó, E., Jakabóczki, G., Szemes, P.T.: Proposal of a secure modbus RTU communication with adi shamir's secret sharing method. Intl. J. Electron. Telecommun. **64**(2), 107–114 (2018). https://doi.org/10.24425/119357
18. Stopakevich, A.O.: Analysis of the shortcomings of the typical practice of developing computer human-machine interfaces in industrial automation systems. In: 72 Scientific and Technical Conference Odessa National Academy of Telecommunications by O.S. Popov, pp. 45–47. Odessa (2017). [in Ukrainian]
19. Rybak, O.V.: Mathematical modeling, analysis and optimization in CAD of the technological process of grinding plasma coatings. Control Navig. Commun. Syst. **4**(66), 33–37 (2021). https://doi.org/10.26906/SUNZ.2021.4.033. [in Ukrainian]

Parametric Synthesis of Electrohydraulic Control System for Variable Displacement Pump

Leonid Kozlov(✉), Viktor Bilichenko, Andrii Kashkanov, Artem Tovkach, and Vadym Kovalchuk

Vinnytsia National Technical University, 95, Khmelnytsky Highway, Vinnytsia 21021, Ukraine
osna3030@vntu.edu.ua

Abstract. The article presents a scheme of an electric-hydraulic control system for a pump. The control system enables the pump to operate in one of four modes: constant flow, constant pressure, constant power, flow, and pressure compensation. The choice of mode depends on the program implemented by the controller. Experimental studies of the controller and the electromagnetic amplifier have been carried out. The transfer functions of the electromagnet amplifier were determined. Work processes in the system were studied based on the developed mathematical model in the MATLAB-Simulink environment. A comprehensive criterion for evaluating the efficiency of the system has been developed. The criterion includes adjustment time, the amount of pressure overshoot, and power losses in the control system. At the determined value of the efficiency criterion, the pump has an adjustment time of $t_p = 0.44$ s, pressure overshoot $\sigma = 22\%$, and power losses in the control system do not exceed $N_p = 1.82$ kW.

Keywords: Hydraulic System · Proportional Control · Controller · Mathematical Model · Performance Indicator · R&D Investment · Energy Efficiency

1 Introduction

One of the main trends in the development of mobile equipment is the transition to electrohydraulic systems with LS-regulation. Such systems are built, as a rule, based on variable pumps, electrohydraulic proportional pressure valves, distributors, and programmable controllers. Regulating pumps in such hydraulic systems are equipped with regulators, providing the operation of pumps in decal modes. It enables to secure the necessary static and dynamic characteristics and high energy efficiency. Developing a complete control system for regulating pumps is an urgent problem today.

2 Literature Review

The paper [1] considers the classification of concepts for the construction of mobile equipment hydraulic systems based on different types of pumping facilities. The efficiency of hydraulic systems based on variable and fixed pumps was compared regarding

V. Tonkonogyi et al. (Eds.): InterPartner 2023, LNME, pp. 48–57, 2024.
https://doi.org/10.1007/978-3-031-42778-7_5

energy and dynamic characteristics. It was noted [2] that the most effective option is to use a variable pump in combination with a variable engine equipped with a controller, but such a system is the most expansive of all the revised versions. The paper notes that using LS-hydraulic systems for excavators of small and medium-sized groups provides improved energy characteristics and the ability to regulate motion parameters in wide ranges, reducing harmful emissions into the atmosphere. The article [3] discusses the hydraulic system of the excavator rotation. More accurate coordination of the pump delivery and the flow consumed by the swing hydraulic motor was achieved. It ensures the improvement of energy characteristics and the reduction of oscillation. The paper [4] is devoted to the study of electrohydraulic system with LS-regulation. The system provides good energy characteristics and a small error in tracking mode when working with oncoming and accompanying loads. The paper [5] considers the possibilities of improving the energy efficiency of the excavator hydraulic system using electrohydraulic control of a variable pump capacity and adjustable throttles that control the flows simultaneously at the inlet and outlet of the hydraulic cylinder. The paper [7] studies a hydraulic system with LS-regulation based on a variable pump and controller. In the applied PID controller, the selection of coefficients ensured the minimization of the integral accuracy assessment. The article [8] presents a hydraulic system with LS-regulation based on a variable pump, proportional directional valves, and a controller. The hydraulic system ensures effective coordination of the flows of the pump and hydraulic cylinders. The work [9] is devoted to improving the PID controller of the 3rd order for managing an electrohydraulic gear. Optimization of its parameters and comparison of the performance with four other controllers were carried out. The accuracy of the electrohydraulic gear has been increased.

A number of works are devoted to the development of new schemes of pump regulators and choosing their parameters that provide the necessary characteristics. The paper [10] presents the enhanced construction of a one-spool regulator for an axial-piston pump, which is controlled with load sensitivity. It is proved that the regulator provides the necessary sensitivity and speed compared to two-spool regulators for LS-controlled hydraulic systems produced by industry. The article [11] presents a mathematical model of an automatic two-spool regulator to supply an axial piston pump. Recommendations for selecting gaps and overlaps of the slide valves and the values of orifices holes have been elaborated. The paper [12] presents an LS-hydraulic system with an improved pump regulator. The regulator's modernization leads to reduced power losses and oscillation. The article [13] considers a hydraulic system with a variable pump and an electrohydraulic regulator. The comparison was made between using a PID controller and a FUZZY controller to control the hydraulic system. Currently, intensive work is underway to develop new schemes of electrohydraulic regulators for hydraulic systems with LS-regulation to increase accuracy and energy efficiency and improve static and dynamic characteristics.

Nonlinear mathematical models of electrohydraulic hydraulic systems and their components have been widely used for determining ways of increasing their efficiency. The paper [15] presents a model of the LS-hydraulic system. The mathematical model adequately reflects the working processes in the hydraulic system. The solution of the equations of the model needs a little time. It can be used to develop complexes that test control facilities and diagnostics. The article [16] is devoted to developing a mathematical model of an axial piston pump equipped with an automatic regulator, which ensures the operation of the pump with a load-sensitive hydraulic system. The model adequately reflects the working processes in the pump and allows to determine the dynamic characteristics. The work [17] contains the developed mathematical model of a hydraulic system with LS-regulation. The possibility of simplifying the mathematical model on the base of the adequacy index, which provides a balance between the results' accuracy and the model's complexity, is shown. The paper [18] presents the study of the PID controller for the electrohydraulic hydraulic system of the excavator. The suggested PID controller is automatically set to work when the external conditions change.

3 Research Methodology

Figure 1 contains the control circuit of the pump with variable working volume. The pump with such a control system operates in one of four possible modes: constant flow; constant pressure; constant power; pressure, and flow compensation. The program executed by the controller determines the mode selection. The control system operation in pressure and flow compensation mode is described below. The pump control system 1 contains a regulator 2 electromagnet 3. Regulator 2 consists of case 4, where there is a valve 5 with a spring 6, a servo valve 7, and two throttles 8 and 9. The servo valve 7 contacts with the valve seat 10. Two adjustable throttles 11 and 12, are installed in series at pump outlet 1. A pressure sensor 13 is installed between throttles 11 and 12. Controller 14 and amplifier 15 are connected to the sensor 13. The amplifier output 15 is connected to electromagnet 3. The working volume of pump 1 is changed by circuit plate 16, the position of which is determined by the servo cylinders 17 and 18. The servo cylinder 18 has a damper 19. A mathematical model was described in the paper to determine the characteristics of the given electrohydraulic control system and the choice of parameters of the pump regulator [19]. The high-order mathematical model includes nonlinear differential equations which highly accurately describe the working processes in the developed hydraulic system. The model is presented in the MATLAB-Simulink environment, using the Rosenbrok numerical method designed for solving systems of rigid differential equations. The absolute accuracy when solving the equations was 10^{-6}, the relative accuracy was 10^{-3}. The time to find the solutions to the equations does not exceed 10 min. In a mathematical model of this type, it is convenient to use experimentally determined dependencies that describe the operation of hydraulic system elements.

Figure 2 shows the installation's electrical circuit for studying the controller's characteristics and the electromagnet amplifier.

The installation comprises signal generator 1 of G3-112/1 type, controller of 2 Arduino UNO type, amplifier of 3 E-MI-AC type, electromagnet of 4 ZOR-A type, filter 5 and analog-to-digital converter L-CARD-140MD. 2 series of studies were carried out on the plant.

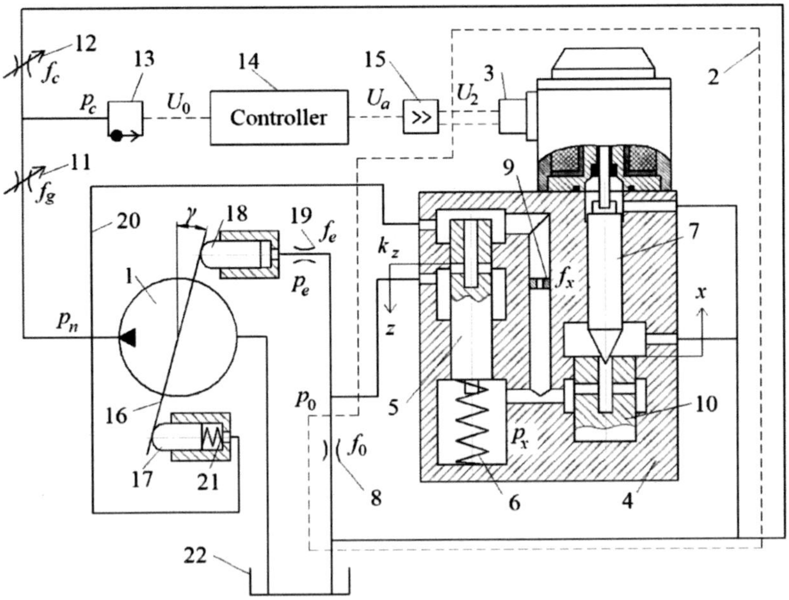

Fig. 1. Control system of a variable working volume pump.

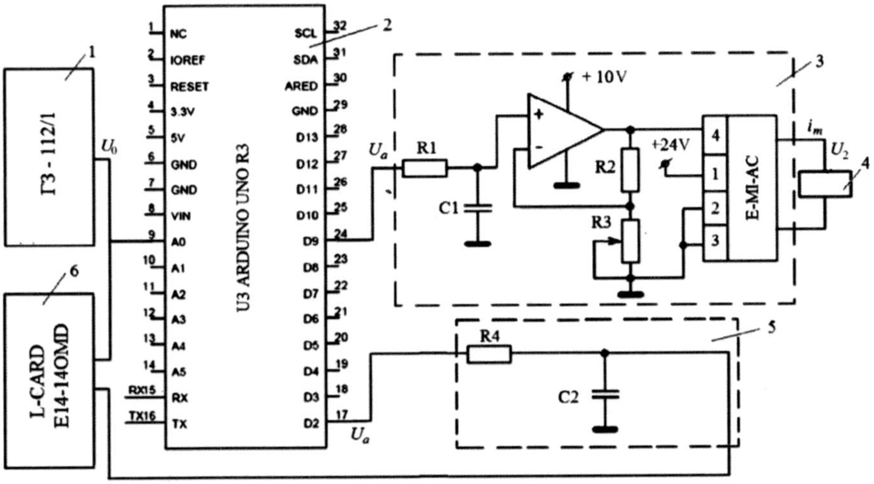

Fig. 2. Scheme for studying the control system characteristics.

4 Results and Discussion

In the first series of studies, a sine wave signal U_0 with a frequency of 8 Hz and an amplitude of 1.25 V was given to the input of controller 2. The program was executed in the controller that changed the amplitude of the sine wave signal by 1.2 times without changing its frequency. A voltage signal U_a was recorded at the controller's output, which passed through filter 5. Figure 3 shows an oscillogram on which the time change of the signals U_0 and U_a is registered. The controller does not change the input signal frequency; the output signal amplitude corresponds to the program's results. When converting a sine wave signal U_0 to a signal U_a, there is a delay in time $\Delta T_1 = 0.02$ s. Changing the program volume from 1.0 to 3.5 Kbyte does not affect the delay value of ΔT_1. The input signal U_0 in the experiments varied in the range from 6 to 18 Hz, corresponding to the frequencies of natural oscillations in the electrohydraulic control system in various combinations of design parameters.

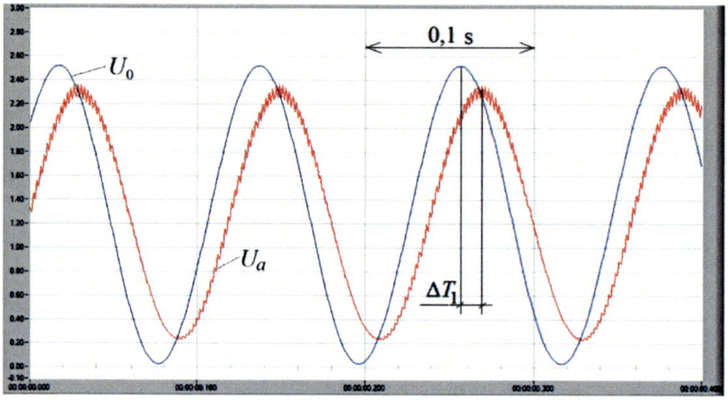

Fig. 3. Oscillogram of the signals changes in time at the input of the controller U_0 and the output of the controller U_a.

In the considered frequency range of the signal U_0, no deviation of the amplitude and frequency of the output signal U_a was observed. Considering that the sampling rate of the controller's inputs is about 9 kHz and also the results of experimental studies of the controller, the conclusion was that it is possible to model the controller during operation in an electrohydraulic proportional link control system with a transmission coefficient, that is set programmatically.

The study was conducted to determine the transfer function of the electromagnet amplifier using the installation, the scheme of which is shown in Fig. 2. The analog-to-digital converter 6, connected to amplifier 3, measured the voltage U_2 and the current i_m at its output. A step signal was applied to the input of the amplifier in the form of a voltage change U_a. Figure 4, a shows the oscillogram of voltage U_2 and current i_m changes at the output of the amplifier with a step voltage $U_a = 1.5$ V and $U_a = 4.5$ V, respectively.

As the signal at the output of the amplifier has pulse-width modulation, the average values of the current i_m (see Fig. 4, c) and voltage U_2 (see Fig. 4, b) for the period were determined. The conducted studies allowed us to conclude that the operation of the amplifier in the range of voltage $U_a = (0.2\ldots4.5)$ V and the output current of the controller up to 40 mA when the amplifier is supplied with a voltage of up to 10 V can

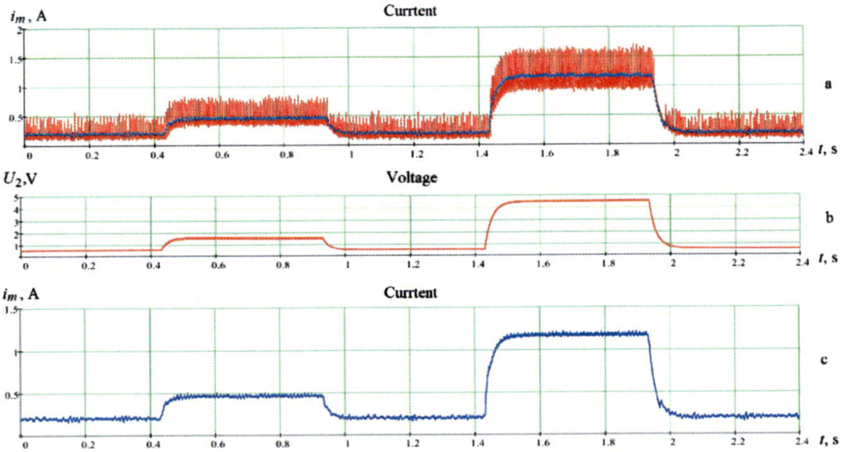

Fig. 4. Change of voltage U_2 and current i_m at the output of the electromagnetic amplifier at the step signal U_a on the controller output.

be modeled by a periodic link with a transfer function

$$F_n(s) = \frac{k_{ng}}{T_n \cdot s + 1}, \tag{1}$$

wherein $k_{ng} = 1, 0$ A/V is the transmission coefficient; $T_n = 0.02$ s is a time constant.

The mathematical model presented in [19] was improved by using the experimentally determined transfer function of the amplifier (Eq. 1). This transfer function is used in the mathematical description of the control line operation from sensor 13 to electromagnet 3 (Fig. 1). The improved mathematical model calculates the time of adjustment, pressure overshoot, and power losses in the developed hydraulic system.

The variable pump with the proposed control system is designed for use in mobile equipment: excavators, loaders, hydraulic manipulators, and cranes. Basic requirements for a variable displacement pump with an electrohydraulic control system, which must be ensured during design: minimization of regulation time; minimization of the amount of pressure overshoot; minimization of power losses. The work [6] presents a hydraulic system with LS-regulation. The hydraulic system includes a pump of a variable working volume with an electric-hydraulic pumping system, a proportional directional valve, a hydraulic cylinder, and a controller. The results of modeling and experimental studies of the hydraulic system are presented. It was established that increasing the pressure reserve in the hydraulic system improves its speed but reduces energy efficiency. The work [14] considered a hydraulic system with electro-hydraulic pumping. The hydraulic system comprises a control pump with an electric-hydraulic pumping system, two hydraulic cylinders, throttle control, and a MELS-controller. The MELS-controller ensures the ability to operate the hydraulic system in several modes: flow control, load sensing, power limitation, and pressure control. An algorithm for the operation of a MELS-controller when switching modes is presented. It is shown that significant pressure fluctuations are avoided when switching from one mode to another. To enhance the elaboration processes of electrohydraulic pump control systems similar to those presented in works [6, 14], we proposed using the complex criterion kk to evaluate the efficiency of a hydraulic system of this type. The criterion includes, with appropriate weight factors, such efficiency indicators as regulation time t_p, pressure overshoot σ, and power loss N_p. Such indicators are among the most important when designing hydraulic systems. The application of the proposed criterion allows for determining the design parameters of the electrohydraulic pump control system at the design stage, depending on the priority of the consumer of the performance indicators of the hydraulic system.

The pump control system is quite a complex dynamic system. Figure 5, a represents an unstable process in the control system with the stepwise change in the pressure value from the inlet to the throttle. Unstable operation of the control system occurs due to a sharp increase in the pressures of p_c and p_n (Fig. 1).

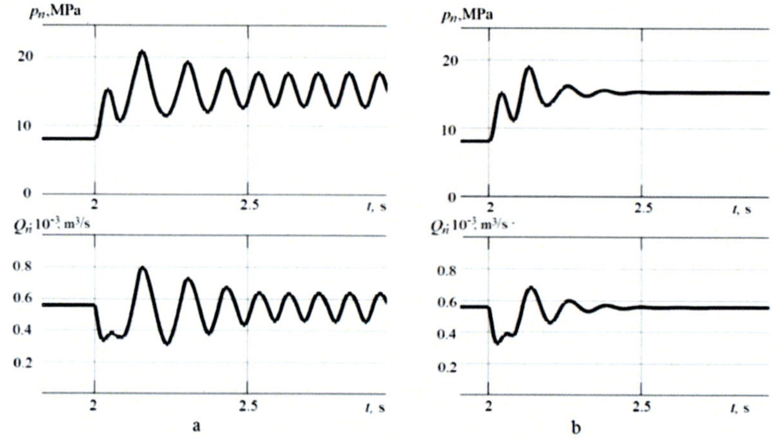

Fig. 5. Unstable process in the electro-hydraulic control system (a) and transient process in the control system at parameter values $f_0 = 1.8{\cdot}10^{-6}$ m^2, $k_z = 8.0{\cdot}10^{-3}$ m, $f_x = 1.8{\cdot}10^{-6}$ m^2, $k_x = 5.0{\cdot}10^{-3}$ m (b).

The paper [20] presents the research results of the influence of the design parameters of the electrohydraulic control system on the regulation time t_p and the value of overshoot σ. The following design parameters have the most significant influence on the dynamic characteristics (Fig. 1): f_0 – throttle area 8; f_x – throttle area 9; k_z – the gain ratio of the operating window of the spool 5; k_x – the gain ratio of the operating window of the servo valve 7. These parameters also have a certain effect on power losses during pump operation. The choice of the values of these parameters is complicated since the increase in the area of the throttles f_0 and f_x in a certain range reduces t_p and σ, but increases power losses. Complex criterion was used to synthesize the coupling of the design parameters f_0, f_x, k_z, k_x, which minimize the time of regulation t_p, pressure overshoot σ, and power loss N_p. The criterion was determined with the specified weight coefficients for the values t_p, σ, and N_p according to the formula:

$$k_k = k_1 \frac{t_p}{t_{p,max}} + k_2 \frac{\sigma}{\sigma_{max}} + k_3 \frac{N_p}{N_{p,max}},$$

wherein t_p, σ, N_p are the current values of the time of regulation, overshoot, and power loss, determined by a particular combination of design parameters;

$t_{p,max}$, σ_{max}, $N_{p,max}$ are the maximum values in a series of experiments;

$k_1 = 0.3$, $k_2 = 0.3$, $k_3 = 0.4$ are weight coefficients to account for the influence of regulation time, pressure overshoot, and power loss.

The design parameters of the control system were changed in the following ranges: $f_0 = (0.8\ldots2.0)\cdot10^{-6}$ m^2; $k_z = (2\ldots8)\cdot10^{-3}$ m; $f_x = (1.6\ldots2.6)\cdot10^{-6}$ m^2; $k_x = (1\ldots5)\cdot10^{-3}$ m.

Tree values from the specified range were taken into account for each of the parameters. In total, 81 combinations of parameters f_0, f_x, k_z, k_x were calculated, with a stepwise change in the pressure value of the p_c from 8.0 MPa to 14 MPa. The pump supply setting was $Q_n = 0.85\cdot10^{-3}$ m^3/s. For each of the 81 conjugation design parameters, the indicators t_p, σ, N_p were defined, and the value of the criterion k_k was calculated. The minimum value of the efficiency factor of the pump control system, with $k_k = 0.51$ was obtained in experiment No. 59. The transition process is illustrated in Fig. 5, b.

Therein, the values of the design parameters synthesized as a result of a series of experiments to investigate the system are: $f_0 = 1.8\cdot10^{-6}$ m^2, $k_z = 8.0\cdot10^{-3}$ m, $f_x = 1.8\cdot10^{-6}$ m^2, $k_x = 5.0\cdot10^{-3}$ m. t At the determined value of the efficiency criterion, the pump has an adjustment time of $t_p = 0.44$ s, pressure overshoot $\sigma = 22\%$, and power losses in the control system do not exceed $N_p = 1.82$ kW.

5 Conclusions

Based on experimental studies, it was determined that the transfer function of the controller could be described by the proportional link $F_k = k_c$, and the transfer function of the electromagnet amplifier by the aperiodic link $F_n(s) = \frac{k_{ng}}{T_n \cdot s + 1}$.

A complex criterion was used to evaluate the efficiency of the variable displacement pump control system, including control time, pressure overshoot, and power loss with corresponding weighting coefficients.

As a result of the parametric synthesis, it was determined that the best value of the criterion for evaluating the efficiency of the electrohydraulic system for controlling a variable displacement pump is achieved with such values of design parameters $f_0 = 1.8\cdot10^{-6}$ m^2, $k_z = 8.0\cdot10^{-3}$ m, $f_x = 1.8\cdot10^{-6}$ m^2, $k_x = 5.0\cdot10^{-3}$ m.

References

1. Lovrec, D., Tič, V.: Speed-controlled hydraulic drive systems for heavy machinery. IMK-14 - Istrazivanje i razvoj **27**(2), 61–72 (2021). https://doi.org/10.5937/IMK2102061L
2. Quan, Z., Ge, L., Wei, Z., Li, Y.W., Quan, L.: A survey of powertrain technologies for energy-efficient heavy-duty machinery. Proc. IEEE **109**(3), 279–308 (2021). https://doi.org/10.1109/JPROC.2021.3051555

3. Huang, W., Quan, L., Huang, J., Yang, J.: Flow matching with combined control of the pump and the valves for the independent metering swing system of a hydraulic excavator. Proc. Inst. Mech. Eng., Part D: J. Automobile Eng. **232**(10), 1310–1322 (2018). https://doi.org/10.1177/0954407017728848
4. Wang, W., Wang, B.: An energy-saving control strategy with load sensing for electro-hydraulic servo systems. J. Mech. Eng. **62**(12), 709–716 (2016). https://doi.org/10.5545/sv-jme.2016.3685
5. Ding, R., Zhang, J., Xu, B., Cheng, M., Pan, M.: Energy efficiency improvement of heavy-load mobile hydraulic manipulator with electronically tunable operating modes. Energy Convers. Manage. **188**, 447–461 (2019). https://doi.org/10.1016/j.enconman.2019.03.023
6. Wu, J., Wang, F., Xu, B., Sun, Z.: Characteristics study of electro-hydraulic load-sensing control system for mobile machines. Proc. Inst. Mech. Eng., Part D: J. Autom. Eng. **237**(9), 2344–2354 (2022). https://doi.org/10.1177/09544070221105386
7. Kumar Hati, S., Pada Mandal, N., Sanyal, D.: Energy-saving design of variable-displacement bi-directional pump-controlled electrohydraulic system. Proc. the Instit. Mech. Eng., Part I: J. Syst. Control Eng. **235**(7), 1218–1236 (2021). https://doi.org/10.1177/0959651820973898
8. Yuan, W., Wu, T.: Review of flow-matching technology for hydraulic systems. Processes **10**, 2482 (2022). https://doi.org/10.3390/pr10122482
9. Ghani, M.F., Ghazali, R., Jaafar, H.I., Soon, C.C., Sam, Y.M., Has, Z.: Improved third order pid sliding mode controller for electrohydraulic actuator tracking control. J. Robot. Control **3**(2), 219–226 (2022). https://doi.org/10.18196/jrc.v3i2.14236
10. Mondal, N., Saha, R., Sanyal, D.: A single stage spool valve for the pressure compensator of a variable displacement pump: design, dynamic simulation and comparative study with a real pump. Sādhanā **47**(4), 279 (2022). https://doi.org/10.1007/s12046-022-02038-3
11. Mondal, N., Saha, R., Mookherjee, S., Sanyal, D.: A novel method to design pressure compensator for variable displacement axial piston pump. Proc. Inst. Mech. Eng., Part E: J. Proc. Mech. Eng. **233**(2), 314–334 (2019). https://doi.org/10.1177/0954408918783409
12. Quanyi, H., Hong, Z., Shujun, T., Xuxin, Q.: Performances analysis of a novel load-sensing hydraulic system with overriding differential pressure control. Proc. Inst. Mech. Eng., Part C: J. Mech. Eng. Sci. **231**(23), 4331–4343 (2017). https://doi.org/10.1177/0954406216667760
13. Lovrec, D., Detiček, E.: Improvement of the statical behaviour of pressure controlled axial piston pumps. J. Mech. Eng. **55**(12), 766–774 (2009)
14. Cheng, M., Sun, B., Ding, R., Xu, B: A multi-mode electronic load sensing control scheme with power limitation and pressure cut-off for mobile machinery. Chin. Mech. Eng. Soc. **10**, 2–10 (2021). https://doi.org/10.21203/rs.3.rs-602646/v1
15. Casoli, P., Pompini, N., Riccò, L.: Simulation of an excavator hydraulic system using nonlinear mathematical models. Strojniški vestnik-J. Mech. Eng. **61**(10), 583–593 (2015). https://doi.org/10.5545/sv-jme.2015.2570
16. Bedotti, A., Pastori, M., Scolari, F., Casoli, P.: Dynamic modelling of the swash plate of a hydraulic axial piston pump for condition monitoring applications. Energy Procedia **148**, 366–373 (2018). https://doi.org/10.1016/j.egypro.2014.01.090
17. Hu, Q., Zhang, H., Tian, S., Qin, X.: Model reduction of a load-sensing hydraulic system via activity index analysis. J. Mech. Eng. **63**(1), 65–78 (2017). https://doi.org/10.5545/sv-jme.2016.3450
18. Ma, W., Ma, S., Qiao, W., Cao, D., Yin, C.: Research on PID controller of excavator electro-hydraulic system based on improved differential evolution. Machines **11**(2), 17 (2023)

19. Bogachuk, V., Kozlov, L., Tovkach, A., Badakh, V., Tarasenko, T., Kobylianskyi Y.: Influence of Electrohydraulic Controller Parameters on the Dynamic Characteristics of a Hudrosystem with Adjustable Pump. «Mehatronics», vol. II. Taylor & Francis Group, CRC Press, London (2020)
20. Buriennikov, Y., et al.: Optimization of parameters of the mobile machine adaptive hydraulic circuit. Int. J. Modern Manuf. Technol. **13**(3), 14–21 (2021). https://doi.org/10.54684/ijmmt.2021.13.3.14

Mathematical Modeling of Thermomechanical Phenomena in Machining of Products Made of Functionally Graded Materials

Maksym Kunitsyn(✉), Anatoly Usov, and Yulia Sikirash

Odessa Polytechnic National University, 1, Shevchenko Ave, Odessa 65044, Ukraine
m.v.kunitsyn@op.edu.ua

Abstract. Mathematical modeling of thermomechanical processes that accompany the mechanical processing of products to control them in technological systems is one reserve for improving the quality of products and their performance in mechanisms. Deterministic modeling of thermomechanical phenomena in the mechanical processing of structurally homogeneous materials using equations based on continuous functions allows us to obtain solutions that are represented as analytical relations in closed form and are convenient for analyzing these processes and based on them to make a rational choice of technological parameters to ensure the required characteristics of the processed surfaces of products. The article presents a numerical and analytical model for determining the thermomechanical state during the mechanical processing of structurally inhomogeneous materials containing inhomogeneities such as interfacial cracks and inclusions. Based on this model, functional dependencies of surface layer quality criteria with technological control parameters are determined to ensure products' processed surfaces' required characteristics.

Keywords: Processing · Layer · Structural · Technological · Defects · Surface · Parameters · Temperature · Stress · Thermal State · Process Innovation · Industrial Growth

1 Introduction

Among the processes characteristic of mechanical processing, the principal place belongs to thermomechanical phenomena occurring in the processed surfaces of products. They significantly affect the formation of the thermomechanical state of the working surfaces of the parts. They are associated with the formation of defects such as cracks, burns, and structural changes in the treated surfaces, which contribute to a decrease in the operational properties of parts. Mechanical processing of modern materials, characterized by heterogeneity, is accompanied by high thermal stress and the formation of a thermoelastic state of the processed surface of the products, which increases the processing error and changes the quality of the surface layer due to the appearance of defects in it. Therefore, mathematical modeling of thermomechanical processes accompanying the mechanical processing of products to control them in technological systems is one of the reserves

V. Tonkonogyi et al. (Eds.): InterPartner 2023, LNME, pp. 58–71, 2024.
https://doi.org/10.1007/978-3-031-42778-7_6

for identifying and forecasting process features that affect the quality of products and their performance in mechanisms. Thus, tooling is the most expensive and vulnerable regarding uptime in producing fiber optic cables operating at elevated temperatures and intense wear. Therefore, its working surfaces are subject to increased requirements for roughness, hardness, dimensional stability, and the absence of defects such as burns and cracks. Modeling the formation of thermomechanical fields during the processing of forming elements of tooling and the establishment of functional relationships between technological parameters, critical temperatures and stresses, and the tool's durability allows you to create optimal conditions for their operation.

Modern materials' mechanical processing, characterized by heterogeneity, is accompanied by high heat stress. The formation of a thermopile state of the processed surface of products, which increases the processing error, changes the surface layer's quality because of the appearance of defects in it. Therefore, mathematical modeling of thermomechanical processes that accompany the mechanical processing of products to control them in technological systems is one reserve for improving the quality of products and their performance mechanisms.

2 Literature Review

Modeling thermomechanical processes in machining products can be divided into two types: modeling processes using continuous functional dependencies and modeling thermomechanical processes with discontinuous functional dependencies. Deterministic modeling of thermomechanical phenomena using equations based on continuous functions [1] allows us to obtain solutions that are represented as analytical relations in closed form and are convenient for analyzing these processes and based on them to make a rational choice of technological parameters to ensure the required characteristics of the processed surfaces of products [2].

Thus, the most expensive and vulnerable uptime in fiber-optic cables is tooling that works in high temperatures and intense wear conditions. Therefore, its working surfaces are subject to increased requirements for roughness, hardness, dimensional stability, and the absence of defects such as cauterization and cracks.

Modeling the formation of thermomechanical fields informing tooling elements and establishing functional relationships between technological parameters, critical values of temperatures and stresses, and tool life allows optimal production conditions.

Mechanical processing of parts made of composite materials characterized by structural heterogeneity associated with functional gradient properties is accompanied by thermomechanical phenomena, which are not always described by continuous dependencies. Thus, the temperature and thermoelastic fields formed in the surface layer of products made of composite materials are subjected to discontinuities in the areas of accumulation of inhomogeneities during operation [3].

The most effective methods for modeling physical processes occurring in environments of inhomogeneous structure and electromagnetic signals in environments with variable characteristics, and the formation of micro-cracks in structural materials that have various types of heterogeneities of ancestral origin, until this time, remain methods using Cauchy integrals [4], boundary value problems of the theory of analytical functions, and the method of singular integral equations [5].

The functionality of products depends on the defects in the structure of their material. In natural materials, there are always numerous different micro-defects, the development of which, under the influence of load [6], leads to the appearance of cracks and their increase and, as a result, local or complete destruction [7]. In this paper, based on the method of singular integral equations, a unified approach to modeling thermomechanical processes in mechanical processing in the surface layer of products [8] whose materials have structural and technological inhomogeneities is presented [9].

The choice of a method for modeling thermomechanical phenomena that accompany the mechanical processing of structural elements depends on the object's size under study. For example, micro-studies [10] are associated with inhomogeneities formed in the surface layer at the workpiece's preparation stage during the manufacturing of structural elements. However, considering defects allows us to adequately consider the loss of functional properties of working surfaces of objects on their performance indicators.

The quality of the surface layer of structural elements during their manufacture is formed under the influence of thermomechanical phenomena accompanying mechanical processing [11]. The presence of stress concentrators in various types of heterogeneities of ancestral origin, introduced in obtaining the workpiece and subsequent types of mechanical processing, are the leading indicators of working surfaces' bearing capacity. The formation of technological origin defects occurs because of heat stress during structural elements' mechanical processing [11, 12]. Based on models of temperature fields, stress fields, and fracture mechanics, the regularities of the formation of defects such as structural changes, micro-cracks, and technological possibilities of their elimination are studied depending on the thermal properties of the processed materials, processing modes, design, and characteristics of the tools used [7, 8]. The problem of stress concentration in defects is solved using material mechanics, which consider the micro-uniformity and defect of their materials when calculating structural elements' load-bearing capacity. Finally, considering defects allows for a more adequately representing mechanism of loss of products' functional properties during machining.

Available models of thermomechanical processing processes are obtained under the assumption of uniformity of materials of structural elements. They do not consider defects in the technological inheritance of products [11]. There are studies of the influence of structural transformations in steels during their mechanical processing on the formation of cracks, according to which the presence of a large amount of austenite in the subsurface layer of parts leads to the formation of tensile stresses, which are realized in the form of brittle cracks [6, 7]. In some cases, structural elements' physical and technical processing is characterized by short duration and high heating and cooling rates. Structural changes are insignificant, and thermomechanical stresses reach limit values [8].

Models of the stress-strain state of parts with coatings have been developed that consider the piecewise heterogeneity of the product coating matrix [8, 9]. However, the lack of studies on the influence of inhomogeneities formed in the surface layer of products during mechanical processing on their functional properties and, in particular, on the bearing capacity or wear resistance determines the relevance of constructing a mathematical model of defect formation in the physical and technical processing of structural elements using the criteria of fracture mechanics. The simulation results allow us to assess the impact of structural inhomogeneities formed during machining in products' working surfaces on the loss of the required properties.

The research aims to develop a numerical-analytical model for determining the thermomechanical state of structurally inhomogeneous materials containing inhomogeneities such as interfacial cracks and inclusions during machining. This model determines the functional relationship of the surface layer quality criteria with the technological control parameters to ensure the required characteristics of the products' processed surfaces.

3 Research Methodology

When choosing and justifying the mathematical model, it was considered that thermal and mechanical phenomena accompany manufacturing parts. However, the combined effect on the stress-strain state of the surface layer is caused by temperature fields. Since the bulk of the surface layer of metal during physical and technical processing is elastic, it can use a thermoelastic body model that reflects the relationship of mechanical and thermal phenomena at finite values of heat flows. For studies of the thermomechanical state of the working surfaces of structural elements, information about the propagation of temperatures and stresses and the depth of the material, taking into account its inhomogeneities, is essential.

For further studies of the kinetics of the formation of thermomechanical processes in the processed material, we will use the following system of differential equations [11, 13] as the central theoretical premise, describing the interaction of the deformation field and the temperature field:

$$G\Delta\overrightarrow{U_j} + (\lambda_t + G)\text{grad div}\overrightarrow{U_j} - \rho\frac{\partial^2\overrightarrow{U_j}}{\partial\tau^2} + P_j = \alpha_t\beta_t\text{grad }T, \tag{1}$$

$$\Delta T - \frac{1}{a}\frac{\partial T}{\partial\tau} - \eta l\frac{\partial}{\partial\tau}\text{div}\overrightarrow{U_j} = -\frac{W}{\lambda} + C_q^{-2}\frac{\partial^2 T}{\partial t^2}, \tag{2}$$

$$\beta_t = 3\lambda_t + 2G;\ l = \frac{1+\tau_r\delta}{\delta};\ \eta = \frac{\alpha_t\beta_t T(\Phi,\tau)}{\lambda}$$

where λ_t, G are Lamé constants; ρ is the density of the processed material; α_t is the temperature coefficient of linear expansion of metal; $a = \frac{\lambda}{C_v}$ is the thermal diffusivity; λ is the thermal conductivity; C_v is the volume heat capacity; $\vec{U}(\Phi,\tau)$ is the total vector of movement of the inner point $\Phi(x, y, z)$ of the surface layer under the influence of thermomechanical forces accompanying the processing process; τ_r is the relaxation time; W is the power of the heat source; C_q is the speed of heat propagation in the processed material; τ is the time; P_j is the cutting force; $\text{grad }T(x, y, z) = \frac{\partial T}{\partial x}\vec{i} + \frac{\partial T}{\partial y}\vec{j} + \frac{\partial T}{\partial z}\vec{k}$, $\text{div}\overrightarrow{U_j} = \frac{\partial U_x}{\partial x} + \frac{\partial U_y}{\partial y} + \frac{\partial U_z}{\partial z}$, $(j \rightleftarrows x, y, z)$.

Since the thermal phenomena prevail over the power ones in the final processing methods, we can ignore the term that considers mechanical energy conversion into thermal energy in the heat equation. Therefore, we will come to a parabolic heat equation.

We will neglect the influence of inertial terms and heat propagation speed limitation for the explicit solvability of the specified system (1)–(2). Moreover, to overcome the analytical difficulties associated with solving spatial problems of thermoelasticity, we will consider the planar problem. This transition is justified because, for studying the thermomechanical state of the treated surfaces, information about the propagation of temperatures and deformations and the depth and direction of the source's movement is essential.

When drawing up the design scheme, we assume the processed product is modeled as a piecewise homogeneous half-plane. It allows us to study thermomechanical processes with several types of coatings with a thickness applied to the main matrix. This scheme determines the thermal and deformation conditions for the interface of layers along their section boundaries a_k.

The influence of structural inhomogeneities in the material during smelting and the technological process will be considered in the model by inclusions and defects such as microcracks in the surface layer.

The system of equations that determine the thermal and stress-strain state when processing the surface of parts with coatings, the upper layer of which has inhomogeneities such as inclusions and cracks, contains:

Equation of non-stationary thermal conductivity:

$$\frac{\partial T}{\partial \tau} = a^2\left(\frac{\partial^2 T}{\partial x^2} + \frac{\partial^2 T}{\partial y^2}\right) \begin{array}{l} 0 \le x < \infty; \\ -\infty < y < \infty. \end{array} \tag{3}$$

The equation of the Lamé elasticity in displacements:

$$\frac{\partial \Theta}{\partial x}\frac{1}{1-2\mu} + \Delta\bar{u} = B^T\frac{\partial T}{\partial x}; \bar{u}(x, y) = \frac{u}{2G}; \bar{v}(x, y) = \frac{v}{2G}; \tag{4}$$

$$\frac{\partial \Theta}{\partial y}\frac{1}{1-2\mu} + \Delta\bar{v} = B^T\frac{\partial T}{\partial y}; B^T = \frac{4G(1+\mu)}{1-2\mu}a_k; \Delta = \frac{\partial^2}{dx^2} + \frac{\partial^2}{dy^2}, \tag{5}$$

where $T(x, y, \tau)$ is the temperature at a point with coordinates (x, y) and at any time τ; a – thermal conductivity of the material; α is the temperature coefficient of linear expansion; μ, G are Lamé constants; u, v are components of the point displacement vector (x, y); $\Delta = \frac{\partial^2}{\partial x^2} + \frac{\partial^2}{\partial y^2}$ is the Laplace operator.

The initial conditions for this task can be taken as follows:

$$T(x, y, 0) = 0. \tag{6}$$

Boundary conditions for temperature and deformation fields that take into account heat transfer from the surface outside the contact zone of the tool with the part and intense heat generation in the processing zone have the form:

$$\frac{\partial T}{\partial x} = -\frac{q(y, \tau)}{\lambda}, |y| < a^*, -\lambda\frac{\partial T}{\partial x} + \gamma T = 0, |y| > a^*, \tag{7}$$

$$\sigma_x(x, y, \tau)|_{x=0} = \tau_{\tau y}(x, y, \tau)|_{x=0} = 0, \tag{8}$$

where $q(y, \tau)$ is the intensity of the heat flow generated by the interaction of the tool with the part; λ is the coefficient of thermal conductivity of the processed material; $2a^*$ is the length of the tool contact zone with the work surface; γ is the coefficient of heat exchange with the environment; σ_x, τ_{xy} are normal and tangent stresses.

Conditions for interfacing layers (coatings) [13]:

$$\begin{array}{ll} \text{for temperature fields} & \text{for deformation fields} \\ \overset{k-1}{T}(a_k - 0, y, \tau) = \overset{k}{T}(a_k + 0, y, \tau) & \overset{k-1}{\upsilon}_j(a_k - 0, y) = \overset{k}{\upsilon}_j(a_k + 0, y) \\ \lambda_{k-q} \frac{\partial \overset{k-1}{T}}{\partial x}(a_k - 0, y, \tau) = \lambda_k \frac{\partial \overset{\sigma}{T}}{\partial x}(a_k + 0, y, \tau) & \overset{k-1}{\sigma}_x(a_k - 0, y) = \overset{k}{\sigma}_x(a_k + 0, y) \\ & \overset{k-1}{\tau}_{xy}(a_k - 0, y) = \overset{k}{\tau}_{xy}(a_k + 0, y) \end{array}; \quad (9)$$

where λ_k is the thermal conductivity of the k-th layer; α_k is the thickness of the k-th layer; $\overset{k}{\upsilon}_j$ is the displacement components in the k-th layer.

For surface layers with structural and technological inhomogeneities, the discontinuity conditions of the solution, depending on the defect, will be [13]:

$$\begin{array}{ll} \text{for inclusion} & \text{on crack} - \text{like defects} \\ \langle \overline{\upsilon} \rangle = 0, \langle \sigma_x \rangle \neq 0 & \langle \sigma_x \rangle = 0, \langle \overline{\upsilon} \rangle \neq 0 \\ \langle \overline{\nu} \rangle = 0, \langle \tau_{xy} \rangle \neq 0 & \langle \tau_{xy} \rangle = 0, \langle \overline{\nu} \rangle \neq 0 \end{array} \quad (10)$$

where $\langle \overline{\upsilon} \rangle$, $\langle \overline{\nu} \rangle$, $\langle \sigma_x \rangle$, $\langle \tau_{xy} \rangle$ are jumps in the displacement and stress components.

The deformable surface layer's maximum equilibrium state was tested by classical strength criteria [14, 15]. Of the fracture criteria that consider local physical and mechanical properties of inhomogeneous materials, the most acceptable for this case are the criteria of the force approach [4, 5], associated with the use of the stress intensity factor (SIF). When loading causes the stress intensity to equal the limit value of K_{1c}, the crack-like defect turns into the main crack.

Modeling the effect of the initial piecewise homogeneity of the processed materials (parts with coatings) on thermomechanical processes is carried out using discontinuous solutions [13]. They are solutions that satisfy the Fourier thermal conductivity and Lamé elasticity equations everywhere except for the defect boundaries. When crossing the boundary, the fields of displacement and stress suffer discontinuities of the first kind, i.e., their jumps $\langle \overline{\upsilon} \rangle$, $\langle \overline{\nu} \rangle$, $\langle \sigma_x \rangle$, $\langle \tau_{xy} \rangle$ appear.

The solution of the thermal problem (3)–(10) is carried out using integral Fourier transforms for the variable y and Laplace transformations for τ to the function $T(x, y, \tau)$ in the first ($k = 0$) layer, which is described in integral form as [13]:

$$T_0(x, y, \tau) = \int_{-a}^{a} d\tau \int_{0}^{\tau} \chi(t - \tau, x, y - \eta) dt, \quad (11)$$

where $q(t, x, y - \eta) = \frac{1}{2\pi i} \int_r K_p^m(y - \eta, x) e^{pt} dp$, $\chi(y, \tau) = \sum_{m=0}^{\infty} \chi_m(y) 2e^{-\tau} L_m(2\tau)$ are polynomials of the Laguerre; $K_p^m(y - \eta) = \frac{1}{2\pi} \int_{-\infty}^{\infty} \frac{e^{-t\beta(y-\eta)}}{l_{m_1 \beta p}} d\beta$; $l_{m_1 \beta p}$ is the expression that considers the thermal properties of layers $k = 0 - m$, their thickness, and boundary conditions.

The stress-strain state of the layered half-plane is also estimated by the method of discontinuous solutions. The boundaries of the section $x = a_k (k = 0)$ are defects when passing through which the displacement and stress fields suffer discontinuities.

The construction of discontinuous solutions of the Lamé equations with prescribed jumps is done using the functions of Trefftz [16]:

$$\begin{gathered} \overline{\upsilon} = \psi_1 + (x - a)\psi_0', \\ \overline{\nu} = \psi_2 + (x - a)\psi_0', \\ \psi_0' = \frac{\psi_1' + \psi_2'}{3 - 4\eta}, \\ e = \psi_1' + \psi_2'' + \psi_0', \end{gathered} \tag{12}$$

where $\Delta\psi_0(x, y) = 0$, $\Delta\psi_j(x, y) = b_k^t T^{(j)}$, $(j = 1, 2)$.

Stresses are found by the equations:

$$\begin{gathered} \sigma_x = (1 - \mu)\psi_0' + \psi_1' + (x - a_k)\psi_0''; \\ \sigma_y = -\mu\psi_0' + \psi_2' + (x - a_k)\psi_0''; \\ \tau_{xy} = \psi_1^2 + 2(x - a_k)\psi_0' + \psi_2' + \psi_0'. \end{gathered} \tag{13}$$

where $\Delta\psi_0(x, y) = 0$, $\Delta\psi_j(x, y) = b_k^t T^{(j)}$, $(j = 1, 2)$.

Application of generalized Fourier transforms for variables x, y to Eqs. (2), (3), (6), (7), taking into account (10), allows us to obtain recurrent relations that link the displacements and stresses in the k-th layer with the stresses and displacements formed in the first layer under the action of non-stationary temperature fields [17].

Under conditions of uneven heating, thermal deformations occur in the surface layer, which causes temperature stresses. Crack formation occurs under the influence of these stresses, concentrated in defects [18].

The most interesting is the behavior of stresses near the vertices of defects such as cracks, sharp inclusions, and structural imperfections, i.e., stress features at $y \to \pm l_k$. The stress field's nature near the end of the defect obtained in the classical's framework elasticity theory is determined by the stress intensity coefficients $K_I + iK_{II}$.

The influence of the tool design parameters on the thermomechanical state of the surface layer was determined using the model problem (1), (4) and boundary conditions in the form of:

$$q(y, \tau) = \frac{c\sqrt{\tau}}{\lambda}\left[H(y) - H\left(y - 2a^*\right)\right]\sum_{k=0}^{n} \sigma\left(y + kl - v_{kp}\tau\right) \tag{14}$$

where $H(y)$ is the function of Heaviside; $\sigma(y)$ is the Dirac Delta function; n is the number of cutting edges of the tool that pass through the contact zone during $\tau = \frac{\sqrt{\pi t_{gr}}}{\upsilon_{kp}}$; λ is the thermal conductivity of the product material; $c\sqrt{\tau}$ is the heat flow from a single edge; v_g, v_{kp}, t_{gr} are processing modes; $2a^*$ is the arc length of the tool contact with the part; l^* is the distance between the cutting edges.

The maximum values of the instantaneous temperature T_M, from the individual grains to the constant component $-T_K$, were obtained theoretically and confirmed experimentally. These values were later used as criteria for predicting the conditions for forming cauterization-type defects and their depth.

The study of the role of heterogeneity of the coating structure in the mechanism of reducing crack resistance was carried out using the local fracture criterion established by the model in the form of the following inequality:

$$l_0 < \frac{Da\lambda^2 \upsilon_q^u K_c^2}{\pi^2 \left[c\upsilon_{kp} G(1+\nu) a_t \left(1 - exp\left(\frac{\upsilon_q \sqrt{Dt}}{a\tau}\right)\right)\right]}, \tag{15}$$

where υ_{kp}, υ_q, t are processing modes; D, C are tool parameters; λ, a are thermophysical characteristics of the treated coating; K_C is the crack resistance of this coating; G is the elastic modulus; ν is the Poison's ratio; a_t is the temperature coefficient of linear expansion; l_0 is the characteristic linear size of the structural parameter (structure defect).

The technique of a choice of tools and modes of processing wear-resistant coatings provides no cracks or chips on the treated surfaces based on the conditions (14), inequalities (15), and constraints on stresses of the detachment of the coating from the matrix:

$$\sigma_{max}\left(M, a_t, E, \nu, a, \lambda, \upsilon_{kp}, \upsilon_q, t_{gr}\right) \leq [\sigma]_{SI} \tag{16}$$

The model allows optimal performance technological parameters for defect-free parts processing with coatings based on the coating's initial thermal and mechanical properties—detail system.

4 Results and Discussion

The quality of the processed surfaces will be ensured using the technological control parameters to choose such processing modes. The tool's characteristics are that the current values of the processing temperature and heat flow $q(y, \tau)$, stress $\sigma(M)$, and cutting forces P_y, P_z will not exceed their limit values.

Implementing a system of limiting inequalities in the values of the temperature itself and the depth of its propagation in the form of:

$$T(x, y, \tau) = \frac{C}{2\pi\lambda} \sum_{k=0}^{n} H\left(\tau - \frac{kl}{\upsilon_{kp}}\right) \times H\left(\frac{L+kl}{\upsilon_{kp}}\right) \int_{\gamma_1}^{\gamma_2} f\left(x, y, \tau, \tau'\right) d\tau' \leq [T]_M \tag{17}$$

$$\begin{gathered} T_k(0, y, \tau) = \frac{C\upsilon_{kp}}{\pi\lambda l\sqrt{\upsilon_g}} \int_a^\tau \int_{-e}^{e} \frac{\chi(\eta, t) e^{\frac{(y-\eta)^2}{4(\tau-t)}}}{2\sqrt{\pi(\tau-t)}} \times \\ \times \left\{ \frac{1}{\sqrt{\pi(\tau-1)}} + \gamma e^{\gamma^2(\tau-t)}\left[1 + \Phi\left(\gamma\sqrt{\tau-t}\right)\right]\right\} d\eta dt \leq [T] \end{gathered} \tag{18}$$

$$T_k^{\max}(L, 0) = \frac{C\upsilon_{kp}\alpha}{\lambda l \upsilon_q^2} \sqrt{\frac{\alpha}{\pi}} \left[1 - exp\left(-\frac{\upsilon_q\sqrt{Dt_{gr}}}{\alpha}\right)\right] \leq [T] \tag{19}$$

avoids the formation of structural changes. It can serve as a basis for the design of processing cycles based on the thermal criterion.

Processing of materials and alloys without cracks can be provided if the stresses formed in the intensive cooling zone are limited to the maximum values:

$$\sigma_{max}(x,\tau) = 2G\frac{1+\nu}{1-\nu}\alpha_t T_k erf\left(\frac{x}{2\sqrt{\alpha\tau}}\right) \leq [\sigma_i] \tag{20}$$

With a dominant influence of hereditary inhomogeneity on the intensity of crack-like defects, it is necessary to use criteria that include deterministic relationships of technological parameters and properties of the inhomogeneities themselves. As such, it can use the restriction of the stress intensity coefficient:

$$K = \frac{1}{\pi\sqrt{l}}\int_{-e}^{e}\sqrt{\frac{l+t}{l-t}}\sigma_x, \sigma_y dt \leq K_{Ic} \tag{21}$$

Alternatively, providing with the help of control technological parameters of the limit value of the heat flow, at which the balance of structural defects is maintained:

$$q^* = \frac{P_z \upsilon_{kp} \alpha_s}{\sqrt{Dt_{gr}}} \leq \frac{\sqrt{3}\lambda K_{Ic}}{Hl\sqrt{\pi l}\sigma} \tag{22}$$

Conditions for defect-free processing can be implemented using information about the structure of the processed material. So, with the general nature of structural imperfections of length $2l$, their regular location relative to the contact zone of the tool with the part, it can use the condition of defect equilibrium as a criterion ratio in the form of:

$$l_0 < \frac{K_C^2}{x[GT_k(1+\nu)\alpha_t]^2} \tag{23}$$

In this formula, the process part is contained in the contact's relation temperature value Tк to the processing conditions.

These inequalities link the limiting characteristics of the temperature and force fields with the control and technological parameters. They specify the range of combinations of these parameters that meet the obtained thermomechanical criteria. Finally, it considers the properties of the processed material and guarantees the required quality of products.

Based on the obtained criteria, an algorithm for ensuring the surface layer of parts during machining, considering the maximum performance, is constructed.

It is known that the quality of processing the working surfaces of tooling for producing fiber-optic cables significantly affects their functionality [19]. Reducing the roughness of processing, all other things being equal, increases static strength, especially brittle, and significantly limits endurance. It was found [19] that samples with low surface roughness have a higher impact strength. The influence of the micro geometry of working surfaces on technological stresses explains these facts. It is believed that the irregularities formed during surface treatment are effective concentrators, leading to surface cracks on the form-forming surfaces of tooling. Optical fibers that are used in communication systems are made of quartz glass. This material provides a low loss level in a fairly broad frequency range, allowing large distances between regenerators. An optical fiber is a two-layer dielectric waveguide through which one or more modes (types of electromagnetic waves) propagate.

The main reason leading to the scattering of light energy in optical fibers (OF) is microscopic inhomogeneities in the OF material that occur during their manufacture and the fiber-coating interface's roughness. Therefore, one of the main requirements for the tool in producing fiber-optic cables is the low roughness of its working surfaces, the high geometric accuracy of the calibration part, and compliance with special conditions in OF production. Using the correlation between the roughness of the working area of the die and the defect value $(-\alpha, \alpha)$ in various finishing operations: finish grinding and super finish grinding, polishing, the SIF dependence $K_{III} = f(\alpha)$ is found (Fig. 1).

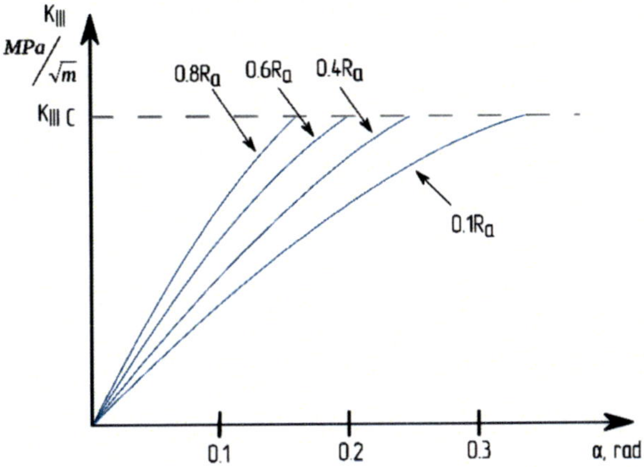

Fig. 1. Dependence of SIF on the size of the defect and roughness of the working area of the tool R_a.

At existing operating stresses on the forming surfaces, tooling is necessary to choose the type and processing modes of the working surfaces, which provide the necessary roughness for preserving the functional properties of fiber-optic products (Fig. 2).

Thus, to ensure the required quality characteristics of the fiber-optic cables, it is necessary to ensure a roughness of $0.8 \leq R_a \leq 1.2$ microns on their working surfaces when manufacturing tools. This roughness can be achieved through grinding operations and subsequent finishing polishing.

Based on the nature of mandrel molds' operation and dies for their manufacture, the most acceptable are hard alloys. Therefore, it is essential that when manufacturing these tools, there are no micro defects on their working surfaces, the roughness corresponds to the required values, and the error of the calibration part is within the tolerance.

The determination of rational technological parameters in the manufacture of mandrel dies should begin from the conditions of their choice in the range that will avoid the development of micro defects in cracks when machining the tool's working surfaces. Existing methods for evaluating the strength of tools obtained by sintering according to the level of the ultimate stress state formed during machining could be more effective. Such calculations ignore the presence of hereditary defects in the processed product, as well as the presence of growing subcritical microcracks, which leads to a significant overestimation of the allowable stresses and, as a result, to an incorrect assessment of

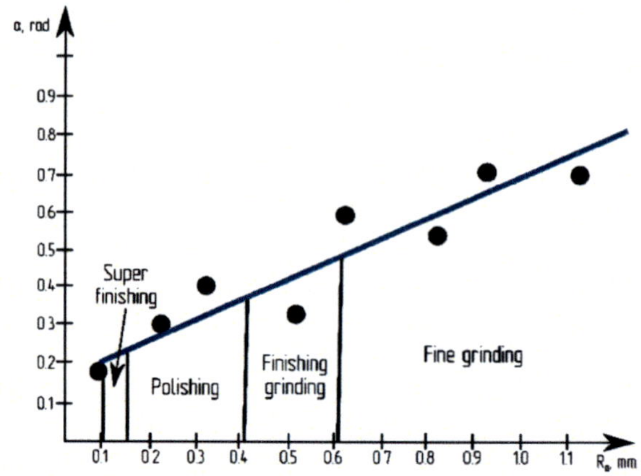

Fig. 2. Influence of working surface treatment modes of tool equipment on its functional properties.

the tool resource, which can be represented as:

$$N_p = N_1 + N_2 \tag{24}$$

where N_1 is the percentage of durability due to the growth of micro defects to a critical size during operation, N_2 is the percentage of tool durability associated with the growth of a macrocrack to the limit state during its manufacture.

Given that tools made of superhard materials are inherently highly defective, the N_2 component can be the central part of their durability, i.e., it ultimately determines the tool's life.

To determine the tool life, the characteristics of the material's resistance to the growth of a fatigue crack formed at the stage of mechanical processing of forming surfaces are used.

To avoid structural transformations in the tool's surface layer, it is necessary to limit the contact temperature, which is formed when processing the working surface of the tool, to the temperature of structural changes (18). Given that the current values of processing temperatures are determined by the processing modes and characteristics of the tool being processed, consistent assignment of modes, and rational selection of characteristics of the processing tool, it is possible to approach the performance of defect-free processing to T_{opt} iteratively – the optimal processing temperature, which should not exceed the temperature of structural transformations in the material of tools for manufacturing fiber-optic cables.

Fractographic analysis of die and mandrel matrix fractures showed that thermal fatigue characterizes their working surfaces' cracks. A significant localization of deformation is observed under the influence of variable stresses formed during the machining of these tools due to uneven heating and cooling. Internal defects formed at the stage of sintering dies and mandrel matrices, being stress concentrators, often serve as a place of origin for cracks. Therefore, when choosing the material for obtaining die blanks, and mandrel matrices, the low coefficient of thermal expansion α_t, high thermal conductivity

λ_m, and low elastic modulus E_m were considered. This approach to choosing the physical and mechanical properties of the material for tool blanks in producing fiber-optic cables also creates a reserve of durability.

The nature of cracks in the machining of working surfaces of tools for fiber-optic cable production indicates that the grain boundaries of the tool material are more damaged than the grain body. The grid of cracks forming on the tool surface indicates the locality of technological stresses and, most importantly, their relatively rapid relaxation. Theoretical studies show that the most significant temperature stresses are formed when grinding the dies' working surfaces. Moreover, their level increases with increasing heat transfer coefficient. The time of maximum stress formation is equal to the time of contact of the processing tool with the processed surface of dies, mandrel molds, and other equipment for fiber-optic cable production.

When sintering dies blanks, mandrel dies, and other tools for producing signal transmission equipment, it is essential to ensure their material is minimally defective. Otherwise, in the presence of crack-like defects and depending on their density, "hot" prominent cracks appear on their forming surfaces under the influence of temperature stresses.

Based on the modeling of the thermomechanical state of the working surfaces of tooling during its manufacture and operating conditions, it is established that the optimal material for forming and calibration tools is tungsten-cobalt hard alloys. Hard alloys' physical and mechanical characteristics and wear resistance are mainly determined by the ratio of tungsten carbide and cobalt volume fractions, whose properties differ significantly.

In cases where wear-resistant coatings are applied to the tooling's calibration surface, the problem arises of preserving its functionality during operation. Ensuring the fiber-optic cables' quality characteristics significantly depends on the coating condition, which is prone to cracking during finishing operations under the influence of thermomechanical processes due to its partial detachment from the primary material of the calibration tool (Fig. 3).

Therefore, based on the developed model and the obtained criterion ratio (23), it is possible to prevent cracks and chips on the forming surfaces of tooling by the established functional relationships of the crack resistance criterion (21) with the control technological parameters of finishing operations.

The designations used in Fig. 3: 1 is wear-resistant coating; 2 is the matrix of the primary material; q is heat flow entering the coating-matrix system when processing the calibration surface of tooling; $(-l, l)$ is the partial separation of the coating from the matrix; σ^t is the tensile stresses formed in the matrix; σ^c – compressive stresses in the coating; Δ_p is the coating thickness; $q_1 > q_2$.

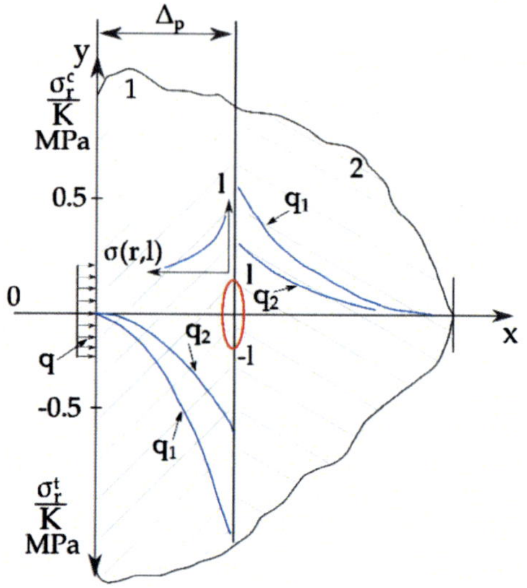

Fig. 3. Calculation scheme for determining the stress-strain state of the coating-matrix system at the boundary, which partially separates the coating from the matrix, size 2l.

5 Conclusions

A mathematical model has been developed that describes thermomechanical processes in the surface layer when processing parts made of materials and alloys, taking into account their inhomogeneities that affect the formation of performance characteristics. The calculated dependencies between the stress intensity factor and the primary control technological parameters are obtained. According to the known characteristics of inherited defects, the limit values of heat flow during machining are determined to ensure the processed surfaces' required quality.

Modeling of thermomechanical processes that form in the surface layer of processed products makes it possible to obtain criteria ratios, the implementation of which, by selecting technological parameters of processing on their basis, makes it possible to provide the required quality indicators for the working surfaces of products, in particular, to avoid defects such as burns and cracks.

The practical use of the obtained results is that there is information about the distribution of hereditary defects in the presence of electron microscopy of the surface layer of products before finishing operations. Choosing among them the defect of the most significant dimensions l_0 – the characteristic linear size of the structural parameter (structure defect) and using the obtained criteria relations, tooling, and processing modes are assigned to link the limiting characteristics of the temperature and force fields with control and technological parameters. They set the range of combinations of these parameters that satisfy the obtained thermomechanical criteria. At the same time, the processed material's properties are considered, and the required quality of products is guaranteed.

References

1. Shahzad, A.: Thermophysical Properties of Complex Materials. BoD – Books on Demand (2020)
2. Kishawy, H.A., Hosseini, A.: Machining Difficult-to-Cut Materials: Basic Principles and Challenges. Springer International Publishing, Cham (2019)
3. Shyha, I., Huo, D. (eds.): Advances in Machining of Composite Materials: Conventional and Non-conventional Processes. Springer International Publishing, Cham (2021)
4. Stephenson, D.A., Agapiou, J.S.: Metal Cutting Theory and Practice. CRC Press (2018)
5. Raja, V.S., Shoji, T.: Stress Corrosion Cracking. Elsevier (2011). https://doi.org/10.1533/9780857093769
6. Doi, T.K., Ohnishi, O., Uhlmann, E., Dethlefs, A.: Lapping and Polishing. In: Handbook of Ceramics Grinding and Polishing, pp. 263–325. Elsevier (2015). https://doi.org/10.1016/B978-1-4557-7858-4.00006-6
7. Caggiano, A., Teti, R.: CBN grinding performance improvement in aircraft engine components manufacture. Proc. CIRP **9**, 109–114 (2013)
8. Kumar, V., Ram, M.: Predictive Analytics: Modeling and Optimization. CRC Press (2021)
9. Frémond, M., Maceri, F., Vairo, G. (eds.): Models, simulation, and experimental issues in structural mechanics. SSSSM, vol. 8. Springer, Cham (2017). https://doi.org/10.1007/978-3-319-48884-4
10. Kuna, M.: Finite Elements in Fracture Mechanics: Theory – Numerics – Applications. Springer, Netherlands (2015)
11. Nowacki, W.: Thermoelasticity. Elsevier Science (2013)
12. Povstenko, Y.: Fractional Thermoelasticity. Springer (2015)
13. Bradt, R.C., Hasselman, D.P.H., Lange, F.F.: Fracture Mechanics of Ceramics: Volume 2 Microstructure, Materials, and Applications. Springer US (2013)
14. Andrianov, I., Gluzman, S., Mityushev, V.: Mechanics and Physics of Structured Media: Asymptotic and Integral Equations Methods of Leonid Filshtinsky. Academic Press (2022)
15. Ebrahimi, F.: Mechanics of functionally graded materials and structures. BoD – Books on Demand (2020)
16. Muskhelishvili, N.I.: Singular Integral Equations: Boundary Problems of Function Theory and Their Application to Mathematical Physics. Dover Publications (2013)
17. Ladopoulos, E.G.: Singular Integral Equations: Linear and Non-linear Theory and Its Applications in Science and Engineering. Springer, Berlin Heidelberg (2014)
18. Kheir, N.: Systems Modeling and Computer Simulation. Routledge (2018)
19. Al-Azzawi, A.: Fiber Optics: Principles and Advanced Practices, 2nd edn. CRC Press (2019)

A Control Configured Mechatronic Mechanism

Vasily Larshin[1](✉), Anatoly Gushchin[1], Volodymyr Marchenko[2], Alina Tselikova[3], and Igor Dudarev[3]

[1] Odessa Polytechnic National University, 1, Shevchenko Ave., Odessa 65044, Ukraine
vasilylarshin@gmail.com

[2] Odessa Military Academy, 10, Fontanskaya Doroga Street, Odessa 65009, Ukraine

[3] Odessa State Academy of Civil Engineering and Architecture, 4, Didrihson Street, Odessa 65029, Ukraine

Abstract. The design features of a mechatronic mechanism (MM) configured by the control were given. MM provides automatic control of the torque on the shaft by adjusting the pressing force of the working body to the object to be machined, for example, the force of pressing the cutting tool to the workpiece. The MM relevance, for example, in motor-spindles of CNC machines was determined by the need to improve (based on a new machining technology paradigm) the technology for machining parts from modern superhard and difficult-to-machine materials, as well as materials with pronounced anisotropic properties (materials with coatings, reinforced fibrous materials), precious stones, and porous materials. Design of a MM sensitive element was developed for an automatic control system "by disturbance" and based on a ball-bearing screw converter with the necessary range of linear movement of the working body of the technological machine. The developed MM mathematical model connects its geometric and electromagnetic parameters and shows the influence of the helical groove angle in the ball-bearing screw converter on the force of pressing a cutting tool to the machining workpiece. It made it possible to develop recommendations on the value of the helical groove angle for implementing various technological machining modes in nano-, micro-, and macro-technologies conditions.

Keywords: Product Innovation · Machining Technology · Electromagnetic Force · Electrodynamic Force · Action and Counteraction · Pressing Force · Torque Control

1 Introduction

The increasing volume of creation of competitive technology in industrial production involves using new structural materials with improved physical and mechanical properties or a strictly organized structure. Examples of such materials are numerous. These are artificially grown crystals with a hardness that is commensurate with the hardness of the machining (e.g., cutting) tool working part. These are polymer composite materials with solid inclusions adjacent to a soft polymer component. These are aluminum alloys with increased adhesion to the cutting tool. New materials require a significant novelty

V. Tonkonogyi et al. (Eds.): InterPartner 2023, LNME, pp. 72–81, 2024.
https://doi.org/10.1007/978-3-031-42778-7_7

in approaches to their mechanical machining technology. An equally important aspect is the technological problems of high-quality machining of materials with a pronounced anisotropy of properties throughout their volume. A representative of this kind of material is, for example, syntegran (synthetic granite). It contains solid inclusions and a binder component – synthetic resins. These resins have high adhesion not only to solid fractions but also to the cutting tool. This is especially evident at elevated cutting temperatures, as it leads to various machining defects if it is performed without temperature control based on cutting force and torque control.

The above materials are difficult to process with high quality because, for example, it is difficult to maintain the required trajectory of the drill due to the possible withdrawal of its axis from a given direction. The reason is the occurrence of uneven lateral forces when the geometric dimensions of the voids of the material being processed are commensurate with the dimensions of the cutting part of the machining tool.

Opportunities for quality machining are provided by a mechatronic technological system based on a mechatronic mechanism (MM), which allows at the initial stage providing the required force parameters of machining (for example, cutting force and torque) and at the final stage of machining – the necessary accuracy. Such MM can perform its functions only due to the joint action of the programmable forces (from the MM control system) and the oppositely directed forces from the external environment [1]. These two forces (action and reaction) "configure" MM; give it the very ability to function. The conventional machines (designed not on the control configured MM concept) have inherent self-stability in their designs. Therefore, they can operate in the self-braking mode (due to the action of friction forces) and have a static transfer function. In the case of control-configured MM, however, there is a structurally inherent mechanical instability of MM. Without an automatic control system and environmental resistance, such MM will not work [2]. The specified mechanical instability often turns out to be necessary to ensure the MM required dynamic characteristics, with an instantaneous change in loads on the machining tool. MM developed in the paper is connected with creating and practically implementing a new paradigm in machining the above materials, for example, on CNC machines.

The paper aims to develop a control-configured mechanism, i.e., configured by control, to provide kinematic (displacement, speed, acceleration) and power (cutting force and torque) machining parameters, for example, on CNC machines. The object of research is modern mechatronic and intelligent technological machines, for example, CNC machines. The subject of research is MM, which is configured by automatic control.

2 Literature Review

This paper proposes a new technology-oriented approach, the basics of which have been described previously with two [1] and one [3] control windings. The ball-bearing screw converter described below is part of MM, designed and implemented based on the well-known concept of a control-configured vehicle or, in other words, a vehicle configured by control [4]. The set of control operations forms a control process. By analogy, a set of work operations forms a work (physical) process. Automatic devices can also partially or entirely perform control operations [5]. However, nothing is said here on the control configured vehicle concept.

The literature notes that the control-configured vehicle concept can be achieved through intelligent control [6] and full intelligent automation [7].

Implementing the control-configured vehicle concept in systems operating in the multilevel control mode is most effective in adaptive [8] and intelligent [9] technological machines. Therefore, it is necessary to rely on the well-known provisions in this area: hierarchical control [10] and intelligent machining systems [11]. At the same time, the principle of hierarchical (multilevel) control makes it possible to optimize the ratio between the control ranges of power parameters for systems of the upper and lower levels of control, on the one hand [12], and the choice of the principle of automatic control (by deviation, by disturbance), on the other hand [13].

Conventional systems for monitoring [14] and controlling [15] production processes are not automatic control systems. These are information support systems for the operator to control tool conditions [16] and machining processes [17]. Therefore, unlike automatic control systems, they reach the commercial level in general for machining [18] and particularly with an active disturbance rejection control [19]. Some examples are spindle monitoring and power monitoring are presented in the research works [20] and [21], respectively.

3 Research Methodology

3.1 Machining Technology Influence on the MM Design

The following is a description and principle of operation of the latest ball-bearing screw converter design, which has shown reliable results as a comparator and force converter in mechatronic "action-counteraction" systems in machining. It includes a non-magnetic (textolite) body 1, in which a kinematic ball-bearing pair "screw-nut" (ball-bearing screw) with screw 2 (aka spindle) and nut 3 is installed (Fig. 1).

This pair is not self-braking, i.e., screw 2 can perform angular and reciprocating movements (concerning nut 3) in both clockwise directions, while the rotational movement of "screw" 2 is set by "nut" 3. A pair of "screw-nut" is fixed in housing 1 on two ball bearings 4 which allows both the "nut" 3 to rotate freely about its longitudinal axis and the "screw" 2 to carry out linear movements along the same axis.

The following are marked in Fig. 1: Q is the weight (gravity) of MM moving parts, N; F_{ed} is the electrodynamic force acting on the spindle, N; F_{ed} is the electromagnetic force acting on the spindle, N. The rotational movement on the ball-bearing screw converter nut is transmitted through bypass coupling 5 from motor 6. The linear movement of spindle 2 is carried out due to the oppositely directed forces F_{em} and F_{ed}, applied to its axis. This allows the spindle to smooth helical movements inside the "nut" without backlash and jamming.

Nut 3 is a source of rotational motion from the drive motor 6. It is made of non-magnetic material (brass, bronze, titanium alloy) and is a hollow cylindrical body (Fig. 1, a, b, c), on the side surface of which a two-start slotted groove is made with step t at the angle of inclination α. This groove performs the function of a two-start thread with

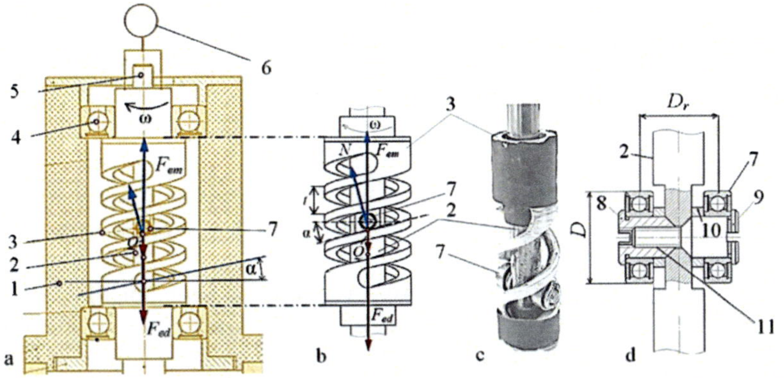

Fig. 1. Designing the ball-bearing screw converter.

the same pitch and angle. From the ends of the "nut" 3, it is equipped with two guide bushings (upper and lower) rigidly fixed inside it (not shown in Fig. 1). The bushings are made of non-magnetic material (fluoroplastic), which has a low coefficient of sliding friction. They provide the possibility of a smooth multidirectional helical movement of the spindle ("screw" 2) relative to the "nut" 3.

Screw 2, also known as the spindle, installed inside nut 3, is made of titanium alloy. This allows for reducing the MM moving part weight. On screw 2 (spindle), two support ball bearings 7 are fixed on the same axis. This axis is perpendicular to the longitudinal axis of the spindle. The outer rings of ball bearings 7 act as a bearing surface for the external thread of the "screw" 3 in a non-self-locking "screw-nut" connection. They provide the required mutual axial position of screw 2 relative to nut 3. Bearings 7 are fastened together by threaded tightening elements 8 and 9 (Fig. 1, d). To eliminate the error in the relative position of the bearings 7 and the spindle 2, spring washers (compensators) 10 are provided (Fig. 1, d), and the gaps in the threaded connection 11 are filled with sealant.

Such a design of the ball-bearing screw converter makes it possible to implement the control-configured MM with a dynamic balance of oppositely directed forces and torques that inevitably arise during machining.

When assembled (Fig. 1, c), the ball-bearing screw converter transforms the rotation of "nut" 3 and the reciprocating movement of "screw" 2 into the screw movement of the latter relative to nut 3. This is necessary to ensure a dynamic balance of forces and torques arising during system operation.

3.2 Control Configured Elements

The rotation of the spindle with the tool is provided by the drive of the main movement, and its reciprocating movements are provided by the joint action of three forces (Fig. 1): Q, F_{ed}, and F_{em}, N. Moreover, $Q = mg$, where Q and m are the weight (in N) and mass (in kg) of MM moving parts; g is free-fall acceleration, м/s^{-2}.

The electrodynamic force F_{ed} (in Newtons) acting on a current-carrying conductor of length l (in meters) in a magnetic field with induction B (in Tesla) can be found based on the well-known dependence for the Ampère force [1]: $F_{ed} = k_{ed}BIl$, where k_{ed} is an experimentally determined coefficient.

The electromagnetic force F_{em} (in Newtons) created by the coil with current (solenoid) can be found by the equation [1]

$$F_{em} = k_{em}B^2S, \tag{1}$$

where k_{em} is the coefficient determined experimentally; S is the effective cross-sectional area of the solenoid, m^2.

At the beginning stage of the cycle of the technological operation of processing, in which $F_{ed} = 0$, the forces Q and F_{em} during the execution of this cycle remain constant, and for this stage of the cycle, the condition is fulfilled: $F_{em} \geq Q$ which is close to the equality $F_{em} = Q$.

The force F_{ed} is a variable value and can be programmatically changed in the range: $0 \leq F_{ed} \leq [P]$, where $[P]$ is the allowable axial load on the drill (in Newtons), while simultaneously applying a torque to it, N·m.

A wide range of force change F_{ed} allows implementing three modes of MM operation: (1) accelerated approach of the tool to the machining zone; (2) working feed which provides the required value of power parameters (axial force and torque); (3) removal of the tool from the cutting zone (periodic during drilling and final at its completion).

Let us take a closer look at these three modes.

1. The approach of the tool to the cutting zone is provided by software with two counteracting forces F_{ed} and F_{em}, on the spindle, the ratio between which allows selecting the necessary feeds of the tool with its accelerated approach and working stroke, which begins when the drill initially touches the surface to be machined.
2. The optimal power mode of the machining process (for example, cutting while drilling) is carried out by the forces F_{ed} and F_{em}, while the value of F_{ed} along the length of the processing is changed programmatically at $F_{em} =$ const, which ensures the optimization of cutting modes. The cutting resistance torque T_{cut} that occurs in the machining zone is one of the power components involved in the dynamic balance of the spindle in its axial direction. This torque acts through the tool and the spindle on the ball-bearing screw converter, ensuring optimal machining process power parameters.
3. The removal of a tool (for example, a drill) from the cutting zone is realized by the joint action of two oppositely directed forces F_{ed} and F_{em}. Periodic and final removal of the tool during drilling is performed to clean it from chips and after the end of processing, respectively. By changing the ratio between the forces F_{ed} and F_{em}, it is possible to control the withdrawal tool speed from the hole. The maximum withdrawal tool speed is reached with $F_{ed} = 0$.

The described principle of operation of the ball-bearing screw converter corresponds to the control-configured MM for which maneuverability is a priority property. This is achieved by including in the balanced system of forces the mechanical resistance of the external environment as a necessary and active component of this system. In this system, oppositely directed forces are F_{ed} and F_{em} in "dynamic equilibrium", which is stabilized by the automatic control system "by disturbance".

Thus, a software change in the magnitude of one force F_{ed} allows implementing many options for performing technological strokes and operations, including intermittent, vibrational, or pulsed machining. For example, when drilling, the axial force and torque take on the values and directions determined by this technological process requirements.

3.3 A Mechatronic Mechanism Mathematical Model

As a power and kinematic mechanism, the ball-bearing screw converter participates in dosing the forces and torques applied through the spindle to the tool, and each "dose" must be sufficient for the high-quality implementation of the process at any moment of its course. It is known, for example, that when drilling through deep holes, the axial force and cutting torque should not be constant over the entire drilling depth. So, for example, at a depth of $0.05L$ (drill plunge), the forces should essentially differ in magnitude from those power parameters of the cutting mode that are realized at a depth of $0.5L$, and completely different cutting forces are assumed at the drill exit from the hole being drilled, i.e., at a drilling depth of $0.95L$, where L is the drilling depth.

Thus, the body for dosing forces and torques in the system of adaptive control of cutting processes must be programmable and flexible. In addition, any automatic control (regulation) system assumes the presence in it of a body for comparing the value of the actual state of the controlled parameter with a given (preset) value. Such a role in the considered MM is played by the ball-bearing screw converter, which provides not only a comparison of the force load in the cutting zone with a programmed value but also acts as a dispenser of normalized forces supplied to the cutting zone.

Let us call a "dose" a part of the sum of forces Q,F_{ed}, F_{em}, and trace how the "dose" which is necessary for the implementation of the programmed forces in the cutting zone is "separated" from the "whole". This can be traced by analyzing successively the schemes of action of the forces presented in Fig. 1, a, b, d, and Fig. 2.

Let us replace the set of forces $\overline{Q}$, $\overline{F}_{ed}$, and $\overline{F}_{em}$ (Fig. 2) with one resultant force $2\overline{T}$, i.e.

$$2\overline{T} = \overline{Q} + \overline{F}_{ed} + \overline{F}_{em}. \quad (2)$$

Let us transfer the resultant force $2\overline{T}$ to the point C, formed by the intersection of the axes $S - S\ and\ C - C$ (Fig. 2).

Forces $\overline{Q}$ and $\overline{F}_{ed}$ are directed to the surface to be machined, and the force F_{em} is directed "from" the same surface.

Since the vectors $\overline{Q}$, $\overline{F}_{ed}$, and $\overline{F}_{em}$ are located on the same straight line, the modulus of the resultant force can be found in the equation

$$2T = F_{ed} + Q - F_{em}. \quad (3)$$

Since the force $2T$ is distributed through two ball bearings, it is possible to operate with the force T, with the transfer of its action to the points lying on the axis of these ball bearings.

The force vector T must be transferred to the point O (taking into account the parallel transfer rule), in which the outer ring of the support ball bearing is forced to close with the inclined plane of the "nut" turn.

As a result of such a transfer, the reaction $\overline{N}$ occurs (Fig. 2) at the point O, and $N = T\cos\alpha$. Further, it is necessary to decompose the reaction N into two components (N_x, horizontal) and N_z, vertical), as well as determine their values as follows (Fig. 2, b): $N_x = N\sin\alpha$ and $N_z = N\cos\alpha$.

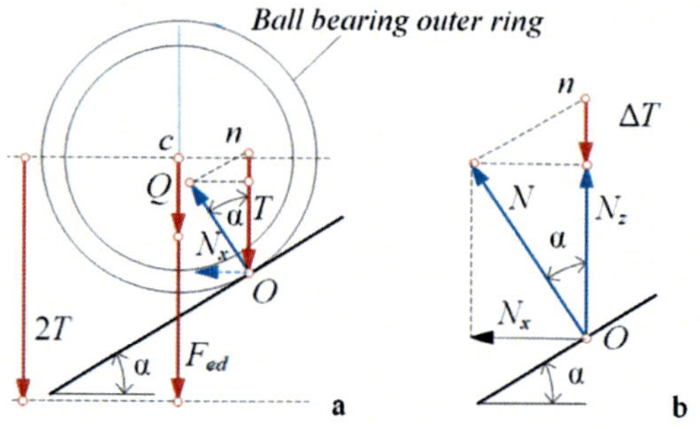

Fig. 2. Force interacting of a screw-spindle ball bearing with the nut threaded surface (a) and the same on a scale of 2:1 (b).

That is why

$$N_z = T\cos\alpha\cos\alpha = T\cos^2\alpha. \tag{4}$$

In turn

$$\Delta T = T - N_z. \tag{5}$$

That is

$$\Delta T = T - T\cos^2\alpha = T\left(1 - \cos^2\alpha\right) = T\sin^2\alpha. \tag{6}$$

Thus

$$\Delta T = 0.5(Q + F_{ed} - F_{em})\sin^2\alpha \tag{7}$$

Focusing the two components of ΔT on one central axis, we get

$$2\Delta T = \left(mg + k_{ed}BIl - k_{em}B^2S\right)\sin^2\alpha. \tag{8}$$

Equation (8) can be used to perform practical calculations for the created MM, configured by control, including the linear impulse definition for the resultant force $2\Delta T$.

4 Results and Discussion

4.1 Influence of the Helical Groove Angle on the Axial Machining Force

Representing Eq. (8) in dimensionless form, we get

$$\eta = \frac{2\Delta T}{\left(mg + k_{ed}BIl - k_{em}B^2S\right)} = \sin^2\alpha. \tag{9}$$

Now it is possible to build a graph (Fig. 3), which allows getting visual recommendations on the value of the helical groove inclination angle α for the implementation of various technological processing modes: zone *A* – nanotechnology; zone *B* – microtechnology; zone *C* – macrotechnology.

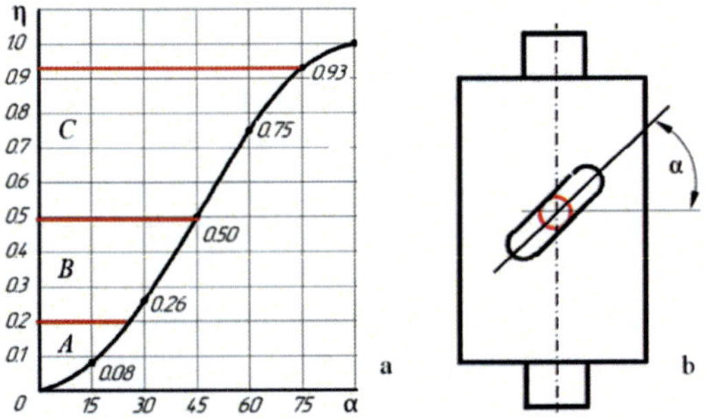

Fig. 3. Influence of the helical groove angle α on the dimensionless axial force η (a) and the location of the angular dimension α (b).

4.2 Results of Experimental MM Studies

Experimental studies to establish the MM technological capabilities and areas of its effective application were carried out in Germany (Stuttgart Technical University in cooperation with HUGO RECKERTH GMBH) and Ukraine (National University Odessa Polytechnic and Odessa State Academy of Civil Engineering and Architecture). Research has been carried out in many areas, including: 1) achieving dimensional accuracy and quality characteristics of the surface layer of parts made of difficult-to-machine materials (research in Germany); 2) establishing a nomenclature of materials to be machined with unusual physical and mechanical properties, as well as improving the MM design in the direction of improving its quality indicators (research in Ukraine).

Another direction of successful experimental work is engraving inscriptions on granite slabs with a diamond tool with a working part diameter of 0.7 mm (rotation speed of 20,000 rpm, and water as coolant). Results: the absence of macro- and micro-chipping of solid mineral inclusions in the material to be processed.

Studies of the developed MM in machining superhard and anisotropic materials showed the expediency of endowing the MM with sensory organs, which made it possible to identify the opportunity for new product innovation. The MM peculiarity, established in previous studies, associated with tactile sensitivity of the working body of a technological machine (analogous to a living organism's properties), is a prerequisite for creating intelligent technological machines of various natures and purposes. This opens the way for the production of mechatronic technological systems of a new generation that cover the entire technosphere of human activity, for example, medical, military, and industrial equipment.

5 Conclusions

The following features of modern difficult-to-machine constructional materials, including ceramic and composite ones, have been established: increased hardness, pronounced anisotropy of properties, and susceptibility to defects. These features make it challenging to apply existing machining technologies and lead to the need to develop fundamentally new technologies for their high performance and defect-free machining on CNC mechatronic machines.

A new paradigm of the control-configured mechatronic mechanism has been formulated based on a technology-oriented approach, according to which the functioning of this mechanism is determined by features of high performance and defect-free machining, for example, on CNC machines. Such a technology can be realized due to the combined action of a programmable electrodynamic force and the counteraction of the environment. These two forces (action and counteraction) "configure" the mechatronic mechanism, i.e., give it the ability to function according to the machining technology needed. These are the main advantages of this mechatronic mechanism and the results' practical significance.

The design of the mechatronic mechanism and its mathematical model have been developed, which show the possibility of regulating the power machining parameters (cutting force and torque) in a wide range, i.e., from nano- and micro- to macro-values. Recommendations are given on the value of the helical groove angle of the ball-bearing screw converter for implementing various technological operations connected with the machining of the above materials.

References

1. Larshin, V.P., Gushchin, A.M.: Mechatronic technological system information support. Appl. Aspe. Info. Technol. **4**(2), 153–167 (2021). https://doi.org/10.15276/aait.02.2021.3
2. Suomalainen, M., Karayiannidis, Y., Kyrki, V.: A survey of robot manipulation in contact. Robot. Auton. Syst. **156**, 104224 (2022). https://doi.org/10.1016/j.robot.2022.104224
3. Gushchin, A.M., Larshin, V.P., Lysyi, O.V.: Adaptive mechatronic mechanism information model. Appl. Aspe. Info. Technol. **5**(2), 105–119 (2022). https://doi.org/10.15276/aait.05.2022.8
4. Isermann, R.: Mechatronic systems – innovative products with embedded control. Control. Eng. Pract. **16**(1), 14–29 (2008). https://doi.org/10.1016/j.conengprac.2007.03.010
5. Taylor, A.N., Berrueta, T.A., Murphey, T.D.: Active learning in robotics: a review of control principles. Mechatronics **77**, 102576 (2021). https://doi.org/10.1016/j.mechatronics.2021.102576

6. Yildirim, S.: Intelligent mechatronic systems applications in engineering. In: Ivanov, V., Trojanowska, J., Pavlenko, I., Rauch, E., Peraković, D. (eds.) Advances in Design, Simulation and Manufacturing V. DSMIE 2022. Lecture Notes in Mechanical Engineering, pp. 123–134. Springer, Cham (2022). https://doi.org/10.1007/978-3-031-06025-0_13
7. Karabegović, I., Karabegović, E., Mahmić, M., Hak, E., Dašić, P.: The implementation of industry 4.0 supported by service robots in production processes. In: Ivanov, V., Trojanowska, J., Pavlenko, I., Zajac, J., Peraković, D. (eds.) Advances in Design, Simulation and Manufacturing IV. DSMIE 2021. Lecture Notes in Mechanical Engineering, pp. 193–202. Springer, Cham (2021). https://doi.org/10.1007/978-3-030-77719-7_20
8. Dobrinski, A., Möhring, H.-C., Stehle, T.: Development of an adaptronic spindle for a faultless machining of homogeneous and inhomogeneous materials. Journal of Machine Engineering **21**(2), 5–23 (2021). https://doi.org/10.36897/jme/136277
9. Dobrinski, A., Dudarev, A.: Intelligent automated drilling in the laminate composites and hybrid materials. Materials Today: Proceedings **38**(4), 1980–1983 (2021). https://doi.org/10.1016/j.matpr.2020.09.723
10. Lima, P.U., Saridis, G.N.: Design of intelligent control systems based on hierarchical stochastic automata. World Scientific Publishing Co. Pte. Ltd, Singapore (1996)
11. Kuppuswamy, R., Jani, F., Naidoo, S., Jongh, Q.: A study on intelligent grinding systems with industrial perspective. Int J Adv Manuf Technol **115**, 3811–3827 (2021). https://doi.org/10.1007/s00170-021-07315-9
12. Mitrishkin, Y., Rodolfo, E.: Hierarchical control system for complex dynamical plants. In: Proceedings of the 6th International Conference on Informatics in Control, Automation and Robotics, Intelligent Control Systems and Optimization, ICINCO 2009, pp. 56–65. Milan, Italy (2009)
13. Klocke, F.: Manufacturing Process 2. Grinding, Honning, Lapping. Springer, Verlad Berlin Heidrlberg (2009)
14. Stavropoulos, P., Chantzis, D., Doukas, C., Papacharalampopoulos, A., Chryssolouris, G.: Monitoring and control of manufacturing processes: a review. Procedia CIRP **8**, 421–425 (2013). https://doi.org/10.1016/j.procir.2013.06.127
15. Goyal, D., Mongia, C., Sehgal, S.: Applications of digital signal processing in monitoring machining processes and rotary components: a review. IEEE Sens. J. **21**(7), 8780–8804 (2021). https://doi.org/10.1109/JSEN.2021.3050718
16. Rehorn, A., Jiang, J., Orban, P.: State-of-the-art methods and results in tool condition monitoring: a review. Int. J. Adv. Manuf. Technol. **26**, 693–710 (2005). https://doi.org/10.1007/s00170-004-2038-2
17. Ambhore, N., Kamble, D., Chinchanikar, S., Wayal, V.: Tool condition monitoring system: a review. Materials Today: Proceedings **2**(4–5), 3419–3428 (2015). https://doi.org/10.1016/j.matpr.2015.07.317
18. Byrne, G., Dornfeld, D., Inasaki, I., Ketteler, G., König, W., Teti, R.: Tool condition monitoring (TCM) – The status of research and industrial application. CIRP Ann. **44**(2), 541–567 (1995). https://doi.org/10.1016/S0007-8506(07)60503-4
19. Fareh, R., Khadraoui, S., Abdallah, M.Y., Baziyad, M., Bettayeb, M.: Active disturbance rejection control for robotic systems: A review. Mechatronics **80**, 102671 (2021). https://doi.org/10.1016/j.mechatronics.2021.102671
20. Process control system for grinding machines, https://www.balancesystems.com/process-control-systems-for-grinding-machines-touch-detection-balancing-gauging/, last accessed 24 June 2023
21. Tian, Y.B., Liu, F., Wang, Y., Wu, H.: Development of portable power monitoring system and grinding analytical tool. J. Manuf. Process. **27**, 188–197 (2017). https://doi.org/10.1016/j.jmapro.2017.05.002

Determination of an Effective Supply Chain: Case Study for Delivering Products from the USA to Ukraine

Olexiy Pavlenko[1], Dmitriy Muzylyov[2], and Vitalii Ivanov[3,4](✉)

[1] Kharkiv National Automobile and Highway University, 25, Yaroslava Mudrogo Street, Kharkiv 61002, Ukraine
[2] State Biotechnological University, 44, Alchevskyh Street, Kharkiv 61002, Ukraine
[3] Sumy State University, 2, Rymskogo-Korsakova Street, Sumy 40007, Ukraine
ivanov@tmvi.sumdu.edu.ua
[4] Technical University of Kosice, 1, Bayerova Street, Presov 08001, Slovak Republic

Abstract. The solution to finding optimal product delivery variants is always relevant. The best supply chain scheme choice is essential for long-distance transportation of specific cargoes, such as vehicles and spare parts. Therefore, the purpose of the study was to establish the dependence between the cost of delivery of automotive products on the technological and economic parameters of the provision of transport services. This approach made it possible to select the best supply chain option based on the minimum unit cost value during shipping vehicles and spare parts in containers from the USA to Ukraine. Improvement of operations determining the interaction of different kinds of transport, as well as achieving the optimal level of distribution of goods between them, allows the development of timely rational management decisions, significantly increasing the quality and efficiency of the supply chains. The study used mathematical modeling as the primary tool to achieve goals. The designed model includes several cost parameters that characterize specific conditions of certain transportation schemes. Unit cost values were compared to determine the most efficient option from the proposed ones. The most effective option is Supply Chain 3 because it has minimal values of unit costs in most tests of the experiment. The newly developed regression models give a universal mathematical tool by which researchers can choose an effective option of supply chains for delivering vehicles and their repair parts in containers. This approach can assess other distribution channels with the same or analog conceptions of route designing.

Keywords: Sustainable Supply Chain · Containers · Simulation · Regression Analysis

1 Introduction

The goods supply process on world markets is connected with performing a number of works, operations, and services, the complex of which will ensure effective goods distribution. Companies supplying goods from abroad must solve tasks that arise of

V. Tonkonogyi et al. (Eds.): InterPartner 2023, LNME, pp. 82–93, 2024.
https://doi.org/10.1007/978-3-031-42778-7_8

preparing effective supply chains for servicing orders [1] in this situation. It is important for any customer of logistics services with significant demand for quality services at lower prices.

A reliable supply chain organization is formed using existing infrastructure elements of the transport network, considering risks and failures [2]. The resources must use more effectively, and the logistical system must build with the simultaneous use of various supplying sources and with using such delivery principles as "flexibility" and "just-in-time" [3] because this help to reduce failure risks and problems in supply chains.

The development of Ukraine's economic ties with world trade leaders directly affects the creation of reliable supply chains. The United States is one of the promising markets for economic recovery and a reliable partner of our country. We can say that the United States ranks fifth [4] according to results analyzing importing goods cost by countries. The imports' percentage from the United States reaches 11% of the total import volume of other countries. The most significant amount of goods are imported from the United States according to the Ukrainian commodity nomenclature of foreign economic activity – mineral fuel, oil, and its distillation products (30.4% of total imports from the United States to Ukraine), and in second place – land transport, except for rail, accounting for 21.5% of the total volume. The volume of vehicle imports increased by 36.4% [4] compared to the last year, 2021. All this suggests that demand will be restored in the post-war period, and this perspective direction for determining options of reliable supply chains will function more efficiently.

The need to develop new reliable systems for goods supply between countries should be based on the principles of Industry 4.0 [5], SMART solutions [6, 7], sustainable development of all kinds of transport [8], as well as on the definition of rational conditions for all participants in the supply chain [9]. Therefore, a promising object of research using mathematical simulation methods is determining an effective supply chain for automotive spare parts in containers from the United States to Ukraine.

The research aim is to determine the effective logistics chain for supplying automobile spare parts and automotive equipment in containers from the USA to Ukraine, considering possible options for organizing work and the advantages of existing routes when using various means of transport. The scientific novelty of the proposed research is concluded in new parameter sets that are used for designing regression models. This aspect allows the creation of more rational routing for delivery vehicles and their repair parts in containers.

2 Literature Review

It is difficult to underestimate the importance of supply chain innovation for effective operational management practices [10]. New supply chains designed using digital technologies give the most outstanding results [11]. Sustainability is fundamental in creating new logistical chains [12]. It creates an opportunity to form effective supply chain options based on a statistical analysis of demand data [13]. An integrated supply chain structure based on the hierarchical clustering triplet is also proposed to achieve sustainability, which is implemented by a leading furniture company [14] and in textile manufacturing [15].

Flexibility, resilience, and reliability as supply chain trends have become relevant strategies to address the current challenges of the digital world [16]. However, digitizing or adopting digital technologies in supply chain operations poses significant challenges such as coordination, tracking, and flexibility. On the other hand, initiatives such as Industry 4.0 are becoming industry-level trends, improving supply chain operations through high levels of transparency, control, dynamic network reconfiguration, and modernization provided by intelligent technology integration [17].

The actual supply chain management tasks are very complex, so the procedure for optimizing them requires significant computational resources due to the extensive size and uncertainty of their critical variables [18]. Simulation optimization is widely used to determine optimal variables because the problem is too complex [19, 20]. It is also worth considering the reliability of optimal solutions due to the uncertainty of real systems. The prediction characteristics of artificial neural networks compare with different structural settings in the article [21], and the authors propose to select the most appropriate networks for obtaining reliable solutions.

Designing a reliable multimodal supply network for products with various uncertainty sources is important while minimizing costs and risks [22]. The main aspect is achieving contradictory costs and risk estimates of significance for all supply chain participants. It is often necessary to create a system for product supply with numerous participants and constantly improve management [23, 24]. The flexibility of designed supply chains is highlighted by database preparation of the whole system for possible risk situations.

Planning the efficiency of container use has become more complicated due to the rapid development of the container transportation market in the world. It is particularly evident with an increase in the number of mega-container carriers operated by major container operators [25]. The authors of this article developed a model of a container transportation network that covers both the main route and a segment of coaster shipping, considering the sea container port as a hub. Ports have a crucial role in the container transportation system, especially in the main hubs of the continents – Asia (ports of China), North America (ports of the USA), and Europe (ports of the Netherlands and Germany) [26].

3 Research Methodology

The current, reliable conditions for functioning the supply system, which Ukraine seeks, require more attention to logistics, fast growth, and improvement. The efficiency and quality of supply chains significantly depend on actions optimization to coordinate the functioning of different kinds of transport, the optimal distribution of consignment sizes between them, and the timely formation of rational management decisions. Particular attention should be paid first to the two most important indicators of the transport process – the transportation cost and the timing of order implementation for the cargo delivery.

The United States ranks second in the world after China in terms of seaport quantity and first in terms of the total length of highways and railways. At the same time, real assets are not considered an advantage. Exclusive importance is attached to the management of information flows because industry leaders' efforts are concentrated on the technology of organizing collections, analysis, and decision-making. That's why it gives good results. For example, Baltimore receives almost one and a half million containers yearly as the most efficient US port. The port handles ships of 14,000 containers (TEUs) and up to 9,000 trucks [27].

Some essential features of the American transportation market should be considered when planning and organizing the delivery of containers from the United States to Ukraine because ignorance of them can lead to curious but expensive situations. There are, for example, other norms of lengths, weights, and another norm of load on the axis (all about 20% more than in Ukraine or Europe) in the USA. There are also no height restrictions on the railway and, basically, "53 ft" containers for transporting goods are used, and not the usual "20 ft" and "40 ft" (through the use of powerful diesel traction on railways, which allows you to transport "53 ft" containers in 2 rows).

The port of New York and the port of Houston are the main ports for designing supply chains of cargo in containers, including automobiles, from the United States. Ports of destination are in Ukraine - the port of Odesa, and in the European Union - the port of Klaipeda (Lithuania) (Fig. 1). Thus, interaction schemes of participants in the cargo delivery process in containers are formed based on existing lines of maritime transport (Table 1).

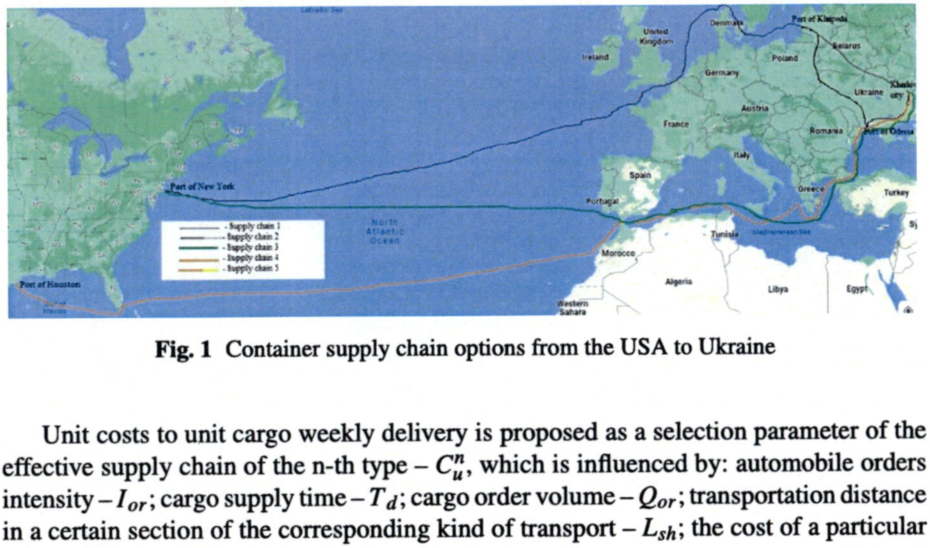

Fig. 1 Container supply chain options from the USA to Ukraine

Unit costs to unit cargo weekly delivery is proposed as a selection parameter of the effective supply chain of the n-th type – C_u^n, which is influenced by: automobile orders intensity – I_{or}; cargo supply time – T_d; cargo order volume – Q_{or}; transportation distance in a certain section of the corresponding kind of transport – L_{sh}; the cost of a particular service – S_c.

$$C_u^n = f(I_{or}, T_d, Q_{or}, L_{sh}, S_c), n = 1, ..., N \quad (1)$$

The mathematical model for determining the effective supply chain has been formed to achieve the goal set in the article

$$C_u^n = \frac{\sum_{i \in K} C_{l(ul)i}^{cont} + \sum_{j=1}^{J} (C_{i(ul)j}^{car} + C_{shj} + C_{\exp j}^{sh}) + C_{doc}^{\inf}}{I_{or} \cdot T_w} \quad (2)$$

Table 1 Supply Chain Designation

Supply chain type	Departure area	Dispatching port	Destination port	Railway	Road carrier	Recipients' area
"Supply Chain 1"	northern part of the United States	the port of New York (USA)	the port of Klaipeda (Lithuania)	the railway station of Klaipeda (Lithuania)- the railway station of Odesa (Ukraine)	the carrier by road in Kharkiv (Ukraine)	the company's warehouse in Kharkiv (Ukraine)
"Supply Chain 2"	northern part of the United States	the port of New York (USA)	the port of Klaipeda (Lithuania)	-	the carrier by road in Kharkiv (Ukraine)	the company's warehouse in Kharkiv (Ukraine)
"Supply Chain 3"	northern part of the United States	the port of New York (USA)	the port of Odesa (Ukraine)	-	the carrier by road in Kharkiv (Ukraine)	the company's warehouse in Kharkiv (Ukraine)
"Supply chain 4"	southern part of the USA	the port of Houston (USA)	the port of Odesa (Ukraine)	-	the carrier by road in Kharkiv (Ukraine)	the company's warehouse in Kharkiv (Ukraine)
"Supply Chain 5"	southern part of the USA	the port of Houston (USA)	the port of Odesa (Ukraine)	the railway station of Odesa - the railway station of Kharkiv		the company's warehouse in Kharkiv (Ukraine)

This model takes into account the next factors such as are: the costs of loading (unloading) the container with the i-th cargo type; costs for loading (unloading) of the container on the vehicle (truck, railway wagon, ship) of the j-type of transport ($C^{cont}_{l(ul)i}$); container transportation costs by a vehicle (truck, railway wagon, ship) of j-th kind transport ($C^{car}_{i(ul)j}$); costs of container waiting to be sent by j-th kind transport ($C^{sh}_{\exp j}$); cost of documentation and information support ($C^{\inf}_{doc}$).

The cost components take into account the cost of performing the relevant types of work, the time of their execution, the risk components (idle time, service waiting, failures), work scopes (transportation distances, loading and unloading time, storage, etc.) at all logistics areas of the container supply chain.

4 Results and Discussion

Statistical assessment of experimental study data results is necessary to solve problems according to the proposed method. Experimental studies were conducted according to the data of 10 companies operating in the Kharkiv market (Ukraine). Order flows in 2021 will be used as input. Flow characteristics include cargo order volume, vehicle order intensity, and delivery time. The value was obtained by analyzing orders for the delivery of vehicles and spare parts of automobiles from the United States to Ukraine. The total number of orders (observations) is one hundred.

The required number of observations is determined (Table 2) to obtain the most reliable data on the value changes in the technological process for container delivery by certain supply chains.

Table 2 Results of sample size calculations for specific parameters

Indicators	Cargo order volume (t)	Delivery time (days)	Vehicle order intensity (unit/time)
Expected value	10.05	31.5	6.5
Standard deviation	35.28	35.72	26.10
Measurement error	0.50	1.58	0.33
Sample size	59	65	80

The Chi-square or Pearson square test was used to test correlation significances between two variables. Using Statistica, calculations were made to identify the distribution laws of specific parameters of incoming order flows. It was found that the values of cargo order volumes, vehicle order intensities, and cargo delivery time are distributed according to normal distribution laws of random variables. It is confirmed by the appropriate level of confidence (more than 5%).

Three factors were found using the model data (2) affecting the estimated indicator, and minimum and maximum values were determined - by evaluating selected values of three parameters (Table 3). An experimental plan was designed according to the previously described data set. The results of the obtained data are presented in Table 4.

Table 3 Exposure factor variation levels

Value level	Cargo order volume (t)	Delivery time (days)	Vehicle order intensity (unit/time)
Minimum	0.1	25	1
Maximum	20	38	12

Regression analysis was performed by functions of two types: linear and power functions. The experiment used MS Excel, particularly the Data Analysis -Regression

Table 4 Experiment plan

Observation series	Variation levels		
	Cargo order volume (t)	Delivery time (days)	Vehicle order intensity (unit/time)
Test 1	0.1	1	25
Test 2	0.1	1	38
Test 3	0.1	12	25
Test 4	20	1	25
Test 5	20	1	38
Test 6	20	12	25
Test 7	0.1	12	38
Test 8	20	12	38

function packages. This software helps to find values of such parameters as equation coefficients under variables, variance, and regression statistics. The power function model was found to be most adequate since the value of the "R-square" index is close to one and is 0.96. The values of the regression model coefficients were also checked, which are adequate only for vehicle order intensities according to the values of standard error, "t-statistics," "P-values," and minimal and maximal values. The regression models were designed for five supply chain options of vehicles in containers from the USA to Ukraine (Table 5).

Table 5 Regression models for relevant supply chain options

Supply chain options	Regression model
"Supply chain 1"	$C_u^n = e^{9.38} \cdot I_{or}^{-0.42}$
"Supply chain 2"	$C_u^n = e^{9.39} \cdot I_{or}^{-0.419}$
"Supply chain 3"	$C_u^n = e^{9.19} \cdot I_{or}^{-0.39}$
"Supply chain 4"	$C_u^n = e^{9.63} \cdot I_{or}^{-0.46}$
"Supply chain 5"	$C_u^n = e^{9.6} \cdot I_{or}^{-0.45}$

The parameter was calculated based on obtained regression models to find an effective option. The parameter values are accepted as vehicle order intensity from minimum to maximum rates $\{1.12\}$. Based on the modeling results, the designed graphs show the dependences of container unit cargo delivery costs from the change in vehicle order intensities (Fig. 2).

The plot curves show that Supply Chain 3 is less costly. At the same time, unit costs have minimum values when ordering volumes reach twelve vehicles per week. The most expensive are Supply Chain 4 and Supply Chain 5 since the distance of transportation

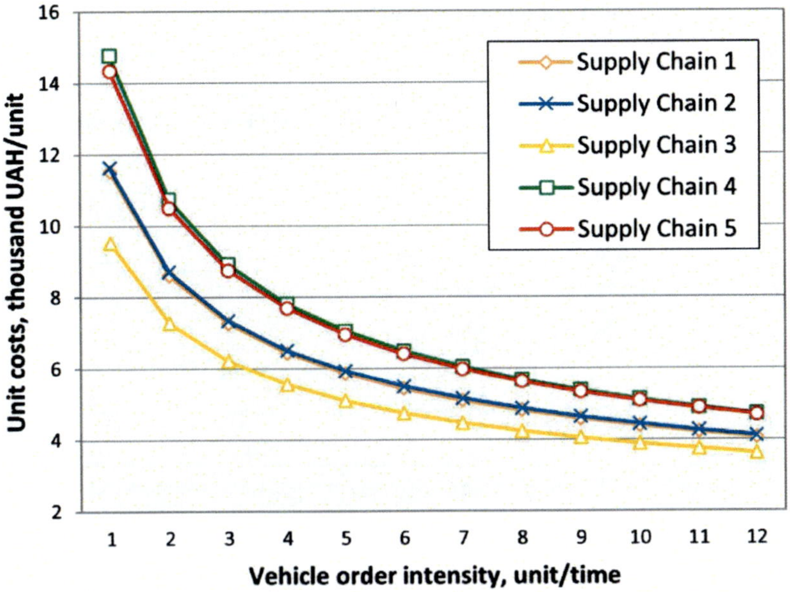

Fig. 2 Dependence plot of unit costs for goods delivery in containers from values changes of vehicle orders intensity

by sea between the port of Houston and the port of Odesa is much longer than the first three options.

Unit cost values were compared (Table 6) to determine the most efficient option from the proposed ones.

The most effective option is Supply Chain 3: unit costs are less by an average of 820 UAH compared to Supply Chain 1, and an average of 890 UAH also reduces expenses compared to Supply Chain 2. Comparing Supply Chain 1 and Supply Chain 2 results in an average cost savings of 69 UAH using Supply Chain 1. Comparing Supply Chain 4 and Supply Chain 5 shows that unit costs give an average savings of 115 UAH using the first option.

Using the developed methodology, an economical or more practical option for product delivery can be chosen under specific conditions. This aspect is important in such studies [28] because here, not only choice issues solving, but also focusing on participant satisfaction.

Table 6 Results of comparative effect determination

Measure values	Effect (UAH/unit)			
	Supply Chain Comparison 1 and 2	Supply Chain Comparison 1 and 3	Supply Chain Comparison 2 and 3	Supply Chain Comparison 4 and 5
1	-115.37	1986.99	2102.36	435.32
2	-92.26	1335.49	1427.75	243.97
3	-80.79	1051.33	1132.11	167.07
4	-73.46	883.99	957.45	124.29
5	-68.21	770.97	839.19	96.67
6	-64.19	688.3	752.48	77.23
7	-60.95	624.57	685.52	62.74
8	-58.28	573.58	631.86	51.5
9	-56.01	531.63	587.64	42.5
10	-54.06	496.37	550.42	35.14
11	-52.34	466.21	518.55	28.99
12	-50.82	440.06	490.88	23.77

5 Conclusions

This paper presents the conditions determining which researcher can choose an effective option for the supply chain of automotive equipment in containers from the United States to Ukraine. The current methodology developed to do this included the following elements: development of alternative supply chain options; defining of parameters for choice evaluating of effective options; designing a mathematical model for determining values of estimated parameters; modeling and evaluation of results.

It was proposed to consider this process through five options for their organization with the participation of automobile, railway, and maritime transport to determine an effective supply chain for cars in containers from the United States to Ukraine. The data of ten companies that organize the delivery of containers (cars and spare parts) from the United States through New York and Houston ports were used to form these schemes. The recipients' main ports are Odesa (Ukraine) and Klaipeda (Lithuania). The designed options consider the possibility of using them in Ukraine to transport containers for automobile and railway transport.

Mathematical modeling is the most promising for describing the process of moving loads in containers according to the options for use. This approach reduces the number of difficulties, the main reason for which is the complication of cause-and-effect relations in complex interacting components. The unit delivery costs are selected as the selection parameter for effective supply chains of vehicles in containers considering the restriction system.

A full-factor experiment plan was developed for three parameters, consisting of eight series of tests. The various combinations of external impact parameters were simultaneously used with varying intervals of cargo order volumes, intensities of truck orders, and cargo delivery times. Regression analysis was performed by functions of two types: linear and power functions. The power function model was found to be most adequate since the value of the "R-square" parameter is close to 1 and is 0.96.

The corresponding regression models were designed for five options of vehicle supply chains in containers from the USA to Ukraine. The final effect showed that the minimum values were obtained for "Supply Chain 3" by the entire range of the values change for vehicle order intensities. But the largest value of the effect was 2102.36 UAH when using "Supply Chain 3" - organizing the delivery of cars from the sender, who is located in the northern part of the United States, to the port of New York (USA), then by sea to the port in the city of Odesa. Then the carrier delivers by road to Kharkiv and the recipient.

Acknowledgements. The research was partially supported by the The NAWA Ulam Programme (grant number BPN/ULM/2022/1/00045) and International Association for Technological Development and Innovations.

References

1. Galkin, A., Yemchenko, I., Lysa, S., Tarasiuk, M., Chortok, Y., Khvesyk, Y.: Exploring the relationships between demand attitudes and the supply amount in consumer-driven supply chain for FMCG. Acta Logistica **9**(1), 1–12 (2022)
2. Abbasi, S., Saboury, A., Jabalameli, M.S.: Reliable supply chain network design for 3PL providers using consolidation hubs under disruption risks considering product perishability: An application to a pharmaceutical distribution network. Comp. Indu. Eng. **152**, 107019 (2021)
3. Trojanowski, P., Trusz, A., Stupin, B.: Correlation Between Accidents on Selected Roads as Fundamental for Determining the Safety Level of Road Infrastructure. In: Ivanov, V., Trojanowska, J., Pavlenko, I., Rauch, E., Peraković, D. (eds.) Advances in Design, Simulation and Manufacturing V. DSMIE 2022. Lecture Notes in Mechanical Engineering, pp. 104–113. Springer, Cham (2022). https://doi.org/10.1007/978-3-031-06025-0_11
4. State Statistics Service, http://www.ukrstat.gov.ua, last accessed 07 December 2022
5. Muzylyov, D., Shramenko, N.: Mathematical Model of Reverse Loading Advisability for Trucks Considering Idle Times. In: Karabegović I. (ed.) New Technologies, Development and Application III. NT 2020. Lecture Notes in Networks and Systems, vol. 128, pp. 612–620. Springer, Cham (2020)
6. Mesjasz-Lech, A., Włodarczyk, A.: The role of logistics infrastructure in development of sustainable road transport in Poland. Research in Transportation Business & Management 100841 (2022).
7. Pavlenko, I., et al.: Using regression analysis for automated material selection in smart manufacturing. Mathematics **10**(11), 1888 (2022). https://doi.org/10.3390/math10111888
8. Rahman, T., Paul, S.K., Shukla, N., Agarwal, R., Taghikhah, F.: Supply chain resilience initiatives and strategies: A systematic review. Comp. Indu. Eng. **170**, 108317 (2022)

9. Pavlenko, O., Muzylyov, D., Shramenko, N., Cagáňová, D., Ivanov, V.: Mathematical Modeling as a Tool for Selecting a Rational Logistical Route in Multimodal Transport Systems. In: Cagáňová, D., Horňáková, N. (eds.) Industry 4.0 Challenges in Smart Cities. EAI/Springer Innovations in Communication and Computing, pp. 23–37. Springer, Cham (2023)
10. Malacina, I., Teplov, R.: Supply chain innovation research: A bibliometric network analysis and literature review. Int. J. Produ. Eco. **251**, 108540 (2022)
11. Bigliardi, B., Filippelli, S., Petroni, A., Tagliente, L.: The digitalization of supply chain: a review. Procedia Comp. Sci. **200**, 1806–1815 (2022)
12. Joshi, S.: A review on sustainable supply chain network design: Dimensions, paradigms, concepts, framework and future directions. Sustai. Oper. Comp. **3**, 136–148 (2022)
13. Rossolov, A., Lobashov, O., Kopytkov, D., Naumov, V.: Sustainable suburban supply chain. Transportation Research Procedia **45**, 795–802 (2020)
14. Dalalah, D., Khan, S.A., Al-Ashram, Y., Albeetar, S., Abou Ali, Y., Alkhouli, E.: An integrated framework for the assessment of environmental sustainability in wood supply chains. Envir. Technol. Inno. **27**, 102429 (2022)
15. Tseng, M.-L., Bui, T.-D., Lim, M.K., Fujii, M., Mishra, U.: Assessing data-driven sustainable supply chain management indicators for the textile industry under industrial disruption and ambidexterity. Int. J. Produ. Eco. **245**, 108401 (2022)
16. García-Reyes, H., Avilés-González, J., Avilés-Sacoto, S.V.: A Model to Become a Supply Chain 4.0 Based on a Digital Maturity Perspective. Procedia Computer Science **200**, 1058–1067 (2022)
17. Rad, F.F., et al.: Industry 4.0 and supply chain performance: a systematic literature review of the benefits, challenges, and critical success factors of 11 core technologies. Industrial Marketing Management **105**, 268–293 (2022)
18. Mediavilla, M.A., Dietrich, F., Palm, D.: Review and analysis of artificial intelligence methods for demand forecasting in supply chain management. Procedia CIRP **107**, 1126–1131 (2022)
19. Volkov, V., Taran, I., Volkova, T., Pavlenko, O., Berezhnaja, N.: Determining the efficient management system for a specialized transport enterprise. Naukovyi Visnyk Natsionalnoho Hirnychoho Universytetu **4**, 185–191 (2020)
20. Ivanov, V., Vashchenko, S., Rong, Y.: Information support of the computer-aided fixture design system. Proc. of 12th Int. Conf. (ICTERI'2016) 1614, pp. 73–86. CEUR-WS (2016)
21. Sharifnia, S.M.E., Biyouki, S.A., Sawhney, R., Hwangbo, H.: Robust simulation optimization for supply chain problem under uncertainty via neural network metamodeling. Comp. Indu. Eng. **162**, 107693 (2021)
22. Goerlandt, F., Montewka, J.: Maritime transportation risk analysis: Review and analysis in light of some foundational issues. Reliab. Eng. & System Safety **138**, 115–134 (2015)
23. Haidabrus, B., Grabis, J., Protsenko, S.: Agile project management based on data analysis for information management systems. In: Ivanov, V., Trojanowska, J., Pavlenko, I., Zajac, J., Peraković, D. (eds.) Advances in Design, Simulation and Manufacturing IV. DSMIE 2021. Lecture Notes in Mechanical Engineering, pp. 174–182. Springer, Cham (2021)
24. Prokopovych, I.V., Kokhanov, A.B., Khamitov, V.M., Tikhenko, V.M., Dašić, P.: Standardizing life cycle organization: A synergetic quality management approach. J. Eng. Sci. (Ukraine) **10**(1), B1–B7 (2023). https://doi.org/10.21272/jes.2023.10(1).b1
25. Kurt, I., Aymelek, M., Boulougouris, E., Turan, O.: Operational cost analysis for a container shipping network integrated with offshore container port system: A case study on the West Coast of North America. Marine Policy **126**, 104400 (2021)
26. Sugawara, J.: Port and hinterland network: a case study of the Crescent Corridor intermodal freight program in the US. Transportation Research Procedia **25**, 916–927 (2017)

27. Delivery of containers from the USA, https://stoles.com.ua/services/delivery-containers-usa/, last accessed 12 July 2022
28. Qiu, R., et al.: A model-experience-driven method for the planning of refined product primary logistics. Chemical Engineering Science **254**, 117607 (2022)

Check for updates

Optimization of the Lifting Machines' Hoisting Mechanism Design Scheme

Volodymyr Semenyuk, Oleksandr Vudvud(✉), and Valeriy Lingur

Odessa Polytechnic National University, 1, Shevchenko Ave, Odessa 65044, Ukraine
o.m.vudvud@op.edu.ua

Abstract. Increasing the productivity of lifting machines and their reliability is possible by reducing the dynamic loads that occur at the lifting mechanism starting and braking. Carried out is the analysis of methods to reduce the lifting machines' hoisting mechanism dynamic loads during the acceleration period when load lifting. The method chosen for our study is reducing the mass inertia moments of parts located on the hoisting mechanism's slow-speed shafts. The study aimed to reduce the hoisting mechanism dynamic loads at cargo lifting by optimizing the gear drive mechanism design scheme. One of the promising methods to reduce dynamic loads in machine drives consists of the multithreading principle use. A new design scheme for the gearbox has been developed. This structure represents a multi-threaded two-stage gearbox, where each stage consists of a central gear, intermediate wheels, which axes are fixed in the housing, and a gear wheel with internal gearing. It has been established that using intermediate wheels of different diameters can obtain different gear ratios of the gearbox. It is determined that the minimum gear ratio for a two-stage multithreaded gearbox is 9. With such a gear ratio of the gearbox, the dynamic loads that occur during the lifting mechanism drive start-up are minimal. Accordingly, the maximum efficiency of the lifting mechanism per its operation cycle was obtained. Also, the optimal gear ratio ensures the minimum dimensions of a multi-threaded two-stage gearbox.

Keywords: Gear Ratio · Moment of Inertia · Dynamic Loads · Product Innovation

1 Introduction

In modern mechanical engineering, there are two trends in improving the machine drive:

– ensuring the working body's smooth acceleration and braking by reducing the inertia moment of the drive's moving mechanical parts: clutches, gears, drums, and other parts. It allows to improve the dynamic characteristics and significantly reduce the dynamic load during acceleration and braking and thus decrease the drive's efficiency factor. For example, in [1], the issues of ensuring the crane lifting mechanism reliability are considered. It is especially important when high explosive risk cargo transporting. The reliability of such mechanisms can be improved by slowing down their wear process, eliminating the source of wear, and strictly observing the operation and maintenance rules. The paper does not provide quantitative values on the probability of the lifting mechanism's failure-free operation;

V. Tonkonogyi et al. (Eds.): InterPartner 2023, LNME, pp. 94–104, 2024.
https://doi.org/10.1007/978-3-031-42778-7_9

– the transition from a drive operated with a mechanical gearbox to a gearless drive increases the efficiency factor due to the rejection of parts characterized by a significant inertia moment

To implement the second trend, i.e., the gearbox rejection embodying structure, work is underway to create low-speed electric motors with acceptable mass-dimensional indicators. Synchronous machines with permanent magnets are used at gearless drives in the valve motor mode. Such low-speed motors are not widely used due to the scarcity and relatively high cost of permanent magnets needed to create those machines. In addition, such motors' repair and disposal costs are approaching the cost of a new drive. Also, gearless drives for machine mechanisms are created based on asynchronous electric motors with short-circuited rotors using frequency converters. Frequency converters ensure their torque constancy when the speed of the controlled electric drive decreases. Still, at the same time, there is a decrease in the useful power with a corresponding decrease in the efficiency factor value. Artificially reducing the rated speed of the motor in a wide range lead to a significant decrease in the efficiency factor since at the control range equal to 8, the efficiency factor decreases up to 53%. To increase the energy efficiency of machine drives, it is advisable to engage the method of minimizing the electric motor nominal speed artificial reduction range and, at the same time to provide the required speed of the machine's working body due to the additional use of a gearbox with a high-efficiency factor [2]. Thus, the research and development of an energy-efficient hoisting mechanism for lifting machines is highly relevant, being of actual scientific interest, and has practical value.

This study aims to reduce the dynamic loads of lifting mechanisms when lifting cargo by optimizing the gearbox design scheme.

This study's objectives: development of a gearbox design scheme with improved dynamic properties; determination of optimal gear ratios for the gearbox with improved dynamic properties.

2 Literature Review

Given that the problem of creating highly efficient lifting machines is urgent, researchers and engineers from different countries worldwide are searching for a solution.

The method for increasing the efficiency of the tower crane mechanisms' joint operation is proposed in [2]. A methodology of selecting the mechanism's optimal structure from a variety of possible solutions by introducing a quality criterion for each structure is developed, and a set of parameters optimized for the movements full scope is determined. The method was illustrated for a crane with both lever mechanisms and rope mechanisms.

Large dynamic loads during the bridge-type cranes' lifting mechanism start-up led to breakdowns of the main parts and sometimes to accidents since the method used to prevent these loads does not provide the necessary accuracy and reliability of overload protection. Developed is a bridge-type crane protection system [3], which allows the implementation of a new method of overload protection, namely, lifting mechanism moving parts disconnecting translationally with simultaneous disconnection of this mechanism drive electric motor. To ensure the required safe operation of the crane, it is necessary to develop constructive schemes of the lifting mechanism capable of excluding so-called systematic overloads efficiently.

It is proposed to decrease peak loads when starting the crane lifting mechanism by reducing the inertia moment of coupling with the brake pulley [4]. Namely, the point is to manufacture a coupling made of composite materials. It is proposed to reduce the inertia moments by increasing the start-up and acceleration time of the lifting mechanism drive. However, reducing the start-up and acceleration time leads to a decrease in the crane's performance.

The influence of rope stiffness on dynamic loads arising in the metal structure of overhead cranes is considered in [5]. When simulating the dynamics of the cargo lifting process with a bridge crane, two dynamic crane models are proposed, rendering it possible to clarify the magnitude of dynamic loads on the metal structure during the lifting mechanism operation.

For all lifting cranes, of key relevance is to ensure high reliability. Therefore, their emergency braking is used in addition to the normal braking of lifting mechanisms [6]. Emergency braking is unavoidable in several situations, for example, when the electrical power is suddenly cut off. The study analyzes the emerging dynamic loads during emergency braking and when braking with a mechanical brake.

To assess the bridge-type crane dynamic loading, a system for crane load monitoring and diagnostics is proposed [7]. This system is developed based on experimental data obtained in various laboratory tests.

In [8], a new system for cargo lifting process control was developed; a controller proposed for this system uses two signals: this one of angular travel and this displacement integral, combining them into a new signal of coupled dissipation. The proposed controller effectively suppresses dynamic load fluctuations at the expense of increasing passivity.

The mathematical model of the bridge crane lifting transmission system is analyzed in [9]. The relationship between the dynamic load stiffness coefficients and the crane drive parameters is established. The bridge crane dynamic model is analyzed for non-stationary modes of lifting mechanism operation [10]. The bridge crane oscillation amplitude is established for these modes, and recommendations on dynamic load reduction are given.

Work [11] is devoted to the study of planetary gears dynamics. Models of the basic structures of planetary gears have been developed using computer software; graphs of carrier angular acceleration have been constructed without account and taking into account the gears' gaps; it is shown that the smallest acceleration of the carrier increases by 1.5 times if taking into account the gaps at the gears.

Studies of the gear-roller planetary gear structure and dynamics have been carried out in [12]. A replacement mechanism was used for this study. The parameters of the replacement mechanism have been considered. The paper [13] presents a methodology for building a mathematical model of a planetary gearbox with rigid links of its planetary series, an example of modeling a dynamic engine – torque converter – planetary gearbox – automobile – road system is given. The simulation results show that the planetary transmission system has three modes of operation [14]. The main frequencies include the motor rotation speed and the frequency of gearing.

The machine drives use planetary gears, which are created on the principle of multithreading, therefore allowing to significantly reduce the size and weight of gearboxes using planetary gear. The principle of multithreading is used not only for transmissions with rotational motion but also for mechanisms implementing the translational one. The roller transfer mechanism [15] is based on the principle of multithreading; in this mechanism, the power flow is divided because it consists of two rows of rollers arranged in staggered order. The analysis of the works under consideration shows that these studies aimed to reduce dynamic loads in machine drives. Still, simultaneously, the problem of creating highly efficient machine drives requires continued research and development of new drive design schemes for various machines, including cargo lifting machines.

3 Research Methodology

3.1 Analysis of Methods for Reducing the Lifting Machines' Hoisting Mechanism Dynamic Loads During Acceleration When Cargo Lifting

With the lifting mechanism's steady movement characteristic for the repeated-short-term operation of lifting machines, the torque on the electric motor shaft must comply with the condition:

$$M_{dv} = M_{st} + M_d \leq \psi_{av} \cdot M_n, \tag{1}$$

where M_{dv} is the motor shaft torque, Nm; M_{st}- the static torque on the motor shaft when lifting the nominal load, Nm; M_d- the dynamic torque on the motor shaft determined from the condition of providing the acceleration necessary for speeding up, Nm; M_H- the nominal torque of the electric motor, Nm; ψ_{av}- the average starting torque multiplicity.

The dynamic torque on the motor shaft, determined from the condition of providing the acceleration necessary for speeding up, is determined by the dependence:

$$M_d = I_{pr} \frac{\omega}{[t_n]^2}, \tag{2}$$

where is the I_{pr} inertia moment of the lifting mechanism moving masses and the lifted load brought to the motor shaft during the start-up period when cargo lifting, $kg{\cdot}m^2$; ω - the current angular velocity of the motor rotor *rad/s*; $[t_n]$ – the permissible duration of the start, s, equals to:

$$[t_n] = \frac{V_f}{[a_n]}, \tag{3}$$

where $[a_n]$ is the permissible acceleration at start-up, m/s^2; V_f is the actual speed of the load-bearing element, m/s.

The design scheme of the lifting machines' hoisting mechanism includes an electric motor, a clutch with a brake pulley, a gearbox, a drum, and a chain hoist. Taking into account the lifting mechanism design scheme and based on the analysis of dependencies (2) and (3), the following methods for reducing dynamic loads can be formulated: method of reducing the motor rotor's moment of inertia; method of reducing the inertia moment of the clutch with the brake pulley; method for reducing the masses inertia moments of parts located on the slow-speed shafts of the lifting mechanism; method of regulating in transients the angular velocity of the lifting mechanism drive's electric motor rotor during the start-up period, that is, thought the use of controlled transient processes.

The motor rotor's inertia moment can be reduced by replacing one motor with two or more while maintaining the total power level. As a rule, the total inertia moment of two half-power motors turns out to be less than the inertia moment of one full-power motor within the same motors series.

The inertia moment of the clutch with the brake pulley can be reduced by using composite materials.

The rational design of the mechanical transmission (selection of the optimal gear ratio) reduces the inertia moments of the masses of parts located on the lifting mechanism's slow-speed shafts.

3.2 Development of a Design Scheme for a Gearbox with Improved Dynamic Properties

One of the promising methods of reducing dynamic loads in machine drives is using the multithreading principle in their design elaboration. The multithreading principle allows to divide the power flow into several parallel branches. As the experience of using gears with power flow separation shows, the mass of multi-threaded gears is 1.5–3 times less than the mass of single-threaded gears. Accordingly, the moment of inertia at such gears is also smaller, and this allows the reduction of the dynamic loads in the machine drives. Multithreaded principles can be arranged as so-called simple gears and planetary gears.

Planetary gears are widely used in mechanical engineering as they have many advantages: small dimensions and less weight; large gear ratios in one stage; it is possible to use gears with internal gearing characterized with increased load-bearing capacity; less noise producing compared to simple gears.

However, planetary gears require increased manufacturing accuracy and a larger number of parts and are more difficult to assemble than gears with fixed axes. The efficiency factor drops sharply with the increase in the gear ratio.

A significant disadvantage of planetary gears is that they have a large moment of inertia at launch because, at drive acceleration, it is necessary to overcome the resistances associated with both the carrier's moment of inertia and the satellites' moment of inertia. To eliminate this drawback, it is proposed to use a simple multithreaded gear, which can be represented as a planetary gear with a stopped carrier.

Figure 1 shows a design diagram of a multithreaded two-stage gearbox, each consisting of a central gear 1, intermediate gears 2, which axes are fixed in the housing, and a gear wheel 3 with internal engagement. Similarly, the second stage is made with a central gear 5, intermediate wheels 4, and a wheel with an internal gearing 6.

In a simple multithreaded gear, the central gear dimensions are smaller than in a simple single-threaded one since each gearing transmits only a part of the entire load.

A simple multithreaded gear, having almost the same advantages as a planetary type one, is easier to assemble, does not require increased manufacturing accuracy, and due to design features, the reduced inertia moment of rotating gears is significantly less than that of a planetary transmission gear.

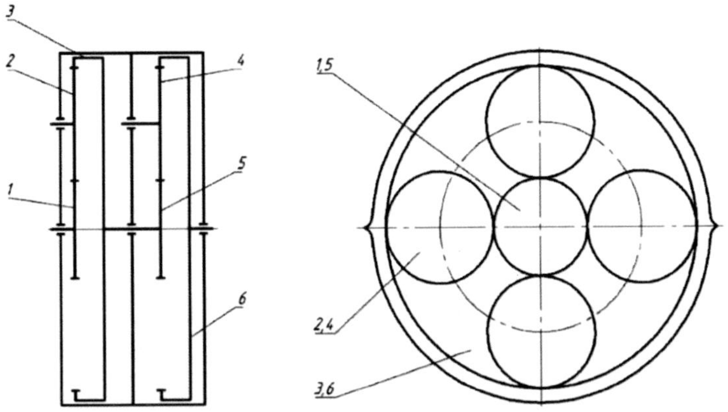

Fig. 1. Design scheme of a multithreaded two-stage gearbox.

3.3 Determining the Optimal Gear Ratios of the Gearbox Design Scheme with Improved Dynamic Properties

An important direction of lifting machines improvement is to increase their productivity, on which the overall success of industry, transport, and construction depends to a certain extent.

To increase the productivity of lifting machines, it is necessary to reduce, all other things being equal, their start-up time as well as the braking time. Reducing the start-up and braking time increases the dynamic loads of the machine mechanisms since, to achieve the machine's working speed in this case, it is necessary to increase acceleration (deceleration) during the specified periods.

An increase in dynamic loads at start-up leads to a decrease in the machine's efficiency during this period and, accordingly, a decrease in the machine's efficiency during the work cycle.

By applying the method of reducing the masses inertia moments of parts located on the lifting mechanism's slow-speed shafts, it is possible to reduce dynamic loads during start-up. Since the dynamic moment on the lifting mechanism drive electric motor shaft depends on the inertia moment of the lifting mechanism and the lifted load moving masses brought to the motor shaft during the start-up period, to determine this moment, we consider the calculation scheme shown in Fig. 2.

The equation for the inertia moment reduced to the motor shaft has the form:

$$I_{pr}\frac{\omega_1^2}{2} = I_1\frac{\omega_1^2}{2} + I_2\frac{\omega_1^2}{2} + I_3\frac{\omega_1^2}{2} + I_4\frac{\omega_2^2}{2\cdot\eta_1} + I_5\frac{\omega_3^2}{2\cdot\eta_2} +$$

$$+I_6\frac{\omega_3^2}{2} + I_7\frac{\omega_4^2}{2\cdot\eta_3} + I_8\frac{\omega_5^2}{2\cdot\eta_4} + I_9\frac{\omega_5^2}{2\cdot\eta_5} + m_c\frac{v_c^2}{2\cdot\eta_6}, \tag{4}$$

where, I_{pr}– the inertia moment of the mechanism brought to the motor shaft at the start–up; ω_1, ω_2, и ω_3– angular velocities, respectively, of the motor shaft, the first

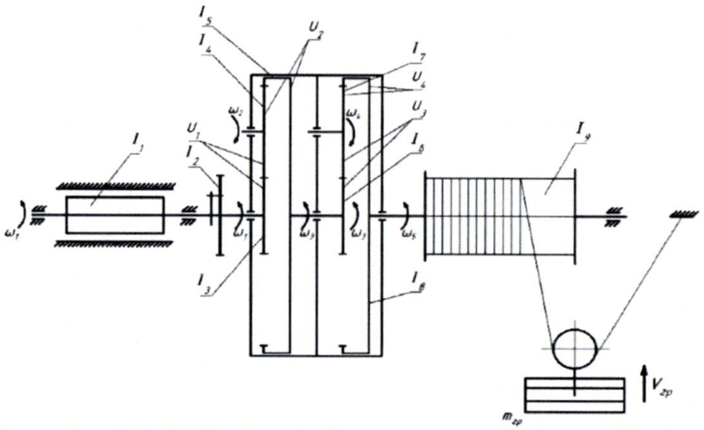

Fig. 2. The cargo lifting mechanism's scheme for bringing moving masses.

and second stage shafts of the gearbox, and the drum shaft; $I_1, I_2, I_3, I_4, I_5, I_6, I_7 I_8, I_9,$– inertia moments, respectively, of the motor rotor, clutch with brake pulley, the first and second stage gears of the gearbox and drum; v_c, m_c- respectively, the speed and weight of the cargo being lifted; $\eta_1, \eta_2, \eta_3, \eta_4, \eta_5, \eta_6$– respectively, the efficiency of the first gear engagement and the second, third, and fourth gear engagements, the efficiency factor of the mechanism from the drum to the motor shaft; the efficiency factor of the lifting mechanism, including the efficiency factor of the chain hoist. Substituting in (4) $\mathrm{v_c} = \frac{\omega_5 \cdot r_6}{\mathrm{a \cdot u_1 \cdot u_2 \cdot u_3 \cdot u_4}}$ $\omega_2 = \frac{\omega_1}{u_1}$; $\omega_3 = \frac{\omega_1}{u_1 \cdot u_2}$; $\omega_4 = \frac{\omega_1}{u_1 \cdot u_2 \cdot u_3}$; $\omega_5 = \frac{\omega_1}{u_1 \cdot u_2 \cdot u_3 \cdot u_4}$, and dividing the left and right parts of expression (4) into $\frac{\omega_1^2}{2}$, we get:

$$I_{pr} = I_1 + I_2 + I_3 + I_4 \frac{1}{u_1^2 \cdot \eta_1} + I_5 \frac{1}{(u_1 \cdot u_2)^2 \cdot \eta_2} +$$
$$+ I_6 \frac{1}{(u_1 \cdot u_2)^2} + I_7 \frac{1}{(u_1 \cdot u_2 \cdot u_3)^2 \cdot \eta_3} + I_8 \frac{1}{(u_1 \cdot u_2 \cdot u_3 \cdot u_4)^2 \cdot \eta_4} +$$
$$+ I_9 \frac{1}{(u_1 \cdot u_2 \cdot u_3 \cdot u_4)^2 \cdot \eta_5} + m_c \frac{r_b^2}{(a \cdot u_1 \cdot u_2 \cdot u_3 \cdot u_4)^2 \cdot \eta_6}, \quad (5)$$

where, $u_1, u_2, u_3, u_4,$ – respectively the gear ratios of the first, second, third, and fourth gears; r_b– the drum radius; a – the chain hoist multiplicity.

Considering the power flow from the motor to the drum and taking $\eta_1 = \eta_2 = \eta_3 = \eta_4 = \eta_5 = 1$, after the transformations, we write:

$$I_{pr} = I_1 + I_2 + I_3 + I_4 \frac{1}{u_1^2} + (I_5 + I_6) \frac{1}{(u_1 \cdot u_2)^2} +$$
$$+ I_7 \frac{1}{(u_1 \cdot u_2 \cdot u_3)^2} + I_8 \frac{1}{(u_1 \cdot u_2 \cdot u_3 \cdot u_4)^2}, \quad (6)$$

The reduced inertia moment of the lifting mechanism, according to expression (6), will have the smallest value, all other things being equal, when using a multithreaded two-stage gearbox made according to the design scheme represented in Fig. 1 with gear ratios $u_1,\ u_2,\ u_3,\ u_4,$ if the following condition is met:

$$\left\{I_3 + I_4 n \frac{1}{u_1^2} + (I_5 + I_6)\,\frac{1}{(u_1 \cdot u_2)^2} + I_7 n \frac{1}{(u_1 \cdot u_2 \cdot u_3)^2} + I_8 \frac{1}{(u_1 \cdot u_2 \cdot u_3 \cdot u_4)^2}\right\} \Rightarrow min, \tag{7}$$

where n is the number of intermediate gears in one stage.

Since the first and second stages of the gearbox are similar, we consider condition (7) for the first stage:

$$\left\{I_{pr} = I_3 + nI_4 \frac{1}{u_1^2} + I_5\,\frac{1}{(u_1 \cdot u_2)^2}\right\} \Rightarrow min, \tag{8}$$

Let us transform the expression (8) by assuming that the formula determines the inertia moments of each of the gear wheels in this gearbox:

$$I = \frac{G \cdot D^2}{7 \cdot g} \tag{9}$$

where I is the inertia moment of the gear mass, kgm·s^2; D – the gear diameter, m; G – the rotating gear weight, kg; g– free fall acceleration, m/s^2.

The weight of a rotating gear wheel can be found as follows:

$$G = V \cdot \gamma \tag{10}$$

where, $V = \frac{\pi \cdot D^2}{4} \cdot h$– the cylindrical gear volume, m3; h– the gear width, m; $\lambda\gamma$– the specific weight of its material, kg/m^3.

Taking for the central gear $h = D_3$, and for the intermediate wheels $h = 0,25D_3$ and considering that $\frac{D_4}{D_3} = u_1$; $\frac{D_5}{D_4} = u_2$; $D_4 = D_3 \cdot u_1$; $D_5 = D_3 \cdot u_1 \cdot u_2$, we find expressions for moments of inertia:

$$I_3 = \frac{\pi}{4 \cdot 7g} \cdot \gamma \cdot D_3^5;\ n \cdot I_4 = \frac{\pi}{4 \cdot 7g} \cdot \gamma \cdot D_3^5 \cdot u_1^5;\ I_5 = \frac{\pi}{4 \cdot 7g} \cdot \gamma \cdot D_3^5 \cdot (u_1 + u_2)^5, \tag{11}$$

Substituting the corresponding values I_3, nI_4, I_5 into expression (8), after the transformations, we get:

$$\left\{\frac{\pi}{4 \cdot 7 \cdot g} \cdot \gamma \cdot D_3^5\left[1 + u_1^3 + (u_1 \cdot u_2)^3\right]\right\} \Rightarrow min \tag{12}$$

Taking into account the fact that the central wheel diameter D_3 can be found based on the strength of the teeth by contact stresses, expression (12) takes the form:

$$\left[1 + u_1^3 + (u_1 \cdot u_2)^3\right] \Rightarrow min \tag{13}$$

By expression (13), we can determine at what gear ratios of the multithreaded gearbox stages u_1, u_2 there will be minimal dynamic loads of the cargo hoisting mechanism at lifting machines.

4 Results and Discussion

By replacing one engine with two or more, while maintaining the total power of the drive of the lifting mechanism, a decrease in the moment of inertia of the engine rotor is achieved. Using composite materials can reduce the moment of inertia of the clutch with a brake pulley.

Multi-threading allows you to divide the power flow into several parallel branches; therefore, the mass of multi-thread gears is 1.5–3 times less than that of single-thread gears. One of the options for a simple multi-threaded transmission is proposed in the form of a multi-threaded two-stage gearbox (Fig. 1).

Using intermediate wheels of different diameters, obtaining different values of the gear ratios at the gearbox first stage is possible. Table 1 shows the possible options, taking into account that $u_1 = \frac{D_4}{D_3}$; $u_2 = \frac{D_5}{D_4}$; $D_5 = D_3 + 2D_4$; $D_3 = 34\,mm$.

Table 1. The possible options.

Options	u_1	u_2	$u_1 \cdot u_2$	D_5, *mm*
$D_4 = D_3$	1	3	3	$D_5 = 3D_3 = 102$
$D_4 = 1{,}5D_3$	1.5	4/15	4	$D_5 = 4D_3 = 136$
$D_4 = 2D_3$	2	5/2	5	$D_5 = 3D_3 = 170$
$D_4 = 2{,}5D_3$	2.5	6/2,5	6	$D_5 = 6D_3 = 204$
$D_4 = 3D_3$	3	7/3	7	$D_5 = 7D_3 = 238$

Analysis of the data given in Table 1 shows that the diametrical dimensions (D5) of the gearbox are relatively small and increase with increasing gear ratio $u_1 \cdot u_2$.

Analyzing the expressions (13) and taking into account expressions (11), we state that the smallest value of the reduced inertia moment and the smallest dynamic loads for the first stage of a multithreaded reducer will be identified at $u_1 = 1$, $u_2 = 3$. The values of the reduced inertia moment and the minimum dynamic loads will also be relevant for the gearbox second stage. For a two-stage multithreaded gearbox, dynamic loads will be minimal at a gear ratio equal to 9.

The studies in this paper offer a new method for designing lifting mechanisms for hoisting machines compared to the method for reducing inertial masses proposed in [4] and the method for reducing dynamic loads in planetary gear [11]. Still, at the same time, the results obtained do not contradict earlier studies.

5 Conclusions

An efficient method to reduce dynamic loads during the start-up period at cargo lifting will be the method of reducing the inertia moment of the masses of the parts located on the lifting mechanism's slow-speed shafts.

One of the promising methods of reducing the machine drives' dynamic loads is using the multithreading principle in the design.

The principle of multithreading has been applied when creating planetary gears. These gears have many advantages, but a significant disadvantage of planetary gears is that they have a significant inertia moment at launch because, during the drive acceleration, it is necessary to overcome the resistance associated with both the inertia moments of satellites and the inertia moment of the carrier.

A new constructive scheme of the hoist mechanism reducer is proposed, which is a simple multi-threaded gear train.

It is established that the reduced inertia moment of the two-stage multithreaded gearbox gears depends on the sum of the squares of the first-stage gear ratios and the gearbox as a whole. The value of the reduced inertia moment and dynamic loads for a two-stage multithreaded gearbox will be minimal when the gearbox gear ratios are equal to 9.

References

1. Faltinová, E., Mantič, M., Kuľka, J., Kopas, M.: Reliability analysis of crane lifting mechanism. Scient. J. Silesian Univ. Technol. Seri. Transp. **98**, 15–26 (2018). https://doi.org/10.20858/sjsutst.2018.98.2
2. Chwastek, S.: Finding the globally optimal correlation of cranes drive mechanisms. Mech. Based Des. Struct. Mach. **51**(6), 3230–3241 (2023). https://doi.org/10.1080/15397734.2021.1920978
3. Volodymyr, T., Volodymyr, S., Iihor, S., Voleriy, L., Oleksandr, V.: Features of overload protection for bridge type cranes. In: Karabegović, I. (ed.) New Technologies, Development and Application IV. NT 2021. Lecture Notes in Networks and Systems, vol 233, pp. 162–168. Springer, Cham (2021). https://doi.org/10.1007/978-3-030-75275-0_19
4. Mudrov, A.: Reduction of peak loads during the start of the lifting mechanism in cranes. Construction technologies. Bulletin of Kazan State University **2**, 239–245 (2018)
5. Haniszewski, T.: Modeling the dynamics of cargo lifting process by overhead crane for dynamic overload factor estimation. Journal of Vibroengineering **19**(1), 75–86 (2017). https://doi.org/10.21595/jve.2016.17310
6. Niu, C., Ouyang, H.: Nonlinear dynamic analysis of lifting mechanism of an electric overhead crane during emergency braking. Appl. Sci. **10**(23), 8334 (2020). https://doi.org/10.3390/app10238334
7. Smoczek, J., Hyla, P., Kusznir, T.: Machine learning based approach to a crane load estimation. Journal of KONBiN **51**(4), 1 (2021). https://doi.org/10.2478/jok-2021-0040
8. Zhang, S., et al.: Passivity-based coupling control for underactuated three-dimensional overhead cranes. ISA Trans. **126**, 352–360 (2022). https://doi.org/10.1016/j.isatra.2021.07.040
9. Zhang, Z., Jin, H., Li, X., Tian, J.: Rigid dynamic load of the crane lifting mechanism when the series resistance starts. J. Phys.: Conf. Ser. **1578**, 012195 (2020). https://doi.org/10.1088/1742-6596/1578/1/012195
10. Čolić, M., et al.: Mathematical modelling of bridge crane dynamics for the time of non-stationary regimes of working hoist mechanism. Archive of Mechanical Engineering **69**(2), 189–202 (2022). https://doi.org/10.24425/ame.2022.140415
11. Dzhomartov, A., Ualiev, G.: Investigation of the dynamics of planetary gears. Rep. Nati. Acad. Sci. Rep. Kazakhstan **3**(301), 10–18 (2015)
12. Jianjun, T., et al.: Dynamic modeling and analysis of planetary gear train system considering structural flexibility and dynamic multi-teeth mesh process. Mech. Mach. Theory **186**, 105348 (2023). https://doi.org/10.1016/j.mechmachtheory.2023.105348

13. Moshrefzadeh, A., Fasana, A.: Planetary gearbox with localised bearings and gears faults: simulation and time/frequency analysis. Meccanica **52**, 3759–3779 (2017). https://doi.org/10.1007/s11012-017-0680-7
14. Cui, T., Li, Y., Zan, C., Chen, Y.: Dynamic modeling and analysis of nonlinear compound planetary system. Machines **10**(1), 31 (2022). https://doi.org/10.3390/machines10010031
15. Semenyuk, V., Lingur, V., Punchenko, N., Falat, P.: Roller function-generating mechanism preventing the crank-drive machines' overloads. In: Zawiślak, S., Rysiński, J. (eds.) Engineer of the XXI Century. EngineerXXI 2018. Mechanisms and Machine Science, vol. 70, pp. 29–38. Springer, Cham (2020). https://doi.org/10.1007/978-3-030-13321-4_3

3D Reconstruction of a Virtual Building Environment

Ihor Tytarenko[1], Ivan Pavlenko[2,3](✉), and Stella Hrehova[3]

[1] Beketov National University of Urban Economy in Kharkiv, 17, Marshal Bazhanov Street, O.M, Kharkiv 61002, Ukraine

[2] Sumy State University, 2, Rymskogo-Korsakova Street, Sumy 40007, Ukraine
i.pavlenko@cm.sumdu.edu.ua

[3] Technical University of Košice, 1, Bayerova Street, Prešov 080 01, Slovak Republic

Abstract. After years of the russian's military invasion of Ukraine, preserving the architectural heritage has become more critical. Moreover, the world's increased interest in the historical heritage of Eastern Europe countries prompts the scientific community to restore the building environment with an up-to-date computer quickly means for 3D modeling. Therefore, the research aims at the virtual reproduction of historical heritage objects with the extensive use of existing computational software. The research methodology includes collecting information from historical, archaeological, cartographic, and other sources. It was compared with archaeological and historical maps. Further superimposing of schemes with the related features on a working map was also realized. A detailed study of the object's current state for the case study of virtual 3D reconstruction was carried out based on the Google Maps services and Geographic Information System (GIS) mapping. Moreover, a relief for the studied territory was analyzed using available topography and geodetic data. As a result of the research methodology implementation, the relief model allowed the reproduction of the structure of architectural objects. This allowed creating the initial 3D model, designing its layout, and realizing the final texturing and rendering. The developed virtual model makes it possible to reproduce essential architectural, cultural, and historical heritage objects.

Keywords: Sustainable Land Management · Infrastructural Development · Smart Cities · 3D Reconstruction · Building Rendering · Sustainable Development

1 Introduction

The modern history of Eastern Europe has a very significant and rapid development of historical heritage research. The active national identification of Ukrainian people contributed to the greater activation of the scientific community in the fields of local history, archeology, and the improvement of historical authenticity [1].

European scientists use advanced computer technologies, especially 3D modeling, to achieve more effective results in scientific research [2]. However, Ukraine is just beginning to actively develop in using Building Information Model (BIM) technologies

V. Tonkonogyi et al. (Eds.): InterPartner 2023, LNME, pp. 105–114, 2024.
https://doi.org/10.1007/978-3-031-42778-7_10

based on scientific archaeological research with an implementation in actual projects [3].

Creating a modeling methodology to provide specialists with relevant knowledge to help research and popularize historical objects remains relevant. The result of the modeling will be a 3D model that can be used for future BIMs.

Therefore, the research aims to reproduce a historical heritage object using modern 3D technologies. The following objectives have been formulated to achieve this goal. Firstly, information collection should be realized to study historical, archaeological, cartographic, and other scientific sources. Secondly, analyzing archival, archeological, and historical maps will allow further plotting schemes on a working archaeological map, considering the main features.

Also, using Google Maps services and Geographic Information System (GIS) mapping [4] will allow a detailed study of the object's current state. Finally, research on territory relief using topography and geodetic data can be realized by laser scanning technology [5] and photogrammetry [6]. Overall, the result of analytical experiments is modeling, followed by creating a 3D model. It should make it possible to recreate Pontic Olbia as an object of Ukraine's cultural and historical heritage.

2 Literature Review

Due to a high level of up-to-date software means developments, many research works in creating 3D virtual building environments have rapidly increased recently. Particularly, Yue and Limao [7] mainly presented an integrated approach based on BIM and Artificial Intelligence (AI) for smart construction management. Doula et al. [8] studied the effect of the detailing level on city landmarks recognition in a virtual environment. Also, Chernyshev et al. [9] implemented digital administration of the project-based smart construction concept. As a result of introducing a life cycle management system for building objects using BIM was proposed.

Bevilacqua et al. [10] applied digital software to implement digital twins and virtual reality (VR) in the 3D reconstruction of architectural heritage objects. As a result, digital prototypes of interior spaces for the First Italian Parliament, the Charterhouse of Pisa in Calci, and the Cathedral of Carpi (Italy) were developed by applying modern VR software. Also, Lang et al. [11] reconstructed an extinct settlement using VR software for the Advent City (Norway) case based on a series of survived historical photographs.

Another successful experience in 3D virtual reconstruction using VR technologies was demonstrated by Shanthini et al. [12] in the case of recreating the ancient Poompuhar (India). Also, Croce et a. [13] developed a virtual model for reconstructing the Globe Theatre (United Kingdom).

The following research proposed entirely different approaches to the 3D reconstruction of virtual environments. Particularly, Ou [14] implemented digital reconstruction of a building environment. For this purpose, automated decision-making algorithms [15, 16], deep learning [17], and information management systems [18] can be applied. Moreover, Zhou et al. [19] researched key technology for the virtual restoration of ancient buildings. For this purpose, Convolutional Neural Networks [20] were applied. The obtained results allow for improving the quality of 3D virtual reconstruction of architectural and cultural heritage as a fundamental problem for sustainable development.

3 Research Methodology

The following methods have been used during the research (Fig. 1): observation (research and analysis of the territory for the studied object), analytics (data collection, use and analysis of archival, library, and cartographic sources), and 3D modeling (reproducing the research object in a virtual space using appropriate software).

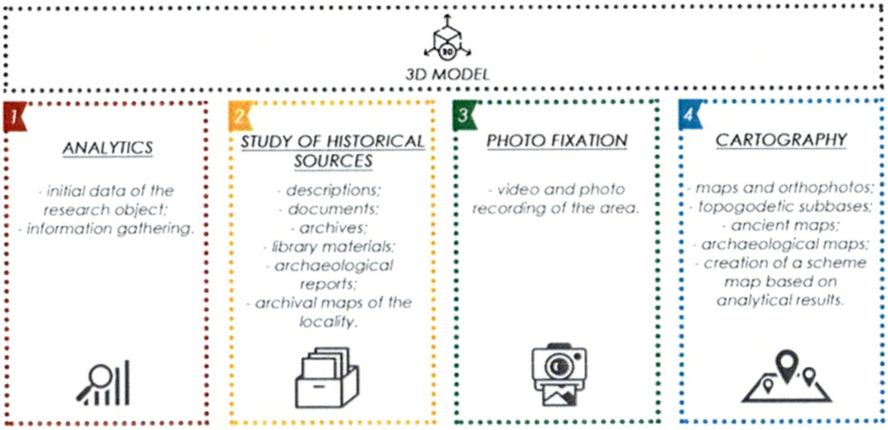

Fig. 1. Research methodology.

The first analytical stage includes collecting initial data on the research object and other necessary information. The second stage deals with the research of historical sources. It mainly provides analysis of descriptions, documents, archives, library materials, archaeological reports, s well as archival maps of the area.

The third stage is video and photo recording of the area. The following stage concerns cartography. It consists of research on modern maps and orthophotos, topogeodetic subbase, and ancient maps of different periods. It also includes the study of archaeological patterns of the area and the analysis and comparison of maps by combining them with different software. Creating a spatial map based on the research on geographic differences in heights (horizons) in the appropriate software is also an integral part of the fourth stage. It also includes creating a 3D model of several relief feature variants. All these activities are needed to reproduce the lost parts of the relief due to water corrosion of the plateau, analysis of the placement of known, studied by archaeologists, and recorded on the diagrams, quarters, and individual objects of the studied object. As a result, maps of the street grid location and city structure will be built based on the analytical results.

The final stage allows creating a 3D model of the object under study with all its elements. It is realized using the proposed stages (Fig. 2).

The following software can be used while modeling and further processing: SketchUp, Blender, and Autodesk 3ds Max – for modeling; Photoshop – to form a mapping scheme for building the structure of the city; 3DF Zephyr, RealityCapture, Agisoft Metashape, and Meshroom – to create photogrammetry; Quixel Megascan – for

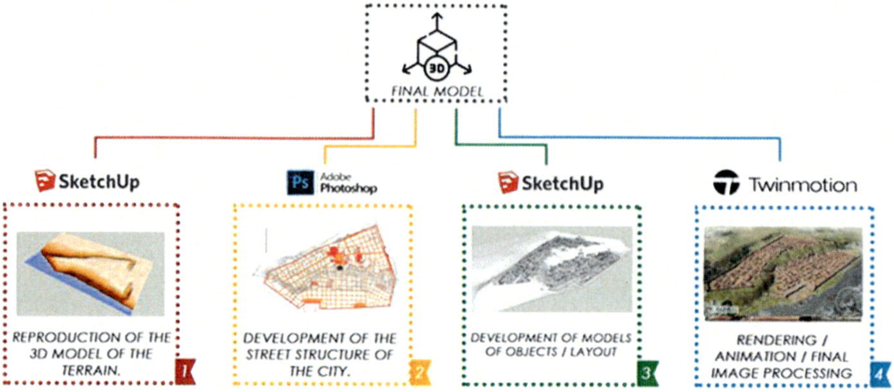

Fig. 2. Stages of 3D modeling.

using an available library of textures, models of photo scans; Twinmotion, Lumion – for rendering images and animations with library support Quixel.

4 Results and Discussion

The following computational means have been used to reproduce the 3D virtual model of the terrain. Firstly, the relief reproduction has been realized using the topogeodetic subbase data. Also, SketchUp has allowed for reproducing horizons and their heights to form a 3D model. After, the relief was reproduced using GIS data, Google Maps, and subsequent import to SketchUp.

As a result, the created relief model is presented in Fig. 3.

Fig. 3. The terrain model.

The development of the street structure of the city is as follows. Firstly, a mapping scheme has been formed using Photoshop software. On it, directions of the existing streets and exact locations of districts and other objects have been analyzed based on the archaeological data.

After, the mapping diagram of the researched object was superimposed on the created relief, forming the street and road network of the city. Thus, the placement of quarters in the structure of the 3D model has been revealed. Overall, the studied areas have been placed relative to the designed grid of streets.

The developed street structure of the city is shown in Fig. 4.

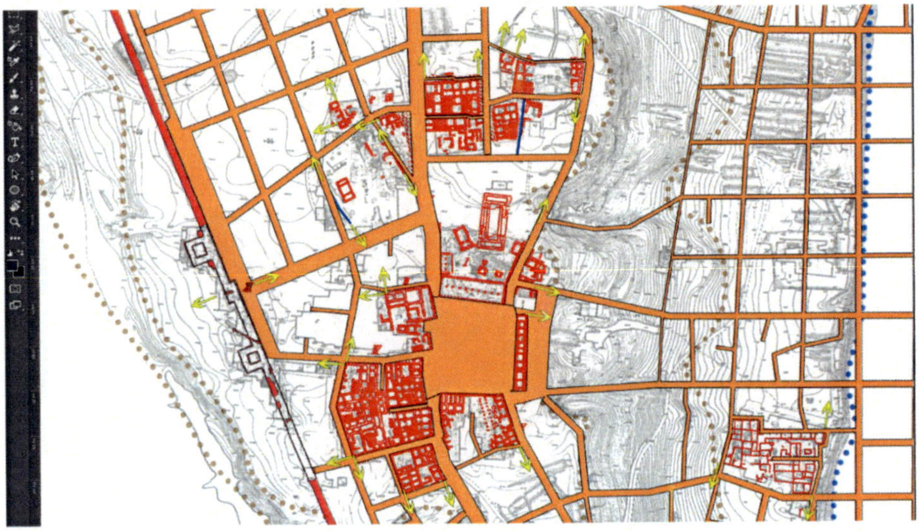

Fig. 4. Reproduction of the city structure.

The development of object models has been carried out as follows. Firstly, models of buildings and structures, defensive walls, and individual objects of the historic environment have been developed using the SketchUp software. After, a search for additional models has been realized. They should have an authentic look at the relevant era. The corresponding object library has also been created.

As a result, an initial 3D model has been built (Fig. 5).

The next stage deals with the 3D model layout. For this purpose, 3D objects have been placed according to the archaeological schemes and the created mapping scheme. As a result, a complete model of the city has been formed (Fig. 6).

The last stage includes rendering, animation, and final image processing. Firstly, a realistic virtual environment has been created using the Twinmotion software and Quixel Megascan library of materials. These computational means allow for further image and video rendering, considering constructional materials and historical environment features.

The rendering results have been processed and adjusted with Photoshop to create the final virtual model (Fig. 7).

Fig. 5. The initial 3D model.

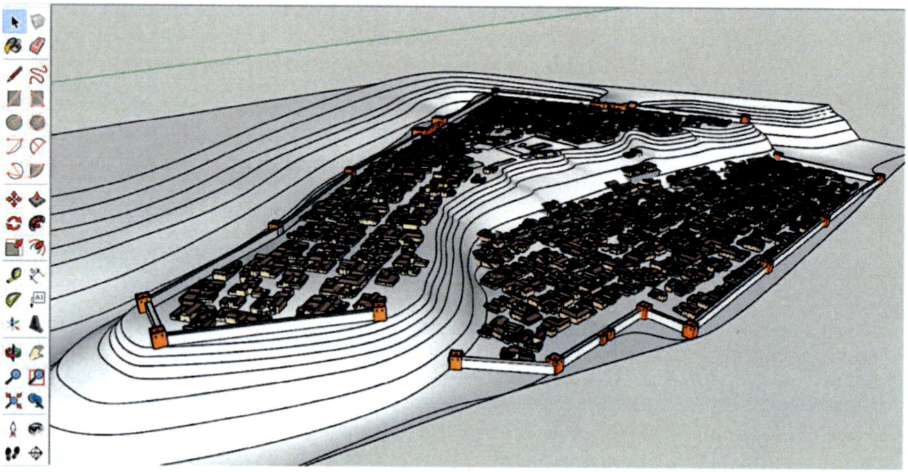

Fig. 6. Design layout of the model.

The developed 3D virtual model contains parts of the upper and lower city with quarters and streets, defense walls, river port, residential quarters, teminos, and central agora.

The obtained results are valuable for science, culture, and society. Particularly, the methodology of modeling the object under study can be applied to further work with the object at different stages. E.g., it can be supplemented by using BIM and CAD technologies for a promising reproduction or reconstruction of historical and architectural objects. Moreover, in today's realities, Ukraine is experiencing a growth of national identity. This encourages knowledge of own history, reconstruction, and research of partially lost objects of heritage. The proposed approach allows for enabling these processes.

Fig. 7. The final texturing and rendering.

The accuracy of the relief creation data can be improved using technologies of extended [21] and mixed reality [22], photogrammetry [23], or laser scanning [24] of a locality by the stereotopographic method as the most accurate but resource-consuming method of modeling and obtaining an information model. These tasks require applying the following software, e.g., Agisoft Metashape, Digitals, AutoCAD, and teleoperation by an unmanned aerial vehicle (UAV) [25, 26].

The proposed approach improves the results obtained by the previous study [3]. Also, architectural objects created based on archaeological research and foundation schemes can be imported into Revit [27] and extended using BIM technology and embedded vision equipment [28] to develop a comprehensive information model [29].

In the nearest future, it is also planned to investigate the following historical objects: the Konotop fortress of the period of 1659, the ancient Vorgol settlement in the Sumy region, the settlement Zelenyi Hai near Sumy, and other objects of architectural and historical heritage, as well as create their 3D virtual models of these objects using the developed methodology.

5 Conclusions

As a result of the research, a methodology of 3D virtual reconstruction for the building environment was proposed based on the comprehensive analysis of historical, archaeological, cartographic, and scientific sources.

A detailed study of the object's current state using the example of Pontic Olbia was carried out using Google Maps and GIS mapping services. Also, the study of the relief for the territory was carried out using topography and geodetic data. Laser scanning and photogrammetry technologies can also be applied to increase the accuracy of the obtained results.

Overall, the results of an analytical experiment are virtual modeling followed by creating a 3D model. It should make it possible to reconstruct the object of cultural and historical heritage.

As a result of implementing the proposed methodology, a 3D model of the architectural environment was created using this approach and close interaction with archaeological, historical, and cartographic materials. This model contains parts of the upper and lower city with quarters and streets, defense walls (Northern and Western Gates), river port, residential quarters of all available archaeological sites, Western and Eastern Temenos (Temples of Apollo Yatros, Apollo Dolphin, and Zeus), and Central Agora (Eastern Trade Row, Eastern and Western Stands, Dicasterium, and Gymnasium).

The model can be used in further experiments and its adjustment. Due to geodetic, topographical, and archaeological data, the areas of the model investigated by archaeologists can be considered as a historically grounded version of the 3D virtual reconstruction.

Acknowledgment. The research was partially realized within the Ulam NAWA Programme project BPN/ULM/2022/1/00042, ordered by the Polish National Agency for Academic Exchange. Also, it was partially supported by the Slovak Research and Development Agency under contract No. APVV-15-0602, and by the projects KEGA 022TUKE-4-2023, VEGA 1/0704/22 granted by the Ministry of Education, Science, Research and Sport of the Slovak Republic. Moreover, the authors appreciate the support of the International Association for Technological Development and Innovations.

References

1. Martynenko, N.M.: History of Ukrainian Culture. KNMU, Kharkiv (2015)
2. Xu, C.F., Zhou, X.C.: Three-dimensional animation technology in showing the application of the art of design. MATEC Web of Conferences **44**, 02044 (2016). https://doi.org/10.1051/matecconf/20164402044
3. Tytarenko, I., Dreval, I.: 3D modeling of a virtual built environment using digital tools: Kilburun fortress case study. Applied Sciences (Switzerland) **13**(3), 1577 (2023). https://doi.org/10.3390/app13031577
4. Fiorini, G., Friso, I., Balletti, C.: A geomatic approach to the preservation and 3D communication of urban cultural heritage for the history of the city: The journey of Napoleon in Venice. Remote Sensing **14**(14), 3242 (2022). https://doi.org/10.3390/rs14143242
5. Tan, Y., Chen, L., Wang, Q., Li, S., Deng, T., Tang, D.: Geometric quality assessment of prefabricated steel box girder components using 3D laser scanning and building information model. Remote Sensing **15**(3), 556 (2023). https://doi.org/10.3390/rs15030556
6. Meyer, T., Brunn, A., Stilla, U.: Geometric BIM verification of indoor construction sites by photogrammetric point clouds and evidence theory. ISPRS J. Photogramm. Remote. Sens. **195**, 432–445 (2023). https://doi.org/10.1016/j.isprsjprs.2022.12.014
7. Pan, Y., Zhang, L.: Integrating BIM and AI for smart construction management: current status and future directions. Archives of Computational Methods in Engineering **30**(2), 1081–1110 (2023). https://doi.org/10.1007/s11831-022-09830-8
8. Doula, A., Kaufmann, P., Guinea, A.S., Muhlhauser, M.: Effects of the level of detail on the recognition of city landmarks in virtual environments. In: Proceedings of the 2022 IEEE Conference on Virtual Reality and 3D User Interfaces Abstracts and Workshops, VRW 2022, pp. 860–861. IEEE, Piscataway (2022). https://doi.org/10.1109/VRW55335.2022.00281
9. Chernyshev, D., Ryzhakova, G., Honcharenko, T., Petrenko, H., Chupryna, I., Reznik, N.: Digital administration of the project based on the concept of smart construction. In: Alareeni,

B., Hamdan, A. (eds.) Explore Business, Technology Opportunities and Challenges After the Covid-19 Pandemic. ICBT 2022. Lecture Notes in Networks and Systems, vol. 495, pp. 1316–1331. Springer, Cham (2023). https://doi.org/10.1007/978-3-031-08954-1_114
10. Bevilacqua, M.G., Russo, M., Giordano, A., Spallone, R.: 3D reconstruction, digital twinning, and virtual reality: Architectural heritage applications. In: Proceedings of the 2022 IEEE Conference on Virtual Reality and 3D User Interfaces Abstracts and Workshops, VRW 2022, pp. 92–96. IEEE, Piscataway (2022). https://doi.org/10.1109/VRW55335.2022.00031
11. Lang, M., Kluge, P., Kruse, F., Rosenkranz, V.: Advent City: The reconstruction of an extinct settlement in virtual reality. Polarforschung **90**(2), 49–61 (2022). https://doi.org/10.5194/polf-90-49-2022
12. Shanthini, E., Sangeetha, V., Vaishnavi, V., Aisvariya, V., Lingadharshini, G., Sakthi Surya, M.L.: Recreating Poompuhar ancient history using virtual reality. In: Rajakumar, G., Du, KL., Vuppalapati, C., Beligiannis, G.N. (eds.) Intelligent Communication Technologies and Virtual Mobile Networks. Lecture Notes on Data Engineering and Communications Technologies, vol. 131, pp. 135–144. Springer, Singapore (2023). https://doi.org/10.1007/978-981-19-1844-5_12
13. Croce, P., Leccese, F., Mordagà, M., Salvadori, G.: Towards reconstructing the Shakespeare's first Globe Theatre: A virtual model for research and development. Frontiers of Architectural Research **11**(6), 1104–1120 (2022). https://doi.org/10.1016/j.foar.2022.05.004
14. Ou, W.: Interactive garden landscape digital reconstruction system based on Particle Swarm Algorithm. In: Sugumaran, V., Sreedevi, A.G., Xu , Z. (eds.) Application of Intelligent Systems in Multi-modal Information Analytics. ICMMIA 2022. Lecture Notes on Data Engineering and Communications Technologies, vol. 138. Springer, Cham (2022). https://doi.org/10.1007/978-3-031-05484-6_55
15. Javanbakht, T.: Optimization of graphene oxide's characteristics with TOPSIS using an automated decision-making process. J. Eng. Sci. **10**(1), E1–E7 (2023). https://doi.org/10.21272/jes.2023.10(1).e1
16. Javanbakht, T.: A novel automated decision-making process for analysis of ions and organic materials in drinking water. J. Eng. Sci. **10**(1), H1–H7 (2023). https://doi.org/10.21272/jes.2023.10(1).h1
17. Kuric, I., Kandera, M., Klarák, J., Ivanov, V., Więcek D.: Visual product inspection based on deep learning methods. In: Tonkonogyi, V., et al. (ed.) Advanced Manufacturing Processes. InterPartner-2019. Lecture Notes in Mechanical Engineering, pp. 148–156. Springer, Cham (2020). https://doi.org/10.1007/978-3-030-40724-7_15
18. Haidabrus, B., Grabis, J., Protsenko, S.: Agile Project Management Based on Data Analysis for Information Management Systems. In: Ivanov, V., Trojanowska, J., Pavlenko, I., Zajac, J., Peraković, D. (eds.) Advances in Design, Simulation and Manu-facturing IV. DSMIE 2021. Lecture Notes in Mechanical Engineering, pp. 174–182. Springer, Cham (2021). https://doi.org/10.1007/978-3-030-77719-7_18
19. Zhou, X., Liu, Y., Lian, X., Guo, H., Lee, S.: Research on key technology of virtual restoration of ancient buildings based on convolutional neural network. In: Proceedings of SPIE - The International Society for Optical Engineering 12348, 123480Q. SPIE, Bellingham (2022). https://doi.org/10.1117/12.2641425
20. Zidek, K., Lazorik, P., Pitel, J., Pavlenko, I., Hosovsky, A.: Automated training of convolutional networks by virtual 3D models for parts recognition in assembly process. In: Trojanowska, J., Ciszak, O., Machado, J., Pavlenko, I. (eds.) Advances in Manufacturing II, vol. 1 - Solutions for Industry 4.0. Lecture Notes in Mechanical Engineering, pp. 287–297. Springer, Cham (2019). https://doi.org/10.1007/978-3-030-18715-6_24
21. Buń, P., Husár, J., Kaščak, J.: Hand Tracking in Extended Reality Educational Applications. In: Trojanowska, J., Kujawińska, A., Machado, J., Pavlenko, I. (eds.) Advances in Manufacturing

III. MANUFACTURING 2022. Lecture Notes in Mechanical Engineering, pp. 317–325. Springer, Cham (2022). https://doi.org/10.1007/978-3-030-99310-8_25
22. Buń, P., Gapsa, J., Husár, J., Kaščak, J.: Mixed reality training in electrical equipment operating procedures. In: Trojanowska, J., Kujawińska, A., Machado, J., Pavlenko, I. (eds.) Advances in Manufacturing III. MANUFACTURING 2022. Lecture Notes in Mechanical Engineering, pp. 306–316. Springer, Cham (2022). https://doi.org/10.1007/978-3-030-99310-8_24
23. Stellacci, S., Condorelli, F.: Remote survey of traditional dwellings using advanced photogrammetry integrated with archival data: the case of Lisbon. In: The International Archives of the Photogrammetry, Remote Sensing and Spatial Information Sciences - ISPRS Archives, vol. 43(B2–2022), pp. 893–899. ISPRS, Hannover (2022). https://doi.org/10.5194/isprs-archives-XLIII-B2-2022-893-2022
24. Cai, H., Wang, Y., Lin, Y., Li, S., Wang, M., Teng, F.: Systematic comparison of objects classification methods based on ALS and optical remote sensing images in urban areas. Electronics (Switzerland) **11**(19), 3041 (2022). https://doi.org/10.3390/electronics11193041
25. Stedman, H., Kocer, B.B., Kovac, M., Pawar, V.M.: VRTAB-map: A configurable immersive teleoperation framework with online 3D reconstruction. In: Proceedings of the 2022 IEEE International Symposium on Mixed and Augmented Reality Adjunct, ISMAR-Adjunct 2022, pp. 104–110. IEEE, Piscataway (2022). https://doi.org/10.1109/ISMAR-Adjunct57072.2022.00029
26. Pérez-Portugal, A., Atencio, E., Muñoz-La Rivera, F., Herrera, R.F.: Calibration of UAV flight parameters to inspect the deterioration of heritage façades using orthogonal arrays. Sustainability (Switzerland) **15**(1), 232 (2023). https://doi.org/10.3390/su15010232
27. Barros, B.A.F.S., Sotelino, E.D.: Constructability and sustainability studies in conceptual projects: a BIM-based approach. J. Constr. Eng. Manag. **149**(4), 04023012–04023021 (2023). https://doi.org/10.1061/JCEMD4.COENG-12767
28. Židek, K., Maxim, V., Piteľ, J., Hošovský, A.: Embedded vision equipment of industrial robot for inline detection of product errors by clustering-classification algorithms. Int. J. Adv. Rob. Syst. **13**(5), 1 (2016). https://doi.org/10.1177/1729881416664901
29. Luis Lugo, J., Gallegos Reyes, J.C., Weidner, J., Raheem, A., Cheu, RL: Evaluation of 3D reconstruction methods from terrestrial LiDAR point cloud data to create digital twins of civil infrastructure projects. In: The International Conference on Transportation and Development 2022: Application of Emerging Technologies - Selected Papers from the Proceedings of the International Conference on Transportation and Development 2022, vol. 1, pp. 81–92. ASCE, Reston (2022). https://doi.org/10.1061/9780784484319.008

Manufacturing Technology and Machining Processes

The Multifractal Analysis of Periodic Surface Relief of Parts After Face Milling

Nataliia Balytska(✉), Vyacheslav Kryzhanivskyy, Petro Melnychuk, Heorhii Vyhovskyi, and Pavel Moskvin

Zhytomyr Polytechnic State University, 103, Chudnivska Str., Zhytomyr 10005, Ukraine
tmkts_nno@ztu.edu.ua

Abstract. The multifractal (MF) analysis was applied to prove the existence of self-similarity and fractal symmetry of special microforms formed on the surface of structural carbon steel C35 and aluminum alloy AA2024 after their face milling. The micro images of surface specimens previously milled at various cutting conditions were used as the input information to implement this approach to describe the state of surfaces. The paper calculates generalized statistical sums for the area and volumes of spatial microforms to prove the presence of linear dependences between their values and sizes of elementary particles in the method of coarse partitions. This result was used as a proof of the presence of fractal symmetry among the indicated surface parameters for the case of a periodic spatial relief formation on the surface as an outcome of cyclic action of face mill blade. The MF spectra and their basic parameters are calculated for the surfaces of samples formed under different cutting conditions. The correspondence of the characteristic functions of the performed MF analysis to their canonical forms was discussed. This is considered as a proof that the indicated geometric surface parameters of the material, even after machining, form a system with self-similarity and fractal symmetry properties. The quantitative relationships between the MF spectrum parameters of surface area and volumes of microforms formed with face milling and cutting conditions were discussed.

Keywords: Surface Microrelief · Self-Similarity · Renyi's Numbers · Multifractal Spectra

1 Introduction

The relief, which is formed on the surface of parts applying various machining methods, is one of the factors that determine their functional properties at micro level.

Therefore, the development of experimental research methods and mathematical description of surface relief within small spatial scales is considered to be an important theoretical and practical task.

By this time, a whole range of quantitative characteristics has been developed, substantiated and widely implemented in practice to describe micro relief of sample surface. The documents [1, 2] specify terms, definitions, parameters and complete specification

V. Tonkonogyi et al. (Eds.): InterPartner 2023, LNME, pp. 117–126, 2024.
https://doi.org/10.1007/978-3-031-42778-7_11

operator for surface texture by profile methods. Typically, different measuring devices can be used to measure these characteristics in a relatively easy way.

The harmonic analysis approach [3], namely the Fourier series [4], should be also recognized as an application of spatial relief linear characteristics [5]. Despite the considerable progress in implementation of such mathematical approach while solving the stated problem, it should be recognized that the application of harmonic function rows (the Fourier series) featured by a low convergence rate makes the task rather mathematically complex and cumbersome. The artificial limitation of the number members in the Fourier series which is equivalent to disregard high-order members in respective series results in both the significant errors in final outcomes and the distortion of interpretation of the physical picture of the process. This is why the approach of using areal methods [6] to determine the surface texture should be considered as one that can be improved.

The methods using power series theory and fractal system structure concepts have to be considered as unusual ones. Such methods are used to describe surface condition in which the spatial features of surface relief are taken into account and they rely on the principles of self-similarity and more general forms of symmetry. The approach based on introduction of fractional dimensions to describe micro space geometry could make it possible to provide further theoretical description of many phenomena which are difficult to interpret within classic physical concepts. Definitely theory of material cutting makes an extensive use of mathematical expressions derived from statistical processing of experimental results and these expressions include power functions with fractional indicators as well. The latter can intrinsically reflect the appearance of fractal properties within the interacting system workpiece-tool blade. This situation stimulates the use of fractal representations to describe the processes occurring at machining.

2 Literature Review

One of the directions solving the problem of mathematical description of surface state is associated with application of the fractal analysis (FA) methods. As a rule, the description of complex geometric forms that are experimentally observed on the surface of specimens tends to be, in point of fact, unpromising. At the same time, it does not seem justified to ignore the influence of appearance of other forms of symmetry, namely the fractal ones, on formation of final mechanical and thermodynamic properties at the precision level of their studies.

Presently, there are an insufficient number of publications devoted to the development and application of the MFA to the description of part surfaces after machining. The main discussion in the works [7, 8] is concentrated both on equipment issues and features of obtaining high-quality micro images of the surface after machining, and on the analysis of various methods of preliminary processing the initial information for its subsequent correct application in the FA or MFA methods.

It is known about the works on influence of cutting conditions on roughness and the corresponding fractal characteristics of the surface. In [9–11], as the initial parameter for fractal parameterization was chosen the length of the surface profile line. It is not absolutely clear in which physical models of the cutting process this parameter can be utilized. This may be an explanation for the difficult-to-interpret effect found in [9], namely an increase in the part surface roughness [10], while the fractal dimension of the surface profile simultaneously decreases. In the mentioned studies, the fractal dimension was calculated on the base of the analysis of two-dimensional profile images and did not take into account the entire spatial pattern of the relief. In addition, the authors of [11] did not manage to find a significant relationship between the fractal dimension and the surface roughness of the part after turning.

Besides, the use of fractal analysis to describe the surface area after turn-milling [12] made it possible to detect the relationship between the fractal dimension of the surface area and the machining conditions. There is a similar situation with the use of the fractal method to describe the part surfaces from various materials [13] after end milling [14].

The importance of choosing a geometric parameter of a system for its fractal parameterization is pointed out by many authors. In this case, the priority in choosing the geometric parameter of the system is given to the surface area of the part after machining. In addition, the fractal dimension of the machined surface area is even recommended for use as an additional parameter when describing the surface relief of a composite material after milling [15].

The FA method was used by us in [16] to solve the problem of describing periodic surface relief formed after face milling. The use of standard mathematical programs made it possible to perform calculations for fractal characteristics of steel and aluminum alloy surface. However, it is unlikely that the use of Housdorff dimension alone will be sufficient to fully describe the complex shape of surface relief. In this respect, it might be preferable to use multifractal analysis to describe such a complex system where the system is characterized by the Renyi's numbers.

Thus, the purpose of the given paper is to quantitatively prove the presence of fractal symmetry and self-similarity phenomena for the surface areas and volumes of microforms which are naturally formed on specimen surface as a result of face milling by applying the MFA mathematical tool. Self-similarity phenomenon detection will provide a physical basis for obtaining quantitative data on the MF surface spectrum parameters and volumes of microforms formed on specimens of C35 steel and AA2024 aluminum alloy as a result of face milling.

3 Research Methodology

3.1 The MFA Implementation and Self-Similarity of Microforms on Specimen Surfaces

The particular attention was paid to the selection of the physical parameter that completely characterizes surface condition of specimens and is subjected to fractal parameterization when applying the MFA. The surface area and volume of surface elements were selected among the possible geometric parameters of spatial shapes that are formed on part surface. These parameters basically determine the surface energy of the system and form the micro surface relief of the sample.

The MFA implementation was carried out in accordance with the standard procedure [17–20]. The input information for MFA surface area and volume of microforms on the surface was obtained by microphotograph. The generation of measure of each cell of the space was performed by dividing the space covering the base measure set into N

cells. The relative value of surface area and microform volume was taken as a measure of surface cell $\nu_{S,i}$ and volume $\nu_{V,i}$. This value entered into the given partition cell:

$$\nu_{S,i} = S_i/S \quad \nu_{V,i} = V_i/V, \tag{1}$$

where S_i and V_i are correspondingly surface and volume of small selected element of surface relief, $S = \sum_{i=1}^{N} S_i$ и $V = \sum_{i=1}^{N} V_i$ are surface and volume of entire analyzed surface that were found using the data of its image. The basic sites were calculated by triangulating the digital picture of the actual surface.

The MF spectrum parameter search was conducted by using the method of coarse partitions in accordance with the standard procedure [17–20]. At the same time, general statistical sum was compiled for a cell of the given size for both surface area and volumes:

$$Z_S(q, l_k) = \sum_{i=1}^{K} \nu_{S,i}^{q}, \quad Z_V(q, l_k) = \sum_{i=1}^{K} \nu_{V,i}^{q}, \tag{2}$$

where l_k is the normalized current length of cube edge used at the current stage in coarse partitioning method. Calculation changes (cube size aggregations) of cell scales were performed following the dependency $l_{k+1} = l_k \cdot 2$ (k = 1, 2, 3,...).

If there is fractal symmetry in the analyzed physical system, all the obtained calculation dependences $\ln Z(q, l_k)$ upon $\ln l_k$ shall consist of a set of points grouped along straight lines. Therefore, accumulation and analysis of such information for each of the selected geometric surface parameter and for each MFA exponent is considered to be of fundamental nature. The calculation of linear parameter regression between the specified parameters of the system was performed by the method of least squares for each among the selected values of magnification number q. The linear regression coefficient data served as the basis for calculating all the functions that make up MF spectrum of the system: $\tau(q), f(\alpha)$ and $D(q)$. All the mentioned functions were calculated by numeric methods.

3.2 Specimens and Surface Imaging Features

The specimens for the study were obtained by milling on the machine model 6P12. The cutting tool was a milling cutter with the diameter of 40 mm with four pentagonal inserts made of carbide alloy with a titanium nitride coating. The specimens were the plates of steel C35 and aluminum alloy AA2024. The cutting speed of 130 m/min (125 rpm) and the feed rate of 100 mm/min were chosen as constant parameters for the study. In this case, the depth of cut was a variable parameter.

Digital camera was mounted directly on the XS-2610 MICROmed microscope. The surface was photographed in reflected light. The uniformity of illumination was ensured by using a set of LEDs which were evenly arranged on the ring holder around the specimen. The surface relief conditions were chosen so that the photograph would focus on the full extent of the surface relief and each part of it would be photographed with the same microscope zoom.

Figure 1 demonstrates surface photographs of specimens made of C35 steel and AA2024 aluminum alloy after face milling. The presented images are the typical ones. They were processed by the MFA methods using the developed software.

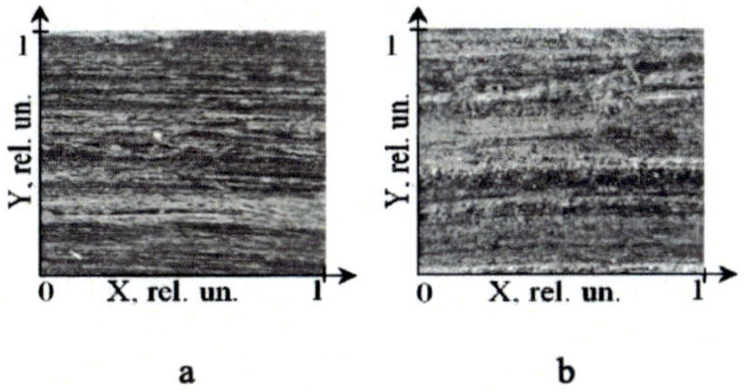

Fig. 1. The typical surface microphotographs of specimens made of steel C35 (a) and AA22024 aluminum alloy (b) after face milling (feed rate $f = 100$ mm/min, depth of cut $a_p = 1{,}0$ mm, cutting speed $v = 130$ m/min). The size of the analyzed area is 300 × 300 pixels.

The data presented in Fig. 1 alongside the photographic images of the surface from [16], produce the complete set of experimental results taken into account when implementing the MFA.

The most important calculation outcomes, which should be regarded as the most significant quantitative proof of the existence of fractal symmetry between geometrical parameters of surface microforms are demonstrated in Fig. 2. The graphs show a typical form of statistical summation for changes of surface areas of microforms and their volumes on the given length of an edge of space unit cell in the method of coarse partitioning for the set of numbers q.

Figure 3 suggests, the experimental data are distinctly grouped along straight lines. The typical value of correlation coefficients for the values $q < 4$ in the method of least squares, as a rule exceeded 0.92, naturally varying from specimen to specimen. The very presence of linear relationship among the analyzed in logarithmic coordinates is known to be the main proof of self-similarity existence in the system [20]. At the same time, it is obvious that the volumes and surface areas of microforms are related to their linear scales and it is due to power function. This result leads to assertion that relief obtains fractal symmetry in the system of microforms that form such surface relief. This occurs even if a longwave periodic relief appears on the surface due to periodic action of cutting tool blade.

Sufficiently high correlation coefficients for different values of q parameter make it possible to assert that fractal symmetry and self-similarity exist within the system. Moreover, that is the MF spectrum that is typical for the surfaces under consideration and is a promising method for describing such systems. Therefore, it is the MF system parameters that are recommended to be applied when describing the surface condition of the specimens obtained under the specified conditions.

Figure 3 demonstrates the typical calculation outcomes of the MFA characteristic functions $\tau(q)$,$f(\alpha)$, $D(q)$ according to [19, 20]. Figure 3 makes it possible to conclude that they correspond to their canonical forms. This means that the sequence of the Renyi's numbers is decreasing, the function sequence $f(\alpha)$ has its specific maximum and the distinctive for the MFA fracture is observed on the dependency $\tau(q)$. No single sample with any deviation in the behavior of the characteristic functions from that predicted by theory [19, 20] was found among those which were studied. This result is an equally

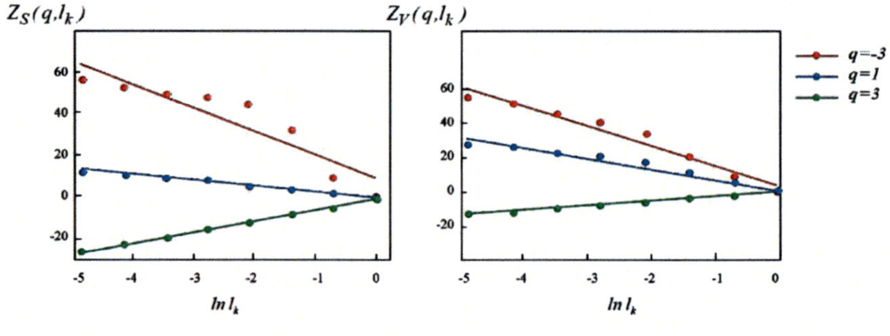

Fig. 2. Generalized statistical summations $Z_S(q, l_k)$ (a) and $Z_V(q, l_k)$ (b) in dependence on l_k for the micro image demonstrated in Fig. 1, b.

essential proof of self-similarity and fractal symmetry properties within the system of relief forming microforms.

The Renyi's numbers $D_{S,0}$ and $D_{V,0}$ as well as fractal order parameters $\Delta_S = D_{S,1} - D_{S,q\to\infty}$ $\Delta_V = D_{V,1} - D_{V,q\to\infty}$ (the degree of fractal symmetry disruption) were chosen as the most informative MF parameters capable of describing surface structure in accordance with the recommendations [18–20]. The typical value of q parameter usually did not exceed $q = 80$ in the calculations of the given paper. The calculations performed demonstrated that an increase of the parameter q over 80 led to a change of each subsequent Renyi's number D_{q+1} by less than 10^{-4}. It was due to the fact that the obtained images contained the number of pixels that did not exceed 300.

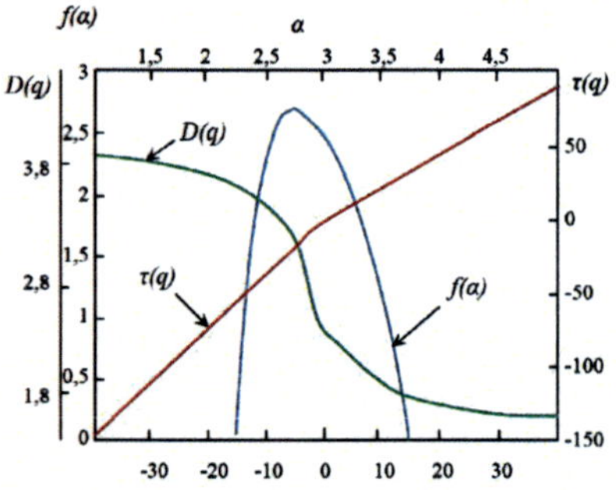

Fig. 3. Characteristic function of MFA $\tau(q), f(\alpha)$ and $D(q)$ for the image in Fig. 1, b.

4 Results and Discussion

The obtained quantitative data on the MF surface parameters of specimen surface enable searching for their correlation with the conditions of surface machining. The totality of such correlations between the Renyi's number D_0, the order parameter Δ_{80} for different machining conditions are presented in Fig. 4. In order to interpret the obtained data, it should be remembered that the value of the numbers $D_{S,1}D_{S,0}$, which is equal to 2 corresponds to ideal flat surface within the fractal formalism. It is natural, that the volume next to the surface is filled with material and it corresponds to the value of the numbers $D_{V,0}$,$D_{V,0}$ is equal to 3. The Renyi's number distinctive feature from the indicated values for a planar surface corresponds to the solution to the task of finding the dimensionality of a surface that is non-planar, porous, and developed at a micro level.

The data in Fig. 4 show that the given study found the relationship among the Renyi's numbers for surface area and microform volumes and cut depth at constant feed rate (f = 100 mm/min) and cutting speed (v = 130 m/min). The straight lines in the graphs were drawn by averaging the experimental dependencies of considered MF parameters on the relevant cutting conditions using the method of least squares.

As for the general trends in the dependencies presented in Fig. 4, their global interpretation seems natural. Definitely, there is a transition from the "soft" cutting conditions typical for the finishing to the "hard" ones observed at roughing as cutting depth increases. The surface becomes more non-planar and its Renyi's numbers for the surface area ($D_{S,0}$) increase deviating more and more from "two". Such a conclusion is also confirmed by the Renyi's numbers dependence for microform volume ($D_{V,0}$), which tends to decrease at cutting depth increasing. This course of dependence corresponds to an increase of overall non-planarity of the final surface, when $D_{V,0}$ demonstrates an increasing deviation from "three", reflecting the appearance of unfilled volumes near the surface specimen.

The emergence of such a clear dependency has a fairly simple interpretation. Cutting depth increase at constant feed rate does not change the long-wave oscillation parameters of surface relief and, therefore, has no significant effect on surface configuration entropy. It should be reminded that the Renyi's numbers are closely related to the entropic component of the energy system [19, 20]. At the same time, depth of cut increase causes an increase of surface area formed at the micro level and produces deeper void areas, which appeared when the material was removed by the cutter. These factors produce an increase in surface roughness as a consequence of the increased magnitude of energy impact on the workpiece during cutting process. The latter is presented in Fig. 4 demonstrating the mentioned trends in the behavior of the corresponding Renyi's numbers as cut depth increases.

The fractal order parameter of the area Δ_S for microform volumes Δ_V is an important parameter of fractal spectrum. The data on the dependence of these parameters on the feed rate are shown in Fig. 4. In general, the width of the fractal spectrum (parameters Δ_S and Δ_V) increases with the transition to more energy intensive machining modes, which are typical for roughing. Here the MF spectrum absolute width from the elements of its surface becomes slightly larger after machining (Fig. 4). It occurs when less hard material (aluminum alloy) is used. The "sensitivity" of these surface parameters $((\partial D_{S,0}/\partial t,\ \partial D_{V,0}/\partial t))$ to an increase in cutting depth, even for materials with significantly different physical and mechanical characteristics, is approximately the same. The latter statement follows from Fig. 4, which demonstrates an almost parallel pairwise course of the dependences under consideration with their almost identical slope.

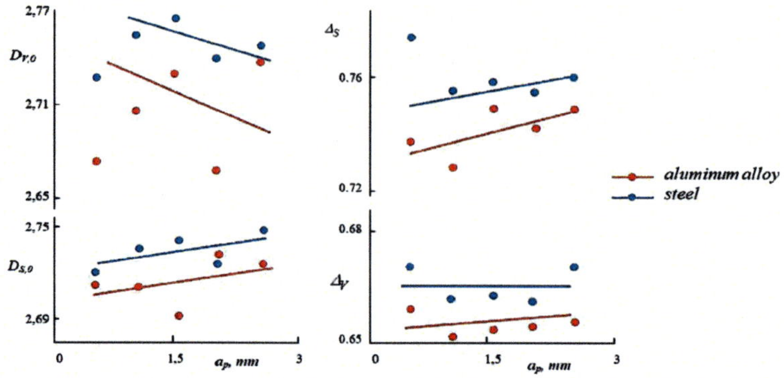

Fig. 4. The dependence of the MF parameters on the depth of cut (the experimental data of the present study).

The significant deviation of experimental results from the general linear trend should also be noted. It occurs with the given interdependences by parameter values of Δ_S and Δ_V at the small cutting depth (0,5 mm). A similar effect is also observed when considering the dependence of the Renyi's numbers on the cutting depth at its small values in Fig. 4. The authors tend to believe that it is also a reflection of far more substantial contribution of the machining system vibrations to the overall periodic pattern formed on the specimen surface.

The low enough functional dependences of MF width of spectra on cutting depth for each of the materials should be taken in account as well. Such data can be explained within geometric representations of surface relief as follows. The main periodic field remains practically unchanged and the changes mainly affect only surface relief components at the variations of cutting depth when the feed rate is constant. This is reflected in the slight variation of the Renyi's numbers depending on depth of cut.

The relatively high typical value of the fractal order parameter Δ_S has to be mentioned as well. It corresponds to significant width of the fractal spectrum from the surface microelements. The data obtained by the specified parameter demonstrate its typical value at the level of 0.65. In comparison, the same parameter does not exceed 0.35 [17] for significantly high quality and homogeneous polycrystalline epitaxial film surfaces. Since such high values of the order parameter are observed for all the studied machining modes, this effect is likely to demonstrate the features of the used milling machine.

5 Conclusions

The MFA method was applied to image processing of the surface of specimens made of C35 steel and AA2024 aluminum alloy after face milling. The existence of self-similarity phenomenon and fractal symmetry between microforms that form specimen surface relief was quantitatively proved. The MF spectra of the surface area and volumes of microforms formed on the specimen surface when cutting were calculated and analyzed. It was shown that fractal symmetry was present even in a system whose surface is periodic due to the impact of a tool blade.

The data were obtained on the dependence of MF spectra for the specimen surface on the milling conditions. It was demonstrated that the Renyi's numbers and ordering parameters are in close correlation with the depth of cut.

References

1. ISO 21920-2:2021 Geometrical product specifications (GPS) – Surface texture: Profile – Part 2: Terms, definitions and surface texture parameters (2021)
2. ISO 21920-3:2021 Geometrical product specifications (GPS) – Surface texture: Profile – Part 3: Specification operators (2021)
3. Harte, D.: Multifractals: Theory and Applications. Chapman and Hall. CRC, New York (2001)
4. Constantoudis, V., Ioannou-Sougleridis, I., Dimou, A., Ninou, A., Chatzichristidi, M., Makarona, E.: A symmetry-based approach to the characterization of complex surface morphologies: application in CuO and NiO nanostructures. Micro Nano Eng. **16**, 100148 (2022). https://doi.org/10.1016/j.mne.2022.100148
5. Godreche, C., Luck, J.M., Godreche, C.: Multifractal analysis in reciprocal space and the nature of Fourier transform of self-similar structures. J. Phys. A: Math. Gen. **23**(16), 3769–3797 (1990). https://doi.org/10.1088/0305-4470/23/16/024
6. ISO 25178-2:2021 Geometrical product specifications (GPS) – Surface texture: Areal – Part 2: Terms, definitions and surface texture parameters (2021)
7. He, W., Wang, Y., Zhao, F., Zhou, P.: Evaluation method for 3D topography of CFRP cutting surfaces based on multifractals spectrum and wavelet. MACE **2010**, 2354–2357 (2010)
8. Talu, S., Guzzo, P.L., Astinchap, B., Ghanbaripour, H.: Multifractal analysis of ultrasonically machined surfaces of cylindrical quartz crystals: the effect of the abrasive grits. Surf. Topogr.: Metrol. Prop. **9**, 045051 (2021)
9. Zuo, X., Zhu, H., Zhou, Y., Ding, C., Sun, G.: Development of fractal dimension and characteristic roughness models for turned surface of carbon steels. Fractals **24**(4), 1650042 (2016). https://doi.org/10.1142/S0218348X16500420
10. Mukhopadhyay, A., Barman, M., Sahoo, P.: Optimization of fractal dimension of turned AISI 1040 steel surface considering different cutting conditions: fractal dimension of turned steel surface. IJSEIMS **7**(2), 19–33 (2019)
11. Petropoulos, G., Bouzid, W., Pandazaras, C., Dramalis, D.: Fractal geometry of metal surfaces obtained by turning. Mater. Technol. **21**(3), 163–169 (2006)
12. Niu, Z., Jiao, L., Chen, S., Yan, P., Wang, X.: Surface quality evaluation in orthogonal turn-milling based on box-counting method for image fractal dimension estimation. Nanomanuf. Metrol. **1**(2), 125–130 (2018). https://doi.org/10.1007/s41871-018-0015-x
13. Mao, H., et al.: Surface quality evaluation in meso-scale end-milling operation based on fractal theory and the Taguchi method. Int. J. Adv. Manuf. Technol. **91**(1–4), 657–665 (2017)
14. Zuo, X., Zhu, H., Zhou, Y., Yang, J.: Estimation of fractal dimension and surface roughness based on material characteristics and cutting conditions in the end milling of carbon steels. Proc. Inst. Mech. Eng. Part B: J. Eng. Manuf. **231**(8), 1423–1437 (2017)
15. Liu, J., Cheng, K., Ding, H., Chen, S.: An investigation of influence of cutting parameters on three-dimensional surface topography in micromilling SiCp/Al composites. Proc. Inst. Mech. Eng. Part B: J. Eng. Manuf. **235**(5), 829–838 (2021). https://doi.org/10.1177/0954405420971098
16. Moskvin, P., Balytska, N., Melnychuk, P., Rudnitskyi, V., Kyrylovych, V.: Special features in the application of fractal analysis for examining the surface microrelief formed at face milling. EEJET **2**(1(86)), 9–15 (2017). https://doi.org/10.15587/1729-4061.2017.96403

17. Moskvin, P., Kryzhanivskyy, V., Rashkovetskyi, L., Lytvyn, P.: Multifractal spectrums for volumes of spatial forms on surface of ZnxCd1−xTe–Si(111) heterostructures and estimation of the fractal surface energy. J. Cryst. Growth **450**, 28–33 (2016). https://doi.org/10.1016/j.jcrysgro.2016.05.035
18. Feder, J.: Fractals. Springer, New York (1988). https://doi.org/10.1007/978-1-4899-2124-6
19. Klymenko, S.A., Melnijchuk, Yu.O., Vstovskyj, G.V.: Fractal parameterization of the structure of materials, their processing by cutting and wear resistance of cutting tools. INM im. V.M. Bakulya, Kyiv (2009)
20. Falconer, K.J.: Fractal Geometry Mathematical Foundation and Application. Wiley, New York (1990)

Quaternion Model of Workpieces Orienting Movements in Manufacturing Engineering and Tool Production

Irina Cherepanska[1(✉)], Artem Sazonov[1], Dmytro Melnychuk[2], Petro Melnychuk[2], and Yuriy Khazanovych[1]

[1] National Technical University of Ukraine "Igor Sikorsky Kyiv Polytechnic Institute", 37, Beresteiskyi Ave., Kyiv 03056, Ukraine
cherepanskairina@gmail.com

[2] Zhytomyr Polytechnic State University, 103, Chudnivska St., Zhytomyr 10005, Ukraine

Abstract. One of the most significant requirements in modern mechanical engineering and instrument-making productions is workpieces' orientation systems productivity and their simulation speed. Workpieces' orientation systems simulation, in particular their orientation movements, is one of the most challenging and essential pre-production tasks. The complexity of the mathematical model equations, its accuracy, and computational complexity affect simulation effectiveness. The article describes the mathematical model of the workpiece orienting movements, which relies upon the mathematical apparatus of quaternions. Moreover, the article shows the principal properties of quaternions and the rules for working with them. Theoretical and experimental studies of the workpiece orienting movements quaternion model are presented, particularly the sequence and results of determining the geometric parameters of these movements, i.e., rotation angle and direction of the vector, which is collinear to the workpiece's axis of rotation. The model has been studied for an arbitrary workpiece. The results show that the performed works' productivity and speed increased by approximately 40%. At the same time, computational costs and intellectual efforts are significantly reduced compared to the traditional methods of the workpiece's movement description.

Keywords: Quaternions · Automatic Orientation · Workpieces · Circular Motion · Linear Motion · Orienting Movements · Rotation Matrices · Euler Angles · R&D Innovation · Industrial Growth

1 Introduction

Modern production rates of mechanical engineering and instrument-making productions impose new higher requirements of the workpieces' orientation systems (WOS) productivity and provide for their simulation. The task of WOS simulation is one of the most challenging, solved during the flexible manufacturing systems (FMS) design, particularly mechanical and instrument-making engineering. For this reason, it requires the use of efficient, universal, and efficient methods and mathematical apparatus for the simulation of orientation movements, which are performed by workpieces during automatic

V. Tonkonogyi et al. (Eds.): InterPartner 2023, LNME, pp. 127–135, 2024.
https://doi.org/10.1007/978-3-031-42778-7_12

orientation. The process of automatic orientation means the workpiece's transferring from the initial non-oriented position (INP) to the predetermined final oriented position (FOP) in the absolute Cartesian coordinate system *XYZ* by performing a set of circular and linear motions that make up the composition of orienting movements. The complexity of the mathematical model equations, its accuracy, and computational complexity affect the effectiveness of the modeling of the WOS as a whole.

Traditionally, rotation matrixes [1, 2], Newton-Euler equations, Euler-Krylov angles, or Bryant angles [3] describe circular motions and movements of rigid bodies. However, all these models are rather complicated and cumbersome, requiring a lot of computing costs and intellectual efforts, as noted in the article [3]. It significantly complicates and slows down the modeling process. In addition, mentioned models do not provide any possibilities of a mathematical description of the linear motion of the OM during their automatic orientation. Therefore, the obtained mathematical model does not fully reflect the structure of the orienting movements of the workpiece. That is why it is necessary to use new approaches and mathematical apparatuses quaternions to build mathematical models of the orienting movements of the OM in the simulation of the WOS, which will help solve the existing problem.

An excellent alternative to the mentioned models can be the so-called quaternion model of workpiece orienting movements, which allows describing both circular and linear motions of workpieces. It should be noted that the idea of using quaternions to describe rigid bodies' actual circular motions is not new. Quaternions are used for tasks of dynamics and spatial measurements of angles because they simplify the equations of the model of circular motions, increasing computational efficiency and accuracy, described in the literature [4–11]. However, there is no information regarding using quaternions to describe workpiece linear motions of, and accordingly, the entire composition of their orienting movements during automatic orientation.

2 Literature Review

Scientists pay attention to traditional and well-studied models that only describe circular motion during rigid bodies' circular and linear movement simulations.

For example, in the article [1] authors describe the rotation and movement of rigid bodies in a generalized space using dual orthogonal matrices. The article [2] showed simple rotation matrices used to describe the robot's movements. At the same time, the authors note the disadvantages of this approach. It is indicated that even though the system can precisely describe workpieces' movements, it requires significant computational effort. In the article [3], various models of the motion of rigid bodies are given, matrix models of guide cosines, Newton-Euler equations, Euler-Krylov angles, or Bryant angles. The authors note the significant complexity of these models and point to the need to compile alternative equations with a more straightforward form and greater computational efficiency.

As an alternative to traditional models for describing the dynamics of rigid bodies, particularly their circular movements, researchers suggest using quaternions. The literature [4(15] shows that the mathematical apparatus of quaternions can be successfully applied to the problems of rigid body dynamics, particularly during its rotations, as well as the spatial measurement of angles. Using quaternions for solving these problems increases computing speed, accuracy, and productivity because their distinctive feature is compactness. Compared with the other standard methods, quaternions are represented by only one equation with four parameters, which provides models of circular motions of rigid bodies by systems of equations with six parameters. For example, the article [4] presents the quaternion calculation of angle values in goniometric measurements. The authors show that the quaternion method of angle calculations can increase the accuracy and speed of measurements. An increase in accuracy is achieved by performing fewer mathematical operations, which allows for reducing the rounding error of calculation results, which accumulates during repeated measurements and can reach large values. The performance of calculations during multiple measurements is increased due to the usage of a smaller amount of random access memory (RAM) of a computer. However, the authors do not consider the question of the usage of quaternions in the description of the orienting movements of workpieces in FMS of mechanical and instrument-making engineering.

The article [5] compares the matrix and quaternion algorithms for the spatial rotation of the object during autocollimation measurements. The quaternion algorithm is shown to be more effective in reducing the measurement error. The authors do not consider the issue of using quaternions to describe the orienting movements of workpieces in FMS of mechanical and instrument-making engineering.

The article [6] presents a new quaternion-based filter developed for gyroscopes and accelerometers for spatial angle measurement. The article's authors claim that quaternions increase the speed of calculations. The authors do not consider the possibility of using quaternions when describing the orienting movements of workpieces in the FMS of mechanical and instrument-making engineering.

The article [7] presents a quaternion algorithm with high computational efficiency and is used to calculate turning angles at any position of the measurement object. The article does not cover the possibility of using quaternions in describing the orienting movements of Workpieces in FMS of mechanical and instrument-making engineering.

Article [8] is considered the concept of quaternion coupling for rotations of a rigid body in the form of a standard ideal mechanical coupling, which is helpful for the construction of energetically consistent integrators. However, the authors do not consider the issue of quaternions used to describe orienting movements of workpieces in FMS of mechanical and instrument-making productions.

A new quadcopter orientation control scheme based on quaternions is discussed in [11]. The general structure of the quaternion is described as a hypercomplex number with rank 4, which can be used to avoid the geometric singularity characteristic of other models (Euler angles, matrix representations with direction cosines) when representing the dynamics of a solid body. However, the authors do not consider the possibility of using quaternions when describing the orienting movements of workpieces in the FMS of mechanical and instrument-making engineering.

The article [10] indicates that it is known that the usage of quaternions in the dynamic modeling of solids allows for avoiding the singularity caused by Euler rotations. This article shows that the dynamic response of a rigid body modeled by quaternions can become unbounded when a torque is applied to the body. The authors ignore the question of using quaternions to describe the orienting movements of workpieces in FMS of mechanical and instrument-making engineering.

The article [11] describes the motion and rotation of flexible space vehicles uniformly with double quaternions. And the article [12] discusses the usefulness of quaternions in describing the kinematics of a solid body in three-dimensional (3D) space in biomechanics. The quaternions are compared with other options for describing circular motions of a rigid body, including Cardan and Euler angles. However, the potential usage of quaternions to describe the orienting movements of workpieces in FMS of mechanical and instrument-making engineering when arranging the production environment is not discussed.

The article [13] is discussed that the formalization of the rotation of a body by such traditional methods as the set of Euler angles or the description of the angular axis has at least two possible ways of representing the orientation of the rigid body, which means that in some configurations these formulations will give an infinite number of possible representations, leading to a simulation singularity or a non-defined governing influence. For these reasons, the rotation quaternion is the preferred orientation formulation in computer control and simulation applications. The article does not consider the question of the description of the orientational movements of workpieces in FMS of mechanical and instrument-making engineering, which are carried out during the arrangement of the production environment.

The article [14] presents work using two quaternion representations regarding the dynamics of space vehicles. The validity of quaternions compared to their Cartesian equivalents, mainly matrices of direction cosines, is discussed. However, the authors do not consider the question of using quaternions to describe the orienting movements of workpieces in FMS of mechanical and instrument-making engineering when modeling the automatic orientation system at the stage of technological production preparation.

The article [15] shows that the quaternion is a simple and effective mathematical apparatus for describing an ordinary solid body's translational and rotational motion. However, the presented information is declarative. The problems of the practical application of quaternions, particularly for describing the orienting movements of workpieces in the simulation of WOS in FMS, are not solved.

Thus, doubt the effectiveness and perspective of using quaternions to describe the full composition of the orienting movements of the OM in the simulation of the WOS is not in doubt.

3 Research Methodology

Quaternions q are powerful and effective mathematical apparatus for modeling processes of the orientation of rigid bodies in 3D space. Graphically, the quaternion is represented as some directed line v in space C, the vector v, around which the object rotates (Fig. 1). Quaternions consist of scalar s and vector $v = a \cdot i + b \cdot j + c \cdot k$ parts. The scalar s and vector v parts of the quaternion q are formed by an ordered sequence of real numbers s, a, b, c, associated with four basic elements 1, i, j, k. At the same time, the basic elements i, j, k of the quaternion q make up a trio of orthogonal unit vectors that are pairwise collinear with the corresponding axes X, Y, Z of the Cartesian coordinate system: i $\uparrow\uparrow$ x; j $\uparrow\uparrow$ y; k $\uparrow\uparrow$ z (Fig. 1).

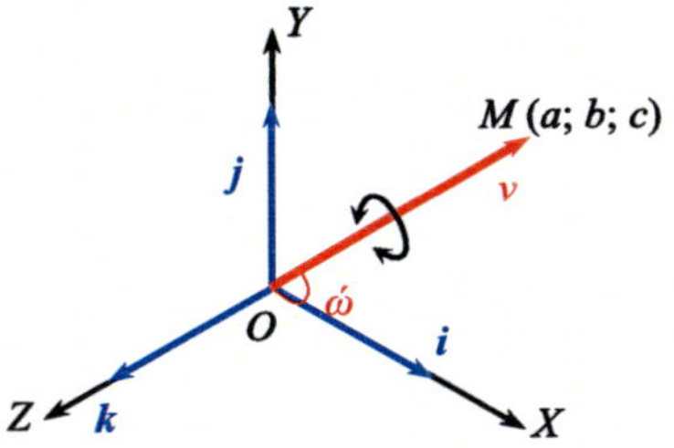

Fig. 1. Example of the location of the essential elements of a quaternion in 3D space.

Depending on the task condition, it is convenient to represent the quaternion q in the form:

- scalar number s and 3D vector v: $q(s;\ v) = s \cdot 1 + [a \cdot i + b \cdot j + c \cdot k]$;
- hypercomplex number: $q(s;\ ai;\ bj;\ ck) = s \cdot 1 + a \cdot i + b \cdot j + c \cdot k$;
- 3D vector of Euclidean space, at $s = 0$: $q(ai;\ bj;\ ck) = 0 \cdot 1 + [a \cdot i + b \cdot j + c \cdot k]$;

the trigonometric sum of a scalar number s and a vector v: $q(v, \omega) = \cos \frac{\omega}{2} + v \cdot \sin \frac{\omega}{2}$, where v – unit vector, the direction of which coincides with the body's axis of rotation, ω – angle of rotation of the body.

The absence of a singularity, mathematical operations' relative simplicity, and equations' compactness are positive features of quaternions q. That is why there is no doubt about the usefulness of quaternions q as a model of orienting movements of workpieces in the FMS of machine and instrument engineering.

The usage of quaternions q to describe the movements of the workpiece and determine their position in the 3D space of the Cartesian coordinate system, which is used in mechanical and instrument-making engineering, is based on the rules of multiplication of basic elements: i, j, k: $i^2 = j^2 = \mathrm{k}^2 = -1$; $i{\cdot}j = k$; $j{\cdot}k = i$; $k{\cdot}i = j$; $j{\cdot}i = -k$; $k{\cdot}j = -i$; $i{\cdot}k = -j$ and the following properties of quaternions:

- commutativeness, that is, the possibility of permuting arguments $[q_1 + q_2 = q_2 + q_1]$, and associativity, that is, the possibility of performing operations in an arbitrary order $[(q_1 + q_2) + q_3 = q_2 + (q_1 + q_3)]$ when adding;
- non commutativeness, that is, not the possibility of permuting arguments $[q_1 \cdot q_2 \neq q_2 \cdot q_1]$, and associativity, that is, the possibility of performing operations in an arbitrary order $[(q_1 \cdot q_2) \cdot q_3 = q_1 \cdot (q_2 \cdot q_3)]$ during multiplication;
- distributiveness $[q_1 \cdot (q_2 + q_3) = q_1 \cdot q_2 + q_1 \cdot q_3]$.

Operations with quaternions involve a combination of scalar and vector algebra rules. Thus, the length of the vector v of the quaternion q in 3D space is determined by the formula (1):

$$|v| = \sqrt{a^2 + b^2 + c^2}, \tag{1}$$

where v – vector part of the quaternion q, $v = a \cdot i + b \cdot j + c \cdot k$. The vector v sets the direction of the object's rotation axis and goes from the point O of the beginning of the Cartesian coordinate system to some point M with coordinates (a; b; c) (Fig. 1).

Thus, it is obvious that quaternions allow to describe all types of orienting movements of workpiece – circular and linear motions. During automatic orientation of the workpiece in the 3D space C, it is necessary to perform the transformation $\Omega\ (v_k;\ \varpi_k)$: $C \to (v_m;\ \varpi_m) \mid k = \overline{1;K};\ m = \overline{1;M}$ of the finite set of circular and linear motion so that the workpiece occupies the final predetermined position in the space C, $(v_m;\ \varpi_m)$, where v, ϖ – the vector of rotations and the angular coordinates of the workpiece, respectively. In other words, during automatic orientation, the workpiece must perform a specific set of circular and linear motions relative to one, two, or three axes of the Cartesian coordinate system *XYZ* of the 3D space *C*.

In quaternion concepts, the above can be interpreted as a displacement of the local Cartesian coordinate system *XYZ*, connected to the workpiece relative to the absolute Cartesian coordinate system *XYZ* of the 3D space *C* with the basis *i, j, k* by some angle $\varpi \in (\alpha,\ \beta,\ \gamma)$ relative to the axis, the direction of which is given by the unit vector v (Fig. 2).

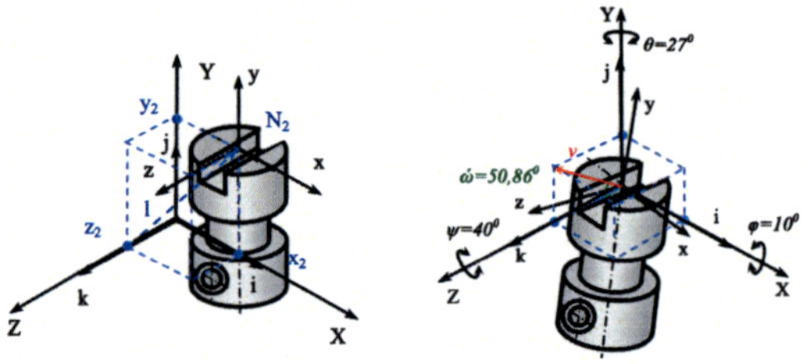

a b

Fig. 2. Examples of linear (a) and circular (b) orienting workpiece movements in 3D space.

Using the trigonometric form of the representation of quaternions, the geometric properties of the orienting movements of the workpiece relative to the axes *OX*, *OY*, and *OZ* of the absolute Cartesian coordinate system *XYZ* of the 3D space *C* can be expressed in quaternion form by expression (2):

$$
\begin{aligned}
q_1(v;\ \alpha) &= \cos\frac{\alpha}{2} + i \cdot \sin\frac{\alpha}{2},\\
q_2(v;\ \beta) &= \cos\frac{\beta}{2} + j \cdot \sin\frac{\beta}{2},\\
q_3(v;\ \gamma) &= \cos\frac{\gamma}{2} + k \cdot \sin\frac{\gamma}{2},
\end{aligned}
\tag{2}
$$

where α, β, γ – the angles of rotation of the workpiece relative to the corresponding axes OX, OY, OZ of the absolute Cartesian coordinate system of the 3D space C. Moreover, in the case of circular motions, the workpiece$(\alpha, \beta, \gamma) \in [0°; 360°]$, in the case of linear motions $(\alpha, \beta, \gamma) \in 0°$.

Accordingly, the oriented position obtained by the workpiece as a result of circular motions in the 3D space *C* of the absolute Cartesian coordinate system *XYZ*, at some angle $\varpi \in (\alpha,\ \beta,\ \gamma)$ relative to the axis, the direction of which is given by the unit vector v (Fig. 2, b), can be determined as the product of the corresponding quaternions according to expression (3). The oriented position obtained by the workpiece as a result of linear motions in the 3D space of the absolute Cartesian coordinate system *XYZ* can be defined as the sum of the corresponding quaternions according to expression (4).

$$q(v;\ \omega) = q_1(v;\ \alpha) \cdot q_2(v;\ \beta) \cdot q_3(v;\ \gamma) = \\ = (\cos\frac{\alpha}{2} + i \cdot \sin\frac{\alpha}{2}) \cdot (\cos\frac{\beta}{2} + j \cdot \sin\frac{\beta}{2}) \cdot (\cos\frac{\gamma}{2} + k \cdot \sin\frac{\gamma}{2}). \tag{3}$$

$$q(v) = q_1(v_x) + q_2(v_y) + q_3(v_z) = x \cdot i + y \cdot j + z \cdot k. \tag{4}$$

The quaternion model of the orienting movements of the workpiece in its general form can be presented as follows:

$$q = \{q(v;\ \omega) \wedge q(v)\} \tag{5}$$

where q – the general quaternion of the orienting movements of the workpiece;
$q(v;\omega)$ – quaternion of circular motions workpiece;
$q(v)$ – quaternion of linear motions of the workpiece;
Λ – notation of the logical conjunction operation.

4 Results and Discussion

Applying the quaternion model to determine the geometric parameters of the orienting movements of the workpiece involves the performance of only 11 mathematical operations in 3 stages.

For the first stage, it is necessary to submit the magnitudes of circular motions in trigonometric form. In the second stage, it is necessary to calculate the product of quaternions according to the above rules. At the third stage, it is necessary to determine the arcos of the angle of rotation in degrees from the scalar part of the resulting quaternion and from its vector part to select the coordinates of the point *M* to which the vector *v*, which sets the direction of the axis of rotation of the workpiece.

It emphasizes the ease of usage of the proposed model, the reduction of computing costs and intellectual efforts when working with it, and the improvement of productivity and speed of the performed work due to the reduced time spent on performing calculations.

For example, the results below show calculations of the geometric parameters of the orienting movements of the arbitrary workpiece, which illustrates the workability and efficiency of the developed quaternion model. So, for example, with the automatic orientation of the workpiece, it is necessary to transfer from the INP to the FOP relative to the absolute Cartesian coordinate system *XYZ* by performing circular motions relative to the axes *OX*, *OY*, *OZ* to the corners $\alpha = 45°$, $\beta = 30°$, $\gamma = 0°$ respectively. The geometric parameters of this process are defined as the product of the corresponding quaternions according to expression (3) as follows:

$$q(v;\omega) = q_1(v;90) \cdot q_2(v;60) \cdot q_3(v;0) = (\cos\tfrac{90}{2} + i \cdot (\sin\tfrac{90}{2}) \cdot (\cos\tfrac{60}{2} + j \cdot (\sin\tfrac{60}{2})$$
$$= (\cos 45 + i \cdot \sin 45) \cdot (\cos 30 + j \cdot \sin 30) = (\tfrac{\sqrt{2}}{2} + i \cdot \tfrac{\sqrt{2}}{2}) \cdot (\tfrac{\sqrt{3}}{2} + j \cdot \tfrac{1}{2})$$
$$\tfrac{\sqrt{2}}{2} \cdot \tfrac{\sqrt{3}}{2} + j \cdot \tfrac{\sqrt{2}}{2} \cdot \tfrac{1}{2} + i \cdot \tfrac{\sqrt{2}}{2} \cdot \tfrac{\sqrt{3}}{2} + i \cdot j \cdot \tfrac{\sqrt{2}}{2} \cdot \tfrac{1}{2}$$
$$= \tfrac{\sqrt{6}}{4} + (i \cdot \tfrac{\sqrt{6}}{4} + j \cdot \tfrac{\sqrt{2}}{4} + k \cdot \tfrac{\sqrt{2}}{4}) = 52,24 + (0,61i + 0,35j + 0.35k).$$

The obtained calculation results can be interpreted as follows. During automatic orientation, the workpiece in the 3Dspace *C* of the absolute Cartesian coordinate system *XYZ* is rotated by an angle 52,24° relative to the axis, the direction of which is specified by the unit vector *v*, which goes from the origin of the coordinates of the point *O*, (0,00; 0,00; 0,00) to the point *M* (0,61; 0,35; 0,35) (Fig. 3).

When the quaternion model of workpiece orientation movements is used, it is necessary to perform 11 mathematical operations, particularly 4 multiplication operations, 3 - division, 3 - finding the root of a number, and 1 - trigonometric operation for determining the arcos angle in degrees. For comparison, it should be noted that the usage of matrix models of guiding cosines or Euler angles for similar conditions involves the performance of a significantly larger number of mathematical operations.

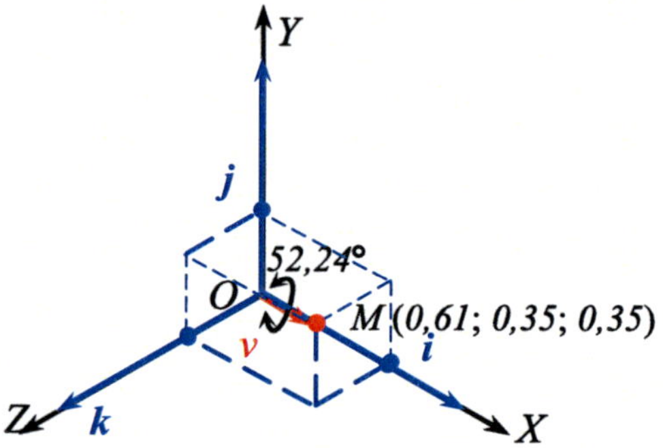

Fig. 3. Example of an illustration of the geometric parameters of the resulting angle and the rotation vector of the OM calculated according to the quaternion model.

The above points to the significant effectiveness of the quaternion model, increasing productivity and reducing the complexity of the procedure for determining the geometric parameters of the orienting movements of the workpiece. In addition, taking into account that in production conditions, automatic orientation is carried out, as a rule, for large batches of workpieces, saving computing resources and reducing the time for performing mathematical operations can be quite large. At the same time, a significant increase in performance and speed of calculations and modeling, in general, is obvious. Therefore, using the quaternion model will reduce the labor intensity of the performed tasks, reduce time and increase the productivity of modeling and the process of automatic orientation of the workpiece.

5 Conclusions

The research and numerical simulations confirmed the possibility of the pioneered quaternion model use for industrial purposes in mechanical and instrument-making engineering and WOS simulation. The obtained results are essential and show that the performed works' productivity and speed increased by approximately 40%. At the same time, computational costs and intellectual efforts are significantly reduced compared with traditional methods of describing the movements of the workpiece, which are based on rotation matrices. After all, quaternions provide the information necessary to orient the workpiece using only 4 numbers instead of 9, as in traditional rotation matrices. It stimulates using the developed quaternion model at the pre-production stage in mechanical engineering and instrument-making.

References

1. Savcı, Ü.Z.: Screw motion via matrix algebra in three-dimensional generalized space. Symmetry **14**, 22–35 (2022). https://doi.org/10.3390/sym14112235
2. Rusydi, M.I., Okamoto, T., Ito, S., Sasaki, M.: Rotation matrix to operate a robot manipulator for 2D analog tracking objects using electrooculography. Robotics **3**, 289–309 (2014). https://doi.org/10.3390/robotics3030289
3. Holzinger, S., Schöberl, J., Gerstmayr, J.: The equations of motion for a rigid body using non-redundant unified local velocity coordinates. Multibody Syst. Dyn. **48**, 283–309 (2020). https://doi.org/10.1007/s11044-019-09700-5
4. Cherepanska, I.Y., Sazonov, A.Y., Krushynska, N.I., Priadko, V.A., Lukiniuk, M.V.: Quaternion method of calculating angles while measuring via goniometric precision instrument system. Bull. Karaganda Univ. "Phys." Ser. **1**(101), 46–56 (2021). https://doi.org/10.31489/2021Ph1/46-56
5. Van Hoang, P., Konyakhin, I.A.: Error analysis of object rotation parameters measurement with autocollimation method using computer models on the base of quaternions. J. Instrum. Eng. **60**(12), 1157–1160 (2017)
6. Valenti, R.G., Dryanovski, I., Xiao, J.: A quaternion-based orientation filter for IMUs and MARGs. Sensors (Basel) **15**(8), 19302–19330 (2015). https://doi.org/10.3390/s150819302
7. Xu, X., Tian, X., Zhou, L.: A robust incremental-quaternion-based angle and axis estimation algorithm of a single-axis rotation using MARG. Sensors **6**, 42605–42615 (2018)
8. Möller, M., Glocker, C.: Rigid body dynamics with a scalable body, quaternions and perfect constraints. Multibody Syst. Dyn. **27**, 437–454 (2012). https://doi.org/10.1007/s11044-011-9276-5
9. Fresk, E., Nikolakopoulos, G.: Full quaternion based attitude control for a quadrotor. In: 2013 European Control Conference (ECC), pp. 3864−3869. IEEE (2013). https://doi.org/10.23919/ECC.2013.6669617
10. Choi, H., Yang, B.: On singularity of rigid-body dynamics using quaternion-based models. ASME. J. Appl. Mech. **79**(2), 024502 (2012). https://doi.org/10.1115/1.4005575
11. Zhang, W., Zhang, J., Zhang, C., Chen, D., He, Y.: Modeling and simulation of flexible spacecrafts based on dual quaternions. J. Phys: Conf. Ser. **2203**, 012018 (2022). https://doi.org/10.1088/1742-6596/2203/1/012018
12. Challis, J.H.: Quaternions as a solution to determining the angular kinematics of human movement. BMC Biomed. Eng. **2**, 5 (2020). https://doi.org/10.1186/s42490-020-00039-z
13. Carvalho, A.S., Botto, M.A., Martins, J.M.: Reflections and the geometry of quaternion extraction methods. Mech. Mach. Theory **176**, 105006 (2022). https://doi.org/10.1016/j.mechmachtheory.2022.105006
14. Stanfield, K., Bani Younes, A.: Dual-quaternion analytic LQR control design for spacecraft proximity operations. Sensors **21**(11), 3597 (2021). https://doi.org/10.3390/s21113597
15. Li, J.: Research on the rigid body pose estimation using dual quaternions. Adv. Mech. Eng. **11**(1) (2019). https://doi.org/10.1177/1687814018823115

Regularities of Oscillations During Turning and End Milling

Serhiy Dyadya, Yuriy Vnukov, Olena Kozlova(✉), and Pavlo Trishyn

National University "Zaporizhzhia Polytechnic", 64, Zhukovsky Str., Zaporizhzhia 69063, Ukraine

kozlova@zp.edu.ua

Abstract. The intensity of vibrations during machining negatively affects the quality of machined parts, tool life, and productivity. Research is being conducted to reduce it, and various measures are being implemented. They relate to the choice of cutting modes, machining strategy, tool geometry, application of damping media, and vibration control. But at the same time, it is essential to determine which types of oscillations need to be suppressed. Therefore, the work aimed to identify the pattern of oscillations during turning and final milling. The devices that allow you to adjust the tool's and parts' dynamic characteristics and record their oscillations during cutting were used for conducting experiments. During intermittent turning, the regularity of occurrence of oscillations in the following sequence was obtained. When the cutter cuts in, forced oscillations occur, which are superimposed on the accompanying free oscillations of the technological system "tool–part". After their damping, self-oscillations are superimposed on the forced oscillations. Each oscillation type has distinctive features and operates for a specific time. Unlike turning, the end milling process is short-term. Therefore, during cutting, only forced oscillations act, on which the accompanying free oscillations of the "tool–part". As the spindle speed increases, the cutting time decreases, and if it is shorter than the period of the accompanying free oscillations, only forced oscillations are effective. Determining the characteristic features of oscillations during cutting makes it possible to prescribe measures that suppress their intensity purposefully.

Keywords: Turning · End-Milling · Technological System · Forced Oscillations · Self-Oscillations · Accompanying Free Oscillations · Manufacturing Innovation

1 Introduction

The dynamic application of an external force to an elastic system brings it out of equilibrium and causes the appearance of vibrations. In blade processing, such a force is the cutting force. At the same time, the intensity of vibrations depends on such properties of the technological system "tool–part" (hereinafter TS), such as mass and rigidity. It is to reduce the intensity of vibrations or to control it that research and various proposed measures are aimed at. The main task of these works is to ensure surface quality during

V. Tonkonogyi et al. (Eds.): InterPartner 2023, LNME, pp. 136–144, 2024.
https://doi.org/10.1007/978-3-031-42778-7_13

cutting [1] and processing productivity [2]. At the same time, the main attention is paid to studying the action of self-oscillations. But it is known that forced and free oscillations also occur during cutting. Therefore, the issue of determining which types of oscillations are prioritized for different processing methods is relevant.

The novelty of the work is the identification of patterns of oscillations during mechanical processing and the dependence of their action on the cutting time.

2 Literature Review

Regardless of the type of oscillations during cutting, the main focus is on limiting their intensity. To control them, vibration sensors are used, which are installed on the spindle [3], or they monitor the current of the spindle drive [4]. It allows monitoring of the level of oscillation. When choosing cutting modes for stable milling, frequency analysis of the tool [5] or part according to petal diagrams is used [6]. To predict the negative impact of vibrations in real conditions, with the help of a digital double, correlation interactions are built in machine structures during cutting [7], making choosing productive processing modes possible. The introduction into the production of milling cutters with a variable pitch of teeth [8], with a different angle of inclination of the helical cutting edges [9], with a comb [10] and toothed [11] cutting edges or variation in the cutting speed [12] allows to destroy the subsetting of regenerative self-oscillations to permanent [13]. A decrease in the intensity of oscillations is also achieved by increasing the tool's rigidity [14], using tool mandrels with damping properties [15]. But the latter is expensive, so the possibility of using the geometric dimensions of the tool following the modes of their oscillations and the mode of operation of the machine is being considered [8]. With the help of the RCSA matrix, it becomes possible to create and investigate different dynamic conditions when connecting the tool with the holder and the spindle [16] and implement those that ensure vibration resistance [17].

It should be noted that the research works consider either the general effect of vibrations on the cutting process without dividing it into types or the effect of self-oscillations.

In general, forced oscillations, free and self-oscillations, are known [13]. Each species has its characteristics that distinguish them from each other. Forced oscillations arise from the action of periodically changing external forces. Free oscillations of an elastic system occur under the action of internal forces. They are always damped due to the effect of friction. The amplitude of free oscillations depends on the initial conditions. Self-oscillations are constant, undamped oscillations that operate without a variable external force. The amplitude of self-oscillations does not depend on the initial conditions.

Because the main task in cutting is to ensure its constancy, and vibrations interfere with this, it is essential to know precisely which type of vibrations should be suppressed. Studying the regularity of their action will allow us to move away from uncertainty in the recommendations regarding the selection of anti-vibration measures.

3 Research Methodology

To determine the regularity of the action of oscillations during cutting, a study was performed during intermittent turning of a cylindrical workpiece with a longitudinal groove (Fig. 1). The turning operation was chosen because of the possibility of obtaining a long cutting time.

Fig. 1. A workpiece with a longitudinal groove for researching the regularity of the action of oscillations.

The design of the device used in this case allows you to adjust the dynamic characteristics of the cutter, record its oscillations during cutting, and determine the cutting time [19]. Table 1 shows the initial parameters for the research when turning a part made from steel 45 with a cutter with a VK8 carbide plate.

Table 1. Output parameters during turning.

Frequency of free oscillations, f_{fo}, Hz	Spindle rotation frequency, n, rpm	Cutting depth, t, mm	Feed, S, mm/rev
463	1000	0.5	0.1

Studies of the patterns of the occurrence of oscillations during milling were performed on a special stand [20], which allows for simulating the processing of parts with different dynamic characteristics due to the ability to adjust the departure of the elastic element with the sample being processed. Table 2 shows the initial data for which the experiments were carried out when milling samples from St. 3, end-mill d = 16 mm.

The frequencies of oscillations that occur during cutting were determined in the MatLab program by fast Fourier digestion.

Table 2. Output parameters during milling.

Frequency of free oscillations, ffo, Hz	Mill rotation frequency, n, rpm	Radial depth, a_e, mm	Axial depth, a_p, mm	Feed per tooth, S_z, mm/rev
512	120 200 700	0.3	4	0.1

4 Results and Discussion

Figure 2 shows a fragment of the oscillogram of the longitudinal oscillations of the cutting tool when turning a cylindrical workpiece with a groove.

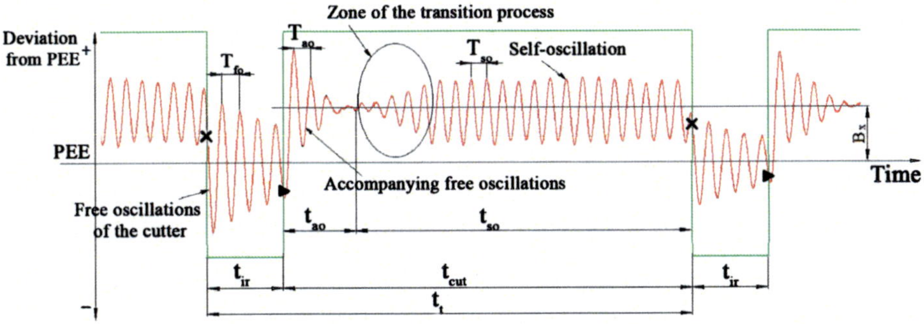

Fig. 2. A fragment of the oscillogram of cutter oscillations when turning a workpiece with a longitudinal groove: PEE - position of elastic equilibrium of the incisor; T_{fo} - the period of free oscillations of the incisor; T_{ao} - the period of accompanying free oscillations of the vehicle; T_{so} – self-oscillation period; t_{fo} – time of action of accompanying free oscillations; t_{so} – time of action of self-oscillations; t_{ir} – idling running time; tcut – cutting time; t_t – time of one turn of the workpiece; B_x - static deviation of the cutter from the PEE during turning; ► - the cutting point of the cutter in the workpiece; × - point of exit of the cutter from the workpiece

The presence of a groove on the surface of the workpiece leads to the cutting of the cutter with an impact at each turn in the time $t_t = 5.89 \cdot 10^{-2}$ s), which includes the cutting time $t_{cut} = 4.96 \cdot 10^{-2}$ s and the idle running time $t_{ir} = 0.93 \cdot 10^{-2}$ s.

Oscillations recorded when turning a workpiece with a groove have characteristic features of forced oscillations, free oscillations, and self-oscillations. Forced oscillations occur under the action of the cutting force, which presses the workpiece by the amount of B_x. On top of them are superimposed the damping accompanying free oscillations of the TS with a period of $T_{ao} = 2.05 \cdot 10^{-3}$ s ($f_{ao} = 488$ Hz) and undamped steady self-oscillations with a period of $T_{so} = 1.86 \cdot 10^{-3}$ s ($f_{so} = 537$ Hz) [19]. During the movement of the workpiece in the absence of cutting, there are damping-free oscillations of the cutter relative to the position of elastic equilibrium (PEE) with a period $T_{fo} = 2.16 \cdot 10^{-3}$ s ($f_{fo} = 463$ Hz). Based on the obtained result, it can be stated that the regularity of the action of oscillations during cutting has the following sequence: forced oscillations, which are superimposed on the accompanying free oscillations of the TS and self-oscillations, free oscillations of the cutter relative to the PEE after the end of the cutting. At the same time, each type of oscillation has its duration. Forced oscillations are active during the entire cutting time t_{cut} (in research $t_{cut} = 4.96 \cdot 10^{-2}$ s). Accompanying free oscillations of TS act during the time of t_{so} (in research $t_{so} = 1.1 \cdot 10^{-2}$ s). After their attenuation and until the end of the cutting, self-oscillations take place during the time t_{so} (in research $t_{so} = 3.86 \cdot 10^{-2}$ s). The time of free oscillations of the cutter is determined by the idle running time t_{ir} (in research $t_{ir} = 0.93 \cdot 10^{-2}$ s) from the end of cutting to the subsequent cutting of the tool.

The use of the term "accompanying free oscillations" of the TS superimposed on the forced oscillations was used by Ya.H. Panovko [18], which better reflects the physical process of oscillations during cutting than the term "free oscillations" of TS. It also separates them from the free oscillations of the cutter when idling.

Determining that when cutting, oscillations act in a certain sequence and for a certain time allows each type of blade processing to be considered considering their characteristics. At the same time, it should be noted that the accompanying free oscillations of the vehicle and self-oscillations occur under certain conditions, while forced oscillations always operate. Firstly, it is necessary to create initial conditions under which the amplitude of oscillations will arise. The others require a wavy mark on the cutting surface after the previous pass of the tool.

If there are reasons for self-oscillations in continuous turning, when the cutting time is extended, then intermittent cutting is short-lived in final milling. Therefore, the question regarding the current fluctuations is relevant.

To answer it, research was carried out during the milling of samples from Steel 3 end-mill d = 16 mm with mechanical attachment of cutting plates. The stand on which the experiments were carried out allows you to adjust the protrusion of the elastic plate with the processed material [19]. Due to this, the necessary dynamic characteristics of it are established. When the plate flew out by h = 80 mm, the frequency of free oscillations was $f_{fo} = 512$ Hz ($T_{fo} = 1.95 \cdot 10^{-3}$ s). Counter and parallel milling occurred at an axial cutting depth of $a_p = 4$ mm, a radial depth of $a_e = 0.3$ mm, and feed per tooth of $S_z =$ 0.1 mm, with spindle rotation frequencies of n = 120, 200, and 700 rpm.

Figure 3 shows a fragment of oscillograms of sample oscillations during up and down milling with a spindle rotation frequency of n = 120 rpm.

When milling with a spindle rotation frequency of n = 120 rpm, only forced oscillations are active during the cutting time $t_{cut} = 4.36 \cdot 10^{-2}$ s. Due to the peculiarities of the change in the thickness of the layer to be cut, during up-milling, the indentation of the sample B_{max} under the action of the cutting force is less than during down-milling.

Figure 4 shows fragments of oscillograms of sample oscillations during up- and down milling with a spindle rotation frequency of n = 200 rpm.

Increased spindle rotation frequency decreases the cutting time ($t_{cut} = 2.69 \cdot 10^{-2}$ s). During up-milling, oscillations occur, which are superimposed on the forced ones. These are the accompanying free oscillations of the TS with the T_{fo} period ($t_{fo} = 1.16 \cdot 10^{-3}$ s, $f_{fo} = 862$ Hz), the amplitude of which depends on the initial conditions, particularly the initial speed. They act during the t_{ao} time ($t_{ao} = 0.08 \cdot 10^{3}$ s) and decay. During the down final milling, the accompanying free oscillations of the TS when cutting with

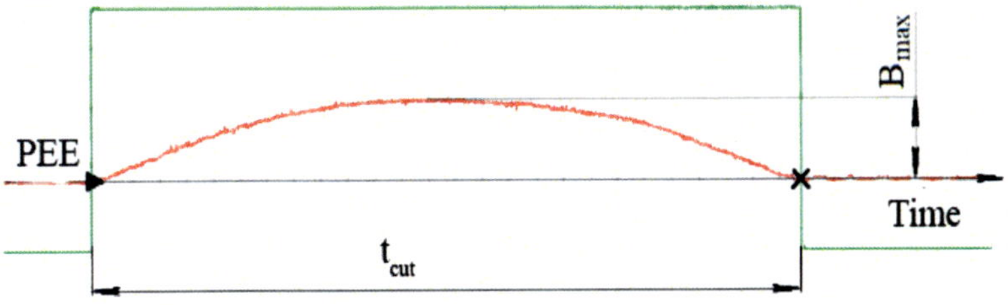

a

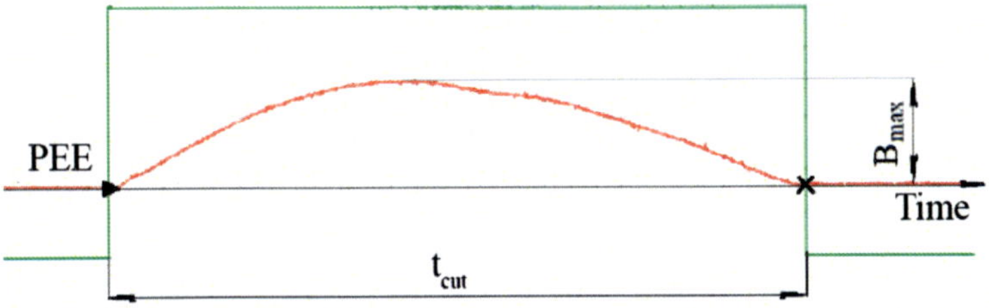

b

Fig. 3. Fragments of oscillograms of sample oscillations during up- (a) and down (b) end-milling with a spindle rotation frequency of n = 120 rpm.

a spindle rotation frequency of n = 200 rpm do not occur. When cutting, the most significant thickness of the sheared layer creates such properties of the TC that dampen these oscillations. After the cutter exits the cutting zone, the free oscillations of the sample with the period of T_{ao} ($t_{ao} = 1.95 \cdot 10^{-3}$ s, $f_{ao} = 512$ Hz) act, which also decays. The cutter cuts into the sample, which does not oscillate.

Figure 5 shows fragments of oscillograms of sample oscillations during up- and down milling with a spindle rotation frequency of n = 700 rpm.

An increase in the initial cutting speed increases the amplitude of the accompanying free oscillations. When milling with a spindle rotation frequency of n = 700 rpm, these oscillations are superimposed on forced oscillations, both during up-milling at the beginning of cutting (Fig. 5, a) and during down-milling at the end of cutting (Fig. 5, b). The occurrence of free oscillations and their damping during end-milling is associated with a change in the thickness of the cut layer. The fundamental thing here is that there is not enough cutting time for self-oscillations characterized by stability.

The obtained results allow a different look at the speed zones of oscillations, which were considered in the paper [20]. Without changing the boundaries of these zones, it is necessary to correct the use of the term "self-oscillation". Instead, the term "accompanying free oscillations of the TS" should be used when considering speed zones.

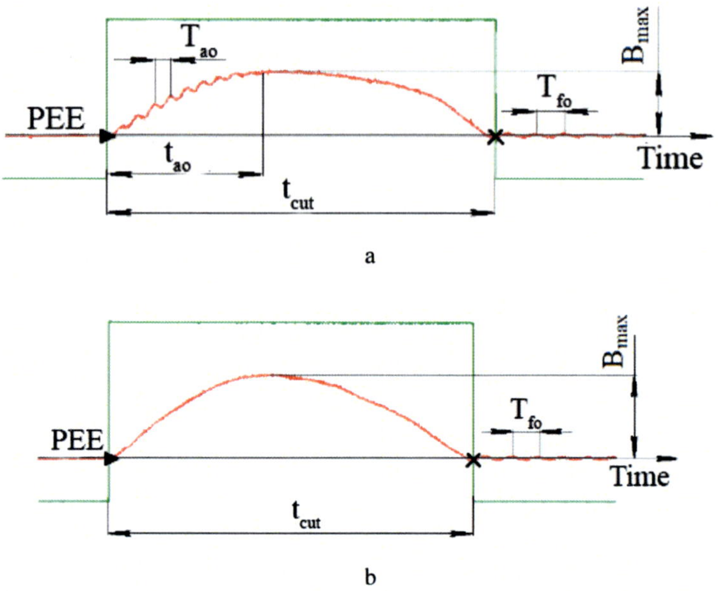

Fig. 4. Fragments of oscillograms of sample oscillations during up- (a) and down (b) end-milling with a spindle rotation frequency of n = 200 rpm.

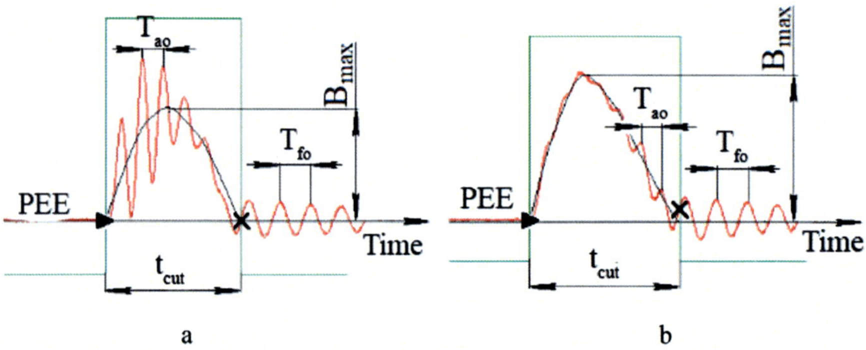

Fig. 5. Fragments of oscillograms of sample oscillations during up- (a) and down (b) end-milling with a spindle rotation frequency of n = 700 rpm.

5 Conclusions

The regularities of the action of various types of oscillations during cutting have been determined. First, forced oscillations act, and the accompanying free oscillations of the vehicle are superimposed on them. After their damping, the sub-adjustment of the TS occurs, and stable self-oscillations occur.

The possibility of different types of oscillations depends on the cutting time.

When turning, when the cutting time is extended, there are conditions for the action of all kinds of oscillations.

The cutting time limitation does not allow stable self-oscillations in the end-milling of thin slices. During end-milling, the amplitude of the accompanying free oscillations of the TS depends on the initial conditions. Therefore, as the cutting speed increases, it also increases.

The obtained results allow a different look at the speed zones of oscillations, which were considered in the paper [20]. Without changing the boundaries of these zones, it is necessary to correct the use of the term "self-oscillation". Instead, the term "accompanying free oscillations of the TS" should be used when considering speed zones.

Further research should be planned to determine the influence on the dynamic properties of the TS and the amplitude of the accompanying free oscillations of the cutting modes and the geometry of the tool.

References

1. Wojciechowski, S., Wiackiewicz, M., Krolczyk, G.M.: Study on metrological relations between instant tool displacements and surface roughness during precise ball end milling. Measurement **129**, 686–694 (2018). https://doi.org/10.1016/j.measurement.2018.07.058
2. Neslony, P., Krolczyk, G.M., Wojciechowski, S., Chudy, R., Zak, K., Maruda, R.W.: Surface quality and topographic inspection of variable compliance part after precise turning. Appl. Surf. Sci. **434**, 91–101 (2018). https://doi.org/10.1016/j.apsusc.2017.10.158
3. Postel, M., Aslan, D., Wegener, K., Altintas, Y.: Monitoring of vibrations and cutting forces with spindle mounted vibration sensors. CIRP Ann. **68**(1), 413–416 (2019). https://doi.org/10.1016/j.cirp.2019.03.019
4. Aslan, D., Altintas, Y.: On-line chatter detection in milling using drive motor current commands extracted from CNC. Int. J. Mach. Tools Manuf. **132**, 64–80 (2018). https://doi.org/10.1016/j.ijmachtools.2018.04.007
5. Munoa, J., et al.: Chatter suppression techniques in metal cutting. Manuf. Technol. **65**, 785–808 (2016). https://doi.org/10.1016/j.cirp.2016.06.004
6. Li, Z., Wang, Z., Shi, X.: Fast prediction of chatter stability lobe diagram for milling process using frequency response function or modal parameters. Int. J. Adv. Manuf. Technol. **89**, 2603–2612 (2017). https://doi.org/10.1007/s00170-016-9959-4
7. Liang, Z., et al.: The process correlation interaction construction of Digital Twin for dynamic characteristics of machine tool structures with multi-dimensional variables. J. Manuf. Syst. **63**, 78–94 (2022). https://doi.org/10.1016/j.jmsy.2022.03.002
8. Comak, A., Budak, E.: Modeling dynamics and stability of variable pitch and helix milling tools for development of a design method to maximize chatter stability. Precis. Eng. **47**, 459–468 (2017). https://doi.org/10.1016/j.precisioneng.2016.09.021
9. Comak, A., Ozsahin, O., Altintas, Y.: Stability of milling operations with asymmetric cutter dynamics in rotating coordinates. J. Manuf. Sci. Eng. **138**(8), 081004–081004-7 (2016). https://doi.org/10.1115/1.4032585
10. Tehranizadeh, F., Rahimzadeh Berenji, K., Yıldız, S., Budak, E.: Chatter stability of thin-walled part machining using special end mills. CIRP Ann. **71**(1), 365–368 (2022). https://doi.org/10.1016/j.cirp.2022.04.057
11. Tehranizadeh, F., Rahimzadeh Berenji, K., Budak, E.: Dynamics and chatter stability of crest-cut end mills. Int. J. Mach. Tools Manuf. **171**, 103813 (2021). https://doi.org/10.1016/j.ijmachtools.2021.103813

12. Nam, S., Eren, B., Hayasaka, T., Sencer, B., Shamoto, E.: Analytical prediction of chatter stability for modulated turning. Int. J. Mach. Tools Manuf. **165**, 103739 (2021). https://doi.org/10.1016/j.ijmachtools.2021.103739
13. Schmitz, T.L., Smith, K.S.: Machining Dynamics: Frequency Response to Improved Productivity. Springer, Cham (2009). https://doi.org/10.1007/978-3-319-93707-6
14. Yue, C., Gao, H., Liu, X., Liang, S.Y., Wang, L.: A review of chatter vibration research in milling. Chin. J. Aeronaut. **32**(2), 215–242 (2019). https://doi.org/10.1016/j.cja.2018.11.007
15. Xia, Y., et al.: Chatter suppression in large overhang face milling using a toolholder with high dynamic performance. Int. J. Adv. Manuf. Technol. **108**, 1713–1724 (2020). https://doi.org/10.1007/s00170-020-05515-3
16. Ji, Y., Bi, Q., Zhang, S., Wang, Y.: A new receptance coupling substructure analysis methodology to predict tool tip dynamics. Int. J. Mach. Tools Manuf. **126**, 18–26 (2018). https://doi.org/10.1016/j.ijmachtools.2017.12.002
17. Schmitz, T., Betters, E., Budak, E., Yüksel, E., Park, S., Altintas, Y.: Review and status of tool tip frequency response function prediction using receptance coupling. Precis. Eng. **79**, 60–77 (2023). https://doi.org/10.1016/j.precisioneng.2022.09.008
18. Panovko, Ya.G.: Fundamentals of the Applied Theory of Vibration and Impact. Mechanical Engineering, Leningrad (1976)
19. Vnukov, Y., Dyadya, S.I., Kozlova, O.B., Trishin, P.R., Zubarev, A.E.: Influence of cutting time on types of oscillations during blade processing. Ukr. J. Mech. Eng. Mater. Sci. **9**(1), 53–66 (2023)
20. Vnukov, Yu.N., Dyadya, S.I., Kozlova, Ye.B., Logominov, V.A., Chernovol, N.N.: Self-oscillations during milling of thin-walled elements of parts. ZNTU, Zaporozhye (2017)

An Impact of the Cutting Fluid Supply on Contact Processes During Drilling AISI 321 Stainless Steel

Eshreb Dzhemilov[1], Eskender Bekirov[1], Ruslan Dzhemalyadinov[1], and Alper Uysal[2(✉)]

[1] Crimean Engineering and Pedagogical University, Simferopol 295015, Republic of Crimea, Ukraine

[2] Yildiz Technical University, 34349 Istanbul, Turkey

auysal@yildiz.edu.tr

Abstract. The article describes processing blind holes when drilling stainless steel AISI 321. Helical drills made of high-speed steel with channels for the internal supply of a lubricating and cooling technological medium were used as a cutting tool. The tool's geometry did not change, corresponding to actual production conditions. Drilling in stainless steel is always done with cutting fluids. The use of minimum lubrication technology in this operation, even if they contain various additives, is impractical due to the lack of a guarantee that the lubricant will enter the chip formation zone. An increase in the volume of process fluids requires special attention to the selection of compositions due to the tightening of requirements for the regeneration and disposal of waste media. The data obtained indicate the expediency of using vegetable-based cutting fluids in connection with their effect on reducing contact pressures and the height of the micro-roughness of the treated surface.

Keywords: Drilling · Stainless Steel · Contact Pressures · Cutting Temperatures · Surface Roughness · Cutting Fluids · Sustainable Manufacturing

1 Introduction

Obtaining holes in solid material, without a pre-prepared hole, is performed in drilling operations. When processing body parts, drilling is performed at the final stages of the technological process. Maintaining the hole's quality requirements of the part is very important. At modern enterprises, the drilling of body parts is often performed with twist drills equipped with carbide inserts. However, the share of high-speed steel tools is relatively high, especially in material processing enterprises with tool production. It is how complex-profile tools and those for which it is advisable to regrind are made.

Drilling operations are carried out when supplying lubricating-cooling technological media to the cutting zone. An essential aspect is the need for a guaranteed lubricant supply to the chip formation zone. Constant chip removal and high specific pressures in the machining zone limit the use of minimum lubrication technology. Drilling stainless

V. Tonkonogyi et al. (Eds.): InterPartner 2023, LNME, pp. 145–153, 2024.
https://doi.org/10.1007/978-3-031-42778-7_14

steel is also accompanied by high-temperature loads. In the absence of sufficient cooling of the cutting zone, the tool life of high-speed steel is sharply reduced. As a result, there is a significant increase in the micro-roughness of the machined surface. The combination of these factors leads to the need to supply a large volume of technological media when drilling holes. At the same time, most of the technological media compositions used in the production are mineral oils or emulsions based on them. Also, lubricant manufacturers add a package of additives to their formulations to increase efficiency, such as surfactants, extreme pressure and antifoam additives, corrosion inhibitors, and more. With prolonged exposure to the operator's body, such substances lead to occupational diseases. In addition, it is essential to properly dispose of used process fluids, which entails additional costs. Therefore, the search for alternative, environmentally friendly lubricants has become widespread, allowing maintaining the required surface quality without compromising machining performance.

2 Literature Review

At this stage of product development, there is a tendency to reduce negative factors [1] in manufacturing machine parts associated with cutting fluids based on oil products. One of the directions is the complete rejection of process media and the transition to dry cutting [2]. However, this method is unsuitable for several machining operations and often requires improving the quality of holes when drilling holes in stainless steel, have been widely considered by many researchers in special tools or switching to high-speed machining. So in work [3], the authors proposed a method for applying microstructures with a linear groove on the rake face surface of the tool to reduce the components of the cutting forces during dry machining. When machining stainless steels, an active Built-up Edge formation occurs on the rake face surface of the tool due to the intense adhesive interaction between the workpiece and the tool. Study [4] shows the effect of built-up layers on tool wear, cutting force, and surface finish. High-speed machining is often used in the processing of materials [5].

In [6, 7], the authors formulated the models for predicting surface roughness during machining of parts. In [8], an automated approach to rational material selection was developed. Optimizing the drill geometry can lead to lower cutting and machining forces without internal cooling of the working area [9]. In the study [10], the authors examined chip combustion during the high-speed turning of titanium alloys. However, this method has a limitation due to increased tool wear [11].

The second way to reduce the negative impact is the use minimum quantity lubrication technology. The authors in [12] propose using the MQL combined with Arduino Uno control. In [13], the efficiency of using a process media with the addition of nanofluid supplied by the MQL method for turning austenitic steel was experimentally shown. The use of additives in the form of nanofluid provides a sustainable approach when using MQL. The authors in [14] showed the positive effect of MQL technology when turning AISI 304 stainless steel on reducing the roughness of the resulting surface. The supplied liquid contained nanoparticles of two sizes, 30 and 40 nm. The authors proposed a new approach to predicting the resulting surface roughness by changing the size of the supplied particles in the composition of the cutting fluid and using machine learning. In [15], the authors showed the effect of MQL and the addition of nano-additives (nanoMQL) to the applied liquid on reducing the height of microroughness of the machined surface and tool wear when turning ER7 steel. According to the presented results, using MQL technology reduces roughness by more than 24% and wear by 34%. Adding nanoparticles to the composition leads to an improvement of 34% and 37.6%, respectively. Aslan and al. [16] presented a comparative assessment of dry cutting when using minimal lubrication technology. They noted that when machining Strenx 900 structural steel, oil supplied in fine droplets leads to a decrease in roughness by more than 30%, tool life increase by more than 90%, and cutting forces decrease by 28%.

In [17] shows the effect of a hybrid nanofluid based on palm oil by adding graphene nanoparticles and hexagonal boron nitride fed into the treatment zone using MQL technology. The authors note a decrease in surface roughness when machining Inconel 718 by more than 40% compared to dry cutting. Tool life increases by more than 50%.

Despite all the advantages of MQL, this technology is poorly suited for processes occurring in closed-cutting conditions, such as drilling blind holes. Constant chip removal makes it difficult for small doses of lubricants to enter the treatment area.

The third way to reduce the negative impact is by using environmentally friendly cutting fluids [18]. The basis of such formulations is usually fatty acids of plant and animal origin. The work [19] shows the effect of emulsions based on vegetable oils, which reduce the wear of the cutting insert and surface roughness.

Ecological compositions of lubricants are also created based on non-edible oils of vegetable origin. In [20], the authors developed two formulations based on Jatropha and Pongamia oil. The authors note the positive effect of vegetable oils on reducing cutting forces and surface roughness. The authors present similar results in a review article [21]. The effectiveness of biodegradable oils in metalworking processes compared to mineral oils has been noted.

This article discusses the impact of various process media on the quality of holes after drilling AISI321 stainless steel.

3 Research Methodology

Experimental studies were carried out on a radial drilling machine model 2K522. The tool used was a twist drill with channels for the internal supply of cutting fluid made from HSSCo. Drilling was carried out for stainless steel AISI321 (ASTM - A 213 TP321H; DIN - X6 Cr Ni Ti 18 10). The drilling diameter was 12 mm. The following process media were selected for cooling the working area: water-based - Aquafreeze 6, oil-based - MOBILMET 423, and sunflower oil. Coolant was supplied to the cutting zone through the internal channels of the tool at a working pressure of 5 bar.

A specially designed experimental setup simultaneously records contact pressures and temperatures along the workpiece generatrix in one pass of the tool (Fig. 1).

Estimating the cutting temperature is difficult due to recording values directly in the contact zone. Because of the peculiarities of the drilling process, the most rational would be measurements using thermocouples. The values will be more approximate when using a natural thermocouple. However, the errors in taring and the existing interference during the equipment operation will also distort the actual values. In our studies, it was decided to use artificial thermocouples. In this case, three thermocouples were assembled into one circuit. The thermocouple junctions were located opposite the strain gauges in three pre-prepared holes with the same pitch from the end of the cylindrical billet. The depth of the hole with the placed thermocouple was calculated so that the distance between the junction and the generatrix wall was 250 μm when the main hole was obtained. The remaining space of the hole where the thermocouples were placed was filled with thermal paste to reduce the measurement error. Temperature data were recorded when the drill edge passed opposite each thermocouple.

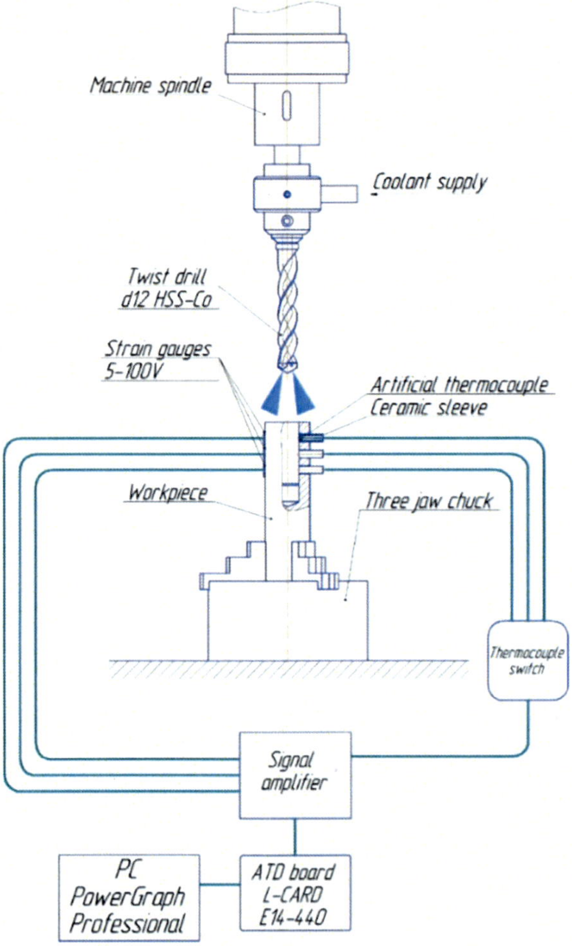

Fig. 1. Experimental setup.

As the drill was deepened, the circuit was switched between each thermocouple to avoid overlapping graphs and to facilitate the processing of the obtained data. The data recording rate was fifty readings per second. This method does not allow estimating the actual temperature in the plastic deformation zone, but this is sufficient to evaluate the qualitative effect of various technological media.

Processing modes are selected according to average production values for two types of materials: cutting speed – 9.42 m/min (255 r/min), feed – 0.06 mm/rev.

Roughness parameters after drilling were measured using a profilometer TR-200 model.

4 Results and Discussion

The cutting process's power characteristics largely determine the chip formation's nature. The effect of the influence of cutting fluids used in the machining process can be assessed precisely by the force indicators of the cutting process. Applying the method for evaluating power characteristics using strain gauges makes it possible to evaluate dynamically changing indicators on parts of any configuration. Figure 2 shows the averaged data after 6 measurements on three strain gauges located along the generatrix of the workpiece.

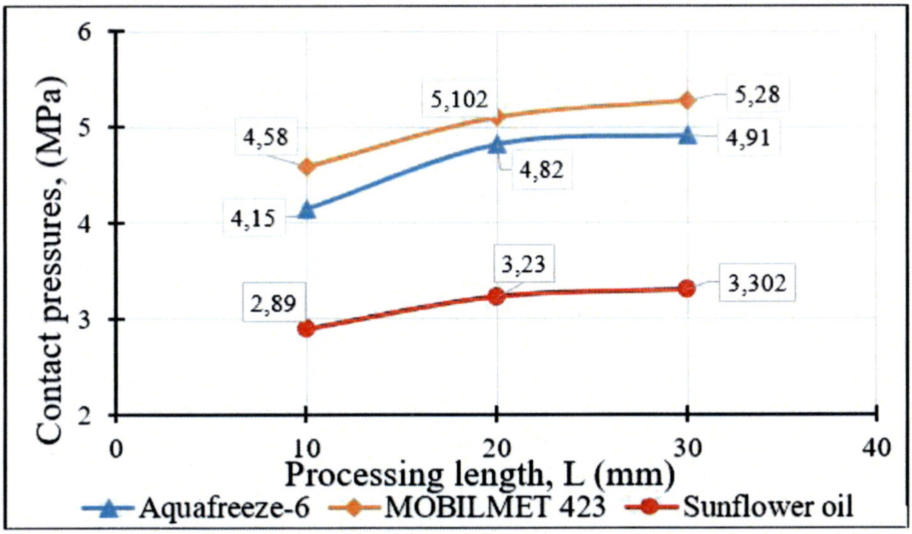

Fig. 2. The influence of cutting fluids on the change in contact pressure along the length of the generatrix of the workpiece when drilling AISI321 stainless steel.

As can be seen from the graphs, the use of sunflower oil as a process medium leads to a decrease in contact pressure values by more than 60% compared to mineral coolant and emulsion.

A decrease in the values of contact pressure indicates a decrease in the resulting cutting force due to a change in the angle of the plane of the actual chip shear.

Another characteristic of the cutting process that affects the tool's life is the temperature in the contact zone. According to the proposed method, measurements and graphs were drawn. They show the qualitative characteristics of the applied lubricating-cooling technological media for temperature changes in the contact zone. The values shown show the average data after six measurements (Fig. 3).

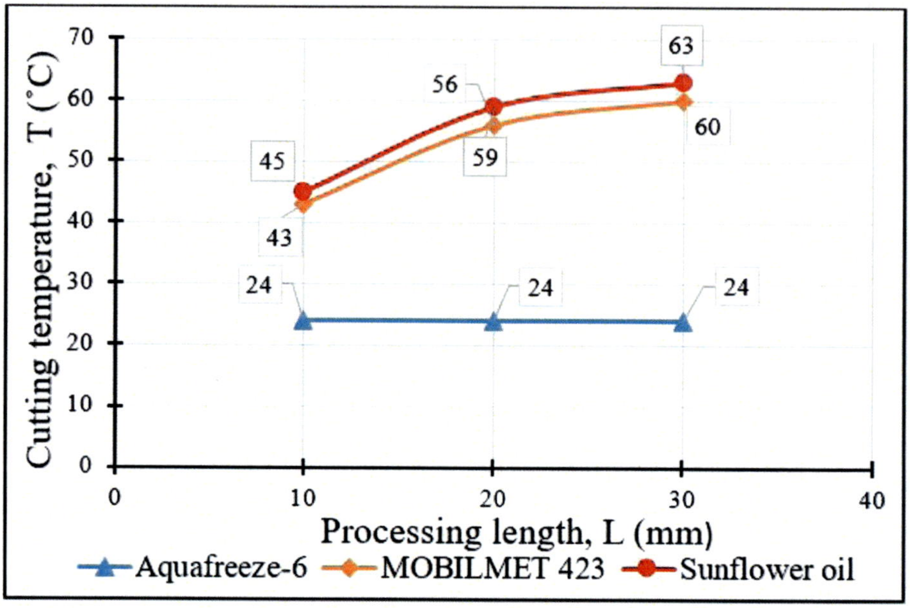

Fig. 3. Effect of cutting fluids on cutting temperature when drilling AISI321 stainless steel.

As can be seen from the graphs, the most pronounced effect on the cooling of the working area is the use of a water-miscible emulsion. Due to the high pressure, heat is actively removed from the cutting zone. The use of oil lubricating-cooling technological media leads to an almost twofold increase in temperature compared to the emulsion.

The nature of the graphs indicates that the actual temperature in the cutting zone is above 250 °C, which exceeds the smoke point of the oils. At such a temperature and high specific pressures, the destruction of carbon chains occurs.

The main characteristic of the drilling process is the quality of the resulting hole. In the study, the control of the roughness of the resulting surface was carried out as a quality parameter.

It can be seen from the histograms (Fig. 4) that the most significant values of the arithmetic mean deviation of the surface microroughness are recorded after treatment with the use of MOBILMET 423 cutting fluid based on mineral oil. Due to active heat removal, using a water-miscible emulsion Aquafriz-6 makes it possible to maintain the effect of the surfactants included in the composition, reducing the coefficient of friction of the removed chips.

Polar molecules can explain a significant decrease in the roughness index when using sunflower oil in the composition of fatty acids (e.g., oleic, linoleic) in the oil. Due to their ordered arrangement, polar molecules create a film on the surface of the tool and workpiece, which helps reduce friction (Fig. 5).

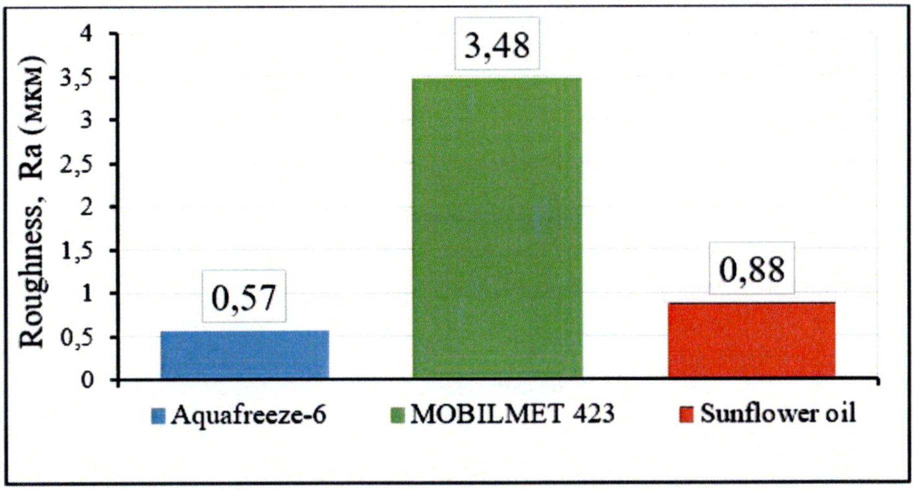

Fig. 4. Influence of cutting fluids on cutting roughness when drilling AISI321 stainless steel.

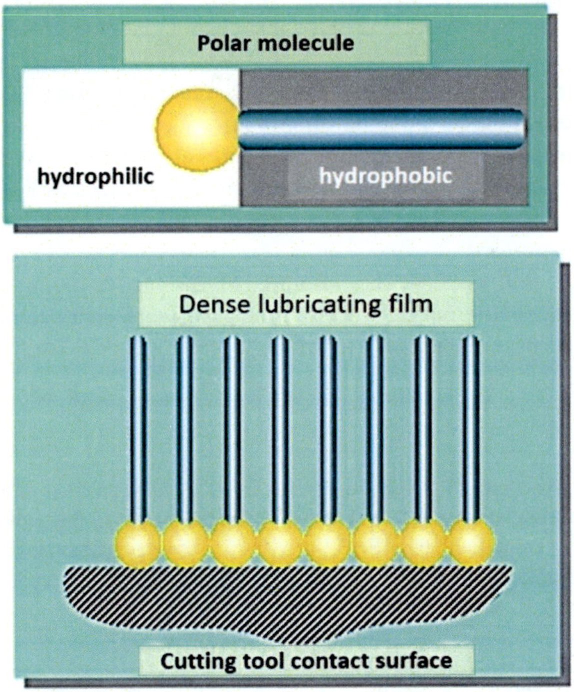

Fig. 5. Creation of a film due to polar molecules of vegetable oil.

Sunflower oil also contains about 38% of the monounsaturated fatty oleic acid as a surfactant. Due to the high values of chemisorption of polar molecules, the elastoplastic contact zone shifts towards the cutting edge due to the Ribender effect. As a result, there is a decrease in the zone of plastic contact and a change like build-up formation. This leads to a change in the geometry of the rake angle of the tool. Increasing the rake angle value changes the position of the actual chip cut plane. All these factors contribute to the facilitation of the chip formation process and, as a result, to a decrease in the roughness value.

5 Conclusions

The blind hole drilling process in stainless steels has several features that necessitate the correct selection of the applied lubricating-cooling technological medium to obtain the required surface quality and increase tool life. Based on the data obtained, it can be concluded that it is more rational to use water-soluble emulsions to reduce the temperature in the cutting zone. However, high contact pressure values will lead to intensive tool wear and subsequent surface quality deterioration.

Due to their low smoke point and shielding of the contact zone, cutting fluids based on mineral oils lead to an increase in cutting temperature and contact loads. All this leads to a significant deterioration in the quality of the resulting surface.

The most significant effect was obtained by using sunflower oil as a cutting fluid. Due to the presence in the composition of fatty acids having a polar structure of molecules, after processing, a roughness decrease is observed. This is due to the change in the angle of the actual shear plane. In this case, a decrease in contact pressure indicates a significant decrease in the resulting cutting force.

Vegetable oils have a low smoke point, like mineral oils. However, due to destruction (thermal and mechanical), sunflower oil fatty acids form free radicals that passivate juvenile surfaces, thereby reducing surface energy and facilitating the cutting process due to the Ribender effect.

Further studies will evaluate the influence of various technological media on the power characteristics of the cutting process, contact pressures, temperature, and vibration of the system in one pass of the tool to optimize processing modes for specific conditions.

References

1. Haidabrus, B., Protsenko, S., Rosenberger, P., Grabis, J.: Data analysis of readiness programs of machine-building enterprises. In: Ivanov, V., Trojanowska, J., Pavlenko, I., Zajac, J., Peraković, D. (eds.) DSMIE 2020. LNME, pp. 128–136. Springer, Cham (2020). https://doi.org/10.1007/978-3-030-50794-7_13
2. Zhang, N., Yang, F., Liu, F., Liu, G.: Cutting performance of micro-textured WC/Co tools in the dry cutting of Ti-6Al-4V alloy. Int. J. Adv. Manuf. Technol. **107**, 3967–3979 (2020). https://doi.org/10.1007/s00170-020-05161-9
3. Song, X., Takahashi, Y., He, W., Ihara, T.: Study on the protective effect of built-up layer in dry cutting of stainless steel SUS30. In: 34th ASPE Annual Meeting, American Society of Precision Engineering, pp. 138–148. American Society for Precision Engineering, Pittsburgh (2019)
4. Duan, R., Deng, J., Ai, X., Liu, Y., Chen, H.: Experimental assessment of derivative cutting of micro-textured tools in dry cutting of medium carbon steels. Int. J. Adv. Manuf. Technol. **92**, 3531–3540 (2017). https://doi.org/10.1007/s00170-017-0360-8

5. Halchuk, T.N., Povstyanoy, O., Bembenek, M., Redko, R.G., Chetverzhuk, T.I., Polinkevych, R.M.: Impact of technological system's characteristics on the machining accuracy of bearing rings. J. Eng. Sci. **10**(1), A22–A30 (2023). https://doi.org/10.21272/jes.2023.10(1).a4
6. Karpus, V.E., Ivanov, V.A.: Locating accuracy of shafts in V-blocks. Eng. Res. **32**(2), 144–150 (2012). https://doi.org/10.3103/S1068798X1202013X
7. Song, Y., Cao, H., Wang, Q., Zhang, J., Yan, C.: Surface roughness prediction model in high-speed dry milling CFRP considering carbon fiber distribution. Compos. B Eng. **245**, 110230 (2022). https://doi.org/10.1016/j.compositesb.2022.110230
8. Pavlenko, I., et al.: Using regression analysis for automated material selection in Smart Manufacturing. Mathematics **10**(11), 1888 (2022). https://doi.org/10.3390/math10111888
9. Sedlak, J., Zouhar, J., Kolomy, S., Slany, M., Necesanek, E.: Effect of high-speed steel screw drill geometry on cutting performance when machining austenitic stainless steel. Sci. Rep. **13**, 9233 (2023). https://doi.org/10.1038/s41598-023-36448-y
10. Takemoto, S., Sato, M., Matsuno, T., Yamamoto, K.: Chip combustion in high-speed dry cutting of titanium alloy. In: Proceedings of International Conference on Leading Edge Manufacturing in 21st century (LEM21), vol. 2017.9, p. 9:116. The Japan Society of Mechanical Engineers, Tokyo, Japan (2017). https://doi.org/10.1299/jsmelem.2017.9.116
11. Hovorun, T.P., et al.: Physical-mechanical properties and structural-phase state of nanostructured wear-resistant coatings based on nitrides of refractory metals Ti and Zr. Funct. Mater. **26**(3), 548–555 (2019). https://doi.org/10.15407/fm26.03.548
12. Muttaqin, A.Z., Darsin, M., Kharisma, Y.R., Syuhri, A., Trfiananto, M.: Development of Minimum Quantity Lubricant (MQL) automation in applying cutting fluid on lathes. J. POLIMESIN **20**(2), 83–86 (2022). https://doi.org/10.30811/jpl.v20i2.2618
13. Touggui, Y., Uysal, A., Emiroglu, U., Dzhemilov, E.: An experimental and statistical investigation on cutting forces in turning of AISI 304 stainless steel under dry, MQL and nanofluid MQL conditions. In: Ivanov, V., Trojanowska, J., Pavlenko, I., Zajac, J., Peraković, D. (eds.) DSMIE 2021. LNME, pp. 513–522. Springer, Cham (2021). https://doi.org/10.1007/978-3-030-77719-7_51
14. Dubey, V., Sharma, A.K., Pimenov, D.Y.: Prediction of surface roughness using machine learning approach in MQL turning of AISI 304 steel by varying nanoparticle size in the cutting fluid. Lubricants **10**(5), 81 (2022). https://doi.org/10.3390/lubricants10050081
15. Çamlı, K.Y., et al.: Performance of MQL and nano-MQL lubrication in machining ER7 steel for train wheel applications. Lubricants **10**(4), 48 (2022). https://doi.org/10.3390/lubricants10040048
16. Aslan, A., Salur, E., Kuntoğlu, M.: Evaluation of the role of dry and MQL regimes on machining and sustainability index of Strenx 900 steel. Lubricants **10**(11), 301 (2022). https://doi.org/10.3390/lubricants10110301
17. Makhesana, M.A., Patel, K.M., Bagga, P.J.: Evaluation of surface roughness, tool wear and chip morphology during machining of nickel-based alloy under sustainable hybrid nanofluid-MQL strategy. Lubricants **10**(11), 315 (2022). https://doi.org/10.3390/lubricants10110315
18. Ivchenko, O., et al.: Method for an effective selection of tools and cutting conditions during precise turning of non-alloy quality steel C45. Materials **15**(2), 505 (2022). https://doi.org/10.3390/ma15020505
19. Bacak, S., Karabıyık, Ö.: Effects of various vegetable oil and water mixtures used as cutting fluids on tool wear and surface roughness. Int. J. 3D Print. Technol. Digit. Ind. **6**, 392–398 (2022). https://doi.org/10.46519/ij3dptdi.1132387
20. Jeevan, T.P., Jayaram, S.R.: Performance evaluation of Jatropha and Pongamia oil based environmentally friendly cutting fluids for turning AA 6061. Adv. Tribol. **1**, 2425619 (2018). https://doi.org/10.1155/2018/2425619
21. Deshpande, V., Jyothi, P.N.: A review on sustainable eco-friendly cutting fluids. J. Sustain. Environ. Manag. **1**(2), 306–320 (2022). https://doi.org/10.3126/josem.v1i2.45383

Load Parameters of the Gear Machining by Power Skiving and Their Influence on the Machining System

Ihor Hrytsay[1], Vadym Stupnytskyy[1](✉), Andrii Slipchuk[1], and Jan Ziobro[2]

[1] Lviv Polytechnic National University, 12, Bandera St., Lviv 79013, Ukraine
Vadym.v.stupnytskyi@lpnu.ua

[2] Technical Institute of Vocational Academy in Sanok, 21, Mickiewicza Str., 38-500 Sanok, Poland

Abstract. The study results for cutting gears by the Power Skiving method was presented based on a graphical and analytical model of chips and an analytical model of cutting force and torque. A methodology for modeling 3D chips and analyzing the cut layers of a disk cutter at the level of certain edges was developed. Based on the example under certain initial conditions, the capabilities of the developed models were shown to predict the load on a single tooth and the entire machine tool system during multi-tooth continuous cutting and the influence of power factors on the machining process. Two factors significantly influence the cutting force: the cross-sectional area and the cutting thickness. An increase in the cutting area proportionally increases the cutting force and the load on the system. The shear thickness affects the cutting force through its effect on the chip compression ratio: when the shear thickness decreases, the intensity of shear plastic deformation and cutting force increase. Shear intensity increases sharply at low chip thicknesses, characteristic of force shearing, so high cutting forces accompany this process.

Keywords: Gear Machining · Power Skiving · Simulation · Chip Parameters · Cutting Force · Shear Deformation · Process Innovation

1 Introduction

In recent years, the power skiving method, formerly known as teeth lathe turning, has been widely used to produce spur and helical gears for internal and external gearing and spline surfaces. The equipment is multi-purpose five-axis CNC machines and specialized gear milling machines. Positive features of this process include high chip removal rates, no idle strokes, and short machining times, as well as the ability to combine it with other shaping methods in one operation on CNC machines.

Due to the latest technological developments, in particular, achievements in the field of machine tool engineering, their software, and tool production, this technology has become flexible and efficient, as well as an alternative to traditional gear-cutting methods with an expansion of the range of cut gears up to a module of 10 mm.

V. Tonkonogyi et al. (Eds.): InterPartner 2023, LNME, pp. 154–166, 2024.
https://doi.org/10.1007/978-3-031-42778-7_15

The features of the power skiving method are significant forces and moments on the axes of the tool and blank combined with high cutting speed. Significant machine rigidity and a high level of synchronization of tool and work gear rotation are required to achieve high machining accuracy. Failure to meet these requirements leads to increased vibrations and gear profile errors. When such a gear operates in the transmission, these deviations cause the kinematic error, i.e., the instability of the transmission ratio, as well as the occurrence of high-frequency oscillations in machine drives.

Cutting force is one of the main physical parameters characterizing the cutting process. Elastic strains and vibrations, machining accuracy and surface roughness, machine tool durability, cutting tool life, machining performance, and energy consumption depend on this force. The importance of cutting forces increases in multi-tooth cutting when the load on the system results from a combination of elementary forces that occur on single teeth and cutting tool edges. The complexity of modeling the cutting force increases for continuous processes during the continuous rolling of the tool and workpiece (e.g., power skiving, hobbing, hard scudding, and grinding gears with worm grinding wheels).

To avoid these negative phenomena that can occur in machining, it is necessary to consider the combination of various factors: cutting parameters - speed, feed, depth, and number of passes; machine tool power and rigidity, in particular, the tool spindle, especially when cutting internal gears; tool geometry and coatings. To solve such problems, it is necessary to have a methodology that considers a large number of initial factors and describes their influence on a gear-cutting process. These tasks can be solved based on mathematical modeling and simulation.

2 Literature Review

The analysis of the state of the art in the field of research and modeling of the power skiving process showed the following. Based on the importance of cut parameters and chip shape and size in cutting processes, several authors pay attention to this problem. Thus, papers [1–6] investigate the cutting process and develop models and software for chip modeling, calculating chip thickness and length, and cutting tool design. The author of [1] develops the design of universal high-speed assembly heads with insertable edges and software for this process on 6-axis CNC bevel gear cutting machines a six-axis. The high efficiency and stability of tools with replaceable inserts of stick blade cutters are shown, and skiving is proposed to machine both internal and external gears instead of being [2].

F. Klocke [3] investigates the effect of cut thickness on cutting force and tool wear, the change in tool geometry with process conditions, and the effect of cutting conditions on the chip welding phenomenon. Paper [4] studied the cutting power parameters. Papers [5, 6] analyze tool design and the influence of geometric parameters on power performance during gear machining. The paper [7] examines how tool geometry relates to technological parameters and cutting conditions. The effect of coatings and changes in the geometric parameters of the cutting tool on the cutting process is analyzed in [8]. A paper on chip thickness and sliding speed is presented in [9], and a study of tool wear involves chip welding to tool surfaces during machining.

Using standard systems with a commercial CAD environment, the results simulate the chip formation process and calculate its thickness.

An approach for simulating the power skiving method to predict non-deformed chip geometry and the morphology of the produced gear using a CAD system paper [10] is presented. Based on simulation analysis, a chip thickness can be calculated.

The following features are characteristic of the noted works. At first, it is an inaccurate reproduction of the process of kinematics. The cutting motion is formed by a combination of axial travel and the revolution movement (cutting speed) when a cutting speed is imparted to the tool, not the workpiece, as is customary in these works. Accordingly, the resulting velocity vector will be directed at a different angle than the base surfaces and axes. This is important for correctly determining the tool's cutting forces, friction, and geometric parameters.

Secondly, the chip formation process simulation considers traces of the tool teeth in the axial feed. This allows only a partial reproduction of the chip since it is finally formed in continuous rolling (angular change of positions) of a formed gap and a cutting tooth.

Thirdly, the research results are largely descriptive, and the chip parameters are not used as a basis for the further and deeper analytical study of this process. Simultaneously, most power, contact, tribological, and thermal phenomena depend on their values and laws of change along the tool movement.

In these conditions, researching and developing adequate chip formation models, quantifying the cuts' parameters, and calculating the cutting force and torque in power skiving is relevant.

Based on the above, this study aims to create a computer system for modeling the cutting force in the power skiving process based on a 3D model of undeformed chips and to analyze the effect of this force on the cutting process.

3 Research Methodology

In this work, the methods of analytical calculation, graphical and analytical modeling, and computer simulation at standard CAD systems are used.

This paper uses the following logical research scheme based on the basic (classical) provisions of cutting theory: analysis of the kinematic scheme of the process => chip modeling => chip cross-sectional parameters => intensity of shear and contact deformations on the surfaces => single (elementary) cutting forces on the edges => total cutting force on all simultaneously working teeth => fluctuations of the cutting force in a continuous cutting cycle. This data can then be used to solve various problems that arise in the design of the machining process: predicting machining accuracy, friction, and heat flows, tool temperature and wear, vibrations and violations of the machine's elastic system, as well as improving the tool geometry and selecting appropriate protective coatings. This approach ensures comprehensive and systematic research.

4 Results and Discussion

4.1 Process Kinematics

In Power skiving, the main cutting motion is the cutter's rotation combined with the cutter's movement along the gear blank axis (for spur gear), which is caused by the inclination of the axis. Auxiliary motions are the axial feed rate f of the tool's axial feed rate f, and the workpiece's circular feed rate movements are shown in Fig. 1: V_{cut} is the speed of rotational movement of the cup cutter; V_f is the speed of the tool in the axial feed, which coincides with the speed of the tooth reduction along a gear axis due to the crossing of the axes; V_g is the speed of the gear rotation; V_Σ is the speed of the resulting cutting motion.

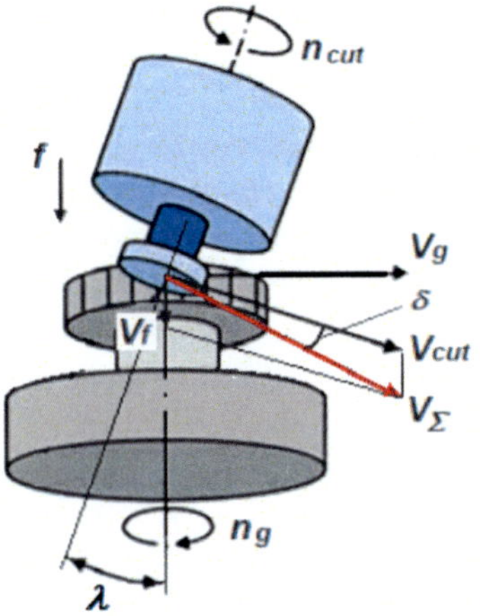

Fig. 1. Kinematic diagram of Power skiving.

According to this scheme, the tooth trajectory is the geometric sum of the vector of rotational motion around the tool axis and the straight-line motion along the wheel axis in the axial feed.

Let us establish the relationship between the parameters V_{cut} and V_f. If we bring these values to a common dimension, then $V_f = \frac{f \cdot n_c}{1000}$, m/min. Based on elementary calculations, we can obtain the following expression for the angle δ between these vectors:

$$\delta \approx actg \frac{f \cdot n_c \cdot \cos \omega}{1000 \cdot V_{cut}} \quad (1)$$

Taking into account the values of the parameters in this equation that are used in practice, it can be argued that the angle δ between the vectors of the linear, rotational speed of the cutter V_{cut} and the total cutting speed V_S is in the range of 1–5°, that is, for practical calculations, it can be assumed that they practically coincide. This means that the cutting process, i.e., oblique shear, occurs in the direction of the vector V_{cut}, not in the direction of axial feed, as follows from the schemes presented in the works mentioned above. This conclusion changes approaches to the calculation of the cutting force and its components in the power skiving method and is also essential for determining the kinematic angles of the tool.

4.2 Method of Computer Simulation of Undeformed Sheared Layers

To form a 3D model of an undeformed chip, we used a technique similar to that in [11–13]. This technique, developed for hobbing, represents a continuous cutting motion during the rolling process by a sequence of discrete cuts. Spatial geometric chip models were created for all the teeth of the active hob helical surface, each of which operates in a specific zone and eliminates chips inherent only to it. These models were used to determine all the necessary parameters of the cross-sections of the cuts and the patterns of their change in different gear-cutting phases following the kinematics.

Based on this approach and the kinematics described above, an undeformed chip is modeled and cut by a single tooth of a cup cutter. The shape and size of this chip on all cutter teeth will be the same for certain cutting conditions, which is the difference from worm hobbing.

This article presents the results of the study for the following initial data: modulus $m = 5$ mm; axial feed rate $f = 0.5$ mm/rev; the number of wheel teeth $Zg = 33$, number of cutter teeth $Z_{cut} = 24$; helical cutter angle $\lambda = 20°$; cutting speed 200 m/min; workpiece material - alloy chromium steel, tensile strength 650 MPa; cutting to full profile depth. The shape of the tool and its parameters correspond to an involute gear of the same parameters.

Figure 2 shows partially formed gaps (a) from sources available on the Internet, the contour of the transition surface in a partially formed gap (b), and the geometric model of the undeformed chip (c).

4.3 Parameters of Cuts

With single-pass cutting, the cut parameters are large; thus, significant cutting forces and elastic deformations will occur. As a result, in practice, multi-pass cutting is used to reduce these parameters. Machining is performed at high cutting speeds to compensate for the time lost.

Adapting the developed methodology to the conditions of multi-pass gear cutting made it possible to model chips and calculate their parameters when cutting in three passes. Cross-sections of the cuts in the passes in discrete angular positions of the cutter tooth are shown in Fig. 3.

The graphs of thickness, chip cross-sectional area, and chip compression ratio in successive cutting passes are shown in Fig. 4. The cutting is carried out with a depth of 0.7 m and 0.6 m, respectively, and the height of the gear tooth is 2 m.

Compared to machining a gear to the full profile height, the maximum cutting area and cut thickness on different passes are reduced by 75% to 200%. Their values by paths (Fig. 4, a, b. Figure 4, c) show the value of the chip compression ratio depending on the thickness of the cut according to the modeling data of this parameter in the Deform 2D system.

4.4 Modeling of the Cutting Force

Let us define the cutting force component that coincides with the cutting speed vector and is commonly referred to in cutting theory as the main component R_o. This force at zero rake angle coincides with the tangential component of the cutting force P_z in the technological coordinate system, and the effective cutting power depends on its value. To calculate the force R_o, we turn to the shear force P_τ. The magnitude of this force

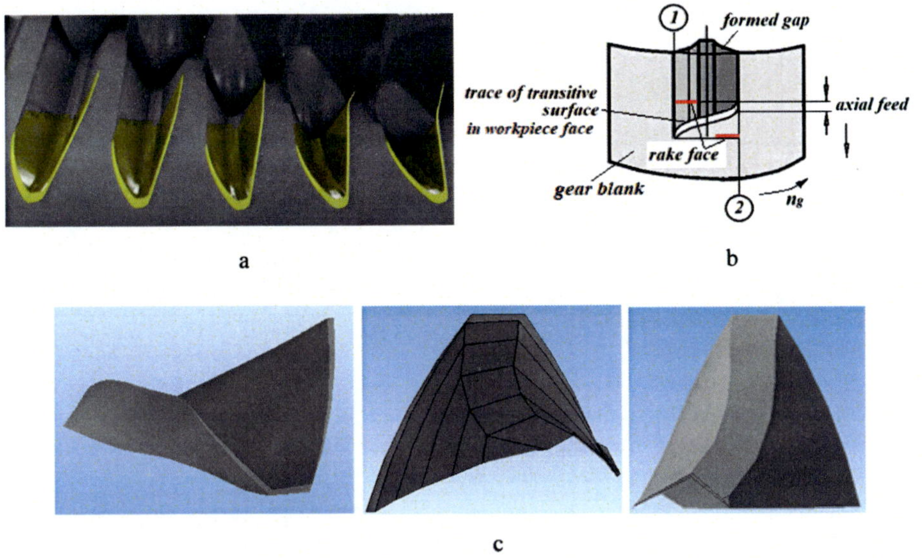

Fig. 2. The partially formed gaps (a), the contour of the transition surface (b), the 3D geometric model of the undeformed chip (c).

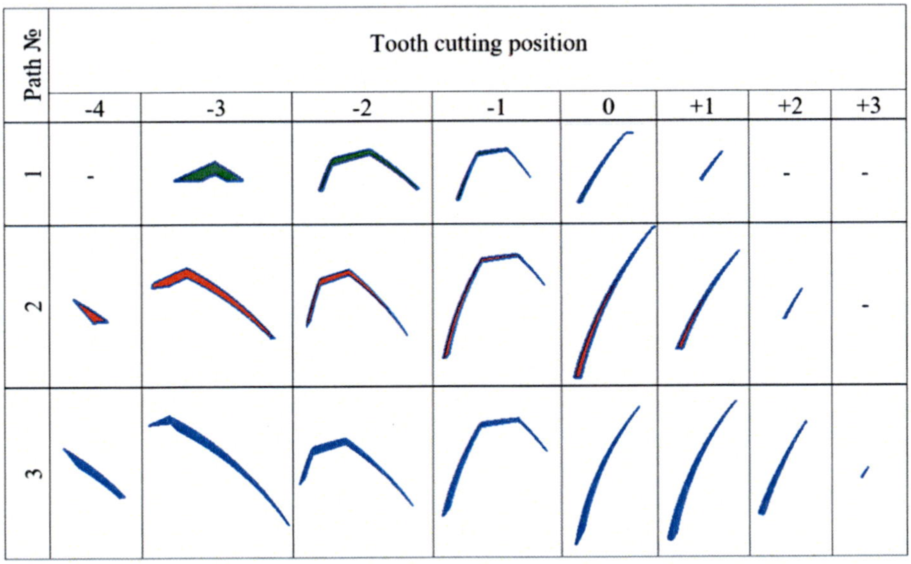

Fig. 3. Cross-sections of slices by paths.

is determined by the shear area S_{shr} and the ultimate shear strength of the workpiece material, MPa: $P_{\tau} = [\tau] \cdot S_{shr}$.

The following geometric relationship exists between these forces: $P_o = P_{\tau} \cdot \cos \Phi$. In our case, we operate with the cut's cross-sectional area; it can replace the shear area:

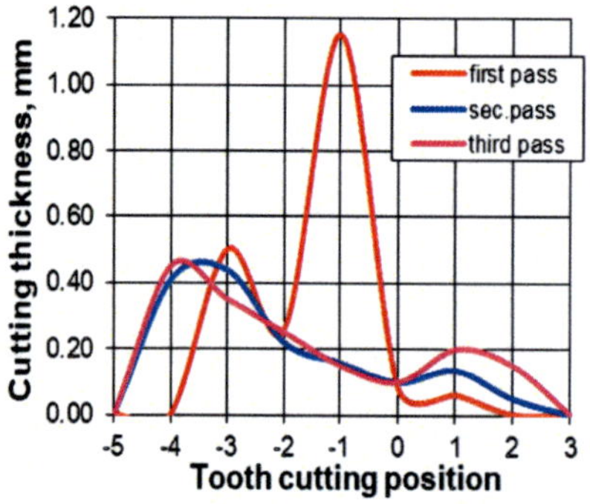

a

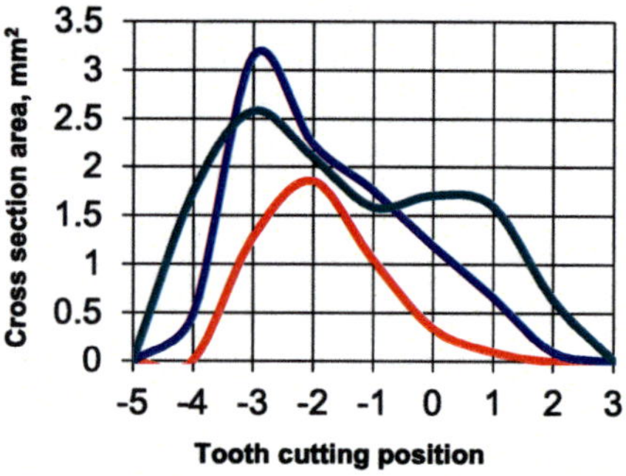

b

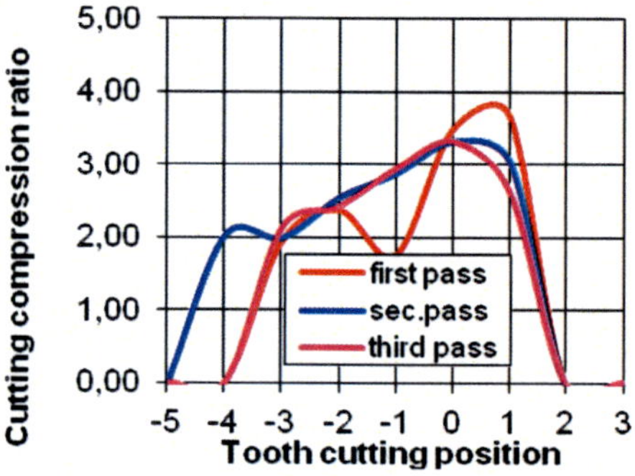

c

Fig. 4. Chip thickness (a), cross-sectional area (b), and chip compression ratio (c) at successive angular cutting positions of a simple tooth during three passes.

$S_{shr} = \frac{S_{crs}}{\sin \Phi}$. Considering these dependencies, the equation for the force R_o is presented in the following form:

$$P_o = P_\tau \cdot \cos \Phi = [\tau] \cdot S_{crs} \cdot \frac{\cos \Phi}{\sin \Phi} \tag{2}$$

The resulting expression contains the chip deposition intensity coefficient ξ, which is defined as the following ratio:

$$\xi = ctg\Phi \cdot \cos \gamma + \sin \gamma \xi \,, \tag{3}$$

where γ is the front angle of the cutter.

Since a rake angle of a gear cutter is zero, then $\xi = ctg\Phi$, the formula (2) will take the form:

$$P_o = P_\tau \cdot ctg\Phi = [\tau] \cdot S_{crs} \cdot \xi \tag{4}$$

According to the third theory of material strength, the shear strength is equal to half the endurance limit of the material since alloy steels are used for gears, with strengths ranging from 600 to 850 MPa, [τ] = 300 to 430 MPa.

The methodology for determining the chip compression ratio is based on rheological computer modeling of the cutting process in the Deform 2D system, depending on the thickness of the cross-section of the cuts [14].

Using the values of the cut parameters above (Fig. 4), Fig. 5 shows how the R_o force changes on individual passes. Based on the research conducted when cutting to the full height of a gear tooth profile, it can be stated that when cutting in three passes, the maximum cutting force decreases by 2.5 to 3 times on each pass.

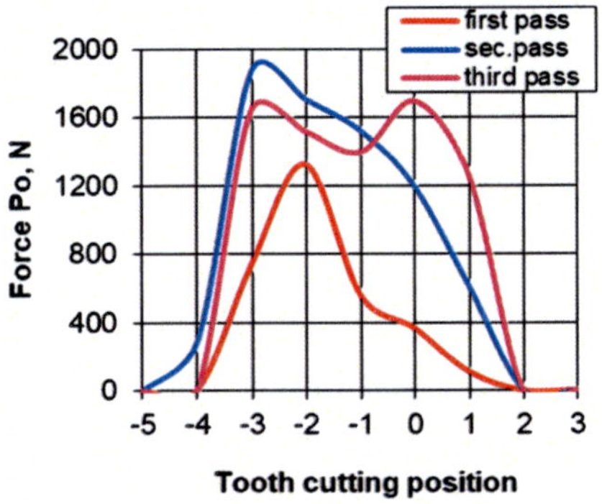

Fig. 5. Cutting forces in three consecutive passes and at full profile depth.

4.5 Determination of Friction Force, Circumferential Force, and Torque

According to the definition of the R_o force, its direction coincides with the cutting speed vector (Fig. 1). The conditions of intense oblique cutting, especially at small angles of inclination of the cutting tool, cause significant friction on the front surface during chip flow. The direction of friction on the leading edge coincides with the cutting speed, while it acts in the opposite direction on the output edge. That is, these forces are directed oppositely. To investigate their influence, it is necessary to decompose the parameters of the cuts along the edges of the cutting tool tooth. The developed methodology makes this possible. Figure 6 shows the parameters of the cuts along the leading, top, and trailing edges.

Using the results of determining the coefficient of friction for similar conditions given [15]. Figure 7 shows the friction forces on the edges of a disc cutter tooth in its successive angular positions during cutting, considering the direction of the corresponding forces.

Given the signs of the friction forces on the opposing edges, we can find the resulting force acting on the gear axis as the difference of these forces for each angular position of the tool tooth. The moment on the gear axis due to this force is shown in Fig. 8, assuming that only one tooth is cutting.

Taking into account the values of a tool and a gear blank angles inclination from 15° to 45°, it can be noted that with an increase in this angle, the torque acting on the axis of the workpiece decreases, but the axial load in the direction of axial feed on the machine components increases; the torque on the tool axis remains unchanged. The increase

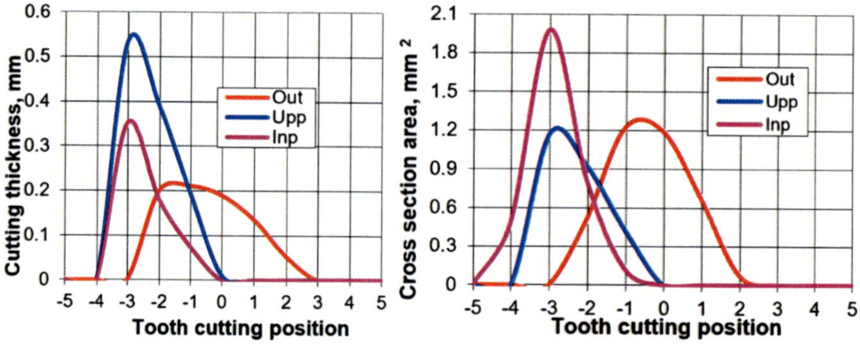

a b

Fig. 6. Graphs of the thickness (a) and cross-sectional area (b) of chips along the edges in the second pass.

in productivity that can be achieved in this case is associated with a deterioration in machining accuracy and surface quality of the gear teeth.

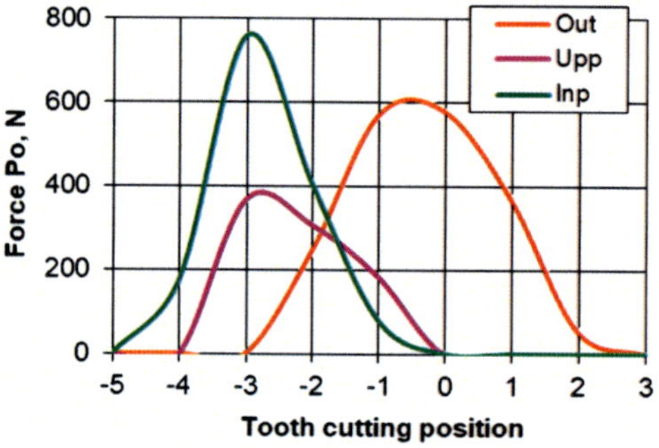

Fig. 7. Friction forces on tool tooth blades in single-tooth cutting.

For the initial conditions of this article, the number of tool teeth that are simultaneously cut is 3.25. In other words, this is the coefficient of end overlap between the machine tool meshing of the cutter and the wheel being cut. This overlap leads to an increase in the forces and moments acting in the system and a cyclical change in the value of these parameters along the wheel's rotation angle. Based on the results of force modeling, Fig. 9 shows a graph of torque changes on the disk cutter's axis.

The frequency of force and torque changes must be determined to evaluate the impact of this oscillatory process on the machine and tool. For a tool revolution, the frequency equals the number of cutter teeth - Z_{cut}. The second frequency will be: $\nu = n \cdot Z_{cut} = \frac{1000 \cdot V \cdot Z_{cut}}{\pi \cdot D_{cut}}$. For 200 m/min cutting speed and 130 mm outer diameter of the cutter, the frequency of change of the cutting torque and force will equal 192 Hz.

A higher oscillation frequency at higher cutting speeds carries less excitation energy, i.e., it results in a lower amplitude of force and torque oscillations. This positively affects

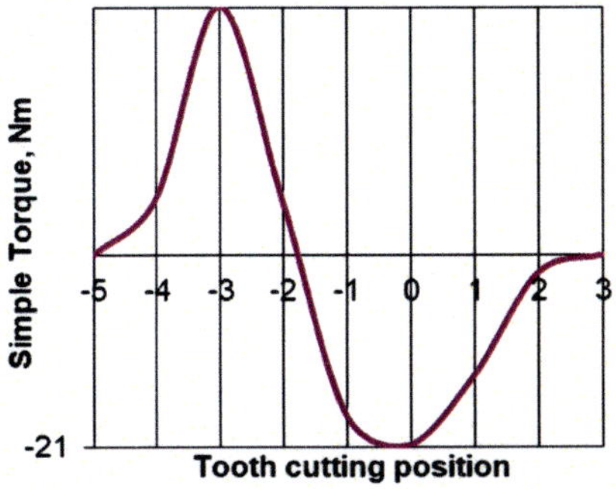

Fig. 8. Torque on the axis of a circular cutter for single-tooth cutting.

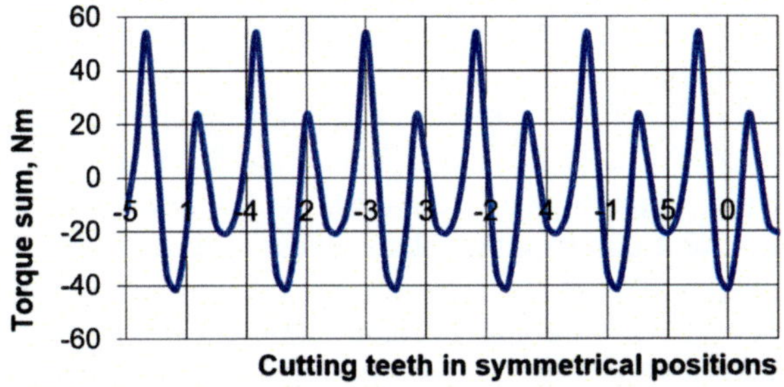

Fig. 9. Change in torque on the cutter axis due to multi-tooth cutting.

the surface quality of the gear teeth being cut. However, on the other hand, increased vibrations and oscillations increase the cutting path, which leads to a decrease in the stability of the cutting tool. As the number of tool teeth increases, the average value of these parameters and the quasi-static load on the machine's elastic system will also increase.

As follows from the results obtained, under the influence of periodic factors, the maximum torque value acting on the axis of the cutter and the tool spindle of the machine in the direction of rotation of the workpiece is 54 Nm, and in the opposite direction 42 Nm, the range is 94 Nm, and the average torque value is 4 Nm. Such data indicate that cyclic loads with alternating signs pose the greatest danger in this process. These disturbances can lead to significant errors in gear machining due to insufficient torsional rigidity of the machine tool and loss of synchronization in the rotational movements of the tool and workpiece. These factors also negatively affect the condition of the machine tool and can lead to cyclic fatigue of the tool, which can result in the breakage of its blades and teeth.

5 Conclusions

A system for computer simulation of the parameters of a power skiving gear-cutting process has been developed. The system is based on a three-dimensional model of undeformed chips and makes it possible to calculate and predict the power factors of this process - cutting force, its components, and torques. The system operates at the level of individual edges and allows for analyzing these parameters at any angle of the cutting and forming area.

In multi-pass cutting, the cut thickness is small (for these initial conditions, from 0.2 mm to 0.6 mm). In these conditions, a nose radius of cutting teeth must be selected with this thickness in mind to reduce the negative impact of cutting replacement by allowance crushing. Depending on the cutting conditions, the minimum radius can be determined using the method described in this article.

Two factors significantly influence the cutting force: the cross-sectional area and the thickness of the cut. An increase in the cutting area proportionally increases the cutting force and load on the system. The thickness of the cut affects the cutting force through its effect on a chip compression ratio: as the thickness of the cut decreases, the intensity of shear plastic deformation and the cutting force increase. The shear intensity increases sharply at low chip thicknesses, typical for power skiving, so high cutting forces accompany this process.

The blades of the disk cutter teeth in power skiving are loaded unevenly: the top blades are the least loaded, and the leading blades are the most loaded. In the range of initial factors considered in this paper, an increase in a chip cross-section dominates the effect on the cutting force compared to the thickness of the cuts.

The solution to the problem of p.2 involves the creation of a sharp cutting wedge on the teeth, which reduces their strength. In addition, for machining wheels with increased physical and mechanical properties (high strength and hardness), it is necessary to ensure the increased hardness of the tool teeth, leading to increased blades' fragility. In such conditions, tooth strength can be increased by selecting appropriate protective coatings that reduce surface friction forces.

The simulation results show that when cutting in three passes, the cutting force can be reduced by up to three times compared to cutting to the full high of the gear tooth profile. However, an increase in the number of passes leads to a decrease in the thickness of the cuts. This will cause an exponential increase in shear intensity and cutting force when certain critical values of the number of passes are exceeded. In this case, increasing the number of passes will lead to the opposite effect and reduce the efficiency of the gearing process.

Because the rake face of the teeth is positioned at a relatively small angle to the cutting speed direction, the friction conditions between the chips and this surface change: on the leading edges, the friction vector coincides with the cutting speed vector, while on the trailing edges, these vectors are opposite.

Multi-tooth cutting in power skiving results in a cyclical change in the cutting force and torque in a wide range of amplitudes with a cyclical change in their sign (direction of action). The frequency of fluctuations in their magnitude depends on the number of cutter teeth and cutting speed. It is necessary to increase the number of cutter teeth to reduce the intensity of these oscillations and their impact on the vibration occurrence and oscillations in the elastic system of the machine tool. However, it is necessary to consider the simultaneous increase in quasi-static (average) force and torque and their negative impact on the gear-cutting process.

For wheels that are cut in several passes, using different tools for roughing and finishing passes is recommended to reduce tooling costs and improve machining accuracy and gear surface quality. When using CNC machines, this task is not difficult.

Accordingly, a tool with straight side blades and a sharp blade edge with positive front angles is sufficient for the roughing pass, as the shape of the tooth face does not affect the accuracy at this stage. The finishing pass will be performed with a precision cutter with the appropriate tooth profile and a zero leading angle to obtain the desired wheel tooth profile.

References

1. Stadtfeld, H.J.: Power skiving of cylindrical gears on different machine platforms. Gear Technol. **31**(1), 52–62 (2014)
2. Brecher, C., Brumm, M., Krömer, M.: Design of gear hobbing processes using simulations and empirical data. Proc. CIRP **33**, 484–489 (2015). https://doi.org/10.1016/j.procir.2015.06.059
3. Klocke, F., Brecher, C., Löpenhaus, C., Krömer, M.: Influence of tolerances on characteristic manufacturing deviations in soft gear machining. In: International Conference on Gears, vol. 1, pp. 1–15. Springer, Garching (2015). https://doi.org/10.13140/RG.2.1.2874.5362
4. Guo, E., Hong, R., Huang, X., Fang, C.: Research on the cutting mechanism of cylindrical gear power skiving. Int. J. Adv. Manuf. Technol. **79**(1), 541–550 (2015). https://doi.org/10.1007/s00170-015-6816-9
5. Guo, E., Hong, R., Huang, X., Fang, C.: A novel power skiving method using the common shaper cutter. Int. J. Adv. Manuf. Technol. **83**(1), 157–165 (2016). https://doi.org/10.1007/s00170-015-7559-3
6. Guo, E., Hong, R., Huang, X., Fang, C.: Research on the design of skiving tool for machining involute gears. J. Mech. Sci. Technol. **28**(12), 5107–5115 (2014). https://doi.org/10.1007/s12206-014-1133-z
7. Antoniadis, A., Vidakis, N., Bilalis, N.: A simulation model of gear skiving. J. Mater. Process. Technol. **146**(2), 213–220 (2004). https://doi.org/10.1016/j.jmatprotec.2003.10.019
8. Spath, D., Hühsam, A.: Skiving for high-performance machining of periodic structures. CIRP Ann. Manuf. Technol. **51**(1), 91–94 (2002). https://doi.org/10.1016/S0007-8506(07)61473-5
9. Klocke, F., Brecher, C., Löpenhaus, C., Ganser, P., Staudt, J., Krömer, M.: Technological and simulative analysis of power skiving. Proc. CIRP **50**, 773–778 (2016). https://doi.org/10.1016/j.procir.2016.05.052
10. Tapoglou, N.: Calculation of non-deformed chip and gear geometry in power skiving using a CAD-based simulation. Int. J. Adv. Manuf. Technol. **100**(5), 1779–1785 (2019). https://doi.org/10.1007/s00170-018-2790-3
11. Bergsa, T., Brimmersa, J., Georgoussisa, A., Krömer, M.: Investigation of the chip formation during hobbing by means of an analytical approach. Proc. CIRP **99**, 226–231 (2021). https://doi.org/10.1016/j.procir.2021.03.033
12. Hrytsay, I., Stupnytskyy, V., Topchii, V.: Improved method of gear hobbing computer aided simulation. Arch. Mechan. Eng. **66**(4), 475–494 (2019). https://doi.org/10.24425/ame.2019.131358
13. Hrytsay, I., Stupnytskyy, V.: Advanced computerized simulation and analysis of dynamic processes during the gear hobbing. In: Tonkonogyi, V., Ivanov, V., Trojanowska, J., Oborskyi, G., Edl, M., Kuric, I., Pavlenko, I., Dasic, P. (eds.) InterPartner 2019. LNME, pp. 85–97. Springer, Cham (2020). https://doi.org/10.1007/978-3-030-40724-7_9

14. Özel, T., Altan, T.: Process simulation using finite element method—Prediction of cutting forces, tool stresses and temperatures in high-speed flat end milling. Int. J. Mach. Tools Manuf. **40**(5), 713–738 (2000). https://doi.org/10.1016/S0890-6955(99)00080-2
15. Bouzakis, K.-D., Friderikos, O., Tsiafis, I.: FEM-supported simulation of chip formation and flow in gear hobbing of spur and helical gears. CIRP J. Manuf. Sci. Technol. **1**(1), 18–26 (2008). https://doi.org/10.1016/j.cirpj.2008.06.004

Optimization of Cutting Modes During Sustainable Machining of Products Based on Economic Criteria

Yaroslav Kusyi[1(✉)], Olha Kostiuk[1], Andrii Kuk[1], Aldo Attanasio[2], and Paola Cocca[2]

[1] Lviv Polytechnic National University, 12, Bandera St., Lviv 79013, Ukraine
jarkym@ukr.net
[2] University of Brescia, 38, Via Branze, 25123 Brescia, Italy

Abstract. The sustainable manufacturing and machining concept using functionally-oriented technologies involves design and technological, economic, social, and environmental dimensions. Currently, rational approaches in the environmental direction are primarily based on effectively recycling cutting tools and chips to minimize the consumption of cutting fluids and energy. However, developing functionally-oriented technologies using economic criteria improves product competitiveness, increases machining productivity, and ensures the choice of rational cutting modes while manufacturing machine parts. The optimization technique of cutting modes during parts machining according to economic criteria using Markov chains was suggested for the first time. The target function is the maximum machining time in specified technological operations or certain technological steps during part manufacturing. It is determined according to regulated reliability indicators, e.g., gamma-percentile operating times to failure as a primary parameter of dependability. Adopting the developed technique in mechanical engineering practice will allow the optimization of cutting modes of parts machining according to regulated reliability indicators due to the operational conditions.

Keywords: Sustainable Manufacturing · Functionally-Oriented Technology · Markov Chains

1 Introduction

The concept of sustainable manufacturing and machining provides complex consideration of design and technological regularities of provision of regulated technical requirements [1], economic, social, and environmental dimensions [2].

Much attention is paid to compliance with environmental standards [3, 4]. This is fair since the machining of products involves significant energy consumption of various types [5], the use of lubricating and cooling fluids, the disposal of which is a global problem of the world economy [6]. Recycling devices for blank manufacturing, metal-cutting tools, and chips is another essential task for sustainable manufacturing [7] and machining [8, 9].

V. Tonkonogyi et al. (Eds.): InterPartner 2023, LNME, pp. 167–181, 2024.
https://doi.org/10.1007/978-3-031-42778-7_16

However, the efficiency of any physical process or phenomenon, the competitiveness of machined products, and the validity of the design and technological solutions are evaluated by economic calculations. Besides, the limit values of economic criteria [10] are often used as target functions in optimization tasks of varying complexity in mechanical engineering [11].

2 Literature Review

Machining accuracy indicators, parameters of microrelief of the functional surfaces for machine parts, their operational characteristics, and reliability indicators are provided by cutting, methods of surface plastic deformation, and heat treatment according to the principles of object-oriented [12] and functionally-oriented technological process planning [13].

Object-oriented technologies are implemented in conventional computerized systems of technological preparation of mechanical engineering production [12, 13]. These technologies involve using the typical technological processes of product manufacturing with the step-by-step implementation of interconnected stages. The main task for object-oriented technologies is to provide the product quality parameters by providing the regulated requirements for the design and technological documentation. The minimum technological cost of product manufacturing, the maximum productivity of the equipment [14], and the rational loading of the technological equipment are optimization criteria during the technological process planning [15]. Functional analysis of the operational characteristics of products is not carried out, relying on the designer's experience, qualifications, and technical literacy [13, 16].

Technological cost $C_{techn.i}$ is calculated by the following formula:

$$C_{techn.i} = C_{standi} \cdot T_{p.-c..i}\left(T_{p..i}\right), \quad (1)$$

where $C_{stand.i}$ – cost standard for i-th technological step/operation [€/min]; $T_{p.-c.i}$, $T_{p.i}$ – piece–calculation time or piece time for i-th technological step/operation, respectively [min].

The cost standard consists in its most general form of the cost price standards per 1 min of equipment operation C_{EOi}, additional costs due to capital investments C_{ACi}, and the costs of live labor for society C_{SCi} (which are not included in the cost price):

$$C_{standi} = C_{EOi} + C_{ACi} + C_{SCi}. \quad (2)$$

Piece–calculation time $T_{p.-c.i}$ calculated by [17]:

$$T_{p.-c.i} = T_{p.i} + T_{p.-f.i}/n, \quad (3)$$

where $T_{p.-f.i}$ – preparation and finishing-up time [min]; n – the number of parts in the batch.

Piece time is determined by [17]:

$$T_{p.i} = T_{mach.i} + T_{add.i} + T_{serv.i} + T_{r.i}, \quad (4)$$

where $T_{mach.i}$, $T_{add.i}$, $T_{serv.i}$, $T_{r.i}$ – machining time, additional time, time for the equipment service, and time for the rest and worker's private purposes, respectively.

A minimum of the piece achieves the technological cost reduction–calculation time or piece time according to the rational choice of the technological equipment (metal-cutting machines).

The machining time, which is calculated, and the additional time chosen according to the standards significantly influence the change of the piece–calculation time [17] or piece time [18].

The reduction of the technological cost is ensured by the minimum value of the machining time at the defined technological step/operation in the general case for the optimal construction of the device (additional time = const) [13].

The minimum machining time for object-oriented technologies on a defined technological operation is the target function of the optimization task and a function of the synthesized parameters [18]:

$$t_{0\,min}(t, S, n, V) \to min, \tag{5}$$

where t, S, n, and V are elements of cutting modes: cutting depth, feed rate, rotary speed, and cutting speed, respectively.

The main advantage of the principle of object-oriented technologies is the proven algorithmic construction of technological design for making rational technological decisions in connection with a specialized theoretical knowledge base. However, neglecting the product's functional properties in its further operation hinders the development of the Product Life Cycle Support/Product Life Cycle Management concepts [19] for object-oriented technologies. Implementation of the principle of parallel engineering is impossible due to the lack of recurrent and iterative relationships with other stages of the integrated design system and technological production preparation. Technological support of product shaping by simulated rheological modeling and other research methods is not implemented. The change in the material characteristics of the part is not considered during the analysis of the technological inheritability of its properties [15, 20].

While machining the machine parts' functional surfaces, the main and auxiliary bases require implementation according to functionally-oriented technologies. A characteristic feature of functionally-oriented technologies is the technological provision of the most effective operational characteristics and reliability indicators of the products in compliance with the accuracy and quality parameters of the surface layer of the product specified by the designer [1, 12].

The general methodology of the synthesis of functionally-oriented technologies for the production of mechanical engineering products is determined by the implementation of a set of technological influences of different quality and functional properties; the general structure and parameters of each are provided under the functional features of the operation of individual elements of the product module in the machine [12, 13].

The principle of functionally-oriented technologies is implemented according to the opposite of object-oriented technologies. A comprehensive analysis of the material of the mechanical engineering product, accuracy standards, and quality parameters of its executive surfaces from the standpoint of ensuring operational characteristics with the selection of rational input data is carried out. At the same time, the input data are adapted into the technological chain "input blank - final part" during the technological inheritability of the product quality parameters with the possibility of prediction of its main parameters in a certain period at the main stages of their Life Cycles [14, 15].

Currently, economic criteria are rarely used to evaluate the effectiveness of implementing functionally-oriented technologies. Therefore, developing techniques for optimizing cutting modes during machine parts machining using economic criteria is an urgent mechanical engineering task.

The paper aims to develop a technique for optimizing cutting modes during parts machining through economic criteria.

3 Research Methodology

3.1 General Provisions

The object of research for modern functionally-oriented technologies using the concepts of Industry 4.0 is the Life Cycles of Parts or Products, and their essential phases and stages should be described by a set of quantitative indicators of operational safety according to the requirements of Sustainable Manufacturing and Machining. The parameters of accuracy, the quality of the microrelief of the functional surfaces of machine parts, their operational characteristics, and reliability indicators are closely related to the structure of the technological processes of parts manufacturing. The most challenging task during the design of the functionally-oriented technologies is the provision of the relationships between reliability indicators (in particular, dependability and durability) and technological parameters of the manufacturing process: cutting modes and elements of a piece–calculation time or piece time [18].

Probable failure-free indicators (reliability factor P(t), gamma-percentage failure time) require the involvement of an adequate mathematical apparatus for modeling actual physical processes during the manufacture of machine parts in conditions of Sustainable Manufacturing/Machining and Industry 4.0 [15].

Calculation of the reliability of the technical system, in general, and its elements, in particular, is calculated by computer simulation of physical and structural models. Physical element reliability models are used at lower levels, and structural models are used at higher ones [16].

Reliability theory is widely used in the automotive industry [21, 22], electric vehicle manufacturing [23, 24], and information technologies [25, 26] to estimate the reliability of specific systems that consist of standard elements with known reliability characteristics, in particular, the intensity of failures. Since typical elements of mechanical systems (shafts, gears, and bearings) work in different machines with different operating loads, their operational characteristics and reliability indicators significantly change. In addition, the elements of mechanical systems, as a rule, are interconnected, which does not allow the use of simplified algorithms for calculating the probability of failure-free operation of technical systems.

Currently, the provision of the product quality parameters is carried out using a technological system. The execution of technological steps in the specified technological operation is implemented by the components of the technological system: machine tool - clamping device - metal-cutting tool (s) - workpiece, and the machining of the workpiece is ensured by the elements of the technological media: machine tool - clamping device - metal-cutting tool (s). The metal-cutting machine and the clamping device provide the product quality parameters regulated by technical requirements for certain operations. At the same time, metal-cutting tools are flexible elements of the technological media, which allow for managing the product parameters using a rational choice of their structural and geometric parameters for a specified machined material and type of machining [18].

However, the influence of the elements of the technological media on the provision of regulated product quality parameters for functionally-oriented technologies and taking into account technological inheritability has not been investigated in detail.

The necessary operational characteristics of mechanical engineering products are provided during manufacture according to the principle of functionally-oriented technologies according to their official purpose. However, ISO standards regulate reliability indicators, the most important of which are dependability and durability parameters. Therefore, it is necessary to provide the dependability and durability indicators of machine parts regulated by technical conditions to effectively use functionally-oriented information technologies [27, 28] in mechanical engineering production [29].

Actual reliability indicators are determined at the second stage of implementing the principles of the functionally-oriented technologies based on the results of the mathematical simulation to provide the quality parameters of parts' functional surfaces. The parameters of the technological operation are corrected according to the regulated values. In addition, the optimization problem of synthesizing processing modes has been formed and solved [18].

Therefore, predicting the reliability indicators of machine parts at the phases and stages of their Life Cycles is one of the essential tasks in mechanical engineering [13, 18].

Regulated input indicators of the reliability of parts, established by calculating dimensional chains in machines, are analyzed by estimating input data for the previous technological processes planning of parts manufacturing. The number of subsystems (operations and steps) of the technical system (technological process) is established, and the possible states of their functioning are analyzed based on structural analysis, synthesis of competitive variants of the technological process, and the choice of their rational variant. The technological graph of reliability using Markov chains is developed after processing important information for each subsystem. The system of differential equations develops the obtained technological graph. The system of differential equations is solved according to the established boundary (initial) conditions. Reliability indicators, determined according to the necessary service functions of mechanical engineering parts, are consistent with the regulated parameters of dependability and durability (gamma-percentile operation time to failure and resource) according to the simulation results. The structure of the specified technological operations, in particular, and the technological process, in general, is corrected according to the results of mathematical simulation [15, 18].

Along with that, the calculation of the reliability indicators of modern technical systems is based on the prediction of the limit state of the element of the technological system [18].

The limit state of the machine is determined at the system level by the regulated parameters of failure and durability set by the customer. Limit indicators of reliability from the aggregate/component unit (a subsystem of the first order) and node (a subsystem of the n-th order) to the level of the part (a subsystem of the k-th order) are alternately determined using the theory of dimensional chains. The provision of reliability indicators is carried out in the reverse sequence: the limit states of nodes are described through the limit states of their parts, the limit states of aggregates are described through the limit states of their nodes, and the limit state of the machine is described through the limit states of its units [15, 18].

The limit state for the elements of the technological system has a different interpretation during the production of the product. The limit state of the workpiece is determined by the actual value of at least one of its parameters near the lower (external surfaces) or upper (internal surfaces) tolerance limit according to technical requirements and unacceptable technological defects in its material in the specified technological operation. The limit state for a metal-cutting machine, clamping device, and metal-cutting tool are determined by the deterioration of their initial parameters during operation due to wear and fatigue, as well as the actual value of at least one of their regulated initial parameters reaching values near the lower or upper limits of tolerances [18, 29].

Prediction of the probability for ensuring the regulated quality parameters of mechanical engineering products under the condition that the metal-cutting machine, clamping device, and metal-cutting tools do not reach the limit state during machining in a separate technological step, a separate technological operation prevents the elements of the technological media (first of all, the metal-cutting tool (s)) from approaching the limit state. This approach allows optimizing the technological process by correcting its structure based on the results of mathematical simulation [18, 29].

At the same time, full-scale tests of products after providing the regulated parameters of their quality ensure the determination of product reliability indicators in conditions as close as possible to operational ones [18].

3.2 Development of an Algorithm for the Formation of Regulated Product Quality Parameters, Taking into Account the Influence of the Elements of the Technological System "Metal-Cutting Machine - Clamping Device - Metal-Cutting Tool(S) – Workpiece"

The technological process as a technical system has been investigated by analyzing the parameters of first-order subsystems (technological operations) and second-order subsystems (technological steps).

At the initial stage of research, the regulated parameters of the reliability of the e-elements of the technological system during the machining of specified products at different levels of subsystems of the technical system (technological process) are determined (e.g., technological operation, technological step). After that, a block diagram of reliability is developed for the specified technological operation, taking into account the influence of the elements of the technological media on the formation of the regulated parameters of the product. Based on the block diagram of reliability, a technological graph of reliability using Markov chains is developed for each technological operation of product manufacturing. The developed technological graphs of reliability determine the failure rate for the elements of the technological system for each technological operation [18, 26].

A system of differential equations is developed to predict the provision of regulated parameters of the quality of the part workpiece during its machining at the i-th technological operation, taking into account the influence of the elements of the technological system. Solutions of differential equations are found in parametric and numerical forms considering the boundary conditions [18].

The optimization task of synthesizing cutting modes is solved based on the obtained values of product reliability indicators. The optimization criterion is the limit value of the machining time in a specified technological step or technological operation, established according to the calculated and regulated parameters of dependability or durability (probability of failure-free operation, gamma-percentage working before failure, and resource) [18, 26].

A system of differential equations is developed to predict the provision of regulated parameters of the quality of the part workpiece during its machining at the ith technological operation, taking into account the influence of the elements of the technological system. Solutions of differential equations are found in parametric and numerical forms considering the boundary conditions [18].

The optimization task of synthesizing cutting modes is solved based on the obtained values of product reliability indicators. The optimization criterion is the limit value of the machining time at a specified technological step or technological operation, established according to the calculated and regulated parameters of dependability or durability (probability of failure-free operation, gamma-percentage working before failure, and resource) [18, 26].

The machining time in certain technological steps or technological operations is a function of the cutting modes of product machining [17].

The limit values of the machining time for the regulated parameters of product reliability are determined based on the solution of the optimization problem. The structure of the technological route of surface treatment, cutting modes, the structure of the technological operation in particular, and the structure of the technological process, in general, are optimized.

3.3 Using Markov Chains to Predict Quality Parameters During Parts Manufacturing

For the first time, the mathematical apparatus of Markov chains has successfully been implemented to develop mathematical models for predicting the influence of the elements of the technological system to ensure the regulated parameters of the quality of mechanical engineering parts [18].

Technological operation 005 (horizontal milling) during reducing-gear housing manufacturing was analyzed [18]. The workpiece is machined by horizontal milling machine 6H81 using a face milling cutter and universal clamping device [18].

A system of Chapman-Kolmogorov differential equations was developed for operation 005 of the technological process of manufacturing the reduction-gear housing for the technological graph of reliability (Fig. 1) [18]:

$$\begin{cases} \frac{dP_1}{dt} = -\lambda_{MCM\_005} \cdot P_1(t) - \lambda_{MCMD\_005} \cdot P_1(t) - \lambda_{MCT\_005} \cdot P_1(t), \\ \frac{dP_2}{dt} = \lambda_{MCM\_005} \cdot P_1(t), \\ \frac{dP_3}{dt} = \lambda_{MCMD\_005} \cdot P_1(t), \\ \frac{dP_4}{dt} = \lambda_{MCT\_005} \cdot P_1(t), \\ P_1(t) + P_2(t) + P_3(t) + P_4(t) = 1, \quad t \in [0; t] \end{cases} \tag{6}$$

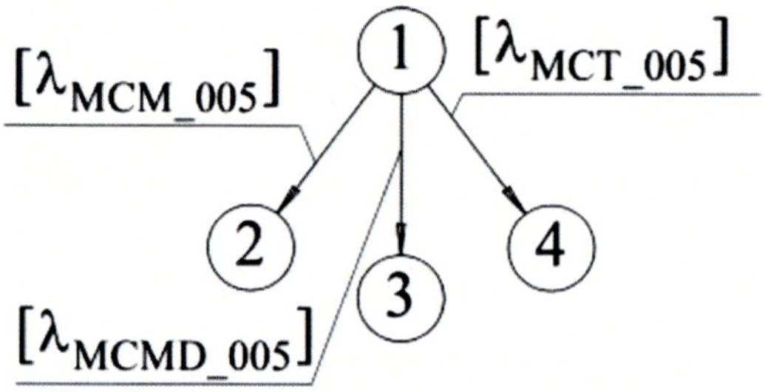

Fig. 1. Technological graph of reliability for technological operation 005 for reducing-gear housing manufacturing.

In Fig. 1, λ_{MCM_005}, λ_{MCCD_005}, λ_{MCT_005} are failure rates of the metal-cutting machine, the clamping device on the metal-cutting machine, and the metal-cutting tool (face milling cutter) or their elements in the technological operation 005 during machining of reduction-gear housing.

The solutions of the system of differential equations (6) for $\lambda_{MCM\_005} = 5.48 \cdot 10^{-6}$ min^{-1}; $\lambda_{MCMD\_005} = 1.111 \cdot 10^{-5}$ min^{-1}; $\lambda_{MCT\_005} = 2.5 \cdot 10^{-3}$ min^{-1} will have the form, taking into account the initial conditions [18]:

$$P_1(t) = e^{-2.517 \cdot 10^{-3} \cdot t}, \tag{7}$$

$$F_2(t) = \frac{\lambda_{MCM\_005} \cdot e^{-\left(\lambda_{MCM_{005}} + \lambda_{MCMD_{005}} + \lambda_{MCT\_005}\right) \cdot t}}{-\left(\lambda_{MCM_{005}} + \lambda_{MCMD_{005}} + \lambda_{MCT\_005}\right)} + 2.178 \cdot 10^{-3}, \tag{8}$$

$$F_3(t) = \frac{\lambda_{MCMD\_005} \cdot e^{-\left(\lambda_{MCM_{005}} + \lambda_{MCMD_{005}} + \lambda_{MCT\_005}\right) \cdot t}}{-\left(\lambda_{MCM_{005}} + \lambda_{MCMD_{005}} + \lambda_{MCT\_005}\right)} + 4.415 \cdot 10^{-3}, \tag{9}$$

$$F_4(t) = \frac{\lambda_{MCT\_005} \cdot e^{-\left(\lambda_{MCM_{005}} + \lambda_{MCMD_{005}} + \lambda_{MCT\_005}\right) \cdot t}}{-\left(\lambda_{MCM_{005}} + \lambda_{MCMD_{005}} + \lambda_{MCT\_005}\right)} + 0.993, \tag{10}$$

where P_1 (t) is the probability of providing regulated quality parameters of the workpiece in the technological operation 005 of the technological process of the reduction-gear housing machining in case of failure to reach the limit state of the metal-cutting machine, clamping device located on the metal-cutting machine and metal-cutting tool (or tools) in general and their elements in particular; $F_2(t)$ is the probability of failure to provide regulated quality parameters in the technological operation 005 of the technological process of the reduction-gear housing machining, when the limit state is reached by the metal-cutting machine in general or its elements in particular; F_3 (t) is the probability of failure to provide the regulated quality parameters in the technological operation 005 of the technological process of the reduction-gear housing machining, when the limit state is reached by the clamping device on a metal-cutting machine in general or its elements in particular; F_{4i} (t) is the probability of failure to provide regulated quality parameters in the technological operation 005 of the technological process of the reduction-gear housing machining when the limit state is reached by the metal-cutting tool (or tools) in general or its (their) elements in particular [18].

Numerical values of probabilities: $P_1(t) = 0.996$; $F_2(t) = 8.09 \cdot 10^{-6}$; $F_3(t) = 1{,}64 \cdot 10^{-5}$; $F_4(t) = 3.69 \cdot 10^{-3}$ determined after substituting the values $\lambda_{MCM\_005} = 5.48 \cdot 10^{-6}\ \text{min}^{-1}$; $\lambda_{MCMD\_005} = 1.111 \cdot 10^{-5}\ \text{min}^{-1}$; $\lambda_{MCT\_005} = 2.5 \cdot 10^{-3}\ \text{min}^{-1}$, $t_0 = 1.48$ min (obtained from technological documentation) for machining the reduction-gear housing at technological operation 005 [18].

4 Results and Discussion

The limit value of the machining time for certain technological steps (i.e., technological operations) is calculated during the design and technological preparation of part production according to the specified reliability parameters (including dependability or durability) and the laws of their distribution.

The maximum machining time in the specified technological operation or certain technological step during mechanical engineering part manufacturing is the target function of the optimization task for the choice of its rational parameters [18]:

$$t_{0\,max}(t, S, n, V) \rightarrow max, \tag{11}$$

The system of equations for the optimization task in general for specified gamma-percentile operating time to failure is determined by [18]:

$$P(t) = e^{-(A \cdot t)}; P(t) = P(t_{\gamma 1}); \ldots\ldots..; P(t) = P(t_{\gamma 6}), \tag{12}$$

where: $P(t) = e^{-(A \cdot t)}$ is a reliability function obtained during the solution of the system of Chapman-Kolmogorov differential equations for a particular technological operation or a particular technological step; $P(t_{\gamma 1})$, … $P(t_{\gamma 6})$ are regulated gamma-percentile operating times to failure.

The solution to the optimization task for the technological process planning of parts manufacturing is determined by [18]:

$$t_{0\,k} \leq [t_0], \tag{13}$$

where: t_{0k} is the machining time for implemented k-th technological step ($k = 1 \ldots n$), $[t_0]$ is the limit value of the machining time according to the specified laws of reliability indicator change for a specific workpiece.

The machining time for technological operations realized by cutting is determined by [17]:

$$t_0 = L_{mach.st.} / \left(S_{0\,pasp..} \cdot n_{pasp.}\right), \tag{14}$$

$$t_0 = L_{mach.st.} / S_{min\,pasp..}, \tag{15}$$

where $L_{mach.st.}$ is the estimated cutting length during certain machining by a specified metal-cutting tool on a certain technological step; $S_{0pasp.}$, $S_{min.pasp.}$, $n_{pasp.}$ are values of cutting modes elements (feed per spindle revolution, feed per minute, and rotary speed) according to regulated parameters of the metal-cutting machine, respectively.

$$L_{mach.st.} = l + l_1 + l_2, \tag{16}$$

where l is the length of the machined surface, mm; l_1 is the cut-in length of the metal-cutting tool, mm; l_2 is the run-out length of the metal-cutting tool, mm.

The calculated length of the machining step of the metal-cutting tool remains constant while processing a specific surface of the workpiece in the technological step of a particular technological operation [17].

So, the relevant conditions serve the optimization criteria for a constant estimated cutting length during certain machining by the specified metal-cutting tool in a certain technological step ($L_{mach.st.}$ = const) [18]:

$$S_0 \geq [S_0] \text{ or } S_{min} \geq [S_{min}], n \geq [n], V_{cut.} \geq [V_{cut.}], \tag{17}$$

where $[S_0]$, $[S_{min}]$, [n], and $[V_{cut}]$ are the limit values of feed per spindle revolution, feed per minute, rotary speed, and cutting speed for particular machining by cutting and provision the regulated accuracy, quality of the microrelief for functional surfaces of the workpieces, and their reliability indicators.

The constructive characteristics of the face milling cutter at technological operation 005 during the manufacturing of reduction-gear housing are external diameter D = 250 mm, diameter under the mandrel d = 60H7 mm, height H = 75 mm, the number of teeth z = 14, $\angle\varphi = 45°$ [17].

The results of solving the optimization task (12) for all values of gamma-percentile operating time to failure (t_{γ_i}) during the manufacture of the reduction-gear housing in technological operation 005 is presented in Fig. 2. The gamma-percentile time to failure for a specified mechanical engineering part is calculated based on the investigated dimensional chains for a certain machine according to its initial gamma-percentile operating time to failure.

The optimization task is solved for all regulated values of the gamma-percentile time to failure (Fig. 2), although one value of the failure rate is set for a specific machine in actual production.

The system of Eqs. (12) for operation 005 under machining of the reduction-gear housing is presented in general by [18]:

$$P(t) = e^{-\left(2.517 \cdot 10^{-3} \cdot t\right)}; P(t_\gamma) = 1.00; P(t_\gamma) = 0.99; P(t_\gamma) = 0.95; P(t_\gamma) = 0.90; P(t_\gamma) = 0.80; P(t_\gamma) = 0.50. \tag{18}$$

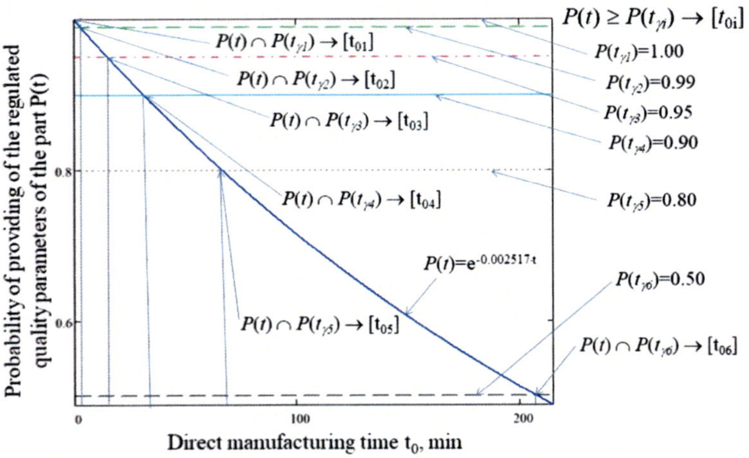

Fig. 2. Checking the condition of ensuring workability of a workpiece according to the regulated reliability parameters (gamma-percentile operation time to failure (t_γ)) in technological operation 005 of machining of the reduction-gear housing.

The limit values of the machining time for technological operation 005, inversely proportional to the elements of cutting modes, are obtained by the results of calculations in Mathcad: for $P(t_\gamma) = 1$ $[t_0] = 0$ min; for $P(t_\gamma) = 0.99$ $[t_0] = 3.99$ min; for $P(t_\gamma) = 0.95$ $[t_0] = 20.38$ min; for $P(t_\gamma) = 0.9$ $[t_0] = 41.87$ min; for $P(t_\gamma) = 0.8$ $[t_0] = 88.67$ min; for $P(t_\gamma) = 0.5$ $[t_0] = 275.43$ min (Fig. 2) [18].

Increasing the rotary speed and feed during workpieces machining reduces the machining time at certain technological steps or general technological operations. The metal-cutting tool life also decreases under these conditions [18].

Calculations are made for $P(t_\gamma) = 0.99$ and $[t_0] = 3.99$ min in the framework of this paper (Fig. 2).

Condition (13) is provided according to Fig. 2 and technological requirements of technological documentation [18]:

$$t_{0\,005} = 1.48 < [t_0] = 3.99 \text{ min}.$$

The cutting modes are determined for the specified variant of the technological step during milling of a basic plane of the reduction-gear housing at technological operation 005 due to the technical documentation: t = 3 mm; $S_Z = 0.2$ mm/t; $S_{min.pasp.} = 180$ mm/min; $n_{pasp.} = 120$ min^{-1} [18]. The length of the machined surface l = 225 mm.

The cut-in length of the metal-cutting tool is determined by [17]:

$$l_1 = 0.5 \cdot \left(D - \sqrt{D^2 - B}\right) + t/tg(\varphi), \tag{19}$$

$$l_1 = 0.5 \cdot \left(250 - \sqrt{250^2 - 225^2}\right) + 3/tg\left(45^\circ\right) = 73.514 \approx 74 \text{ mm}.$$

Run-out length of the metal-cutting tool is calculated by [17]:

$$l_1 = l_2' + D, \tag{20}$$

where $l_2' = 4$ mm – [17].

$$l_1 = 4 + 250 = 254 \text{ mm}. \quad (21)$$

So, the estimated cutting length during milling of reduction-gear housing by face milling cutter Ø 250 mm at technological operation 005 is determined by:

$$L_{mach.st.} = 225 + 74 + 254 = 553 \text{ mm}.$$

To ensure the regulated gamma-percentile operating time to failure $P(t_\gamma) = 0.99$ for the limit machining time $[t_0] = 3.99$ min at technological operation 005, the limit value of the feed per minute is calculated by (12):

$$[S_{min\ pasp.}] = L_{mach.st.}/[t_0], \quad (22)$$

$$[S_{min\ pasp.}] = 553/3.99 = 138.569 \approx 139 \text{ mm/min for } P(t_\gamma) = 0.99.$$

Fulfillment of condition (17) according to the results (22) is provided by the nearest value $[S_{min.pasp.}] = 145$ mm/min according to the passport characteristics of the metal-cutting machine with a stepped drive of the cutting motion and feed motion and the following (ascending) value of a feed rate for the horizontal milling machine at operation 005. So,

$$S_{min\ pasp.} = 180 \text{ mm/min} > [S_{min\ pasp.}] = 145 \text{ mm/min}.$$

Feed per minute S_{min} is determined by [29]:

$$S_{min} = S_Z \cdot z \cdot n, \quad (23)$$

So,

$$[n_{pasp.}] = [S_{min\ pasp.}]/(S_Z \cdot z), \quad (24)$$

$$[n_{pasp.}] = 180./(0.2 \cdot 14) = 64.286 \approx 65 \ \text{min}^{-1} \text{ for } P(t_\gamma) = 0.99.$$

Fulfillment of condition (14) according to the results (19) is provided by the nearest value $[n_{pasp.}] = 71$ min^{-1} according to the passport characteristics of the metal-cutting machine with a stepped drive of the cutting motion and feed motion and the following (ascending) value of a feed rate for the horizontal milling machine at operation 005.

$$n_{pasp.} = 120 \text{ min}^{-1} > [n_{pasp.}] = 71 \text{ min}^{-1}.$$

Cutting speed is calculated by:

$$V_{cut.} = (\pi \cdot D \cdot n)/1000, \quad (25)$$

$$[V_{cut.}] = \left(\pi \cdot D \cdot \left[n_{pasp.}\right]\right)/1000, \quad (26)$$

$$V_{cut.} = \left(\pi \cdot D \cdot n_{pasp.}\right)/1000, \tag{27}$$

$$[V_{cut.}] = (\pi \cdot 250 \cdot 71)/1000 = 55.76 \text{ m/min},$$

$$V_{cut.} = (\pi \cdot 250 \cdot 120)/1000 = 94.25 \text{ m/min},$$

$$V_{cut.} = 94.25 \text{ m/min} > [V_{cut.}] = 55.76 \text{ m/min for P}(t) = 0.99.$$

The failure rate of elements of technological media is considered to ensure the reliability of the metal-machining technological system at a specified technological operation or technological step during workpiece machining according to the principles of the functionally-oriented technologies (metal-cutting machine, clamping device, and metal-cutting tool). The limit value of machining time in a certain technological operation or technological step is determined using the obtained value of the probability of providing regulated quality parameters of the workpiece in the technological operation of the technological process of the reduction-gear housing machining in case of failure to reach the limit state of the metal-cutting machine, a clamping device located on the metal-cutting machine and metal-cutting tool (or tools) in general and their elements in particular and gamma-percentile operating time to failure regulated by technical requirements according to the operational conditions of products. An increase in the rotary speed, cutting speed, and feed rate reduce the machining time and increases the probability of regulated quality parameters of the workpiece. However, increasing the cutting modes during workpiece machining reduces the tool's life and increases the probability of failure to ensure the necessary quality parameters of the workpiece according to the metal-cutting tool reaching its limit state.

5 Conclusions

The following main conclusions have been drawn based on the research results. For the first time, the technique of optimization of cutting modes during parts machining according to the economic criteria and using Markov chains is suggested. The target function serves the maximum machining time at a specified technological operation or particular technological step during part manufacturing determined according to regulated reliability indicators, for example, gamma-percentile operating times to failure as a primary parameter of dependability.

Solutions of the system of Kolmogorov-Chapman differential equations according to the regulated values of gamma-percentile operating times to failure make it possible to establish the limit value of the machining time, which is a target function from one side and a function of cutting modes from the other one. The optimization of cutting modes (feed per minute $S_{min.}$, feed rate n, and cutting speed $V_{cut.}$), which are inversely proportional to the limit values of the machining time, ensures an increase in labor productivity, taking into account the recommendations of the tool manufacturer for functionally-oriented technologies. The average value of the probability of ensuring the regulated quality parameters under the condition that the elements of the technological media do not reach the limit state is 92.4%, while the limit value is 78.6% during the machining of the reduction-gear housing.

Adopting the developed technique in mechanical engineering enterprises will provide the selection of rational cutting modes of parts machining according to operating conditions and regulated reliability indicators.

Acknowledgment. This research has been conducted as part of the ongoing project "Comprehensive system of functional-oriented planning of machining difficult-to-cut materials for the military-industrial complex (Kompleksys)". It has been funded by the Research Council of Lithuania and the Ministry of Education and Science of Ukraine under the Lithuanian–Ukrainian Cooperation Programme in the Fields of Research and Technologies (Grant No. S-LU-22-6).

References

1. Davim, J.P.: Surface Integrity in Machining. Springer, London (2010). https://doi.org/10.1007/978-1-84882-874-2
2. Davim, P.J.: Manufacturing in Digital Industries: Prospects for Industry 4.0. Advanced Mechanical Engineering, Walter de Gruyter Gmbh, Berlin/Boston (2020)
3. Kopei, V.B., Onysko, O.R., Panchuk, V.G.: Component-oriented acausal modeling of the dynamical systems in Python language on the example of the model of the sucker rod string. PeerJ Comput. Sci. **5**, e227 (2019)
4. Cerdas, F., Thiede, S., Herrmann, C.: Integrated computational life cycle engineering: application to the case of electric vehicles. CIRP Ann. **67**(1), 25–28 (2018)
5. Liao, Z., et al.: Surface integrity in metal machining - part I: fundamentals of surface characteristics and formation mechanisms. Int. J. Mach. Tools Manuf. **162**, 103687 (2021)
6. M'Saoubi, R., Outeiro, J.C., Chandrasekaran, H., Dillon, O.W., Jr., Jawahir, I.S.: A review of surface integrity in machining and its impact on functional performance and life of machined products. Int. J. Sustain. Manuf. **1**, 203–236 (2008)
7. Davim, J.P.: Sustainable Machining: Materials Forming, Machining and Tribology, 1st edn. Springer, Cham (2018). https://doi.org/10.1007/978-3-319-51961-6
8. La Monaca, A., et al.: Surface integrity in metal machining - part II: functional performance. Int. J. Mach. Tools Manuf. **164**, 103718 (2021)
9. Luscinski, S., Ivanov, V.: A simulation study of Industry 4.0 factories based on the ontology on flexibility with using Flexsim® software. Manag. Prod. Eng. Rev. **11**(3), 74–83 (2020). https://doi.org/10.24425/mper.2020.134934
10. De Man, J.C., Strandhagen, J.O.: An Industry 4.0 research agenda for sustainable business models. Procedia CIRP **63**, 721–726 (2017)
11. Gavrysh, O., Boiarynova, K.: The methodological approach to monitoring of the economic and functional state of innovation-oriented machinery engineering enterprises at the modern technological modes. Econ. Annals-XXI **164**, 65–70 (2017)
12. Stupnytskyy, V.: Features of functionally-oriented engineering technologies in concurrent environment. Int. J. Eng. Res. Technol. (IJERT) **2**(9), 1181–1186 (2013)
13. Stupnytskyy, V., Hrytsay, I.: Computer-aided conception for planning and researching of the functional-oriented manufacturing process. In: Tonkonogyi, V., et al. (eds.) Advanced Manufacturing Processes: Selected Papers from the Grabchenko's International Conference on Advanced Manufacturing Processes (InterPartner-2019), September 10-13, 2019, Odessa, Ukraine, pp. 309–320. Springer, Cham (2020). https://doi.org/10.1007/978-3-030-40724-7_32

14. Onysko, O., Panchuk, V., Kopei, V., Havryliv, Y., Schuliar, I.: Investigation of the influence of the cutter-tool rake angle on the accuracy of the conical helix in the tapered thread machining. J. Phys. Conf. Ser. **1781**(1), 012028 (2021)
15. Kusyi, Y.M., Stupnytskyy, V.V., Kuk, A.M., Topilnytskyy, V.G.: Development of the fundamental diagram of the formation and transformation of the products properties during their manufacturing. J. Phys. Conf. Ser. **1781**(1), 012027 (2021)
16. Kopei, V.B., Onysko, O.R., Panchuk, V.G.: Principles of development of product lifecycle management system for threaded connections based on the Python programming language. J. Phys. Conf. Ser. **1426**(1), 012033 (2020)
17. Yurchyshyn, I.I., Lytvynyak, Y.M., Hrytsay, I.Y., et al.: Mechanical Engineering Technology: Handbook for the Performance of Qualification Works. Lviv Polytechnic National University, Lviv (2009)
18. Kusyi, Y., Stupnytskyy, V., Onysko, O., Dragašius, E., Baskutis, S., Chatys, R.: Optimization synthesis of technological parameters during manufacturing of the parts. Maint. Reliab. **24**(4), 655–667 (2022)
19. Prokopovych, I.V., Kokhanov, A.B., Khamitov, V.M., Tikhenko, V.M., Dašić, P.: Standardizing life cycle organization: a synergetic quality management approach. J. Eng. Sci. **10**(1), B1–B7 (2023). https://doi.org/10.21272/jes.2023.10(1).b1
20. Ivchenko, O., et al.: Method for an effective selection of tools and cutting conditions during precise turning of non-alloy quality steel C45. Materials **15**(2), 505 (2022). https://doi.org/10.3390/ma15020505
21. Klocke, F.: Manufacturing Processes 1: Cutting. Springer, Berlin (2011). https://doi.org/10.1007/978-3-642-11979-8
22. Nyberg, P., Frisk, E., Nielsen, L.: Generation of equivalent driving cycles using Markov chains and mean tractive force components. IFAC Proc. Vol. **47**(3), 8787–8792 (2014)
23. Gruosso, G., Mion, A., Gajani, G.S.: Forecasting of electrical vehicle impact on infrastructure: Markov chains model of charging stations occupation. eTransportation **6**, 100083 (2020)
24. Sorlei, I.-S., Bizon, N., Thounthong, P., et al.: Fuel cell electric vehicles—a brief review of current topologies and energy management strategies. Energies **14**, 251 (2021)
25. Yakovyna, V., Seniv, M., Symets, I., Sambir, N.: Algorithms and software suite for reliability assessment of complex technical systems. Radio Electron. Comput. Sci. Control (4), 163–177 (2020)
26. Pavlenko, I., et al.: Using regression analysis for automated material selection in smart manufacturing. Mathematics **10**(11), 1888 (2022). https://doi.org/10.3390/math10111888
27. Gaydabrus, B.V., Druzhinin, E.A., Kiyko, S.G.: Main aspects of project management and development programs of IT-availability of the manufacturing enterprises. Metall. Min. Ind. **7**(12), 326–330 (2015)
28. Zaleta, O.M., Povstyanoy, O.Y., Ribeiro, L.F., Redko, R.G., Bozhko, T.Y., Chetverzhuk, T.I.: Automation of optimization synthesis for modular technological equipment. J. Eng. Sci. **10**(1), A6–A14 (2023). https://doi.org/10.21272/jes.2023.10(1).a2
29. Kopei, V., Onysko, O., Odosii, Z., Pituley, L., Goroshko, A.: Investigation of the influence of tapered thread profile accuracy on the mechanical stress, fatigue safety factor and contact pressure. In: Karabegović, I. (ed.) New Technologies, Development and Application IV, pp. 177–185. Springer, Cham (2021). https://doi.org/10.1007/978-3-030-75275-0_21

Calculation of the Accuracy of the Drill-String NC13 Thread Profile Turned from Difficult-to-Machine Steel

Oleh Onysko(✉), Volodymyr Kopei, Vasyl Vytvytskyi, Viktor Vriukalo, and Tetiana Lukan

Ivano-Frankivsk National Technical University of Oil and Gas, 15, Karpatska St., Ivano-Frankivsk 76019, Ukraine
onysko.oleg@nung.edu.ua

Abstract. The process of drilling wells is accompanied by significant environmental pollution. From time to time, it is proposed to make threaded connectors in drill strings from high-strength and stainless steel to reduce emissions. Such steels are difficult to machine, requiring negative rake angles of threading lathe cutters. The article proposes an algorithm for calculating the accuracy of the thread profile to determine the possibility of turning it using such cutters. The result of the predictive calculation proved that using a tool with a rake angle of −7° for turning an NC13 drill thread can lead to a deviation from the nominal value of the half-profile angle, which is 40% of the size tolerance. It was shown that such a deviation could be avoided if a tool with a zero rake angle was used on the lust finish infeed of 0.035 mm.

Keywords: Drill-String Thread · Lathe Tool · Machining · Smart Manufacturing · Sustainable Manufacturing

1 Introduction

Drilling wells is the cause of emissions into the atmosphere. Even more environmental problems are caused by significant drilling waste at the site of drilled and exploited wells. To reduce environmental damage, many measures are proposed that improve the technological condition of drilling equipment [1] and offer entirely new approaches using the fuzzy analytical hierarchy process [2], thanks to which oil pollution on the ground can be cleaned [3]. As is well known, the most fundamental way to preserve the environment is to abandon the extraction of fossil raw materials [4]. However, the growing needs for drinking water require cleaning surface groundwater and increasing the use of underground water resources [5]. In each case, a need to apply drilling, including inclined and horizontal drilling, remains. For arranging water wells and drilling in complex, aggressive environments of underground formations, the possibility of using stainless and high-strength steels is increasingly being considered [6]. Such proposals require significant innovative processing in university communities [7], as well as the involvement of scientific research in the field of mechanical metalworking engineering

V. Tonkonogyi et al. (Eds.): InterPartner 2023, LNME, pp. 182–192, 2024.
https://doi.org/10.1007/978-3-031-42778-7_17

[8], which should be quickly tested for viability in innovative educational and scientific centers [9].

2 Literature Review

Among the proposed works on the engineering calculation of strength estimates in drill pipe joints, there are developed analytical models of contact interaction in pipes with holes [10]. The improvement of control of the strength of the connection in drill and casing pipes is presented in [11], and technical innovations regarding the thread itself with a spring collet insert are studied in [12]. The study of high torque with simultaneous deviation from vertical drilling led to the emergence of new effective design solutions [13]. However, these works require significant design changes, which is economically unattainable in the mass use of drill strings. Scientific proposals for improving existing threaded connections by improving their quality in the manufacturing process and eliminating machining errors appear to be more effective [14, 15]. However, the proposed development does not reference using stainless and high-strength steel in the drill and casing strings joints. There are interesting solutions regarding specialized vibration damping [16], which would perhaps somewhat improve the situation with environmental emissions. There are also proposals for self-centering during screwing [17], which increases the efficiency of drilling operations. However, these works require detailed technological implementation, which requires much time.

The study of the machinability of rust-resistant steels is presented in [18]. However, it is only about general turning, not about the thread machining process. The technological processes and the functionality of the life cycle of products are studied in works [19, 20], and with the disclosure of the influence of vibration strengthening of the part. The work [21] shows the possibilities of increasing resistance against destruction due to detailed study and new approaches to the modernization of the process of casting blanks. Ways for forecasting the machining process are shown in the articles [22–25]. However, they do not specify the issue of the formation of threads, but only the force analysis of metal cutting. The dependence of the intergranular damage of the casting on the method of technological processing was studied in [26]. This work is essential when processing difficult-to-machine steels, but it should be refined concerning threaded parts that are difficult to process. In the works [27, 28], the product's entire life cycle is studied in terms of heredity as a function of the elements of the technological manufacturing process. The works [29] studied the mechanical processing of drill-string threads, where a force analysis of chip formation during tapping is introduced, although without conclusions regarding the accuracy of the obtained thread. In the article [30], deviations from accuracy are considered, but deformation processes cause them and not by the cutter's geometric parameters, which are considered when it comes to threading parts made of hard-to-machine steel [6, 31].

Research [32] also considers the geometric parameter of the rake angle of the tool as an argument. However, as in works [6, 31], there is no analysis of the accuracy of the obtained cutting. The work [33] presents the theory of the study of the influence of the rake angle on the accuracy of the helix line lead angle and not the thread itself. When inspecting inserts from well-known manufacturers of thread-turning tools, no offer with different values of rake angles is found [34]. Papers [35, 36] touch on the research of methods of feeding the cut to the process of tapping, namely: oscillating [35], which contributes to the improvement of chip formation; flank and increment infeed [36] – to improve the wear resistance [37].

Finally, the article [38] proposes a regression model for automated material selection in Smart Manufacturing. The article [39] considers the accuracy and is methodologically fundamental for this article in terms of the application of the algorithm, which determines the profile of the obtained thread as a function of the front angle of the cutter. However, the article does not touch on the cutting process. It does not apply another argument - the amount of radial infeed of cutting and its effect together with a rake angle on the accuracy of the obtained profile.

3 Research Methodology

Study [6] deals with the positive values of rake angles (12) on the accuracy of the thread profile used in the oil and gas production industry (Fig. 1a). This angle is chosen for the efficiency of processing the stainless steel.

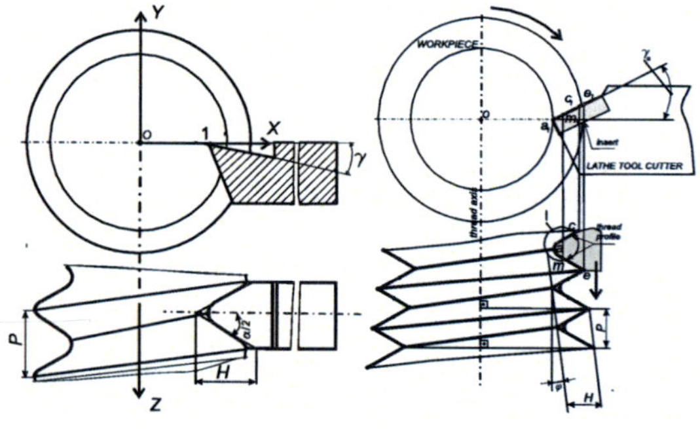

a b

Fig. 1. Scheme of the threading by lathe tool cutter: A – positive Rake Angle (γ), B – negative Rake Angle (γ_a).

However, the work [39] substantiates the choice of the negative rake angle of the carbide insert (−7°), which provides increased stability when tapping on a steel workpiece (steel 35X3NM) with a hardness of 51–55 HRc (Fig. 1b). For the study [39], the M39x2 thread was chosen, for which 12 passes of infeed with a cutting feed of 0.150 mm at the beginning and 0.035 mm at the final were applied (Table 1). The closest drill-string thread to this metric M39x2 thread in terms of its dimensions is the NC13 tapered thread, the profile of which, according to the API7 standard, corresponds to the shape of V - 0.05 (Fig. 2).

According to the API 7 standard, the side profile of the thread is rectilinear. In the ZX coordinate system with the origin at point A (Fig. 3), this profile is described by the formula:

$$Z = tg\frac{\alpha}{2}x \tag{1}$$

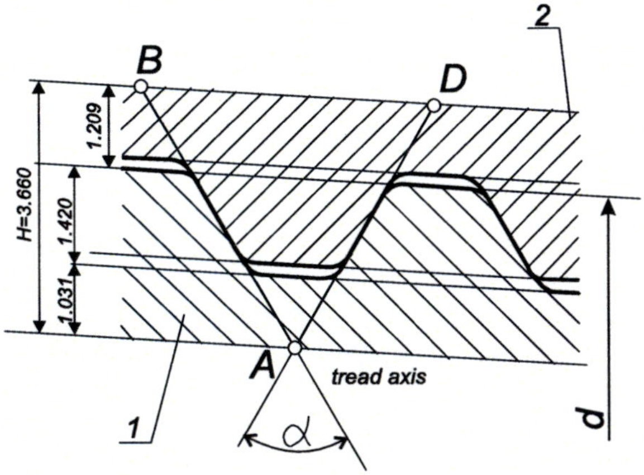

Fig. 2. The scheme of the tool-joint tapered thread profile (V - 0,05) according to the standard API 7: *1* - pin, *2* - box, *P*- pitch, *d* - major diameter, *H* - the height of the fundamental triangle, α -profile angle, points *A*, *B*, *D* are the vertexes of thread fundamental triangle.

Table 1. Selected cutting feeds for the study of turning the M39x2 thread and for research of turning the NC13 thread.

Pass number	Infeed	Cutting depth
1	0.150	0.150
2	0.150	0.300
3	0.150	0.450
4	0.150	0.600
5	0.150	0.750
6	0.150	0.900
7	0.150	1.050
8	0.150	1.200
9	0.075	1.275
10	0.075	1.350
11	0.035	1.385
12	0.035	1.420

If the rake angle of the cutter is not equal to 0 (Fig. 1a, Fig. 1b), the AD profile is curvilinear and is described by a transcendental expression in ZX coordinates (Fig. 3):

$$Z(x) = tg\left(\frac{\alpha}{2}\right)x\frac{\sin\tau}{\sin\gamma} - \frac{P}{2\pi}\tau \tag{2}$$

where:

$$\tau = \gamma - arcsin\left(\frac{r \cdot sin\,\gamma}{x}\right);$$

P – thread pitch; R – minor radius.

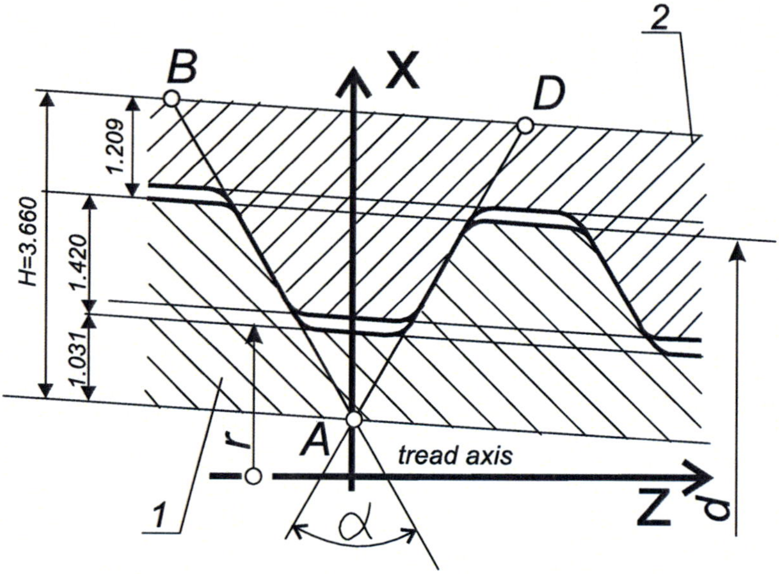

Fig. 3. Profile V050 of a tapered drill bit NC13 and a rectangular coordinate system established with the origin at point A XAZ: r – minor radius, d – major diameter, 1– pin, 2– box.

Since thread turning involves a certain number of passes with the same cutter, in this way, after every cutting pass position, there is a deviation from the actual profile, which is marked with a dotted line (Fig. 4). Conventionally, Fig. 4 shows only the correct deviations from the nominal profile obtained as a result of three thread-final infeed pass of cutting No. 10, 11, and 12.

An enlarged version of the scheme for obtaining lateral deviations of the thread in the turning process indicates the value of b. This value is the maximum radial deviation from the nominal value of the predicted curvilinear profile of the thread obtained in one cutting infeed pass (Fig. 5).

Therefore, the task of the research is as follows:

- based on the algorithm according to formula (2), there is a need to determine the value b, which functionally depends on the parameters of the NC13 thread and the front angle of the cutter $-8°$.
- based on the obtained value of b, predict the number of infeed passes that should be made with a tool with a zero value of the rake angle to ensure the accuracy of the profile.

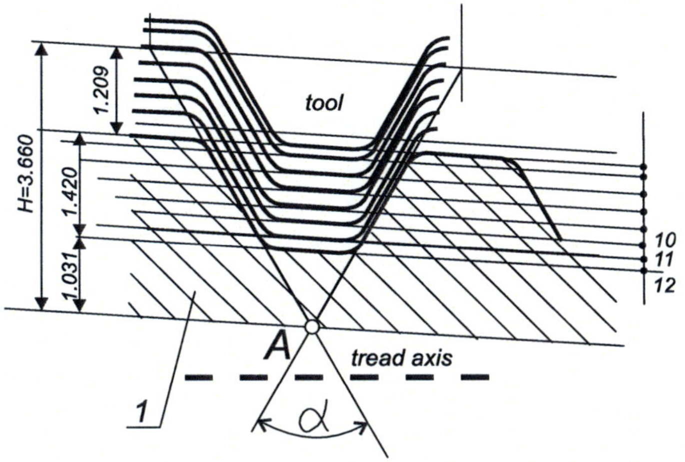

Fig. 4. The scheme of formation of the NC13 thread using a tool for the 12th pass of infeed: 1 – the body of the workpiece; dashed lines show deviations from the nominal profile.

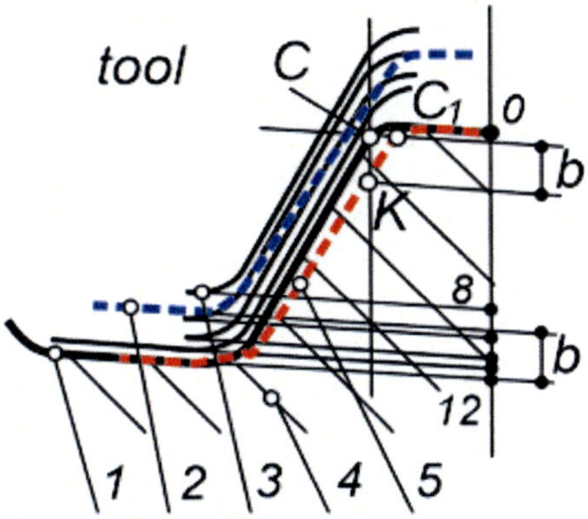

Fig. 5. Scheme for explaining the critical value b: the upper dashed line 2 has not reached the limit of the profile of the thread; the lower dashed line 5 exceeded the limit of the profile by the amount b; 1 – the nominal profile, 3 – the profile of the cutting edge when cutting No. 8; 4 – workpiece.

4 Results and Discussion

Based on formulas (1), (2), a visual algorithm was performed and, applying it with an iteration step of 0.01 mm along the X axis (Fig. 2), the corresponding Z1 coordinates of the profile points according to formula (1) and the corresponding Z2 coordinates have calculated the intersection of the profile according to formula (2) and the difference a = Z1–Z2 is obtained.

The values *a* and *b* are the legs of a right triangle (Fig. 6). Therefore, the value *b* is determined using the equation:

$$b = a/\mathrm{tg}(\alpha/2) \quad (3)$$

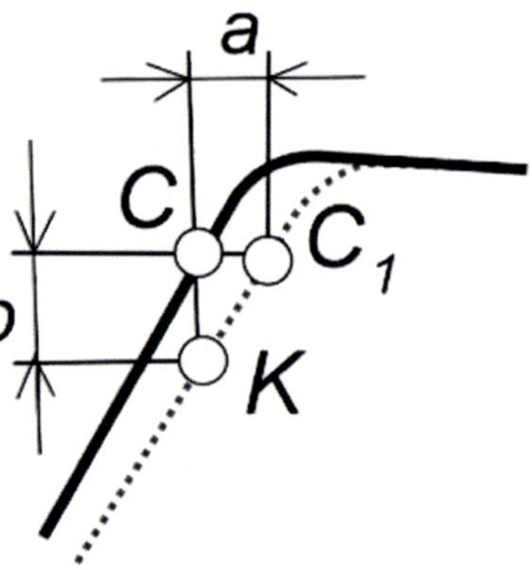

Fig. 6. The scheme for determining the value of b as a function of the parameter $\boldsymbol{a} = Z1–Z2$: bold line – nominal profile according to API7 standard, dashed line – obtained profile using a non-zero value of the rake angle.

The diagram in Fig. 2 shows that the tapered tool-joint thread is asymmetric relative to the X-axis. According to the standard, the profile angle α is 60°. It is divided by the X-axis into two identical half-profile angles α/2 = 30°. However, the left side of AB is greater than the right side of AD. Therefore, the calculation of z coordinates should obviously be carried out separately for both sides of the profile. Standard coordinates are only coordinates Z1 and Z2 for point A, as is evident from the diagram in Fig. 3.

Therefore, the X coordinate of point A is equal to 14.090 mm, and its coordinates are Z1 = Z2 = 0 mm, which is shown in Fig. 7. The Z-axis is directed in both directions relative to A in order not to use inconvenient negative coordinate values in calculations.

Algorithmic calculation includes all points of segments AD and AB. However, according to Figs. 4 and 5, sections GF and EC are required for this study (Fig. 7).

The X coordinates of points G, F, E, and C are calculated due to the X coordinates of the extreme points of the threaded profile A, B, and D. For example, $Xc = X_B - 1.209$ (mm), $Xe = Xa + 1.031$(mm), $Xg = Xb - 1.209$ (mm), $Xf = Xg - 1.420$ (mm). The coordinates Z1 and Z2 of points G, F, E, and C determined by the visual algorithm are listed in Table 2 and schematically shown in Fig. 7.

Table 2 and Fig. 8 data indicate that the most considerable deviations $a = Z2 - Z1$ are at points C and D. Point D lies on the original theoretical triangle, so only real point C is considered, where the $a = 0.020$ mm. Figure 7 corresponds to the distance between points C and C1 in the direction of the Z axis. One of the tasks is to determine the radial (along the X-axis) displacement of the thread profile in the case of using a cutter with a rake angle of −7°. So, applying formula (3), we get b = 0.035 mm. This value equals the final cutting infeed pass (Table 1).

Thus, during the final feeding of No. 12, the deviation from the nominal value of the thread profile is predicted. According to API7 Standard, the tolerance for the half-profile angle is 40′, i.e. 0.67°. From the triangle AC1J, the half-profile angle α1/2 is calculated as follows:

$$^{\alpha_1}/_2 = arctg\frac{|AJ|}{|JC_1|} = arctg\frac{1.43}{16.54 - 14.09}, \tag{4}$$

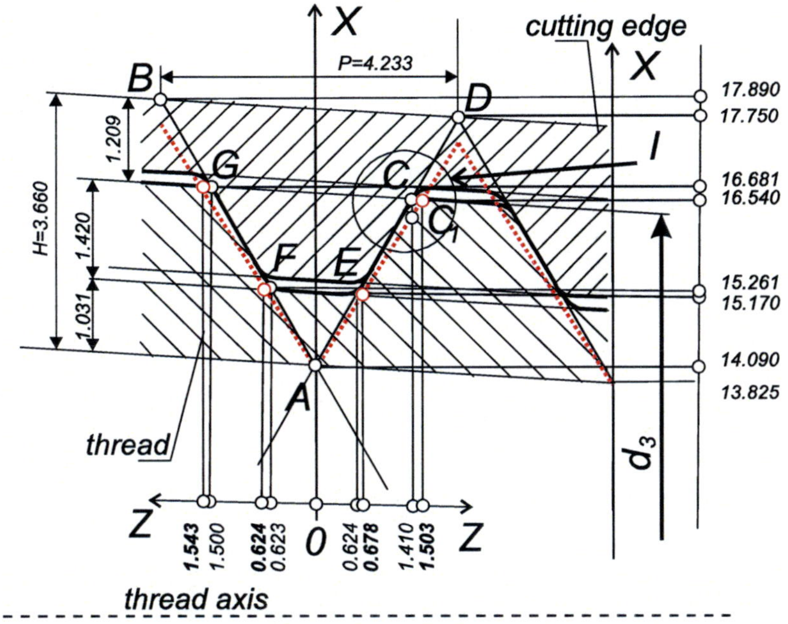

Fig. 7. Projected thread profile NC23 long side.

Table 2. Coordinates of the nodal points of the nominal profile of the NC23 thread and the predicted profile turned with a cutter with a rake angle of −7°.

Point	X	Z1	Z2	$a = Z2 - Z1$
A	14.090	0	0	0
B	17.890	2.180	2.190	0.010
C	16.541	1.410	1.430	0.020
D	17.750	2.110	2.140	0.030
E	15.121	0.589	0.597	0.008
F	15.261	0.675	0.685	0.010
G	16.681	1.500	1.510	0.010

Therefore, according to formula (4), the half-profile angle $\alpha 1/2 = 30.27°$. Thus, a deviation of 0.27° is 40% of the tolerance of 0.67°.

5 Conclusions

Using a cutter with a rake angle of −7° for turning an NC13 drill-string tool-joint thread can lead to a deviation from the nominal value of the half-profile angle of this thread – 0.27°, which is 40% of the size tolerance.

Avoidance of such a deviation is possible due to using a cutter with a zero value of the rake angle on the final infeed pass of 0.035 mm.

We will study a double-point tool for turning threads in the nearest feature.

Acknowledgment. The authors are grateful to the Ministry of Science and Education of Ukraine for the grant to implement project D-2-22-P (RK 0122U002082). The team of authors expresses their gratitude to the reviewers for valuable recommendations that have been taken into account to improve significantly the quality of this paper.

References

1. Liakh, M., et al.: Environmentally-efficient approaches to oil and gas producing sites. In: E3S Web of Conferences, vol. 280, p. 09002 (2021). https://doi.org/10.1051/e3sconf/202128009002
2. Lyu, H.M., Sun, W.J., Shen, S.L., Zhou, A.N.: Risk assessment using a new consulting process in fuzzy AHP. J. Constr. Eng. Manag. **146**(3), 04019112 (2020). https://doi.org/10.1061/(ASCE)CO.1943-7862.0001757
3. Poberezhna, L., Shkitsa, L., Poberezhnyi, L., Popovych, P., Brych, V., Shevchuk, O.: Assessment of environmental risks using elements of fuzzy logic. In: 2022 12th International Conference on Advanced Computer Information Technologies, ACIT 2022, pp. 219–222 (2022). https://doi.org/10.1109/ACIT54803.2022.9913179
4. Arkhypova, L.M., Mandryk, O.M., Moskalchuk, N.M., Prykhodko, M.M., Radlovska, K.O.: Renewable energy resources in the system of sustainable development of Carpathian region of Ukraine. J. Phys. Conf. Ser. **1781**, 012010 (2021). https://doi.org/10.1088/1742-6596/1781/1/012010
5. Kravchynskyi, R., Korchemlyuk, M., Khilchevskyi, V., Arkhypova, L., Mykhailyuk, I., Mykhailyuk, J.: Spatial-factorial analysis of background status of the Danube River basin state on the northeastern slops of the Ukrainian Carpathians. J. Phys. Conf. Ser. **1781**(1), 012011 (2021). https://doi.org/10.1088/1742-6596/1781/1/012011
6. An, Q.L., Guo, G.G., Zheng, X.H., Chen, M., Liu, G., Zhang, Y.S.: Experimental study on cutting characteristics for buttress thread turning of 13%Cr stainless steel. Key Eng. Mater. **443**, 262–267 (2010). https://doi.org/10.4028/www.scientific.net/KEM.443.262
7. Shkitsa, L.Y., Kornuta, V.A., Kornuta, O.V., Bekish, I.O.: The model of informational space for innovation and design activities in the university. Sci. Innov. **15**(6), 14–22 (2019). https://doi.org/10.15407/scine15.06.014
8. Kiyko, S., Druzhinin, E., Prokhorov, O., Haidabrus, B.: Multi-agent model of energy consumption at the metallurgical enterprise. In: Ivanov, V., Trojanowska, J., Pavlenko, I., Zajac, J., Peraković, D. (eds.) DSMIE 2020. LNME, pp. 156–165. Springer, Cham (2020). https://doi.org/10.1007/978-3-030-50794-7_16
9. Shkitsa, L., Kornuta, V., Kornuta, O., Bekish, I., Bui, V.: Information support of design innovation activity of the technical university. Manag. Syst. Prod. Eng. **28**(2), 127–132 (2020). https://doi.org/10.2478/mspe-2020-0019
10. Dubei, O.Y., Tutko, T.F., Ropyak, L.Y., Shovkoplias, M.V.: Development of analytical model of threaded connection of tubular parts of chrome-plated metal structures. Metallofiz. Noveishie Tekhnologii **44**, 251–272 (2022). https://doi.org/10.15407/mfint.44.02.0251
11. Shats'kyi, I.P., Lyskanych, O.M., Kornuta, V.A.: Combined deformation conditions for fatigue damage indicator and well-drilling tool joint. Strength Mater. **48**, 469–472 (2016). https://doi.org/10.1007/s11223-016-9786-8

12. Shatskyi, I., Ropyak, L., Velychkovych, A.: Model of contact interaction in threaded joint equipped with spring-loaded collet. Eng. Solid Mech. **8**(4), 301–312 (2020). https://doi.org/10.5267/j.esm.2020.4.002
13. Zhu, X., Zhang, Z.: Design of an ultra-high torque double shoulder drill-pipe tool joint for extended reach wells. Nat. Gas. Ind. **4**(5), 374–381 (2017). https://doi.org/10.1016/j.ngib.2017.10.002
14. Pryhorovska, T., Ropyak, L.: Machining error influence on stress state of conical thread joint details. In: Proceedings of the International Conference on Advanced Optoelectronics and Lasers, CAOL 2019, pp. 493–497 (2019). https://doi.org/10.1109/CAOL46282.2019.9019544
15. Ivanov, V.: Process-oriented approach to fixture design. In: Ivanov, V., et al. (eds.) Advances in Design, Simulation and Manufacturing, pp. 42–50. Springer, Cham (2019). https://doi.org/10.1007/978-3-319-93587-4_5
16. Shats'kyi, I.P., Shopa, V.M., Velychkovych, A.S.: Development of full-strength elastic element section with open shell. Strength Mater. **53**, 277–282 (2021). https://doi.org/10.1007/s11223-021-00286-y
17. Dutkiewicz, M., Velychkovych, A., Shatskyi, I., Shopa, V.: Efficient model of the interaction of elastomeric filler with an open shell and a chrome-plated shaft in a dry friction damper. Materials **15**, 4671 (2022). https://doi.org/10.3390/ma15134671
18. Sonawane, G.D., Sargade, V.G.: Machinability study of duplex stainless steel 2205 during dry turning. Int. J. Precis. Eng. Manuf. **21**, 969–981 (2020). https://doi.org/10.1007/s12541-019-00305-8
19. Kusyi, Y.M., Stupnytskyy, V.V., Kuk, A.M., Topilnytskyy, V.G.: Development of the fundamental diagram of the formation and transformation of the products properties during their manufacturing. J. Phys. Conf. Ser. **1781**(1), 012027 (2021)
20. Kusyj, J., Kuk, A.: A centrifugal vibration strengthening method devised to improve technological reliability of machine parts. Eastern-Eur. J. Enterp. Technol. **1**(7), 41–51 (2015)
21. Kusyi, Y., Stupnytskyy, V.: Optimization of the technological process based on analysis of technological damageability of castings. In: Ivanov, V., Trojanowska, J., Pavlenko, I., Zajac, J., Peraković, D. (eds.) DSMIE 2020. LNME, pp. 276–284. Springer, Cham (2020). https://doi.org/10.1007/978-3-030-50794-7_27
22. Zhang, G., Guo, C.: Modeling flank wear progression based on cutting force and energy prediction in turning process. Procedia Manuf. **5**, 536–545 (2016). https://doi.org/10.1016/j.promfg.2016.08.044
23. Zaleta, O.M., Povstyanoy, O.Y., Ribeiro, L.F., Redko, R.G., Bozhko, T.Y., Chetverzhuk, T.I.: Automation of optimization synthesis for modular technological equipment. J. Eng. Sci. **10**(1), A6–A14 (2023). https://doi.org/10.21272/jes.2023.10(1).a2
24. Ivchenko, O., et al.: Method for an effective selection of tools and cutting conditions during precise turning of non-alloy quality steel C45. Materials **15**(2), 505 (2022). https://doi.org/10.3390/ma15020505
25. Chalyj, V., Moroz, S., Tkachuk, A., Zablotskyi, V., Trokhymchuk, I., Stelmakh, A.: Formation of bearings parts waviness in centerless mortise grinding on rigid supports. J. Eng. Sci. **10**(1), A15–A21 (2023). https://doi.org/10.21272/jes.2023.10(1).a3
26. Kusyi, Y., Kuzin, O., Kuzin, N.: The dependence of intergrain damageability of casting on the technological treatment route. Eastern-Eur. J. Enterp. Technol. **1**(5(79)), 39–47 (2016). https://doi.org/10.15587/1729-4061.2016.59845
27. Kusyi, Y.: Technological inheritability of product material using the criterion of technological damageability. In: Ivanov, V., Trojanowska, J., Pavlenko, I., Zajac, J., Peraković, D. (eds.) DSMIE 2021. LNME, pp. 52–61. Springer, Cham (2021). https://doi.org/10.1007/978-3-030-77719-7_6

28. Kusyi, Y.M., Kuk, A.M.: Investigation of the technological damageability of castings at the stage of design and technological preparation of the machine Life Cycle. J. Phys. Conf. Ser. **1426**(1), 012034 (2020)
29. Khoshdarregi, M.R., Altintas, Y.: Generalized modeling of chip geometry and cutting forces in multi-point thread turning. Int. J. Mach. Tools Manuf. **98**, 21–32 (2015). https://doi.org/10.1016/j.ijmachtools.2015.08.005
30. Koleva, S., Enchev, M., Szecsi, T.: Compensation of the deviations caused by mechanical deformations during machining of threads. Procedia Manuf. **13**, 480–486 (2017). https://doi.org/10.1016/j.promfg.2017.09.066
31. Baizeau, T., Campocasso, S., Fromentin, G., Rossi, F., Poulachon, G.: Effect of rake angle on strain field during orthogonal cutting of hardened steel with c-BN tools. Procedia CIRP **31**, 166–171 (2015). https://doi.org/10.1016/j.procir.2015.03.089
32. Yanda, H., Ghani, J.A., Haron, C.H.C.: Effect of rake angle on stress, strain and temperature on the edge of carbide cutting tool in orthogonal cutting using FEM simulation. ITB J. Eng. Technol. Sci. **42**(2), 179–194 (2010). https://doi.org/10.5614/itbj.eng.sci.2010.42.2.6
33. Onysko, O., Panchuk, V., Kopey, V.B., Havryliv, Y., Schuliar, I.: Investigation of the influence of the cutter-tool rake angle on the accuracy of the conical helix in the tapered thread machining. In: International Conference of Applied Sciences. Journal of Physics: Conference Series, vol. 1781, p. 012028 (2021). https://doi.org/10.1088/1742-6596/1781/1/012028
34. Threading. Tread turning tools. Sundwik Coromant, https://www.sandvik.coromant.com/ru-ru/tools/threading-tools/thread-turning-tools. Accessed 13 Feb 2023
35. Berglind, L., Ziegert, J.: Modulated tool path (MTP) machining for threading applications. Procedia Manuf. **1**, 546–555 (2015). https://doi.org/10.1016/j.promfg.2015.09.029
36. Costa, C.E., Polli, M.L.: Effects of the infeed method on thread turning of AISI 304L stainless steel. J. Braz. Soc. Mech. Sci. Eng. **43**, 253 (2021). https://doi.org/10.1007/s40430-021-02978-7
37. Hovorun, T.P., et al.: Physical-mechanical properties and structural-phase state of nanostructured wear-resistant coatings based on nitrides of refractory metals Ti and Zr. Funct. Mater. **26**(3), 548–555 (2019). https://doi.org/10.15407/fm26.03.548
38. Pavlenko, I., et al.: Using regression analysis for automated material selection in smart manufacturing. Mathematics **10**(11), 1888 (2022). https://doi.org/10.3390/math10111888
39. Onysko, O., Medvid, I., Panchuk, V., Rodic, V., Barz, C.: Geometric modeling of lathe cutters for turning high-precision stainless steel tapered threads. In: Ivanov, V., Trojanowska, J., Pavlenko, I., Zajac, J., Peraković, D. (eds.) Advances in Design, Simulation and Manufacturing IV, pp. 472–480. Springer, Cham (2021). https://doi.org/10.1007/978-3-030-77719-7_47

Integrated Process Model for Development and Manufacturing of Customized Orthopedic Implants

Vitalii Pasichnyk[1(✉)], Svitlana Burburska[1,2], Yuliia Lashyna[1], and Volodymyr Korenkov[1]

[1] Igor Sikorsky Kyiv Polytechnic Institute, 37, Beresteiskyi Ave., Kyiv 03056, Ukraine
v.pasichnyk@kpi.ua

[2] LLC "Osteonica", 23, Preobrazhenska str., Kyiv 03110, Ukraine

Abstract. Customized orthopedic implants are one of the first successful applications of additive technologies in the medical field, and their usage provides significant advantages for patient treatment. The design of customized orthopedic implants and their manufacturing based on advanced additive and subtractive technologies and quality assurance necessitate flexible production management that requires careful process planning for cost and time management. The paper uses the business process model and notation to discuss computer-integrated manufacturing modeling for customized implants. The study highlights the key business roles and their respective tasks and identifies five main stages of customized implant development and production: initial data acquisition, reverse engineering, design, prototyping, and manufacturing. The required human, information, and material resources for each stage were analyzed, and the dependencies of time and cost were obtained. Optimization tasks based on time or cost minimization criteria were formulated for development and manufacturing. The results obtained in this study can provide a foundation for efficient time and cost management.

Keywords: Additive Technologies · 3D Printing · Computer Integrated Manufacturing · Endoprosthesis · Customized Implants · Process Innovation

1 Introduction

The use of customized orthopedic implants is highly desirable in many cases, compared to standard products with similar purposes, and offers significant advantages in terms of patient treatment. This approach can notably shorten operation duration and reduce the risk of postoperative complications related to aseptic loosening, i.e., the separation between the implant and the bone, which often occurs due to differences in geometry [1]. However, the widespread use of individual orthopedic implants has been limited for various reasons until recently. Firstly, there have been insufficient technological capabilities of the available equipment. Secondly, there has been a lack of understanding of the features and interconnection of all stages of production process planning and manufacturing of parts [2]. These limitations have increased the time and cost required

V. Tonkonogyi et al. (Eds.): InterPartner 2023, LNME, pp. 193–208, 2024.
https://doi.org/10.1007/978-3-031-42778-7_18

to manufacture such products. With the emergence and development of additive technologies, their popularity in biomedicine has been overgrowing [3]. These technologies offer a significant advantage over conventional ones by allowing for unit production of products with complex geometric shapes using universal manufacturing equipment [4]. This approach is either impossible or extremely expensive using conventional technologies. Such additive process features offer broad product customization opportunities, including products used for medical purposes [5]. From a practical standpoint, information management and manufacturing products with a high degree of uniqueness are two crucial tasks of orthopedic surgery [6].

The production of individual orthopedic implants is a technically complex task that involves processing a wide range of information, such as medical, design, production, and management data. Increasing the degree of customization can improve the reliability of surgery while reducing the duration and overall cost of treatment [7]. However, this degree of customization adds a high level of complexity and variability to the product, complicating design and manufacturing process planning processes, and necessitates the development of specialized management procedures to ensure product quality and safety [6]. Two critical tasks are resource planning during individual implant production [1] and process and data modeling [8].

2 Literature Review

Design and manufacturing processes are an integral part of the value chain. Therefore, a business process model can be the fundamentals for managing product development and optimizing production processes.

Several methods are available for modeling business processes, and among the most popular today is the Business Process Model and Notation (BPMN) system [9]. This system represents business processes through diagrams using basic elements to describe complex semantic constructs. Work operations, key participants and performers, and data flow, including those occurring in the additive manufacturing process, could be integrated into a unified system using BPMN, making it an effective tool for process modeling [10].

The Unified Modeling Language (UML), a graphical description language for object-oriented modeling, was initially created for designing and documenting software systems [11]. However, it can also be effectively used for modeling business processes and system design. In additive manufacturing, the UML graphical description can be utilized to depict the hierarchy of processes and their interactions [8].

In modeling complex systems, the IDEF methodologies can be applied [12]. The IDEF0 functional modeling methodology can be used to study the system structure. Each system function can be portrayed as a separate process with IDEF3 (Process Description Capture), a methodology for documenting the processes occurring in the system. It can describe the sequence of operations for each process and the logic of their implementation. These methodologies are extensively used for describing production systems using additive technologies [13].

To the best of the author's knowledge, there is currently no comprehensive model available for developing a customized implant from the initial patient contact to the delivery of the final product. The expected impact of the study is to reduce the production time of custom orthopedic implants through rational resource planning and the elimination of unnecessary iterations and, as a result, their cost reduction.

3 Research Methodology

This study aims to examine and evaluate the various stages, components, and parameters involved in the development and manufacturing of customized implants using additive technologies. Based on this analysis, a comprehensive computer-integrated model was developed to optimize the entire process in terms of time and cost, from the initial data preparation to the start of surgery. The following steps were taken to accomplish this goal: review and analysis of existing practices for implementing the production process, selection of a suitable business process modeling methodology, identification of key factors that affect resource consumption during the production process, and investigation of the nature of interrelationships between main processes.

Based on the analysis of production practices at the Laboratory of Biomedical Engineering of the Institute of Traumatology and Orthopaedics of the National Academy of Medical Sciences of Ukraine [2] and LLC "Osteonica" [14], a comprehensive model of the production process was developed.

Since the study covered different aspects of the production process, from customer contact to shipment of the packaged product, the decision was made to use the BPMN business process modeling methodology [10].

The process involves three key business roles, which are described below.

Medical Service is responsible for all tasks with a medical focus, such as patient consultations, prescription of medical tests, and planning of surgeries. As the product's customer, this business role is responsible for model verification and accepting the manufacturing process results.

Development is responsible for implementing all stages of product design, including prototyping, and manages the entire product life cycle. In addition, the developer acts as a customer for manufacturing processes.

Manufacturing is a generalized business role that encompasses all contractors involved in manufacturing. In practice, it can be represented by several companies specializing in additive manufacturing, machining, heat treatment, cleaning, and coating.

The resources (R) required to produce a custom implant have been categorized into Human Resources, IT Resources, and Material Resources. In the following sections, each group will be discussed in detail. The amount of use (Resource.amount) and the cost per unit of amount (Resource.rate) are the characteristics attributed to each resource. The total resource usage cost is calculated as follows:

$$\text{Resource.cost} = \text{Resource.amount} \times \text{Resource.rate}.$$

Human Resources (H). This paper focuses on the critical specialists responsible for carrying out Medical (H_{MD}), Development (H_D), and Manufacturing tasks (H_M). A detailed description of these resources is provided in Table 1. Resources needed for support processes, such as project management, accounting, and supply chain management, are not considered as a part of this study.

Table 1. Human resources categories.

Medics	Development Engineers	Manufacturing Engineers
H_{MD}^{A} – Attending Doctor	H_{D}^{RE} – Reverse Engineer	H_{M}^{PP} – Process Planner
H_{MD}^{S} – Surgeon Doctor	H_{D}^{CAD} – Design Engineer	H_{M}^{AE} – AM Engineer
H_{MD}^{R} – Radiologist	H_{D}^{FEA} – FEA Engineer	H_{M}^{AO} – AM Operator
	H_{D}^{RP} – Rapid Prototyping Specialist	H_{M}^{AP} – Post-processing operator
		H_{M}^{CAM} – CAM Engineer
		H_{M}^{CNC} – CNC-machine operator
		H_{M}^{Gr} – Grinding machine operator
		H_{M}^{Cl} – Cleaning machine operator
		H_{M}^{Ct} – Coating specialist
		H_{M}^{St} – Sterilizer operator
		H_{M}^{P} – Packing and shipment responsible person

IT Resources (I) can be categorized into three distinct groups: Software (I_{SW}), Knowledge (I_{Kn}), and Computing (I_{C}). Specifically, the study examines the necessary IT resources for two process categories, namely Development and Manufacturing. Table 2 provides a detailed description of these resources.

Table 2. IT resources categories.

Development		Manufacturing
I_{SW}^{Im} – 3D Imaging Software	I_{Sw}^{RP} – Slicing Software	I_{SW}^{AM} – AM software tool
I_{SW}^{CAD} – CAD Software	I_{Kn}^{DB} – Database of best practices	I_{SW}^{CAM} – CAM Software
I_{SW}^{CAE} – CAE Software	I_{C}^{FEA} – Computing resource for FEA	

Material Resources (M). This category considers the following types of resources: equipment (M_{Eq}) and materials (M_{M}). A detailed description of these resources is presented in Table 3.

The present paper explores the content, characteristics, impact, and limitations of the abovementioned elements.

The final focus will be on identifying the time characteristics by utilizing the objective function of minimizing the development cycle and manufacturing of a customized implant before the surgical operation. Additionally, this study seeks to determine the cost components and their corresponding amounts.

Table 3. Material resources categories.

Medics	Development Engineers	Manufacturing Engineers
M_{Eq}^{CT} – CT scanner M_{Eq}^{MRI} – MRI scanner	M_{Eq}^{RP} – 3D printer for rapid prototyping (FDM) M_{Eq}^{F} – 3D printer for fixture (SLA/DLP/LCD) M_{M}^{RP} – prototype material M_{M}^{F} – fixture material	M_{Eq}^{AM} – Additive Machine M_{Eq}^{F} – Furnace for heat treatment M_{Eq}^{CNC} – CNC Machining Center M_{CT}^{CNC} – Cutting tools for machining M_{F}^{CNC} – Fixture for machining M_{Eq}^{Gr} – Grinding machine M_{Eq}^{Cl} – Cleaning equipment M_{Eq}^{St} – Sterilizer; M_{M}^{Pr} – product material M_{M}^{Ct} – coating material

4 Results and Discussion

The product development and manufacturing process can be represented in five consecutive stages: 1 – obtaining initial data; 2 – reverse engineering; 3 – design; 4 – prototyping; 5 – manufacturing. At each stage, specific business roles perform certain tasks and interact by exchanging data and prototyping/manufacturing results with one another. The tasks and characteristics of the business roles at each stage are defined below.

Stage I. Obtaining initial data commences with a visit by the patient to a medical facility, as depicted in Fig. 1.

The medical advice provided (task T_{MD1}^{I}) entails analyzing the patient's health status, prescribing any necessary medical tests, such as computed tomography (CT) and/or magnetic resonance imaging (MRI), and carrying out the required tests. Subsequently, the need for a custom implant is evaluated (gateway, G_{MD1}^{I}), and if such a need exists, a request, together with initial data comprising a set of CT/MRI scans, is sent to an organization (task T_{MD2}^{I}) responsible for designing and manufacturing the implant and surgical guide. The development process begins with the acquisition of the CT/MRI scans (task T_{D1}^{I}).

Resources:

$$R_{MD1}^{I} = \left\{H_{MD}^{A} \wedge H_{MD}^{S} \wedge H_{MD}^{R}, M_{Eq}^{CT} \wedge M_{Eq}^{MRI}\right\} \tag{1}$$

Stage completion time:

$$T_{MD1}^{I}.time = H_{MD}^{A}.time + H_{MD}^{S}.time + H_{MD}^{R}.time + M_{Eq}^{CT}.time + M_{Eq}^{MRI}.time - t' - t'', \tag{2}$$

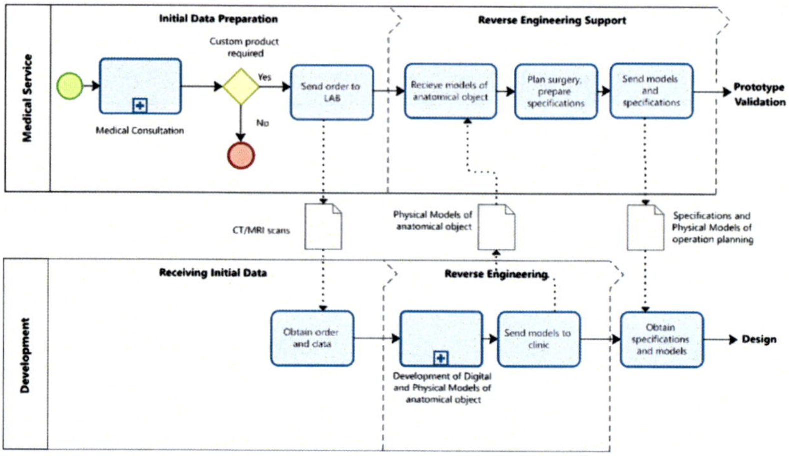

Fig. 1. Interaction of business roles at the initial data preparation and reverse engineering.

where t' is the duration of simultaneous use of resources H_{MD}^{R} and M_{Eq}^{CT}, and t'' is the duration of simultaneous use of resources H_{MD}^{R} and M_{Eq}^{MRI}.

Cost of resource usage:

$$T_{MD1}^{I}.cost = H_{MD}^{A}.cost + H_{MD}^{S}.cost + H_{MD}^{R}.cost + M_{Eq}^{CT}.cost + M_{Eq}^{MRI}.cost \quad (3)$$

Result: decision regarding the development of a customized implant, set of CT/MRI scans (data, D_{MD}^{I}). Since Stage I is not divided into substages, its integral characteristics align fully with those described above, namely:

Stage I resources: $R^{I} = \{R_{D1}^{I}\}$. Stage I duration: $T^{I}.time = T_{MD1}^{I}.time$.

Stage I cost of resource usage: $T^{I}.cost = T_{MD1}^{I}.cost$.

Stage II. Reverse engineering begins with the creation of a virtual model of an anatomical object through processing a set of images gathered via CT/MRI diagnostics (task T_{D1}^{II}).

Resources:

$$R_{D1}^{II} = \left\{H_{D}^{RE}, I_{SW}^{Im} \wedge I_{SW}^{CAD}\right\} \quad (4)$$

Task completion time: $T_{D1}^{II}.time = H_{D}^{RE}.time$.

Cost of resource usage:

$$T_{D1}^{II}.cost = H_{D}^{RE}.cost + I_{SW}^{Im}.cost + I_{SW}^{CAD}.cost \quad (5)$$

The result of this task is a digital model (CAD model, M_{AO}) of the anatomical object, presented in the form of an STL, VRML, PLY, or DXF file.

The next task entails producing a full-scale model (task T_{D2}^{II}), utilizing the STL file obtained in the previous step as input data. The full-scale model (physical object, O_{D}^{II}) is created from plastic using additive FDM/SLS technologies or photopolymer resin using SLA technology. Subsequently, the final product is transferred to the medical institution (task T_{D3}^{II}).

Resources:

$$R_{D2}^{II} = \left\{H_{D}^{RP}, I_{SW}^{RP}, M_{Eq}^{RP}, M_{M}^{Pr}\right\} \tag{6}$$

Task completion time:

$$T_{D2}^{II}.time = H_{D}^{RP}.time + M_{Eq}^{RP}.time - t' \tag{7}$$

where t' is the duration of simultaneous use of resources H_{D}^{RP} and M_{Eq}^{RP}

Cost of resource usage:

$$T_{D2}^{II}.cost = H_{D}^{RP}.cost + M_{Eq}^{RP}.cost + M_{M}^{Pr}.cost + I_{SW}^{RP}.cost \tag{8}$$

Result: a full-scale model of the anatomical object (physical object, O_{MD}^{II}) in 1:1 scale.

For a medical institution, the first task of the second stage involves obtaining a full-scale model (task T_{MD1}^{II}), followed by the task of planning the surgical procedure while preparing the product requirements document (task T_{MD2}^{II}). This stage is crucial for developing a customized implant as it involves preparing necessary input data. This includes determining the location of the surgical target, identifying relevant surrounding structures, locating blood vessels and nerves in proximity to the implant site, determining the surgical access point and path to the designated target structure, and identifying the position of surgical instruments and probes required for treatment delivery and/or implant placement. The stage concludes with the transfer of the planning results (task T_{MD3}^{II}) to the development contractor in the form of a full-scale model marked with appropriate attachment sites and the product requirements document for the customized implant.

Resources: $R_{MD2}^{II} = \left\{H_{MD}^{S}\right\}$. Task completion time: $T_{MD2}^{II}.time = H_{MD}^{S}.time$.

Cost of resource usage: $T_{MD2}^{II}.cost = H_{MD}^{S}.cost$.

Result: a full-scale model of the anatomical object with the results of the surgery simulation (physical object, O_{MD}^{II}) and the product requirements document (data, D_{MD}^{II}).

An integral characteristics of Stage II:

Stage II resources:

$$R^{II} = \left\{R_{D1}^{II}, R_{D2}^{II}, R_{MD2}^{II}\right\} \tag{9}$$

Stage II duration:

$$T^{II}.time = T_{D1}^{II}.time + T_{D2}^{II}.time + T_{MD2}^{II}.time \tag{10}$$

Stage II cost of resource usage:

$$T^{II}.cost = T_{D1}^{I}.cost + T_{D2}^{II}.cost + T_{MD2}^{II}.cost \tag{11}$$

Stage III. Design (Fig. 2) begins after the initial data is received from the medical institution (task T_{D1}^{III}).

In addition to the digital model of the anatomical object and the product requirements document, the design must also consider the constraints and specifications of the manufacturing process. In order to obtain this information, the executor of the *Development* process is required to send a request to prospective manufacturers (task T_{D2}^{III}).

Resources: $R^{III}_{D1-2} = \{H^{CAD}_{D}\}$.
Task completion time: $T^{III}_{D1-2}.time = H^{CAD}_{D}.time$.
Cost of resource usage: $T^{III}_{D1-2}.cost = H^{CAD}_{D}.cost$.
Result: production request (data, D^{III}_{D1}).

The executor of the *Manufacturing* process processes the request (task T^{III}_{Mf1}), prepare a proposal (task T^{III}_{Mf2}), and send it to the developer (task T^{III}_{Mf3}).
Resources: $R^{III}_{Mf1-3} = \{H^{PP}_{\mathrm{M}}\}$.
Task completion time: $T^{III}_{Mf1-3}.time = H^{PP}_{\mathrm{M}}.time$.
Cost of resource usage: $T^{III}_{Mf1-3}.cost = H^{PP}_{\mathrm{M}}.cost$.

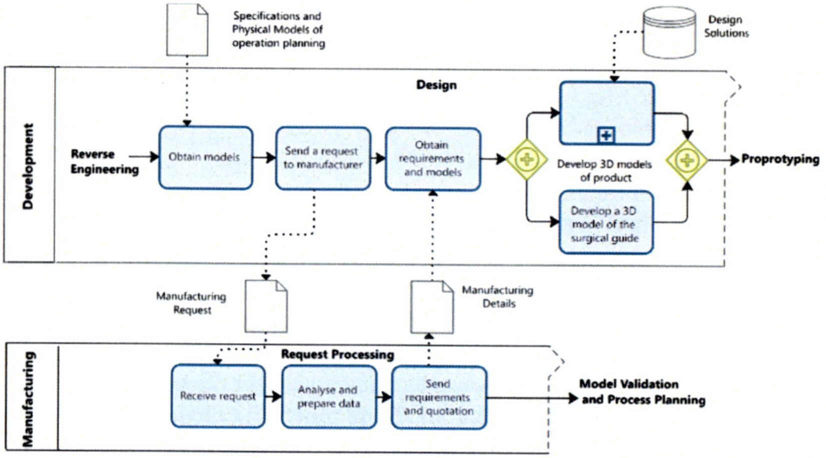

Fig. 2. Interaction of business roles at the design stage.

Result: proposal containing estimated timeframes for completing the work, preliminary cost estimation, as well as technical specifications of the equipment to be used, and recommendations on product design (data, D^{III}_{Mf}).

The executor of the Development process receives manufacturing data (task $\mathrm{T}^{\mathrm{III}}_{\mathrm{D3}}$) begins the tasks associated with the development of a customized implant (task $\mathrm{T}^{\mathrm{III}}_{\mathrm{D4}}$). The development involves using a database of standard solutions and design guidelines.

Resources:

$$R^{III}_{D3-4} = \left\{H^{CAD}_{D}, H^{FEA}_{D}, I^{CAD}_{SW}, I^{CAE}_{SW}, I^{DB}_{Kn}\right\} \quad (12)$$

Task completion time:

$$T^{III}_{D3-4}.time = H^{CAD}_{D}.time + H^{FEA}_{D}.time + I^{FEA}_{C}.time - t', \quad (13)$$

where $t\prime$ is the duration of simultaneous use of resources H^{FEA}_{D} and I^{FEA}_{C}

Cost of resource usage:

$$T^{III}_{D3-4}.cost = H^{CAD}_{D}.cost + H^{FEA}_{D}.cost + I^{CAD}_{SW}.cost + I^{CAE}_{SW}.cost + I^{FEA}_{C}.cost \quad (14)$$

Result: a comprehensive implant model (implant CAD model, M_{i1}) that will be used for prototyping and medical validation; a computational model (implant CAE model, M_{i2}) optimized for finite element calculations; and an additive manufacturing model (implant CAD model, M_{i3}) which may contain the necessary elements for subtractive machining and allowances for cutting, and do not include elements obtained at subsequent manufacturing stages. A model for the surgical guide is being developed simultaneously (task T_{D5}^{III}). Resources:

$$R_{D5}^{III} = \left\{ H_D^{CAD}, I_{SW}^{CAD}, I_{Kn}^{DB} \right\} \tag{15}$$

Task completion time: $T_{D5}^{III}.time = H_D^{CAD}.time$
Cost of resource usage:

$$T_{D5}^{III}.\text{cost} = H_D^{CAD}.cost + I_{SW}^{CAD}.cost + I_{Kn}^{DB}.cost \tag{16}$$

Result: a digital model of the surgical guide (CAD model, M_{sg}).
Integral characteristics of Stage III:
Stage III resources:

$$R^{III} = \left\{ R_{D1-2}^{III}, R_{Mf1-3}^{III}, R_{D3-4}^{III}, R_{D5}^{III} \right\} \tag{17}$$

Stage III duration:

$$T^{III}.time = T_{D1-2}^{III}.time + T_{Mf1-3}^{III}.time + max\left(T_{D3-4}^{III}.time, T_{D5}^{III}.time \right) \tag{18}$$

Stage III cost of resource usage:

$$T^{III}.cost = T_{D1-2}^{III}.cost + T_{Mf1-3}^{III}.cost + T_{D3-4}^{III}.cost + T_{D5}^{III}.cost \tag{19}$$

Stage IV. Prototyping (Fig. 3) begins with the additive manufacturing of the implant and the surgical guide prototypes utilizing one of the available FDM/SLS/SLA technologies. The resulting prototypes are subsequently transferred to the medical institution for validation (task T_{D1}^{IV}).

Resources:

$$R_{D1}^{IV} = \left\{ H_D^{RP}, I_{SW}^{RP}, M_{Eq}^{RP}, M_M^{RP}, M_{Eq}^F, M_M^F \right\} \tag{20}$$

Task completion time:

$$T_{D1}^{IV}.time = H_D^{RP}.time + M_{Eq}^{RP}.time + H_D^F.time + M_{Eq}^F.time - t' - t'', \quad (21)$$

where t' is the duration of simultaneous use of resources H_D^{RP} and M_{Eq}^{RP}, and t'' is the duration of simultaneous use of resources H_D^F and M_{Eq}^F.

Cost of resource usage:

$$T_{D1}^{IV}.cost = H_D^{RP}.cost + M_{Eq}^{RP}.cost + M_M^{RP}.cost + M_{Eq}^F.cost$$
$$+ M_M^F.cost + I_{SW}^{RP}.cost \quad (22)$$

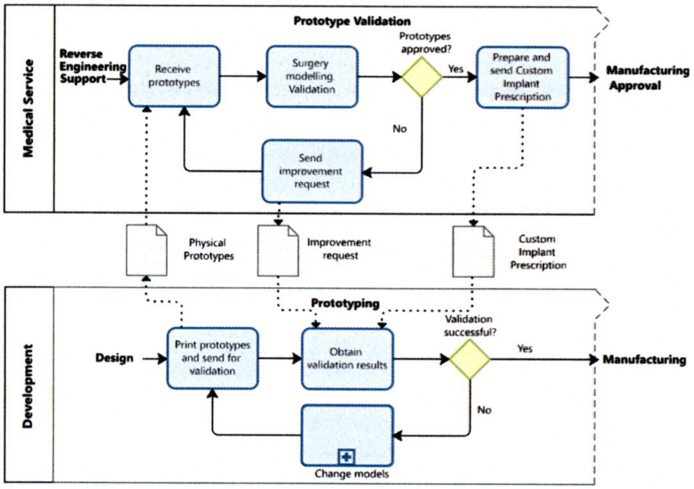

Fig. 3. Interaction of business roles at the prototyping stage.

Result: the implant prototype (physical object, O_{D1}^{IV}) and the surgical guide (physical object, O_{D2}^{IV}).

The medical institution receives the models (task T_{MD1}^{IV}), performs the simulation of the surgical procedure (task T_{MD2}^{IV}) and analyzes the results (gateway, G_{MD1}^{IV}).

Resources: $R_{MD1-2}^{IV} = \{H_{MD}^{S}\}$. Task completion time: $T_{MD1-2}^{IV}.time = H_{MD}^{S}.time$.

Cost of resource usage: $T_{MD1-2}^{IV}.cost = H_{MD}^{S}.cost$.

Result: depending on the simulation results, two potential scenarios exist for the subsequent course of the process.

The simulation result is not satisfactory $G_{MD1}^{IV} = false$. If there is a need to modify the design, the medical institution will communicate its recommended changes to the developer (task T_{MD3}^{IV}).

Resources: $R_{MD3}^{IV} = \{H_{MD}^{S}\}$. Task completion time: $T_{MD3}^{IV}.time = H_{MD}^{S}.time$.

Cost of resource usage: $T_{MD3}^{IV}.cost = H_{MD}^{S}.cost$.

The developer makes changes to the model (task T_{D2}^{IV}), reprints it, and submits it for re-validation.

Resources: $R_{D2}^{IV} = \{H_{D}^{CAD}\}$. Task completion time: $T_{D2}^{IV}.time = H_{D}^{CAD}.time$.

Cost of resource usage: $\mathrm{T}^{\mathrm{IV}}_{\mathrm{D2}}.\mathrm{cost} = \mathrm{H}^{\mathrm{CAD}}_{\mathrm{D}}.\mathrm{cost}$.

In the event of reprinting, all resources with their respective time and cost correspond to the task $\mathrm{T}^{\mathrm{IV}}_{\mathrm{D1}}$. Re-simulation of a surgical procedure: task $\mathrm{T}^{\mathrm{IV}}_{\mathrm{MD1}}$, task $\mathrm{T}^{\mathrm{IV}}_{\mathrm{MD2}}$, gateway, $\mathrm{G}^{\mathrm{IV}}_{\mathrm{MD1}}$.

The simulation result is satisfactory $\mathrm{G}^{\mathrm{IV}}_{\mathrm{MD1}} = \mathrm{true}$. In this case, the doctor will provide a Custom Implant Prescription ($\mathrm{D}^{\mathrm{IV}}_{\mathrm{MD1}}$) which includes the patient's information, the physician's information, and the technical specifications of the designed product (task $\mathrm{T}^{\mathrm{IV}}_{\mathrm{MD4}}$). Subsequently, the next stage of the process commences.

Resources: $\mathrm{R}^{\mathrm{IV}}_{\mathrm{MD4}} = \{\mathrm{H}^{\mathrm{S}}_{\mathrm{MD}}\}$. Task completion time: $\mathrm{T}^{\mathrm{IV}}_{\mathrm{MD4}}.\mathrm{time} = \mathrm{H}^{\mathrm{S}}_{\mathrm{MD}}.\mathrm{time}$.

Cost of resource usage: $\mathrm{T}^{\mathrm{IV}}_{\mathrm{MD4}}.\mathrm{cost} = \mathrm{H}^{\mathrm{S}}_{\mathrm{MD}}.\mathrm{cost}$.

Result: Custom Implant Prescription, $\mathrm{D}^{\mathrm{IV}}_{\mathrm{MD1}}$.

Integral characteristics of Stage IV:

Stage IV resources:

$$R^{IV} = \left\{R^{IV}_{D1}, R^{IV}_{MD1-2}, \left[R^{IV}_{MD3}, R^{IV}_{D2}\right], R^{IV}_{MD4}\right\} \tag{23}$$

Stage IV duration:

$$T^{IV}.time = T^{IV}_{D1}.time + T^{IV}_{MD1-2}.time + \left[T^{IV}_{MD1-3}.time + T^{IV}_{D1}.time + T^{IV}_{D2}.time\right] + T^{IV}_{MD4}.time \tag{24}$$

Stage IV cost of resource usage:

$$T^{IV}.cost = T^{IV}_{D1}.cost + T^{IV}_{MD1-2}.cost + \left[T^{IV}_{MD1-3}.cost + T^{IV}_{D1}.cost + T^{IV}_{D2}.cost\right] + T^{IV}_{MD4}.cost \tag{25}$$

Stage V. Manufacturing of a customized implant (Fig. 4) begins with the transfer of models and the product requirements document to the manufacturer (task T^{V}_{D1}).

Resources: $R^{V}_{D1} = \{H^{CAD}_{D}\}$. Task completion time: $T^{V}_{D1}.time = H^{CAD}_{D}.time$.

Cost of resource usage: $T^{V}_{D1}.cost = H^{CAD}_{D}.cost$.

The manufacturer validates the models and approves the product requirements document (task T^{V}_{Mf1}, T^{V}_{Mf2}). Following this, an evaluation report is sent to the developer. The report includes the results of the analysis, which may indicate any necessary design changes that should be implemented (T^{V}_{Mf3}).

Resources: $R^{V}_{Mf1-3} = \{H^{PP}_{M}\}$. Task completion time: $T^{V}_{Mf1-3}.time = H^{PP}_{M}.time$.

Cost of resource usage: $T^{V}_{Mf1-3}.cost = H^{PP}_{M}.cost$.

Result: evaluation report, D^{V}_{Mf1}.

The developer analyzes the manufacturer's recommendations (task T^{V}_{D2}) and, upon a positive decision (gateway, G^{V}_{D1}), makes changes to the design (task T^{V}_{D3}), approved by the medical institution.

Resources: $R^{V}_{D2-3} = \{H^{CAD}_{D}\}$. Task completion time: $T^{V}_{D2-3}.time = H^{CAD}_{D}.time$.

Cost of resource usage: $T^{V}_{D2-3}.cost = H^{CAD}_{D}.cost$. Result: digital model, O^{V}_{D1}.

If changes are requested to the design, the medical institution provides authorization for the proposed modifications (task T^{V}_{MD1}).

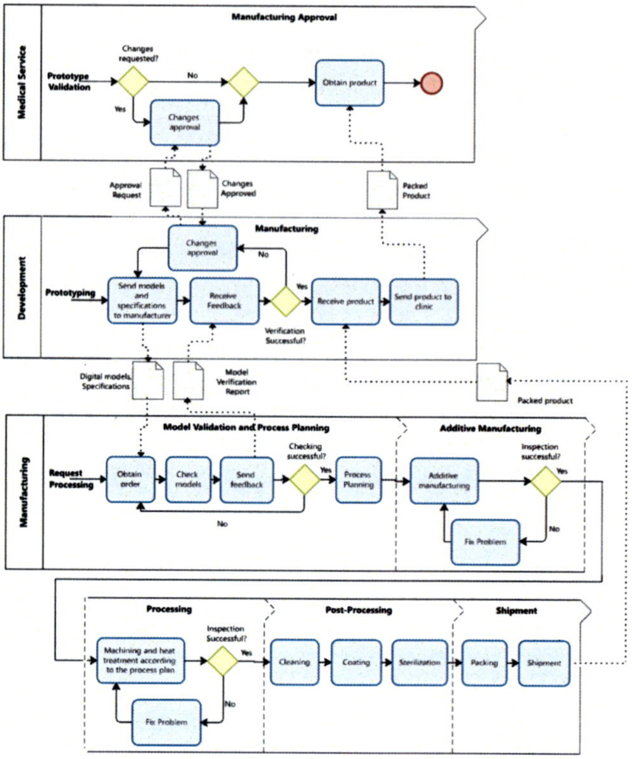

Fig. 4. Interaction of business roles at the manufacturing stage.

Resources: $R_{MD1}^{V} = \{H_{MD}^{S}\}$.

Task completion time: $T_{MD1}^{V}.time = H_{MD}^{S}.time$.

Cost of resource usage: $T_{MD1}^{V}.cost = H_{MD}^{S}.cost$.

Result: protocol of changes approval, D_{MD1}^{V}.

If the results of model validation are successful (gateway, G_{Mf1}^{V}), the manufacturer proceeds to the development of a process plan for the customized implant (task T_{Mf4}^{V}).

Resources: $R_{Mf4}^{V} = \{H_{M}^{PP}\}$. Task completion time: $T_{Mf4}^{V}.time = H_{M}^{PP}.time$.

Cost of resource usage: $T_{Mf4}^{V}.cost = H_{M}^{PP}.cost$. Result: process plan, D_{Mf2}^{V}.

The fabrication process initiates with the additive manufacturing of the component, encompassing tasks such as developing the control program, machine setup, additive build, and post-processing operations (task, T_{Mf5}^{V}). Subsequently, the fabricated component is assessed for conformance with the technical specifications and, if it meets the requirements (gateway, G_{Mf2}^{V}), proceeds to the next stage.

Resources:

$$R_{Mf5}^{V} = \left\{H_{M}^{AM}, H_{M}^{AO}, H_{M}^{AP}, I_{SW}^{AM}, M_{Eq}^{AM}, M_{Eq}^{F}, M_{M}^{Pr}\right\} \tag{26}$$

Task completion time:

$$T_{Mf5}^{V}.time = H_{M}^{AM}.time + H_{M}^{AO}.time + H_{M}^{AP}.time + M_{Eq}^{AM}.time + M_{Eq}^{F}.time$$

$$- t' - t'', \tag{27}$$

where $t\prime$ is the duration of simultaneous use of resources H_{Mf}^{AO} and M_{Eq}^{AM}, and t'' is the duration of simultaneous use of resources H_{Mf}^{AP} and M_{Eq}^{F}.

Cost of resource usage:

$$T_{Mf5}^{V}.cost = H_{M}^{AM}.cost + I_{SW}^{AM}.\text{cost} + H_{M}^{AO}.cost + H_{M}^{AP}.cost \\ + M_{Eq}^{AM}.cost + M_{Eq}^{F}.cost + M_{M}^{Pr}.cost \tag{28}$$

Result: fabricated component, O_{Mf1}^{V}.

Following the established manufacturing process, control programs for CNC machines are prepared and subtractive operations are implemented (task, T_{Mf6}^{V}). These operations may include milling, drilling, and grinding. The machined component is inspected for adherence to the technical specifications and, if compliant (gateway, G_{Mf3}^{V}), advances to the next phase.

Resources:

$$R_{Mf6}^{V} = \left\{ H_{M}^{CAM}, I_{SW}^{CAM}, H_{M}^{CNC}, M_{Eq}^{CNC}, M_{CT}^{CNC}, M_{F}^{CNC} \right\} \tag{29}$$

Task completion time:

$$T_{Mf6}^{V}.time = H_{M}^{CAM}.time + H_{M}^{CNC}.time + M_{Eq}^{CNC}.time + H_{M}^{Gr}.time + M_{Eq}^{Gr}.time \\ - t' - t'', \tag{30}$$

where $t\prime$ is the duration of simultaneous use of resources H_{Mf}^{CNC} and M_{Eq}^{CNC}, and t'' is the duration of simultaneous use of resources H_{Mf}^{Gr} and M_{Eq}^{Gr}.

Cost of resource usage:

$$T_{Mf6}^{V}.cost = H_{M}^{CAM}.cost + I_{SW}^{CAM}.cost + H_{M}^{CNC}.cost + M_{Eq}^{CNC}.cost + M_{CT}^{CNC}.cost \\ + M_{F}^{CNC}.cost + H_{M}^{Gr}.cost + M_{Eq}^{Gr}.cost \tag{31}$$

Result: machined component, O_{Mf2}^{V}.

Upon completion of cutting, the component undergoes cleaning (task, T_{Mf7}^{V}) and coating (task, T_{Mf8}^{V}) operations. Subsequently, the sterilization of the product may be performed either by the manufacturer or the medical institution (task, T_{Mf9}^{V}).

Resources:

$$R_{Mf7-9}^{V} = \left\{ H_{M}^{Cl}, H_{M}^{Ct}, H_{M}^{St}, M_{Eq}^{Cl}, M_{Eq}^{Ct}, M_{Eq}^{St}, M_{M}^{Ct} \right\} \tag{32}$$

Task completion time:

$$T_{Mf7-9}^{V}.time = H_{M}^{Cl}.time + H_{M}^{Ct}.time + H_{M}^{St}.time + M_{Eq}^{Cl}.time + M_{Eq}^{Ct}.time \\ + M_{Eq}^{St}.time - t' - t'' - t''' \tag{33}$$

where $t\prime$ is the duration of simultaneous use of resources H_M^{Cl} and M_{Eq}^{Cl}, t'' is the duration of simultaneous use of resources H_M^{Ct} and M_{Eq}^{Ct}, and t''' is the duration of simultaneous use of resources H_M^{St} and M_{Eq}^{St}.

Cost of resource usage:

$$T_{Mf7-9}^{V}.cost = H_M^{Cl}.cost + H_M^{Ct}.cost + H_M^{St}.cost + M_{Eq}^{Cl}.cost + M_{Eq}^{Ct}.cost + M_{Eq}^{St}.cost + M_M^{Ct}.cost \quad (34)$$

Result: finished product, O_{Mf3}^{V}.

At the final stage, the product is packaged and sent to the developer or a medical institution.

Resources: $R_{Mf10}^{V} = \{H_M^P\}$. Task completion time: $T_{Mf10}^{V}.time = H_M^P.time$.

Cost of resource usage: $T_{Mf10}^{V}.cost = H_M^P.cost$.

Result: packaged product, D_{Mf4}^{V}.

An integral characteristics of Stage V:

Stage V resources:

$$R^V = \left\{R_{D1}^V, R_{Mf1-3}^V, \left[R_{D2-3}^V\right], \left[R_{MD1}^V\right], R_{Mf4}^V, R_{Mf5}^V, R_{Mf6}^V, R_{Mf7-9}^V, R_{Mf10}^V\right\} \quad (35)$$

Duration of the stage V:

$$T^V.time = T_{D1}^V.time + T_{Mf1-3}^V.time + \left[T_{D2-3}^V.time\right] + \left[T_{MD1}^V.time\right] + T_{Mf4}^V.time + T_{Mf5}^V.time + T_{Mf6}^V.time + T_{Mf7-9}^V.time + T_{Mf10}^V.time \quad (36)$$

Stage V cost of resource usage:

$$T^V.cost = T_{D1}^V.cost + T_{Mf1-3}^V.cost + \left[T_{D2-3}^V.\text{cost}\right] + \left[T_{MD1}^V.cost\right] + T_{Mf4}^V.cost + T_{Mf5}^V.cost + T_{Mf6}^V.cost + T_{Mf7-9}^V.cost + T_{Mf10}^V.cost \quad (37)$$

The study has therefore identified the primary resources that directly impact the time required to manufacture a customized implant, from the initial patient request to the delivery of the final product to the medical facility. The optimization of the development and manufacturing process can be aimed at two objectives:

1. Minimizing time

$$\sum_{i=1}^{5} T_i.time \rightarrow min,$$

2. Minimizing cost

$$\sum_{i=1}^{5} T_i.cost \rightarrow min,$$

where T_i is the ith stage of the development and manufacturing process.

The selected BPMN method allows for presenting thoroughly the main stages of creating custom implants, key processes, such as additive manufacturing [8], key business roles, and their interaction. The developed model comprises the key resources necessary for the implementation of each stage, which determine the cost and duration of the processes, and therefore the optimization problem can be formulated. The limitation of the developed model is that it does not fully consider the implementation of the quality control and assurance system [6]. Future improvement of the model is possible by supplementing it with additional resources - information and human, as well as processes related to quality management. In addition, the model can be further advanced for the parallel development and production of several different custom implants.

5 Conclusions

This paper presents a computer-integrated manufacturing model for creating customized implants using additive technologies, which employs the BPMN system for modeling the business process. The process involves three main business roles: a medical institution, a customized implant developer, and a manufacturer. To complete this complex technical task, these roles interact through the exchange of information and the results of their activities throughout five stages: obtaining initial data, reverse engineering, design, prototyping, and manufacturing. The paper analyses the key human, information, and material resources required for the successful completion of each task and presents the duration and cost of each task dependent on the required time and cost of resources at each stage. The optimization task is formulated for two criteria: time and cost of development and production. Future research will focus on improving and developing the proposed model, particularly in terms of parallel design and manufacturing process planning for several customized implants.

References

1. Hauser, M., King, R., Wysk, R., Harrysson, O.: Resource planning for direct fabrication of customized orthopedic implants using EBM technology. J. Manuf. Syst. **60**, 500–511 (2021). https://doi.org/10.1016/j.jmsy.2021.07.003
2. Pasichnyk, V., Kryvenko, M., Burburska, S., Haluzynskyi, O.: Design and engineering assurance for the customized implants production using additive technologies. In: Ivanov, V., Trojanowska, J., Pavlenko, I., Zajac, J., Peraković, D. (eds.) DSMIE 2021. LNME, pp. 81–94. Springer, Cham (2021). https://doi.org/10.1007/978-3-030-77719-7_9
3. Jakus, A.E.: An introduction to 3D printing - past, present, and future promise. In: Dipaola, M., Wodajo, F. M. (eds.) 3D Printing in Orthopaedic Surgery, pp 1−15. Elsevier (2019). https://doi.org/10.1016/B978-0-323-58118-9.00001-4
4. Cronskär, M., Bäckström, M., Rännar, L.: Production of customized hip stem prostheses – a comparison between conventional machining and electron beam melting (EBM). Rapid Prototyping J. **19**(5), 365–372 (2013). https://doi.org/10.1108/RPJ-07-2011-0067
5. Saptarshi, S.M., Zhou, Ch.: Basic of 3D printing: engineering aspect. In: Dipaola, M., Wodajo, F.M. (eds.) 3D Printing in Orthopaedic Surgery, pp. 17−30. Elsevier (2019). https://doi.org/10.1016/B978-0-323-58118-9.00002-6

6. Martinez-Marquez, D., Mirnajafizadeh, A., Carty, C.P., Stewart, R.A.: Facilitating industry translation of custom 3D printed bone prostheses and scaffolds through Quality by Design. Procedia Manuf. **30**, 284–291 (2019). https://doi.org/10.1016/j.promfg.2019.02.041
7. Zdravković, M.M., Stojković, M.S., Mišić, D.T., Trajanović, M.D.: Towards semantic interoperability framework for custom orthopaedic implants manufacturing. IFAC Proc. Vol. **45**(6), 1327–1332 (2012). https://doi.org/10.3182/20120523-3-RO-2023.00368
8. Belkadi, F., Vidal, L.M., Bernard, A., Pei, E., Sanfilippo, E.M.: Towards an unified additive manufacturing product-process model for digital chain management purpose. Procedia CIRP **70**, 428–433 (2018). https://doi.org/10.1016/j.procir.2018.03.146
9. Wang, W., Ding, H., Dong, J., Ren, C.: IBM research report. A comparison of business process modeling methods (2006). https://dominoweb.draco.res.ibm.com/f8f910dcdf33d3008525719d0053e950.html. Accessed 26 Feb 2023
10. Business Process Model and Notation (BPMN) Version 2.0 (2011). http://www.omg.org/spec/BPMN/2.0. Accessed 26 Feb 2023
11. OMG Unified Modeling Language Version 2.5 (2015). https://www.omg.org/spec/UML/2.5. Accessed 26 Feb 2023
12. Serifi, V., Dašić, P., Ječmenica, R., Labović, D.: Functional and information modeling of production using IDEF methods. Strojniski Vest. – J. Mech. Eng. **55**(2), 131–140 (2013)
13. Feng, S.C., Witherell, P., Ameta, G., Kim, D.B.: Activity model for homogenization of data sets in laser based powder bed fusion. Rapid Prototyping J. **23**(01), 137–148 (2017). https://doi.org/10.1108/RPJ-11-2015-0160
14. Laboratory of Biomedical Engineering "Osteonica" Homepage. https://osteonica.com/en. Accessed 26 Feb 2023
15. Lantada, A.D., Morgado, P.L.: Enhancing product development through CT images, computer-aided design and rapid manufacturing: present capabilities, main applications and challenges. In: Homma, N. (eds.) Theory and Applications of CT Imaging and Analysis, pp. 269–290. InTech, Rijeka (2011). https://doi.org/10.5772/14134

Creation of a Combined Technology for Processing Parts Based on the Application of an Antifriction Coating and Deforming Broaching

Ihor Shepelenko[1(✉)], Yakiv Nemyrovskyi[2], Yaroslav Stepchyn[2], Sergii Mahopets[1], and Oleksandr Melnyk[2]

[1] Central Ukrainian National Technical University, 7, Universytetskyi Ave., Kropyvnytskyi 25006, Ukraine
kntucpfzk@gmail.com

[2] Zhytomyr Polytechnic State University, 103, Chudnivska Str., Zhytomyr 10005, Ukraine

Abstract. This paper presents the research results on developing a combined technology for processing parts, which integrates methods of applying antifriction coatings and surface plastic deformation. The study revealed that the existing technologies for finishing antifriction non-abrasive treatments aimed at enhancing the tribological characteristics of the surface layer do not effectively strengthen it. For this issue, the proposal was to employ deforming broaching, which can harden the surface layer of the part and enhance the bond strength between the coating and the base material. Experimental investigations were conducted to examine the feasibility of the combined technology using samples made of cast iron EN-GJL-200. The study focused on analyzing geometrical changes resulting from deforming broaching and the alteration of physical-mechanical properties associated with implementing finishing antifriction abrasion-free machining operations. These investigations facilitated the development of various combined processing technologies, encompassing finishing antifriction abrasion-free processing and deforming broaching. Based on the obtained results, a technological process for restoring cylinder liners of internal combustion engines was devised. A comparison between the existing and proposed technological processes demonstrated the advantages and prospects offered by the proposed approach.

Keywords: Combined Technology · Finishing Antifriction Non-Abrasive Treatment · Antifriction Coatings · Deforming Broaching · Surface Quality · Roughness · Strengthening · Manufacturing Innovation

1 Introduction

Improving the quality of engineering products depends on the surface layer properties of the parts working surfaces included in the product. When choosing the material of the workpiece and its processing technology, the core, and the part's surface layer's functions should be distinguished. Such a design-technological concept of creating products

V. Tonkonogyi et al. (Eds.): InterPartner 2023, LNME, pp. 209–218, 2024.
https://doi.org/10.1007/978-3-031-42778-7_19

is strategic and universal because it dominates throughout their entire life cycle: during design, production, and operation, as well as during the restoration of individual components and parts [1].

Numerous methods have been developed to modify the operational properties of working surfaces by leveraging a combination of geometric and physical-mechanical characteristics within the surface layer. Through the application of these methods, it becomes feasible to achieve the desired levels of accuracy, roughness, supporting surface area, and microrelief, as well as enhance the overall strengthening effect, residual stresses, microstructure, texture, adhesive properties of coatings and their bond strength with the substrate, and the longevity of the employed plasticity.

The combination of physical-mechanical and geometric characteristics in the surface layer obtained during the processing process enables the enhancement of various properties, such as wear resistance, fatigue strength, and antifriction properties, as well as the strength of tension seats, among others.

Based on the observations above, it can be confidently stated that the primary focus of contemporary mechanical engineering lies in developing novel engineering approaches for the surfaces of machine parts. Additionally, there is a growing emphasis on advancing combined technologies that facilitate control over the composition, structure, and properties of the working surface layer of these parts.

Consequently, it can be concluded that combined (hybrid) technologies are highly effective processes for the surface engineering of machine and mechanism parts, applicable not only in primary production but also in secondary operations such as repair and restoration.

2 Literature Review

To enhance reliability and longevity, it is imperative that every part, irrespective of its material of manufacture, possesses a protective coating that aligns with its intended purpose and operating conditions [2]. In this context, coatings represent intentionally formed surface layers that exhibit distinct properties differing from the base material, thereby significantly altering its characteristics. Concurrently, the overall dimensions of the part increase due to the thickness of the coating layer. This aspect enables the utilization of coating application methods during the production of new parts and facilitates the restoration of worn components and structures [3].

There are numerous existing methods of applying coatings, which are typically classified from various perspectives. In mechanical engineering, the authors categorize the coatings [4] into different types based on their intended purposes. These categories include wear-resistant, corrosion-resistant, antifriction, heat-resistant, heat-protective, sealing, and other types.

Antifriction coatings, characterized by a low coefficient of friction, hold a special significance among coatings due to their favorable antifriction properties when applied to friction surfaces [5]. Moreover, these coatings can serve various purposes, such as restoration, running-in, hard lubrication, and multifunctionality [6]. Among the various methods available for obtaining antifriction coatings, producing multi-functional coatings is particularly advantageous. This approach enables their use in dimensional restoration, surface run-in reduction, and as solid lubricants [7].

Finishing antifriction non-abrasive treatment (FANT) should be included among the methods of obtaining multi-functional coatings, the essence of which is the frictional interaction of the processed tool made of antifriction material (brass, copper, and bronze) with the surface of the part. As a result of this treatment, the friction surfaces of the parts are covered with a thin layer (1…5 μm) of brass, copper, bronze, or other antifriction solid lubricating materials [8].

At the same time, it should be noted that the existing FANT technologies do not provide sufficient parts surface strengthening and, therefore, wear resistance for a longer time. This, of course, limits the scope of using FANT as a technological method of processing parts.

The strength of antifriction coatings obtained through the Friction-Assisted Nanostructuring Technique (FANT) can be enhanced by incorporating surface plastic deformation (SPD) methods, which simultaneously strengthen the surface layer of the part. However, the lack of comprehensive research on the impact of SPD on the coating quality and the base material's surface hinders the successful integration of these processes. This issue is particularly significant when dealing with products composed of low-plasticity materials, such as cylinder liners of internal combustion engines (ICE) made of modified cast iron. The operational stability of these liners is heavily reliant on the condition of the surface layer, which is determined by technological processing methods.

Among the various methods of SPD, the method of deforming broaching (DB) has received wide application in the metalworking industry. This method's essence consists of successive graded plastic deformation of the part's inner surface during translational movement of the deforming elements through the hole being processed. DB successfully solves the issues of strengthening the surface layer and obtaining residual compressive stresses, ensuring high parts' operational properties [9].

Combining FANT and DB in a single technological process is a prerequisite for creating a combined technology that would improve the surface layer quality of the processed product. Implementing such technology requires conducting experimental studies and establishing the fundamental laws of the proposed technology.

This research aims to study the influence of FANT and DB on the geometric and physical-mechanical characteristics of the working surface when creating a hybrid technology.

The following tasks should be solved to achieve this goal:

- investigate the influence of DB and FANT on the geometrical and physical-mechanical properties of the processed surface;
- determine the effect of operations sequence on the quality characteristics of the processed surface;
- develop a technological process for processing holes in cast iron products using FANT and DB.

3 Research Methodology

Experiments were performed on bushings made of gray cast iron EN-GJL-200, the chemical composition and mechanical properties of which are presented in [10].

The internal surface of each bushing underwent the boring process, followed by the application of antifriction coating using Friction-Assisted Nanostructuring Technique

(FANT) per the prescribed guidelines [8]. The FANT technology was executed at the Central Ukrainian National Technical University (Kropyvnytskyi, Ukraine) using a specialized device designed for applying antifriction coatings through the friction-mechanical method (see Fig. 1) (Table 1).

Table 1. Chemical composition and mechanical properties of gray cast iron EN-GJL-200 [10].

Chemical Composition %					Mechanical Properties			
C	Cr	Si	Mn	Ti	Hardness, HB MPa	Strength limit, MPa	Poisson's ratio, μ	Elastic modulus, E, GPa
2.65	0.10	0.10	0.5	0.08	1700–1900	200	0.27	160

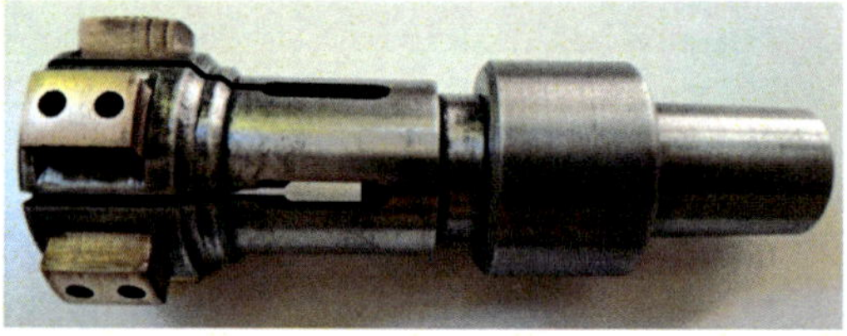

Fig. 1. Design of a tool for friction-mechanical application of an antifriction coating.

The tool (Fig. 1) has 5 brass inserts, which are brought into contact with the processed surface during simultaneous reciprocating movement. This ensures the formation of a brass coating in the contact areas. Next, the tool is turned to an angle that ensures the application of the coating on the areas that connect with the areas of the already applied coating. During the reciprocating movement of the assembly tool, the following areas with an antifriction coating are formed. This cycle is repeated until a solid antifriction coating covers the entire working surface. At the same time, the processing zone is constantly wetted with a technological environment, the composition of which is selected based on recommendations [11]: 12% solution of HCl in glycerin in a ratio of 1:3. The use of such a composition of the technological environment provides softening and dissolution of oxide films on the surfaces of the processed part and antifriction bars.

DB was carried out on a vertical-broaching machine mod. MA7U-750 uses the prefab constructions of broaches (Fig. 2), which allow the broaching process both with single working elements and groups of elements in different combinations [12].

These experiments were performed at the Institute for Superhard Materials named after VM Bakul (Kyiv, Ukraine).

Note that the DB was carried out according to two schemes:

- before applying the antifriction coating;
- after applying the antifriction coating.

At the same time, different sets of deforming elements made of HG420 hard alloys were used. This made it possible to perform DB according to the following options:

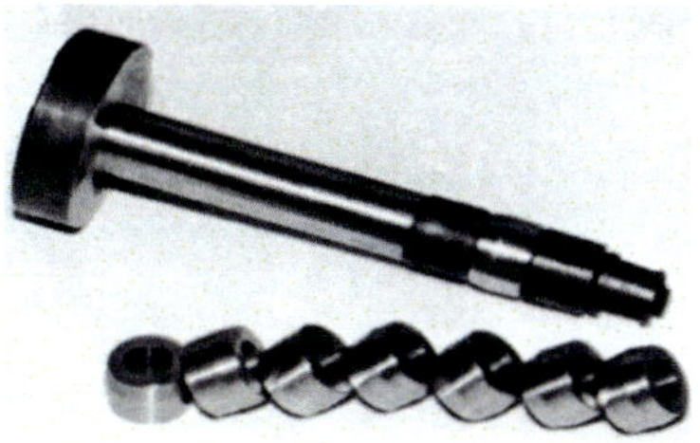

Fig. 2. General view of the deforming broaching.

- by one deforming element with the working cone angle $\alpha = 2°$ and tension $a = 0{,}1$;
- by two deforming elements, the working cone angle $\alpha = 4°$ and tension $a = 0.1$ on each element, which means the total tension – $\Sigma a = 0.2$.

The roughness of the surface, following the processes of boring, FANT, and DB, is evaluated using a Mahr XR20 profilograph. Additionally, the hardness of the surface layer is measured using the Vickers method on an HPO device with a load of 50N.

4 Results and Discussion

The conducted research made it possible to study the change in the height parameter of the roughness R_a and the microrelief of the working surface when processing the bushings according to the considered technologies (Figs. 3, 4).

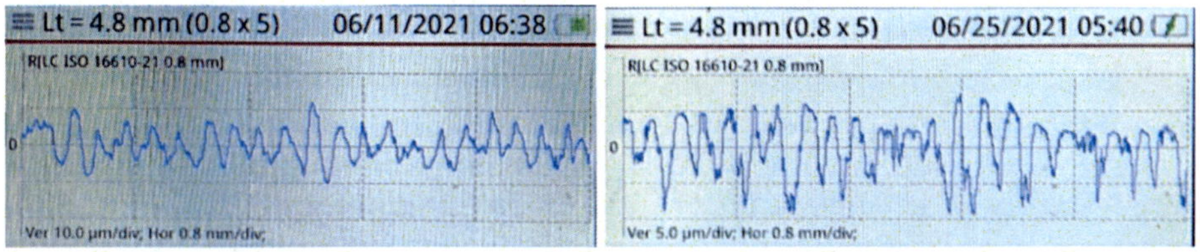

a) b)

Fig. 3. Microrelief of the processed inner surface of the bushing after operations: a) FANT; b) DB ($\alpha = 2°$, $a = 0{,}1$).

As follows from Fig. 3(a) and 4(a), after applying antifriction coatings by FANT to the inner surface of the bushings, the surface roughness changes insignificantly. The height parameter of the roughness decreased to the value of $R_a \approx 3.7$ µm. The peaks of the microroughness were somewhat smoothed, and the size of the support surface t_p increased by approximately 12%. The change in the height parameter R_a, during processing using the considered operations, is presented in Fig. 5.

After the DB operation, there were changes with the microrelief. Moreover, for a bushing broached by a deforming element with an angle $\alpha = 2°$, they are negligible (Fig. 3, b). This should be explained by the fact that the contact pressures reach the value $q = 0.3$ GPa. Therefore, in this case, according to recommendations [13], the antifriction

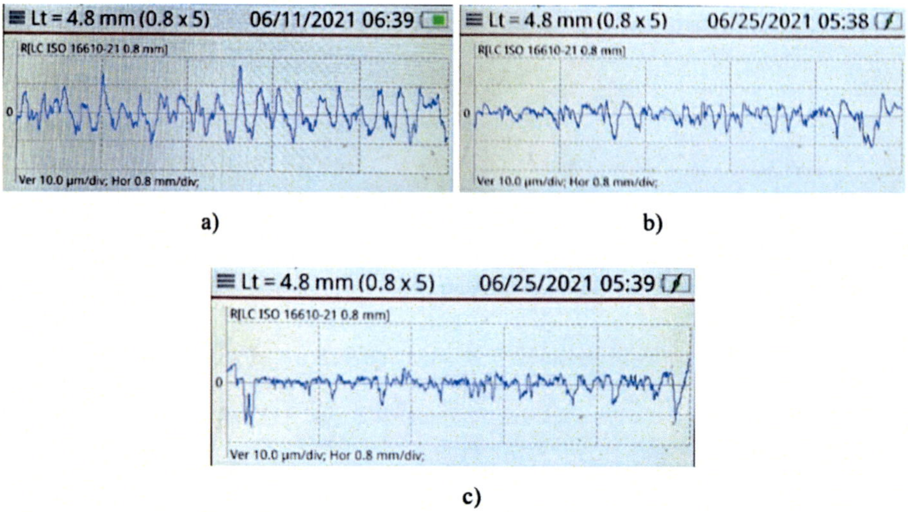

Fig. 4. Microrelief of the processed inner surface of the bushing after operations: a) FANT; b) DB ($\alpha = 4°$, $a = 0{,}1$); c) DB ($\alpha = 4°$, $\Sigma a = 0{,}2$).

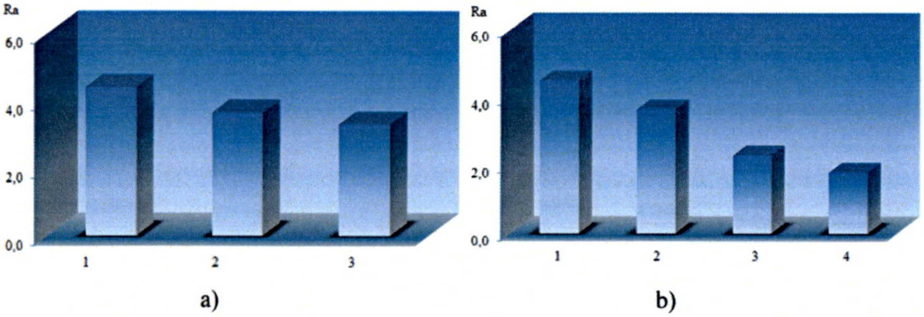

Fig. 5. Changing the R_a parameter during bushing processing using operations: a) 1 – boring; 2 – FANT; 3 – DB by one element ($\alpha = 2°$, $a = 0{,}1$); b) 1 – boring; 2 – FANT; 3 – DB with one deforming element ($\alpha = 4°$, $a = 0{,}1$); 4 – DB with two deforming elements ($\alpha = 4°$, $\Sigma a = 0{,}2$).

layer plays the role of solid lubrication, localizing shear deformations in the lubrication layer and changing the roughness slightly – $R_a \approx 3.7$ µm (Fig. 5, a, b). The microrelief of the processed by DB surface (Figs. 3, b, 4, b) does not significantly differ from the microrelief obtained by FANT.

The picture is somewhat different when the bushing is deformed by two elements with an angle $\alpha = 4°$ (Fig. 4, c). After the first element's passage, the contact pressure reaches the value $q = 0.85$ GPa, which exceeds the value of the critical contact pressure for this material [14]. Simultaneously, the microrelief corresponds to the microrelief inherent to the DB operation (Fig. 4, c). It represents flat areas with hollows that serve as reservoirs for lubrication. The value of the step parameter S_m decreases, which favorably affects wear resistance. The value of the support surface increases to $t_p \approx 60\%$, which also favorably affects wear resistance. The height parameter of the roughness has decreased and is $R_a = 2.3$ μm (Fig. 5, b).

After DB with two deforming elements, the character of the microrelief does not change, and the height parameter of the roughness decreases further, reaching a value of $R_a = 1.8$ μm (Fig. 5, b). Parameters S_m and t_p also increase. It should be noted that the obtained antifriction coating is characterized by high integrity (Fig. 6), which indicates its quality. The thickness of the antifriction layer was 5 μ.

The results indicate that for optimizing the surface layer, it is possible to devise a technological process by incorporating the DB operation after FANT, considering the specific operational requirements. It is essential to consider the potential use of an antifriction coating in the form of solid lubrication [15], which occurs under low contact pressures typically associated with thin-walled bushings processed by deforming elements featuring working cone tilt angles of $\alpha \leq 2°$.

Fig. 6. The cast iron bushing processed surface appeared after FANT and deforming broaching.

The second treatment option involves high contact pressures, where the shielding properties of the antifriction coating on the working surface are ineffective. In such cases, the original microrelief is transformed into a microrelief suitable for the DB process using liquid lubrication.

The research has facilitated the development of different variations of combined technology for processing parts, incorporating FANT and DB operations. Depending on the technical requirements of the product, each of these options can be utilized for processing holes.

The first option involves processing the hole by DB after the FANT operation. In this scenario, the antifriction coating is applied to the hole using the friction-mechanical method to create adhesive coating base regions [16]. Consequently, a grid of brass particles, securely attached through adhesion, forms on the hole's surface. Following this, the primary brass coating is applied, interacting with the adhesively fixed brass areas, thereby enhancing the quality of the coating applied through FANT. Subsequently, the DB treatment leads to improved adhesion between the coating and the base material and strengthening of the base itself [17].

Let us consider the II variant of the technological process, when the pre-bored surface (Fig. 7, a) is deformed by a working element with an angle of $\alpha \leq 2°$ to create platforms (Fig. 7, b), which serve as a place for adhesion of glued brass and, in general, for intensification of the adhesive grid application process. The next FANT operation fills the microrelief hollows with antifriction material (Fig. 7, c). It should be noted that after applying the antifriction coating, a repeat DB operation is possible, which is assigned depending on the technical requirements of the product.

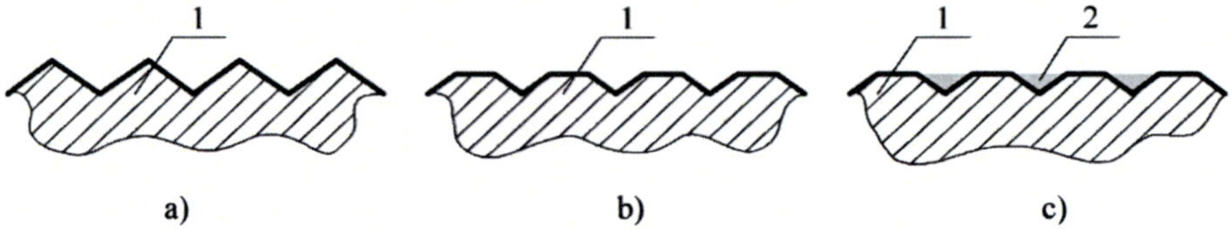

Fig. 7. Formation of the surface layer: a) boring; b) DB; c) FANT: 1 – base; 2 – antifriction coating.

Given that the parameters of the rough layer remain relatively unchanged after the FANT operation, particular attention should be given to the preceding operations before applying antifriction coatings. Consequently, when designing a technological process where FANT serves as the final operation, it is crucial to select a preliminary operation that enhances the microrelief and physical-mechanical characteristics of the surface layer, specifically in terms of wear resistance.

Considering the above, the following technological process of restoring internal combustion engine cylinder liners to the first repair size was developed and recommended (Fig. 9).

Fig. 8. Scheme of the technological process for restoration of ICE cylinder liners.

Following the outlined procedure, once the engine is disassembled, the cylinder liners undergo a thorough cleaning process to remove any dirt, scale, and defects present. During the inspection stage, liners exhibiting wear beyond the third repair size and those with cracks and chips are deemed unsuitable and consequently rejected.

Combined broaching is performed to create an optimal combination of mechanical and geometric parameters of the working surface.

FANT of the surface ensures the application of an antifriction coating.

A comparison of the physical-mechanical characteristics of the surface layer processed by the existing and proposed (Fig. 8) technological processes showed a significant difference between them (Fig. 9).

The DB operation substantially strengthens the liner material's surface layer, increasing by up to 25%. Additionally, the strengthening depth reaches approximately 0.3 mm,

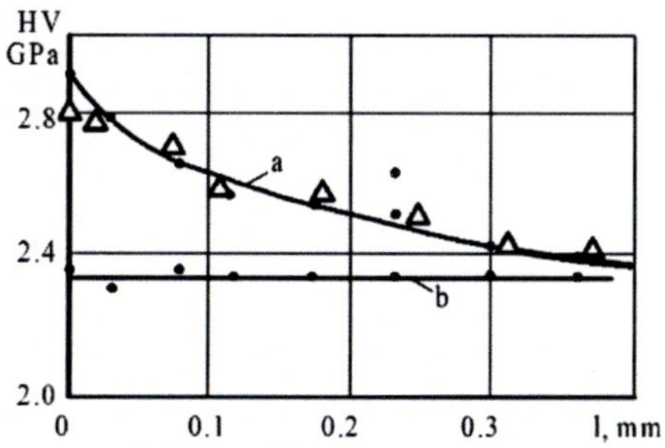

Fig. 9. Hardness distribution by the thickness of the ICE cylinder liner wall during processing: a – according to the new technological process (• – after broaching; Δ – after broaching and FANT); b – according to the existing technological process (boring and honing).

ensuring the presence of reinforced material within the friction pair, even during extended periods of operation. Therefore, the combination of DB and FANT has successfully led to the development of an effective technological process for processing the inner surface of the cylinder liner in internal combustion engines (ICE) made of gray-modified cast iron.

5 Conclusions

The studies showed the prospects of creating combined hole processing technologies based on DB and FANT.

Depending on the technical requirements for the part, the creation of technologies is implemented in the following two directions: DB operation forms the geometric and physical-mechanical properties of the surface layer and can be used before and after FANT operation.

Using FANT operation after DB increases the antifriction properties of the surface layer.

The implementation of DB following FANT generates the essential microrelief and enhances the surface layer's physical and mechanical properties. This combined process results in a substantial strengthening effect, with an increase of up to 25% in the surface layer's strength. The hardening depth achieved reaches 0.3 mm, enhancing the quality of the antifriction coating and improving the overall durability of the part.

A combined technological process for processing holes in cast-iron internal combustion engine liners has been developed, incorporating DB and FANT. In this process, DB plays a crucial role in forming the required microrelief and enhancing the physical-mechanical properties of the surface layer. On the other hand, FANT acts as a solid lubricant, improving the antifriction properties of the treated surface.

References

1. Huang, S., Xu, Z., Wang, G., et al.: Reconfigurable machine tools design for multi-part families. Int. J. Adv. Manuf. Technol. **105**, 813–829 (2019). https://doi.org/10.1007/s00170-019-04236-6

2. Galedari, S.A., Mahdavi, A., Azarmi, F., et al.: A comprehensive review of corrosion resistance of thermally-sprayed and thermally-diffused protective coatings on steel structures. J. Therm. Spray Tech. **28**, 645–677 (2019). https://doi.org/10.1007/s11666-019-00855-3
3. Meng, Y., Xu, J., Ma, L., et al.: A review of advances in tribology in 2020–2021. Friction **10**, 1443–1595 (2022). https://doi.org/10.1007/s40544-022-0685-7
4. Pashkov, D.M., Belyak, O.A., Guda, A.A., et al.: Reverse engineering of mechanical and tribological properties of coatings: results of machine learning algorithms. Phys. Mesomech. **25**, 296–305 (2022). https://doi.org/10.1134/S1029959922040038
5. Jha, S., Chen, Y., Renner, P., et al.: Design of anti-frictional ceramic-based composite coatings. J. Mater Eng. Perform. **31**, 3076–3093 (2022). https://doi.org/10.1007/s11665-021-06416-6
6. Bolotov, A., Novikova, O., Novikov, V.: Triboengineering properties of oxide coatings with antifriction fillers. In: Radionov, A.A., Gasiyarov, V.R. (eds.) ICIE 2022. LNME, pp. 534–543. Springer, Cham (2023). https://doi.org/10.1007/978-3-031-14125-6_53
7. Singh, S., Rai, H., Pandey, K.K., et al.: Improving tribological properties of al alloys via robust one step graphene coatings using plasma spraying. Tribol. Lett. **71**, 42 (2023). https://doi.org/10.1007/s11249-023-01713-8
8. Shepelenko, I., Nemyrovskyi, Y., Tsekhanov, Y., Mahopets, S., Bevz, O.: Peculiarities of interaction of micro-roughnesses of contacting surfaces at FANT. In: Ivanov, V., Trojanowska, J., Pavlenko, I., Zajac, J., Peraković, D. (eds.) DSMIE 2020. LNME, pp. 452–461. Springer, Cham (2020). https://doi.org/10.1007/978-3-030-50794-7_44
9. Sheykin, S.Y., Rostotskyi, I.Y., Protsyshyn, V.T., et al.: Improving the performance of hard-alloy deforming broaches using modified technological lubricants. J. Superhard Mater. **42**, 276–282 (2020). https://doi.org/10.3103/S1063457620040085
10. Álvarez, L., Luis, C.J., Puertas, I.: Analysis of the influence of chemical composition on the mechanical and metallurgical properties of engine cylinder blocks in grey cast iron. J. Mater. Process. Technol. **153–154**, 1039–1044 (2004)
11. Kuksenova, L.I., Rybakova, L.M., Nazarov, Y.A.: Cast iron surface structure and wear after an antifriction non-abrasive finishing treatment. Met. Sci. Heat. Treat. **35**, 499–504 (1993). https://doi.org/10.1007/BF00774916
12. Ortiz-De-Zarate, G., et al.: Experimental and FEM analysis of surface integrity when broaching Ti64. Procedia CIRP **71**, 466–471 (2018)
13. Wenlong, D., Xuefeng, Y., Fei, S., et al.: Antifriction and wear resistance analysis of cemented carbide coatings. Int. J. Adv. Manuf. Technol. **122**, 2795–2821 (2022). https://doi.org/10.1007/s00170-022-10092-8
14. Gavrilov, K.V., Morozov, A.V., Seleznev, M.V., et al.: Evaluation of anti-friction properties of solid lubricant coatings for a piston skirt of a high-force diesel. J. Frict. Wear **41**, 480–485 (2020). https://doi.org/10.3103/S1068366620050104
15. Tang, H., Ren, Y., Zhang, X.: Tribological performance of MoS2 coating on slipper pair in axial piston pump. J. Cent. South Univ. **27**, 1515–1529 (2020). https://doi.org/10.1007/s11771-020-4387-x
16. Pavlov, A.A., Soboleva, E.S.: Anti-frictional copper-fluoroplast coating for parts of power equipment. Chem. Petrol. Eng. **57**, 56–59 (2021). https://doi.org/10.1007/s10556-021-00894-0
17. Sheykin, S.Y., Grushko, O.V., Melnichenko, V.V., et al.: On the contact interaction between hard-alloy deforming broaches and a workpiece during the shaping of grooves in the holes of tubular products. J. Superhard Mater. **43**, 222–230 (2021)

Wear of Oval and Round Calibers Rolls of High-Speed Wire Block

Maksym Shtoda(✉)

LLC "Technical University "Metinvest Polytechnic", 80, Pivdenne Hwy, Zaporizhzhia 69008, Ukraine

maksym.shtoda@mipolytech.education

Abstract. It was proposed to use a new measurement method for studying the research of wear of round and oval calibers of hard-alloy rolls of a high-speed wire block. The method allows for assessing the extent of wear of the caliber groove over its complete width. With the traditional adjustment of the finished block, during rolling wire with a diameter of 5.5 mm, the fourth and fifth stands were most subject to wear. It was shown that due to the use of fluid bearings as supports and the cantilever position of the rolls at an angle of 45° to the horizon, the upper and lower streams wear unevenly in width. The wear of the lower roll is greater than the upper one. The non-linear regression formulas for calculating the value of wear grooves carbide rolls were obtained. It was proposed to consider a wear model of an oval caliber consisting of four parts for modeling character wears near the actual process. New formulas were obtained for calculating the wear value during hot rolling in the system of calibers "oval–round" in carbide rolls. The obtained results correspond to the existing ideas of the theory of destruction and exploitation of rolling rolls.

Keywords: Optical-Light Method Study Wear · High-Speed Wire Block · Oval Calibers · Round Calibers · R&D Investment

1 Introduction

Wire rod is one of the most important and popular types of long products. From the point of view of the breadth of the circle consumers, wire rod is a universal type of metal product. According to the OECD, its share in all long products produced worldwide is about 12%. Economic (low productivity) and technological (billet enter speed in the first stand of a continuous mill) are why wire rods must be rolled at high speed. The rolling speed in the finishing pass on modern continuous mills is more than 100 m/s for the smallest wire rod diameters of 5.0…6.5 mm. There is intensive wear of the calibers because of high rolling speeds in the last passes. The normative durability of caliber hard-alloy rolls of a wire finishing block is 1200…1400 tons or 16…18 h of continuous operation. The actual durability can be 700…800 t due to the peculiarities of operation.

It should be noted that it is not only the nature and amount of wear in each caliber essential but also the determination of the maximum wear among all calibers of the

V. Tonkonogyi et al. (Eds.): InterPartner 2023, LNME, pp. 219–230, 2024.
https://doi.org/10.1007/978-3-031-42778-7_20

finishing block for the campaign. This value determines the diameter of the carbide rolls after regrinding since the diameters in each module of the finishing block must be the same. If this condition is not observed, rolling will be impossible due to violating the calibration constant in the finishing block.

One major factor that adversely affects the strip's quality during hot rolling is roll wear [1]. Improving rolls' lifespan in hot rolling processes is of primary concern for metal forming industries. Replacement and early breaking represent 15% of the production costs [2]. In reality, the cross-section of a wire rod with perfect roundness does not exist. This is called"ovality", calculated as the subtracted length of the short axis from the long axis of the cross-section. In rod mill, ovality is usually in the range of 0–±0.4 mm. The wear-on roll is the biggest obstacle for us to have zero ovality [3].

This work aims at designing a method for studying wear rolls of the wire block mill 400/200 PJSC "KAMET-STEEL". Typically, caliber wear is measured using a clock indicator. Its accuracy is ±0.01 mm. We must say that other methods and devices for the experimental study of wear were also published in the technical literature [4]. The characteristic of roll grooves wire blocks is that they have relatively small cross-sections. In this case, the clock indicator succeeds measuring wear only in the middle of the caliber. Research shows that in some cases maximum wear exists aside from the center of the roll groove. There is a problem with studying wear of caliber on all wide its cross-section in these conditions.

2 Literature Review

The paper [5] studies the influence of lubrication on the friction and wear of car rolling bearings. In paper [6], the effect of quenching and different tempering on abrasive and shock-abrasive wear of low-carbon manganese (10–24% Mn) steels, phase composition, and mechanical properties was studied. In work [7] is to evaluate the performance of some wear models, either with different mathematical formulations or definitions of the unknown wear coefficients, on predicting the work-roll wear amplitude in Hot Strip Mills (HSM). The paper [8] demonstrates that the rate of roll wear accelerates after reaching a critical production level. Work [9] calculated and measured through simulation wear of four specimens and shown that with increasing axial force of the mandrel, the wear value has increased on the mandrel. A direct relationship is shown between wear and the force acting on the working tool.

The theory of adhesion-deformation contact of surfaces of bodies and their changes, which can be caused by friction and wear processes, was developed in work [10]. The theory of friction and wear of solids under various conditions is summarized in the handbook [11]. The wear process is significantly affected by loads, sliding and rolling speeds, roughness of contact surfaces, and other factors [12]. If previous works consider wear in general and focus on the study of the wear of machine parts, the work [13] is already focused on generalizing research and friction during rolling. The approaches based on probability theory to develop mathematical models of complex processes, such as contact and wear between the rolled strip and rolls of a rolling mill, are described in the work [14]. The rich industrial experience in the operation and wear of rolling rolls of section mills is summarized in [15]. It provides experimental data on the wear of various shapes' calibers and defines the basic concepts associated with the wear of rolling rolls. The book [16] presents the results of industrial studies of the wear of section rolls, shows the dependence of the wear rate on the main technological parameters of the rolling section process, and gives recommendations for improving the durability of rolls of section rolling mills. The results of [17] show that the main types of roll wear are thermal, oxidative, frictional, and abrasive.

It is known that the wear of rolls in the deformation zone during rolling depends on physical, chemical, and mechanical conditions. It also depends on the temperature of the metal on contact. When estimating wear, we need to take into account dependent the intensity of deterioration of rolls from physical and chemical action environments (water, slag, lubricant), significant gradients of stress, and deformation fields from the kinematic process, including sliding in a contact zone between bar and rolls. In addition, we need to consider the emergence of defects in surface layers and the heterogeneity and anisotropy of their properties. Wearing a working tool during rolling includes a part such as burring of the surface, seizure, crumbling, and abrasion. The most serious form of surface damage to a working tool during hot rolling is abrasive wear. It includes all components of wear which action getting worse high-temperature deformation of metal and cooling water supply. Although rolls of wire block were made from carbide material, the roll grooves exposed significant wear during his work. Also, previous researchers did not consider the effect of high rolling speeds on the wear of carbide rolls. Which, of course, leads to the intensification of the processes of destruction of the contact surface of the rolls.

3 Research Methodology

For study wear of caliber on all its wide achieved by developing optical and light methods of studying wear roll grooves and by creation special measuring device. The optical and light stand was used as a measurement device (Fig. 1). Stand include an illuminant, magnifying lenses, a screen, and a camera for fixing pictures. During the study, the roll put between illuminant and magnifying lenses on distance focus. Then the template is put inside the rolling groove, and the special device strictly fixes its positions. The larger image of the gap between the template and roll groove arises on the stand's screen when the light turns on (Fig. 1).

The measurement of the wear caliber process consists of two stages. The first stage is an inverted reverse picture of the gap between the new roll groove (before work), and the template got on the screen. Then got an on-screen picture photographed with benchmark length (special measuring scale for scaling) and on computer determined sizes of actual gaps between the rolling groove of new caliber and template. The second stage made the same actions but for rolling after the campaign. The difference between the end and initial sizes gap between the roll groove and template is a picture of wear caliber in the cross-section roll. Photos of the gap between threadbare caliber and template for ten stands of wire block and micro-relief surface thrown-out rolls are in Fig. 2 and 3. Sizes gaps on specified parts of the cross-section of calibers are shown with figures on photos. We notice that rolls of the first seven stands of wire block have two identical calibers. Each roll in the rest three stands have four calibers. Only one caliber in the first seven

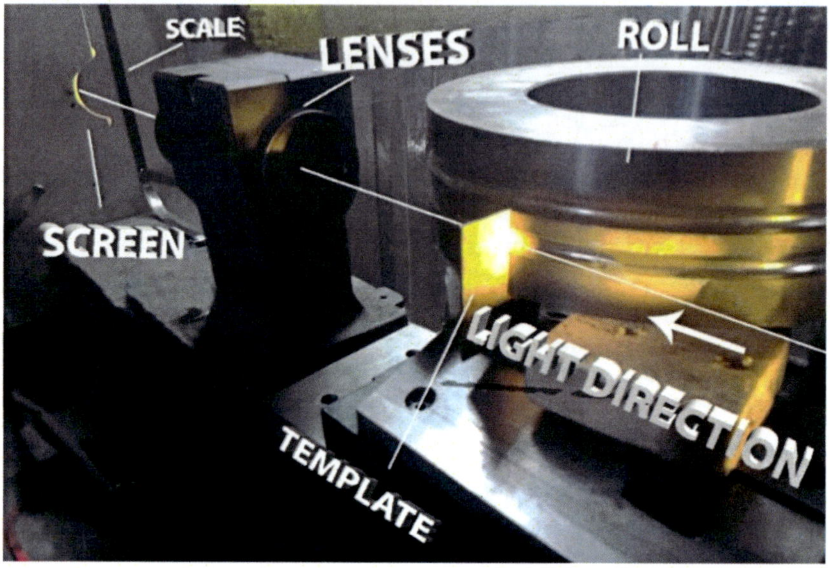

Fig. 1. Stand with roll, template, and picture of the gap.

stands was used during the research. The 2756 t of metal was rolled in these rolls. The rolling in 8–10 stands was realized in two calibers with 1350 t and 1406 t tonnage.

There are conditions of deformation, caliber sizes, mean pressure, and separating force at the stands of the wire block in Table 1.

4 Results and Discussion

We estimate precision measuring wear rolls using an optical light stand compared to clock indicator readings before analyzing research results, which are given in Fig. 2 and 3. In this aim, measurements of wear grooves caliber after two methods produced campaign work rolls. In both cases, measurements were performed at the center of the grooves caliber. There are results measurements of wear rolls in all stands of wire block in Tables 2 and 3.

Statistical analysis of the data was obtained. To do this, the t-statistic checks whether the sample from one set in Table 2 belongs to the other in Table 3.

Critical values of t with the reliability of P = 0.99 and degrees of freedom $k = n_c + n_u - 2 = 102$ is $t_{cr} = 2.62$. Comparing the calculated and critical values, we conclude that with high reliability, measurements made using an optical light stand do not differ significantly from readings of clock indicators. We obtained similar results when both samples' dispersion was compared using the Fisher criterion. It must also note that the readings of the optical light stand were checked for normal distribution, and their correspondence to the Gaussian curve was established.

Now we analyze the results of measurement wear rolls shown in Fig. 2 and 3 and Table 4. It received at conditions given in Table 1. As can see, there is almost equable wear of calibers from 0.2 mm to 0.3 mm in the first three stands. Abrasive wear is evident,

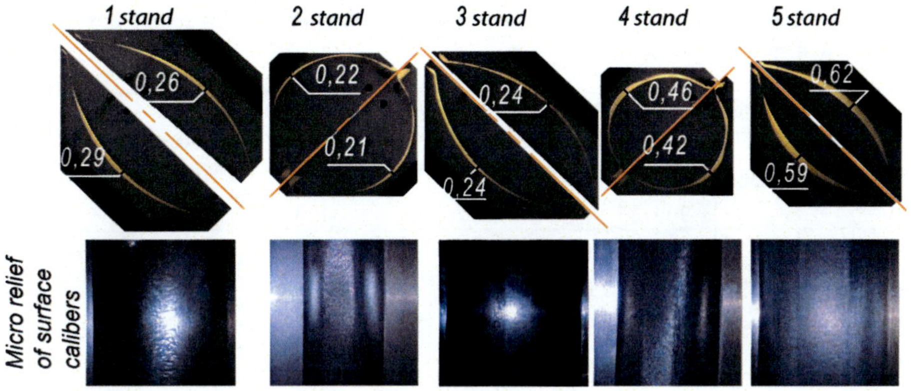

Fig. 2. Photographs of the gaps between the upper and lower grooves rolls and the templates of the first five stands of the wire block after the campaign work.

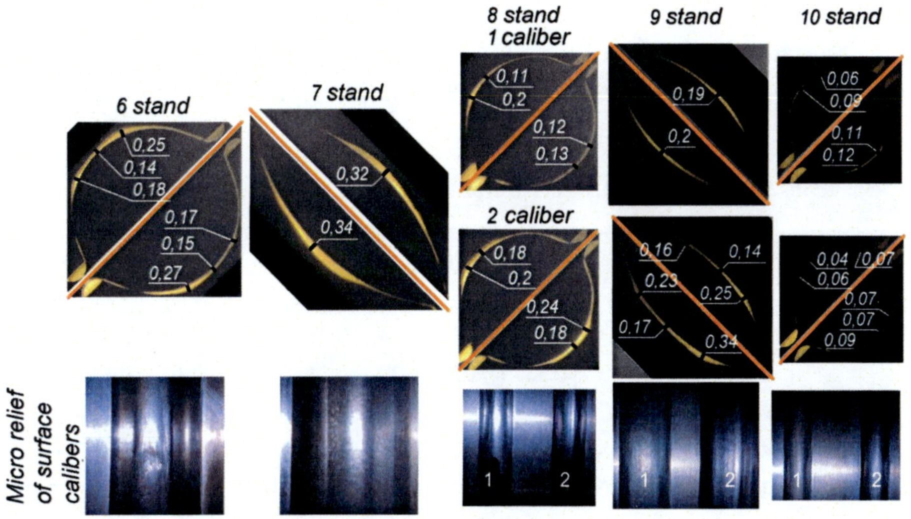

Fig. 3. Photographs of the gaps between the upper and lower grooves rolls and the templates of the last five stands of the wire block after the campaign work.

especially on the surface grooves rolls of the first two stands. There is an initial thermal grid on the surface of the threadbare groove rolls of these stands. There is an abrupt transition from a smooth surface to a rough one with outlines of alternate protrusions and depressions on the rolls of the second stand, which is associated with incomplete caliber filling.

The rolls of the fourth stand have intense wear along the bottom of the caliber. The most significant wear of the surface rolls is observed in the fifth stand, which is 0.5–0.6 mm. Although the surface remains smooth, there is practically no spotty. The micro-relief is degraded near the gaps of the caliber. Significant wear of the surface of

Table 1. Calibers sizes, deformations regime, and separating force during rolling round profile Ø 5.5 mm in wire block.

No. Stand	Calibers sizes, mm		Gaps, mm	Metal dimensions, mm				Mean pressure, MPa	Separating force, kN
	Width	Groove depth		h_0	h_1	b_0	b_1		
1	23.65	4.6	1.9	17.2	11.1	17.2	20.7	280	155
2	13.83	6.14	1.6	20.7	13.9	11.1	13.2	286	85
3	18.89	3.5	1.83	13.2	8.83	13.9	16.0	305	121
4	11.45	4.95	1.3	16.0	11.2	8.83	9.86	305	63
5	16.85	2.7	1.2	9.86	6.6	11.2	13.4	318	96
6	8.97	3.7	1.42	13.4	7.76	6.6	7.76	308	40
7	13.56	2.1	1.2	7.76	5.4	7.76	10.9	305	62
8	7.27	2.92	1.3	10.9	7.14	5.4	6.18	300	27
9	10.24	1.6	1.1	6.18	4.3	7.14	8.47	300	46
10	5.76	2.18	1.1	8.47	5.46	4.3	5.37	286	20

Table 2. Results measurements of wear caliber rolls by optical light stand, mm.

Roll	Caliber	Number stand of wire block									
		1	2	3	4	5	6	7	8	9	10
Up-per	No: 1	0.32	0.23	0.53	0.10	0.24	0.05	0.24	0.09	0.11	0.11
	No: 2	0.32	0.28	0.33	0.07	0.27	0.14	0.28	0.10	0.13	0.13
	No: 3	–	–	–	–	–	–	–	0.12	0.12	0.10
	No: 4	–	–	–	–	–	–	–	0.05	0.11	0.11
Lower	No: 1	0.36	0.30	0.44	0.14	0.22	0.13	0.31	0.12	0.12	0.10
	No: 2	0.32	0.29	0.40	0.07	0.32	0.13	0.34	0.12	0.13	0.10
	No: 3	–	–	–	–	–	–	–	0.11	0.11	0.10
	No: 4	–	–	–	–	–	–	–	0.08	0.13	0.10

the rolls of the fifth stand is associated with a significant mean pressure of the metal on the rolls in this pass (Table 1) and the slide of the metal in the deformation zone.

In the sixth stand, the minimum wear is located near the top of the caliber on either side. The maximum wear of the rolls of this stand is observed along the diagonal of the caliber and amounts to 0.25–0.27 mm, which indicates the tipping of the bar in the caliber.

Two identical calibers worked on each roll of 8–10 stands in Fig. 3 are shown under the numbers 1 and 2. Initially, caliber number 1 worked, and 1350 tons were rolled on

Table 3. Results measurements of wear caliber rolls by clock indicator, mm.

Roll	Caliber	Number stand of wire block									
		1	2	3	4	5	6	7	8	9	10
Up-per	No: 1	0.34	0.33	0.36	0.11	0.28	0.07	0.21	0.06	0.09	0.05
	No: 2	0.32	0.28	0.54	0.22	0.18	0.08	0.29	0.07	0.12	0.06
	No: 3	–	–	–	–	–	–	–	0.09	0.10	0.02
	No: 4	–	–	–	–	–	–	–	0.06	0.10	0.03
Lower	No: 1	0.32	0.34	0.35	0.11	0.33	0.14	0.31	0.06	0.08	0.01
	No: 2	0.38	0.33	0.44	0.19	0.28	0.10	0.44	0.10	0.11	0.00
	No: 3	–	–	–	–	–	–	–	0.08	0.05	0.01
	No: 4	–	–	–	–	–	–	–	0.03	0.09	0.02

Table 4. The results study wear calibers of wire block rolling mill rolls 400/200, mm.

Rough calibers							
Roll	No stand						
	1	2	3	4	5	6	7
Upper	0.26	0.24	0.24	0.46	0.62	0.14(0.25)	0.32
Lower	0.29	0.21	0.24	0.53	0.59	0.15(0.27)	0.34

Finish calibers				
No caliber	Roll	No stand		
		8	9	10
1	Upper	0.11(0.2)	0.19	0.09(006)
	Lowe	0.11(0.12)	0.2	0.11(0.12)
2	Upper	0.18(0.2)	0.16 (0.25)	0.04(0.07)
	Lowe	0.18(0.2)	0.17 (0.34)	0.07(0.09)

Note: the maximum wear is shown in parentheses (not at the bottom of the caliber)

them. Then the metal was deformed in caliber number 2, and 1406 tons were produced. Although the volume of metal rolled in both calibers was approximately the same, the wear of second caliber numbers was significantly higher than the first. Thus, the maximum wear of the second number of the caliber of the 9th stand was 0.34 mm, and that of the first number was only 0.2 mm. An increase in the rolled bar can explain such a difference in the character of their wear from rough calibers of the block. The metal rolled on relatively new rolls of the first seven stands when rolling in the first numbers (Fig. 3), and to the second numbers of the finishing calibers, the metal arrived after deformation in rough calibers with a slightly distorted geometry due to their wear.

Given the wear grooves, round and oval calibers in the finished wire block are visibly affected by the console position of the rolls and the use of fluid bearings for their supports. This design peculiarity leads to the skewness of the rolls in the bearings and the roll axes' deformation under the rolling load's influence (Fig. 4).

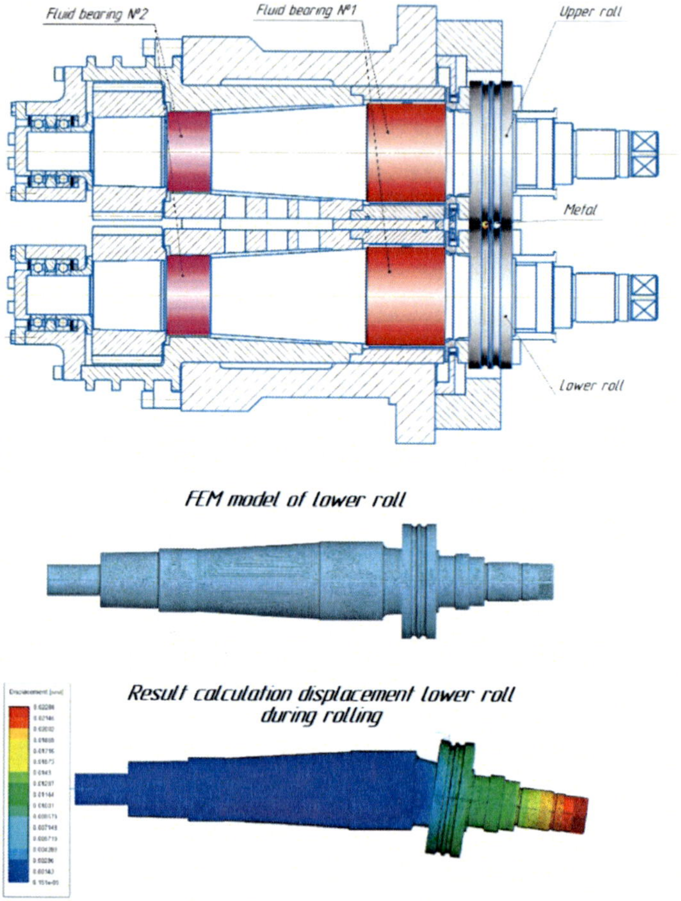

Fig. 4. The design position of the fluid rolling bearings and carbide rolls in the finishing block module. Simulation results of static loading of the roll under a rolling force of 67 kN.

As a result, the shape of the calibers and billets for all modules becomes asymmetric. Templates of oval profiles after rolling in finishing block calibers are shown in Fig. 5, which most clearly illustrate the asymmetry of the process. It leads to lopsided reduction and force action metal to the rolls in the width of each groove.

The location of the rolls at an angle of 45° to the horizon leads to the fact that during rolling, a slightly larger amount of water enters the deformation zone from the side of the lower groove under the action of gravity, which cools the equipment of the finishing block. Water entering the deformation zone instantly turns into steam, which is

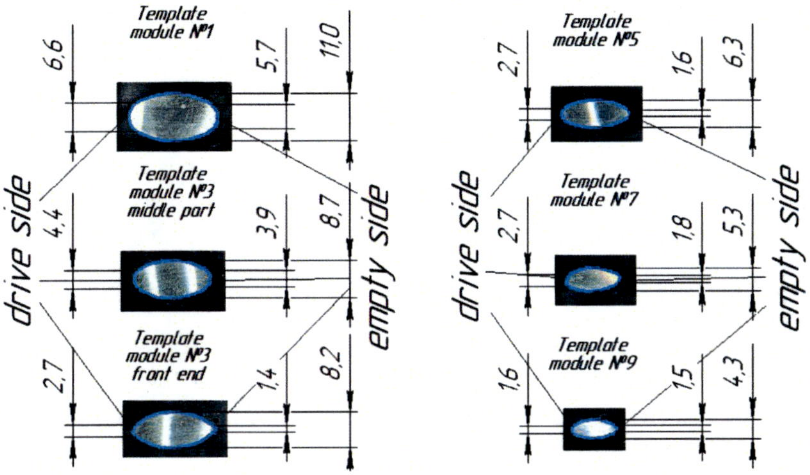

Fig. 5. Templates of oval profiles after rolling in finishing block calibers.

compressed between the billet's hot metal and the pass's cold surface, causing extreme destruction to the lower groove roll.

Thus, the design characteristic of the modules in wire block leads to asymmetric wear of grooves of round and oval calibers. The subsequent analysis aims to elaborate a statistical model for the wear of carbide rolls with oval and round passes for hot rolling wire rods. Based on the analysis of the character of wear oval and round calibers of the high-speed finishing block with cantilevered rolls, it is necessary to consider the asymmetry of the rolling process associated with the design particularities of the equipment. This paper proposes a wear model of an oval caliber consisting of four parts: 1 – upper roll, drive side; 2 – upper roll, empty side; 3 – lower roll, drive side; 4 – lower roll, empty side. Wear models of round passes have also been developed for similar four half (Fig. 6).

An analysis of the technical literature shows that the material of the rolls depends the most significant impact on the value of rolls wear during hot rolling, the shape of billet and caliber, the deformation parameters of the process, the size of metal sliding at the contact with the rolls, the force rolling and the quantity of rolled products from one installation of the caliber. Was choose the following dimensionless parameters as independent factors and their borders of values: $m_1 = \frac{l_d}{D}$; $m_2 = \frac{V_0}{V_r}$; $m_3 = \frac{N}{F}$, where l_d - length of deformation zone in study section; D - diameter of roll in study section; V_0 - speed of metal before deformation zone; V_r - linear speed of rotation roll in study section; N - quantity of metal rolled in study caliber, $[\text{t} \cdot 10^4] = [\text{N}]$; F - rolling force, [N].

The formulas for calculating the value of wear grooves carbide rolls were obtained on the results of the analysis by the method of multiple non-linear regression after choosing the most appropriate type of model for each part of the oval and round caliber, mm:

For oval caliber:

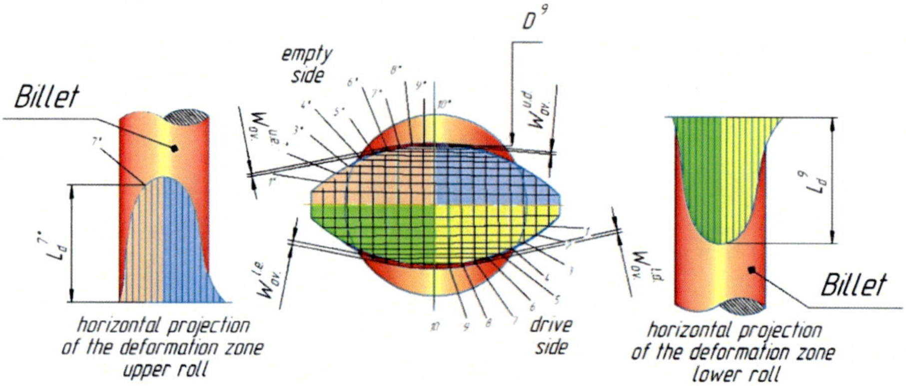

Fig. 6. An Example of fragmentation the deformation zone into sections when developing a mathematical wear oval or round caliber model.

- lower roll empty side

$$W_{ov.}^{l.e.} = 1,888 + 2,975 \cdot m_1 - 6,053 \cdot m_2 + 4,97 \cdot 10^{-3} m_3 + 3,869(m_2)^2 - 1,14 \cdot 10^{-5}(m_3)^2; \quad (1)$$

- lower roll drive side

$$W_{ov.}^{l.d.} = 0,334 + 2,751 \cdot m_1 - 0,864 \cdot m_2 + 3,37 \cdot 10^{-3} m_3 - 6,96 \cdot 10^{-6}(m_3)^2; \quad (2)$$

- upper roll empty side

$$W_{ov.}^{u.e.} = 3,065 + 2,721 \cdot m_1 - 12,775 \cdot m_2 + 1,17 \cdot 10^{-2} m_3 + 8,881(m_2)^2 - 2,29 \cdot 10^{-5}(m_3)^2; \quad (3)$$

- upper roll drive side

$$W_{ov.}^{u.d.} = 0,096 + 2,519 \cdot m_1 - 1,075 \cdot m_2 + 5,99 \cdot 10^{-3} m_3 - 1,15 \cdot 10^{-5}(m_3)^2. \quad (4)$$

For round caliber:

- lower roll empty side

$$W_{r.}^{l.e.} = 0,223 + 62,533 \cdot (m_1)^3 - 4,98 \cdot 10^{-10}(m_3)^3; \quad (5)$$

- lower roll drive side

$$W_{r.}^{l.d.} = 0,214 + 39,598 \cdot (m_1)^3 - 2,94 \cdot 10^{-10}(m_3)^3; \quad (6)$$

- upper roll empty side

$$W_{r.}^{u.e.} = 0,207 + 69,64 \cdot (m_1)^3 - 2,04 \cdot 10^{-4} m_3; \quad (7)$$

- upper roll drive side

$$W_{ov.}^{u.d.} = 0,187 + 50,742 \cdot (m_1)^3 - 3,22 \cdot 10^{-10}(m_3)^3. \quad (8)$$

5 Conclusions

Research showed that the maximum depth of wear of calibers is unevenly distributed at stands of wire block. When the rolling wire is 5.5 mm in diameter, the fourth and fifth stands, where abrasive wear prevails, are most susceptible to wear.

The non-linear regression formulas for calculating the value of wear grooves carbide rolls were obtained. It is proposed to consider a wear model of an oval caliber consisting of four parts for modeling character wears near the actual process. As can be seen from the statistical analysis results, the wear of oval and round calibers depends on the length contact line. A decreased rolling speed of the metal round strips at the entrance to the deformation zone of oval calibers, which corresponds to increased sliding, leads to an increase in wear. For all the wear models obtained, it is characteristic that an increase in dimensionless parameters m_3 leads to a reduction in wear. This relationship is explained by the fact that the growth of the parameter m_3 occurs due to a decrease in the rolling force and with a conditionally constant quantity of metal per roll campaign. The models can be used in practice to develop material-saving technologies for rolling wire rods on modern high-speed rolling mills.

The results are supposed to be used in a mathematical model of the high-speed rolling of wire rods in a finishing block, considering the wear of round and oval calibers.

References

1. John, S., Sikdar, S., Mukhopadhyay, A., Pandit, A.: Roll wear prediction model for finishing stands of hot strip mill. Ironmaking Steelmaking **33**(2), 169–175 (2006)
2. Bataillea, C., Lucb, E., Bigerellea, M., Deltombea, R., Dubara, M.: Rolls wear characterization in hot rolling process. Tribol. Int. **100**, 328–337 (2016)
3. Byon, S.M., Park, H.S., Lee, Y.: Experimental study for roll gap adjustment due to roll wear in single-stand rolling and multi-stand rolling test. J. Mech. Sci. Technol. **22**, 937–945 (2008)
4. Moore, D.F.: Principles and Applications of Tribology. Pergamon International Library of Science, Technology, Engineering and Social Studies: International Series in Materials Science and Technology (1975)
5. Dykha, O., Makovkin, O., Posonsky, S.: Influence of lubrication on the friction and wear of car rolling bearings. Probl. Tribol. **26**(3/101), 81–88 (2021)
6. Malinov, L.S., et al.: Effect of particular combinations of quenching, tempering and carburization on abrasive wear of low-carbon manganese steels with metastable austenite. Mater. Sci. Forum **945**, 574–578 (2019)
7. Souto, N., Marchand, E., Gay, A., Koont, Z., Legrand, N.: Performance analysis of work-roll wear models on hot rolling. Key Eng. Mater. **926**, 621–631 (2022)
8. Pathak, P., Das, G., Jha, S.K.: Influence of roll wear in hot rolling of steel at hot strip mills. In: Kumari, R., Majumdar, J.D., Behera, A. (eds.) Recent Advances in Manufacturing Processes. LNME, pp. 153–169. Springer, Singapore (2022). https://doi.org/10.1007/978-981-16-3686-8_13
9. Darki, S., Raskatov, E.Y.: Study of the rollers geometry effects in rotary tube piercing and wear analyses of mandrel according to the finite element method. Proc. Inst. Mech. Eng. Part E J. Process Mech. Eng. **235**(5), 1676–1684 (2021)
10. Hutchings, I., Gee, M., Santner, E.: Friction and wear. In: Czichos, H., Saito, T., Smith, L. (eds.) Springer Handbook of Materials Measurement Methods. Springer Handbooks, pp. 685–710. Springer, Berlin, Heidelberg (2006). https://doi.org/10.1007/978-3-540-30300-8_13

11. Jeyaprakash, N., Yang, C.-H.: Friction, Lubrication, and Wear. IntechOpen (2020). https://doi.org/10.5772/intechopen.93796
12. Zhang, X., Zhang, L., Yu, X., Zhu, X., Tu, Y., Kang, X.: A novel and quantitative determination method for the running-in process through dimensionless real contact area. Wear **514–515**, 204554 (2023). https://doi.org/10.1016/j.wear.2022.204554
13. Sun, J., Chen, S.-Z., Han, H.-H., Chen, X.-H., Chen, Q.-J., Zhang, D.-H.: Identification and optimization for hydraulic roll gap control in strip rolling mill. J. Cent. South Univ. **22**(6), 2183–2191 (2015). https://doi.org/10.1007/s11771-015-2742-0
14. Hacken, H.: Synergetics: an overview. Rep. Prog. Phys. **52**(5), 515 (1989). https://doi.org/10.1088/0034-4885/52/5/001
15. Spuzic, S., Strafford, K.N., Subramanian, C., Savage, G.: Wear of hot rolling mill rolls: an overview. Wear **176**(2), 261–271 (1994). https://doi.org/10.1016/0043-1648(94)90155-4
16. Zhang, J.-Y., Yang, J.-X., Ma, G.-T., Wang, J.-S., Feng, W.: Research on wear of roll for section steel. J. Iron Steel Res. **15**(4), 29–32 (2003)
17. Kumar, S.S.S., Sathiyanarayanan, S., Rao, A.S., Raghu, T.: Roll bonding behaviour of ti-6Al-4V sheets during pack rolling. Trans. Indian Inst. Met. **62**(2), 129–133 (2009). https://doi.org/10.1007/s12666-009-0017-x

Modeling of Vibrational-Centrifugal Strengthening for Functional Surfaces of Machine Parts

Vadym Stupnytskyy[1], Yaroslav Kusyi[1](✉), Egidijus Dragašius[2], Saulius Baskutis[2], and Rafal Chatys[3]

[1] Lviv Polytechnic National University, 12, Bandera St., Lviv 79013, Ukraine
jarkym@ukr.net
[2] Kaunas University of Technology, 56, Studentu St., 51424 Kaunas, Lithuania
[3] Kielce University of Technology, 7, al. Tysiąclecia Panstwa Polskiego, 25-314 Kielce, Poland

Abstract. Microrelief quality indicators and operational characteristics of machine parts are implemented by using rational finishing treatment (conventional cutting methods) and finishing strengthening (chemical and thermal treatment and methods of surface plastic deformation) during their manufacturing. The main difference between vibration technologies and traditional finishing methods of machining and static deformation of SPD is the possibility of forming regular microreliefs, which allows for achieving the geometric parameters of the quality of the surface layer for machine parts. In this paper, dynamic processes are modeled using electromagnetic vibration-centrifugal strengthening devices. The Poincare and Lyapunov method of small parameters was used to develop a system of differential equations to study the laws of motion of the main elements of a vibration-centrifugal strengthening device with an electromagnetic drive. The dynamics of the variation of the amplitude spectrum in time and the phase portraits of the executive elements for the non-working and working modes of operation of the vibration-centrifugal amplification device were analyzed. Further research on electromagnetic vibration-centrifugal strengthening devices is suggested.

Keywords: Vibrational-Centrifugal Strengthening · Electromagnetic Device · Mathematical Model · Industrial Innovation

1 Introduction

The diversity and differences in the operating conditions of mechanical engineering machines [1] and equipment [2] require the use of rational materials [3, 4], the development of an optimal structure [5] of the technological processes for parts manufacturing [6], therefore the choice of effective finishing treatment [7, 8] according to sustainable [9] machining requirements [10] and Industry4.0 [11] principles [12], including digitalization [13] of economic approaches in manufacturing [14], mathematical modeling [15] of manufacturing [16] and technological processes [17] become an integral factor in modern production [18].

V. Tonkonogyi et al. (Eds.): InterPartner 2023, LNME, pp. 231–244, 2024.
https://doi.org/10.1007/978-3-031-42778-7_21

This leads to significant difficulties in using the typical manufacturing processes of such products [19] per the principle of object-oriented technologies [20]. In addition, the intensive operation conditions of mechanical engineering products require a comprehensive prediction of adjustable accuracy indicators, microrelief parameters, shape deviations, functional surfaces of machine parts' mutual location, operational characteristics, and reliability indicators [3, 9].

Therefore, the overall probability of ensuring the regulated parameters of the machine-building production object P(t), which is determined by the numerical values of the technological parameters within the tolerance limits throughout the entire technological process of parts manufacturing [21], is formed based on the features of three groups of technological process final parameters (Fig. 1) [22]:

1. A small part of the output parameters of intermediate operations (in particular, the material characteristics) are transferred to the final category (parameters of I group, Fig. 1).
2. Finishing and finishing-strengthening treatment during the final technological operations of parts manufacturing allows for forming accuracy indicators and surface quality parameters regulated by technical requirements (parameters of II group, Fig. 1).
3. Part of the output parameters (parameters of III group, Fig. 1) are functionally related to the parameters of previous intermediate operations according to the technological inheritability of properties and parameters for mechanical engineering parts and machines.
4. The characteristics of workpiece material with unsatisfactory input control or its absence affect the formation of the final parameters of the products.
5. Control operations (C_i, Fig. 1) are intended for the timely detection and prevention of defects at various stages of the implementation of technological processes for parts manufacturing.

Technological support of finishing and finishing-strengthening operations and developing effective equipment for their adoption in mechanical engineering practice are essential tasks in applied mechanics.

2 Literature Review

The choice of operational characteristics, reliability indicators, rational finishing, and finishing-strengthening treatments used in the technological operations of parts manufacturing is essential for forming the quality parameters of the final products. The provision of these indicators by technological methods determines the current level of technology and the progressive development of technologies [8, 21].

The analysis of machining technologies for parts of different classes [9, 23] shows that the formation of their quality parameters takes place during the final operations of parts manufacturing using the finishing and finishing-strengthening treatments implemented by various methods, in particular, conventional cutting methods, chemical and thermal treatment and application of coatings, methods of surface plastic deformation, and associated (coupled) and combined methods of finishing [7, 9].

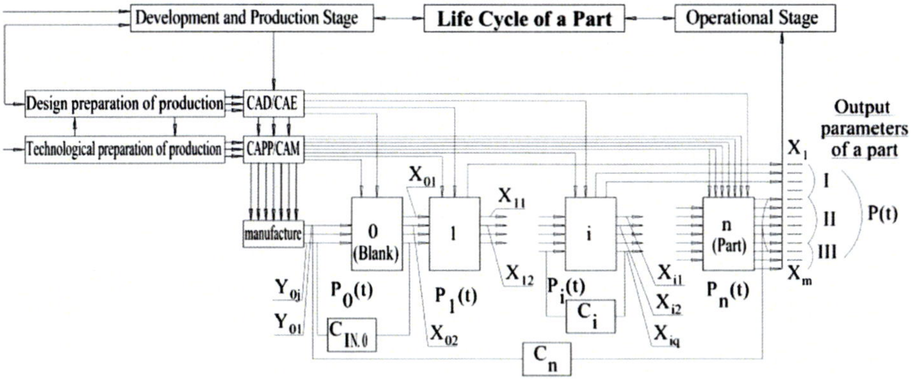

Fig. 1. Schematic diagram of the formation of the output parameters of mechanical engineering parts.

The progressive direction of increasing the products' operational characteristics and reliability indicators is the modification of the surface layers of the machine parts operated with friction pairs, exposed to a corrosive, aggressive environment, and worn out. It is known that the most common metallic coatings [24] are chromium [25]. The most common non-metallic coatings are oxide coatings formed by chemical and electrochemical methods [26] and plasma electrolytic oxidation [27, 28]. The development of coating methods includes wear and corrosion tests, modeling of the stress-strain state of single-layer coatings [29, 30] and multilayer coatings [31], including those with cracks [32], and considering the effects of extra-ordinary operational loads [33], as well even quality control of their surfaces [34].

The vibration-centrifugal strengthening (VCS) method [7, 21], developed at the Lviv Polytechnic National University under the guidance of prof. Dr. Aftanaziv is favorably distinguished from various methods of strengthening parts by surface plastic deformation, for example, ultrasonic finishing [35], friction finishing [36], and diamond burnishing [37, 38]. The VCS method belongs to the group of dynamic strengthening methods due to the strengthening tool's variable-contact interaction with the part's processed surface. The advantages of this method are the provision of a high level of deformation energy, high productivity, simplicity, reliability, compactness, and versatility of strengthening devices, and the possibility of high-quality machining of the internal surfaces of parts. The process of vibrational centrifugal strengthening does not change the geometric shape of the part and does not require a special allowance for treatment. VCH can be used to machine parts made of both non-ferrous metals and alloys, as well as different types of steel deformed in a cold state [7, 8]. With a wide adjustment range, choosing the optimal strengthening modes for parts is easy. Vibrational treatment is particularly effective in strengthening parts exposed to sign-changing cyclic loads during operation.

Classification criteria of vibrational centrifugal strengthening to ensure the quality parameters of long-sized cylindrical parts are the type of surface treatment; location of strengthening devices concerning the processed surface; the shape of deformable bodies; the movement of the executive body of strengthening devices; type of feeds; type of used drive; connection of drive elements with executive bodies of strengthening devices. This classification covers only the main kinematic and structural features of the strengthening equipment that uses VCS to strengthen different parts. In addition, strengthening devices differ in the number and location of deformable elements (i.e., single, double-row).

Using vibration-centrifugal strengthening devices with an electromagnetic drive is a priority direction in developing the VCS method [15]. Therefore, the simulation of dynamic processes under the action of electromagnetic devices is essential in finishing the functional surfaces of mechanical engineering products.

3 Research Methodology

3.1 Methods of Research

The basic scheme of the electromagnetic device for strengthening the internal surface of the cylindrical part is shown in Fig. 2. The device consists of flange 1, electromagnetic drive components: armature 2, and stator 3 with three electromagnetic coils, disc-separators 4 and 5 with deformable elements 8, connected by torsions 6 and 7. Rollers 9 are intended to determine the device's location on the machined surface 10.

An electromagnetic field is generated between the components of the electromagnetic drive by alternating power supply to the electromagnetic coils of the stator 3. The attraction of armature 2 to the electromagnetic coils of stator 3 takes place at a frequency equal to the frequency of the power grid or a multiple of it. Stator 3 oscillates with the armature 2 in antiphase. Disc-separators 4 and 5, attached to stator 3 and armature 2 by torsions 6, 7, start to perform oscillating-rolling movements. These movements are accompanied by contact interaction with the internal surface of the machined part 10 by deformable bodies 8. The contact of part 10 with the separator disc-separators 4 and 5 occurs through a small number of deformable bodies 8 in a discrete time interval. The contact of part 10 with the next group of deformable bodies 8 occurs with an impact, and the contacting elements are the massive part 9 and disc-separators 4 and 5. The presence of contact interaction between part 10 with the disc-separators 4 and 5 due to the small number of deformable elements 8 causes high contact stresses and strengthening of the surface layers of the product (Fig. 2).

In Fig. 2 marked: $y(n)_0(t)$, $z(n)_0(t)$ are linear displacements in the direction of oscillations of flange 1 due to the resulting movements of elastic-oscillating systems: flange 1 – armature 2 – disk-separator 4 and flange 1 – stator 3 – disk-separator 5; $y(n)_1(t)$, $z(n)_1(t)$; $y(n)_3(t)$, $z(n)_3(t)$ are linear displacements in the direction of oscillations of the components of the electromagnetic device (CED): the armature 2 and the stator 3, attached to the flange by means of elastic systems, the stiffness of which is equal to c1 and c3 during supplying power to n-the electromagnetic coils in a certain period of time t; $y(n)_2(t)$, $z(n)_2(t)$; $y(n)_4(t)$, $z(n)_4(t)$ – linear displacements in the coordinate directions of oscillations Y and Z of the disk-separators attached to the respective CED in the specified time interval t; c_0, k_0 – stiffness and coefficient of viscous linear damping of guide rubber rollers 8, respectively, c_1, c_3, k_1, k_3 – stiffness of sections of elastic systems and coefficients of viscous linear damping between the flange and the corresponding CED: armature 2 and stator 3, respectively; c_2, c_4, k_2, k_4 – stiffness of sections of elastic systems and coefficients of viscous linear damping between the corresponding CED and

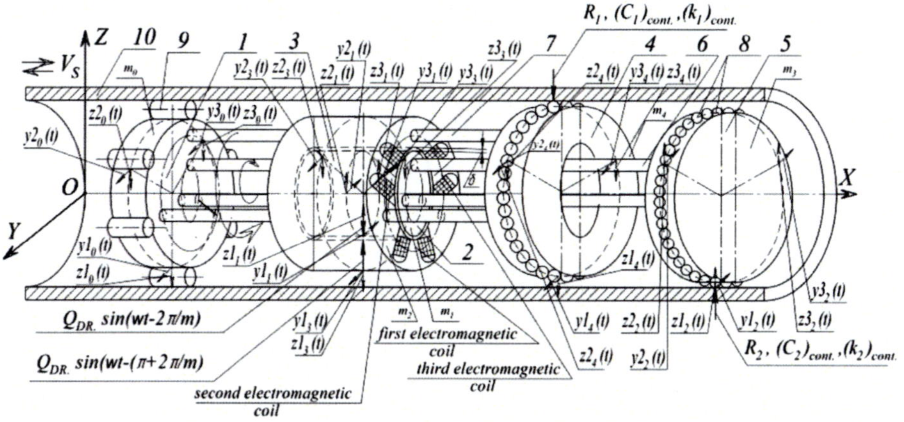

Fig. 2. The basic scheme of the electromagnetic vibration-centrifugal device for strengthening the internal surface of the cylindrical part.

the disc-separators attached to them; $(C_1)_{cont.}$, $(k_1)_{cont.}$, $(C_2)_{cont.}$, $(k_2)_{cont.}$ – contact stiffnesses and coefficients of rolling friction in the places of the interaction of the bodies 8 of the device and the machined surface of the part 10, respectively, δ – the air gap between the corresponding CED; m_0, m_1, m_2, m_3, m_4 – masses of the flange 1, armature 2, stator 3, and disk-separators 4, 5 connected to the respective CED, respectively; $Q_{DR.} \cdot sin(\omega \cdot t + \varphi)$ is the force of the electromagnetic drive, which varies according to the sinusoidal law; $(R_1)_y$, $(R_1)_z$, $(R_2)_y$, $(R_2)_z$ – components of contact forces R_1, R_2 of the oscillating and rolling motion of the disk-separators 4, 5 and the surface of part 10; ω – forced circular frequency of CED oscillations; φ_i, φ_j – phase shift angles between CED; t is an arbitrary time interval.

The schematic diagram (Fig. 2) is replaced by a calculation diagram (Fig. 3) for modeling the laws of motion of the elements of the electromagnetic vibration-centrifugal device with elastic systems.

The following assumptions were accepted when developing the equations of motion of a single-drive electromagnetic device (Fig. 3) [21]:

- movements of attraction between the armature and each electromagnetic coil of the stator during power supply, under the influence of an electromagnetic field, cause parallel oscillations in the planes of the axial sections of the elements of the corresponding strengthening devices, while the general motion is elastic-oscillating or rolling motion around the central axis;
- the strengthening tool is considered a conservative system, regardless of the friction in the joints of its elements; in this case, the total energy remains unchanged;
- the masses of disk separators and the masses of deformable bodies (balls, rollers) form a concentrated mass;
- the masses of electromagnetic drive components with main and auxiliary elements (body, coils, bushings, and threaded connections) form a concentrated mass;
- material recovery factor K = 0 (during machining, the material's properties are not restored to their original state).

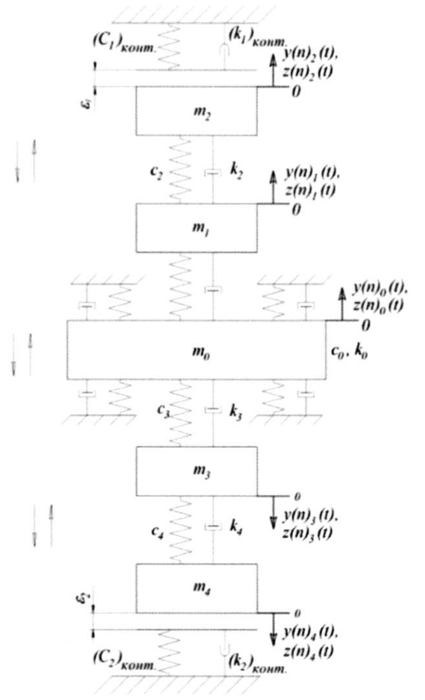

Fig. 3. The calculation diagram for modeling the laws of motion of the elements of the electromagnetic vibration-centrifugal device with elastic systems.

Systems of equations are obtained in the general case. The Poincare and Lyapunov insignificant (small) parameter method was used to create equations and simplify the system solution.

A generating system is applied instead of the original system of equations, which is obtained from the original under certain constraints. The sought solution of the equations of the generating system may be simpler than the solution of the equations of the original system. So:

$$\begin{cases} -c_0 \cdot y_0 - m_0 \cdot \ddot{y}_0 - c_1 \cdot (y_0 - y_1) - c_3 \cdot (y_0 - y_3) - k_0 \\ \quad \times \dot{y}_0 - k_1 \cdot (\dot{y}_0 - \dot{y}_1) - k_3 \cdot (\dot{y}_0 - \dot{y}_3) = 0; \\ c_1 \cdot (y_0 - y_1) - m_1 \cdot \ddot{y}_1 - c_2 \cdot (y_1 - y_2) - c_3 \cdot (y_0 - y_3) + k_1 \\ \quad \times \dot{y}_0 - k_1 \cdot (\dot{y}_0 - \dot{y}_1) - k_2 \cdot (\dot{y}_1 - \dot{y}_2) + Q_{DR.} \cdot \sin(\omega \cdot t + \varphi_1) = 0; \\ c_2 \cdot (y_1 - y_2) - m_2 \cdot \ddot{y}_2 + k_2 \cdot (\dot{y}_1 - \dot{y}_2) - (1 + f_1) \cdot (R_1)_y \cdot \sin(\omega \cdot t + \varphi_2) = 0; \\ c_3 \cdot (y_0 - y_3) - m_3 \cdot \ddot{y}_3 - c_4 \cdot (y_3 - y_4) \\ \quad + k_3 \cdot (\dot{y}_0 - \dot{y}_3) - k_4 \cdot (\dot{y}_3 - \dot{y}_4) + Q_{DR.} \cdot \sin(\omega \cdot t + \varphi_3) = 0 \\ c_4 \cdot (y_3 - y_4) - m_4 \cdot \ddot{y}_4 + k_4 \cdot (\dot{y}_3 - \dot{y}_4) - (1 + f_1) \cdot (R_2)_y \cdot \sin(\omega \cdot t + \varphi_4) = 0. \end{cases} \tag{1}$$

$$\begin{cases} -c_0 \cdot z_0 - m_0 \cdot \ddot{z}_0 - c_1 \cdot (z_0 - z_1) - c_3 \cdot (z_0 - z_3) - k_0 \\ \times \dot{z}_0 - k_1 \cdot (\dot{z}_0 - \dot{z}_1) - k_3 \cdot (\dot{z}_0 - \dot{z}_3) = 0; \\ c_1 \cdot (z_0 - z_1) - m_1 \cdot \ddot{z}_1 - c_2 \cdot (z_1 - z_2) - c_3 \cdot (z_0 - z_3) + k_1 \\ \times \dot{z}_0 - k_1 \cdot (\dot{z}_0 - \dot{z}_1) - k_2 \cdot (\dot{z}_1 - \dot{z}_2) + Q_{DR.} \cdot \sin(\omega \cdot t + \varphi_1) = 0; \\ c_2 \cdot (z_1 - z_2) - m_2 \cdot \ddot{z}_2 + k_2 \cdot (\dot{z}_1 - \dot{z}_2) - (1 + f_1) \cdot (R_1)_y \cdot \sin(\omega \cdot t + \varphi_2) = 0; \\ c_3 \cdot (z_0 - z_3) - m_3 \cdot \ddot{z}_3 - c_4 \cdot (z_3 - z_4) \\ +k_3 \cdot (\dot{z}_0 - \dot{z}_3) - k_4 \cdot (\dot{z}_3 - \dot{z}_4) + Q_{DR.} \cdot \sin(\omega \cdot t + \varphi_3) = 0 \\ c_4 \cdot (z_3 - z_4) - m_4 \cdot \ddot{z}_4 + k_4 \cdot (\dot{z}_3 - \dot{z}_4) - (1 + f_1) \cdot (R_2)_y \cdot \sin(\omega \cdot t + \varphi_4) = 0. \end{cases} \tag{2}$$

$$x = V_S \tag{3}$$

The conditions for the provision of the operation modes of the device serve as limitations (Fig. 3):

- non-working mode:

$$A_2 = \sqrt{y_2^2 + z_2^2} < \varepsilon_1, \tag{4}$$

$$A_4 = \sqrt{y_4^2 + z_4^2} < \varepsilon_2, \tag{5}$$

$$A_1 + A_3 = \sqrt{y_1^2 + z_1^2} + \sqrt{y_3^2 + z_3^2} < (0.63 - 0.72) \cdot \delta, \tag{6}$$

- working mode:

$$A_2 = \sqrt{y_2^2 + z_2^2} \geq \varepsilon_1, \tag{7}$$

$$A_4 = \sqrt{y_4^2 + z_4^2} \geq \varepsilon_2, \tag{8}$$

$$A_1 + A_3 = \sqrt{y_1^2 + z_1^2} + \sqrt{y_3^2 + z_3^2} < (0.63 - 0.72) \cdot \delta, \tag{9}$$

A combination of working and non-working modes is also possible during the machining of products.

3.2 Mathematical Apparatus for Solving a System of Differential Equations

Modern mathematical media provide opportunities to solve systems of second-order differential equations efficiently. The MATLAB system, as an effective representative of computer mathematics systems, equipped with a high-level programming language for performing technical calculations and numerical and simulation experiments, deserves attention among a wide range of software products focused on solving applied mathematics problems, processing the results of mechanics, physics and visualization solutions.

Systems of differential Eqs. (1), (2), regularity (3), and conditions (4–9) are solved in MATLAB media using the classical Runge-Kutta method of the 4th order of accuracy using the functions ode45, ode113. The accuracy of the solution of the system of equations: the permissible value of the relative error is $\pm 1 \cdot 10^{-5}\%$. Permissible values of absolute errors: displacement $\pm 1 \cdot 10^{-6}$ m; speeds of $\pm 1 \cdot 10^{-6}$ m/s.

The established permissible values of errors satisfy the accuracy of solving systems of differential equations using widely known universal methods such as Runge-Kutta or Adams–Bashworth–Moulton while maintaining a sufficiently high performance of the functions ode45, ode113.

4 Results and Discussion

4.1 Verification of the Mathematical Model

The non-working mode of operation of an electromagnetic device with elastic systems is chosen for verification as the simplest because of the mathematical description of physical phenomena and processes. In this case, the forces of contact interaction between the strengthening device's working bodies and the part's machined surface acquire zero values.

The mathematical model verification results are shown in Figs. 4 and 5 with certain parameters and under the condition that flange 1 (Fig. 1) is given initial displacements $(y_1)_0 = 3 \cdot 10^{-3}$ m and $(z_1)_0 = 3 \cdot 10^{-3}$ m.

Other elements of the mechanical oscillating system respond to excitation by a sudden increase in the amplitude of oscillations at the initial moment. However, the disturbances subside very quickly (approximately in a period), the oscillating process stabilizes, and each element performs harmonic oscillations in a given coordinate direction. This behavior of the investigated model, which mathematically describes the process of vibration-centrifugal strengthening by an electromagnetic device with elastic systems, corresponds to the basics of classical mechanics and allows us to conclude its suitability for actual physical processes according to the abovementioned limitations.

4.2 Initial Data for Mathematical Modelling

Mathematical model studies were implemented using the pre-resonance operating mode of the vibrational-strengthening device. The initial data took the following values: $m_0 = 20$ kg, $m_1 = 1.338$ kg, $m_2 = 2$ kg, $m_3 = 4.357$ kg, $m_4 = 2.8$ kg, $c_0 = 0.3 \cdot 10^5$ N/m, $c_1 = 6210113$ N/m, $c_2 = 230004$ N/m, $c_3 = 2794551$ N/m, $c_4 = 349319$ N/m, $(C_1)_{\text{cont.}} = (C_2)_{\text{cont.}} = 5 \cdot 10^7$ N/m, $\varepsilon_1 = 0.001$ m, $\varepsilon_2 = 0.0008$ m.

4.3 Modeling of Non-working and Working Modes of Operation of ESR with Elastic Systems

The elements of the electromagnetic device with elastic systems perform forced harmonic oscillations in the non-working mode of operation due to the action of the electromagnetic drive force.

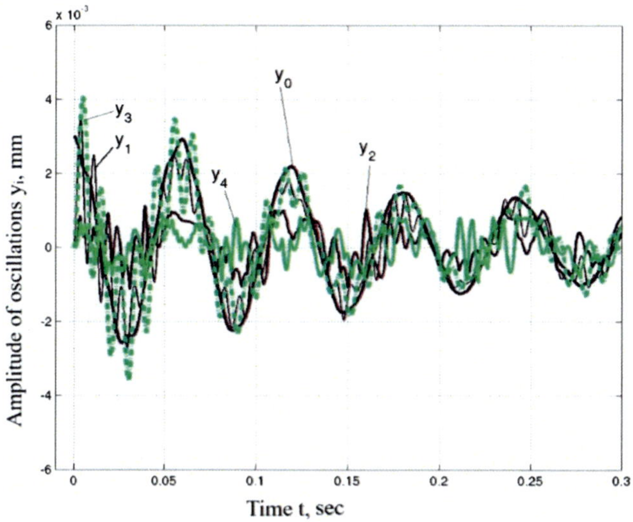

Fig. 4. Dynamics of change $y_0 = f(t)$; ...; $y_4 = f(t)$ with fixed values of parameters c_i, m_i and initial condition $(y_1)_0 = 3 \cdot 10^{-3}$ m.

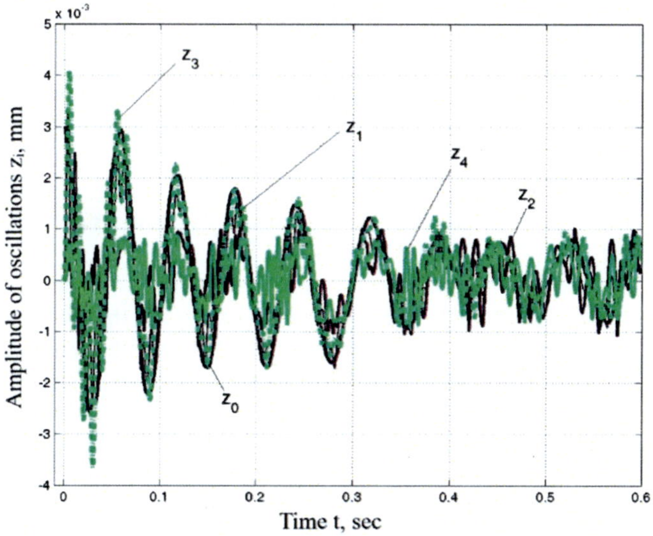

Fig. 5. Dynamics of change $z_0 = f(t)$; ...; $z_4 = f(t)$ with fixed values of parameters c_i, m_i and initial condition $(z_1)_0 = 3 \cdot 10^{-3}$ m.

Contact interaction of the disc-separators 4 and 5 with the internal surface of part 10 (Fig. 1) occurs in working mode because the oscillation amplitudes of the disc-separators 4 and 5 become larger than ε_1, ε_2.

Curves characterized by both modes of operation of the electromagnetic vibration-centrifugal device are presented in Figs. 6, 7 and 8.

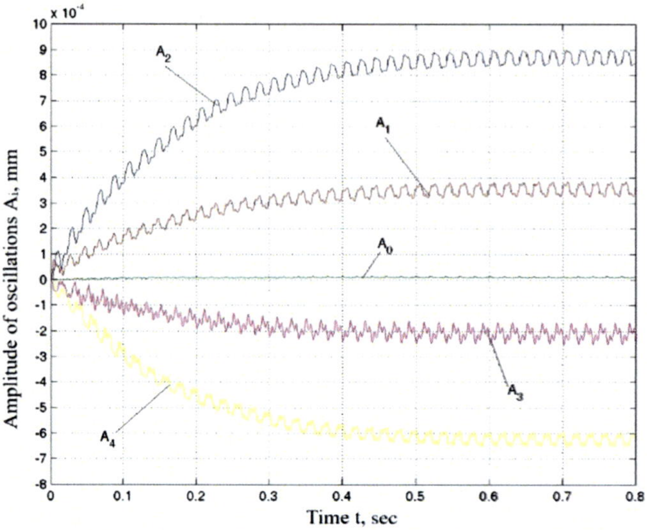

Fig. 6. The dynamics of changes in the amplitude spectrum over time for the non-working mode of operation ($Q_{DR.} = 40$ N).

Transient processes, which are accompanied by an increase in the amplitudes of oscillations of all elements according to (1), (2), determine the operation of the device during the first 0.5 s after the excitement of the system by the force of the electromagnetic drive $Q_{DR.} = 40$ N (Fig. 6). After that, the movement of all components of the five-mass mechanical system occurs with the set amplitudes of oscillations. There is no contact of the deformable bodies 8 of the disk-separators 4, 5 with the internal surface of part 10 (Fig. 1). This is confirmed by the absence of disturbing influences on the trajectories of the disc-separator (Fig. 8, a).

When the driving force is increased to $Q_{DR} = 50$ N after transient processes (t = 0.3 s), the oscillation amplitude increases to values exceeding the values ε_1, ε_2.

The contact of the separator discs 4 and 5 of the device with the internal surface of part 10 (Fig. 1) occurs along the protruding deformable bodies 8, i.e., the mechanical system enters the working mode of operation (Fig. 7). This regime is accompanied by the emergence of resistance forces in the surface layers of part 10 (Fig. 1). Resistance forces are equal to contact interaction forces according to Newton's third law, which is reflected by the appearance of disturbing effects on the phase portrait of the working body (Fig. 8, b).

A further increase in the driving force increases the intensity of interactions of the contacting elements. In addition, the influence of contact interaction, characterized by more chaotic trajectories of movement of working bodies in axial sections, increases.

High-quality machining of the internal surface of part 10 (Fig. 1) is provided when the necessary values of the physical, mechanical, and geometric quality parameters of the surface layer of its material are achieved.

The minimum value of the force of the electromagnetic drive, at which the deformable bodies 8 of the disk-separators 4, 5 and the internal surface of part 10 (Fig. 1) will come

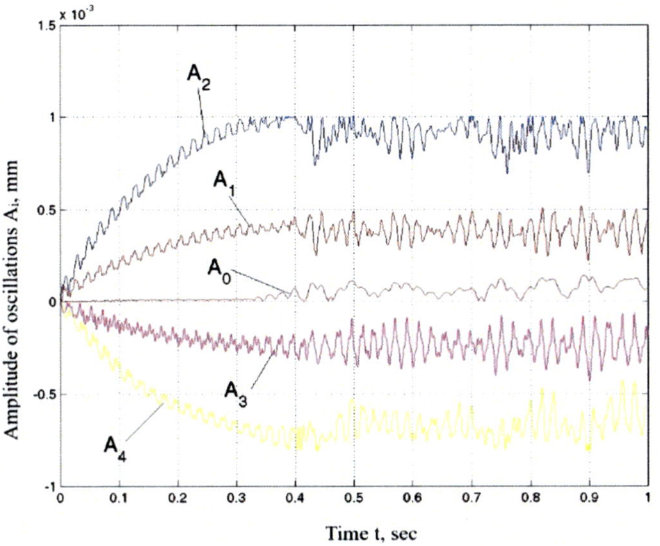

Fig. 7. The dynamics of changes in the amplitude spectrum over time for the working mode of operation ($Q_{DR.} = 50$ N).

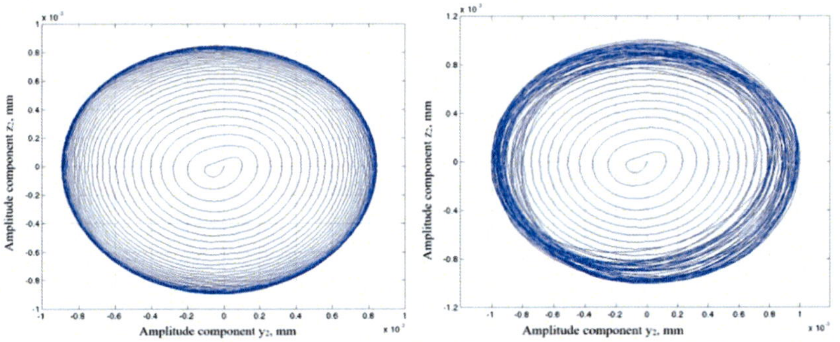

a) b)

Fig. 8. Phase portraits of the disc-separator attached to the anchor for non-working a) and for working b) modes of operation.

into contact, can be set by increasing the amplitudes of oscillations of the working bodies of the electromagnetic vibration-centrifugal device up the values ε_1, ε_2. The technological operation of vibration-centrifugal strengthening is designed after finding the solution of the systems of Eqs. (1), (2), and regularity (3).

5 Conclusions

The main conclusions have been drawn based on the research results.

Technological support of accuracy indicators, quality of surface layers, operational characteristics, reliability indicators, and product competitiveness is essential in compliance with the requirements of the main principles of Industry 4.0 and Sustainable Machining in production, including finishing treatment tools using vibrational-centrifugal strengthening.

The main structural elements of the developed electromagnetic vibration-centrifugal devices for strengthening the functional cylindrical surfaces of machine parts are a flange with a coaxially located armature and a stator with electromagnetic coils connected using elastic elements. Electromagnetic drive components with flange and disk separators with deformable elements form a five-mass oscillating system. The five-mass oscillating system's pre-resonance operation mode ensures the strengthening treatment's stability and reliability.

The mathematical model of the electromagnetic vibration-centrifugal strengthening device with elastic systems was developed based on theoretical research of the finishing treatment using surface plastic strengthening. This model, consisting of systems of differential equations and the regularity of the axial movement of the device, describes the spatial movements of the five-mass components of the device in non-working and working modes.

Solutions of systems of differential equations were performed in MATLAB media using the classical Runge-Kutta method of the 4th order and the multi-step Adams-Bashworth-Moulton method. The research results are the distribution of the amplitude spectrum for all elements of the device in non-working and working modes along two coordinates and the development of a phase portrait of the executive elements of the device. Optimization includes determining the minimum value of the electromagnetic drive force at which plastic deformation of the surface layers of the products will occur and the design of the final technological operation.

Further research will involve modeling the operation of electromagnetic vibration-centrifugal devices using amplitude and force criteria for one-cycle and two-cycle vibration exciters.

Acknowledgment. This research has been conducted as part of the ongoing project "Comprehensive system of functional-oriented planning of machining difficult-to-cut materials for the military-industrial complex (Komplekssys)". It has been funded by the Research Council of Lithuania and the Ministry of Education and Science of Ukraine under the Lithuanian–Ukrainian Cooperation Programme in the Fields of Research and Technologies (Grant No. S-LU-22-6).

References

1. Kopei, V.B., Onysko, O.R., Panchuk, V.G.: Component-oriented acausal modeling of the dynamical systems in Python language on the example of the model of the sucker rod string. PeerJ Comput. Sci. **5**, e227 (2019)
2. Davim, P.J.: Manufacturing in digital industries: prospects for Industry 4.0. In: Advanced Mechanical Engineering, Book 6. Walter de Gruyter Gmbh, Berlin/Boston, Germany/USA (2020)
3. Liao, Z., et al.: Surface integrity in metal machining - Part I: fundamentals of surface characteristics and formation mechanisms. Int. J. Mach. Tools Manuf. **162**, 103687 (2021)

4. Wu, W., et al.: Mechanostructures: rational mechanical design, fabrication, performance evaluation, and industrial application of advanced structures. Progr. Mater. Sci. **131**, 101021 (2023)
5. Babilas, R., Krupińska, B., Szewieczek, D.: The optimization of a technological process forms a competitive position of the factory. J. Achievements Mater. Manuf. Eng. **16**(1–2), 177–183 (2006)
6. Kusyi, Y., Onysko, O., Kuk, A., Solohub, B., Kopei, V.: Development of the technique for designing rational routes of the functional surfaces processing of products. In: Karabegović, I., Kovačević, A., Mandžuka, S. (eds.) NT 2022. LNNS, vol. 472, pp. 135–143. Springer, Cham (2022). https://doi.org/10.1007/978-3-031-05230-9_16
7. Aftanaziv, I.S., Shevchuk, L.I., Strohan, O.I., Kuk, A.M., Samsin, I.L.: Improving reliability of drill pipe by strengthening of thread connections of its elements. Naukovyi Visnyk Natsionalnoho Hirnychoho Universytetu **2019**(4), 22–29 (2019)
8. Aftanaziv, I., Shevchuk, L., Samsin, I., Strutynska, L., Strogan, O.: Development of a technology for the surface strengthening of barrel channels in the large-caliber artillery guns. East.-Eur. J. Enterp. Technol. **3**(1–99), 11–18 (2019)
9. Davim, J.P.: Sustainable Machining. Materials Forming, Machining and Tribology, 1st edn. Springer, Cham (2018). https://doi.org/10.1007/978-3-319-51961-6
10. La Monaca, A., et al.: Surface integrity in metal machining - Part II: functional performance. Int. J. Mach. Tools Manuf. **164**, 103718 (2021)
11. Cocca, P., Marciano, F., Rossi, D., Alberti, M.: Business software offer for Industry 4.0: the SAP case. IFAC-PapersOnLine **51**(11), 1200–1205 (2018)
12. Stefana, E., Marciano, F., Cocca, P., Alberti, M.: Confined space risk management in steel industry: towards the adoption of Industry 4.0 technologies. In: Proceedings of the Summer School Francesco Turco, pp. 94–100, Italy, Brescia (2018)
13. Eklinder-Frick, J., Fremont, V.H.J., Åge, L.-J., Osarenkhoe, A.: Digitalization efforts in liminal space – interorganizational challenges. J. Bus. Ind. Mark. **35**(1), 150–158 (2019)
14. Mnykh, O., Kostiuk, O., Dalyk, V., Zaitseva, A.: Digitalization of economy and assessment of opportunities and implementation. Int. J. Adv. Sci. Technol. **29**(8), 2470–2476 (2020)
15. Jaluria, Y.: Mathematical modeling of manufacturing processes. In: Advanced Materials Processing and Manufacturing. MES, pp. 19—39. Springer, Cham (2018). https://doi.org/10.1007/978-3-319-76983-7_2
16. Bambach, M., Fügenschuh, A., Buhl, J., Jensch, F., Schmidt, J.: Mathematical modeling and optimization for powder-based additive manufacturing. Proc. Manuf. **47**, 1159–1163 (2020)
17. Attanasio, A., Abeni, A., Özel, T., Ceretti, E.: Finite element simulation of high speed micro milling in the presence of tool run-out with experimental validations. Int. J. Adv. Manuf. Technol. **100**(1–4), 25–35 (2019)
18. Abeni, A., Loda, D., Özel, T., Attanasio, A.: Modeling of cutting force in micro-end-milling process with experimental validation on additive manufactured Nickel-based superalloy. Proc. CIRP **102**, 222–227 (2021)
19. Hedmana, R., Sundkvista, R., Almströma, P., Kinnandera, A.: Object-oriented modeling of manufacturing resources using work study inputs. Proc. CIRP **7**, 443–448 (2013)
20. Kotliar, A., Basova, Y., Ivanov, V., Litvynenko, M., Zinchenko, O.: Ensuring the economic efficiency of enterprises by multi-criteria selection of the optimal manufacturing process. Manag. Prod. Eng. Rev. **11**(1), 52–61 (2020)
21. Kusyj, J., Kuk, A.: A centrifugal vibration strengthening method devised to improve technological reliability of machine parts. East.-Eur. J. Enterp. Technol. **1**(7), 41–51 (2015)
22. Kumar, K., Zindani, D., Davim, J.P.: Sustainable Manufacturing and Design. Woodhead Publishing, Cambridge, United Kingdom (2021)

23. Onysko, O., Panchuk, V., Kopei, V., Havryliv, Y., Schuliar, I.: Investigation of the influence of the cutter-tool rake angle on the accuracy of the conical helix in the tapered thread machining. J. Phys.: Conf. Ser. **1781**(1), 012028 (2021)
24. Ropyak, L., Ostapovych, V.: Optimization of process parameters of chrome plating for providing quality indicators of reciprocating pumps parts. East.-Eur. J. Enterp. Technol. **2**(5), 50–62 (2016)
25. Bazaluk, O., Dubei, O., Ropyak, L., Shovkoplias, M., Pryhorovska, T., Lozynskyi, V.: Strategy of compatible use of jet and plunger pump with chrome parts in oil well. Energies **15**(1), 83 (2022)
26. Student, M.M., et al.: Influence of the composition of electrolyte for hard anodizing of aluminum on the characteristics of oxide layer. Mater. Sci. **57**(2), 240–247 (2021)
27. Ropyak, L., Shihab, T., Velychkovych, A., Bilinskyi, V., Malinin, V., Romaniv, M.: Optimization of plasma electrolytic oxidation technological parameters of deformed aluminum alloy D16T in flowing electrolyte. Ceramics **6**(1), 146–167 (2023)
28. Student, M.M., et al.: Influence of the porosity of a plasma-electrolytic coating on the corrosion resistance of D16 alloy. Mater. Sci. **54**(6), 899–906 (2019)
29. Dubei, O.Ya., Tutko, T.F., Ropyak, L.Ya., Shovkoplias, M.V.: Development of analytical model of threaded connection of tubular parts of chrome-plated metal structures. Metallofizika i Noveishie Tekhnologii **44**(2), 251–272 (2022)
30. Dutkiewicz, M., Dalyak, T., Shatskyi, I., Venhrynyuk, T., Velychkovych, A.: Stress analysis in damaged pipeline with composite coating. Appl. Sci. (Switzerland) **11**(22), 10676 (2021)
31. Bembenek, M., et al.: Optical and mechanical properties of layered infrared interference filters. Sensors **22**(21), 8105 (2022)
32. Shats'kyi, I.P., Makoviichuk, M.V., Shcherbii, A.B.: Influence of a flexible coating on the strength of a shallow cylindrical shell with longitudinal crack. J. Math. Sci. (U.S.) **238**(2), 165–173 (2019)
33. Ropyak, L.Y., Velychkovych, A.S., Vytvytskyi, V.S., Shovkoplias, M.V.: Analytical study of "crosshead - Slide rail" wear effect on pump rod stress state. J. Phys.: Conf. Ser. **1741**(1), 012039 (2021)
34. Bembenek, M., et al.: Multiclass level-set segmentation of rust and coating damages in images of metal structures. Sensors **22**(19), 7600 (2022)
35. Pavlenko, D., Kondratiuk, E., Torba, Y., Vyshnepolskyi, Y., Stepanov, D.: Improving the efficiency of finishing-hardeningtreatment of gas turbine engine blades. East.-Eur. J. Enterp. Technol. **1**(12–115), 31–37 (2022)
36. Shepelenko, I., Tsekhanov, Y., Nemyrovskyi, Y., Posviatenko, E.: Improving the efficiency of antifriction coatings by means of finishing the antifriction non-abrasive treatment. In: Tonkonogyi, V., et al. (eds.) InterPartner 2019. LNME, pp. 289–298. Springer, Cham (2020). https://doi.org/10.1007/978-3-030-40724-7_30
37. Koszela, W., Pawlus, P., Rejwer, E., Ochwat, S.: Possibilities of oil pockets creation by the burnishing technique. Arch. Civ. Mech. Eng. **13**(4), 465–471 (2013)
38. Raza, A., Kumar, S.: A critical review of tool design in burnishing process. Tribol. Int. **174**, 107717 (2022)

Advanced Materials

Formation of 2D Copper Oxide Nanostructures on Substrates Exposed to Glow Discharge Plasma

Oleg Baranov(✉)

National Aerospace University, 17, Chkalova St., Kharkiv 61070, Ukraine
O.Baranov@khai.edu

Abstract. A conventional glow discharge was engaged in conducting a process of plasma-enhanced synthesis of 2D CuO nanostructures on a copper sample installed on a cathode exposed to oxidation for 30 min at 360 Pa of oxygen pressure and the temperature of 600 °C. The sample was partially covered with a copper cap with an orifice (1 mm diam.) to remove the ion flux extracted from the glow plasma to the surface area, while other parts of the sample were exposed to the direct action of the plasma. SEM instruments allowed for founding that after 2 min treatment, an array of interconnecting nanodots with an average diameter of 200 nm, formed of substructures of lesser diameters, is grown along the boundaries of the CuO oxide layer. Then the samples treated for 30 min revealed the dense arrays of 2D nanosheets with sizes up to 20 μm. The arrays were similar for the protected and unprotected sample areas. TEM study revealed a periodic structure on the surface of the nanosheets, presented by a set of grooves up to 20 nm in width separated by a similar area. The sizes of the substructure were associated with the sizes of the nanodots, and an assumption of the formation of a frame of the nanosheets of 1D nanowires was made. The proposed growth mechanism was then analyzed and verified using the developed theoretical model of the CuO nanowires growth.

Keywords: Copper Oxide · Nanosheets · Plasma Glow · Ion Bombardment · Manufacturing Innovation

1 Introduction

For the last decade, two-dimensional metal oxide nanostructures have been extensively studied concerning their possible application in energy storage, electronics, optics [1], catalysis [2], and satellite cooling systems [3]. Among the nanostructures, copper oxide nanosheets showed promising ferromagnetic properties [4]. The importance of the structure of the material used in supercapacitors, and the availability of CuO nanosheets concerning this application, was highlighted by Hussain et al. [5]. Nanosheets-built microtubes of copper oxide grown in the microwave-assisted hydrothermal method were used for a sensor developed for blood glucose detection [6]. The hydrothermal method enhanced by applying microwaves resulted in the synthesis of well-developed arrays of 2D and 3D nanostructures with a large surface area of 15 to 20 m^2g^{-1}, which exhibited

V. Tonkonogyi et al. (Eds.): InterPartner 2023, LNME, pp. 247–255, 2024.
https://doi.org/10.1007/978-3-031-42778-7_22

excellent sensing applications towards five gases [7]. The chemical method of growth of 2D CuO nanostructures on foamed nickel substrate in a liquid solution was engaged to obtain the material fit for energy storage applications [8]. Chemical corrosion of copper foils stayed in a liquid solution of sodium hydroxide and ammonium persulfate was necessary to fabricate CuO nanosheets implemented as anode material for batteries. At that, the electrode exhibited good performance in a lithium-ion battery, while its characteristics in the sodium-ion battery were poor [9]. The outstanding performance of copper oxide nanosheets as catalysts for the electrochemical conversion of CO_2 was proved after preparing a complex nanocomposite for more than 20 h in a chemical process conducted in a liquid solution [10]. It can be concluded that the emerging applications of 2D copper oxide nanomaterials are closely related to the production methods and mechanisms of the formation of the nanostructures.

2 Literature Review

Among the large diversity of possible physical [11] and mechanical [12] production methods, hydrothermal chemical solution-based methods are the most applied for producing two-dimensional CuO nanostructures. Thus, Ghosh et al. [13] carried out one pot coprecipitation process in a liquid solution for a few hours to produce 2D copper oxide nanosheets of about 1 μm in width and 3–4 nm in thickness. More than 6 h was necessary to synthesize CuO nanosheets for a gas sensor in a hydrothermal process proposed by Umara et al. [14], which included subsequent stages of stirring, heating, cooling, filtering, washing, and drying. More than 24 h was spent growing CuO nanosheets capable of performing as photocatalyst for degradation of textile dyes using the solution-based chemical method [15]. A much extended time of 120 h was required to synthesize CuO nanosheets on Cu foil in the aqueous ammonia and sodium hydroxide [16]. Similar processes were conducted by Amirtharaj et al. [17] for 12 h. As can be seen, despite the relative cost-effectiveness, the production time is quite extensive, in addition to the use of non-friendly environment chemicals, thus hindering the implementation of this group of methods in industrial applications.

To accelerate the process, plasma was engaged [18] in the production process due to the ability to enrich the production media with many active species like ions, electrons, excited neutrals, and radicals, which is also powered by the ability to conduct a flexible control over the plasma fluxes by use of external magnetic fields [19]. Luan et al. have successfully implemented nitrogen plasma treatment to strengthen the copper oxide 2D structures intended for use for anodes in lithium-ion batteries [20]. Microwave plasma enhanced growth of complex nanostructures composed of CuO-Cu_2O@graphene nanosheets was used in the methane-hydrogen gas mixture at the discharge power of 2000 W and growth temperature of 800 °C to develop an electrode for a supercapacitor [21]. Spark discharges are recognized as a powerful tool to deliver a large quantity of energy into a small volume, thus providing the diversity of the plasma-chemical processes [22]. Pulsed spark discharge was engaged just for 10 min by Hamdan et al. to grow two-dimensional mesoporous CuO agglomerates in water by use of Cu electrodes [23]. Direct application of plasma for the synthesis of 2D nanostructures was applied by Hamdan et al. by use of pin-to-pin configuration to ignite spark discharge in liquid hydrogen [24]. However, the number of publications dedicated to plasma synthesis of 2D CuO nanostructures does not correspond to the fabrication benefits associated with plasma, and the reported researches allow for concluding the critical role of plasma bombardment in the morphology of the growing nanostructures. Thus, implementing atmospheric pressure micro-afterglow plasma discharge in a needle-like configuration of plasma column resulted in an abundant yield of copper oxide nanostructures of various dimensionality achieved by Altaweel et al. on a copper substrate [25]. At that, the higher the nanostructures' dimensionality, the closer a growth location was to the plasma column, i.e., 3D bulky structures were formed close to the plasma, while 0D nanodots were observed on the periphery of the sample. Later a similar experiment confirmed that plasma could be a beneficial factor in the growth of 2D CuO nanostructures if the growth area is located at some 'reasonable' distance from the plasma source [26]. According to the analysis, despite the successful implementation of the gas-plasma methods for the problem of synthesis of one-dimensional copper oxide nanostructures, the plasma-enhanced techniques suitable to obtain 2D CuO nanostructures still needs to be developed.

Moreover, the mechanism of growth of such structures should be proposed. Thus, the experimental part of the present research is dedicated to studying the limiting factors of plasma-enhanced production of 2D CuO nanostructures. The theoretical part considers a mechanism of growth of such structures. Various approaches based on a solution to the heat conduction problem [27] can be used to describe the stages of the synthesis of one- [28] and two-dimensional [29] nanostructures. However, based on the experimental findings, the successful application of the previously reported theoretical model [30, 31] is discussed.

3 Research Methodology

A cylindrical stainless-steel chamber with a diameter of 300 mm and length of 350 mm was engaged to perform the experiments on the growth of copper nanosheets. The simplest method of igniting plasma, DC glow discharge, was chosen to clarify the effect of the ion bombardment. For that, copper electrodes were mounted in the chamber and separated by a gap of 15 mm. The cathode was negatively biased, while the anode and the chamber walls were grounded. A sample made of copper was installed onto the cathode.

The sample was partially covered by a cap with a small orifice (1 mm diam.) to study the effect of the ion bombardment. It was made in the cap's central part, so the sample's central part was protected from the direct action of plasma ions while exposed to the flux of neutral species. Thus, between the edges of the cap and the sample surface. A region, partially shadowed by the plasma ions, was also formed. The chamber was evacuated to the pressure of 10^{-3} Pa and filled with oxygen to the pressure of 360 Pa. The glow discharge was initiated at the voltage of 500 V and the current of 0.1 A. At that, the sample was heated to 600 °C. A schematic of the setup and photograph of the glow discharge can be found elsewhere [32]. After the plasma treatment that lasted for 2 min and 30 min, the samples were studied with scanning electron microscopy (SEM) and transmission electron microscopy (TEM) facilities.

4 Results and Discussion

SEM images of the different surface regions, which were obtained after 2 or 30 min of growth, are shown in Fig. 1. For the samples treated for 2 min, a magnified view of the sample surface that was not covered by the cap after 2 min of the plasma treatment, is shown in Fig. 1, a; the surface under the cap exhibited the same structures.

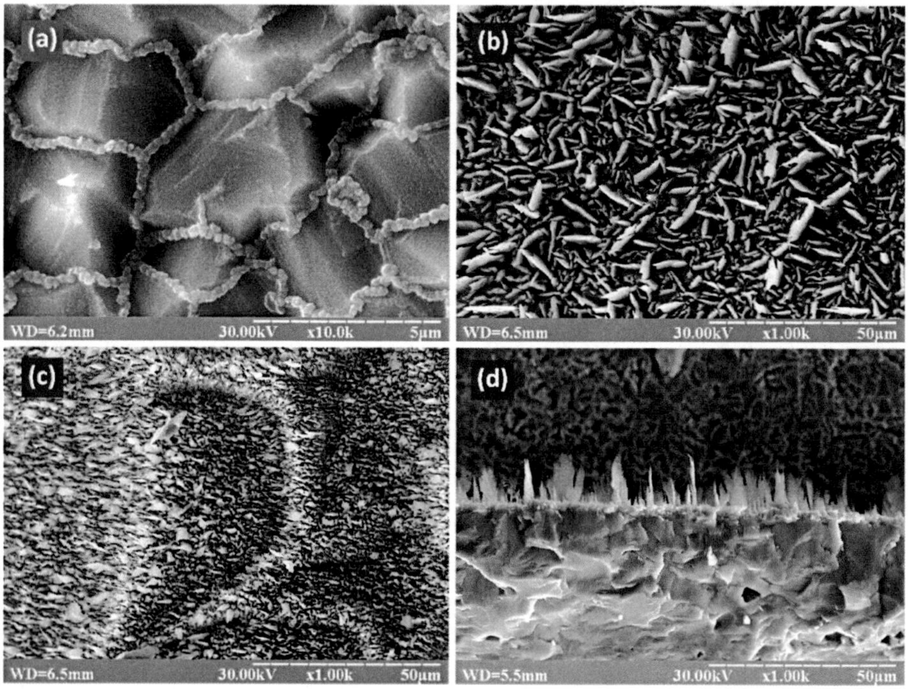

Fig. 1. SEM images of a copper sample treated for 30 min in oxygen plasma (360 Pa, 500 V, and 0.1 A): a – magnified view of the sample surface that was not covered by the cap during 2 min of plasma treatment; b – the same area after 30 min of plasma treatment, where 2D nanostructures with a width of about 15 μm are seen; c – sample area in the proximity to the cap (within 1 mm from the cap walls); d – surface area of the sample that was covered by the cap, which exhibits 2D nanosheets with a maximal height up to 20 μm.

As can be seen, a dense array of connected nanodots with a diameter of about 200 nm was grown on the surface, and the spatial distribution of the nanodots follows the boundaries between the grains of copper oxide. Moreover, the nanodots appear to be formed as a coalescence of lesser nanodots with a few dozen nanometres sizes. Thus, an assumption about the leading role of the boundary diffusion can be made. At that, the nanodots are deformed and show an elliptical shape, which can be treated as a result of the action of relatively strong stress developed in the surface area. The same area of the sample yet treated for 30 min is shown in Fig. 1, b, and 2D nanostructures with a width of about 15 μm are seen with the spatial distribution related to the distribution of the nanodots after 2 min of the treatment. At the same time, the sample area located close to the cap (within 1 mm from the cap walls, partially shadowed from the direct action of plasma ions) exhibits a similar array of the copper oxide nanosheets, which is unaffected by the wrinkles formed on the oxide layer surface.

Moreover, the sample's surface area covered by the cap demonstrates 2D nanosheets with a maximum height of about 20 μm. Thus, the simplest technique of glow discharge allows achieving the large, uniform, and dense arrays of CuO nanosheets just for 30 min, which is in striking contrast with the results typical for the chemical hydrothermal methods; moreover, the distribution of the ion flux over the sample surface does not affect to the formation of the nanostructures.

This finding allows concluding about the ion bombardment's leading role in the fast growth of such structures.

After the SEM studies, the samples treated for 30 min underwent TEM imaging, and the results are shown in Fig. 2.

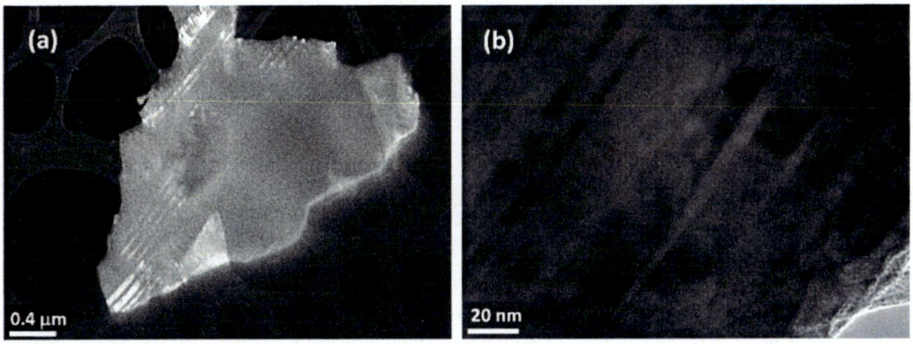

Fig. 2. TEM images of the 2D CuO nanostructures collected from a sample treated for 30 min in oxygen plasma (360 Pa, 500 V, and 0.1 A): a – general view of a nanosheet with parallel grooves that were also found on other nanosheets; b – the enlarged view of the nanosheet where substructures with a width of 5 to 20 nm are seen.

TEM images revealed the interesting substructure of the nanosheets. As can be seen in Fig. 2, a, a set of parallel grooves with well-defined edges was found on the surface of the nanosheets, and this relief feature is common for other nanosheets. The enlarged view in Fig. 2, b demonstrates the substructures' width of 5 to 20 nm. These findings allow an assumption about the possible mechanism of formation of 2D copper oxide nanosheets under strong ion bombardment.

According to the assumption, the nanosheets are formed from a set of nuclei that present the roots of one-dimensional CuO nanostructures known as nanowires. The abundant yields of these 1D structures were observed after the action of the elevated temperatures of 500 °C to 700 °C on the surface of copper samples [30, 31]. At that, the yield was poor for the plasma-enhanced processes conducted at low energy ion bombardment (anodic growth); moreover, disk-shaped microstructures were typical for the experiments [32]. It is likely, that the plasma oxidation under the condition of heavy ion bombardment, which accelerates the oxidation process by order of magnitude [11], results in much faster growth of the copper oxide layer and generation of much higher surface stress, which is beneficial in the formation of a large number of nanometre-sized nuclei of 1D nanostructures formed between the grains of copper oxide and promotes the coalescence of the nuclei into the large nanodots. The distribution of the nanodots along the inter-grain border results in the formation of 2D nanostructures found in the experiment.

To verify the assumption, the growth of two-dimensional nanosheets composed of one-dimensional nanowires was simulated for 10 min using the reported theoretical model [30]. The results of calculations of the dependencies on time of growth of thicknesses of Cu_2O and CuO layers, as well as the dependencies of length and diameter of the nanowires that form a 'skeleton', or basic frame, of the nanosheets, are shown in Fig. 3, a–d, respectively.

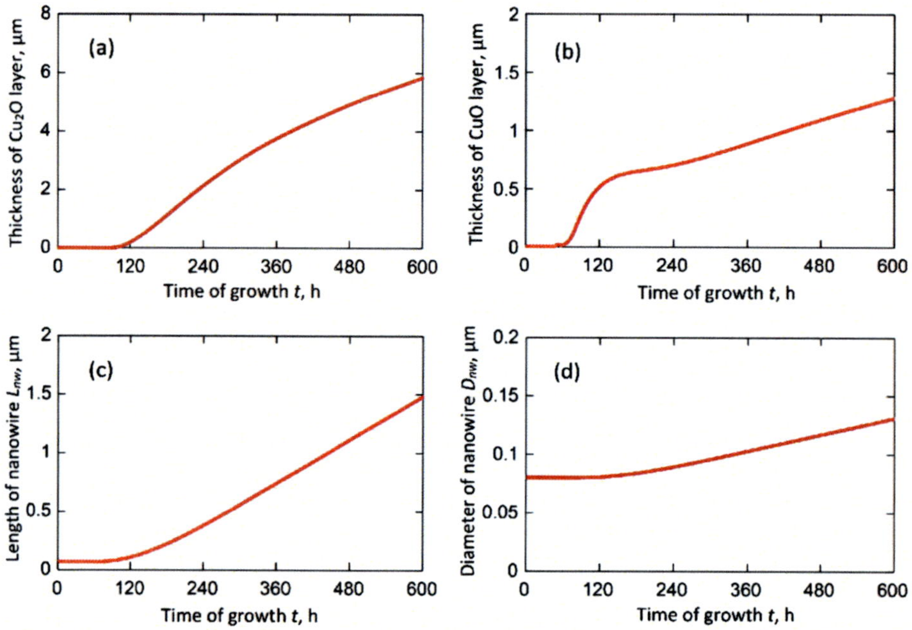

Fig. 3. Results of calculations of the copper oxide system at 600 °C: a, b – dependencies of thicknesses of Cu_2O and CuO layers on time of growth, respectively; c, d – dependencies of length and diameter of nanowires on time of growth, respectively.

To fit the experimental values, the same energies [30] were used, except for the energies of adsorption of oxygen molecules on the side ε_{aO2nws} and top ε_{aO2nwt} surface of the growing nanowires, which were set to 0.87 eV and 1.17 eV, respectively. The increased values of the energies are attributed to the action of the ion flux that causes the generation of a large number of surface defects on the nanowires, thus promoting oxygen adsorption.

The results of simulation of the shape of 2D nanostructures grown at 600 °C for 2, 5, 8, and 10 min are shown in Fig. 4. The initial pattern of the nuclei formed after 2 min on the borders between the hexagonal grains of copper oxide, is shown in Fig. 4, a; at that, the initial diameter of the nuclei used in the system of differential equations [30] was set to 80 nm, which corresponds to the sizes of the substructure observed for the nanodots. In the simulation, an assumption about different starting times for the nanowire growth was used, which explains the different lengths on the nanowires shown in Fig. 4, b after 5 min of growth. Further development of the pattern is shown in Fig. 4, c (8 min) and Fig. 4, d (10 min).

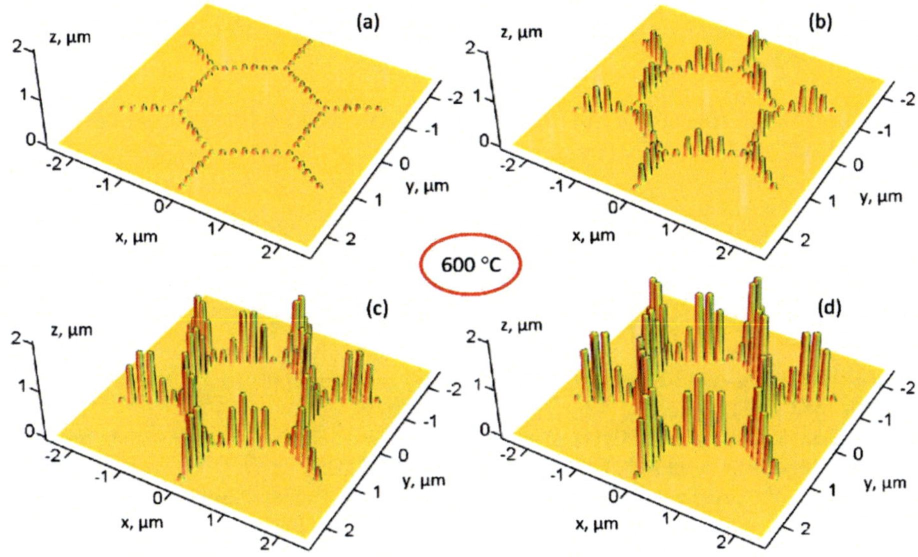

Fig. 4. Results of simulation of the shape of 2D CuO nanostructures grown at 600 °C for different times of growth: a – 2 min; b – 5 min; c – 8 min; d – 10 min.

It should be stressed that the simulation shows only the initial stage of growth of the nanosheets when the skeleton of 2D nanostructures is formed of 1D nanowires; the formation of flat interconnections between the nanowires (like diaphragms) is not considered here. However, based on the experiment, the proposed model allows describing the growth of 2D nanostructures from the array of 1D nanostructures that originated, in turn, from the array of 0D nanostructures aligned along the grains of the oxide.

5 Conclusions

Experimental and theoretical research described in the present paper is dedicated to developing a reliable, highly-productive, environment-friendly, flexible, and cost-effective synthesis of two-dimensional copper oxide nanostructures that can be used for further scientific and industrial applications.

Based on the literature analysis, glow plasma discharge was chosen as the main production tool, and the effect of ion bombardment was studied by screening a part of the sample surface with a cap to remove the ion flux from the flux of the treating particles. At that, the similarity between the nanostructures found under the cap and on the area exposed to the ion flux allows concluding about the neglective effect caused by the ion bombardment concerning the sputtering of the growing nanostructures. Simultaneously, the effect caused by the ions concerning the significant acceleration of the oxidation rates (by order of magnitude) followed by the generation of strong stress in the oxide layer and the plasma etching of the surfaces of the nanowires, which promotes the oxygen adsorption, should be recognized as the main control factors.

In the future, plasma synthesis of 2D CuO nanostructures on substrates with different surface areas will be considered to control the surface stress value for the same plasma operation mode.

Acknowledgment. The author acknowledges the support from the project funded by the National Research Foundation of Ukraine, under grant agreement No. 2020.02/0119, and the support from the project sponsored by the NATO Science for Peace and Security Programme under grant id. G5814 project NOOSE.

References

1. Kumbhakar, P., et al.: Emerging 2D metal oxides and their applications. Mater. Today **45**, 142–168 (2021)
2. Ruzaikin, V., Lukashov, I.: Experimental method of ammonia decomposition study based on thermal-hydraulic approach. Results Eng. **15**, 100600 (2022)
3. Ruzaikin, V., Lukashov, I., Fedorenko, T., Abashin, S.: The equilibrium contact angle of ammonia-stainless steel interface. Results Eng. **16**, 100691 (2022)
4. Zhao, J.G., Liu, S.J., Yang, S.H., Yang, S.G.: Hydrothermal synthesis and ferromagnetism of CuO nanosheets. Appl. Surf. Sci. **257**, 9678–9681 (2011)
5. Hussain, I., et al.: Integration of CuO nanosheets to Zn-Ni-Co oxide nanowire arrays for energy storage applications. Chem. Eng. J. **413**, 127570 (2021)
6. Yang, P., et al.: Fabrication of CuO nanosheets-built microtubes via Kirkendall effect for non-enzymatic glucose sensor. Appl. Surf. Sci. **494**, 484–491 (2019)
7. Yang, C., Xiao, F., Wang, J., Su, X.: 3D flower- and 2D sheet-like CuO nanostructures: microwave-assisted synthesis and application in gas sensors. Sens. Actuators B **207**, 177–185 (2015)
8. Wang, G., Huang, J., Chen, S., Gao, Y., Cao, D.: Preparation and supercapacitance of CuO nanosheet arrays grown on nickel foam. J. Power Sources **196**, 5756–5760 (2011)
9. Liu, Y., et al.: Facile fabrication of CuO nanosheets on Cu substrate as anode materials for electrochemical energy storage. J. Alloy. Compd. **586**, 208–215 (2014)
10. Tang, H., et al.: Rationally designed hierarchical carbon supported CuO nanosheets for highly efficient electroreduction of CO2 to multi-carbon products. J. CO2 Utilization **67**, 102320 (2023)
11. Baranov, O., Filipič, G., Cvelbar, U.: Towards a highly-controllable synthesis of copper oxide nanowires in radio-frequency reactive plasma: fast saturation at the targeted size. Plasma Sources Sci. Technol. **28**, 084002 (2019)
12. Gnytko, O., Kuznetsova, A.: Theoretical research of the chip removal process in milling of the closed profile slots. Arch. Mater. Sci. Eng. **113**(2), 69–76 (2022)

13. Ghosh, A., Miah, M., Bera, A., Saha, S.K., Ghosh, B.: Synthesis of freestanding 2D CuO nanosheets at room temperature through a simple surfactant free coprecipitation process and its application as electrode material in supercapacitors. J. Alloy. Compd. **862**, 158549 (2021)
14. Umara, A., Alshahrani, A.A., Algarni, H., Kumar, R.: CuO nanosheets as potential scaffolds for gas sensing applications. Sens. Actuators B **250**, 24–31 (2017)
15. Rao, M.P., Sathishkumar, P., Mangalaraja, R.V., Asiri, A.M., Sivashanmugam, P., Anandan, S.: Simple and low-cost synthesis of CuO nanosheets for visible-light-driven photocatalytic degradation of textile dyes. J. Environ. Chem. Eng. **6**, 2003–2010 (2018)
16. Li, Y., et al.: CuO nanosheets grown on cupper foil as the catalyst for H2O2 electroreduction in alkaline medium. Int. J. Hydrogen Energy **37**(18), 13611–13615 (2012)
17. Nelson Amirtharaj, S., Mariappan, M.: A facile synthesis of interconnected CuO nanosheets: a promising electrode material for supercapacitor application. Appl. Phys. A **127**, 511 (2021)
18. Breus, A., Abashin, S., Serdiuk, O.: Carbon nanostructure growth: new application of magnetron discharge. J. Achievements Mater. Manuf. Eng. **109**(1), 17–25 (2021)
19. Baranov, O., Romanov, M., Fang, J., Cvelbar, U., Ostrikov, K.: Control of ion density distribution by magnetic traps for plasma electrons. J. Appl. Phys. **112**(7), 073302 (2012)
20. Luan, J., et al.: Plasma-strengthened lithiophilicity of copper oxide nanosheet–decorated Cu foil for stable lithium metal anode. Adv. Sci. **6**(20), 1901433 (2019)
21. Bai, J., et al.: Synthesis of CuO-Cu2O@graphene nanosheet arrays with accurate hybrid nanostructures and tunable electrochemical properties. Appl. Surface Sci. **452**, 259–267 (2018)
22. Korytchenko, K.V., et al.: Numerical simulation of gap length influence on energy deposition in spark discharge. Electr. Eng. Electromech. **1**, 35–43 (2021)
23. Hamdan, A., Agati, M., Boninelli, S.: Selective synthesis of 2D mesoporous CuO agglomerates by pulsed spark discharge in water. Plasma Chem. Plasma Process **41**, 433–445 (2021)
24. Hamdan, A., Kabbara, H., Noël, C., Ghanbaja, J., Redjaimia, A., Belmonte, T.: Synthesis of two-dimensional lead sheets by spark discharge in liquid nitrogen. Particuology **40**, 152–159 (2018)
25. Altaweel, A., Filipič, G., Gries, T., Belmonte, T.: Controlled growth of copper oxide nanostructures by atmospheric pressure micro-afterglow. J. Cryst. Growth **407**, 17–24 (2014)
26. Imam, A., et al.: Nanostructures design by plasma afterglow-assisted oxidation of iron–copper thin films. J. Cryst. Growth **442**, 52–61 (2016)
27. Kantor, B.Ya., Smetankina, N.V., Shupikov, A.N.: Analysis of non-stationary temperature fields in laminated strips and plates. Int. J. Solids Struct. **38**, 8673–8684 (2001)
28. Shypul, O., Myntiuk, V.: Transient thermoelastic analysis of a cylinder having a varied coefficient of thermal expansion. Periodica Polytechnica Mech. Eng. **64**(4), 273–278 (2020)
29. Shupikov, A.N., Smetankina, N.V., Svet, Y.V.: Nonstationary heat conduction in complex-shape laminated plates. J. Heat Transfer **129**(3), 335–341 (2007)
30. Baranov, O., Košiček, M., Filipič, G., Cvelbar, U.: A deterministic approach to the thermal synthesis and growth of 1D metal oxide nanostructures. Appl. Surface Sci. **566**, 150619 (2021)
31. Breus, A., Abashin, S., Serdiuk, O.: Linking dynamics of growth of copper oxide nanostructures in air. In: Nechyporuk, M., Pavlikov, V., Kritskiy, D. (eds.) ICTM 2021. LNNS, vol. 367, pp. 555–564. Springer, Cham (2022). https://doi.org/10.1007/978-3-030-94259-5_47
32. Breus, A., Abashin, S., Lukashov, I., Serdiuk, O.: Anodic growth of copper oxide nanostructures in glow discharge. Arch. Mater. Sci. Eng. **114**(1), 24–33 (2022)

Formation of 2D Carbon Nanosheets and Carbon-Shelled Copper Nanoparticles in Glow Discharge

Andrii Breus(✉), Sergey Abashin, and Oleksii Serdiuk

National Aerospace University, 17, Chkalova Str., Kharkiv 61070, Ukraine
A.Breus@khai.edu

Abstract. A simple and reliable technique of glow discharge ignited at 700 Pa in a mixture of methane with hydrogen (5:1) was used to synthesize 15 min 2D carbon nanosheets with a thickness of about 15 nm and size up to a few μm, as well as composite nanostructures that incorporate copper nanoparticles with a diameter of 15 to 40 nm into the carbon nanosheets. The presence of the copper species was ensured by incorporating a copper anode into the setup. The nanostructures were found in a space between the graphite sample mounted on a graphite cathode and in the dents and craters left after the preliminary stage of ion cleaning of the surfaces. As a result, it was concluded that the necessity of screening the area where the synthesis takes place by a design structure cuts off the flux of ions extracted from plasma but leaves the flux of neutral species containing the carbon and copper precursors.

Keywords: Carbon Nanosheets · Copper Nanoparticles · Glow Discharge · Composite Structures · Manufacturing Innovation

1 Introduction

Nowadays, nanostructures with various morphology [1] and chemical composition [2] were applied in several branches of science and industry because of their outstanding properties acquired at the transition of the materials from the traditional 3D morphology. A synergistic effect originated from the high chemical activity of CuO nanoparticles, and an array of carbon nanosheets with a huge surface-to-volume ratio resulted in the development of a high-performance electrochemical biosensor for glucose detection [3]. Complex hierarchical structures composed of a Pt nanostructured layer and carbon nanowalls with an optional addition of an Au layer were formed on the surface of the SiO2/Ti sample. Adding gold changed the current-voltage relations from the diode-like to the graphene-like [4]. A net of 2D carbon nanostructures was obtained in RF magnetron discharge ignited in an argon atmosphere with a mixture of hydrogen and methane. The nanowalls grown with the addition of the mixture show the increase of the electric field by three orders of magnitude concerning the walls grown just in argon atmosphere, thus making them suitable for application in field emission devices [5]. A surface structure

V. Tonkonogyi et al. (Eds.): InterPartner 2023, LNME, pp. 256–264, 2024.
https://doi.org/10.1007/978-3-031-42778-7_23

with controlled optical properties was obtained by Pierpaoli et al. when doping the carbon nanowalls with boron; the doping is considered the main factor in reaching the transparency of the structures, while the increase in the growth temperature improves the electrical conductivity [6]. The morphology of 2D carbon nanostructures is vital for their optical and electrical properties, such as reflectance, transmittance, absorbance, conductivity, and Hall coefficient [7]. Wearable and highly sensitive pressure sensors were developed on a base of a composite 3D microstructure made of carbon nanowalls embedded in a polydimethylsiloxane matrix [8]. The compressive strength of the typical maze-like structure of carbon nanowalls can reach the value of 50 MPa with Young's modulus of about 28 GPa, which is much higher than for other pyrolytic graphite and graphene-based materials [9]. The mechanical strength and capability to withstand the action of aggressive media are considered the main factors limiting the energy [10] and space [11] applications of nanostructures.

2 Literature Review

To meet the specifications developed for the successful implementation of the nanostructures, many production methods have been developed, including chemical, mechanical [12], and plasma techniques [13]. The later are conditioned by the flexible control of plasma fluxes [14] and the effect caused by the ions on the growing surface layer [15]. Electrodeless methods of plasma synthesis of nanostructures are recognized for the absence of contamination of the nanostructures caused by the materials sputtered from the electrodes [16]. Many techniques have been developed for the last decade based on setups that utilize single-type plasma sources, explained by the relative cost-effectiveness and simplicity [17]. 2D carbon network was grown on a surface of silicon samples in the atmosphere of $Ar/H_2/C_2H_2$ mixture (1050/25/1 sccm) at the gas pressure of 100 Pa when 300 W of capacitively coupled plasma (CCP) discharge power was applied for 10 min [18]. The sample temperature was 700 °C), promoting the growth of relatively thick (500 nm) nanowalls suitable for supercapacitor applications. Inductively coupled plasma (ICP) was used by Shoukat et al. to enhance the CVD process of synthesis of carbon nanowalls on silicon, glass, and pyrex substrates coated with Ni catalyst at the temperature of 650 °C and argon pressure of 2.18 Pa [19]. Chemical vapor deposition, assisted with microwave plasma (MPECVD), was implemented by Kwan et al. to produce a gas sensor with carbon nanowalls engaged as the sensitive element [20]. For the growth of the carbon nanostructures, the mixture of methane and hydrogen at a ratio of 2:1 at the total gas pressure of 65 Pa and the microwave power of 1300 W was applied for 10 min at the substrate temperature of 600 °C. At that, the average height of the nanowalls reached the value of 1 μm at the distance between them of about 200–300 nm. MPECVD was also implemented at 1300 W to grow carbon 2D nanostructures in the mixture of CH4 and hydrogen H2 gases (1:1) at 7 Pa for 10 min for acetone sensing [21]. The ratio of the carbon precursor gas (methane, acetylene, etc.) and discharge power are considered the main control parameters. In the experiments conducted by Batryshev et al., radio-frequency (RF) capacitively coupled plasma (CCP) discharge was ignited to obtain an array of carbon nanowalls on a silicon substrate with nickel catalyst deposited on the substrate surface [22]. For the fixed Ar flow (7 sccm), methane CH4 was introduced into the chamber at the flows of 0.7 to 1 sccm, and the synthesis time varied from 10 to 30 min. At the fixed gas pressure of 250 Pa and substrate temperature of 430 °C, the discharge power was changed independently in the range of 8 to 15 W. It was shown that low power (1–7 W) results in formation of nanotubes; intermediate power (8–15 W) leads to formation of the carbon nanowalls; high power (20–25 W) allows obtaining the multi-layered graphene, while further increasing of the CCP discharge power forms nanoclusters from the nanowalls. At that, the increase in the methane flow rate improves the conditions of the nanowall formation when applied to the low and intermediate power modes. Pulsed filtered cathodic vacuum arc deposition (PFCVAD) was employed by Tüzemena et al. at the pressure of 0.35 Pa to grow the carbon nanowalls that exhibited low reflectance in visible light and are considered for optical applications [23].

However, the characterization and implementation of the developed structures show insufficiency of the existing setups. Thus, bombardment of a substrate surface with energetic ions is the characteristic property of plasma discharges. Implementing large energies like 500–1000 eV leads to drastic thickening of 2D structures, thus turning them into a net of 3D structures, which still can have advanced properties [24]. That is why more sophisticated methods were developed. A review of the growth methods is carried out by Vesel et al., and the problem of developing a reliable method of production of carbon nanowalls is stressed [25]. One of the approaches relies on the combination of different types of plasma sources. Thus, carbon nanowalls were successfully synthesized in a setup that incorporates capacitively coupled (100 MHz) and surface wave (2.45 GHz) sources to generate CH_4/H_2 (2:1) plasma and an inductive energy storage circuit to feed a sample holder with high-voltage pulses (~1 μs) to sustain a temperature of 650 °C of Si sample at the total pressure of 1 Pa in the chamber [26]. For the growth time of 290 s, the nanowalls with a height of 500 nm were obtained with a strong relation of the wall-to-wall distance to the substrate bias voltage. In the experiment, for the input voltage increase from 90 to 150 V, which corresponds to the output voltage of about 500 V and change of power from 60 to 100 W, the average wall-to-wall distance increased from 200 to 700 nm, thus proving the substrate bias being a reliable tool to control the density of 2D carbon nanostructures on the sample surface. It can be concluded from the results that the intensive ion bombardment prevents the nucleation of the nanostructures. Despite the successful implementation, the methods imply an increase in the cost of the equipment.

Another approach is improving the simple methods concerning the modification of the preliminary stages, such as deposition of additional layers or use of catalyst, implementation of screens to remove the negative effect of ion bombardment, change of gases, etc. The effect of a substrate's chemical composition and morphology on the formation of carbon nanowalls was studied by Yerlanuly et al. [27]. It was observed that the height of the carbon nanowalls decreases at the increase of thickness of Al foil. For porous Al_2O_3 membrane, the wider pore size resulted in a perfect reproduction of the membrane surface by the carbon nanowalls. Copper foil is widely applied as a catalyst in synthesizing 2D carbon nanostructures because its presence favors the nucleation at the atomic steps of the Cu(111) surface [28]. Despite the comprehensive utilization of hydrogen in the processes of synthesis of carbon nanosheets, which is conditioned by the ability of this gas to create radicals from the carbon precursor gas, nitrogen was also successfully implemented in electron-cyclotron resonance (ECR) plasma discharge [29].

In the present paper, the utilization of a conventional glow discharge over a graphite cathode with the implementation of a copper anode is described, and the effect of the discharge configuration and material application is discussed. The work is considered a further development of the research [30] to design an optimal configuration of a simple and reliable plasma-based setup for producing carbon nanostructures.

3 Research Methodology

To carry out the research, a stainless-steel vacuum chamber was used. A setup comprising a disk-shaped graphite cathode with a diameter of 35 mm and a thickness of 6 mm and a copper anode with a diameter of 15 mm and 5 mm was designed. The anode was grounded, as well as the vacuum chamber walls. The distance between the electrodes was set to 30 mm. Graphite sample with a diameter of 8 mm and a height of 5 mm was arranged in the center of the cathode. The chamber was filled with a mixture of methane CH_4 and hydrogen H_2 at a ratio of 5:1 at a total gas pressure of 700 Pa. A schematic of the setup is shown in Fig. 1.

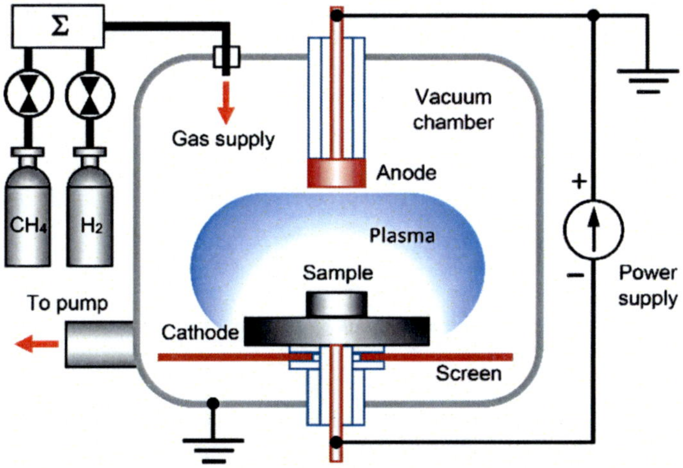

Fig. 1. A schematic of the setup to grow 2D carbon nanostructures on a graphite sample in a glow discharge.

The discharge was sustained at the voltage between the electrodes of 900 V and a discharge current of 0.1 A. At that, the temperature of the sample reached the value of 650 °C. Before achieving the stable mode of the glow discharge, unipolar arcs were generated on the surface of the sample, resulting in craters on the sample surface. After the plasma treatment in the stable operation mode for 15 min, the sample was cooled in the chamber and then studied using scanning electron microscopy (SEM) (Fig. 2).

Fig. 2. A photograph of plasma discharge.

4 Results and Discussion

SEM images obtained for the surfaces of the sample and cathode exposed to the direct action of plasma showed no traces of nanostructures. However, the surfaces of the sample and cathode that contacted each other, i.e., the bottom part of the sample and the part of the cathode covered by the sample, revealed the presence of the nanostructures found in the dent of the surfaces.

SEM images of the surface of the graphite cathode located under the sample are shown in Fig. 3. A general view (Fig. 3, a) of nanostructures found under the edge of the sample in the space that was screened from the plasma exposes the film-like 2D carbon nanostructures and copper microparticles, which are incorporated into a net-like folded microstructure with a thickness of the carbon sheet of about 13.5 nm (Fig. 3, c). In contrast, the surface of the cathode exposed to the direct flux of plasma particles (not covered by the sample) resembles the bulky nanostructures usually observed after the treatment by direct ion flux in ECR plasma setups with a lot of copper droplets probably generated either under the action of the unipolar arcs or as a result of nucleation of the material sputtered from the anode (Fig. 3, d). At the same time, 2D carbon nanostructures are also found in a crater left on the cathode near the edge of the sample (Fig. 3, e) with the thickness of the nanostructures of 15.5 nm and size of about 5 μm (Fig. 3, f).

SEM images of the bottom surface of the edge of the graphite sample, which is contacted with the cathode surface, are shown in Fig. 4.

In this part, the nanostructures differ slightly from those found on the cathode counterpart of the growth area. The size of 2D carbon nanostructures is smaller, and they are sparser, as it is shown in Fig. 4, a. At the same time, the structures are folded with large flat parts of about 1 μm in width and clearly distinguished edges (Fig. 4, b). Moreover, the nanosheets with sizes of about 2 μm (Fig. 4, c) were densely covered (~300 particles/μm^2) with the array of copper nanoparticles of 15 to 40 nm in diameter (Fig. 4, d).

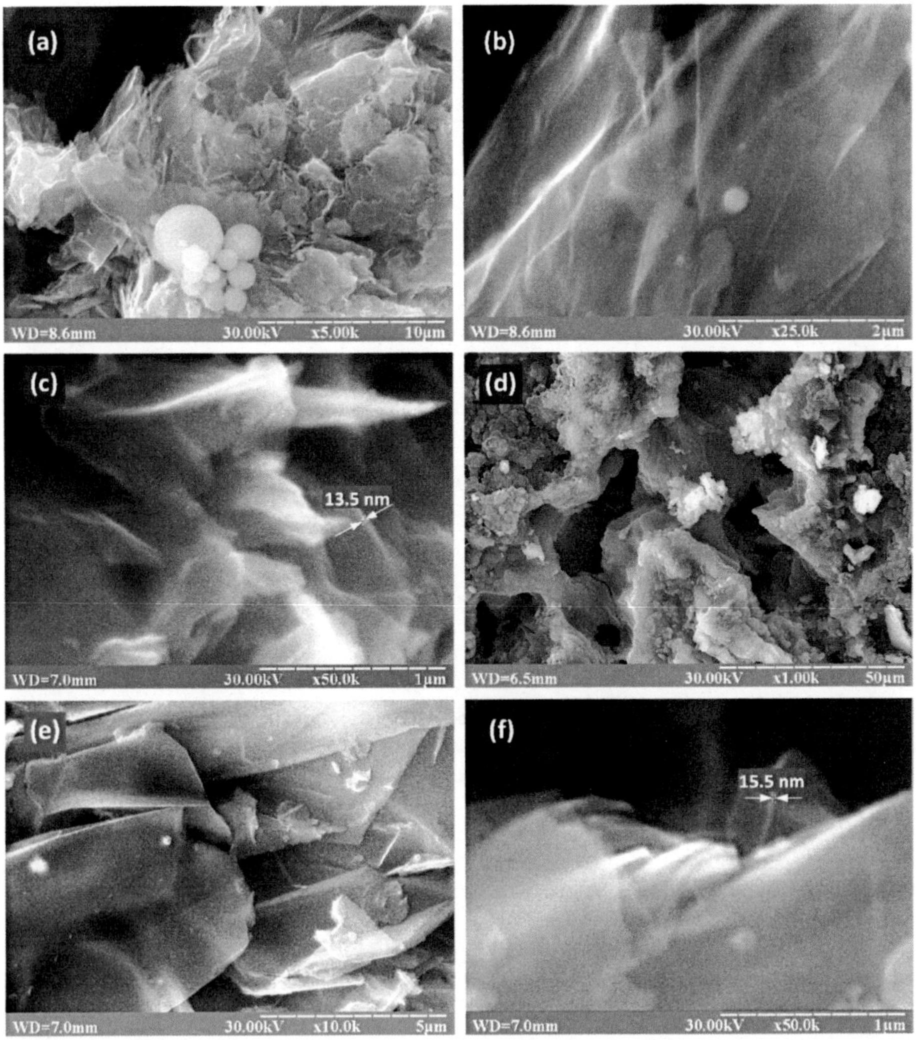

Fig. 3. SEM image of a surface area of the graphite cathode located under the sample that was processed for 15 min in methane-hydrogen (5:1) plasma (700 Pa, 900 V, and 0.1 A): a – general view of nanostructures under the edge of the sample, which exposes film-like 2D carbon nanostructures and copper microparticles; b – magnified view of the nanostructures; c – thickness of the nanostructures of 13.5 nm; d – the surface of the cathode exposed to the direct flux of plasma particles (not covered by the sample); e – 2D carbon nanostructures found in a crater left on the cathode near the edge of the sample; b – a magnified view that reveals the thickness of 15.5 nm of 2D nanostructures.

These findings allow suggesting the following mechanism of formation of the nanostructures. The surfaces of the sample and cathode exposed to the glow discharge plasma undergo the action of plasma ions accelerated in the cathode DC sheath up to the energies of 900 eV, which far exceed the threshold where the sputtering equals the deposition

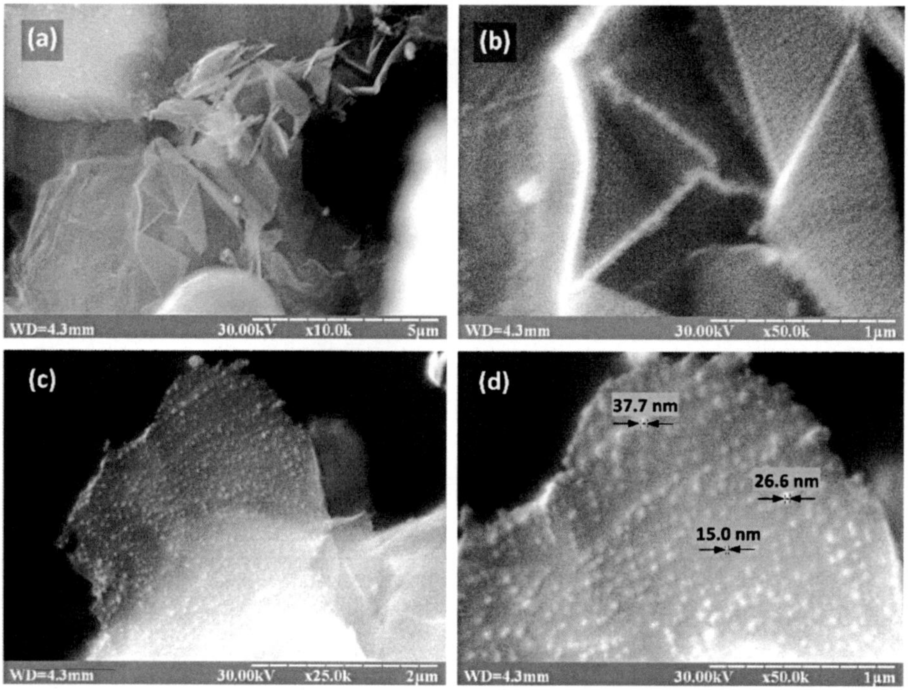

Fig. 4. SEM images of the nanostructures found on the edge of the bottom part of the sample treated for 15 min in methane-hydrogen (5:1) plasma (700 Pa, 900 V, and 0.1 A): a – general view; b – a magnified view, which shows the folded structure of the whole 2D nanosheet with the size of the fragments of about 1 μm; c – enlarged view of 2D carbon nanostructures with a size of about 2 μm covered with copper nanoparticles; d – a magnified view showing the diameters of the copper nanoparticles of 15 to about 40 nm.

rate. Thus, the intense ion bombardment supposedly inhibits the nanostructure growth. However, for the dents and craters on the surface of the negatively biased elements of the setup structure, the shadowing is the factor that cuts off the ion flux, thus leaving only the flux of copper- and carbon-containing neutrals and radicals from plasma. The latter flux can penetrate a few millimeters into the space between the cathode and sample, where the conditions benefit the nanostructure growth.

5 Conclusions

The simple and reliable glow discharge plasma ignition technique proved to be a perspective tool for synthesizing 2D nanostructures. Some limitations caused by the severe ion bombardment should be removed to expand the productivity of the process. For 15 min, dense arrays of 2D carbon nanosheets were grown on the surfaces screened by the details of the surface relief, such as dents and craters left after the preliminary stage of the ion cleaning conducted over the surfaces before the synthesis. At that, the addition of copper anode provided the growth region with a flux of catalyst, which allowed the formation of a composite material comprised of copper nanoparticles incorporated into the carbon sheet. The structure can be considered for further catalytic applications since the copper particles are protected from oxidation with the carbon layer. In the future, the proposed synthesis method is planned to update with a set of graphitic sheets located on the cathode and separated from each other by a few mm spaces to screen the nanostructure growth area from the ion flux.

Acknowledgment. The research was partially sponsored by the NATO Science for Peace and Security Programme under grant id. G5814 project NOOSE. A. Breus and S. Abashin acknowledge the support from the project funded by the National Research Foundation of Ukraine under grant agreement No. 2020.02/0119.

References

1. Baranov, O., Košiček, M., Filipič, G., Cvelbar, U.: A deterministic approach to the thermal synthesis and growth of 1D metal oxide nanostructures. Appl. Surface Sci. **566**, 150619 (2021)
2. Guo, B., et al.: Single-crystalline metal oxide nanostructures synthesized by plasma-enhanced thermal oxidation. Nanomaterials **9**(10), 1405 (2019)
3. Zhai, Z., et al.: Rational construction of 3D-networked carbon nanowalls/diamond supporting CuO architecture for high-performance electrochemical biosensors. Small **15**, 1901527 (2019)
4. Bita, B., et al.: On the structural, morphological, and electrical properties of carbon nanowalls obtained by plasma-enhanced chemical vapor deposition. Hindawi J. Nanomater. **2020**, 8814459 (2020)
5. Guzmán-Olivos, F., Espinoza-González, R., Fuenzalida, V., Morell, G.: Field emission properties of carbon nanowalls prepared by RF magnetron sputtering. Appl. Phys. A **125**, 354 (2019)
6. Pierpaoli, M., et al.: Tailoring electro/optical properties of transparent boron-doped carbon nanowalls grown on quartz. Materials **12**, 547 (2019)
7. Yerlanuly, Y., et al.: Physical properties of carbon nanowalls synthesized by the ICP-PECVD method vs. the growth time. Sci. Rep. **11**, 19287 (2021)
8. Zhou, X., Zhang, Y., Yang, J., Li, J., Luo, S., Wei, D.: Flexible and highly sensitive pressure sensors based on microstructured carbon nanowalls electrodes. Nanomaterials **9**, 496 (2019)
9. Ghodke, S., et al.: Mechanical properties of maze-like carbon nanowalls synthesized by the radial injection plasma enhanced chemical vapor deposition method. Mater. Sci. Eng. A **862**, 144428 (2023)
10. Ruzaikin, V., Lukashov, I.: Experimental method of ammonia decomposition study based on thermal-hydraulic approach. Results Eng. **15**, 100600 (2022)
11. Levchenko, I., et al.: Diversity of physical processes: challenges and opportunities for space electric propulsion. Appl. Sci. **12**(21), 11143 (2022). https://doi.org/10.3390/app122111143
12. Gnytko, O., Kuznetsova, A.: Theoretical research of the chip removal process in milling of the closed profile slots. Arch. Mater. Sci. Eng. **113**(2), 69–76 (2022)
13. Breus, A., Abashin, S., Lukashov, I., Serdiuk, O.: Catalytic growth of carbon nanostructures in glow discharge. In: Ivanov, V., Trojanowska, J., Pavlenko, I., Rauch, E., Peraković, D. (eds.) DSMIE 2022. LNME, pp. 375–383. Springer, Cham (2022). https://doi.org/10.1007/978-3-031-06025-0_37
14. Baranov, O., Romanov, M., Fang, J., Cvelbar, U., Ostrikov, K.: Control of ion density distribution by magnetic traps for plasma electrons. J. Appl. Phys. **112**(7), 073302 (2012)

15. Baranov, O., Fang, J., Rider, A., Kumar, S., Ostrikov, K.: Effect of ion current density on the properties of vacuum arc-deposited TiN coatings. IEEE Trans. Plasma Sci. **41**(12), 3640–3644 (2013)
16. Baranov, O., Filipič, G., Cvelbar, U.: Towards a highly-controllable synthesis of copper oxide nanowires in radio-frequency reactive plasma: fast saturation at the targeted size. Plasma Sources Sci. Technol. **28**, 084002 (2019)
17. Breus, A., Abashin, S., Lukashov, I., Serdiuk, O.: Anodic growth of copper oxide nanostructures in glow discharge. Arch. Mater. Sci. Eng. **114**(1), 24–33 (2022)
18. Guerra, A., et al.: ZnO/Carbon nanowalls shell/core nanostructures as electrodes for supercapacitors. Appl. Surface Sci. **481**, 926–932 (2019)
19. Shoukat, R., Khan, M.I.: Synthesis of nanostructured based carbon nanowalls at low temperature using inductively coupled plasma chemical vapor deposition (ICP-CVD). Microsyst. Technol. **25**, 4439–4444 (2019)
20. Kwon, S., et al.: Room temperature gas sensor application of carbon nanowalls using electrical resistance change by surface adsorption of toxic gases. Mater. Res. Bull. **141**, 111377 (2021)
21. Choi, H., Kwon, S.H., Kang, H., Kim, J.H., Choi, W.: Zinc-oxide-deposited carbon nanowalls for acetone sensing. Thin Solid Films **700**, 137887 (2020)
22. Batryshev, D., Yerlanuly, Y., Alpysbaeva, B., Nemkaeva, R., Ramazanov, T., Gabdullin, M.: Obtaining of carbon nanowalls in the plasma of radio-frequency discharge. Appl. Surface Sci. **503**, 144119 (2020)
23. Tüzemena, E.Ş, et al.: Formation of carbon nanowalls by pulsed filtered cathodic vacuum arc deposition. Diam. Relat. Materials **93**, 200–207 (2019)
24. Çelikel, O., Kavak, H.: Catalyst-free carbon nanowalls grown on glass and silicon substrates by ECR-MPCVD method. Diam. Relat. Mater. **120**, 108610 (2021)
25. Vesel, A., Zaplotnik, R., Primc, G., Mozetič, M.: Synthesis of vertically oriented graphene sheets or carbon nanowalls—Review and challenges. Materials **12**, 2968 (2019)
26. Ichikawa, T., Shimizu, N., Ishikawa, K., Hiramatsu, M., Hori, M.: Synthesis of isolated carbon nanowalls via high-voltage nanosecond pulses in conjunction with CH4/H2 plasma enhanced chemical vapor deposition. Carbon **161**, 403–412 (2020)
27. Yerlanuly, Y., et al.: Synthesis of carbon nanowalls on the surface of nanoporous alumina membranes by RI-PECVD method. Appl. Surface Sci. **523**, 146533 (2020)
28. Wang, M., et al.: Controllable electrodeposition of ordered carbon nanowalls on Cu(111) substrates. Mater. Today **57**, 75–83 (2022)
29. Kar, R., Tripathy, S.P., Keskar, N., Sinha, S.: Effect of processing gas compositions on growth of carbon nanowalls by ECR-CVD process. Mater. Res. Express **6**, 065029 (2019)
30. Breus, A., Abashin, S., Serdiuk, O.: Carbon nanostructure growth: new application of magnetron discharge. J. Achievements Mater. Manuf. Eng. **109**(1), 17–25 (2021)

The Wear Resistance During Oscillating Friction of Steel Specimens with Strengthened Nanocrystalline Layers

Ihor Hurey[1,2], Volodymyr Gurey[1](✉), Tetyana Hurey[1], Marian Bartoszuk[3], and Weronika Wojtowicz[2]

[1] Lviv Polytechnic National University, 12, Bandera Str., Lviv 79013, Ukraine
Volodymyr.I.Hurei@lpnu.ua
[2] Rzeszow University of Technology, 12, Powstancow Warszawy Ave., 35-959 Rzeszow, Poland
[3] Opole University of Technology, 5, Mikolajczyka St., 45-271 Opole, Poland

Abstract. Thermo-deformation treatment refers to methods of surface strengthening using highly concentrated energy sources. In the process of processing, the surface layers of the metal are modified, and strengthened white layers with a nanocrystalline structure are formed. The conducted studies showed that during the thermo-deformation treatment of the working surfaces of specimens made of steel 41Cr4 (quench-hardening and low-temperature tempering) using different technological media (mineral oil (MO), mineral oil with active additives containing polymers (APP) and saturated aqueous solution of mineral salts based on magnesium and calcium chlorides (ASMC)), a strengthened white layer with a thickness of 160 μm to 260 μm and a hardness of 7.6–8.2 GPa was formed. It is shown that the technological medium used during thermo-deformation treatment affects the wear resistance with reversible friction without lubrication. Thus, when using APP and ASMC, the wear resistance of steel-bronze friction pairs increased by 3.1–3.3 times, and when using MO, by 2.1–2.5 times, compared to the non-strengthened pair. During studies of friction pairs of "Steel 41Cr4 – Bronze CuAl10Ni5Fe4", "Steel 41Cr4 – Steel 30HGSA" in oscillating (reversible) friction without lubrication, as well as single-direction sliding friction, strengthened white layers with a nanocrystalline structure significantly increase their wear resistance.

Keywords: Industrial Growth · White Layer · Nanocrystalline Structure · Wear · Process Innovation

1 Introduction

Modern engineering products operate at high speeds and cyclic loads. Their reliability is determined by the quality of manufacturing machine parts, assembly processes, and operating conditions [1]. The operational characteristics of machine parts depend on the parameters of the quality of the working surfaces [2] and the surface layer [3]. The processes of destruction of machine parts begin from their contacting surfaces [4]. In the surface layers of the metal of machine parts [5], the crystal lattice accumulates various

V. Tonkonogyi et al. (Eds.): InterPartner 2023, LNME, pp. 265–275, 2024.
https://doi.org/10.1007/978-3-031-42778-7_24

defects [6], which lead to accumulation of various defects, which lead to the initiation and propagation of cracks [7] and the destruction of the surface layer and the part as a whole [8]. The geometric parameters of the contacting surfaces and the physical, mechanical, and electrochemical properties of the surface layers influence the durability of machine parts during operation [9]. The geometric parameters of the working surfaces of the parts depend on the conditions of their processing in the finishing operations and are characterized by roughness, waviness, and the bearing capacity of the contacting surface [10]. The physical and mechanical properties of the surface layers are determined by the chemical and phase composition, structure [11], grain size of the metal structure components, and stress state [1, 2, 12].

2 Literature Review

During operation, the contact surfaces of the machines are mostly destroyed due to friction processes [13]. One of the effective methods of increasing the durability of machines is the formation of a given surface so that it performs functions that differ from the functions required for the main material. Therefore, it is necessary to provide appropriate parameters for the working surfaces of the friction pair, as well as the quality of the surface layer [1, 11].

A new direction, "Surface Engineering", is used during manufacture to increase the durability of the contacting surfaces of machine parts during operation. During modification, it forms the specified parameters of the treated surfaces' quality and the metal's surface layer [11]. During the modification of the surface layer of metal [14], the process of strengthening [15]. The chemical and phase composition, structure, grain size, hardness, stress state of the metal [16], and crystal lattice dimensions are changing [17]. In this case, the section has no boundary between the strengthened surface layer and the base metal of the part [14]. The methods of modifying the surface layer of metal are processing methods using highly concentrated energy sources (laser [18], plasma [19], electron beam machining [20], and friction strengthening treatment [21, 22]), as well as intensive plastic deformation [23, 24]. In the process of these treatments, strengthened surface layers with a nanocrystalline structure are formed [25] (the grain size of the metal structure is less than 100 nm in at least one direction) [26], which have specific physical, mechanical, and other properties that are significantly different from the properties of the base metal [26, 27].

When applying surface treatment technologies using highly concentrated energy flows in the surface layers of metal under the action of high-temperature gradients [28], structural and phase transformations pass [18, 29]. The surface layers are heated at high speeds to temperatures above the point of phase transformations [30]. After escaping the source of thermal energy, the following rapid cooling of the surface layers due to heat dissipation into the depth of the metal passes [27]. As a result, the structure [31], phase and chemical compositions [29, 32], and physical, mechanical, and chemical characteristics of the metal change in the surface layers [25, 26].

Thermo-deformation treatment (TDT) refers to surface treatment methods using highly concentrated energy flows. Such a flow of energy is created by high-speed friction (60–80 m/s) of a strengthening tool on the work surface of the workpiece. Simultaneous shear deformation occurs during the processing in the contact area "tool-part" [33]. The surface layers of the metal are heated to temperatures above the point of phase transformations (Ac_3) [34]. After moving, the contact zone "tool-part" takes place in high-speed cooling of the metal surface layer. In the surface layer of machine parts, a specific structural-stressed state of metal [35] – a white layer with a nanocrystalline structure is formed [36].

Most studies of the durability of friction pairs, including white layers, were performed during single-directional sliding [13, 35]. Studies [33, 35] have shown that the strengthened white layers with nanocrystalline structure significantly increase the wear resistance of friction pairs during various types of wear.

In practice, there are mechanisms in which friction pairs work in oscillating movements, and the oscillating (reversible) wear-and-tear is realized [37]. An example is the linkage and crank mechanisms, mated parts of the chain conveyors, and other surfaces of mated parts. The course of the wear process with oscillating (reversible) friction of a strengthened white layer with the nanocrystalline structure is unknown since the surface layers constantly change the direction of sliding to the opposite. Therefore, this work aims to influence the obtained strengthened nanocrystalline layers' wear resistance during oscillating (reversible) friction.

3 Research Methodology

Studies of wear in the oscillating (reversible) friction were carried out on the friction and wear testing machine (UMT-1) (Fig. 1) installation according to the scheme "Plane-Plane" (Fig. 2a). Moveable and stationary (fixed) specimens were fixed in the clamps (Fig. 2b). The required load was applied to a fixed specimen by using a load device and the oscillating movements were formed by the crank mechanism and the eccentric for the moveable specimen. The study of wear resistance was carried out on specimens made of steel 41Cr4 (quench-hardening and low-temperature tempering, HRC 52–54). Moveable and stationary (fixed) specimens are in contact with the end (face) surfaces. The chemical composition of the steel 41Cr4 is as follows: mass %: 0.40 C; 0.78 Mn; 0.26 Si; 1.12 Cr; 0.01 S; 0.01 P; Fe: balance. The end surface of the fixed specimen has the shape of a circle with a diameter of 19 mm, and the moveable specimen – has the shape of a ring, an outer diameter is 15.8 mm and an inner – of 11 mm. The area of contact of the specimens is 100 mm^2.

TDT strengthened the end contact surface of the fixed specimens. Counter-specimens (moveable specimens) were made of bronze CuAl10Ni5Fe4 and steel 30HGSA (PL) (quench-hardening and low-temperature tempering, HRC 38–40). The chemical composition of the bronze CuAl10Ni5Fe4 is as follows: mass %: 3.5–5.5 Fe; 0.3 Mn; 0.1 Si; 3.5–5.5 Ni; 0.01 P; 9.5–11 Al; 77.4–83.5Cu; 0.02 Pb; 0.3 Zn; 0.1 Sn and the chemical composition of the steel 30HGSA (PL) is as follows: mass %: 0.28–0.34 C; 0.8–1.1 Mn; 0.9–1.2 Si; 0.8–1.1 Cr; 0.025 S; 0.025 P; 0.3 Cu; 0.3 Ni; Fe: balance. Studies of the wear resistance of friction pairs were carried out without lubrication. For comparison, similar non-strengthened friction pairs were investigated.

Before the test, all specimens of a friction pair were run in before stabilization of the moment of friction and alignment of mating surfaces, which was evaluated in the presence of traces of friction on the area of at least 90% of the friction working surface of each specimen.

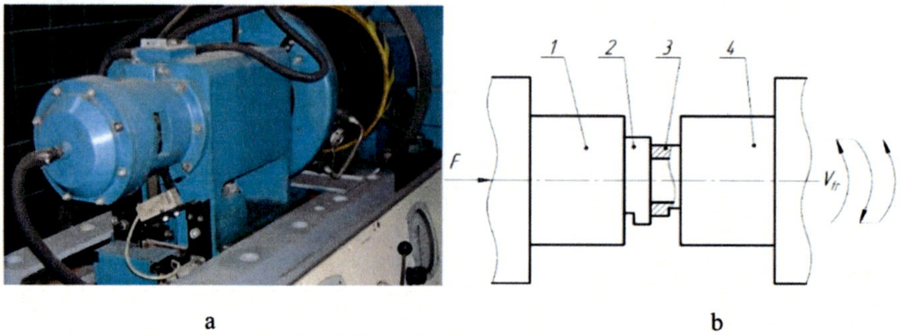

Fig. 1. The friction and wear testing machine (UMT-1) (a) and scheme of study the wear in the oscillating (reversible) friction according to the scheme "Plane-Plane" (b): 1 – fixed specimen; 2 – moveable specimen; 3, 4 – clamps; *F* – the force of pressing; V_{fr} – velocity of friction.

The criterion for the amount of wear was taken from the weight loss of specimens after a certain length of friction, which was determined by weighing on the analytical balance with an accuracy of ±0.2 mg. Also, during the experiments, the amount of linear wear on the working surfaces of the specimens was constantly recorded using the strain gauges, which also recorded the established friction coefficient (factor).

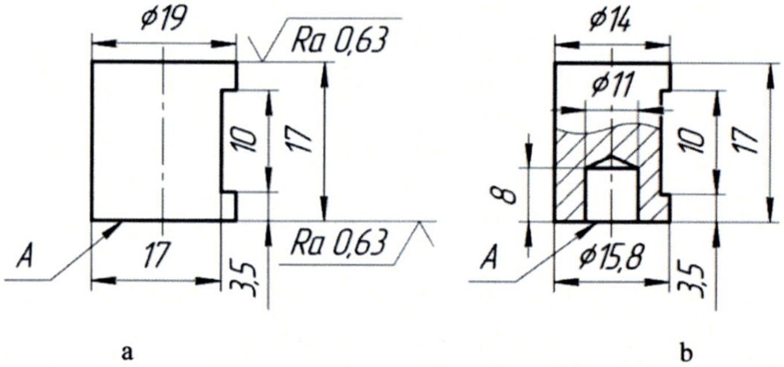

Fig. 2. Specimen (a) and counter-specimen (b) for studying the wear in the oscillating (reversible) friction according to the scheme "Plane-Plane".

After the alignment of mating (contact) surfaces of the friction pair, the drive of friction and wear testing machine was switched on and set the required load. The load was given in the range F = 10…100 MPa and in steps of 10 MPa. The duration of experiments at each load was 100 cycles, and the total duration of the experiment was 1000 cycles. The load on each following range (step) was applied without interrupting the experiment. The angular displacement of the specimen was 90° and the displacement frequency – 5 double strokes per minute.

The TDT of the end (face) contact surfaces of fixed specimens made of steel 41Cr4 were carried out by a modernized surface-grinding machine of the SPC-20a (Jotes, Poland) (Fig. 3). Instead of a grinding wheel, the tool (disk) for TDT made of a stainless steel 1H18N10T (PL) was installed, as well as the modernization of the drive of the machine's main motion. The tool's dimensions correspond to the grinding wheel size used on the surface-grinding machine. The linear velocity on the periphery of the tool was 60–70 m/s. The technological medium was fed to ensure the appropriate quality parameters of the treated surfaces in the tool's contact area with the part during TDT. The media tested were: mineral oil (MO), mineral oil with active additives containing polymers (AAP), and a saturated aqueous solution of mineral salts based on magnesium and calcium chlorides (ASMC).

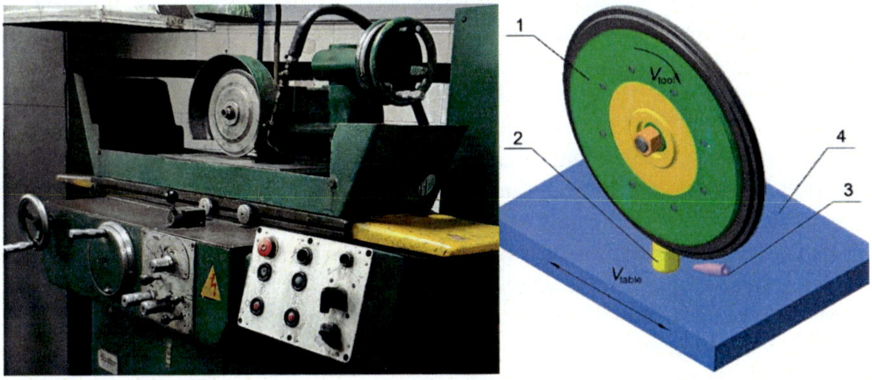

a b

Fig. 3. Surface-grinding machine of the SPC-20a (Jotes, Poland) (a), scheme of the process of strengthening by TDT (b): 1 – tool; 2 – specimen; 3 – nozzle to feed the technological medium; 4 – table of surface-grinding machine; V_{table} – velocity of the displacement of the table; V_{tool} – linear velocity of rotation of the tool.

The phase composition and average grain size L of the steel's surface layer after MPT was determined by X-Ray analysis using the diffractometer DRON-3 with a CuKα X-Ray source (voltage of 30 kV and intensity of 20 mA), spacing of 0.05° and the exposition of 4 s. The diffractograms were post-processed using the software CSD [38]. The X-Ray diffraction patterns were analyzed after the Joint Committee on Powder Diffraction Standards/American Society for Testing and Materials (JCPDS-ASTM) index [39].

The microhardness Hμ was measured using the microhardness testing machine PMT-3 at the load of 50 g on metallographic sections made from flat specimens of steel 41Cr4. Tests were carried out by indentation of a standard 136-degree Vickers diamond pyramid with a square base.

4 Results and Discussion

Metallographic studies have shown that qualitative and solid strengthened surface layers with the nanocrystalline structure are formed after TDT using various technological media on specimens made of steel 41Cr4 (quench-hardening and low-temperature tempering). The thickness of the strengthened white layer was 150–160 μm, and the hardness of the layer was 7.6 GPa, with the hardness of the main metal structure of 4.8 GPa after TDT using the MO as a technological medium. When used AAP as a technological medium, the thickness of the strengthened layer was more significant – 200–210 μm. After TDT with the use of ASMC, the thickness of the strengthened layer was even larger and was 250–260 μm. The hardness of the strengthening layer after processing using AAP and ASMC was almost the same and was 8.2 GPa.

X-ray studies of the obtained strengthened surface layers showed that the grain size of their structure near the surface is 20–50 nm. The structure's size with the layer's depth decreases and changes smoothly to the main structure. Accordingly, the obtained strengthened layers can be attributed to nanocrystalline.

As noted earlier, studies on determining the wear resistance of friction pairs were performed with single-direction sliding friction. During the reversible sliding motion in the friction pair's contact area, the cyclic changes in the load in the local zones of contact of the peaks of contacting surfaces are presented. Friction without lubrication is the severe type of interaction of contact surfaces in which their geometric parameters and surface layer properties are shown.

Experiments have shown that strengthened white layers with nanocrystalline structure significantly increase the wear resistance of oscillating friction (reversing) without lubrication of pair "Steel 41Cr4 (quench-hardening and low-temperature tempering) – Bronze CuAl10Ni5Fe4" (Fig. 4). The technological media used in the process of TDT significantly influenced the wear process. Thus, the increase in the wear resistance of the white nanocrystalline layer obtained by TDT using the technological medium of mineral oil reaches 2.5 times compared to non-strengthened specimens, AAP – 3.3 times, and ASMC – 3.5 times. This also increases the wear resistance of specimens made of bronze, which is paired with strengthened specimens made of steel. The increase in the wear resistance of bronze specimens exceeds 1.2–1.3 times. The magnitude of the established friction coefficient is also reduced.

During the study of the friction pair's non-strengthened specimens, the friction coefficient increases monotonously (Fig. 5). Study specimens have shown that with increasing load to 50 MPa, the friction coefficient at this time also increases monotonously. Its value is smaller than during the friction of non-strengthened pair. The friction coefficient stabilizes and even slightly decreases with the subsequent load increase of more than 50 MPa. It should be noted that the study of friction pairs with strengthened specimens by TDT with AAP and ASMC showed that the friction coefficient is less than in the study of pair, where TDT strengthened the specimen with MO. In the studies of the non-strengthened pair, the coefficient of friction is much higher than in friction pairs with strengthened specimens. Thus, during studies with a unit load of 90 MPa, the friction coefficient decreased by 1.2–1.4 times compared to the non-strengthened pair. The wear resistance during oscillating friction (reversing) without lubricating the pair with strengthened specimens using TDT with technological media AAP and ASMC is much higher than in pairs with non-strengthened specimens, as shown by studies.

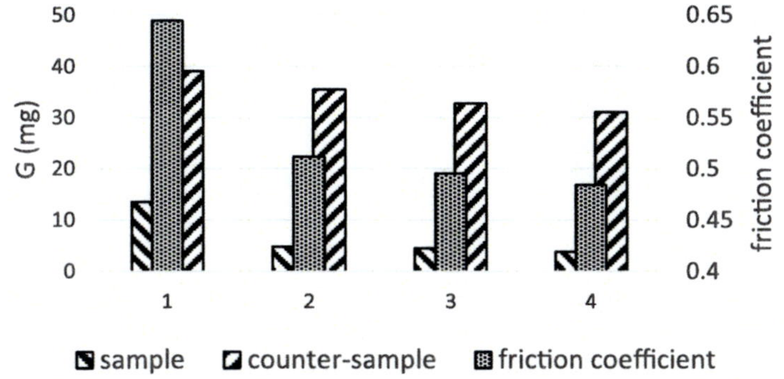

Fig. 4. The magnitude of wear and established friction coefficient of oscillating friction (reversing) without lubrication of the pair "Steel 41Cr4 (quench-hardening and low-temperature tempering) – Bronze CuAl10Ni5Fe4" ($F = 10\ldots100$ MPa; $V_{fr} = 0.005$ m/s; $N = 1000$ cycles): 1 – source (base) metal; 2 – TDT with MO; 3 – TDT with AAP; 4 – TDT with ASMC.

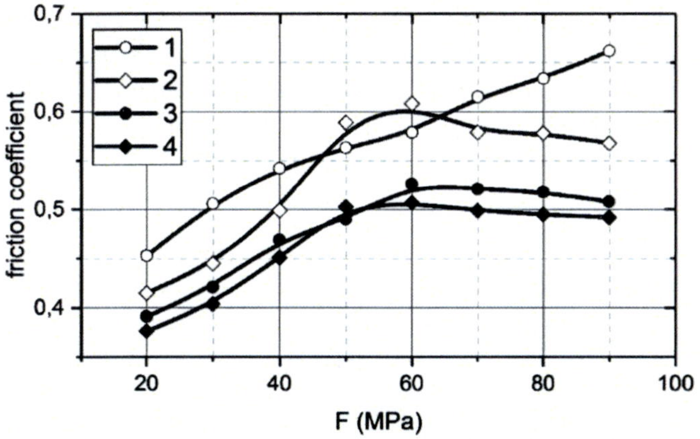

Fig. 5. The dependence of the established friction coefficient on the unit load of oscillating friction (reversing) without lubrication of the pair "Steel 41Cr4 (quench-hardening and low-temperature tempering) – Bronze CuAl10Ni5Fe4" ($V_{fr} = 0.005$ m/s; $N = 1000$ cycles): 1 – source (base) metal; 2 – TDT with MO; 3 – TDT with AAP; 4 – TDT with ASMC.

A similar situation is also observed in the case of oscillating friction (reversing) without lubrication when testing specimens made of steel 30HGSA (PL) (quench-hardening and medium-temperature tempering). In this case, strengthening of steel 41Cr4 (specimens made of steel 30HGSA (PL) were not strengthened, the working surface was only polished) also significantly increases the wear resistance of the friction pair (Fig. 6). So, for example, TDT with MO as a technological medium increases wear resistance by 2.1 times compared to non-strengthened pair, AAP – by 3.1 times, and using ASMC – by 3.3 times.

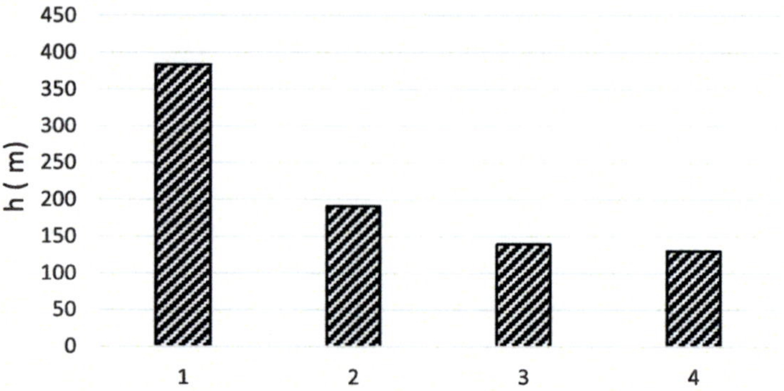

Fig. 6. The magnitude of wear of oscillating friction (reversing) without lubrication of the pair "Steel 41Cr4 (quench-hardening and low-temperature tempering) – Steel 30HGSA (PL)" (Vfr = 0.005 m/s; N = 1000 cycles): 1 – source (base) metal; 2 – TDT with MO; 3 – TDT with AAP; 4 – TDT with ASMC.

The friction coefficient increases monotonically with an increase in the unit load during the wear process of the non-strengthened pair. At the same time, the friction coefficient decreases during the wear process of the friction pair with the specimens after the TDT (Fig. 7). The friction coefficient is much lower after the TDT of specimens using APP and ASMC as a technological medium than the same treatment using MO. Thus, the friction coefficient decreased by almost 1.3 times compared to the non-strengthened pair at a unit load of 100 MPa.

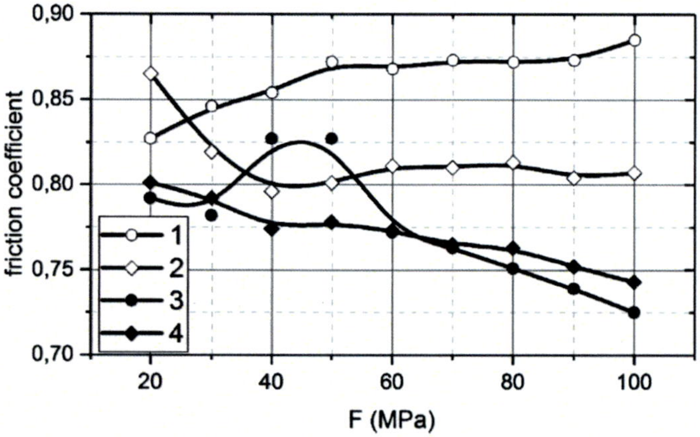

Fig. 7. The dependence of the established friction coefficient on the unit load of oscillating friction (reversing) without lubrication of the pair "Steel 41Cr4 (quench-hardening and low-temperature tempering) – Steel 30HGSA (PL)" ($V_{fr} = 0.005$ m/s; $N = 1000$ cycles): 1 – source (base) metal; 2 – TDT with MO; 3 – TDT with AAP; 4 – TDT with ASMC.

Strengthened surface layers significantly increase the wear resistance of contacting surfaces in oscillating friction. For example, paper [40] shows that applying chrome coatings to the working surfaces of plunger pairs of jet pumps operating in oil wells significantly increased their wear resistance. These parts of the pairs operate under oscillating friction. The thickness of the coatings was 250 μm and 350 μm and the hardness was 8.1 GPa and 11.8 GPa, respectively. The thickness and hardness of the strengthened chromium layers are similar to the parameters of the strengthened layers described in this paper.

5 Conclusions

During TDT with MO, AAP, and ASMC in the surface layers of specimens made of steel 41Cr4 (quench-hardening and low-temperature tempering), the high-quality, uniform, strengthened white layers with nanocrystalline structure are formed. Thus, when using AAP and ASMC, the wear resistance of the friction pair "Steel 41Cr4 – Bronze CuAl10Ni5Fe4" increased by 3.1–3.3 times, and with the use of MO – 2.1…2.5 times, compared to non-strengthened pair. During studies of friction pairs "Steel – Bronze" and "Steel – Steel" with reversible friction without lubrication and single-directional sliding, strengthened white layers with nanocrystalline structure significantly increased their wear resistance. It is advisable to strengthen by using the TDT only one part more technological of friction pair.

References

1. Dwivedi, D.K.: Surface Engineering. Enhancing Life of Tribological Components. Springer, New Delhi (2018). https://doi.org/10.1007/978-81-322-3779-2
2. Davis, J.R.: Surface Engineering for Corrosion and Wear Resistance. ASM International, Detroit (2001)
3. Troshchenko, V.T.: Deformation and destruction of metals under multicycle loading. Naukova Dumka, Kyiv (1981)
4. Troshchenko, V.T., Khamaza, L.A.: Stages of fatigue failure of metals and alloys. Rep. Natl. Acad. Sci. Ukr. **2**, 56–63 (2018). https://doi.org/10.15407/dopovidi2018.10.056
5. Pineau, A., McDowell, D.L., Busso, E.P., Antolovich, S.D.: Failure of metals II: fatigue. Acta Mater. **107**, 484–507 (2016). https://doi.org/10.1016/j.actamat.2015.05.050
6. Santecchia, E., et al.: A review on fatigue life prediction methods for metals. Adv. Mater. Sci. Eng. **2016**, 9573524 (2016). https://doi.org/10.1155/2016/9573524
7. Pedersen, M.M.: Introduction to metal fatigue. Technical report ME-TR-11. Aarhus University (2018). https://doi.org/10.13140/RG.2.2.25216.28163
8. Troshchenko, V.G.: Fatigue of metals at inhomogeneous stress state. Message 1. Methods of stress assessment and research results. Probl. Strengths **2**, 5–24 (2010)
9. Ropyak, L.Ya., Vytvytskyi, V.S., Velychkovych, A.S., Pryhorovska, T.O., Shovkoplias, M.V.: Study on grinding mode effect on external conical thread quality. IOP Conf. Ser. Mater. Sci. Eng. **1018**, 012014 (2021). https://doi.org/10.1088/1757-899X/1018/1/012014
10. Dzyura, V., Maruschak, P.: Optimizing the formation of hydraulic cylinder surfaces, taking into account their microrelief topography analyzed during different operations. Machines **9**(6), 116 (2021). https://doi.org/10.3390/machines9060116
11. Yushchenko, K.A., Borysov, Yu.S., Kuznetsov, V.D., Korzh V.M.: Surface engineering. Naukova dumka, Kyiv (2007)

12. Olugbade, T.O., Lu, J.: Literature review on the mechanical properties of materials after surface mechanical attrition treatment (SMAT). Nano Mater. Sci. **2**, 3–31 (2020). https://doi.org/10.1016/j.nanoms.2020.04.002
13. Hutchings, I., Shipway, P.: Tribology Friction and Wear of Engineering Materials. Butterworth-Heinemann, Oxford (2017)
14. Shaw, L.L.: The surface deformation and mechanical behavior in nanostructured alloys. In: Whang, S.H. (ed.) Woodhead Publishing Series in Metals and Surface Engineering, Nanostructured Metals and Alloys, pp. 481–506. Woodhead Publishing (2011). https://doi.org/10.1533/9780857091123.3.481
15. Quintino, L.: Overview of coating technologies Surface modification by solid state processing. In: Miranda, R. (ed.) Surface Modification by Solid State Processing, pp. 1–24. Woodhead Publishing (2014). https://doi.org/10.1533/9780857094698.1
16. Zhang, F., Duan, C., Sun, W., Ju, K.: Effects of cutting conditions on the microstructure and residual stress of white and dark layers in cutting hardened steel. J. Mater. Process. Technol. **266**, 599–611 (2019). https://doi.org/10.1016/j.jmatprotec.2018.11.038
17. Quintino, L.: Overview of coating technologies. In: Surface Modification by Solid State Processing, vol. 2014, pp. 1–24 (2014). https://doi.org/10.1533/9780857094698.1
18. Bartkowska, A.: Production and properties of FeB-Fe2B-Fe3(B, C). Surface layers formed on tool steel using combination of diffusion and laser processing. Coatings **10**(11), 1130 (2021). https://doi.org/10.3390/coatings10111130
19. Korzhyk, V., Tyurin, Yu., Kolisnichenko, O.: Theory and practice of plasma-detonation technology of surface hardening metal products. PC Technology Center, Kharkiv (2021). https://doi.org/10.15587/978-617-7319-46-6
20. Lukaszkowicz, K.: Review of nanocomposite thin films and coatings deposited by PVD and CVD technology. In: Rahman, M.M. (ed.) Nanomaterials, pp. 145–163. IntechOpen (2011). https://doi.org/10.5772/25799
21. Vilaça, P.: Friction surfacing. In: Miranda, R. (ed.) Surface Modification by Solid State Processing, pp. 25–72. Woodhead Publishing (2014). https://doi.org/10.1533/9780857094698.25
22. Lowe, T.C., Zhu, Y.T., Semiatin, S.L., Berg D.R.: Overview and outlook for materials processed by severe plastic deformation. In: Lowe, T.C., Valiev, R.Z. (eds.) Investigations and Applications of Severe Plastic Deformation. NATO Science Partnership Subseries: 3, pp. 347–356. Springer, Dordrecht (2000). https://doi.org/10.1007/978-94-011-4062-1
23. Segal, V.: Deformation mode and plastic flow in ultra fine grained material. Mater. Sci. Eng. A **406**(1), 205–216 (2005). https://doi.org/10.1016/j.msea.2005.06.035
24. Valiev, R.Z., Islamgaliev, R.K., Alexandrov, I.V.: Bulk nanostructured materials from severe plastic deformation. Progress Mater. Sci. **45**, 103–189 (2000). https://doi.org/10.1016/S0079-6425(99)00007-9
25. Hahn, H.: Unique features and properties of nanostructured materials. In: Zehetbauer, M., Valiev, R.Z. (eds.) Nanomaterials by Severe Plastic Deformation. WILEY-VCH Verlag GmbH & Co. KGaA, Weinheim (2004). https://doi.org/10.1002/3527602461.ch1a
26. Inoue, A., Hashimoto, K.: Amorphous and Nanocrystalline Materials. Preparation, Properties, and Applications. Springer, Heidelberg (2001). https://doi.org/10.1007/978-3-662-04426-1
27. Javaheri, V., et al.: Formation of nanostructured surface layer, the white layer, through solid particles impingement during slurry erosion in a martensitic medium-carbon steel. Wear **204301**, 496–497 (2022). https://doi.org/10.1016/j.wear.2022.204301
28. Zhang, F.-Y., Duan, C., Wang, M., Sun, W.: White and dark layer formation mechanism in hard cutting of AISI52100 steel. J. Manuf. Process. **32**, 878–887 (2018). https://doi.org/10.1016/j.jmapro.2018.04.011

29. Wu, S., Liu, G., Zhang, W., Chen, W., Wang, C.: Formation mechanism of white layer in the high-speed cutting of hardened steel under cryogenic liquid nitrogen cooling. J. Mater. Process. Tech. **302**, 117469 (2022). https://doi.org/10.1016/j.jmatprotec.2021.117469
30. Hosseini, S.B., Klement, U.: A descriptive phenomenological model for white layer formation in hard turning of AISI 52100 bearing steel. CIRP J. Manuf. Sci. Technol. **32**, 299–310 (2021). https://doi.org/10.1016/j.cirpj.2021.01.014
31. Merino, P., et al.: An attempt to generate mechanical white etching layer on rail surface on a new rolling contact test bench. Wear **482–483**, 203945 (2021). https://doi.org/10.1016/j.wear.2021.203945
32. Xu, Y., Fang, L., Cen, Q., Zhu, J.: Nanostructure and transformation mechanism of white layer for AISI1045 steel during impact wear. Wear **258**, 537–544 (2005). https://doi.org/10.1016/j.wear.2004.06.002
33. Kyryliv, V., Maksymiv, O., Gurey, V., Hurey, I., Kyryliv, Y., Zvirko, O.: The mode deformation effect on surface nanocrystalline structure formation and wear resistance of steel 41Cr4. Coatings **13**, 249 (2023). https://doi.org/10.3390/coatings13020249
34. Gurey, V., Shynkarenko, H., Kuzio, I.: Mathematical model of the thermoelasticity of the surface layer of parts during discontinuous friction treatment. In: Ivanov, V., Pavlenko, I., Liaposhchenko, O., Machado, J., Edl, M. (eds.) DSMIE 2021. LNME, pp. 12–22. Springer, Cham (2021). https://doi.org/10.1007/978-3-030-77823-1_2
35. Hurey, I., Hurey, T., Gurey, V.: Wear resistance of hardened nanocrystal-line structures in the course of friction of steel-grey cast iron pair in oil-abrasive medium. In: Ivanov, V., et al. (eds.) DSMIE 2019. LNME, pp. 572–580. Springer, Cham (2019). https://doi.org/10.1007/978-3-030-22365-6_57
36. Gurey, V., Hurey, I.: Influence of surface hardened nanocrystalline layers on the resistance of contact fatigue destruction. In: Ivanov, V., Trojanowska, J., Pavlenko, I., Zajac, J., Peraković, D. (eds.) DSMIE 2020. LNME, pp. 483–491. Springer, Cham (2020). https://doi.org/10.1007/978-3-030-50794-7_47
37. Bai, Z., Ning, Z., Zhou, J.: Study on wear characteristics of revolute clearance joints in mechanical systems. Micromachines **13**, 1018 (2022). https://doi.org/10.3390/mi13071018
38. Krous, W., Nolze, G.: Powder Cell – a program for the representation and manipulation of crystal structures and calculation of the resulting X-ray powder patterns. J. Appl. Cryst. **29**, 301–303 (1996). https://doi.org/10.1107/S0021889895014920
39. Powder Diffraction File Search Manual [inorganic]: Alphabetical Listing and Search Section of Frequently Encountered Phases. Joint Committee on Powder Diffraction Standards, Philadelphia, USA (1976)
40. Bazaluk, O., Dubei, O., Ropyak, L., Shovkoplias, M., Pryhorovska, T., Lozynskyi, V.: Strategy of compatible use of jet and plunger pump with chrome parts in oil well. Energies **15**, 83 (2021). https://doi.org/10.3390/en15010083

Temperature Field Behavior on Plate Width at Thermomechanical Rolling of Low Carbon Microalloyed Steel at the Steckel Mill

Volodymyr Kukhar[1] (✉), Oleksandr Kurpe[2], and Khrystyna Malii[1]

[1] Technical University "Metinvest Polytechnic" LLC, 80, Pivdenne Highway, Zaporizhzhia 69008, Ukraine
kvv.mariupol@gmail.com

[2] Metinvest Engineering LLC, 74, Pivdenne Highway, Zaporizhzhia 69008, Ukraine

Abstract. The metal temperature before the reduction in rolls and before controlled cooling is the most significant factor influencing the formation of rolled products' mechanical properties. The wide sheet chilling of edges effect leads to the appearance of a temperature crown and, accordingly, to an uneven distribution of mechanical properties across the finished product width. Steckel mills (with furnace coilers) can reheat the coils, significantly eliminating the presence of a temperature gradient. However, implementing a thermomechanical controlled process (TMCP) at Steckel mills, which is highly sensitive to uneven temperature distribution, requires forecasting temperature fields to correct thermal conditions. This paper presents the results of developing and implementing a finite-difference mathematical model for calculating the temperature field for flat products concerning the mills with furnace coilers conditions. It is shown that a temperature drop occurs during the passing of the rolled metal from the furnace coiler to the mill stand, which must be compensated to ensure an ordered temperature gradient across the width.

Keywords: Industrial Growth · Temperature Modes · Temperature Gradient · Thermomechanical Controlled Process · Steckel Mill · Furnace Coiler · Finite-Difference Mathematical Model

1 Introduction

Sheet steel manufacturers highly approved rolling mills with reversing stands and furnace coilers (Steckel mills) while producing small batches of various products [1]. The ability to produce a very thin and, at the same time, wider sheet than continuous wide-strip hot mills is considered to be one of the greatest advantages of such mills [2]. In addition to possibly assembling mills with cast and rolling units (Compact Strip Production processes), Steckel mills can contain a roughing universal mill stand. They also provide rolling in a finishing reversing stand at a high and constant speed while maintaining an overall high level of rolling temperature [1]. Usually, synchronization of the operation of stands is required to ensure technological speed, temperature, and power modes [3]. The

V. Tonkonogyi et al. (Eds.): InterPartner 2023, LNME, pp. 276–285, 2024.
https://doi.org/10.1007/978-3-031-42778-7_25

subsequent arrangement of the laminar or accelerated cooling implements the production processes of heat-treated, tempered, and thermomechanical-controlled processed steel of HSS and AHSS classes and the production of pipe grades and DP steels [2]. However, thermomechanical rolling is very sensitive to temperature changes across the width of rolled stock. Therefore, the cooling of edges during the movement of a strip from a furnace coiler to a stand and from the stand to a table of the cooling system must be considered and compensated since the relationship between the temperature gradient and the uneven distribution of mechanical properties across the width of rolled stock is obvious.

2 Literature Review

A thermomechanical rolling process that combines controlled rolling and controlled cooling was used to produce shape [4] and flat [5] rolled products. Simulation of the TMCP process is very time- and resource-intensive because it is a coupled thermo-mechanical-microstructural model split into several independent parts [1, 4–6]: thermo-mechanical processing, austenite microstructure evolution during thermo-mechanical steel treatment, pearlite transformation model, and microstructural-mechanical properties. The modern competitive environment in the flat rolled metal products market requires TMCP on broad-strip (wide-strip) and plate mills that have never been used in this way before by their reconstruction [3, 5]. At the same time, a Coil-Box unit is installed to improve the energy-saving effect and provide better homogeneity of coiled stock temperatures, while accelerated cooling is controlled online (On-Line Accelerated Cooling) [7]. High strength indices for X65-X80 steels (535–827 MPa) are achieved by using different microalloying variations (C-Mn-Mo-Nb, C-Mn-Cr-Nb-Ti, and C-Nb-V-Mo-Ni-Ti), which lead to the formation of ferrite-pearlite and ferrite-bainite structures [8].

The work [9] conducted hot-rolling experiments from the initial temperature of 800–700 °C to the final temperature of 400–300 °C to reveal the effect of the temperature factor obtaining optimal modes with a tensile strength of specimen between 871–851 MPa. The study of the combined action of factors in [10] revealed that the final temperature of 958 °C leads to the best mechanical properties of flat products for high-speed rolling of St60Mn steel. The work [11] studied the effect of TMCP modes on the performance properties of S460 and S690 steel at elevated temperatures.

The work [12] modeled uneven cooling of workpieces at the hot shape and plate rolling, considering the importance of the effect of temperature on the mechanical properties of rolled stocks. At the same time, the work [13] justified the use of induction heating of edges of slotted inductors in the interstand space of a broad-strip mill, which is used for the alignment across the width of the temperature gradient. According to the application of continuous steel teeming [14], descaling [15], and adjustment of the thermal state of mill rolls [16], water and spray cooling are effectively used to equalize the temperature range of plates. Deformation processes, microstructure evolution, and temperature distribution changes can be separately considered for the Steckel mills [17]. Numerical mathematical models have been promoted based on finite difference and finite element methods. The work [18] required using a complex hybrid mathematical and neural network model to predict temperature and power mode changes only for hot rolling conditions. The study of processes of differentiated and non-uniform induction heating [19], together with structural and phase changes in the metal, will unnecessarily complicate the model even more. The study of electrothermal processes will require additional mathematical apparatus for the computational model considering the patterns of changes in electric and physical fields [20]. Therefore, a finite difference approach can be used with sufficient accuracy in modeling the temperature distribution across the width of a wide plate after a furnace coiler and before deformation, which was successfully applied in [6, 8, 12, 21], saving time and material resources.

3 Research Methodology

3.1 Dependencies for Temperature Distribution Calculation

A schematic representation of the Steckel mill with two furnace coilers is shown in Fig. 1. The finite differences method requires conditional discretization of the strip bulk (Fig. 2) and the time interval from the end of heating to the beginning of reduction. Temperature differences were neglected across the thickness. The heat transfer in the longitudinal direction was not considered, nor was the cooling of the front and rear ends because the edges of these ends are designed to be trimmed. The heated strip (see Fig. 2) was conditionally divided into elementary volume (slabs) by length Δb and area $F = h_0 \cdot \Delta l$, where h_0 – initial strip thickness, Δl is the distance covered by the rolled stock in an elementary time interval $\Delta \tau$. Half the width of the strip is $B_0 = 2 \cdot B_{0n}$ (symmetric temperature distribution against the axis y, see Fig. 2).

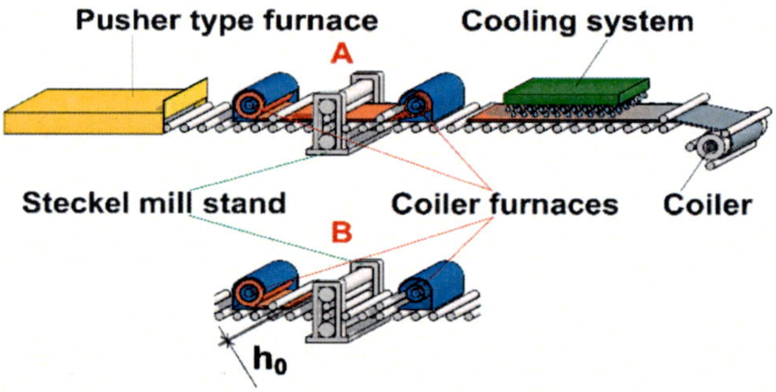

Fig. 1. General view of the Steckel mill, schematic representation: rolling process (A); strip feeding to the rolling stand (B).

The following assumptions are accepted: (i) the average heat flow passing through any slab surface for an elementary period is proportional to the initial value of the temperature gradient during this time; (ii) the heat capacity of an element increases (decreases) in proportion to the increase (decrease) in temperature at the midpoint of its volume, the heat flow spreads from more heated elements to less heated elements.

The element in the middle of the width of the strip has serial number 1, while the slab on the side edge of the pre-rolled stock corresponds to the number N. Temperature

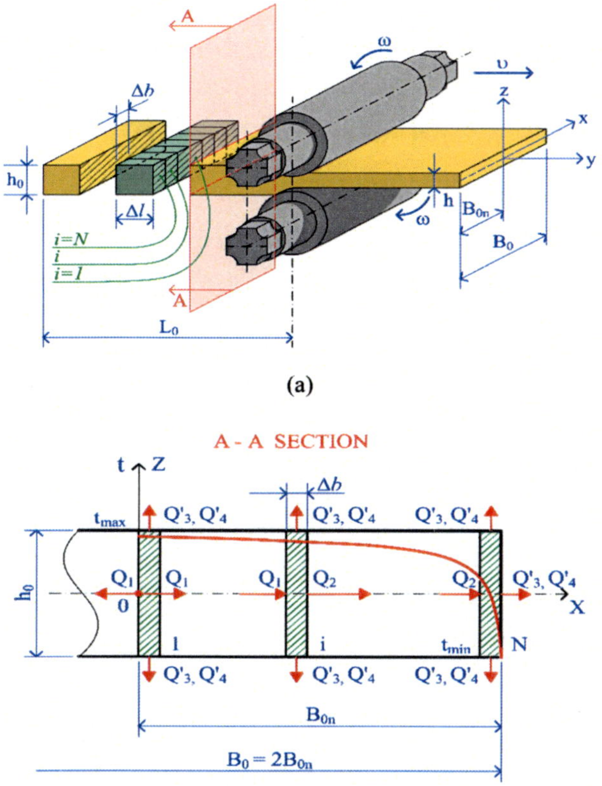

(a)

(b)

Fig. 2. The calculation scheme for the temperature field in the rolled stock and layers (a), the diagram of the distribution of heat flows across the width (b).

marking (in °C): $t_{i;j}$ is the temperature of a random slab in any time interval, where *i* is the slab number, *j* is the time interval number. The temperature was maintained at the level required for the thermomechanical process (830…800 °C).

Equations that determine temperatures in the finite-difference form were obtained by compiling heat balance equations (not given due to their bulkiness) according to the scheme of Fig. 3b and their solution with the implementation of transformations:

– for an intermediate layer (*i*):

$$t_{i;j-1} = t_{i;j} + Fo_{(\Delta b)} \cdot \left\{ \begin{array}{l} \left(t_{i-1;j} - 2t_{i;j} + t_{i+1;j}\right) - \frac{2}{\lambda} \cdot \frac{\Delta b^2}{h_0} \\ \times \left[\sigma \cdot \left(\left(t_{i;j} + 273\right)^4 - 293^4\right) - \alpha \cdot \left(t_{i;j} - 20\right)\right] \end{array} \right\} \quad (1)$$

– for a layer with the maximum heat capacity ($i = 1$):

$$t_{1;j-1} = t_{1;j} + 2 \cdot Fo_{(\Delta b)} \cdot \left\{ \begin{array}{l} (t_{2;j} - t_{1;j}) - \frac{1}{\lambda} \cdot \frac{\Delta b^2}{h_0} \\ \times \left[\sigma \cdot \left((t_{1;j} + 273)^4 - 293^4 \right) - \alpha \cdot (t_{1;j} - 20) \right] \end{array} \right\} \quad (2)$$

– for a layer on the side edge ($i = N$):

$$t_{N;j=1} = t_{N;j} + Fo_{(\Delta b)} \cdot \left\{ \begin{array}{l} (t_{N;j} - t_{N+1;j}) - \frac{\Delta l \cdot (2\Delta b + h_0)}{\lambda \cdot h_0} \\ \times \left[\sigma \cdot \left((t_{N;j} + 273)^4 - 293^4 \right) - \alpha \cdot (t_{N;j} - 20) \right] \end{array} \right\} \quad (3)$$

In Eqs. (1)–(3): $Fo_{(\Delta b)} = a_m \cdot \Delta\tau / \Delta b^2$ is the Fourier criterion for the elementary slab; $a_m = \lambda/\rho \cdot c$ is the material thermal conductivity, m^2/s, were λ is the heat transfer coefficient for the strip material ($W/m \cdot K$), ρ and c are density (kg/m^3) and heat capacity ($J/kg \cdot K$) of the workpiece material at a given temperature; $\sigma = \varepsilon_{st} \cdot \sigma_0$ is the radiation coefficient of the black body, $\varepsilon_{st} = 0.8$, $\sigma_0 = 5.77 \cdot 10^{-8}$ W/(m$^2 \cdot$ K^4); $t = 20$ °C is the ambient temperature; α is the heat exchange coefficient, $W/(m^2 \cdot K)$.

3.2 Dependencies for Continuous Calculation of Physical and Thermophysical Properties of the Material

The strip temperature values $t_{i;j-1}$, $t_{1;j-1}$, and $t_{N;j-1}$ depend on the heating option during hot, standard, or thermomechanical rolling technologies. The dependence for continuous calculation of the density of rolled steel ρ, kg/m^3, is obtained by the equation: $\rho = \rho_o/(1 + 3 \cdot \beta_t)$, where β is thermal expansion coefficient; ρ_O is the steel density at room temperature, kg/m^3:

$$\begin{array}{c} \rho_o = 7876 - 40C - 16Mn - 73Si - 164S + 11Cu \\ +4Ni + Cr + 95W - 120Al + 100As \end{array} \quad (4)$$

where *C, Mn, Si ...* is the content of carbon, manganese, silicon, and other elements in steel, %.

The processing of discrete data (see Fig. 3) for C10E steel, following EN 10084, was performed to determine the dependence of the linear expansion coefficient β on temperature t. According to Fig. 3, the dependence for the coefficient β was considered in 3 ranges, see Table 1. Dependencies of the average heat capacity for carbon steel c for the 2 ranges (see Fig. 3), see Table 2. The data processing results for the dependence of the thermal conductivity coefficient λ on the steel temperature t (see Fig. 3) are also shown in Table 2. The standard deviations (st. dev.) for β, c, and λ are given in the caption of Fig. 3.

The calculation is completed upon reaching the time of the entry of a strip into the roller zone. The following condition should be satisfied to ensure stable calculation according to formulas (1)–(3), minimize and accumulate errors: $Fo_{\Delta x} < 1/2$. The following requirement $\Delta\tau < \frac{1}{2} \cdot \frac{\Delta x^2}{a_m}$ must be met for the value. Calculations were

Table 1. Formulas for calculating the coefficient of linear expansion β.

Temperatures, °C	β, K^{-1}
50…750	$\beta = 8.8385 \cdot t^{0.0772}$
>750…900	$\beta = -0.0174 \cdot \mathrm{t} + 28.403$
>900…1200	$\beta = 9.1222 \cdot \ln(t) - 49.532$

Table 2. Formulas for calculating the capacity (c) and thermal conductivity (λ) coefficients.

Temperatures, °C	c, *kJ/(kg × K)*	Temperatures, °C	λ, *W/(m × K)*
100…800	$c = 0.4556 \cdot e^{0.0005 \cdot t}$	100…900	$\lambda = -0.0363 \cdot t + 59.381$
>800…1300	$c = -3 \cdot 10^{-5} \cdot t + 0.7211$	>900…1200	$\lambda = 20.84 \cdot e^{0.0003 \cdot t}$

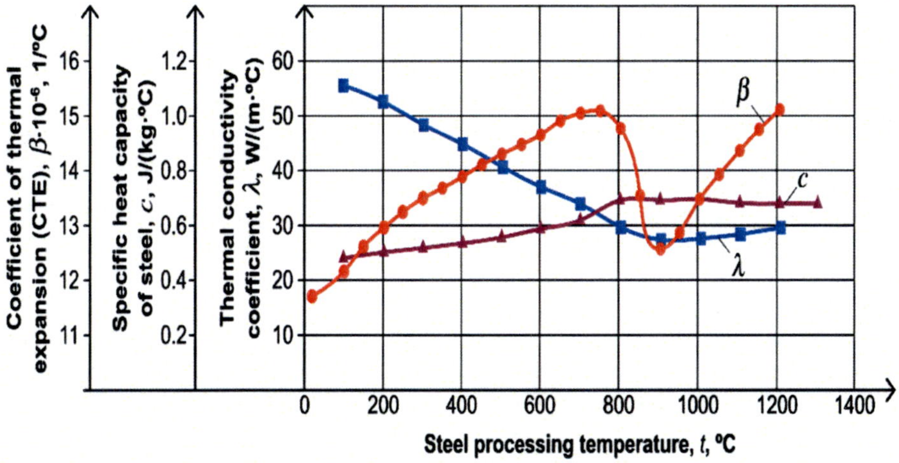

Fig. 3. Results of discrete data processing for the determination of the dependence of the linear expansion coefficient β (st. dev. 0.952), the average heat capacity of carbon steel c (st. dev. 0.078), and the thermal conductivity coefficient λ (st. dev. 9.78) on temperature t.

performed at $\Delta\tau$ and $\Delta\tau' = \Delta\tau/2$ values and the obtained temperatures in all layers were compared with the previous results for optimization. If $t_{i;j-1}$, $\mathrm{t}_{1;j-1}$, and $t_{N;j-1}$ values calculated at $\Delta\tau$ and $\Delta\tau'$, differ by less than 1%, then the calculations are performed at the last $\Delta\tau'$ value; if they differ by more than 1%, then the calculations are repeated with a new $\Delta\tau'' = \Delta\tau'/2$ value, then the same comparison is made, etc.

4 Results and Discussion

The main technological requirement for TMCP requires that a temperature gradient of the rolled stock does not exceed 25 °C across the width in the finishing process before accelerated cooling. This will ensure minimal mechanical properties and anisotropy across the rolled stock's width.

Approbation of the developed mathematical model aiming at estimating the temperature distribution across the width of the semi-finished rolled stock after heating, before thermomechanical rolling on the Steckel mill, was performed under the following conditions: (i) the strip material is X65 steel under the requirements of the API-5L standard; (ii) the strip dimensions: thickness – $h_0 = 40$ mm, width $B_0 = 2B_{0n} = 1510$ mm; (iii) strip heating temperature in a furnace coiler $t = 825$ °C; the time between the strip exit from the furnace coiler and the start of its rolling is 8 s.

The chemical composition of the steel used in the study is given in Table 3.

Table 3. Chemical composition of the steel used in the study.

Steel brand	Chemical composition, %														
	C	Mn	Si	S	P	Al	Nb	V	Cu	Ni	Cr	Mo	Ti	N	B
X65	0.09	1.4	0.23	0.002	0.011	0.036	0.03	0.066	0.02	0.02	0.03	0.01	0.015	0.007	0.0005
Dev*.	0.09-0.11	1.4-1.55	0.07-0.23	≤ 0.004	≤ 0.015	0.020-0.050	0.03-0.04	0.060-0.075	≤ 0.30	≤ 0.30	≤ 0.30	≤ 0.35	0.015-0.035	≤ 0.010	≤ 0.0005

*Chemical deviation (Dev.) under API-5L requirements

The size $\Delta\tau$ that satisfied the requirements of the calculation according to the model was 0.00028 h ($\Delta\tau$ should be less than 0.047 h). The basic temperature distribution of the strip at half the width after heating in the furnace coiler (see Table 4).

Table 4. Basic temperature distribution of the strip at half width after heating in the furnace coiler for thermomechanical modes.

Distance from the edge to the middle of the strip, *mm*	75.5	151	226.5	302	377.5	453	528.5	604	679.5	755	792.75
Temperature*, °C	825	825	826	827	827	829	830	830	831	832	833

*Temperature measurement accuracy ±4 °C

The modeling results are shown in Fig. 4. The calculation results are symmetrically projected on the second half of the width for visualization. According to the modeling results (see Fig. 4), more intense cooling is observed on the side edges of the strip.

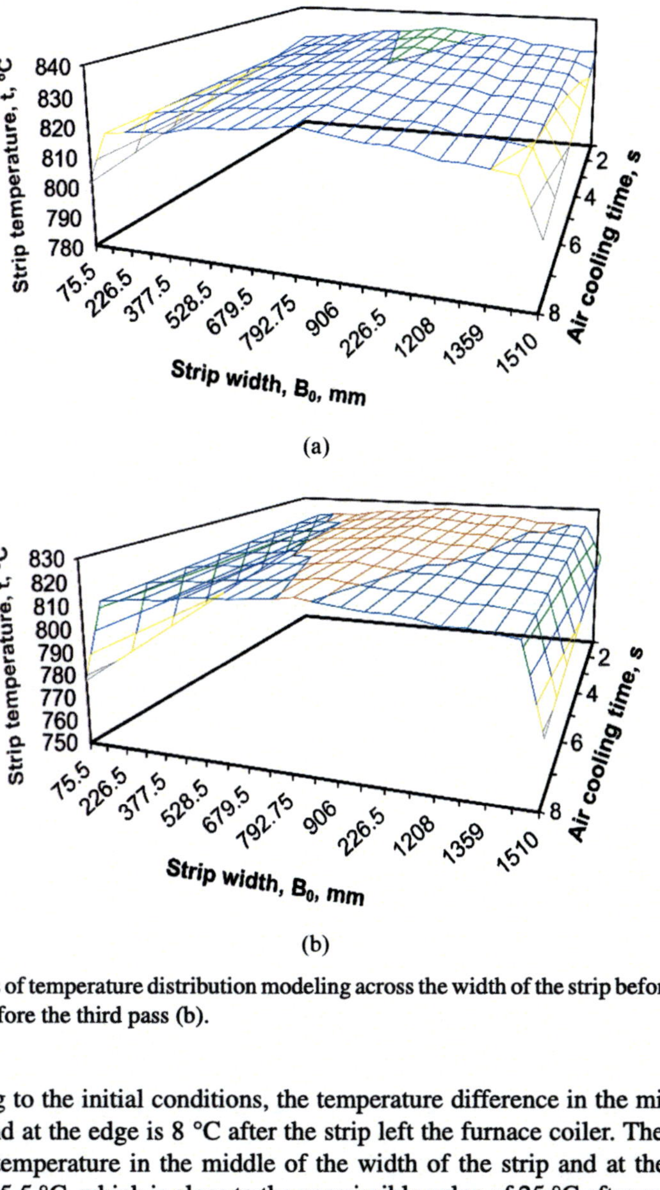

Fig. 4. Results of temperature distribution modeling across the width of the strip before the second pass (a) and before the third pass (b).

According to the initial conditions, the temperature difference in the middle of the strip width and at the edge is 8 °C after the strip left the furnace coiler. The difference between the temperature in the middle of the width of the strip and at the side edge increased to 25.5 °C, which is close to the permissible value of 25 °C after modeling the temperature distribution before the second pass (for the time interval between exiting the furnace coiler and entering the stand for the second pass). Modeling the change in temperature distribution before the third pass established that the difference between the temperature in the middle of the width and at the edge of the strip will increase and amount to 43 °C (see Fig. 4), which does not satisfy the technological conditions.

The current technological regulation of furnace coilers of the Steckel mill is aimed at preserving the established temperature mode without changing the temperature distribution across the width. It should also be noted that in the rolling process, a significant difference in temperature distribution will increase across the width of the strip, even considering deformation heating due to the additional effect of cooling the rolls and during contact with the rolls. It is necessary to adjust the heating mode with a change in the temperature distribution across the section of the furnace coiler to minimize the effect on the mechanical properties of the strip of the cooler side edges. Well-known technical solutions are used in this case: differentiated fuel supply, arrangement of burners, or installation of shielding or heating of the side edges between a furnace coiler and a mill [13, 15, 16].

5 Conclusions

The improved model can be used to calculate changes in the temperature distribution of flat rolled products produced by hot rolling, thermomechanical rolling, and its different implementation options (high temperature and low temperature) and standard rolling. The results of calculations determine the non-uniformity of the temperature distribution across the width of the strip during intermediate cooling on a roller conveyor. The results of calculations are the basis for the design of the strip temperature maintenance technology in furnace coilers of Steckel mills for conditions of the TMCP process. It was established that the difference between the temperature of the central part of the strip and the edge increases from the initial 8 °C to 43 °C after cooling the strip on the roller conveyor during three finishing passes, in the temperature range of the TMCP process, which exceeds the initial drop conditions of 25 °C adopted for TMCP before accelerated cooling. Adjusting the heating mode in the furnace coiler of the Steckel mill was proposed to equalize the side edges' temperature by adjusting the furnace burners' operation near the inlet/outlet window.

References

1. Bulfone, M., Mossutti, M.: Modern plate/Steckel mills: recent developments for new levels of flexibility and competitiveness. In: AISTech - Iron and Steel Technology Conference Proceedings, pp. 2335–2339. AISTech, Philadelphia (2018)
2. Efron, L.I., Ringinen, D.A., Muntin, A.V.: Features of implementing thermomechanical rolling in various types of mills. Metallurgist **66**(3–4), 403–421 (2022)
3. Kurpe, O., Kukhar, V., Puzyr, R., Burko, V., Balalayeva, E., Klimov, E.: Electric motors power modes at synchronization of roughing rolling stands of hot strip mill. In: 25th IEEE International Conference on Problems of Automated Electric Drive. Theory and Practice, pp. 510–513. IEEE, Kremenchuk (2020)
4. He, Q., Sun, J., Yan, C., Zhao, J., Zhang, Z.: Thermo-mechanical modeling and simulation of microstructure evolution in multi-pass H-shape rolling. Finite Elem. Anal. Des. **76**, 13–20 (2013)
5. Roccisano, A., Nafisi, S., Stalheim, D., Ghomashchi, R.: Effect of TMCP rolling schedules on the microstructure and performance of X70 steel. Mater. Charact. **178**, 111207 (2021)
6. Markov, O.E., Gerasimenko, O.V., Kukhar, V.V., Abdulov, O.R., Ragulina, N.V.: Computational and experimental modeling of new forging ingots with a directional solidification: the relative heights of 1.1. J. Braz. Soc. Mech. Sci. Eng. **41**(8), 310 (2019)
7. Endo, S., Nakata, N.: Development of thermo-mechanical control process (TMCP) and high performance steel in JFE steel. JFE Tech. Rep. **20**, 1–7 (2015)

8. Ramirez, M.F.G., Hernández, J.W.C., Ladino, D.H., Masoumi, M., Goldenstein, H.: Effects of different cooling rates on the microstructure, crystallographic features, and hydrogen induced cracking of API X80 pipeline steel. J. Mater. Res. Technol. **14**, 1848–1861 (2021)
9. Kumar, R., Singh, L., Singh, Y., Singh, A.: Influence of metal temperature on strength of rolled steel product. Int. J. Sci. Res. Dev. **1**(9), 1716–1719 (2013)
10. Nwachukwu, P.U., Oluwole, O.O.: Effects of rolling process parameters on the mechanical properties of hot-rolled St60Mn steel. Case Stud. Constr. Mater. **6**, 134–146 (2017)
11. Xiong, M.-X., Pi, P.-W., Gong, W., Yang, M.-F., Ou, Z.-F.: Mechanical properties of TMCP high strength steels with different strength grades at elevated temperatures. J. Build. Eng. **48**, 103874 (2022)
12. Hwang, J.-K.: Thermal behavior of a rod during hot shape rolling and its comparison with a plate during flat rolling. Processes **8**(3), 327 (2020)
13. Zhu, Z.Q., Li, Y., Yang, S.: 3D eddy current and temperature field analysis of edge induction heater. COMPEL – Int. J. Comput. Math. Electr. Electron. Eng. **35**(2), 683–694 (2016)
14. Ramírez-López, A., Dávila-Maldonado, O., Nájera-Bastida, A., Morales, R.D., Rodríguez-Ávila, J., Muñiz-Valdés, C.R.: Analysis of non-symmetrical heat transfers during the casting of steel billets and slabs. Metals **11**, 1380 (2021)
15. Pohanka, M., Kotrbáček, P., Resl, O., Bellerová, H.: Optimal hydraulic descaling. In: Metal – 2020 Proceedings 29th International Conference on Metallurgy and Materials, pp. 118–125, Brno, Czech Republic (2020)
16. Bohacek, J., Raudensky, M., Kotrbacek, P.: Remote cooling of rolls in hot rolling; applicability to other processes. Metals **11**, 1061 (2021)
17. Sosedkova, M.A., Grigorenko, A.S., Radionova, L.V., Lisovskaya, T.A., Lezin, V.D.: Mathematical model of temperature conditions of sheet mills with furnace coilers. Mater. Sci. Forum **989**, 711–718 (2020)
18. Hwang, R., Jo, H., Kim, K.S., Hwang, H.J.: Hybrid model of mathematical and neural network formulations for rolling force and temperature prediction in hot rolling processes. IEEE Access **8**, 153123–153133 (2020)
19. Satywan, K., Swapnil, D., Shubham, K., Paras, K., Pramod, K.: Modelling and simulation of induction heating in heat treatment process. Int. J. Adv. Technol. **13**(2), 1000172 (2022)
20. Yarymbash, D., Kotsur, M., Yarymbash, S., Kylymnyk, I., Divchuk, T.: Electromagnetic properties determination of electrical steels. In: IEEE 15th International Conference on Advanced Trends in Radioelectronics, Telecommunications and Computer Engineering (TCSET), pp. 185–189. IEEE, Lviv-Slavske (2020)
21. Chen, D., Xu, H., Lu, B., Chen, G., Zhang, L.: Solving the heat transfer boundary condition of billet in reheating furnace by combining "black box" test with mathematic model. Case Stud. Therm. Eng. **40**, 102486 (2022)

Mathematical Modeling of Technological Regulations of Furnace Equipment for Carbon Graphite Electrode Production

Serhii Leleka, Anton Karvatskii(✉), Ihor Mikulionok, Olena Ivanenko, and Iryna Omelchuk

National Technical University of Ukraine "Igor Sikorsky Kyiv Polytechnic Institute", 37, Peremohy Avenue, Kyiv 03056, Ukraine
anton@rst.kpi.ua

Abstract. The generalized mathematical model of physical fields, which describes the main technological divisions of producing electrode carbon-graphite products, has been refined in terms of considering the equation of the state of the loose medium and the turbulence of the flows of the working medium. Based on the generalized model, approaches to formulating mathematical models of such individual redistributions of electrode production as calcination of carbon-containing filler, pressing, firing, and graphitization of electrode blanks are shown. Using numerical modeling, resource- and energy-efficient technological regulations for the calcination of carbon-containing fillers in electrocalciners and firing of graphite products in a Riedhammer ring multi-chamber furnace were developed. It was established that the developed regulations for the start-up and operation of the electrocalciner for heat treatment of anthracite ensure the required quality of the final product and long-term operation of the equipment - more than 6 months. It is shown that the developed technological regulations for firing various graphite products in Riedhammer furnaces provide a 7–10% reduction in waste output and technogenic impact on the environment.

Keywords: Energy Efficiency · Calcination · Pressing · Firing · Graphiting · Numerical Modeling

1 Introduction

The main stages of production of carbon-graphite products include calcination of carbon-containing filler, mixing the filler with binding pitch and pressing of "green" blanks of carbon-graphite products, their firing, and graphitization [1].

The specific electricity consumption of electrode production is extremely high and reaches 59400 MJ/t. Therefore, the leading global trends in improving the electrode industry include increasing existing equipment's resource and energy efficiency, developing new innovative technological equipment, and reducing the man-made impact on the environment [1].

V. Tonkonogyi et al. (Eds.): InterPartner 2023, LNME, pp. 286–295, 2024.
https://doi.org/10.1007/978-3-031-42778-7_26

The work aims to refine the generalized mathematical model of the physical fields of technological redistributions of the production of carbon graphite products and its application to build mathematical and numerical models of individual redistributions for the numerical analysis of furnace equipment operation regulations.

2 Literature Review

The analysis of literary sources on the problem of numerical modeling of equipment for the main redistribution of electrode production testifies to the following.

The modeling of temperature and electric fields in electrocalciners for calcinating carbon-containing filler is devoted to studies [2–5].

In works [2, 3], a mathematical model of the thermoelectric and hydrodynamic state of the Elkem-type electrocalciner is given, which was used to minimize the drop in operating temperatures in the radial direction and increase the productivity of the high-temperature unit.

The work [4] is devoted to the problem of optimizing the process of high-temperature calcination of anthracite in electrocalciners, in particular, using a multiphysics numerical model, the influence of the distribution of electromagnetic and temperature fields, gas flow, and kinetics of chemical reactions on the productivity and duration of operation of the electric furnace was investigated.

In work [5], using numerical simulation, a vertical electrocalciner was developed, which makes it possible to obtain high-quality graphitized coke and electroanthracite with a peak temperature of the center of the working space of the furnace of 3000 °C, which was implemented in production. However, judging by the calculation grid, it is difficult to conclude the results' reliability.

The main shortcomings of the reviewed works [2–5] include the lack of consideration of the dynamics of the dense movement of the loose filler [6], as well as data on the verification of the developed numerical models [7, 8].

A limited number of scientific works are devoted to studying the process of pressing "green" billets of carbon graphite products [1]. Since the coke oven mixture during its pressing can behave both as a solid body and as a visco-plastic liquid, the closest approximation of its description can be the Bingham-Papanastasiou liquid [9, 10]. These articles provide mathematical formulations of the specified problem, methods, and results of its numerical solution, which may be useful for numerical analysis of pressing redistribution.

A limited number of scientific papers are also devoted to the theoretical study of the process of firing electrographite blanks [11]. The work [11] proposed a mathematical model of the thermal-hydrodynamic state of the ring multi-chamber Riedhammer furnace, which includes half of the cross-section of the furnace chamber. The results of numerical modeling showed that reducing the size of the windows for the exit of flue gases from the chamber leads to a more uniform temperature field of the electrode blanks. However, this work does not consider the burning process of natural gas.

Scientific works [12–15] are devoted to the mathematical modeling of the graphitization process of electrographite blanks. In the paper [12], theoretical studies of the possibility of using thermal insulation heat recovery during the cooling of Castner's furnaces to increase the energy efficiency of the graphitization process and the productivity of the furnaces were carried out. However, the proposed method is complex, requires significant capital investment for industrial implementation, and does not directly affect the regulation of the electric power supply to the furnace.

The work [13] is devoted to a theoretical study of the influence of contact resistance between electrode blanks on the temperature field of the column during graphitization in Castner's furnace. It is shown that poor thermoelectric contact between the blanks in the column can lead to their deformation and further deterioration of the contact between them.

The work [14] is devoted to the numerical study of the recovery potential of spent heat of graphitization furnaces. It was established that heat dissipation on the upper surface of the coke is 48.5% of the total electricity consumption and 66.9% of the total heat dissipation from the furnace, respectively.

In work [15], the development of a generalized mathematical model of the physical fields of the main technological redistributions of the production of carbon graphite products was probably carried out for the first time. Its application is shown for building or refining mathematical and numerical models of individual redistributions for performing numerical analysis of process and equipment parameters on the examples of pressing "green" electrode blanks and researching the effective thermophysical properties of loose carbon-containing materials. However, the given generalized mathematical model lacks the equation of the state of the fluid medium, the formulation of flow turbulence, etc.

From the conducted literature review of the available sources of information, it follows that the unsolved problem is the refinement of the generalized mathematical model and its application to describe the physical processes of redistribution of production of carbon graphite products since the existing model needs some improvement.

3 Research Methodology

3.1 Problem Statement

The following tasks must be solved to achieve the stated research aim:

- to clarify the generalized mathematical model of physical fields, which describes the technological redistribution of the production of carbon graphite electrode products;
- to apply a generalized mathematical model for formulating mathematical and numerical models of individual redistributions of electrode production to perform a numerical analysis of furnace equipment operation regulations;
- with the use of numerical modeling, develop resource- and energy-efficient regulations for the operation of the furnace equipment for redistribution of calcination of the filler and firing of carbon graphite blanks.

3.2 Generalized Mathematical Model

Solids, liquids/gases, and loose materials are included in the working environments of the technological redistribution of carbon graphite production.

The physical processes that take place in the technological equipment of electrode carbon graphite production include heating due to the flow of electric current, combustion of gas and solid carbon-containing substances, deformation of solid bodies under the action of mechanical and temperature loads, movement of gases and dense movement of loose materials, heat exchange by heat conduction, convection and thermal radiation [15].

The presence of continuous and discrete environments in the production technology of carbon graphite products determines the use of a complex continuous-discrete approach to the mathematical formulation of physical processes using the Euler and Lagrange reference systems.

The continuous formulation of the physical processes of the specified technology is based on the Euler frame of reference. It may include the following equations: conservation of mass, amount of movement and energy, electrical conductivity, and transport of chemical species of combustion reactions [15]:

$$\begin{cases} \frac{d\rho}{dt} + \rho\nabla\cdot\mathbf{v} = 0; \\ \frac{d(\rho\mathbf{v})}{dt} = \nabla\cdot\hat{\boldsymbol{\sigma}} + \rho\mathbf{b}; \\ \frac{d(\rho c_p T)}{dt} = \hat{\boldsymbol{\sigma}} : \nabla\mathbf{v} - \nabla\cdot\mathbf{q} + q_v; \\ \nabla\cdot j = 0; \\ \frac{d(\rho Y_k)}{dt} = \nabla\cdot\mathbf{J}_k + \dot{\omega}_k, k = \overline{1, N}, \end{cases} \tag{1}$$

where $\frac{d()}{dt} = \frac{\partial()}{\partial t} + v_i\nabla_i(), i = 1, 2, 3$ – material derivative; ∇ – operator of Hamilton's; ρ – density; t – time; $(\cdot)$ – operator of the scalar product of tensors; $\mathrm{v} = d\mathrm{u}/dt$ – velocity vector; u, b, q, j, J_k – vectors of displacement, mass forces, heat flux density, electric current density, and diffusion flux density of reaction species, respectively; $\hat{\sigma}$ – stress tensor of the 2nd rank; T – absolute temperature; c_p – mass isobaric heat capacity; $\hat{\sigma} : \nabla\mathrm{v}$ – source of heat of a mechanical nature; (:) – operator of the double scalar product of tensors; $q_v = q_{velec} + q_{vchem} + q_{vrad}$ – a source of heat of a non-mechanical nature; $q_{velec} = \chi(T)|\nabla U|^2$ – the heat source is caused by the flow of electric current; $q_{vchem} = \nabla\cdot\rho\sum_{k=1}^{N}\mathrm{J}_k h_k$ – the heat source is caused by chemical reactions; q_{vrad} – the heat source is related to radiative heat transfer; χ – coefficient of electrical conductivity; U – electric potential; h_k, Y_k – mass enthalpy and mass fraction of species of chemical reactions of combustion, respectively; $\dot{\omega}_k$ – source term due to the average rate of chemical reactions.

The tensors $\hat{\sigma}$, q, j and J_k are determined by the corresponding physical equations (laws). For example, for isotropic homogeneous solid media with linear properties, the physical equation for determining $\hat{\sigma}$ is a generalized Hooke's law [15]

$$\hat{\boldsymbol{\sigma}} = \frac{E}{1+\nu}\left[\hat{\boldsymbol{\varepsilon}} + \frac{\nu}{1-2\nu}\mathrm{tr}(\hat{\boldsymbol{\varepsilon}})\right]\hat{\mathbf{I}}, \tag{2}$$

where E – modulus of elasticity; ν – Poisson's ratio; tr() – tensor trace operator; $\hat{\mathrm{I}}$– unit tensor of the 2nd rank; $\hat{\varepsilon} = \frac{1}{2}(\nabla\mathrm{u} + \mathrm{u}\nabla)$ – tensor of finite deformations of the 2nd rank, provided that the deformations are small values $\ll 1$.

At the same time, the equation of motion and equilibrium of the system of Eqs. (1) is assumed to be independent of time, i.e. $d\mathrm{v}/dt = 0$.

For media with non-linear properties, for example, those in which plastic flow is taken into account [16, 17] by determining the moment of onset of the plastic state of the material using a similar time parameter. At the same time, according to the incremental theory of plasticity, inelastic deformations are considered initial, and the equation of motion and equilibrium (1) is written through increments in the form

$$\nabla \cdot \dot{\hat{\boldsymbol{\sigma}}} + \rho \dot{\mathbf{b}} = 0. \tag{3}$$

The generalized Hooke's law is written in terms of the initial stress increment $\dot{\hat{\sigma}}^{\text{in}}$ and total strain rate $\dot{\hat{\varepsilon}}$ [17]

$$\dot{\hat{\boldsymbol{\sigma}}} = 2G\dot{\hat{\boldsymbol{\varepsilon}}} + \frac{2G\nu}{1-2\nu}\text{tr}\left(\dot{\hat{\boldsymbol{\varepsilon}}}\right)\hat{\mathbf{I}} - \dot{\hat{\boldsymbol{\sigma}}}^{\text{in}}, \tag{4}$$

where G – shear modulus.

At the same time, the onset of the plastic state of the medium is determined by the material plasticity criterion (as a function of the onset of plastic flow) of the form [17]

$$F = \sigma_{\text{eq}} - \sigma_{\text{yield}}, \tag{5}$$

where σ_{eq} – equivalent stress in the material; σ_{yield} – yield strength of the material.

The form of the formulas for determining σ_{eq} and σ_{yield} depends on the choice of plasticity model. In the case of loose material, the Drucker-Prager mechanical model is used, and the equation of state – the fluidity criterion takes the form [16]

$$F = \alpha_{\text{DP}}\text{tr}\left(\hat{\boldsymbol{\sigma}}\right) + \sqrt{\frac{1}{2}\hat{\mathbf{S}} : \hat{\mathbf{S}}} - k_{\text{DP}}, \tag{6}$$

where $\alpha_{\text{DP}} = \frac{2\sin\varphi}{\sqrt{3}(3-\sin\varphi)}$ – bulk material constant; φ – angle of internal friction; $\hat{\text{S}} = \hat{\sigma} - \frac{1}{3}\text{tr}\left(\hat{\sigma}\right)\hat{\text{I}}$ – deviatoric stress tensor; $k_{\text{DP}} = \frac{6c\cos\varphi}{\sqrt{3}(3-\sin\varphi)}$ – yield point of bulk material; c – the amount of cohesion between bulk material particles.

The physical equation for determining the stress tensor $\hat{\boldsymbol{\sigma}}$ in liquids is called the Navier-Stokes law, which is valid for a Newtonian incompressible fluid [9, 10, 15]

$$\hat{\boldsymbol{\sigma}} = 2\mu\dot{\hat{\boldsymbol{\varepsilon}}} - p\hat{\mathbf{I}}, \tag{7}$$

where μ – coefficient of dynamic viscosity; $\dot{\hat{\varepsilon}} = \frac{1}{2}(\nabla \text{v} + \text{v}\nabla)$ – strain rate tensor of the 2nd rank.

In the case of a compressible fluid, law (7) takes the form

$$\hat{\boldsymbol{\sigma}} = \hat{\boldsymbol{\tau}} - p\hat{\mathbf{I}}, \tag{8}$$

where $\hat{\tau} = \mu\left[\nabla \text{v} + \text{v}\nabla - \frac{2}{3}(\nabla \cdot \text{v})\right]\hat{\text{I}}$ – shear stress tensor.

In the case of a nonlinear viscoplastic fluid, a characteristic feature is that such a substance exhibits the properties of a solid body before reaching a certain critical internal shear stress τ_{shear}, and only when this stress value is exceeded it begins to move like an ordinary fluid. For example, for the Bingham-Papanastasiou fluid [9, 10, 15], the shear stress tensor will be determined by the formula

$$\hat{\tau} = \left\{ \mu_{\text{eff}} + \tau_{\text{shear}} / \lfloor \dot{\hat{\gamma}} \rfloor \lfloor 1 - \exp(-m \lfloor \dot{\hat{\gamma}} \rfloor) \rfloor \right\} \dot{\hat{\gamma}}, \tag{9}$$

where μ_{eff} – effective coefficient of dynamic viscosity; $\dot{\hat{\gamma}} = \nabla \text{v} + \text{v} \nabla$ – rate of deformation; m – exponential multiplier.

The physical equations for determining q, j and J_k, described by Fourier's, Ohm's, and Fick's laws, respectively, can be represented in the form of a generalized equation

$$\mathbf{y} = -k \nabla X. \tag{10}$$

The discrete formulation of physical processes in loose materials is based on the Lagrange frame of reference and may include the following equations: translational and rotational motion and energy [6, 7, 15, 19]:

$$\left\{ \begin{array}{c} m_i \frac{d\mathbf{v}_i}{dt} = \sum_j \mathbf{F}_{ij} + m_i \mathbf{b}; \\ I_i \frac{d\boldsymbol{\omega}_i}{dt} = \sum_j \mathbf{M}_{ij}^{\Sigma}; \\ m_i c_{pi} \frac{dT_i}{dt} = \sum_{\text{contacts}\, j-i} \dot{Q}_{i-j} + \dot{Q}_{i,conv} + \dot{Q}_{i,rad} + \dot{Q}_{i,chem}, \end{array} \right. \tag{11}$$

where i – particle index; j – index of the particle interacting with the particle i; m_i – mass; I_i – moment of inertia; ω_i – angular velocity vector; F_{ij} – external force vector; M_{ij}^{Σ} – total moment; $\dot{Q}_{i-j}$, $\dot{Q}_{i,conv}$, $\dot{Q}_{i,rad}$, $\dot{Q}_{i,chem}$ – the power of contact, convective, and radiative heat exchange between particles and the liquid medium and the intensity of heat exchange associated with surface chemical reactions, respectively.

In the presence of turbulent flows of working media, the system of Eqs. (1) is transformed into a system of Reynolds averaged Navier-Stokes equations (RANS), the number of equations of which depends on the choice of the turbulence model. For example, in the case of choosing the k–ε model of turbulence [18], two scalar equations of turbulent kinetic energy and its dissipation rate are added to the transformed system of Eqs. (1).

The connection between the discrete (11) and continuous (1) formulations can, for example, be established by averaging the velocity field [15] obtained from the solution of a discrete problem or by using a discrete phase model (Discrete Phase Models (DPM)) [6, 19]. To close the systems of differential Eqs. (1)–(11), it is necessary to write down the corresponding initial and boundary conditions.

The open-source software LIGGGHTS [19] and OpenFOAM [20] were used for the numerical implementation of the generalized mathematical model (1)–(11).

4 Results and Discussion

4.1 Redistribution of Calcination of Carbon-Containing Filler in an Electrocalciner

The construction of the mathematical model of the mechanical and thermoelectric state of the electrocalciner is based on the generalized mathematical model (1)–(11) and includes the third and fourth equations of the system of Eqs. (1) and the first and second equations of the system (11), physical equations of the type (10) in the form of laws Fourier's and Ohm's.

Its division into several simpler components is used to simplify the mathematical statement of the initial problem of the mechanical and thermoelectric state of the electrocalciner. The first of them includes a discrete description of the movement of the fluid medium, the second – a transition from a discrete to a continuous formulation, with the obtaining of the velocity field of the fluid material as a continuous medium, and the third – a connected thermoelectric problem, in which the dynamics of the fluid medium is taken into account in the convective derivative of the energy equation.

The freely open software codes LIGGGHTS [19] and OpenFOAM [20] were used for the numerical implementation of the described mathematical model of the process of calcination of carbonaceous filler in an electrocalciner. Based on the conducted numerical experiments, a modernized regulation of start-up and operation of the electrocalciner was developed (Fig. 1).

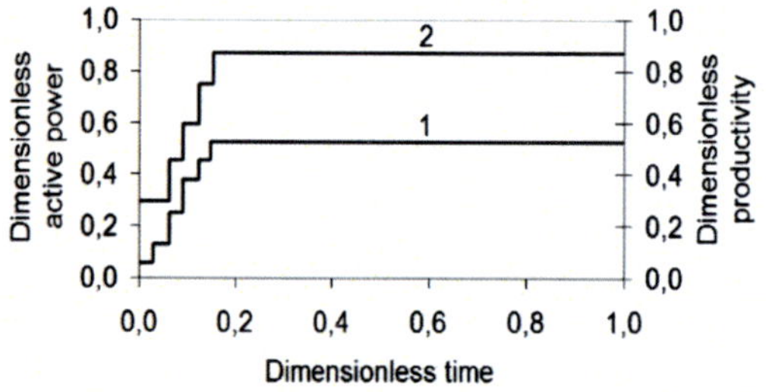

Fig. 1. Modernized regulations for the start-up and operation of the electrocalciner for roasting anthracite: 1 – active electric power supply mode; 2 – performance change mode.

The developed resource- and energy-efficient start-up and operation regulations of the electrocalciner ensure the required quality of the final product (electro-anthracite) and long-term operation of the equipment - more than 6 months.

4.2 Redistribution of the Firing of Graphite Blanks in a Multi-chamber Ring Closed Furnace of the Riedhammer Type

The construction of a mathematical model of the thermal-hydrodynamic state of the Riedhammer-type furnace during the firing of graphite blanks is based on the generalized mathematical model (1)–(11). It includes the following equations: the system of Eqs. (1) in addition to the fourth equation and physical Eqs. (8) and type (10) in the form of Fourier's and Fick's laws. As mentioned above, it is necessary to transform the system of Eqs. (1) into the RANS system and select a specific turbulence model to consider the presence of turbulent flows. The approximation of a "gray" or selective absorbing and emitting medium is used to consider radiation heat transfer.

For the numerical implementation of the described mathematical model of the process of firing electrographite billets of the Riedhammer furnace, the freely open software code OpenFOAM [20] was used. Based on the conducted numerical experiments, a modernized procedure for burning electrode blanks was developed, taking into account the dynamics of the gas evolution of the binder (Fig. 2).

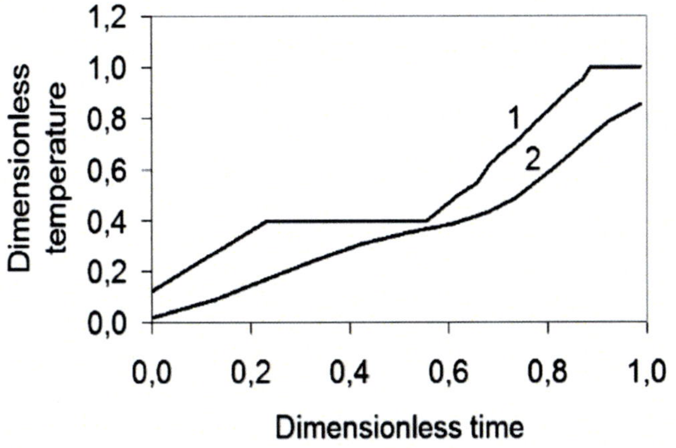

Fig. 2. Temperature parameters of the modernized regulation of the firing process in the Riedhammer furnace: 1 – temperature under the vault of the furnace; 2 – the average volume temperature of the workpieces.

Based on studies of the influence of the technological parameters of the firing process on the thermal and gas-dynamic state of Riedhammer furnaces, scientifically based (taking into account the dynamics of the gas evolution of the binder) technological regulations for the firing of various types of graphite products have been developed, which ensure resource and energy efficiency, a reduction in the yield of defects by 7–10% and man-made impact on the environment. Appropriate tests in industrial conditions confirmed the given quantitative characteristics of the developed regulations.

Using the given generalized mathematical model (1)–(11), it is also easy to formulate such separate mathematical models related to electrode production and describe such models as the pressing of billets of carbon graphite products, gasification of carbon-containing material in a rotary furnace for heat treatment of the filler of electrode products, graphitization electrode blanks in Castner's furnaces, theoretical studies of effective thermophysical properties of loose carbon-containing materials, etc.

For example, the construction of a mathematical model of the thermal-hydrodynamic state of the coke oven mass during its pressing is based on the generalized mathematical model (1)–(11) and includes the following equations: the first three equations of the system of Eqs. (1), physical Eqs. (8), (9) and of type (10) in the form of Fourier's law.

And the construction of a mathematical model of the thermoelectric and mechanical state of the Castner furnace during the process of graphitization of electrode blanks includes the second equation of the system of Eqs. (1) in the form (3), the third and fourth (1) and physical Eqs. (4), type (10) in the form of laws Fourier's and Ohm's and (6) is the flowability criterion of loose material in the form of the Drucker-Prager mechanical model.

5 Conclusions

Based on the results of the analysis of the conducted studies, the following conclusions can be drawn.

The generalized mathematical model of physical fields in producing electrode carbon-graphite products has been refined in terms of considering the equation of the state of the loose medium and turbulence of the working medium flows.

Based on the generalized mathematical model, approaches to formulating mathematical models of separate redistributions of electrode production, such as calcination of carbon-containing filler, pressing, firing, and graphitization of electrode blanks, are shown.

Resource- and energy-efficient technological regulations have been developed for calculating carbon-containing filler in electrocalciners and firing of graphite products in Riedhammer furnaces.

It was established that the developed regulations for the start-up and operation of the electrocalciner ensure the required quality of the final product (electro-anthracite) and increase the equipment's service life to 6 months. And more.

It is shown that the developed technological regulations for firing various graphite products in Riedhammer furnaces provide a 7–10% reduction in waste output and technogenic impact on the environment.

References

1. Jäger, H., Frohs, W. (eds.): Industrial Carbon and Graphite Materials, Volume I: Raw Materials, Production and Applications. WILEY-VCH GmbH, Germany (2021). https://doi.org/10.1002/9783527674046
2. Perron, J., Bouvette, J.-F., Dupuis, M.: Optimization of Anthracite Calcination Process in a Vertical Electric Arc Furnace. Light Metals 597–602 (1996)
3. Hachette, R., Bui, R.T., Simard, G., Perron, J., Bouvtte, J.F.: A CFD dynamic model of the anthracite calciner. Light Metals 677–687 (1997)
4. Gasik, M.M., Gasik, M.I.: Modeling of anthracite treatment in an electrocalcinator. Modern Probl. Metall. **14**, 100–108 (2011)
5. Yang, Y., Gong, S., Ning, Q., Zhou, X., Zhao, H.: Development and application of electrocalciners with increased calcination temperature. Light Metals 1363–1371 (2018). https://doi.org/10.1007/978-3-319-72284-9_178
6. Pöschel, T., Schwager, T.: Computational Granular Dynamics: Models and Algorithms. Springer, Heidelberg (2005)
7. Rao, K.K., Nott, P.R.: An Introduction to Granular Flow. Cambridge University Press, New York (2008)
8. Pazouki, A., et al.: Compliant contact versus rigid contact: a comparison in the context of granular dynamics. Phys. Rev. E **96**, 042905 (2017)
9. Khan, N.A., Sultan, F.: Numerical analysis for the Bingham-Papanastasiou fluid flow over a rotating disk. J. Appl. Mech. Tech. Phys. **59**(4), 638–644 (2018). https://doi.org/10.1134/S0021894418040090
10. Mehmood, A., Khan, W.A., Mahmood, R., Rehman, K.Ur.: Finite element analysis on bingham–papanastasiou viscoplastic flow in a channel with circular/square obstacles: a comparative benchmarking. Processes **8**(7), 779 (2020). https://doi.org/10.3390/pr8070779
11. Hajduk, A., Goede, F.: Innovations in the design and construction of ring pit furnaces. https://www.yumpu.com/en/document/view/14559453/innovations-in-the-design-and-construction-of-ring-pit-sacmi. Accessed 21 Nov 2021
12. Shen, C., Zhang, M., Li, X.: Numerical study on the heat recovery and cooling effect by built-in pipes in a graphitization furnace. Appl. Therm. Eng. **90**, 1021–1031 (2015). https://doi.org/10.1016/j.applthermaleng.2015.04.036

13. Piekło, J., Maj, M.: Analysis of the state of stress in the connection of graphite electrodes. Arch. Foundry Eng. **15**, 85–88 (2015)
14. Lan, Y., Zhao, X., Zhang, W., Mu, L., Wang, S.: Investigation of the waste heat recovery and pollutant emission reduction potential in graphitization furnace. Energy **245**(15), 123292 (2022). https://doi.org/10.1016/j.energy.2022.12329
15. Leleka, S.V.: Generalized mathematical model of physical fields of technological redistributions of manufacturing electrographite products. Energy Technol. Resour. Saving **2**, 28–43 (2021). https://doi.org/10.33070/etars.2.2021.03. (in Ukrainian)
16. Simo, J., Taylor, R.: Consistent tangent operators for rate-independent elastoplasticity. Comput. Methods Appl. Mech. Eng. **48**, 101–118 (1985)
17. Karvatskii, Anton, Mikulionok, Ihor, Leleka, Serhii, Solovei, Vladyslav: Numerical simulation of elasto-plastic behavior of isotropic composite materials. In: Ivanov, Vitalii, Trojanowska, Justyna, Pavlenko, Ivan, Zajac, Jozef, Peraković, Dragan (eds.) DSMIE 2020. Lecture Notes in Mechanical Engineering, pp. 492–501. Springer, Cham (2020). https://doi.org/10.1007/978-3-030-50794-7_48
18. Castro, F.A., Palma, J.M.L.M., Lopes, A.S.: Simulation of the Askervein flow. Part 1: Reynolds averaged Navier–Stokes equations (k ∊ Turbulence Model). Boundary-Layer Meteorol. **107**, 501–530 (2003). https://doi.org/10.1023/A:1022818327584
19. LIGGGHTS Open Source Discrete Element Method Particle Simulation Code. http://www.liggghts.com/. Accessed 20 Oct 2021
20. OpenFOAM – Open Field Operation and Manipulation. http://www.openfoam.org/. Accessed 24 Oct 2020

Effects of Optimized Laser-Ultrasonic Surface Hardening Parameters on Residual Stress and Structure-Phase State of Medium-Carbon Steel

Dmytro Lesyk[1,2,4](✉), Bohdan Mordyuk[2], Silvia Martinez[3], Vitaliy Dzhemelinskyi[1], and Aitzol Lamikiz[3]

[1] National Technical University of Ukraine "Igor Sikorsky Kyiv Polytechnic Institute", 37, Beresteiskyi Ave, Kyiv 03056, Ukraine
lesyk_d@ukr.net

[2] G.V. Kurdyumov Institute for Metal Physics of the NAS of Ukraine, 36, Academician Vernadsky Blvd, Kyiv 03142, Ukraine

[3] University of the Basque Country, 202, Bizkaia Science and Technology Park, 48170 Zamudio, Spain

[4] West Pomeranian University of Technology, 17, Aleja Piastow, 70310 Szczecin, Poland

Abstract. AISI 1045 medium-carbon steel workpieces were selectively hardened by a laser heat treatment (LHT) followed by a multi-pin ultrasonic impact treatment (UIT) to enhance the surface properties. The laser surface hardening was done using a high-power fiber laser and scanning optics mounted in the computer numerical control (CNC) machine. The LHT tests were implemented using a constant temperature strategy. The LHT-processed flat specimens were subsequently peened by the CNC ultrasonic processing with a seven-pin impact head. The paper focuses on the effects of combined laser-ultrasonic surface hardening technology on the hardness, structure-phase, and residual stress state of AISI 1045 steel. Particular attention was paid to analyzing the grain size and residual stress magnitudes under various LHT + UIT conditions. Results have shown that the combined LHT + UIT technique provides a high surface hardness (~60 HRC_5) due to the formation of the fine-grained martensitic microstructure. After UIT treatment, compressive residual stress is formed.

Keywords: Combined Laser-Ultrasonic Surface Processing · Laser Transformation Hardening · Fiber Laser · Ultrasonic Impact Peening · Surface Nanocrystallization · Manufacturing Innovation

1 Introduction

Surface hardening technologies are widely used in manufacturing processes to extend metallic parts' life. At the same time, the prolongation of the operation life of the carbon steel components by preventing their degradation under highly corrosive environments or dynamic and contact loads still is a timely and critical in engineering. Combined

V. Tonkonogyi et al. (Eds.): InterPartner 2023, LNME, pp. 296–306, 2024.
https://doi.org/10.1007/978-3-031-42778-7_27

technologies can significantly enhance the fatigue performance, wear, and corrosion resistance of steel products compared to the mono-technologies.

Most of the malfunction of metal parts occurs on the surface. The main types of destructive processes include friction and wear of metallic products, plastic deformation, fatigue [1], and corrosion [2]. Product improvement is difficult without using new progressive technologies that increase the service life and reliability of parts in difficult operating conditions. One of the most promising directions for solving the issues mentioned above is the improvement of technological processes for the selective surface hardening of metal parts due to structural-phase transformation [3].

2 Literature Review

The surface hardening methods differ significantly in terms of technical indicators and efficiency. Some surface hardening methods are limited for various reasons, such as increased surface roughness, decreased fatigue strength [4], and high porosity [5]. Combining hardening methods makes it possible to achieve the required state of the surface layer and thereby ensure a high level of operational properties. Thermomechanical treatments are very effective processes for the surface modification of steel parts. The thermomechanical surface treatments can be applied simultaneously [6] or separately [7].

One of the most promising decisions for the improvement of the physical and mechanical properties of steel products is the use of a laser heat treatment (LHT) combined with an ultrasonic impact treatment (UIT) [8]. The laser transformation hardening process has a number of specific advantages among thermal methods [9]. Particularly, the LHT treatment leads to fine-grained martensitic structure formation in steels through laser-induced phase transformations. It is well-known that the higher the surface roughness in the laser-processed steel parts is generally demonstrated, the higher the heating temperatures and/or specimen feed rates [10, 11]. It should also be noted that the LHT treatment can be realized selectively and remotely using computer numerical control (CNC) robotic equipment [12] or integrated milling centers [8].

Moreover, the LHT using advanced programmable scanning optics and proportional-integral-derivative (PID) controllers for heat temperature control can be fully controlled as compared to electron-beam hardening [13] or plasma hardening [14]. Muthukumaran et al. [15] confirmed that the residual compressive stress values were achieved in laser-hardened and laser-tempered/overlapped areas. Uysal [16] proposed ways for the treatment of composite materials.

Applying surface plastic deformation techniques after/during LHT can improve the residual stress state and reduce surface roughness. Hu et al. [17] reported that a combined laser-ultrasonic treatment achieved favorable compressive residual stresses (–1200…–500 MPa). The deformation-affected depth can be expanded into the material due to the softening caused by laser heating (hybrid laser-ultrasonic treatment) [18].

This work aims to study the effects of combined laser-ultrasonic surface hardening technology on the hardness, structure-phase state, and residual stress of AISI 1045 steel [19]. Particular attention is paid to studying the grain size and residual stress values under various LHT + UIT conditions.

3 Research Methodology

The steel parts with dimensions of $100 \times 60 \times 20\ mm^3$ were used in this study. The chemical composition of the AISI 1045 carbon steel is the following: 0.45% C, 0.65% Mn, 0.27% Si, 0.24% Cr, 0.12% Ni, 0.19% Cu, 0.03% P, 0.03% S, and balance Fe. The measured values of the average surface hardness are given in Table 1. The surface parts were polished before applying the LHT + UIT treatment. The initial surface roughness was ~ 0.6 μm (Table 2).

Table 1 Average surface hardness of the initial sample

HRC	HRC	HRC	HRC	HRC	HRC	HRC	Average HRC
19.4	17.4	19.1	17.9	18.9	18.4	19.4	18.6

Table 2 Average surface roughness of the initial sample

Ra, μm	Ra, μm	Ra, μm	Ra, μm	Ra, μm	Ra, μm	Ra, μm	Average Ra, μm
0.54	0.62	0.63	0.65	0.57	0.63	0.6	0.61

The scan-based LHT treatment was carried out using a ROFIN SINAR FL010 fiber laser, and SCANLAB HURRY SCAN 25 scanning optics mounted on the CNC milling center (Fig. 1a).

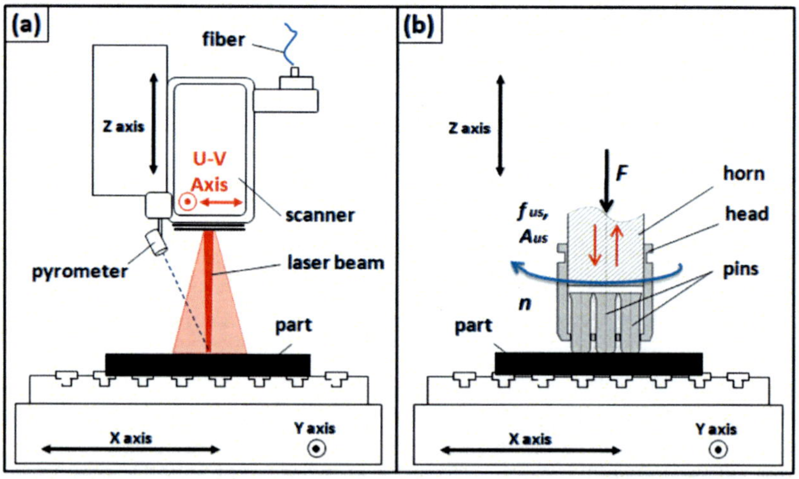

Fig. 1 Laser surface hardening (a) and ultrasonic impact peening (b) processes

The LHT treatment was performed by a single-pass process using a strategy of constant heating temperature. An IMPAC IGAR 12LO pyrometer was applied to measure the surface temperature on the treated part (Fig. 1a).

The UIT treatment was conducted using the CNC milling machine. The ultrasonic vibration system contained a piezoceramic transducer, a step-like horn, and a seven-pin impact head (Fig. 1b). The UIT tests were carried out by an ultrasonic generator with a power output of 0.3 kW and a frequency of 21.6 kHz.

The LHT and UIT parameters are listed in Table 3.

Table 3 Laser-ultrasonic surface hardening parameters

Laser heat treatment						
Heating temperature (°C)	Specimen feed rate (mm/min)	Scanning speed (mm/s)	Laser beam dimension (mm)	Treatment duration (s)	Heat energy density of laser beam (kJ/cm^2)	Power density of laser beam (W/cm^2)
1200 [LHT1]	40	1000	1 × 10	1.50	102	5.4 × 10^3
1200 [LHT2]	90	1000	1 × 10	0.66	46	5.5 × 10^3
1200 [LHT3]	140	1000	1 × 10	0.42	31	5.8 × 10^3
Ultrasonic impact treatment						
Vibration frequency (kHz)	Vibration amplitude (μm)	Rotation speed of impact head (rpm)	Contact pin diameter (mm)	Treatment duration (s)	Mechanical energy density of pins (mJ/cm^2)	Total accumulated energy per area (J/cm^2)
21.6 [UIT1]	18	76	3	60	18	2.2 × 10^2
21.6 [UIT2]	18	76	3	120	26	3.1 × 10^2

The temperature and laser power history recorded on the LHT of AISI 1045 steel at 1200 °C are presented in Fig. 2. It is clear that the measured temperature magnitudes on the surface of the treated parts are constant regardless of the specimen feed rate. The required magnitudes of the laser power (laser beam dimension 1 × 10 mm) are 680 W, 690 W, and 730 W at the specimen feed rate of 40 mm/min, 90 mm/min, and 140 mm/min, respectively.

The surface hardness on the steel parts was measured using a portable digital hardness tester COMPUTEST SC at a load on a diamond cone-shaped Brale indenter of 5 kgf (HRC_5) for a dwell time of 3 s. Seven measurements were carried out in the central area of laser/ultrasonic tracks in each case, and the average magnitude was reported (Table 3).

The combined LHT + UIT-processed parts were mechanically cut and prepared according to a standard metallographic procedure [20]. The heat-affected zone (HAZ) cross-sections were studied using a LEICA DCM3D microscope. The cross-section microstructure observations were conducted using a NIKON OPTIPHOT-100 light optical microscope (LOM). The X-ray diffraction (XRD) analysis was conducted using a

Rigaku Ultima IV diffractometer. The full widths at half maximum (FWHM) of the (310) peak fragments of the XRD patterns were used to estimate the crystallite size and residual macro-stress.

The residual macrostresses in the near-surface layers were determined based on a shift in the diffraction maximum (310) of the LHT + UIT-hardened specimens from their positions registered for the untreated specimens [21].

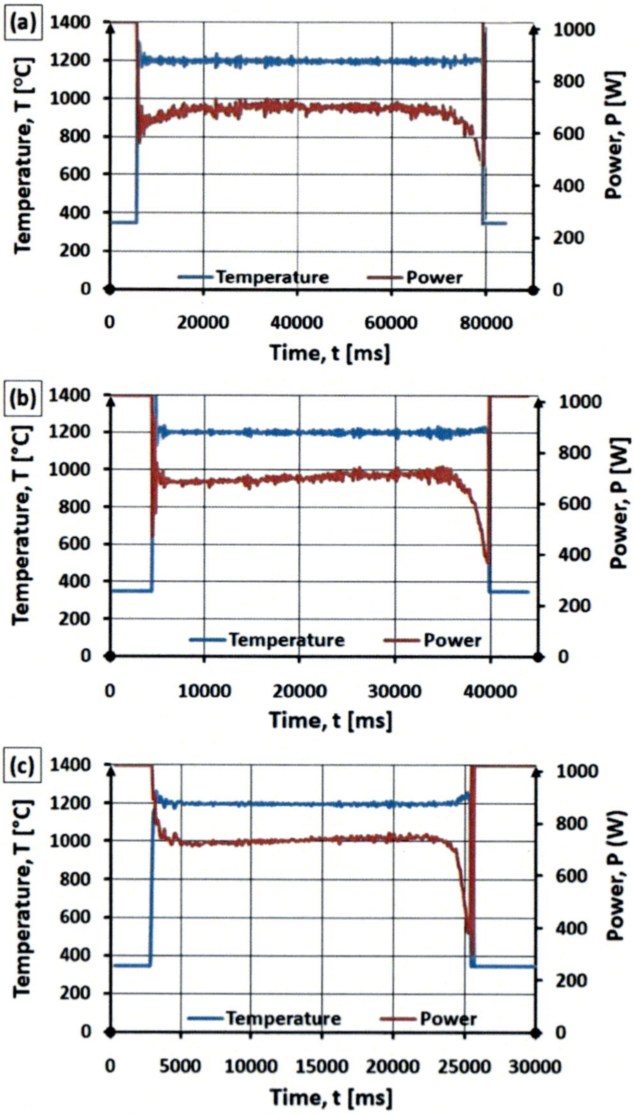

Fig. 2 Temperature-power history after LHT1 (a), LHT2 (b), and LHT3 (c) of AISI 1045 steel

4 Results and Discussions

4.1 Surface Hardness

The measured surface macrohardness values of the studied steel for various LHT and combined LHT + UIT conditions are presented in Fig. 3. The surface hardness values are 18.6 HRC_5 for the as-received sample (Table 1). The LHT and LHT + UIT treatments significantly provide a higher surface hardness due to the formation of the fine-grained martensitic microstructure coupled with a high dislocation density [22, 23]. Compared to the hardness of the untreated sample, the HRC values were increased by about triple after the combined treatment (Fig. 3).

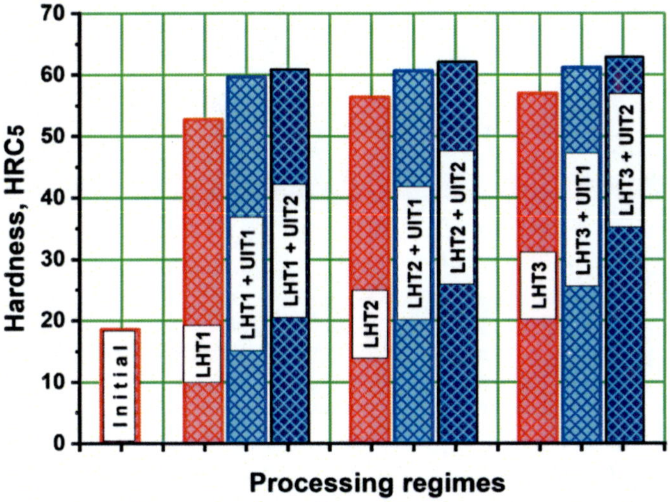

Fig. 3 Average surface hardness after LHT and LHT + UIT-hardened of AISI 1045 steel

4.2 Structure Observation

The cross-sectional micrographs of the LHT + UIT2-hardened medium-carbon steel parts at the heating temperature of 1200 °C are presented in Fig. 4.

The hardening depth at the LHT process depends on the interaction time and surface/interior temperature related to the laser power density (surface heating temperature (Fig. 2)), LHT speed (specimen feed rate), and trace area values. The LHT-treated zone is observed to get thinner with an increase in sample feed rate from 40 mm/s (LHT1 regime) to 140 mm/s (LHT3 regime). The hardening depth of the AISI 1045 steel is about 200–300 μm after LHT1 + UIT2 and LHT2 + UIT2 conditions (Fig. 4a). The LHT traveling speed using the fiber laser and scanning optics was optimized based on the previous experimental and theoretical studies. In this work, the laser transformation hardening process of the studied carbon steel was conducted below the melting temperature to avoid the melting of the surface.

Fig. 4 Subsurface microstructure after LHT1 + UIT2 (a), LHT2 + UIT2 (b), and LHT3 + UIT2-hardened (c) of AISI 1045 steel

Figure 5 shows that the diffraction maximums of α-iron significantly broadened after combined LHT + UIT processes for various LHT speeds. The broadened diffraction peaks relate to dual contributions of the grains/crystallites size and lattice micro-strains formed by the applied process. The combined LHT + UIT processes decreased crystallites with high lattice micro-strains [20]. Additionally, the finishing UIT (applied after LHT) effectively cleans the oxide scale and decreases the intensity of oxides' diffraction peaks.

In our case, high lattice micro-strains also resulted in an increase in carbon contents both in the LHT-induced martensitic needles/laths (having base-centered tetragonal lattice) and in the ferritic grains supersaturated with carbon due to UIT-induced severe plastic deformation of the near-surface layers. However, the broadening extents are different for various LHT speeds used (Fig. 6), and the combined LHT2 + UIT2 treatment can be discriminated as optimal LHT speed (LHT2: 90 mm/s) based on the broadest (310) diffraction peak of α-phase (the peak width is ~ 2.8 degrees), which naturally indicates the lowest crystallites size and the highest lattice micro-strains formed. Assessment of the contributions of crystallites size (D) and lattice micro-strains (η) shows that as compared to the initial sample ($D = 350$ nm, $\eta = 1 \times 10^{-4}$), the combined LHT2 + UIT2 process leads to essential crystallite refinement ($D = 25$ nm) and lattice micro-strains increase by an order of magnitude ($\eta = 2 \times 10^{-3}$). These values are similar to those observed in [24] for combined treatments.

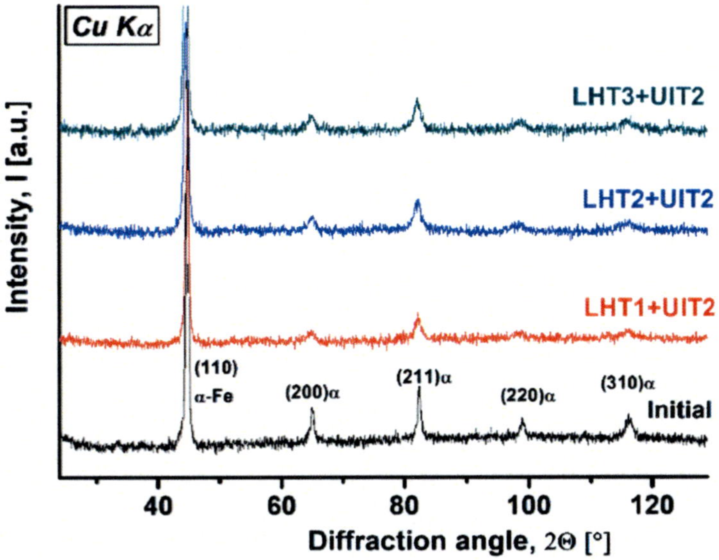

Fig. 5 XRD patterns in the near-surface layer after LHT + UIT-hardened of AISI 1045 steel

4.3 Residual Stress Characterization

Typical positions of diffraction peaks of the complexly processed samples are shown in Fig. 6. Evidently, the (310) diffraction peaks are shifted towards smaller diffraction angles (see the angles nearby the centroid lines of the peaks) that naturally indicate the formation of compressive residual macro-stress in the near-surface layer [25].

The most considerable shift and thus the highest magnitude of the residual stress ($\sigma_R = 1.8$ GPa) were observed for the sample after the combined LHT3 + UIT2 process (LHT3: 140 mm/s), i.e., for the sample underwent laser quenching with the highest cooling rate. Other combined processes led to slightly lower σ_R of similar magnitudes (1.5 GPa). Assessment of the residual stress gives the magnitudes far higher than the yield stress of the initial steel specimen (after annealing $\sigma_Y = 350$ MPa). However, these magnitudes correlate well to the surface hardening (Fig. 3) and meet the well-known relation HV ~ 3.03 σ_Y (3.03 σ_R). In our case, the LHT-processed sample's hardness is similar to the triple residual stress: 53 $HRC_5 = 590$ HV (5.8 GPa) $= 3.03\ \sigma_R = 5.45$ GPa.

Thus, combined laser-ultrasonic surface hardening methods can be used for the operational properties' improvement of hypoeutectoid steels owing to the reduction in the grain size, increasing the dislocation density and lattice micro-strains, as well as the formation of compressive residual macro-stress in the near-surface layer.

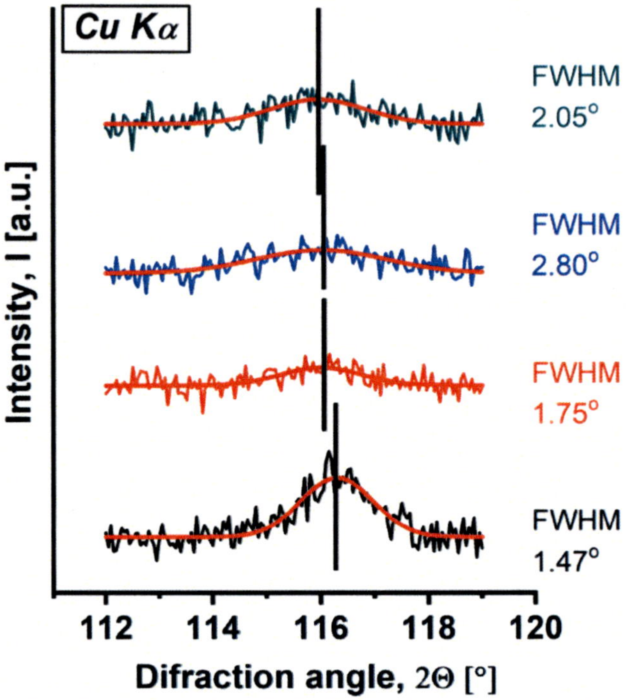

Fig. 6 Diffraction maximum 310 in the subsurface layer of the initial state (black color) and after LHT1 + UIT2 (red color), LHT2 + UIT2 (blue color), and LHT3 + UIT2-hardened (dark cyan color) of AISI 1045 steel (from down upward)

5 Conclusions

The laser-assisted ultrasonic surface modification was implemented to increase the surface properties of AISI 1045 steel. The results of this study can be summarized as follows.

It was established that to maintain a constant temperature (1200 °C) on the surface in the center of the track (10 mm), the required laser power should be at the level 680 W, 690 W, and 730 W at the specimen feed rate of 40 mm/min, 90 mm/min, and 140 mm/min, respectively.

The surface hardness was increased by approximately triple from the base metal level (18.6 HRC_5) after the combined LHT + UIT treatment.

The compressive residual macro-stresses formed in the near-surface layer after the combined LHT + UIT treatment are due to laser-quenching and can achieve up to1.8 GPa that correlate well with the registered surface hardness magnitudes (53 $HRC_5 \approx$ 5.5 GPa (HV)).

Enhanced surface hardness is brought about by crystallite refinement (down to 25 nm) and a high increase in the lattice microstrains owing to martensitic transformation and carbon dissolution in the ferritic lattices.

Acknowledgements. This research was partially supported by the Ulam NAWA program (Grant Number BPN/ULM/2021/1/00153) and the National Academy of Sciences of Ukraine (Grant Number 0119U001167). D.A. Lesyk also thankful the International Association for Technological Development and Innovations.

References

1. Yao, J., Zhang, Q., Liu, R., Wu, G.: Laser surface hardening. In: Laser Applications in Surface Modification. Advanced Topics in Science and Technology in China, vol. 65, pp. 83–102 (2022)
2. Frerichs, F., Lu, Y., Lubben, T., Radel, T.: Process signature for laser hardening. Metals **11**, 465 (2021). https://doi.org/10.3390/met11030465
3. Santhanakrishnan, S., Kong, F., Kovacevic, R.: An experimentally based thermo-kinetic phase transformation model for multi-pass laser heat treatment by using high power direct diode laser. Int. J. Adv. Manuf. Technol. **64**, 219–238 (2013). https://doi.org/10.1007/s00170-012-4029-z
4. Lymarenko, A., Kurgan, V., Bazhanova, A., Khamray, V., Ponomarenko, A., Karabegović, I.: The stress state of a thick-walled hydraulic press cylinder with concentrators. J. Eng. Sci. (Ukraine) **10**(1), D15–D19 (2023). https://doi.org/10.21272/jes.2023.10(1).d3
5. Kennedy, E., Byrne, G., Collins, D.N.: A review of the use of high power diode lasers in surface hardening. J. Mater. Proc. **155–156**, 1855–1860 (2004). https://doi.org/10.1016/j.jmatprotec.2004.04.276
6. Radziejewska, J.: Influence of laser-mechanical treatment on surface topography, erosive wear and contact stiffness. Mater. Des. **32**, 5073–5081 (2011). https://doi.org/10.1016/j.matdes.2011.06.035
7. Wu, B., Wang, P., Pyoun, Y.-S., Zhang, J., Murakami, R.: Effect of ultrasonic nanocrystal surface modification on the fatigue behaviors of plasma-nitrided S45C steel. Surf. Coat. Technol. **213**, 271–277 (2012). https://doi.org/10.1016/j.surfcoat.2012.10.063
8. Lesyk, D., Alnusirat, W., Martinez, S., Mordyuk, B., Dzhemelinskyi, V.: Comparison of the effects of laser, ultrasonic, and combined laser-ultrasonic hardening treatments on surface properties of AISI 1045 steel parts. In: Tonkonogyi, V. (eds) Advanced Manufacturing Processes III. InterPartner 2021. Lecture Notes in Mechanical Engineering, pp. 313–322. Springer, Cham (2022).https://doi.org/10.1007/978-3-030-91327-4_31
9. Sancho, P., Montealegre, M.A., Dominguez, J., Alvarez, P., Isaza, J.: Dynamic control of laser beam shape for heat treatment. J. Laser Appl. **30**, 032507 (2018). https://doi.org/10.2351/1.5040647
10. Yazici, O., Yilmaz, S.: Investigation of effect of various processing temperatures on abrasive wear behaviour of high power diode laser treated R260 grade rail steels. Tribol. Int. **119**, 222–229 (2018). https://doi.org/10.1016/j.triboint.2017.11.006
11. Hovorun, T.P., et al.: Physical-mechanical properties and structural-phase state of nanostructured wear-resistant coatings based on nitrides of refractory metals Ti and Zr. Funct. Mater. **26**(3), 548–555 (2019). https://doi.org/10.15407/fm26.03.548
12. Lee, K.-H., Choi, S.-W., Yoon, T.-J., Kang, C.-Y.: Microstructure and hardness of surface melting hardened zone of mold steel, SM45C using Yb:YAG disk laser. Sci. Technol. Weld. Join. **34**, 75–81 (2016). https://doi.org/10.5781/JWJ.2016.34.1.75
13. Valkov, S., Ormanova, M., Petrov, P.: Electron-beam surface treatment of metals and alloys: Techniques and trends. Metals **10**, 1219 (2020). https://doi.org/10.3390/met10091219

14. Guo, D., Zhang, P., Jiang, Y., Song, C., Tan, D.Q., Yu, D.: Effects of surface texturing and laminar plasma jet surface hardening on the tribological behaviors of GCr15 bearing steel. Tribol. Int. **169**, 107465 (2022). https://doi.org/10.1016/j.triboint.2022.107465
15. Muthukumaran, G., Babu, D.: Analysis of residual stress distribution and corrosion in laser surface hardened low alloy steel with a flat top-hat laser beam, using a high-power diode laser. Arab. J. Sci. Eng. **47**, 8785–8803 (2022). https://doi.org/10.1007/s13369-021-06350-8
16. Uysal, M.U.: Effects of cryogenically treated CFRP composite on the buckling behavior in the adhesively bonded beam. J. Eng. Sci. (Ukraine) **10**(1), D1–D7 (2023). https://doi.org/10.21272/jes.2023.10(1).d1
17. Hu, X., et al.: Rolling contact fatigue behaviors of 25CrNi2MoV steel combined treated by discrete laser surface hardening and ultrasonic surface rolling. Opt. Laser Technol. **155**, 108370 (2022). https://doi.org/10.1016/j.optlastec.2022.108370
18. Liu, J., Suslov, S., Ren, Z., Dong, Y., Ye, C.: Microstructure evolution in Ti64 subjected to laser-assisted ultrasonic nanocrystal surface modification. Int. J. Mach. Tools Manuf. **136**, 19–33 (2019). https://doi.org/10.1016/j.ijmachtools.2018.09.005
19. Ivchenko, O., et al.: Method for an effective selection of tools and cutting conditions during precise turning of non-alloy quality steel C45. Materials **15**(2), 505 (2022). https://doi.org/10.3390/ma15020505
20. Akca, E., Trgo, E.: Metallographic procedures and analysis – A review. Period. Eng. Nat. Sci. **3**(2), 9–11 (2015). https://doi.org/10.21533/pen.v3i2.51
21. Bouras, M., Boumaiza, A., Ji, V., Rouag, N.: XRD peak broadening characterization of deformed microstructures and heterogeneous behavior of carbon steel. Theor. Appl. Fract. Mech. **61**, 51–56 (2012). https://doi.org/10.1016/j.tafmec.2012.08.006
22. Maharjan, N., Zhou, W., Zhou, Y., Wu, N.: Decarburization during laser surface processing of steel. Appl. Phys. A **124**, 682 (2018). https://doi.org/10.1007/s00339-018-2104-5
23. Lin, P., Zhang, Z., Zhou, H., Ren, L.: The mechanical properties of medium carbon steel processed by a biomimetic laser technique. Mater. Sci. Eng. A **560**, 627–632 (2013). https://doi.org/10.1016/j.msea.2012.10.006
24. Lesyk, D.A., Mordyuk, B.N., Martinez, S., Iefimov, M.O., Dzhemelinskyi, V.V., Lamikiz, A.: Influence of combined laser heat treatment and ultrasonic impact treatment on microstructure and corrosion behavior of AISI 1045 steel. Surf. Coat. Technol. **401**, 126275 (2020). https://doi.org/10.1016/j.surfcoat.2020.126275
25. Lee, C., et al.: Temperature dependence of elastic and plastic deformation behavior of a refractory high-entropy alloy. Sci. Adv. **6**, eaaz4748 (2020). https://doi.org/10.1126/sciadv.aaz4748

Numerical Evaluation of the Properties of Highly Efficient Titanium Porous Materials

Oleksandr Povstyanoy(✉), Nataliya Imbirovich, Rostyslav Redko, Olha Redko, and Pavlo Savaryn

Lutsk National Technical University, 75, Lvivska Str., 43018 Lusk, Ukraine
npovstjanoj@ukr.net

Abstract. An exciting combination of such properties as high strength, low density, corrosion resistance, and biocompatibility characterizes titanium. However, the widespread use of titanium at the industrial level has not yet been achieved due to its high extraction and production costs. Therefore, titanium is increasingly used in sectors with high demand, such as the aerospace industry or the production of biomedical devices, where the final high cost is not a major factor. It is believed that processing titanium and its alloys using powder metallurgy (PM) methods is a significant way to reduce the cost of manufacturing titanium products. It also provides the opportunity to develop new alloys that are difficult to obtain using traditional technologies. This work is devoted to processing titanium powder from biomedical production waste using various PM methods. It aims to research the processing of almost pure, chemically homogeneous, and fine-grained titanium-based components. In particular, the main properties that can be achieved (porosity, microstructure, and mechanical properties) and the creation of highly efficient porous materials by advanced methods of isostatic pressing are presented.

Keywords: Titanium · Porous Materials · Isostatic Pressing · Self-Propagating High-Temperature Synthesis · Hardness · Microstructure · Product Innovation

1 Introduction

Titanium is a relatively new engineering material compared to other structural metals such as steel and aluminum, as its industrial use only began in the last century. It is an extraordinary metallic element with many defining characteristics, such as a high melting point (1675 °C), relatively low density (4.5 g/cm^3), high strength, and good fracture resistance.

Titanium and its alloys are significant industrial metals. They are widely used in aerospace, energy, nuclear industry, food industry, chemical, and biomedical engineering, etc. [1]. However, titanium porous, permeable materials (PPM) consists of titanium metal and pores, which is why it not only inherits the inherent characteristics of metal but also provides many operational characteristics (filtering [2], separation [3], sound absorption and thermal insulation [4]).

V. Tonkonogyi et al. (Eds.): InterPartner 2023, LNME, pp. 307–317, 2024.
https://doi.org/10.1007/978-3-031-42778-7_28

Therefore, titanium PPM is used in the chemical industry [5] due to its resistance to corrosion and filtering properties [6].

In recent years, many scientists have conducted many studies using titanium powders as raw materials for obtaining various types of products and ways of processing them using various methods of powder metallurgy, such as pressing and sintering [7], isostatic pressing [8] and uniaxial hot pressing [9].

However, crushed titanium feedstock can also be used as a feedstock in powder metallurgy processes, taking advantage of the fact that it is already in powder form. Titanium powders obtained from biomedical production waste are characterized by chlorides, which create porosity filled with gas during sintering, preventing the production of entirely dense or non-porous titanium products.

The main advantage of using PM methods is that they are end-to-end processes with a higher material yield and almost without additional mechanical processing, which reduces the costs of manufacturing titanium porous products.

2 Literature Review

Today, there are many ways to obtain porous, permeable materials (PPM) from titanium. These are the method of sintering without pressure [10], the technology of pressing in dies [11], the technology of spatial pressing [12], and additive technologies of creating PPM [13]. Among the modern powder metallurgy methods, the simplest and most common is the method of sintering PPM containing a pore former [14, 15]. Porosity, pore shape, and pore size distribution can be adjusted using the concentration of the pore former.

To increase competitiveness and expand the scope of application and assortment of titanium PPMs at the current stage of development of the world industry, it is proposed to modify and simplify the processes of obtaining PPMs, particularly during sintering.

The economic efficiency of such PPM from titanium powders is ensured not only due to operational qualities but also at the production stage due to using cheap raw materials and energy saving at all stages of the technological process [16–18]. The analysis of costs, which are inextricably linked to the technology, is of great importance in determining the cost of titanium porous powder materials.

Thus, the largest share in the cost of PPM production by powder metallurgy methods is occupied by the costs of basic materials and electricity (80–90%). Traditional methods of sintering titanium PPM require sufficiently powerful furnace equipment with protective media. Sintering costs make up 40–50% of the production cost [19, 20]. These circumstances became a prerequisite for developing new methods for obtaining highly efficient porous, permeable materials from titanium powders from biomedical industry waste.

3 Research Methodology

The raw material was selected for the study – MEDGAL® biomedical engineering waste (Fig. 1.). This is pure titanium (99%) waste after processing titanium orthopedic prostheses.

Fig. 1. Titanium waste of biomedical engineering MEDGAL®.

The biomedical industry waste processing technological scheme consisted of two stages – grinding and restorative annealing [21] (Fig. 2).

Grinding was carried out in a vibrating drum mill with an offset axis of rotation. Grinding time was 30, 60, 120, and 180 min. The loading mass of the titanium material was determined from the ratio of the mass of the balls to the mass of the powder and was 0.75:1. The degree of grinding was estimated by the amount of titanium powder of one or another fraction.

Restorative annealing to relieve internal stresses in titanium powder lasted 2–2.5 h at a temperature of 500 °C in a vacuum.

Fig. 2. Particles of titanium powder obtained by technology [21].

In manufacturing titanium PPM that meets modern requirements, the radial pressing scheme is the most rational, and it can be the basis for creating new and improving existing PPM manufacturing technologies, equipment, and tools [22, 23] and new types of products.

To obtain titanium PPM, an installation for pressing various materials was used (Fig. 3).

The developed technology for creating porous materials using radial isostatic pressing followed by sintering using self-propagating high-temperature synthesis (SHS) was

Fig. 3. Installation for pressing various kinds of materials.

developed at the Lutsk National Technical University [6, 7]. For the preparation of mixtures, the initial components are titanium powder obtained from biomedical engineering waste (Ti – 98.18%, C – 0.03%, N_2 – 0.08%, H_2 – 0.32%, Si –0.07%, Ni – 0.14%, Fe – 0.10%) and carbon black (C) – carbon with a bulk density of 0.1 g/cm^3.

Their atomic weights determined the ratio of Ti and C. The reaction of the formation of TiC by the SHS method from a briquette mixture of titanium and carbon black due to the high isothermality of the process belongs to the category of combustion reactions that occur in a narrow zone, moving along the briquette due to heat transfer after local initiation of reactions in a heated mixture of reagents. The optimal ratio of Ti and C of the $TiC_{0.5}$ compound was determined experimentally from the point of view of obtaining high-quality products (absence of cracks, warping, and obtaining isotropic properties by volume).

PPM's SHS sintering technology is based on burning powder mixtures (Ti + C) in the air without preheating (Fig. 4).

Fig. 4. Obtaining PPM based on Ti + C with the help of the SHS – process.

After sintering SHS (t = 30 s), the finished product gradually cools automatically for 30 s.

Figure 5 presents PPMs from titanium powder tubes and cones obtained by radial isostatic pressing and SHS-sintering.

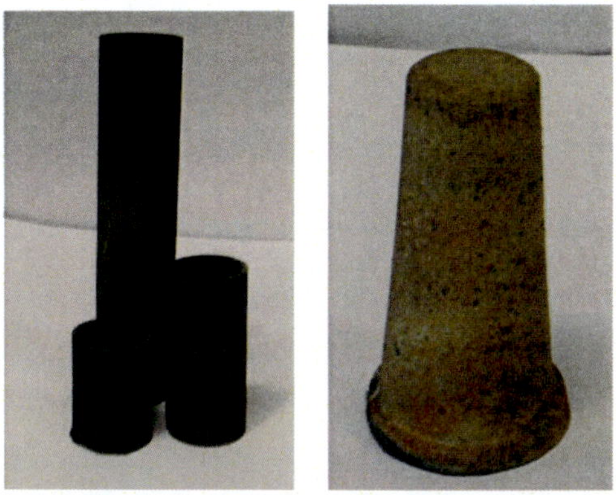

Fig. 5. General view of PPM based on titanium carbide.

The developed and applied technology for obtaining porous materials based on $TiC_{0.5}$ allows to reduce energy consumption both at the pressing stage (by 2 times) and at the sintering stage (by 1.5 times).

4 Results and Discussion

The properties of porous materials based on titanium carbide are significantly superior to those of porous materials based on pure titanium obtained in the SHS-burning mode [24]. Porous materials based on $TiC_{0.5}$ have high chemical resistance.

Figure 6 Illustrates the structure of PPM from titanium. Volumetric porosity due to the packing of particles and microporosity of particles can be seen. The last circumstance allows to increase the specific surface of the porous material (1 g of porous material – 5 m^2) and, as a result, its sorption properties, which is very important for PPM, which can be used for cleaning liquids and gases.

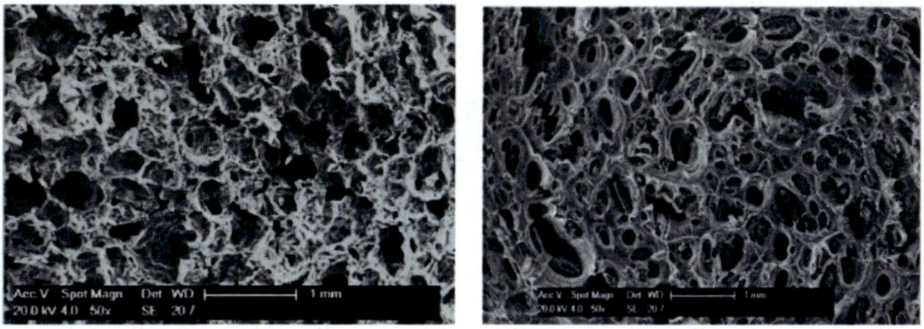

Fig. 6. SEM image of the microstructure of titanium PPMs.

The microstructure of the obtained material (Fig. 6) allows us to state that it is possible to obtain a porous material with a uniform structure during the metal reduction reaction.

One of the most powerful and progressive methods for visualization, obtaining 3D image data of the PPM structure, is direct volume rendering, the parameters of light emission and absorption, which are assigned to each point of the image [25]. Modeling light transmission through a particular image volume allows you to display data without building intermediate polygonal models.

The creation of a 3D image environment is carried out by superimposing flat cross-sections of the appropriate height range of the finished porous powder material (Fig. 7).

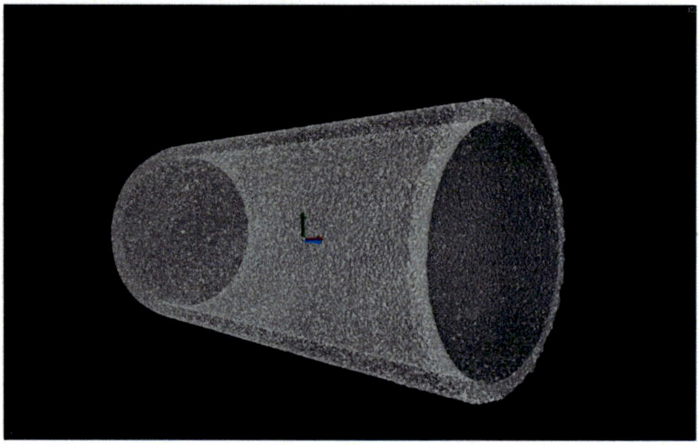

Fig. 7. Visualization of the received image of the cross-section of the titanium PPM.

A combination of various structural components represents metallographic images of PPM grindings. The combination of these structural components (planar and spatial) for PPM is presented in Fig. 8.

The morphology of PPMs determines their filtering characteristics and, therefore, their effectiveness in many application areas. Quantitative and qualitative relationships between the morphology of the porous material and its local and global filtering properties are essential in many fields of application [26–28].

The establishment of quantitative morphological and filtering properties can be based on direct modeling of the porous structure by transfer of the flow of matter in the 3D reconstruction of the material.

The principle of operation is as follows: a section is cut from the PPM sample, generating a clean and smooth surface on the block, which is then displayed on the monitor. After rendering, the new slice is removed from the sample block, giving a new surface to the image. This cycle is repeated until the desired number of images is obtained. Typical slice thickness is in the range of 30–100 nm. A total of 1300 slices were obtained, each 100 nm thick and consisting of 4000 × 4000 pixels.

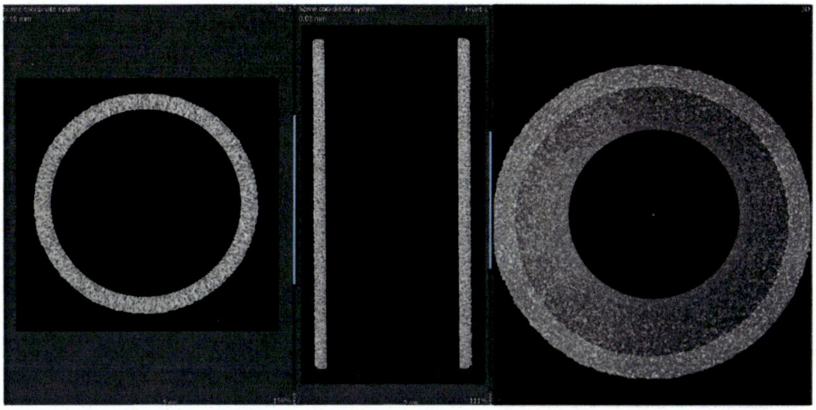

Fig. 8. Determination and analysis of structural components of PPM in 3D image format.

Images were first binarized using an automated sequence consisting of image denoising, filtering, and thresholding [29]. The images were then stitched together to give a representation of the 3D structure of the PPM monolith (Fig. 9).

The studied PPM sample consists of a solid metal structure (monolithic skeleton; opaque) with a porosity of ~ 69% (interstitial voids, shaded).

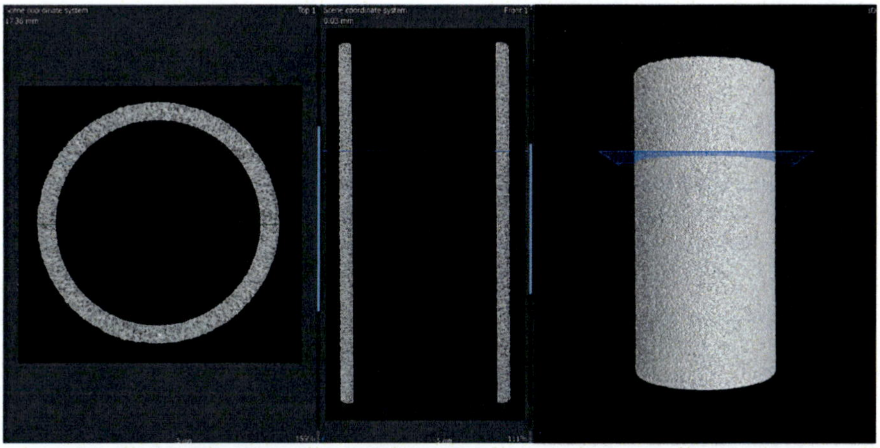

Fig. 9. 3D image of the reconstructed volume of the cylindrical sample of PPM.

Porosity can be determined from the number of black pixels representing hollow space divided by the total number of pixels in the PPM structure. The advantage of the image reconstruction approach is its ability to produce spatial tortuous porosity. It makes it possible to study the radial and axial porosity distributions, namely, perpendicular and parallel to the cylindrically confined PPM monolith axis.

Hardness is an important characteristic of a porous material that reflects the bond energy and symmetry of the structure. The hardness of the sintered sample was measured using a Vickers microhardness tester under a load of 1.96 N (300 g) for 15 s. The test results are presented in Fig. 10.

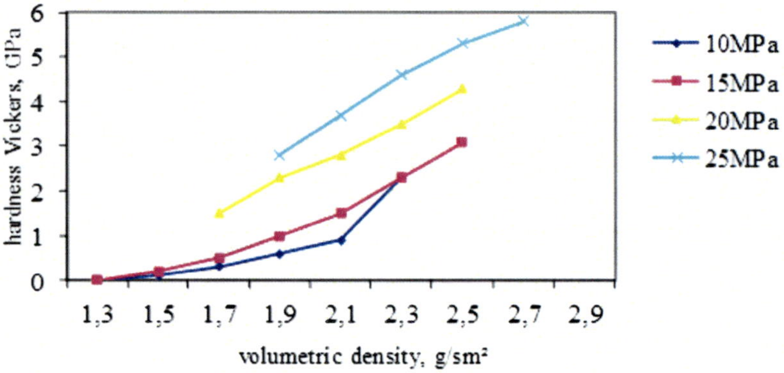

Fig. 10. The relationship between Vickers hardness and bulk density.

The criterion of strength complex is the amount of specific potential energy of shape change accumulated by a deformed object. A dangerous state (flowability) in a stressed state occurs when the specific potential energy of a change in shape reaches its critical value [30, 31].

With the help of ABAQUS and entering all the necessary data, calculations are automatically carried out, and the results of applying pressure from the outside are obtained (Fig. 11).

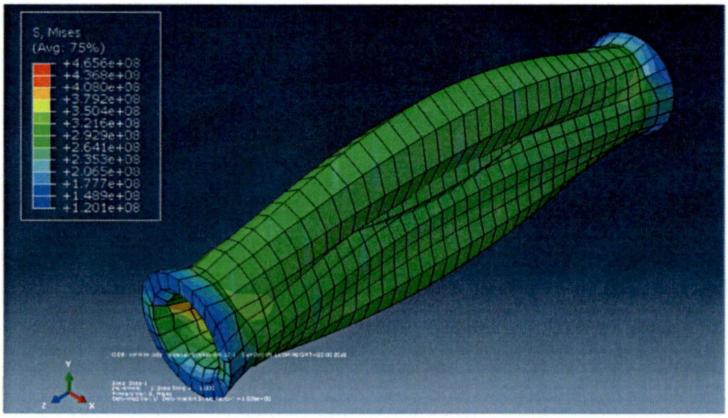

Fig. 11. Distribution of load on PPM according to the criterion of maximum stress from the outside.

In this case, we get a graphic representation of the load distribution on the PPM according to the maximum stress criterion. The detail is displayed in a deformed form.

Permissible loads on the PPM according to the maximum stress criterion are shown in green, and exceeding the maximum permissible loads is shown in red.

Let's change the criterion to the criterion of destruction in ABAQUS and display the pressure distribution (Fig. 12).

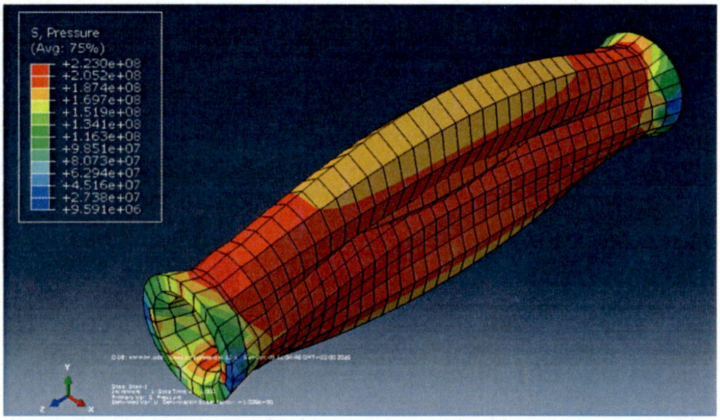

Fig. 12. Distribution of deformations along the PPM under the action of a given pressure.

The modeling environment provides an opportunity to investigate and predict patterns of structure formation and properties of porous materials [32], considering the sizes of structural elements and establishing correlations between structural components.

5 Conclusions

The technology for obtaining highly efficient porous, permeable materials based on titanium powders obtained from biomedical engineering waste has been developed. This technology made it possible to reduce energy consumption at the pressing stage by 2 times and at the sintering stage by 1.5 times due to the absence of external energy consumption for sintering (SHS sintering).

The formation of a 3D image by superimposing flat cross-sections of the appropriate height range of the finished titanium porous, permeable material is shown.

A 3D reconstruction and morphological analysis of the PPM was carried out to determine the average value of porosity in the volume, which is 0.687.

The main structural characteristics of highly efficient porous, permeable materials based on titanium powders were determined - microstructure, porosity, and strength.

References

1. Jung, J.-H.: Study on surface shape control of pure Ti fabricated by electron beam melting using electrolytic polishing. Surf. Coat. Technol. **324**, 106–110 (2017). https://doi.org/10.1016/j.surfcoat.2017.05.061

2. Sklabinskyi, V., Liaposhchenko, O., Pavlenko, I., Lytvynenko, O., Demianenko, M.: Modelling of liquid's distribution and migration in the fibrous filter layer in the process of inertial-filtering separation. In: Ivanov V. et al. (eds) Advances in Design, Simulation and Manufacturing. DSMIE 2018. Lecture Notes in Mechanical Engineering, pp. 489–497. Springer, Cham (2019). https://doi.org/10.1007/978-3-319-93587-4_51
3. Pavlenko, I., Liaposhchenko, A., Ochowiak, M., Demyanenko, M.: Solving the stationary hydroaeroelasticity problem for dynamic deflection elements of separation devices. Vibr. Phys. Syst. **29**, 2018026 (2018)
4. Wang, J.: Energy absorption characteristics and preparation of porous titanium with high porosity. Mater. Today Commun. **34**, 54–60 (2023). https://doi.org/10.1016/j.mtcomm.2022.105003
5. Povstyanoi, O.Y., Sychuk, V.A., McMillan, A., et al.: Metallographic analysis and microstructural image processing of sandblasting nozzles produced by powder metallurgy methods. Powder Metall. Met. Ceram. **54**, 234–240 (2015). https://doi.org/10.1007/s11106-015-9705-8
6. Povstianoi, O.Y., Rud, V.D., Imbirovych, N.Y., et al.: Optimization of the properties of multilayer porous permeable materials. Mater. Sci. **56**, 530–535 (2021). https://doi.org/10.1007/s11003-021-00460-2
7. Zhang, E., Wang, X., Han, Y.: Research status of biomedical porous Ti and its alloy in China. Acta Metall Sin **53**(12), 1555–1567 (2017). https://doi.org/10.11900/0412.1961.2017.00324
8. Zabolotny, O.: Development of processes of pressing of powder materials. Sci. Notes **8**, 135–141 (2001)
9. Bolzoni, L., Ruiz-Navas, E.M., Gordo, E.: Processing of elemental titanium by powder metallurgy techniques. Mater. Sci. Forum **765**, 383–387 (2013). https://doi.org/10.4028/www.scientific.net/msf.765.383
10. Yamanoglu, R., Bahador, A., Kondoh, K.: Fabrication methods of porous titanium implants by powder metallurgy. Trans. Indian Inst. Met. **74**, 2555–2567 (2021). https://doi.org/10.1007/s12666-021-02332-4
11. Liu, Z., Ji, F., Wang, M., Zhu, T.: One-dimensional constitutive model for porous titanium alloy at various strain rates and temperatures. Metals **7**(1), 24–30 (2017). https://doi.org/10.3390/met7010024
12. Shbeh, M.M., Goodall, R.: Open pore titanium foams via metal injection molding of metal powder with a space holder. Met. Powder Rep. **71**(6), 450–455 (2016). https://doi.org/10.1016/j.mprp.2016.06.003
13. Ataee, A., Li, Y., Fraser, D., Song, G., Wen, C.: Anisotropic Ti-6Al-4V gyroid scaffolds manufactured by electron beam melting (EBM) for bone implant applications. Mater. Des. **137**, 345–354 (2018). https://doi.org/10.1016/j.matdes.2017.10.040
14. Stanev, L., Kolev, M., Drenchev, B., Drenchev, L.: Open-cell metallic porous materials obtained through space holders Part II: Structure and properties. A review. J. Manuf. Sci. Eng. **139**(5), 050802 (2017). https://doi.org/10.1115/1.4034440
15. Onishi, T., Azuma, K., Ogasawara, T.: Production of titanium porous material "tiporous" and its evaluation. Mater. Sci. Forum **534–536**(2), 993–996 (2007). https://doi.org/10.4028/0-87849-419-7.993
16. McMillan, A.J., Archer, E., McIlhagger, A., Lelong, G.: Strength knockdown assessment of porosity in composites: modelling, characterising and specimen manufacture. J. Phys. Conf. Ser. **382**, 012027 (2012). https://doi.org/10.1088/1742-6596/382/1/012027
17. Lebedev, V.V., Miroshnichenko, D.V., Nyakuma, B.B., Moiseev, V.F., Shestopalov, O.V., Vyrovets, S.V.: Design of inorganic polymer composites for electromagnetic radiation absorption using potassium titanates. J. Eng. Sci. (Ukraine) **10**(1), C1–C8 (2023). https://doi.org/10.21272/jes.2023.10(1).c1

18. Dzhemelinskyi, V., Lesyk, D., Goncharuk, O., Danyleika, O.: Surface hardening and finishing of metallic products by hybrid laser-ultrasonic treatment. Eastern-Eur. J. Enterp. Technol. **1**(12–91), 35–42 (2018). https://doi.org/10.15587/1729-4061.2018.124031
19. Wang, X.S., Lu, Z.L., Jia, L., Chen, J.X.: Preparation of porous titanium materials by powder sintering process and use of space holder technique. J. Iron. Steel Res. Int. **24**(1), 97–102 (2017). https://doi.org/10.1016/S1006-706X(17)30014-6
20. Reig, L., Amigó, V., Busquets, D., Salvador, M.D., Calero, J.A.: Analysis of sintering of titanium porous material processed by the space holder method. Ceram. Trans. **209**, 273–282 (2010)
21. Rud', V.D., Gal'chuk, T.N., Povstyanoi, A.Y.: Powder metallurgy use of waste from bearing production. Powder Metall. Met. Ceram. **44**, 88–92 (2005). https://doi.org/10.1007/s11106-005-0062-x
22. Zaleta, O.M., Povstyanoy, O.Y., Ribeiro, L.F., Redko, R.G., Bozhko, T.Y., Chetverzhuk, T.I. Automation of optimization synthesis for modular technological equipment. J. Eng. Sci. (Ukraine), **10**(1), A6–A14 (2023). https://doi.org/10.21272/jes.2023.10(1).a2
23. Balasankar, A., et al.: Recent advances in the preparation and performance of porous titanium-based anode materials for sodium-ion batteries. Energies **15**(24), 9495 (2022). https://doi.org/10.3390/en15249495
24. Oh, I.H., Nomura, N., Masahashi, N., Hanada, S.: Mechanical properties of porous titanium compacts prepared by powder sintering. Scripta Mater. **49**, 1197–1202 (2003). https://doi.org/10.1016/j.scriptamat.2003.08.018
25. Korniy, V.: Model and algorithm for processing color metallographic 3D images. Computing **7**(1), 164–170 (2008)
26. Xiao, J., Qiu, G.B.: Research review of space holders of sintered titanium foams with large pores and high porosity. Mater. China **37**(5), 372–378 (2018)
27. Pavlenko, I., Ivanov, V., Gusak, O., Liaposhchenko, O., Sklabinskyi, V.: Parameter identification of technological equipment for ensuring the reliability of the vibration separation process. In: Knapcikova L., Balog M., Perakovic D., Perisa M. (eds) 4th EAI International Conference on Management of Manufacturing Systems. EAI/Springer Innovations in Communication and Computing, pp. 261–272. Springer, Cham (2020). https://doi.org/10.1007/978-3-030-34272-2_24
28. He, G., Liu, P., Tan, Q.: Porous titanium materials with entangled wire structure for load-bearing biomedical applications. J. Mech. Behav. Biomed. Mater. **5**(1), 16–31 (2012). https://doi.org/10.1016/j.jmbbm.2011.09.016
29. Bewerse, C., Emery, A.A., Brinson, L.C., Dunand, D.C.: NiTi porous structure with 3D interconnected microchannels using steel wire spaceholders. Mater. Sci. Eng., A **634**, 153–160 (2015). https://doi.org/10.1016/j.msea.2014.12.088
30. Niespodziana, K.: Synthesis and Properties of Porous Ti-20 wt.% HA Nanocomposites. J. Mater. Eng. Perform. **28**(4), 2245–2255 (2019). https://doi.org/10.1007/s11665-019-03966-8
31. Hovorun, T.P., et al.: Physical-mechanical properties and structural-phase state of nanostructured wear-resistant coatings based on nitrides of refractory metals Ti and Zr. Funct. Mater. **26**(3), 548–555 (2019). https://doi.org/10.15407/fm26.03.548
32. Lytvynenko, A., et al.: Ensuring the reliability of pneumatic classification process for granular material in a rhomb-shaped apparatus. Appl. Sci. (Switzerland) **9**(8), 1604 (2019). https://doi.org/10.3390/app9081604

Check for updates

Structure and Thermal Stability of Vacuum Cu-Mo Condensates

Valentyn Riaboshtan[1](✉), Anatoly Zubkov[1], Maria Zhadko[2], Edward Zozulya[1], and Olena Rebrova[1]

[1] National Technical University "Kharkiv Polytechnic Institute", 2, Kyrpychova Str., Kharkiv 61002, Ukraine
obibobbivalkinobi@gmail.com

[2] Department of Physics and NTIS – European Centre of Excellence, University of West Boemia, 8, Univerzitní Str., Pilsen 30614, Czech Republic

Abstract. The structure of Cu-Mo vacuum condensates was studied by transmission electron microscopy and X-ray diffraction. The components of this system do not form chemical compounds under equilibrium conditions and are mutually insoluble in liquid and solid states. The experimental results indicate that molybdenum atoms are located at the grain boundaries of the copper matrix, predominantly in the form of a monoatomic adsorption layer in the initial condensed state. It has been established that the grain size of the Cu-0.3% Mo condensate is limited by the blocking effect of grain-boundary segregation of molybdenum atoms both during the deposition of the vapor mixture and subsequent annealing. Molybdenum particles formed at the grain boundaries of the copper matrix can have both equilibrium BCC and nonequilibrium FCC crystal lattices. The structure of Cu-0.3% Mo condensates is stable during annealing up to 600 °C. At temperatures above this, degradation of grain boundary segregations, which initiates the formation of BCC Mo particles and the growth of the copper matrix grains, occurs. Kinetic estimates within the Langmuir monolayer adsorption mechanism and the Zinner mechanism of grain boundary pinning correlate well with the experimental results obtained.

Keywords: Manufacturing Innovation · Grain Size · Grain Boundary Segregation · PVD Method · Vacuum Isothermal Annealing · Transmission Electron Microscopy

1 Introduction

In the vapor phase, any substances can be mixed at the atomic and molecular levels without restrictions on composition. Subsequent condensation on different substrates with cooling rates reaching 10^{11}–10^{12} K/s makes it possible to obtain metallic materials that are currently unavailable or at least very difficult to synthesize by melting and casting methods. These are the so-called immiscible systems whose components do not have mutual solubility and do not form chemical compounds in either solid or liquid states. Typical representatives of such objects are thin films, foils, and coatings of the Cu-Mo

V. Tonkonogyi et al. (Eds.): InterPartner 2023, LNME, pp. 318–325, 2024.
https://doi.org/10.1007/978-3-031-42778-7_29

binary system of various compositions with a wide range of structural states, including objects with nano- and submicrometer sizes, anomalous supersaturated solid solutions, not typical for equilibrium conditions phases, etc.

The possibility of more precise control of the condensates structure formation processes than in melting and casting makes them convenient objects of research aimed at solving the existing general problem of increasing the thermal stability of nano- and submicrocrystalline metals.

Moreover, vacuum-plasma technologies make it possible to avoid the veiling effect of impurities on crystallization processes, which have always been present in metals of metallurgical production. Since condensate formation occurs due to the diffusive redistribution of the atoms of the constituent components at the crystallization front, and the crystallization front is a free surface, it becomes possible to involve well-studied phenomena to solve this problem. The possibility of obtaining objects of research, for example, in the form of thin films, facilitates the use of high-resolution electron microscopy methods and eliminates the appearance of artifacts when thinning bulk samples.

2 Literature Review

Composite materials based on copper: Cu-Mo, Cu-Ta, Cu-W, and others, called bulk pseudo-alloys, are obtained mainly by powder technologies [1]. They are also produced as films, foils, and coatings using vacuum plasma technologies [2]. These materials have high physical and mechanical properties [3]. The structure of such pseudoalloys can be stable when heated to high temperatures [4]. Alloying the copper matrix with such refractory metals as Mo, Ta, and W leads to a multiple decrease in the grain size of the copper matrix down to the nanoscale dimension [5]. The degree of refinement of the grain structure depends on the elemental composition [6]. Dependences on the technological conditions of obtaining are also established [7]. Therefore, for these composite materials, the problem of increasing the thermal stability of the initial nanodispersed structure and the corresponding functional properties is an important area of numerous studies [8].

The main methods of thermostabilization are “kinetic” and “thermodynamic”. The “Kinetic” approach involves a reduction in the mobility of the grain boundaries by the second phase particles [9]. The “thermodynamic” method reduces grain boundary energy by segregating alloying elements, which have specific properties to the matrix metal [10]. The first direction is the most studied, theoretically described, and partially used to interpret the experimental results. The second allows for solving several interrelated problems: grain refinement of the matrix metal, increasing the cohesive strength of grain boundaries [11], and increasing the temperature-time intervals of stability of the initial structure and functional properties [12]. This direction is called “grain boundary segregation engineering” [13].

The structural-phase state of grain boundary segregations is very diverse [14]. It depends on many factors: crystallography of grain boundaries [15], a combination of physical and mechanical properties of the matrix metal and segregating substance, elemental composition, etc. [16].

For the Cu-Mo system, the available information indicates the possibility of thermal stabilization of the grain structure by the formation of grain-boundary segregations by molybdenum atoms at the grain boundaries of the copper matrix, both during the formation of the condensate and during subsequent thermal exposure.[12] The Cu-Mo system is characterized by a positive enthalpy of mixing of 28 kJ/mol, a thermodynamic barrier for solution formation, and a stimulus for the concentration of molybdenum atoms at the copper grain boundaries [17]. The interfacial surface energy is 2.8 J/m^2 [18], and the heat of copper adsorption on molybdenum is 3.496 eV [19]. These characteristics indicate the possibility of forming strong bonds between copper and molybdenum atoms at the grain boundaries of the copper matrix, thereby increasing the initial structure's thermal stability and the functional properties of Cu-Mo condensates. The presented information about the role of the second phase particles and equilibrium segregation of atoms of alloying elements in the mechanism of thermal stabilization of the initial nano- and submicrocrystalline state is currently debatable and therefore requires further research. Therefore, this work aims to study the mechanisms of refinement of the copper grain structure upon alloying with molybdenum and its thermal stabilization during subsequent annealing.

3 Research Methodology

Condensates of the Cu-Mo binary system up to 50 μm thick obtained by separate electron-beam evaporation of the constituent components and subsequent crystallization of their vapor mixtures on a non-orienting substrate in a vacuum at a pressure of ~10^{-3} Pa were studied separated from the substrate [20]. Annealing was carried out in a vacuum with a residual pressure of not more than 10^{-2} Pa in the temperature range of 600–900 °C for 30 min. The structure was studied by transmission electron microscopy and X-ray diffraction. The molybdenum content was determined by X-ray fluorescence analysis. The substrate temperature was 450 °C. The sizes of structural parameters were determined by the linear intercept method on the electron-microscopic images.

The choice of technological conditions for obtaining samples for research was due to the desire to create conditions for an equilibrium distribution of molybdenum atoms in the volume of the copper matrix during the formation of condensate during the deposition of the vapor mixture [2]. The maximum dispersing effect on the grain structure of Cu-Mo condensates occurs at a molybdenum concentration of 0.3 at.% [5].

4 Results and Discussion

Figure 1 shows the transmission electron microscopic images of copper and Cu-0.3 at.% Mo condensates obtained under the same technological conditions. The nature of the grain structure of all samples indicates the occurrence of recrystallization processes during the formation of condensates, which corresponds to the selected technological conditions and literature data [2]. According to these images, alloying copper with molybdenum leads to a sharp decrease in the grain size of the copper matrix by more than an order of magnitude.

The intergranular structure of the condensates is unfragmented since the grain sizes determined from bright-field and dark-field images practically coincide. (Table 1; Fig. 1,

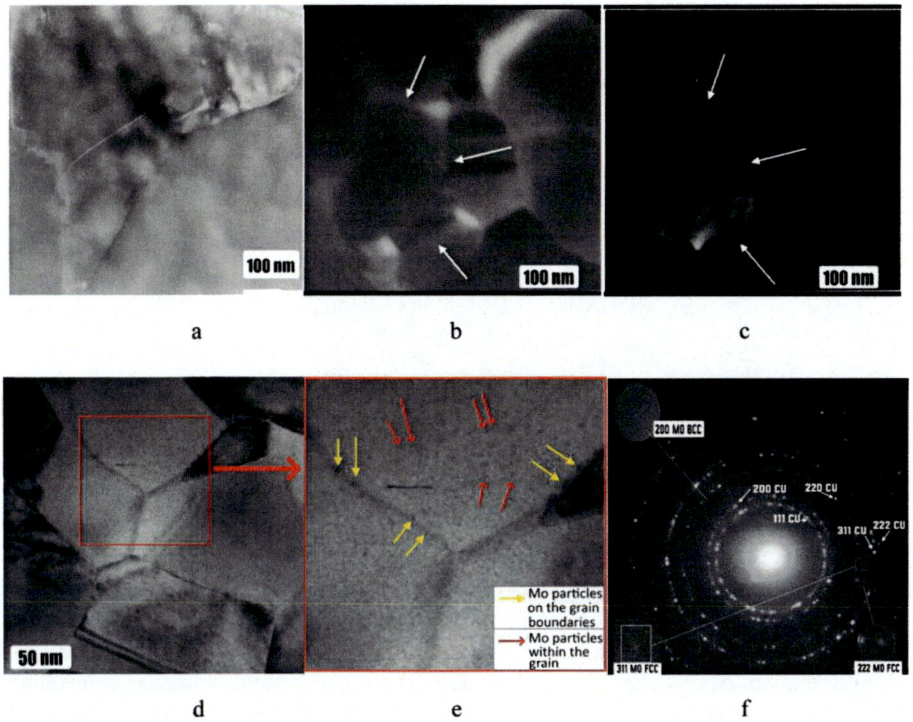

Fig. 1. TEM images of condensates in the initial state taken in different modes: a – copper; b c, d, e, f – Cu-0.3 at.% Mo.

Table 1. Characteristics of structural components determined from electron-microscopic bright and dark field images.

	Initial state	Annealing		
		600 °C	800 °C	900 °C
Grain size *L*, nm bright field mode	250	280	320	510
Grain size *L*, nm dark field mode	240	270	340	500
Mo particles size *D*, nm	2	20	20	30

b, c). Molybdenum particles with an average size of 1 nm and 2 nm, respectively, are observed in the volume of the grains and at their boundaries. (Fig. 1, e).

Besides, the electron diffraction patterns contain single diffraction reflections, which have a low intensity and belong to the equilibrium BCC and nonequilibrium FCC crystal structure of molybdenum. (Fig. 1, g) Moreover, the constancy of the copper crystal lattice parameter indicates the absence of molybdenum dissolution in the copper crystal lattice.

On the images obtained in the registration mode of the characteristic X-ray radiation of molybdenum, there is a distribution of traces of its radiation, which make up certain areas. (Fig. 2) The average size of such Mo-free areas is about 120 nm.

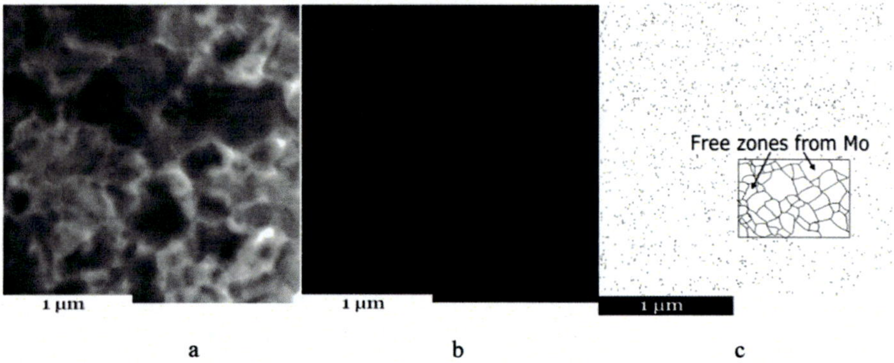

Fig. 2. TEM images of Cu-0.3 at.% Mo condensate: a – bright field mode; b, c – EDS mapping.

The presented results convincingly indicate that molybdenum atoms are located at the grain boundaries of the copper matrix, predominantly in the form of a monoatomic adsorption layer. According to [5], if the achieved grain size of the condensate during the deposition of a two-component vapor is limited by the blocking mechanism of atomic grain boundary segregations, then the amount of molybdenum atoms should be:

$$C = \frac{\pi \cdot d^3}{L \cdot N \cdot n \cdot S} \tag{1}$$

where d – the diameter of the matrix metal atom –0.256 nm; L – the experimentally determined grain size ~250 nm; C – the atomic concentration of molybdenum; S – the area of the adsorption cell; $n = 0.74$ for the FCC lattice, which takes into account its packing density; $N = 2$ coefficient taking into account the ratio of the sizes of adsorption cells and molybdenum atom [5].

If we take the average value of the adsorption cells of crystallographic planes (111), (100), (110) of copper, then it will be [5]:

$$S = 6.2 \cdot 10^{-2}\ \text{nm}^2 \tag{2}$$

If substituting (2) into (1):

$$\text{C} = \frac{3.14 \cdot 0.256^3 nm^3}{2 \cdot 0.74 \cdot 6.2 \times 10^{-2} nm^2 \cdot 250 nm} = 0.002 = 0.2\,\text{at.\%} \tag{3}$$

Excess molybdenum atoms in the amount of ~0.1 at. % are concentrated in the particles observed on electron microscopic images (Fig. 1). In the volume of the copper matrix, the particles are arranged in chains framing free areas. This morphology indicates that they are detached from grain boundaries, not preventing their migration by the Zener mechanism [9] during condensation.

Hence, an important conclusion about the mechanism of dispersing action of molybdenum on the grain structure of copper matrix consists in the blocking action of monatomic adsorption layers formed on the surface of growing grains. Thus, it is energetically advantageous for the molybdenum atoms to form adsorption layers on the grain boundaries instead of forming self-phase particles or anomalous supersaturated solutions in the copper crystal lattice [17]. These conclusions are consistent with literature sources such as [19].

Figure 3 shows images of condensates after isothermal annealing. Up to a temperature of 800 °C, the structure has no noticeable changes.

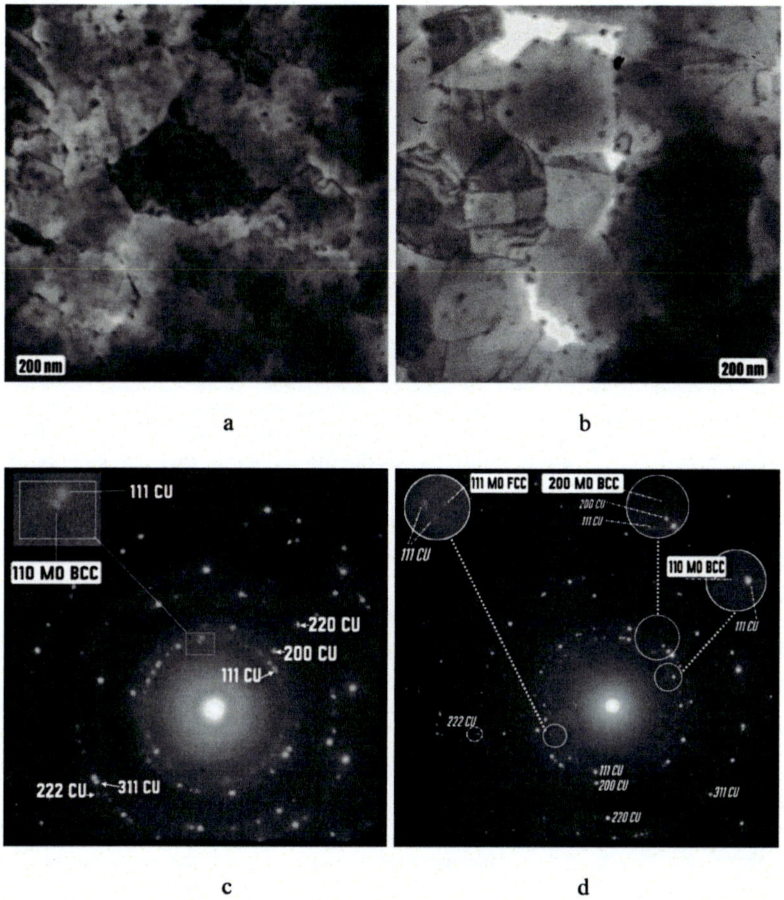

Fig. 3. TEM images of Cu-0.3 at.% Mo condensate after annealing: a) 800 °C, b) 900 °C – bright-field mode; c) 800 °C, d) 900 °C – electron diffraction mode.

After annealing at 800 °C, the grain size slightly increases, and molybdenum particles are still observed at the grain boundaries of the copper matrix, the average size of which is approximately 20 nm. The electron diffraction patterns show more intense diffraction reflections of BCC molybdenum, which have a crystallographic relationship with the

FCC lattice of the copper matrix (110)Mo||(111)Cu, typical for particles coherently connected with the Cu matrix.

Annealing at 900 °C resulted in significant grain growth, and an increase in molybdenum particle size was observed (Table 1). These experimental results also indicate that the temperature of the start of grain growth of the copper matrix is limited not by the mechanical action of molybdenum particles (Zener pinning mechanism) but by the destruction of the adsorption monolayer of molybdenum atoms, which initiates the subsequent growth of particles and grains.

According to [9], for the Zener pinning mechanism to work, the grain size would have to be:

$$L \geq \frac{3D}{2f} \tag{4}$$

where D is the particle size and f – is the volume content of particles, which is 0.45% at a Mo atomic concentration of 0.3%. Then:

$$L \geq \frac{3 \cdot 20nm}{4 \cdot 0,0045} = 3.3\mu m \tag{5}$$

So the theoretical estimate does not correspond to the experimental results in Table 1.

5 Conclusions

The grain size of alloyed copper condensate is determined by the blocking effect of grain-boundary segregations of molybdenum atoms during the vapor mixture's deposition and subsequent thermal exposure.

Molybdenum particles located at the grain boundaries of the copper matrix can have both equilibrium BCC and nonequilibrium FCC crystal lattices. This state results from forming strong bonds between molybdenum and copper atoms at the internal interfaces.

The structure of Cu-0.3%Mo condensates is stable when heated up to 600 °C. At temperatures above this, degradation of grain boundary segregations occurs, which initiates the formation of Mo particles of the bcc phase and the growth of the copper matrix grains.

References

1. Hernández, O.: Effects of Mo concentration on the structural and corrosion properties of Cu–alloy. Metals **9**(12), 1307 (2019). https://doi.org/10.3390/met9121307
2. Derby, B.: Effects of substrate temperature and deposition rate on the phase separated morphology of co-sputtered. Cu-Mo thin films. Thin Solid Films **647**, 50–56 (2018). https://doi.org/10.1016/j.tsf.2017.12.013
3. Spearot, D.E.: Mechanical properties of stabilized nanocrystalline FCC metals. J. Appl. Phys. **126**(11), 110901 (2019). https://doi.org/10.1063/1.5114706
4. Souli, I.: Thermal stability of immiscible sputter-deposited Cu-Mo thin films. J. Alloy. Compd. **783**, 208–218 (2019). https://doi.org/10.1016/j.jallcom.2018.12.250

5. Zhadko, M., Sobol, O., Zelenskaya, G., Zubkov, A.: Methods for calculating the grain boundary adsorption capacity of nanostructured copper based condensates. In: Ivanov, V., et al., (eds.) Advances in Design, Simulation and Manufacturing. DSMIE 2018. Lecture Notes in Mechanical Engineering, pp. 199–206. Springer, Cham (2019). https://doi.org/10.1007/978-3-319-93587-4_21
6. Zhang, P.: Microstructural evolution, mechanical properties and deformation mechanisms of nanocrystalline Cu thin films alloyed with Zr. Acta Mater. **76**, 221–237 (2014). https://doi.org/10.1016/j.actamat.2014.04.041
7. Shang, X.: Effects of ball milling processing conditions and alloy components on the synthesis of Cu-Nb and Cu-Mo alloys. Materials **12**(8), 1224 (2019). https://doi.org/10.3390/ma12081224
8. Schuh, C.A.: Stability of nanocrystalline metals: The role of grain-boundary chemistry and structure. MRS Bull. **46**, 225–235 (2021). https://doi.org/10.1557/s43577-021-00055-x
9. Koju, R.K.: Zener pinning of grain boundaries and structural stability of immiscible alloys. JOM **68**, 1596–1604 (2016). https://doi.org/10.1007/s11837-016-1899-9
10. Borovikov, V.: Effects of grain boundary disorder on dislocation emission. Mater. Lett. **237**, 303–305 (2019). https://doi.org/10.1016/j.matlet.2018.11.114
11. Xing, W.: Preferred nanocrystalline configurations in ternary and multicomponent alloys. Scripta Mater. **127**, 136–140 (2017). https://doi.org/10.1016/j.scriptamat.2016.09.014
12. Wagih, M.: Thermodynamics and design of nanocrystalline alloys using grain boundary segregation spectra. Acta Mater. **217**, 117177 (2021). https://doi.org/10.1016/j.actamat.2021.117177
13. Raabe, D.: Grain boundary segregation engineering in metallic alloys: A pathway to the design of interfaces. Curr. Opin. Solid State Mater. Sci. **18**(4), 253–261 (2014). https://doi.org/10.1016/j.cossms.2014.06.002
14. Kalidindi, A.R.: Stability criteria for nanocrystalline alloys. Acta Mater. **132**, 128–137 (2017). https://doi.org/10.1016/j.actamat.2017.03.029
15. Pan, Z.: Effect of grain boundary character on segregation-induced structural transitions. Phys. Rev. B **93**(13), 134113 (2016). https://doi.org/10.1103/PhysRevB.93.134113
16. Mishin, Y.: Stabilization and Strengthening of Nano-Crystalline Immiscible Alloys. George Mason University Fairfax United States, USA (2018). Technical Report
17. Sabooni, S.: Thermodynamic analysis and characterisation of nanostructured Cu (Mo) compounds prepared by mechanical alloying and subsequent sintering. Powder Metall. **55**(3), 222–227 (2012). https://doi.org/10.1179/1743290111Y.0000000003
18. Wu, K.: An easy way to quantify the adhesion energy of nanostructured cu/X (X= Cr, Ta, Mo, Nb, Zr) multilayer films adherent to polyimide substrates. Acta Metall. Sinica (Engl. Lett.) **29**, 181–187 (2016). https://doi.org/10.1007/s40195-016-0375-4
19. Paunov, M.: An adsorption-desorption study of Cu on Mo (110). Appl. Phys. A **44**, 201–208 (1987). https://doi.org/10.1007/BF00626424
20. Glushchenko, M.: The Influence of copper condensates alloying with Co, Mo, Ta transition metals on the structure and the hall-petch dependence. J. Nano Electron. Phys. **8**(3), 03015 (2016). https://doi.org/10.21272/jnep.8(3).03015

Features of Magnesium Alloy Protection Technologies in Die Casting

Oleg Stalnichenko(✉), Tatiana Lysenko, Oleksii Shamov, Kyryll Kreitser, and Evgeny Kozishkurt

Odessa National Maritime University, 34, Mechnikova Str, Odessa 65029, Ukraine
dakerkir@gmail.com

Abstract. Flux protection for magnesium alloys is now the most common in many industries, but it has many drawbacks. The main one is that it dramatically reduces the quality of casting. The atmosphere's interaction with molten magnesium can lead to the deterioration of casting quality and an emergency. Therefore, since magnesium melting technologies began, its protection methods have been developed. The most advanced method of protection of magnesium alloys is flux-less melting under a protective layer of gases. This technology has been refined, and pulsed gas feeding has been created. For this purpose, a computer-controlled unit with pulsed protective gas feeding was created. This technology was used in the high-pressure casting of critical parts. The casting results using the new technology were compared with flux protection technology casting. Microstructure studies were carried out, and the results were described. And also, measurements of the microhardness of samples were carried out.

Keywords: Magnesium Protection · Magnesium Alloys · Magnesium Microstructure · Inclusions · Manufacturing Innovation

1 Introduction

In today's technology, the development of magnesium production is the most relevant. Over the past decade, magnesium has evolved from a rare material for the space industry to the third most common metal after steel and aluminum [1]. Despite all the advantages of magnesium, its disadvantages hinder its implementation. One of the main problems is its high affinity for oxygen and hydrogen [2].

Magnesium is the lightest structural material in the industry, with a density of only $\rho = 1800–1900$ kg/m^3. It follows that it is 6 times lighter than steel and 1.5 times lighter than aluminum, which reduces the weight of products by up to 30%. But adsorbing oxygen and magnesium inside the alloy releases it to the surface and ignites it [3]. The burning point of magnesium is 2500 °C, and it is not extinguished by water, which causes a particular fire hazard [4].

Magnesium has a high specific strength at such a low density, second only to titanium, as well as specific vibration resistance and stiffness. In addition to its high characteristics, this metal is very competitive in a market economy [5].

V. Tonkonogyi et al. (Eds.): InterPartner 2023, LNME, pp. 326–334, 2024.
https://doi.org/10.1007/978-3-031-42778-7_30

Another problem with magnesium is its low corrosion resistance. Magnesium is the most electronegative metal, and its chemical activity increases with increasing temperature. Because of this, unprotected magnesium castings have low corrosion resistance, and a magnesium hydroxide scale forms on their surface in a couple of hours. A characteristic feature of AZ91D magnesium alloys is their tendency to form microcracks [6]. The interaction of the workshop atmosphere with molten magnesium can lead to a deterioration in casting quality and an emergency. Therefore, since the beginning of the development of magnesium melting technologies, magnesium protection equipment has been developed [7].

This paper investigated the technologies for protecting magnesium alloys in die-casting. To compare and obtain the results, we compared non-metallic inclusions using the new technology of fluxless protection of magnesium alloys in die casting in samples of magnesium alloy AZ91D and samples of a similar alloy produced using the traditional "flux method".

2 Literature Review

Until now, fluxes based on $MgCl_2$, Cl, and F have been used to protect magnesium and its alloys from ignition. However, their use has led to chloride inclusions in castings. Therefore, other compositions and methods were developed and tested.

The next step was to use alkaline earth metal chlorides, such as $BaCl_2$. This substance was porous and did not provide an optimal solution to the problem.

Next, we developed fluxes based on MgO and MgF_2. They created a viscous, dense, protective layer on the melt surface and eliminated the shortcomings of previous experiments. However, sufficient protection requires 20% to 50% of the flux by weight of the alloy to be protected. This method caused substantial economic losses. Therefore, fluxes based on chloride and fluoride salts of metals and some oxides were developed [8].

Currently, fluxes are often used, especially in our country. Therefore, the following requirements are imposed on them:

1. protecting the charge from oxidation and inflammation, as well as from reducing the content of alloying components in the alloy;
2. refining of the melt from non-metallic impurities (oxides, nitrides, etc.) present in the suspended state in the melt;
3. separation of molten metal (and alloy) from the salt phase and slag;
4. obtaining an alloy of a given chemical composition as a result of removing harmful impurities from the melt (e.g., alkali metals of sodium and potassium) and as a result of enhancing the transition of alloying elements - zirconium, yttrium, neodymium, cerium, etc. [9].

Fluxes that contain at least small amounts of magnesium chloride, which partially decomposes at 400 °C, such as VI-2 flux, have the best protective properties. Fluxes not containing magnesium chloride satisfactorily protect the melt at temperatures above 700 °C.

However, these fluxes have characteristic disadvantages. Flux VI-2 is highly hygroscopic. It leads to an increase in gassiness and the formation of additional micro-richness.

Also, its relatively low viscosity and density make it difficult to separate the flux from the metal. The production of special magnesium alloys using this technology is impossible due to the interaction of alloying components with flux components. Removing fluxes from the melt is almost impossible, so their inclusions create additional defects in castings. It sharply reduces their corrosion resistance and reduces the reliability of the castings. As a result, products made using this technology cannot be introduced into the aerospace industry, and their use is limited [10].

From the above, it can be concluded that using fluxes is a past stage in developing magnesium alloy protection technology.

The most advanced method of protecting magnesium alloys is flux-free melting under a layer of protective gases. This technology eliminates most of the above-mentioned negative factors of flux protection. The use of flux-free melting will lead to the following:

- an increase of corrosion resistance of the part by two or more times;
- an increase in the durability of the part by eliminating chlorine ions and, as a result, intergranular corrosion;
- the reduction of bottom sediment by 30%;
- the reduction of irrecoverable alloy losses by 5%;
- the increase in smelting productivity by (10–15)%;
- the reduction of smelting costs by (10–15)%, including electricity consumption;
- the reduction of harmful gas emissions by 20 times - sharp improvement of sanitary and hygienic conditions.

The new technology of flux-free melting of magnesium alloys uses gas protection, which isolates the metal from contact with air better than a flux coating. The gas environment can protect itself if it chemically interacts with the liquid magnesium alloy to form a thin, dense film [11].

The gas environment should delay evaporation during the melting of magnesium vapor and prevent further interaction of the gas environment with the magnesium melt after the film formation. An atmosphere is used for these purposes, consisting of a continuous supply of a mixture of active and inert gases to the alloy mirror.

The most promising protective composition of gases is dried air with active gas. But you need to limit the oxygen content. It is caused by the fact that with low oxygen contents of up to 4%, the ignition of magnesium is slow, at a temperature of 550 - 750 °C with periodic ignition and extinction.

3 Research Methodology

To study and work out research tasks on the impact of technological factors on the protection of magnesium alloys during their melting and preservation when shielding gases are used with their pulsed supply to the units, the design of the melting and distributing electric furnaces with a removable crucible was finalized. This technology will make it possible to produce gas-shielded parts and compare them with parts made with flux shielding.

The die-casting furnaces also had hermetic lids with 200 kg of magnesium alloy capacity. The hermetically sealed cover has an airlock for technological and preventive maintenance. Thermocouples, a movable measuring head for the surface tension of the liquid alloy, and glass for laser sensors to detect the appearance of white smoke (magnesium vapor) were mounted through the lid with sealed components into the magnesium alloy zone. Metal lines were added to supply heated pellets to the level of the metal mirror and pump liquid alloy and a metal line for mixing the alloy.

To improve the metrological accuracy of the research work, the following were used:

- a certified gas mixing apparatus (3 types) was used to regulate the flows of inert and active gases in proportion to the flow of dry air, sulfur hexafluoride, and sulfur dioxide;
- thermocouples, manometers, pressure sensors, and duplicate thermocouples;
- sulfur dioxide SO_2 should be used in a mixture with dry air at temperatures above 800–850 °C.

One of the main issues is the technological process of reducing the oxygen content in dry air to 6%. To solve this problem, a Siemens gas analyzer ULTRAMAT 6 for adsorbing gases was used.

Figure 1 shows a fundamentally new scheme of pulsed gas mixture supply to protect the magnesium melt in the melting unit and the dispensing furnace, where air from a compressor with an operating pressure of up to 1.6 MPa was supplied to the inlet of an adsorption dryer, in the system of which air was cleaned from moisture to the level of the dew point (-70 °C), oil and mechanical impurities. The dried air entered the oxygen reduction system to reduce oxygen to 6%. The mixer (7) supplied the required gas level to the dryer with oxygen reduction and nitrogen compensation. The mixer was supplied in a dynamic mode through a proportional valve (5, 8, 10) with control over a given content, impulsively, taking into account the calculations of the mathematical model of the content of active gases [12, 13].

The system provides for an independent fire extinguishing circuit for the ignition tongue.

To compare the two technologies and to study non-metallic inclusions and the microstructure of magnesium alloy castings, cylindrical aluminum alloy adapters with a diameter of 22 mm were first made, in the center of which conical holes with a diameter of 10 mm were drilled in which the magnesium alloy samples under study were mounted.

The grinds were made: pretreatment to obtain a flat surface was performed on the side surface of the abrasive wheel.

Non-metallic inclusions were examined using a MIM-M microscope equipped with a video camera connected to a computer.

The microscopic examination was performed using a microplanar with F = 40 mm for a magnification of 100 and a planachromat with F = 40 for a magnification of 300.

After examining non-metallic inclusions, the grinds were washed with 96% alcohol and pickled in a 4% alcohol solution of nitric acid.

4 Results and Discussion

The object of study: alloy samples obtained using the old and improved technology (two cone samples with a maximum diameter of 9 mm and 10 mm) and one sample obtained using the new technology in the form of a massive sector with planes at an angle of 90°.

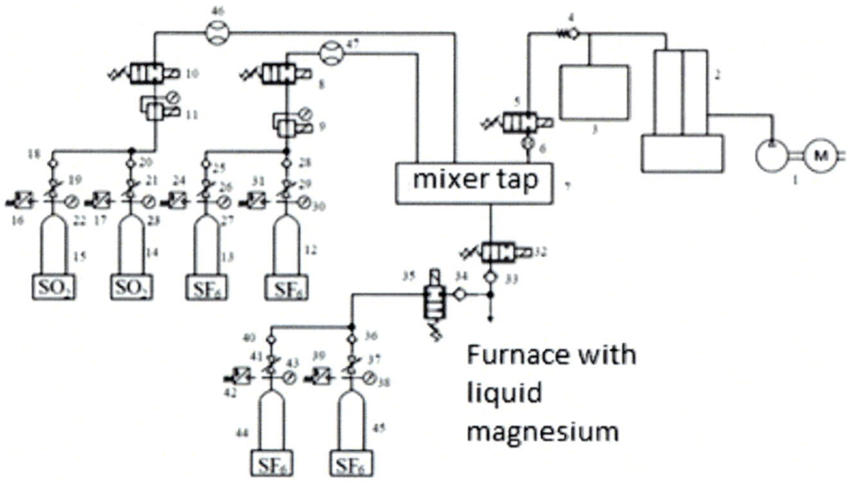

Fig. 1. Supply circuit of pulse fluxless protection.

When the samples were magnified on a microplaner to 100, many dark brown inclusions up to 500–600 microns were found (Fig. 2). The inclusions are unevenly distributed. The large size of the inclusions is of external origin. The exogenous mechanism is the occurrence of gas bubble germs based on defects in the end surface. At a higher magnification, many small satellite inclusions with the same physical state are detected in the immediate vicinity (Fig. 3).

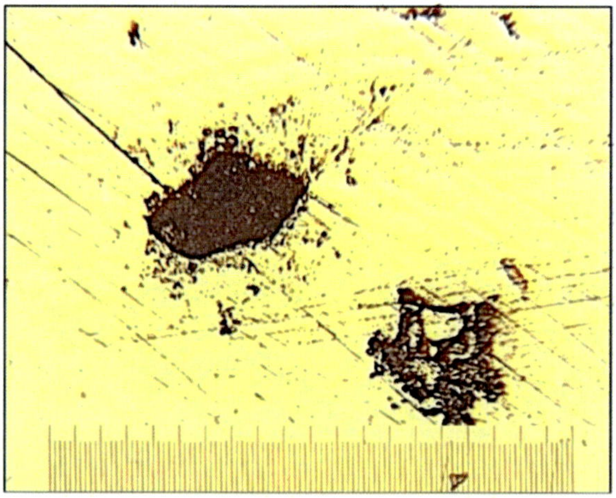

Fig. 2. Accumulation of non-metallic inclusions in metal smelted using the old X 100 technology.

Some areas of the grind contained film inclusions along the grain boundaries (Fig. 4). The area has an acute-angled or elongated shape, affecting the casting quality.

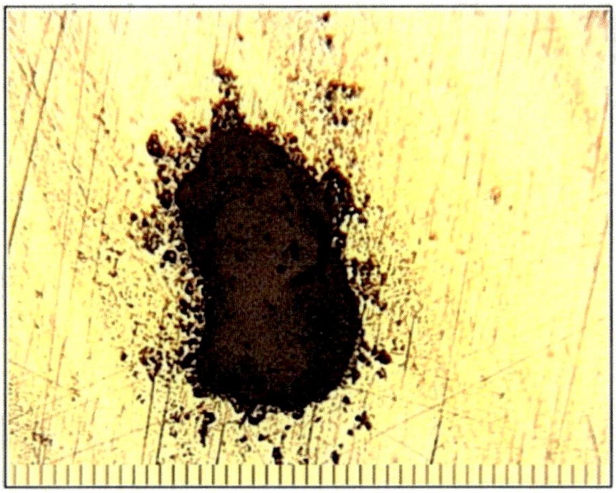

Fig. 3. Satellite inclusions surrounding a large non-metallic inclusion.

The microhardness of non-metallic inclusions was determined using a PMT-3 microhardness tester. The diamond pyramid impressions are shown in Fig. 5, and the microhardness values are shown in Table 1.

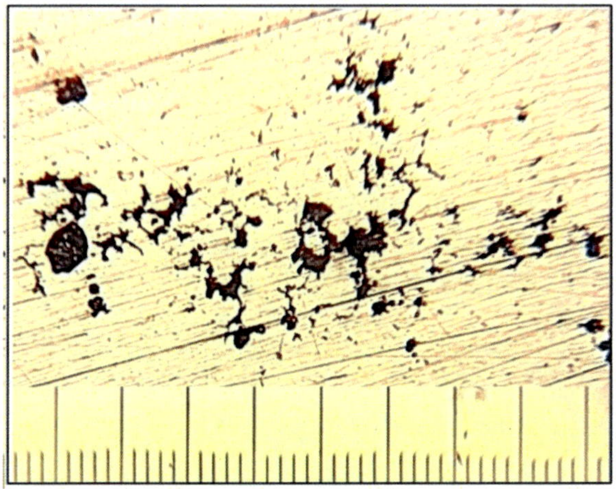

Fig. 4. Non-metallic inclusions oriented along the boundaries of cast grains.

No non-metallic inclusions larger than 0.5 microns were found on the surface of the alloy smelted using the new technology at the above increases (Fig. 5).

The alloy's microstructure smelted using the old technology consisted of small light angular grains of solid aluminum solution in magnesium and dark g-phase inclusions (Fig. 6).

Table 1. Microhardness of the alloy and nonmetallic inclusions.

Place of measurement	The diagonal of the print, microns	Microhardness, MPa
Metal	24,0	161
Incorporated in metal	16,5	387

Fig. 5. Grind the surface of an alloy smelted using the new technology.

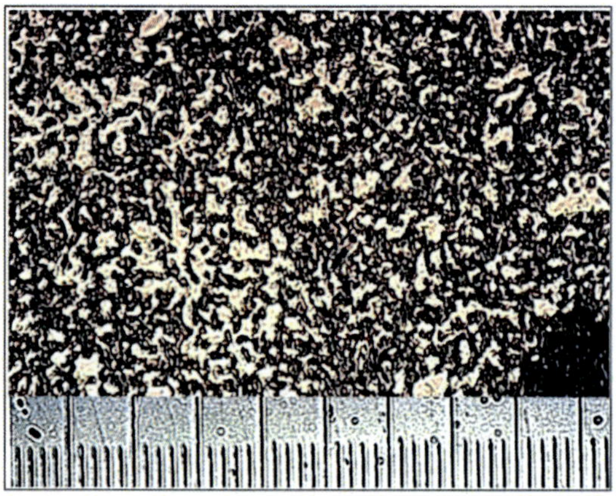

Fig. 6. Microstructure of an alloy smelted using the old technology.

Two indicators were selected to quantify the results of corrosion resistance based on the data obtained in this series of experiments: mass and depth, and the corresponding calculations were performed.

The microstructure of the alloy smelted using the new technology contained light rounded grains surrounded by less contrasting non-metallic rounded inclusions, which are stress concentrators, with sizes of tens of microns. In contrast, the microstructure of the old technology consisted only of angular grains of solid solution, which significantly deteriorated the alloy (Fig. 7).

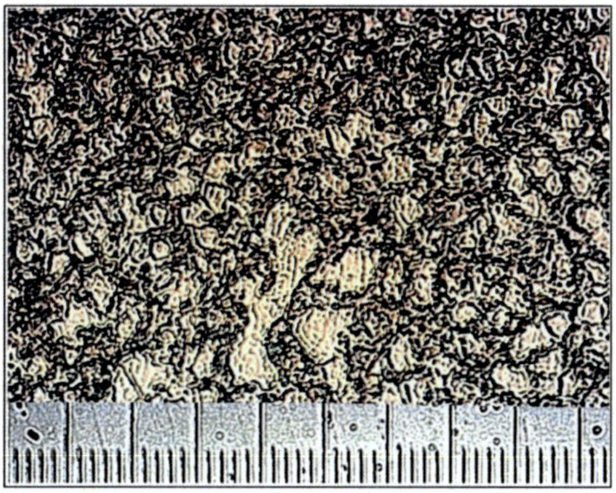

Fig. 7. Microstructure of the alloy smelted using the new technology.

5 Conclusions

The results of the study of non-metallic inclusions by the new technology, non-metallic inclusions with a size more than 0. 5 microns in the samples of magnesium alloy AZ91D and the samples of a similar alloy obtained by the traditional "method of flux" have been compared. Many dark brown inclusions up to 500–600 μm in size were detected, distributed irregularly. It is due to the exogenous mechanism of their emergence from gas bubble germs based on end surface defects. Many small satellite inclusions near large nonmetallic inclusions were also found. The study of the macro- and microstructure of magnesium alloys produced with the use of gas protection of magnesium alloys according to the developed technology of their melting allows for significantly improve the operational and mechanical properties of magnesium alloy castings compared with counterparts using traditional technologies of producing strongly oxidized melt using their protection under a flux layer.

References

1. Jiangfeng, S.: Latest research advances on magnesium and magnesium alloys worldwide. J. Magnes. Alloys **8**(1), 1–41 (2020). https://doi.org/10.1016/j.jma.2020.02.003
2. Brown, R.E.: Magnesium and its alloy. In: Kutz, M. (eds.) Mechanical Engineers Handbook: Materials and Mechanical Design (2015). https://doi.org/10.17798/bitlisfen.502290

3. Easton, M., Beer, A., Barnett, M., et al.: Magnesium alloy applications in automotive structures. JOM **60**, 57–62 (2008). https://doi.org/10.1007/s11837-008-0150-8
4. Mordike, B.L., Ebert, T.: Magnesium: properties applications potential. Mater. Sci. Eng. A **302**(1), 37–45 (2001). https://doi.org/10.1016/S0921-5093(00)01351-4
5. Lysenko, T., Kreitser, K., Kozishkurt, E., Dotsenko, V., Ponomarenko, O.: New technology for producing castings from magnesium alloys with increased corrosion resistance. In: Ivanov, V., Trojanowska, J., Pavlenko, I., Rauch, E., Peraković, D. (eds.) Advances in Design, Simulation and Manufacturing V. DSMIE 2022. Lecture Notes in Mechanical Engineering, pp. 445–454. Springer, Cham (2022). DOI: https://doi.org/10.1007/978-3-031-06025-0_44
6. Hanxue, C.: Research status and prospects of magnesium alloys' melt refining and purification technology. J. Magnes. Alloys **7**(3), 370–380 (2019). https://doi.org/10.1016/j.jma.2019.07.002
7. Luo, A.A., Fu, P., Peng, L., et al.: Solidification microstructure and mechanical properties of cast magnesium-aluminum-tin alloys. Metall. Mater. Trans. A **43**, 360–368 (2012). https://doi.org/10.1007/s11661-011-0820-y
8. Cole, G.S.: Summary of "magnesium vision 2020: A North American automotive strategic vision for magnesium". In: Mathaudhu, S.N., Luo, A.A., Neelameggham, N.R., Nyberg, E.A., Sillekens, W.H. (eds.) Essential Readings in Magnesium Technology, pp. 35–40. Springer, Cham (2016). https://doi.org/10.1007/978-3-319-48099-2_5
9. Friedrich, H.E., Mordike, B.L.: Magnesium Technology: Metallurgy, Design Data Applications. Springer, Cham (2006). https://doi.org/10.1007/3-540-30812-1.pdf
10. Kim, S.K., et al.: Eco-Mg for magnesium future. In: Proceedings of 66th Annual World Magnesium Conference (IMA-2009). San Francisco, IMA, USA (2009)
11. Luo, A.A.: Magnesium casting technology for structural applications. J. Magnes. Alloys **1**(1), 2–22 (2013). https://doi.org/10.1016/j.jma.2013.02.002
12. Callister Jr, W.D.: Materials Science and Engineering, An Introduction. John Wiley& Sons, Inc. (2006)
13. Pan, F.S., Tang, A.T., Long, S.Y., Yang, M.B.: Development and application of wrought magnesium alloys in China. In: Proceedings of the 7th International Conference on Magnesium Alloys and Their Applications, pp. 297–304 (2007)

Organization of the Structure of Composite Construction Materials and the Impact on the Characteristics of Concrete

Hanna Zinchenko[1], Vitaliy Dorofeev[1], Natalia Pushkar[2](✉), Igor Myronenko[3], and Stanislav Fic[4]

[1] Odessa National Polytechnic University, 1, Shevchenko Ave., Odessa 65044, Ukraine
[2] Odessa State Academy of Civil Engineering and Architecture, 4, Didrihson Str, Odessa 65029, Ukraine
push@ogasa.org.ua
[3] Odessa National Maritime University, 34, Mechnikov Str., Odessa 65029, Ukraine
[4] Lublin University of Technology, 38 D, Nadbystrzycka Str., 20-618 Lublin, Poland

Abstract. The organization of the structure of composite construction materials at the micro- and macrolevels and the impact of the structure on the physical and mechanical properties of concrete mixtures were presented. It was established that the structure of materials depends on the initial composition and the conditions for processing raw materials into the final product. It was established that the resulting incipient crack is an intercluster interface that can develop. The introduction of aggregates and the formation of mixed cluster structures can contribute to the early appearance of incipient cracks and restrain their width, total number, and ability to grow. It was established that at the level of structural inhomogeneity, a crack develops stepwise with a microtortuous trajectory along the intercluster interfaces. The formation of internal interfaces between the matrix material and inclusions was determined by the nature of the adhesive and cohesive bond forces at the interfaces, the shrinkage method, and the number of inclusions. A purposeful change in these parameters will make it possible to predict damage to the macrostructure of composite construction materials by hereditary defects. Technological damage's effect on composite construction materials' properties and operational reliability was also presented. The work aims to describe the organization of the structure of composite building materials at the micro- and macro levels, establish the causes of crack initiation, and describe the effect of technological damage on concrete's physical and mechanical properties.

Keywords: Composites · Technological Damage · Microlevel · Macrolevel · Deformations · Cluster · Reliability · Manufacturing Innovation

1 Introduction

Structure formation is an evolutionary process based on chemical [1] and mechanical [2] processes and phenomena. Therefore, to obtain materials with predetermined properties, it is necessary to purposefully organize the initial structure and assign technological methods that make it possible to obtain final structures with the required parameters.

V. Tonkonogyi et al. (Eds.): InterPartner 2023, LNME, pp. 335–345, 2024.
https://doi.org/10.1007/978-3-031-42778-7_31

Composite construction materials are complex systems of the "structure within the structure" type. At the same time, space-time structures of qualitatively different types and scale levels arise in the system [3]. Each subsequent type of changing structure depends on the previous one [4] and the nature of the interaction of individual structural elements and groups [5].

The physical and mechanical properties of composites are determined by the type [6], number, and orientation of defects [7]. Most defects are embedded in the material during receipt [8]. Such defects are called technological [9] or hereditary [10]. Therefore, it is essential to identify the main causes of the origin and development of technological defects at the micro- and macro levels to assess their impact on composite materials' performance characteristics. It is interesting to determine the ways of managing technological damage to obtain materials and structures with standardized indicators.

2 Literature Review

The crack resistance of reinforced concrete structures and its effect on the durability of operated buildings was studied in detail in the works of the Lviv Scientific School, in which the crack resistance of reinforced concrete strengthened beams [11], crack resistance of reinforced concrete columns [12], residual strength of reinforced concrete beams [13] were studied. However, the authors did not consider the initial cracks in concrete while manufacturing structures. Cracks are structural elements that lead to stress concentration in their mouth. They are characterized by length, type of opening, mouth radius, front, morphology, and texture of the banks.

It is known that cracks and damage occur in concrete structures during the technological processing of the material into a product. It has been established that the destruction of reinforced concrete bending elements depends on the presence of technological cracks. A method for determining the initial damage of concrete has been proposed [7]. The resistance of concrete and expanded clay concrete under periodic external influences was also determined [14], and the possibility of using natural composites was studied.

At each structural level, the material's properties are evaluated by an average characteristic, allowing us to consider it a continuous medium. In recent years, works that consider the appearance of technological cracks at the micro-, macro-level, and the level of products and structures are known [15, 16].

Of interest is a more detailed study of cracks that appear in the concrete matrix at the micro- and macrolevels and the establishment of their impact on the strain properties of concrete.

3 Research Methodology

It is advisable to study the mechanism of crack initiation on a representative volume of the system, which includes at least two neighboring K-N clusters. This is since the probability of the appearance of an internal system interface is higher at the boundaries of the interaction of structural blocks.

Cracks that appear in the material at the micro- and macrolevel automatically become cracks in the product or structure, determining its deformability, crack resistance, and load-bearing capacity. Therefore, it is essential to establish the relationship between the damage of the material by technological cracks and its strength and strain characteristics.

Nineteen concrete prisms 10 × 10 × 40 cm in size were tested. The surfaces of the prisms were treated with a tannin solution to reveal surface technological cracks. The initial damage was determined using the area damage coefficients as the ratio of the total length of surface cracks, estimated using a curvimeter L, to the area of the sample S, where the measurements were taken, $K_S = L/S$ (cm/cm^2) is an estimate of the specific length of the crack that arose per unit area. The damage coefficient was also determined along a straight line h = 40 cm, related to the length of surface cracks along it: $K_{40} = L/L_l$ is an estimate of the specific length of the crack surface per unit length, where L is the length of the technological crack. The value of the damage factor of the prisms varied within $K_S = 1.3...2.08$, $K_{40} = 0.95...1.28$. The prisms at the time of testing had the same age.

4 Results and Discussion

It is known that cracks are present at all structural levels of composite construction materials. The reasons for the initiation of cracks are considered to be the intrinsic volumetric strains of the matrix as a whole and its components, the difference in temperature and humidity strains of the material, constrained strain defects, temperature and humidity gradients, osmotic phenomena, corrosive effects of the operating environment. The problem of studying the mechanism of crack initiation during the structure formation of composite construction materials on inorganic binders is of interest.

By an incipient crack, the inner interface of the interface is implied. Such a representation is because for the level of structural inhomogeneity "cluster-cluster" the appearance of an external, concerning the aggregates, the interface is associated with a break in the interparticle bonds of neighboring cluster structures. An increase in the intercluster interface to the aggregate size turns it into a crack which is dangerous for this structure.

The potential ability to form an incipient crack is embedded in a dispersed system at the very initial stages of its formation. It is determined by the composition and concentrations of the dispersed phase, surface interaction between solid particles and the dispersion medium, as well as between themselves. Plastic strains accumulate in the material of the partitions to a critical value, after which the interparticle bonds are broken (Fig. 1), causing an increase in the length of the incipient crack or interface, on which more and more clusters and their groups come out.

Thus, the resulting incipient crack is an intercluster interface that can develop. The introduction of aggregates and the formation of mixed cluster structures can contribute to the early appearance of incipient cracks and restrain their width, total number, and ability to grow.

It has been established that at the level of structural inhomogeneity, a crack develops stepwise with a microtortuous trajectory along the intercluster interfaces.

A two-dimensional stress state can be described using the theory of analytic functions of a complex variable [3, 4]. Using this apparatus makes it possible to obtain numerical results characterizing the initial stage of the operation of composite materials for any characteristics of materials, forms of inclusions, and external influences.

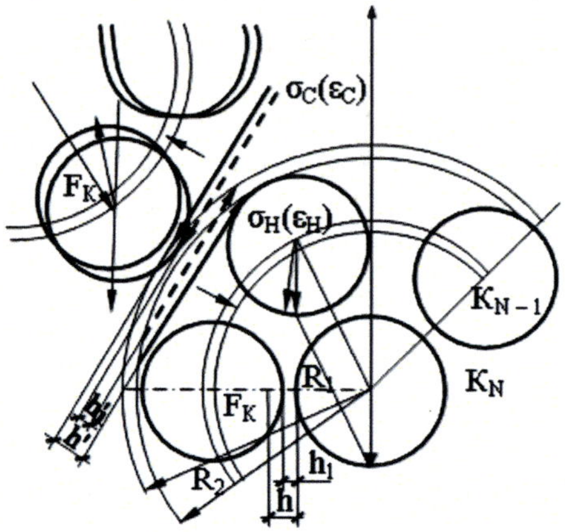

Fig. 1. The mechanism of formation of the intercluster interface: R_1, R_2 – initial and transformed radius of the structural block; $\sigma_C(\varepsilon_C)$ – shear stresses and strains; $\sigma_H(\varepsilon_H)$ – normal stresses and strains; F_K – the strength of intercluster interaction; K_{N-1}, K_N – clusters of different scale levels.

Aggregates are independent structural elements of concrete and dismember the macrostructure into separate structural subblocks. Depending on the cohesive and adhesive bond forces at the interfaces between the matrix material and aggregates, there is a change in shape, delamination, cracking, and the associated redistribution of strains and stresses. Let us consider the influence of aggregates as elements of the structural heterogeneity "mortar part – aggregates" on the distribution mechanism of shrinkage strains and the nature of cracking at the level of the concrete macrostructure.

Three characteristic cases of the formation of adhesive and cohesive bond forces between the matrix and inclusions, which determine the mechanism of formation and distribution of shrinkage strains, can be distinguished:

- adhesion of the matrix material Ra to the aggregate is higher than its cohesive strength Rc, Ra > Rc;
- adhesive and cohesive strengths are equal, Ra = Rc;
- the matrix's adhesion to the aggregates' surface is below its cohesive strength, Ra < Rc.

The model of the concrete structural cell with the distance between the aggregates h = 0.1r is considered. At Ra > Rc, the boundary layers of the matrix material are connected to the aggregate's surface and held on it. It is assumed that the elastic characteristics of the aggregate are much higher than the elastic characteristics of the matrix Ea > Em, and the matrix is considered when it has plastic properties (it is strained irreversibly). Thus, the internal interfaces are determined by the surface of the aggregate and do not change in the process of changing the volume of the matrix material. With the internal interface unchanged, the volumetric strains are redistributed in the matrix material with compression strains εR at the interface, turning into radial tensile strains in the peripheral zones of the matrix (Fig. 2a, b). Under the action of εR in the matrix material, there is a violation of the continuity of the matrix itself without violating the integrity of the interfaces.

The separation of concrete into separate blocks by technological cracks occurs before the application of operational loads. This nature of the distribution of cracks is most dangerous for concrete since the mechanical properties of aggregates in composition with other components are not realized. Aggregates, having good adhesion with the matrix material in the initial periodic volumetric changes in the matrix, are isolated in the block. The resulting structural blocks interact with each other through technological cracks.

Incomplete adhesion of the matrix material to the surface of the inclusions may occur at Ra = Rc. Then destruction is equally probable along the interfaces or in the volume of the material with selective adhesion of the matrix to the active areas of the aggregates' surface due to the interfaces' microrelief. There is a general distribution of strains in the structural cell of concrete and local distribution in the zones of the interfaces (Fig. 2c, d).

The local distribution of strains depends on the areas of zones with imperfect adhesion and their location on the internal interfaces. Cracks occur either in the zone of transition of areas with perfect adhesion to areas with its absence or in the area of sections of the interface without adhesion to the surface of inclusions. The reasons for the appearance of cracks in the transition zones can be the multidirectionality of strains and the localization of shear strains. In areas with weakened adhesion, the adhesion crack is primary. After the appearance of a new interface with a larger area than the initial one, tensile strains develop on it, contributing to the appearance of cracks, which leads to a redistribution of the strain field in the matrix material and allows cracks to connect or quench each other.

A violation of the connection of the mortar part with aggregates characterizes concrete. Sedimentation phenomena, adsorption of the gas component or water films, and volumetric shrinkage can be the reasons for the appearance of adhesion cracks. One of the reasons for the appearance of adhesion cracks is the initial process of volumetric changes in the matrix material at the boundary with inclusions.

In the case of Ra < Rc (adhesion failure), a new internal matrix interface is formed with a decrease in the matrix material volume. At the same time, the boundary of the aggregate interface does not change (Fig. 2e, f). A strain field is formed on the inner interface of the matrix material, which depends on the distance between the inclusions and the shrinkage methods.

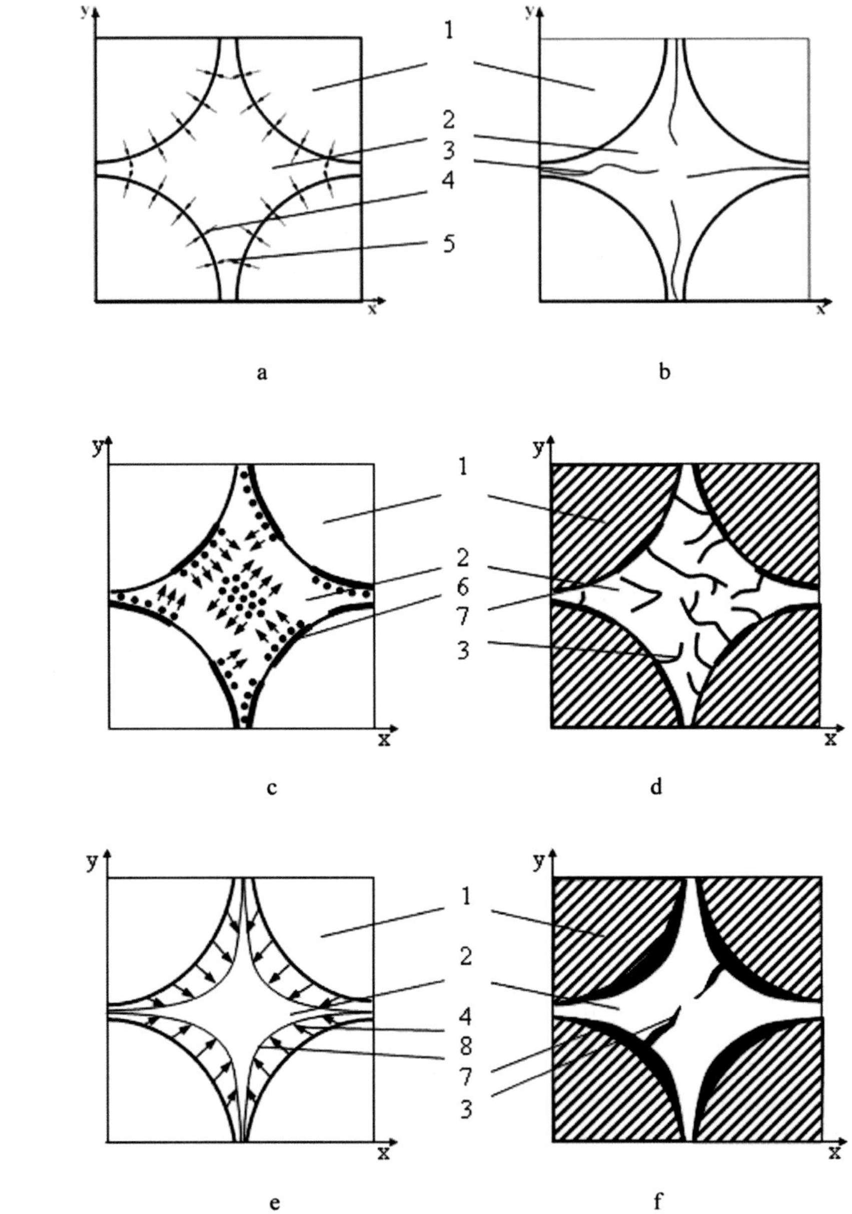

Fig. 2. Strain distribution mechanism and character of crack formation during cubic shrinkage of aggregates at h = 0,1r: a, b – $R_a > R_c$; c, d – $R_a = R_c$; e, f – $R_a < R_c$. 1 – aggregates; 2 – matrix; 3 – cracks in the matrix; 4 – direction of shrinkage strains; 5 – the aggregates' reaction to the matrix's impact; 6 – areas with broken adhesion; 7 – adhesion crack; 8 – new interface.

The system contains tensile strains at the interfaces with a transition to compression strains in the matrix. First, delamination cracks appear in the zone of maximum radial tensile strains. With the subsequent development of adhesion cracks in the matrix material, radial cracks appear depending on the distance between the inclusions. This is due to the uneven distribution of strain gradients at the interfaces of the matrix material. In the zone of multidirectional shrinkage strains, shear strains are localized, which leads to a violation of the integrity of the matrix.

Therefore, the formation of internal interfaces between the matrix material and inclusions is determined by the nature of the adhesive and cohesive bond forces at the interfaces, the shrinkage method, and the number of inclusions. A purposeful change in these parameters will make it possible to predict damage to the macrostructure of composite construction materials by hereditary defects.

Let us consider the compression deformations of concrete. The values of strains of the tested samples within the range of K_S coefficient are unevenly distributed; 69% are concentrated in the first half of the K_S range – from 1.3 to 1.7; 31% are distributed in the second half – from 1.7 to 2.08 (Fig. 3a). Throughout the area of K_{40} change, the values of prism strains are distributed evenly (Fig. 3b).

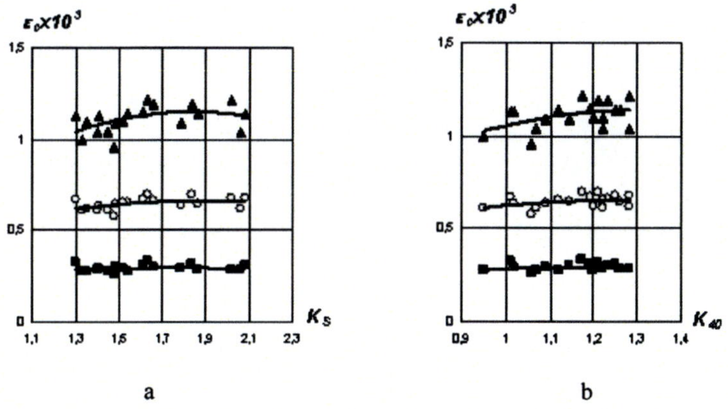

Fig. 3. Prism compression strains depending on the damage coefficients: a – K_S; b – K_{40}.

Strains of concrete at three stress levels have been considered: Level I – (0.259 – 0.333) × 10^{-3}; Level II – (0.581–0.694)×10^{-3}; Level III – (0.959 – 1.219) × 10^{-3}. It is noted that with increased technological damage, the values of total compression strains increase by 8...10%.

Let us consider the tensile deformations of concrete. Tensile strains depending on technological damage, were also considered at three stress levels: 6, 12, and 18 MPa.

Level I: strain values vary from 0.031 × 10^{-3} to 0.082 × 10^{-3} (2.6 times) (Fig. 4a, b). Average values of prism strains do not change: $\varepsilon_c = 0.05 \times 10^{-3}$.

Level II: tensile strain values vary from 0.09 × 10^{-3} to 0.196 × 10^{-3} (2.2 times).

Level III: tensile strain values vary from 0.185 × 10^{-3} to 0.567 × 10^{-3} (3.1 times). It has been established that the values of total tensile strains increase by 25...40% with increased technological damage.

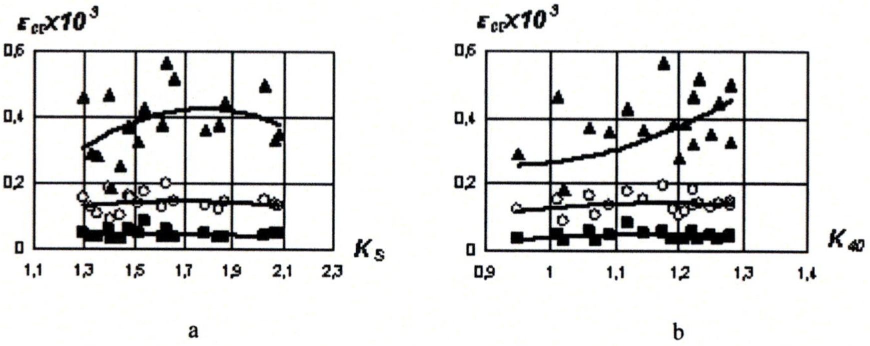

Fig. 4. Tensile strains of prisms depending on damage coefficients: a – K_S; b – K_{40}.

The inelastic compression strains that occur when holding the samples under load, depending on the initial damage to the concrete, are considered.

Level I: experimental values of inelastic compression strains varied from 0.007 × 10^{-3} to 0.023 × 10^{-3}. With an increase in coefficients K_S and K_{40}, the values of inelastic compression strains, on average, did not change and amounted to 0.014 × 10^{-3} (Fig. 5a, b).

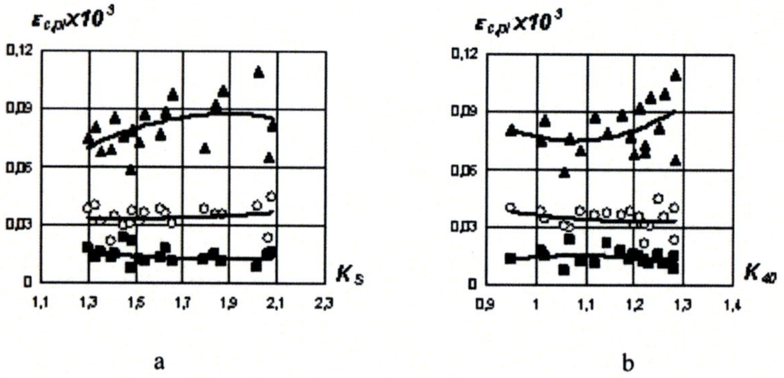

Fig. 5. Inelastic compression strains of prisms depending on the damage coefficients:

Level II: the values of inelastic compression strains varied from 0.021 × 10^{-3} to 0.045 × 10^{-3}. With increased coefficients K_S and K_{40}, the average values of inelastic compression strains are $\varepsilon_{c,pl} = 0.035 \times 10^{-3}$, (Fig. 5a, b).

Level III: average values of inelastic compression strains varied from 0.059 × 10^{-3} to 0.11 × 10^{-3}. The values of inelastic compression strains varied within 9...22%.

Let us consider the inelastic tensile deformations of concrete. Level I: strain values vary from 0.0009×10^{-3} to 0.0063×10^{-3}. The values of average inelastic tensile strains are 0.0028×10^{-3} and 0.0043×10^{-3} (Fig. 6a, b).

Level II: experimental values of inelastic tensile strains vary from 0.007×10^{-3} to 0.028×10^{-3}. With an increase in coefficients K_S and K_{40}, the average values of strains are constant: $\varepsilon_{ct,pl} = 0.014 \times 10^{-3}$.

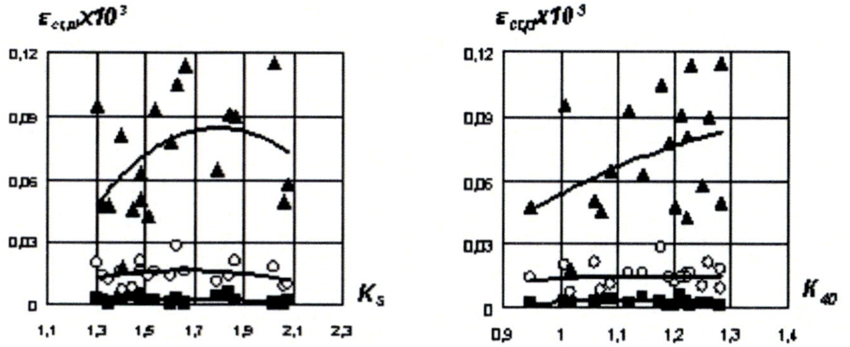

a b

Fig. 6. Inelastic tensile strains of prisms depending on damage coefficients: a – K_S; b – K_{40}.

Level III: values of inelastic tensile strains vary from 0.042×10^{-3} to 0.115×10^{-3}. With an increase in K_S from 1.3 to 1.8, the values of inelastic tensile strains increase from 0.047×10^{-3} to 0.085×10^{-3}. With an increase in damage factor K_{40}, an increase in inelastic tensile strains from 0.046×10^{-3} to 0.082×10^{-3} is observed. It can be concluded that with an increase in the degree of technological damage, the values of inelastic tensile strains increase by 22...60%.

5 Conclusions

The resulting incipient crack is an intercluster interface that can develop. The introduction of aggregates and the formation of mixed cluster structures can contribute to the early appearance of incipient cracks and restrain their width, total number, and ability to grow.

The formation of internal interfaces between the matrix material and inclusions is determined by the nature of the adhesive and cohesive bond forces at the interfaces, the shrinkage method, and the number of inclusions. A purposeful change in these parameters will make it possible to predict damage to the macrostructure of composite construction materials by hereditary defects.

With an increase in technological damage, the values of total compression strains increase by 8...10% and tensile strain values increase by 25...40%. The values of inelastic compression strains vary within 9...22% and inelastic tensile strains increase by 22...60%.

The effect of technological damage on the level of properties and operational reliability of composite building materials has been experimentally established.

In the future, when designing reinforced concrete structures, it is necessary to consider the presence of initial cracks through the damage factor, considering the complete diagram of the σ-ε dependence during compression.

References

1. Mamazhonov, A.U., Nabiev, M.N, Umurzakov, E.K., Abdullaev, I.N., Akbaralizoda, S.: Structure formation of cement stone in the presence of mineral fillers and plasticizing chemical additive ACF-3m. Archit. Constr. Des. 112–117 (2019)
2. Mamazhonov, A.U.: Strength and deformation of concrete with mineral fillers from industrial waste and the chemical additive of ACF resin. Int. J. Discourse Innov. Educ. **4**, 193–198 (2020)
3. Krentowski, J.R., Knyziak, P., Mackiewicz, M.: Durability of interlayer connections in external walls in precast residential buildings. Eng. Fail. Anal. **121**, 105059 (2021). https://doi.org/10.1016/j.engfailanal.2020.105059
4. Mamazhonov, A., Kasimov, L.: Features of the properties of cement systems in the presence of mineral fillers and additives of acetone-formaldehyde resin. Grail Sci. **5**, 102–108 (2021). https://doi.org/10.36074/grail-of-science.04.06.2021.020
5. Oleynyk, N.V., Bychev, I.K., Lukashenko, L.E.: Influence of technological damage of concrete on moments of cracking of reinforced concrete beams. Bull. Odessa State Acad. Constr. Archit. **67**, 48–53 (2017)
6. Vyrovoy, V.N., Korobko, O.A., Panasyuk, V.A.: Multivariance of concrete structure. Concr. Technol. **5–6**, 16–18 (2016)
7. Dorofeev, V., Pushkar, N., Zinchenko H.: The influence of concrete structure on the destruction of reinforced concrete bended elements. In.: II International Scientific Conference EcoComfort and current issues of civil engineering, pp. 103–111. Springer, Cham (2021). https://doi.org/10.1007/978-3-030-57340-9_13
8. Vyrovoy, V.N., Sukhanov, V.G., Vinogradsky, V.M.: Technological events in the structural development of building composites. Bull. Odessa State Acad. Constr. Archit. **62**, 50–59 (2016)
9. Vashpanov, Y., Son, J.-Y., Heo, G., Podousova, T., Kim, Y.S.: Determination of geometric parameters of cracks in concrete by image processing. Adv. Civ. Eng. **2019**, 1–15 (2019). https://doi.org/10.1155/2019/2398124
10. Karpiuk, I., Danilenko, D., Karpiuk, V., Danilenko, A., Lyashenko, T.: Bearing capacity of damaged reinforced concrete beams strengthened with metal casing. Acta Polytechnica **6**, 703–721 (2021). https://doi.org/10.14311/AP.2021.61.0703
11. Kovalchuk, B., Blikharskyy, Y., Selejdak, J., Blikharskyy, Z.: Strength of reinforced concrete beams strengthened under loading with additional reinforcement with different levels of its pretension. In: International Scientific Conference EcoComfort and Current Issues of Civil Engineering, pp. 227–236. Springer, Cham (2021). https://doi.org/10.1007/978-3-030-57340-9_28
12. Selejdak, J., Blikharskyy, Y., Khmil, R., Blikharskyy, Z.: Crack resistance RC columns strengthened by CFPR system. Key Eng. Mater. **878**, 127–133 (2021). https://doi.org/10.4028/www.scientific.net/KEM.878.127
13. Blikharskyy, Y., Vashkevych, R., Kopiika, N., Bobalo, T., Blikharskyy, Z.: Calculation residual strength of reinforced concrete beams with damages, which occurred during loading. Mater. Sci. Eng. **1021**(1), 1–9 (2021). https://doi.org/10.1088/1757-899X/1021/1/012012
14. Vyrovoy, V., Korobko, O., Sukhanov, V., Zakorchemny, Y.: Resistance of concrete and expanded clay concrete under periodic external influences. MATEC Web Conf. **230**(5), 03021 (2018). https://doi.org/10.1051/matecconf/201823003021

15. Tawashi, A., Alamoudi, S.: The thermal effects on the self-compacting concrete beams exposed to the fire "experimental case study." El-Cezeri J. Sci. Eng. **10**(2), 268–283 (2023). https://doi.org/10.31202/ecjse.1209551
16. Babachenko, O.I., Kononenko, H.A., Podolskyi, R.V., Safronova, O.A., Taranenko, A.O.: Structure and fracture resistance of steels in different zones of railway axles. Mater. Sci. **58**(3), 417–421 (2022). https://doi.org/10.1007/s11003-023-00679-1

Structure and Mechanical Properties of V, Nb-Added TRIP-Assisted Steel After Q&P Treatment with Near Ac_3 Austenitization

Vadym Zurnadzhy[1,2](✉), Yuliia Chabak[1,2], Vasily Efremenko[1,2], Alexey Efremenko[1], and Maria Podobova[2]

[1] Pryazovskyi State Technical University, 19, Dmytro Yavornytskyi Ave., Dnipro 49005, Ukraine
vadim.zurnadzhy@gmail.com

[2] Institute of Materials Research, Slovak Academy of Sciences, 47, Watsonova St., Kosice 04001, Slovak Republic

Abstract. The article is devoted to studying the effect of Q&P heat treatment parameters after intercritical annealing near the Ac_3 critical temperature on the phase-structural state and mechanical properties of 0.2 wt. % C-Mn-Si-Cr-Mo TRIP-assisted steel micro-alloyed with Nb and V. The investigations were conducted using optical microscopy (OP) and electron microscopy (SEM, TEM), X-ray diffraction (XRD) analysis, and mechanical properties testing. It was found that intercritical annealing at 900 °C followed by the quenching to 235 °C and the "partitioning" at 350–450 °C promoted the formation of a multiphase structure consisting of martensite, bainite, and minor amounts of proeutectoid ferrite (about 5 vol. %) and retained austenite (5.4–7.4 vol. %). The strength of steel decreased with the holding at the "Partitioning" stage. Retained austenite has a film-like morphology and performs different tendencies to the TRIP effect depending on the "Partitioning" temperature. The optimal was the "Partitioning" at 400 °C for 10–20 min, ensuring the advanced combination of properties: UTS of 1020–1122 MPa; YTS of 908–1020 MPa, TEL of 21.0–23.5%, impact toughness (KCV) of 115 J/cm^2, and the PSE of 25 GPa·%.

Keywords: Quenching · Partitioning · AHSS · TRIP Effect · Retained Austenite · Mechanical Behaviour · Manufacturing Innovation

1 Introduction

The current progress in manufacturing welded structures and requirements for reducing their costs lead to an ever-increasing demand for high-strength sheet steel intended for construction and engineering applications [1]. It is a challenge for the metallurgical industry to develop new high-strength steel grades and improve their production technologies [2]. An improved complex of mechanical properties can be gained through costly heavy alloying of steel or, alternatively, by implementing innovative steel processing technologies at low/moderate alloying. The second approach is more feasible regarding the property/cost ratio.

V. Tonkonogyi et al. (Eds.): InterPartner 2023, LNME, pp. 346–355, 2024.
https://doi.org/10.1007/978-3-031-42778-7_32

2 Literature Review

The TRIP-assisted steels belong to the family of modern advanced high-strength steels (AHSSs) [3]. After heat treatment, a multiphase structure consisting of ferrite (about 50 vol. %), bainite, and retained austenite (RA) is formed in these steels. RA is prone to strain-induced martensite transformation (SIMT), thus enabling the TRIP (Transformation-Induced-Plasticity) effect [4]. The TRIP effect is essential since it significantly contributes to the mechanical [5], fatigue [6], and tribological properties of steels [7, 8]. Due to the TRIP effect and a significant amount of proeutectoid ferrite, TRIP-assisted steels perform higher plasticity and formability [3]. Though, ferrite reduces strength limiting the steel application. The strength of the TRIP-assisted steel can be increased by partially replacing ferrite/bainite with a stronger phase, namely martensite. It can be attained by applying the heat treatment of "Quenching and Partitioning" (Q&P) [9], which was developed to form in low-alloy steel a multiphase structure with an increased amount of metastable RA. Q&P treatment implies quenching the temperature range between the M_S and M_F points to form a "martensite + austenite" structure [10]. After that, the steel is subjected to isothermal holding at the quenching temperature (one-step Q&P) or a temperature above M_S point (two-step Q&P]) to let carbon partitioning from martensite to austenite to stabilize the latter to transformation upon final cooling [11]. At the partitioning stage, the partial transformation of austenite into bainite is possible [12]. The resultant Q&P structure may include martensite (tempered and "fresh"), bainite, ferrite, and an increased amount of RA [13]. The Q&P steels and TRIP-assisted steels contain 1–2 wt. % Mn and 1–2 wt.% Si (or Al) [14]. The latter aims to suppress the cementite precipitation during the carbon partitioning (bainite transformation) to ensure the austenite enrichment with carbon. Also, Nb and V are added to the aforementioned steels to improve their strength through the dispersed precipitates formation [15]. The same effect of nanoscale Cu precipitates on the strength and ductility of quench-partitioned and tempered steel is reported in [16].

It is worth noting that the classical Q&P processing scheme involves austenitization at temperatures above Ac_3 [17]. In contrast, the classical heat treatment of the TRIP-assisted steels includes intercritical annealing (IA) before austempering (i.e., steel is soaked at the temperature between Ac_1 and Ac_3 points) [11]. Incorporating the IA into the Q&P schedule can be feasible to increase the ductility of Q&P-treated steel due to the retention of some amount (5–10 vol.%) of proeutectoid ferrite in the structure [18]. As recently shown in [19], performing IA close to Ac_3 temperature ensures an improved complex of strength, ductility, and impact toughness in the TRIP-assisted steel microadded with strong carbide forming elements (Cr, Mo, Nb, V) due to the limitation of carbon content in austenite before austempering. The latter hinders the cementite precipitation under bainite transformation leading to a significant increase in strength properties without loss of plasticity. This approach to austenitization can also be applied to the Q&P process. However, this issue is not well studied yet, especially regarding the (Nb,V)-micro-alloyed TRIP-assisted steels. Because of the above, the objective of the present work was to investigate the effect of the temperature-time parameters of Q&P treatment on the microstructure and properties of 0.2wt.%C-Mn-Si-Cr-Mo-Nb-V TRIP-assisted steel subjected to near Ac_3 intercritical annealing.

3 Research Methodology

The experimental steel of the following chemical composition (wt. %) 0.20 C; 1.79 Si; 1.73 Mn; 0.55 Cr; 0.20 Mo; 0.11 V; 0.045 Nb; 0.009 S; 0.013 P was used as study material. The steel was smelted in laboratory conditions and eventually hot rolled into a strip 15 mm thick, which was further used for the specimens' preparation. The critical points (Ac_1, Ac_3, Ms) and kinetic of austenite transformation in this steel were studied earlier and presented in [19].

The specimens were subjected to heat treatment according to the Q&P schedule in three stages depicted in Fig. 1.

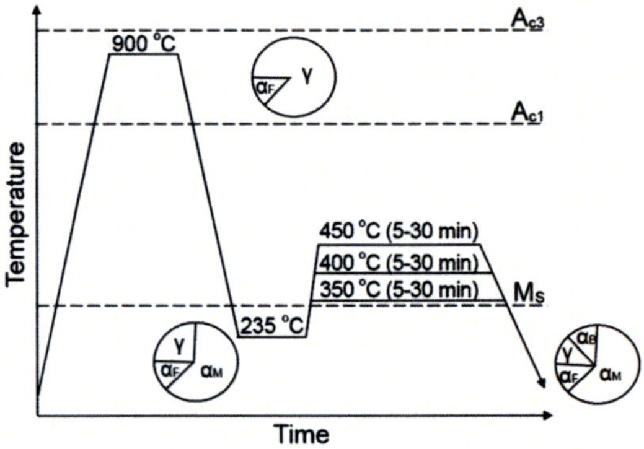

Fig. 1. The scheme of Q&P treatment (γ is austenite; αF, αM, αB are proeutectoid ferrite, martensite, and bainite, accordingly).

In the first stage, the specimens were austenitized at 900 °C for 10 min. The critical points of the steel are $Ac_1 = 760$ °C, $Ac_3 = 930$ °C. Thus 900 °C is relatively close to Ac_3, which allowed forming of an austenite/ferrite structure with a volume fraction of the latter of about 5 vol.%. This temperature was found optimal [19] for the austenitization of studied steel to avoid the precipitation of cementite carbide during the "Austenite → Bainite" transformation. In the second stage, the austenitized specimens were quenched in a molten Wood's alloy to acquire 235 °C, below the Ms point (determined as 349 °C for austenitization at 900 °C [19]). The quenching temperature was calculated according to the "Constrained Paraequilibrium" concept [20] to ensure the maximal content of the retained austenite in Q&P-treated steel. In the third stage, the quenched specimens were isothermally held at 350 °C, 400 °C, and 450 °C to partition carbon between martensite and austenite. The holding durations were 5 min, 10 min, 20 min, and 30 min corresponding to the time "window" of "Austenite → Bainite" transformation in the studied steel [19]. After the "Partitioning" stage, the specimens were cooled to still air.

The Q&P-treated specimens were subjected to tensile and Charpy impact tests performed at room temperature. The tensile specimens were of 5 mm diameter and 30 mm gauge. The impact specimens with a V-notch were 7.5×10×55 (mm) in size. The microstructure was observed using an optical microscope (OM) "Axiovert 40 MAT"

(Carl Zeiss), scanning electron microscope (SEM) "JSM-7000F" (JEOL), and a transmission electron microscope (TEM) "JEM-100-C-XII" (JEOL). The OM/SEM observation specimens were mirror-polished with SiO_2 papers and Al_2O_3-emulsion and eventually etched by the 4 vol.% Nital solution. The specimens for the TEM study were mechanically polished to get the 0.10–0.15 mm thick foils which were then electropolished in 6-vol.% $HClO_4$ solution using a fluid-jet polishing installation. X-ray diffraction analysis was performed using D500 (Siemens) monochromator-equipped diffractometer with CuKα radiation under the following parameters: voltage is 40 kV, a current is 30 mA, step is 0.025 degrees, dwell time is 3 s, scanning speed is 0.0083 degree/s. The volume fraction of retained austenite and the concentration of carbon in RA were calculated according to the methodology described in [21].

4 Results and Discussion

Figure 2 shows the microstructures of the steel Q&P-treated with different temperatures of the "Partitioning" stage. The microstructure has the same pattern consisting of martensite laths segmenting the former austenite grain (the laths are divided by ferritic interlayers). Increasing the "Partitioning" temperature led to coarsening of the martensite with corresponding narrowing of the ferritic interlayers (Fig. 2, d and Fig. 2, f). The microstructural pattern is the same for each temperature of "Partitioning," irrespective of its duration. TEM observation revealed the internal structure of martensitic alpha-phase, featuring a high density of crystal defects resulting from shear deformation under γFe → αFe transformation (Fig. 3a). As well, for all "Partitioning" temperatures structure contained the RA films lying between the alpha-phase laths (Fig. 3b). The film-like RA performs higher stability to SIMT, which is beneficial for the mechanical properties of steel [22]. Cementite inclusions were not found in the structure, being suppressed by 1.79 wt. % Si present in the steel.

The presence of retained austenite in the structure of Q&P-treated steel was confirmed by XRD analysis (Fig. 4). The XRD pattern for each "Partitioning" temperature (duration of 20 min) exhibits all set of diffraction maxima characteristic for γFe with FCC lattice, namely: (111), (200), (220) and (311) indicating the existence of retained austenite in the structure. With that, αFe was a major phase, as follows from the comparison of the intensities of the diffraction peaks. The calculations showed that the volume fraction of RA after "Partitioning" at 350 °C (20 min), 400 °C (20 min), and 450 °C (20 min) was 7.4 vol.%, 7.3 vol. %, and 5.4 vol.%, respectively while the carbon content in RA was 0.60 wt.%, 0.90 wt.% and 0.83 wt.%, respectively.

The latter values are 3–4 times higher than those of the total concentration of carbon in steel, reflecting the enrichment of austenite with carbon during the "Partitioning" stage promoted by the Si-inhibition of cementite precipitation from austenite.

Figure 5 depicts the effect of Q&P parameters on the mechanical properties of the steel. "Partitioning" at 350 °C provided the highest yield tensile strength (YTS) of 1015–1020 MPa after 5–20 min holding (Fig. 5a). At higher "P" temperatures YTS is about 60–80 MPa lower as compared with 350 °C. There is a tendency of decreasing YTS reaching 840–913 MPa after partitioning for 30 min. The same tendency is observed for ultimate tensile strength (UTS) though the decrement in UTS is lower than that

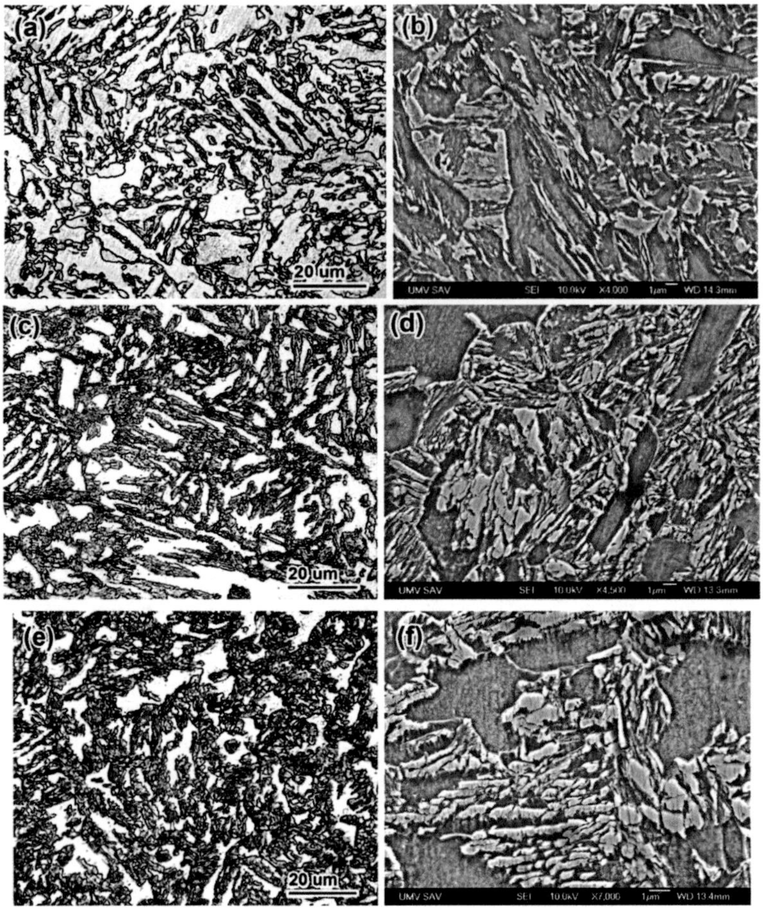

Fig. 2. Microstructure of treated steel with "Partitioning" at (a, b) 350 °C, (c, d) 400 °C and (e, f) 450 °C for 20 min (a, c, e are OM images; b, d, f are SEM images).

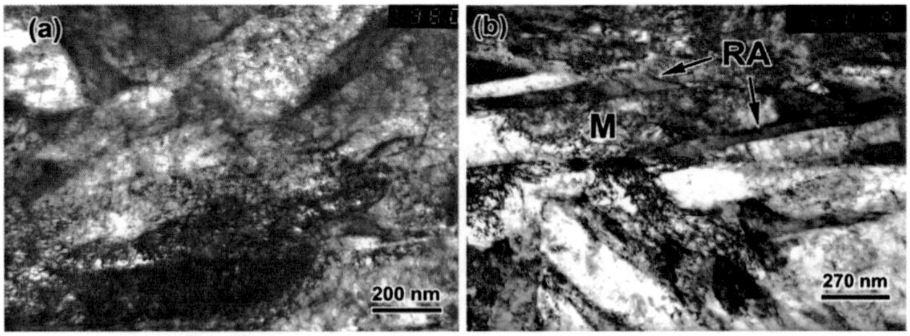

Fig. 3. TEM images of the structure after "Partitioning" at (a) 350 °C (2 min) and (b) 400 °C (20 min) (M is martensite, RA is retained austenite).

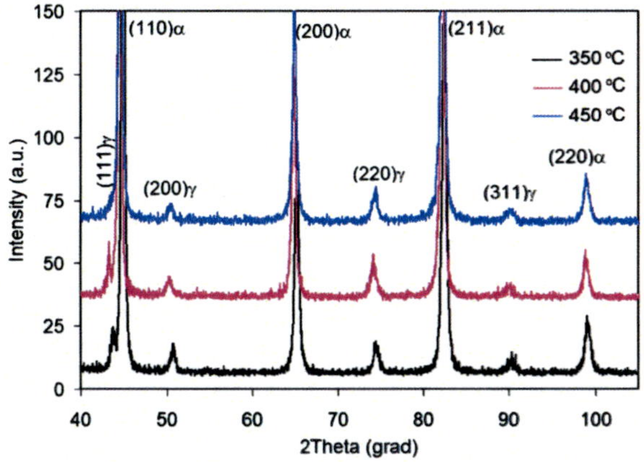

Fig. 4. XRD patterns of the Q&P-treated steel.

of YTS (Fig. 5b). All the "Partitioning" temperatures resulted in about the same UTS values in the range of 1060–1120 MPa (after 5–20 min holding) and in the range of 1010–1050 MPa (after 30 min holding). The highest ductility level was attributed to the "Partitioning" at 400 °C ensuring the total elongation (TEL) of 21.0–23.5% after 10–30 min holding (Fig. 5c). The specimens partitioned at 350 °C and 450 °C performed the lowest TEL level (14–16.5%).

PSE (Product of Strength and Elongation) had similar to the TEL profile (PSE characterizes the combination of different mechanical properties [4]). The maximum PSE level refers to partitioning at 400 °C, reaching maximally 25 GPa·% for 10 min holding (Fig. 5d). Q&P treatment provided the studied steel with a relatively high impact toughness, slightly increasing from 115–120 J/cm^2 at 350–400 °C to 128 J/cm^2 at 450 °C (Fig. 5e).

The observed effects of the "Partitioning" parameters on strength and ductility can be explained by: (a) fuller development of martensite tempering through ε-carbide precipitation and carbon redistribution to austenite, (b) stabilization of RA to SIMT due to carbon enrichment, and (c) the partial transformation of the austenite into bainite, which reduced the proportion of brittle "fresh" martensite in the structure.

Despite the differences in the yield strength after treatment at 350 °C and 400–450 °C, the specimens performed approximately the same tensile strength for all holding temperatures. The TRIP effect presumably compensated the lower YTS after partitioning at 400–450 °C, which led to the strengthening through the deformation-induced transformation of retained austenite.

The optimal combination of strength, ductility, and impact toughness was obtained after Q&P treatment with the partitioning at 400 °C for 10–20 min. This combination includes YTS of 908–997 MPa, UTS of 1020–1092 MPa, TEL of 21.0–23.5% and KCV of 120 J/cm^2. These values provide the PSE of 21.5–25.0 GPa·%, corresponding to the AHSS steels level [1]. The improved PSE value (of 25 GPa·%) is governed mainly by advanced ductility performed at a relatively high strength of >1000 MPa. This feature of the specimens partitioned at 400 °C refers to the tensile behavior of steel illustrated by the strain hardening rate (SHR) curves presented in Fig. 6 (SHR is calculated as dσ/dε, where

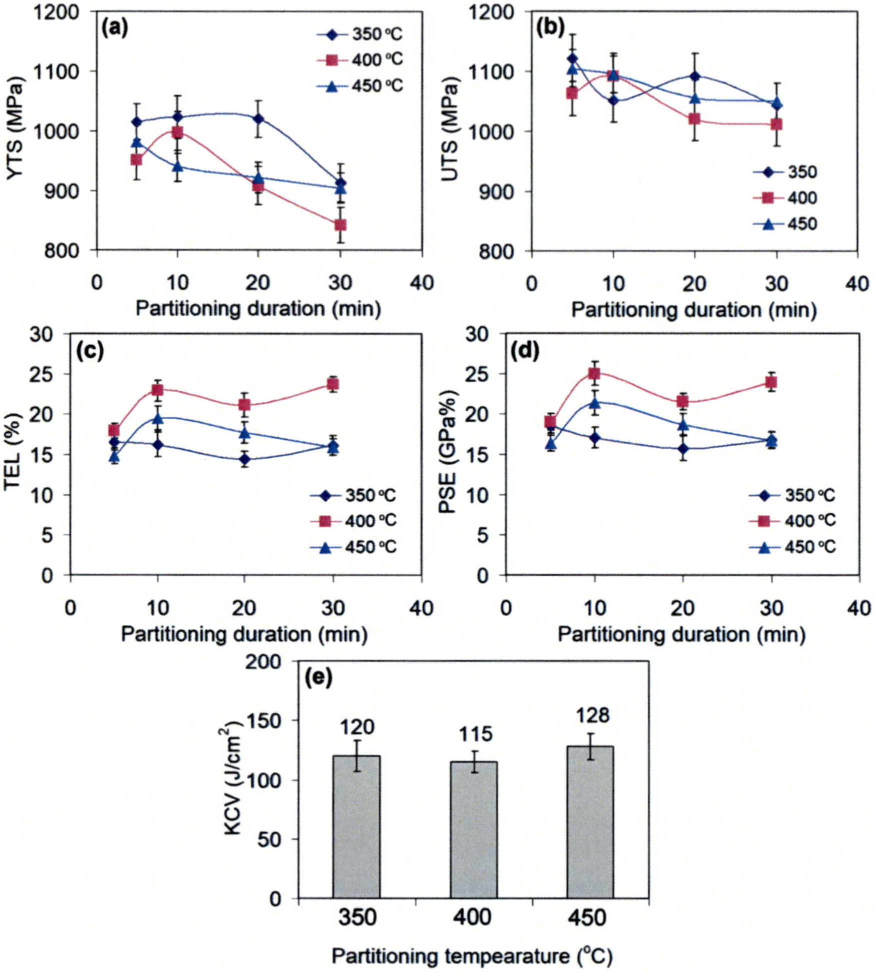

Fig. 5. The effect of Q&P treatment parameters on the mechanical properties of the investigated steel: (a) ultimate tensile strength, (b) yield tensile strength, (c) total elongation, (d) product of strength and elongation, (e) impact toughness.

σ is stress and ε is strain). After reaching the maximum value at the initial deformation ($\varepsilon \leq 0.02$), the SHR significantly decreases in the interval. $\varepsilon = 0.03$–0.04 corresponds to the stage of easy sliding of dislocations [23]. When the dislocation blocked the densely packed slip planes, the second hardening stage started, when SHR increased with a strain of 0.05–0.13.

The second stage is attributed to the TRIP effect manifestation [23]. As can be seen, the SHR curves for different partitioning temperatures perform the same behavior in the ascending section; however, they reach the maximum at different strains: $\varepsilon \approx 0.10$ for 400 °C and $\varepsilon \approx 0.12$–0.13 for 350 °C and 450 °C. That means that the strain hardening (caused by SIMT) lasted longer, exactly at 350 °C and 450 °C leading to the complete

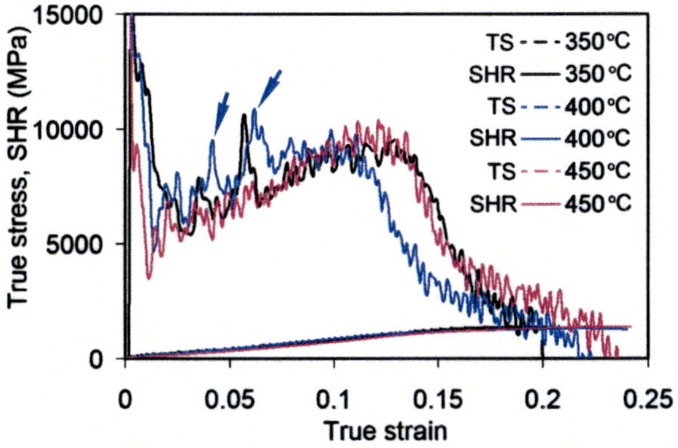

Fig. 6. The true stress (TS) and SHR curves for the specimens partitioned at different temperatures for 20 min.

transformation of retained austenite. In the case of 400 °C, the hardening (SIMT) stopped earlier. Thus more austenite was retained in the structure, providing higher ductility. Such behavior of partitioned at 400 °C structure can be attributed to the formation of RA areas with a different propensity to SIMT: the metastable RA areas transformed at $\varepsilon \leq 0.10$ (see high peaks of SHR shown by the arrows), the stable RA areas (containing 0.90 wt. %C) remained untransformed until the specimen fracture. The specimens partitioned at 350 °C and 450 °C presumably contained metastable RA. Its metastability was caused either by lower carbon content (350 °C, 0.6 wt.%) disabling the chemical stabilization of RA or by martensite tempering (450 °C) that released the stress and eliminated the mechanical stabilization of RA [24].

5 Conclusions

The effect of the Q&P parameters on the structure and mechanical properties of a structural 0.2wt.%C-Mn-Si-Cr-Mo-Nb-V TRIP-assisted steel austenitized near the Ac_3 point has been studied in this work. It was found that after quenching to 235 °C and "partitioning" at 400 °C for 10–20 min, the steel attains a complex structure with an increased (up to 7.3 vol. %) amount of retained austenite. During the "Partitioning" stage, austenite was enriched in carbon up to 0.90 wt.%, which increased its stability to strain-induced martensite transformation under tension. Such structure ensured an improved combination of strength (UTS of 1020–1122 MPa; YTS of 908–1020 MPa), ductility (TEL 21–23.5%), and impact toughness (KCV of 115 J/cm^2) thus reaching the PSE value of 25 GPa·%. It is shown that high ductility can be obtained in TRIP-assisted steel at a low amount of proeutectoid ferrite, which is beneficial for the higher strength of steel.

Acknowledgment. This work was funded by the Ministry of Education and Science of Ukraine under the project No. 0123U100374. V. Efremenko, Yu. Chabak and V. Zurnadzhy appreciate the support in the framework of the "EU Next Generation EU the Recovery and Resilience Plan for Slovakia" under projects No. 09I03-03-V01-00061 and No. 09I03-03-V01-00099. The help of Prof. M. Dabala and M. Franceschi with the XRD study is acknowledged.

References

1. Soleimani, M., Kalhor, A., Hamed, M.: Transformation-induced plasticity (TRIP) in advanced steels: a review. Mater. Sci. Eng. A **795**, 140023 (2020)
2. Karnaukh, S.G., Markov, O.E., Aliieva, L.I., Kukhar, V.V.: Designing and researching of the equipment for cutting by breaking of rolled stock. Int. J. Adv. Manuf. Technol. **109**(9–12), 2457–2464 (2020)
3. Lesch, C., Kwiaton, N., Klose, F.B.: Advanced high strength steels (AHSS) for automotive applications– tailored properties by smart microstructural adjustments. Steel Res. Int. **88**(10), 1700210 (2017)
4. Bleck, W., Guo, X., Ma, Y.: The TRIP effect and its application in cold formable sheet steels. Steel Res. Int. **88**, 1700218 (2017)
5. Franceschi, M., Bertolini, R., Fabrizi, A., Dabalà, M., Pezzato, L.: Effect of ausforming temperature on bainite morphology in a 3.2% Si carbide-free bainitic steel. Mater. Sci. Eng.: A **864**, 144553 (2023)
6. Rykavets, Z.M., et al.: Investigation of the microstructure and properties of TRIP 800 steel subjected to low-cycle fatigue. Uspehi Fiziki Metallov **20**, 620–633 (2019)
7. Hesse, O., Merker, J., Brykov, M., Efremenko, V.: On the strength of low-alloy steels with increased carbon content against abrasive wear. Tribol. Schmierungstech. **60**(6), 37–43 (2013). [in Germany]
8. Stelmakh, A., Kostunik, R., Radzievskyi, V.; Shymchuk, S., Zaichuk, N.: An increase in tribocharacteristics for highly loaded friction units of modern equipment. In: Ivanov, V., Trojanowska, J., Pavlenko, I., Rauch, E., Peraković, D. (eds). Advances in Design, Simulation and Manufacturing V. DSMIE 2022. Lecture Notes in Mechanical Engineering, pp. 504–508. Springer, Cham (2022). https://doi.org/10.1007/978-3-031-06025-0_50
9. Yan, X., Hu, J., Zhang, X., Xu, W.: Obtaining superior low-temperature wear resistance in Q&P-processed medium Mn steel with a low initial hardness. Tribol. Int. **175**, 107803 (2022)
10. Xia, P., et al.: Effect of alloying and microstructure on formability of advanced high-strength steels processed via quenching and partitioning. Mater. Sci. Eng. A **831**, 142217 (2022)
11. Yan, S., et al.: Comparative study on microstructure and mechanical properties of a C-Mn-Si steel treated by quenching and partitioning (Q&P) processes after a full and intercritical austenitization. Mater. Sci. Eng. A **684**, 261–269 (2017)
12. Xu, N., Wang, L., Hu, J., Liu, H., Yu, S., Xu, W.: Batch annealing enhances the yield strength of a high quenching temperature Q&P steel through accelerated bainitic transformation. Mater. Sci. Eng. A **855**, 143888 (2022)
13. Bansal, G.K., et al.: Evolution of microstructure in a low-Si micro-alloyed steel processed through one-step quenching and partitioning. Metall. Mater. Trans. A. **50**, 547–555 (2019)
14. Kaar, S., et al.: Influence of the Al and Mn content on the structure-property relationship in density reduced TRIP-assisted sheet steels. Mater. Sci. Eng. A **735**, 475–486 (2018)
15. Kučerová, L., Bystrianský, M.: Comparison of thermo-mechanical treatment of C-Mn-Si-Nb and C-Mn-Si-Al-Nb TRIP steels. Procedia Eng. **207**, 1856–1861 (2017)
16. Rong, X., et al.: Simultaneous enhancement of strength and ductility via nanoscale Cu precipitates and ultrafine filmy retained austenite in a novel quench-partitioned and tempered steel. Mater. Sci. Eng. A **760**, 47–57 (2019)
17. Zinsaz-Borujerdi, A., et al.: Room temperature mechanical properties and microstructure of a low alloyed TRIP-assisted steel subjected to one-step and two-step quenching and partitioning process. Mater. Sci. Eng. A **725**, 341–349 (2018)
18. Kickinger, C., Suppan, C., Hebesberger, T., Schnitzer, R., Hofer, C.: Microstructure and mechanical properties of partially ferritic Q&P steels. Mater. Sci. Eng. A **815**, 141296 (2021)

19. Zurnadzhy, V., et al.: Alternative approach for the intercritical annealing of (Cr, Mo, V)-alloyed TRIP-assisted steel before austempering. Metals **12**, 1814 (2022)
20. Clarke, A.J., et al.: Carbon partitioning to austenite from martensite or bainite during the quench and partition (Q&P) process: a critical assessment. Acta Mater. **56**(1), 16–22 (2008)
21. Sun, J., Hao, Y.: Microstructure development and mechanical properties of quenching and partitioning (Q&P) steel and an incorporation of hot-dipping galvanization during Q&P process. Mater. Sci. Eng. A **586**, 100–107 (2013)
22. Xu, Y., et al.: Kinetics of carbon partitioning of Q&P steel: considering the morphology of retained austenite. Metals **12**(2), 344 (2022)
23. Zhang, B., et al.: Structure-property relationship in novel low carbon hot-rolled TRIP steels via thermo-mechanical controlled processing and coiling. Mater. Sci. Eng. A **771**, 138643 (2020)
24. Kozłowska, A., et al.: Mechanical and thermal stability of retained austenite in plastically deformed bainite-based TRIP-aided medium-MN steels. Arch. Civ. Mech. Eng. **21**, 133 (2021)

Quality Assurance

Standardization of Scanning Protocols and Measurements for Additive Manufacturing Quality Assurance

Aleksandr Kokhanov[1], Igor Prokopovich[1](✉), Tetiana Sikach[1], Irina Dyadyura[2], and Isak Karabegovich[3]

[1] Odessa Polytechnic National University, 1, Shevchenko Ave., Odessa 65044, Ukraine
igor.prokopovich@gmail.com

[2] Sumy State University, 2, Rymskogo-Korsakova St., Sumy 40007, Ukraine

[3] Academy of Sciences and Arts of Bosnia and Herzegovina, 7, Bistrik St., 71000 Sarajevo, Bosnia and Herzegovina

Abstract. This work is devoted to the (high-throughput) extraction of image biomarkers from acquired, reconstructed, and stored images. The development of new imaging biomarkers involves clearly defined sequential steps. This paper discusses advanced medical imaging data and approaches to processing and optimizing medical imaging data acquisition for accurate, reliable medical models based on real-world data. Segmentation and classification tools provide an approach to feature extraction from images based on nonlinear dynamics methods. In the context of this work, an image is defined as a three-dimensional (3D) set of two-dimensional (2D) digital image fragments. The image fragments are arranged along the Z-axis. Pixels and voxels are represented as rectangles and rectangular parallelepipeds. Pixel and voxel centers coincide with the intersections of a regularly spaced grid. Both views are used in the document. The phase plane method was chosen in this work to analyze 2D imaging. Based on the received metadata, reconstruction is performed using 3D visualization to extract the data of the region of interest. The research results contribute to developing software optimization of medical imaging for additive manufacturing. The issue of harmonizing and standardizing medical image acquisition and reconstruction is being addressed more comprehensively.

Keywords: Standardization · 3D Printing · Additive Manufacturing · AM Medical Devices · AM Standards · Regulations · Quality Assurance · Nonlinear Dynamics · Industrial Innovation

1 Introduction

Additive manufacturing (AM) is a general term for technologies used to connect materials sequentially in creating physical objects according to data from 3D models [1]. Such technologies are currently used for various applications, including those actively used in medicine [2]. The healthcare industry was one of the first to embrace AM technology,

V. Tonkonogyi et al. (Eds.): InterPartner 2023, LNME, pp. 359–368, 2024.
https://doi.org/10.1007/978-3-031-42778-7_33

recognizing its unique potential for mass device personalization and complex structure creation. Standardization in the field of additive manufacturing is essential and necessary in terms of processes, terms, and definitions [3], technological chains (hardware and software), test procedures [4], quality parameters, supply agreements, and all the fundamentals [5]. The ASTM/ISO collaboration (ISO/TC 261 and ASTM F 42) allows the use and implementation of international technical experience in additive manufacturing. Since healthcare is a highly regulated industry, demonstrating AM technology's capabilities alone is insufficient to provide confidence in using this technology's parts. Currently, to effectively apply 3D printing technology in medical applications, it is essential to ensure that medical image data can be consistently and accurately converted into a 3D printed object. The original image data's accuracy and resolution predetermine the final model's accuracy [6]. The main factors affecting the accuracy are the image's resolution and noise level, the contrast between the studied tissues, and artifacts inherent in the imaging system [7]. Segmentation is critical in transforming medical images into a physical model [8]. For example, segmentation of the orbital bone is necessary for orbital wall reconstruction in craniomaxillofacial surgery to support the eye globe position and restore the volume and shape of the orbit. In this regard, for medical 3D printing, segmentation methods must be optimized and combined according to the characteristics of medical images and corresponding body parts to obtain an optimal 3D model. Additional errors may occur when converting DICOM or PACS data to the computational formats used in segmentation editing software and maintaining the STL 3D mesh format when used in additive manufacturing systems. Medical imaging for precision medicine relies on biomarkers that capture patient and disease characteristics accurately, efficiently, reproducibly, and interpretably.

The standard ISO/ASTM/TR 52,916:2022 [9] allows for creating optimized data for medical additive manufacturing (MAM). These data can be generated from static modalities, e.g., computed tomography (CT) and magnetic resonance imaging (MRI). Improving the segmentation algorithm that can extract all region of interest (ROI) boundaries is necessary. Appropriate 3D printing software modeling is required to accurately and consistently render human anatomy.

This work aims to define the basic principles of general procedures for modeling medical images based on methods of nonlinear dynamics. This concerns accurate 3D modeling of medical data using medical image data such as magnetic resonance imaging (MRI), and computed tomography (CT).

2 Literature Review

This article [10] reviews AM's regulatory and standard development activities for medical devices.

The standard ISO/ASTM TR 52,916:2022 [9] allows for improving medical image data, acquisition processing, and optimization techniques. As a result, more accurate solid medical models can be obtained from actual data. The stacked 2D image output from MIS allows for generating solid medical models. The following factors predetermine the final image's accuracy and resolution: a contrast between tissues of interest, image noise, and inherent artifacts.

The standard ISO/ASTM 52,950:2021 [11] allows for applying data exchange for additive manufacturing. It also determines definitions and terms and enables information to be exchanged by describing parts' geometry so that it can be manufactured additively. The data exchange technique outlines the file type and formatting of enclosed data.

The standard ISO/ASTM 52,915:2020 [12] provides the Additive Manufacturing File Format (AMF) specification, an interchange format, to address additive manufacturing technology's current and future needs.

Research into the causes [13] of eliminating errors in medical device additive manufacturing is ongoing. The reason for the accuracy error of AM occurs in the process of converting raw data to medical images and in the process of converting 3D model data.

Currently, to solve the problems of segmentation and classification of objects in images, which are in demand in intelligent information systems for medicine, deep neural networks (NMs) integrated into the architecture of such systems are increasingly used.

The standard ISO/IEC FDIS 3532-1:2023 [14] was developed in response to the need to customize 3D scanning and 3D printing technology within the medical industry, which can be achieved by taking full advantage of information and communication technology (ICT). This document addresses the overview of medical image processing and requirements for image-based modeling.

The standard ISO/IEC FDIS 3532-2:2023 [15] proposes a standardized process for segmentation optimization.

However, despite the effectiveness of projects built on deep neural networks of complex architecture, their developers and researchers faced a natural problem – an increase in computing resources spent on training them (according to OpenAI, there is a three hundred thousandth increase in training costs). The analysis showed that one of the effective ways to solve this problem is to reduce the time spent on training neural networks, which can be achieved by searching for new and improving existing optimization methods used in NM training. More common when training the deepest NMs are many methods based on the gradient optimization method. However, they all have a common drawback in the multi-purpose, multimodal, and noisy function conditions – "slow" convergence, which significantly increases the training time. To eliminate it, a modification of NM learning methods based on gradient descent and applying the nonlinear dynamics theory was proposed, and the corresponding theoretical provisions for implementing the methods were developed. The conducted analysis showed that several methods [16] and algorithms [17] have been presented in recent years which are used to solve these problems. In addition, it is essential to ensure the integrity of the medical image, especially the ROI, before making any diagnostic decisions. ROIs are used to define the area in which the features are calculated. What constitutes an ROI depends on the imaging and research objective. The work [18] uses a template-matching algorithm to find special symbols for locating the ROI.

This paper [19] presents a generalized framework for heterogeneous recurrence analysis of spatial data to investigate recurrence patterns and dynamical properties of complex systems in the spatial domain.

3 Research Methodology

Anatomically matched devices have very complex geometrical contours and shapes. Several challenges exist in the design process between the input data and the final device design. Most of these steps depend on the software-based management of medical images [20]. The images typically consist of a discretized voxel domain, with each voxel (3D pixel) providing some information about that location in space.

The fact that tissue types have different pixel intensities predetermines fundamental segmentation approaches [21]. In the examples of 2D and 3D spatial data, each pixel (or voxel) contains spatial location and attribute information.

This work defines an image as a three-dimensional (3D) stack of two-dimensional (2D) fragments of a digital image. Image fragments are stacked along the Z-axis. In addition, this stack is assumed to have the same coordinate system; the image slices do not rotate or move (in the xy plane) relative to each other. Moreover, digital images usually have a finite resolution. Thus, the intensities in the image are located at equal intervals or intervals [22]. In 2D, such regular positions are called pixels, while 3D uses voxels. Thus, the pixels and voxels appear as an intersection of a regular grid. Alternatively, pixels and voxels can be represented as rectangles and rectangular parallelepipeds (Fig. 1). Then, the centers of the pixels and voxels coincide with the intersections of a regularly spaced grid. Both representations are used in the document.

Pixels and voxels contain the intensity values for each channel of the image. The number of channels depends on the modality of visualization. Most medical imaging generates single-channel images, while the number of channels in microscopy can be higher, e.g., due to different coloring [23]. In such multi-channel cases, features can be extracted for each channel, a subset of channels, or the channels can be combined and converted into a single-channel representation. In the rest of the document, we treat the image as having only one channel. Pixel or voxel intensity is also called gray level or gray tone, especially in single-channel images. Although there is virtually no difference, the terms grayscale or grayscale are often used to denote discrete intensities, including discrete intensities.

Spatial data refer to the data in a spatial domain, either two-dimensional or three-dimensional. Two essential elements of the spatial data are (1) spatial reference (location information), denoted as $x = (x_1; x_2; \ldots x_d)$ and (2) a set of features or attributes that vary over the spatial reference, denoted as $s = (s_1; s_2; \ldots s_m)$. As shown in Fig. 1, a two-dimensional image consists of a set of pixels, in which each pixel i contains the location information $x_i = (x^{(i)}, y^{(i)})$, as well as the pixel value $s_i = (s_R^{(i)}, s_G^{(i)}, s_B^{(i)})$ (the values for R – red, G – green, and B – blue). Similarly, a three-dimensional object is composed of a set of voxels, where a voxel j possesses both the spatial reference, $x_j = (x^{(j)}, y^{(j)}, z^{(j)})$, and the voxel attribute, $s_j = (s_R^{(j)}, s_G^{(j)}, s_B^{(j)})$.

The ROI mask is a regular grid to scan entire rows at a time. However, despite the many image processing methods, there are not many differences in the image attribute information values of the anatomical human structures [9].

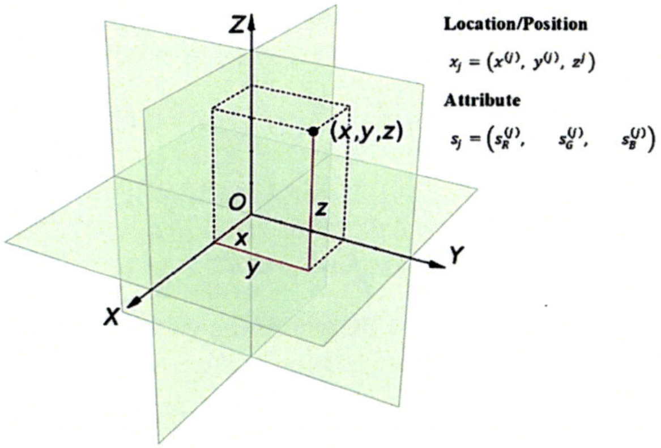

Fig. 1 A point in three-dimensional Euclidean space. 3D spatial data

3.1 Qualitative Solutions of Differential Equations Corresponding to the Pixel Dynamic System

The phase plane method is used to study nonlinear systems [24] described by differential equations of the first and second orders. It consists of the construction and study of the phase portrait of the system in the coordinates of the studied quantity and its derivative. Consider a dynamic system described by a model of two (or more) equations:

$$\begin{cases} \frac{dx}{dt} = P(x, y), \\ \frac{dy}{dt} = Q(x, y), \end{cases} \tag{1}$$

where $P(x, y)$ and $Q(x, y)$ are nonlinear functions with respect to x, y that are continuously differentiable in some region xOy (or in the entire plane).

To simulate such a system, it is necessary to use the phase plane method, which allows considering the system's behavior on the (x, y) plane. Each point of the phase plane defines the system's state at a given moment.

The movement of the configuration points $(x(t), y(t))$ along the phase plane (phase trajectory) in response to changes in the system. The set of phase trajectories sets the phase portrait of the system.

The differential equations of phase trajectories are of the form:

$$\frac{dx}{dy} = \frac{P(x, y)}{Q(x, y)} \text{ either } \frac{dy}{dx} = \frac{Q(x, y)}{P(x, y)} \tag{2}$$

Solving these equations, one can obtain the integral function of phase trajectories. The subsequent procedure is reduced to determining the type of facial points and their durability. For this, decouplings are substituted in the system (1) in the form:

$$x = x_1 + ae^{\lambda t}, y = y_1 + be^{\lambda t}, \tag{3}$$

where $a, b << 1$ are the coefficients; λ is the Lyapunov exponent. After substitution, we obtain the characteristic equations

$$\begin{cases} a \cdot f_1(x_1, y_1, \lambda) + b \cdot f_2(x_1, y_1) = 0, \\ a \cdot f_3(x_1, y_1) + b \cdot f_4(x_1, y_1, \lambda) = 0, \end{cases} \tag{4}$$

$$\begin{vmatrix} f_1(x_1, y_1, \lambda) & f_2(x_1, y_1) \\ f_3(x_1, y_1) & f_4(x_1, y_1, \lambda) \end{vmatrix} = 0. \tag{5}$$

Having solved the determinant, we determine the value of the Lyapunov exponent λ by the sign and value of which one can determine the type of stability of a personal point.

3.2 Qualitative Solutions of Differential Equations Corresponding to the Voxel Dynamic System

The study of the Lorentz system showed that it provides a kinetic picture of the collective behavior of a macroscopic system [25]. The Lorentz system goes to (7) if the dynamic variables X, Y, Z, and time t are replaced according to the relationship:

$$t' \equiv \sigma t, \eta \equiv \frac{X}{\sqrt{b}}, h \equiv \frac{Y}{\sqrt{b}}, S \equiv r - Z. \tag{6}$$

In this case, the Lorentz system is represented in the form

$$\dot{\eta} = -\eta + h, \sigma \dot{h} = -h + \eta S, (\sigma/b)\dot{S} = (r - S) - \eta h, \tag{7}$$

where h is the conjugate field to the synergetic order parameter η; S is the controlling parameter.

Equations (7) show that the parameter σ represents the ratio of the characteristic times of the change of the field h and the synergetic parameter η, the latter of which is taken as the mass scale of the time measurement $t\prime$; accordingly, the parameter b is reduced to the ratio of the characteristic times of the field change h and the control parameter S; finally, the parameter r determines the degree of external action that moves the system away from the equilibrium state.

In turn, relations (6) show that the synergetic order parameter η and the conjugate field h represent the dynamic variables X and Y, referred to the scale $\sqrt{b}$, and the control parameter S is reduced to the variable Z, counted from the threshold r in the opposite direction.

Within the framework of the synergistic approach, further studies of medical images are reduced to a self-consistent description of the time dependences of the conjugate field to the synergetic order parameter to it and the controlling parameter.

Modeling and visualization of numerous results are done in the Python programming language using the Anaconda Distribution software.

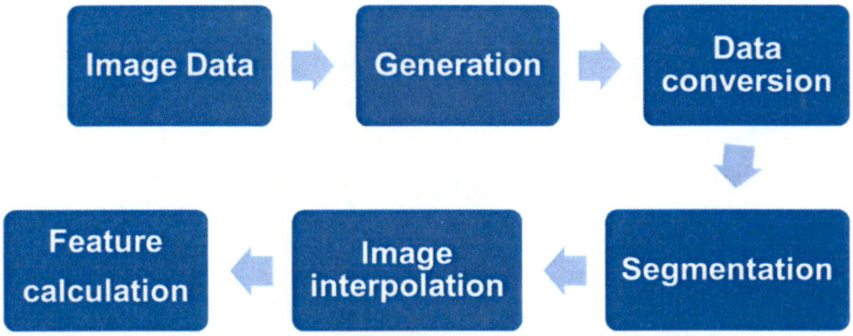

Fig. 2 Image processing scheme for calculating image features

4 Results and Discussion

The scheme of image processing is shown in Fig. 2.

The beginning of image generation (Fig. 3) is the collection of raw image data. It includes analyzing raw information about the inside of the human body and becomes the primary target for all subsequent stages of image processing. Regardless of the image format, the acquisition allows for assessing physical parameters, preprocessing signals, and digitizing.

Fig. 3 The beginning of image generation

After interpolating to the same grid as the interpolated image, the ROI is clearly divided into two masks, namely the intensity mask and the morphology mask. Some rendering techniques require the raw image data to be transformed into a more meaningful representation, e.g., standardized absorbance values. This is done in the data conversion step.

The images are post-processed to enhance the image quality. From a process perspective, segmentation (Fig. 4) results in the creation of an ROI mask R for which every voxel $j \in R(R_j)$ is defined as:

$$R = \begin{cases} 1, j \text{ in } ROI; \\ 0, \text{ otherwise.} \end{cases} \quad (8)$$

The sequence of partitioning an image into multiple labeled regions and locating objects and boundaries in images is considered as image segmentation.

Based on the captured metadata, reconstruction is performed using 3D visualization to extract the data of the region of interest (Fig. 5).

The interpolation algorithms convert the image intensity from the original image grid to the interpolation grid. In such networks, voxels are spatially represented by their

Fig. 4 Segmentation Methods

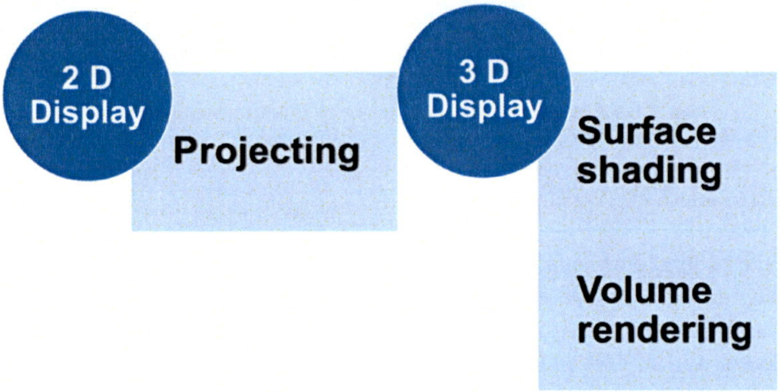

Fig. 5. 3D visualization to extract the data of the region of interest

center. After voxel interpolation, interpolated CT intensities are rounded to the nearest integer. CT matrix size is realized from 2D digital images when each pixel is converted into bits by gray levels. Feature calculation is the final processing step in which feature descriptors are used to quantify ROI characteristics. Once calculated, such characteristics can be used as imaging biomarkers, associating them with interesting physiological and medical outcomes.

5 Conclusions

Standardization of 3D printing processes in medicine is essential for many affected stakeholders. Continuous technological progress and medical imaging hardware and software improvements require constant reevaluation of the quantitative accuracy of medical image evaluation. The development of biomaterials, and medical device components, finished various international and national standards and guidelines regulating medical devices and 3D-printed and regenerative medicines. Quantitative imaging biomarkers have not yet impacted clinical practice due to a lack of comprehensive standardization in terms of technical aspects of imaging, analysis algorithms, processes, and clinical validation. The paper presents essential concepts and computational methods of volumetric visualization. Surface rendering algorithms for volumetric data are briefly described, in which an intermediate representation of the data is used to create an image of the surface contained in the data.

Standardization of image quality as input data for the analysis of different imaging biomarkers will never be fully achieved, and there is some risk of reaching an operational limit. This risk should be identified early enough so that corrective action can be taken, for example, by exploring new ways to reverse the heterogeneity trend by using application-based approaches of nonlinear dynamics in learning using gradient descent algorithms of artificial intelligence to enable complex and deep neural networks to learn from the lack of homogeneity in collected images, both in DICOM metadata and pixel information. The validation of the developed methods should be based on the validation of technical characteristics (precision and reliability assessment) and clinical validation.

The research results allow us to formulate recommendations for reducing the uncertainty in determining the composition of bone tissue substitutes hydroxyapatite and beta-tricalcium phosphate, associated with manifestations of biological variability, reflects the body's response to various environmental factors, and obeys statistical laws.

References

1. ISO/ASTM 52900:2021 Additive manufacturing – General principles – Fundamentals and vocabulary (2021)
2. Sukhodub, L.F., Dyadyura, K.: Design and fabrication of polymer-ceramic nanocomposites materials for bone tissue engineering. J. Nano- Electron. Phys. **10**(6), 06003 (2018)
3. Zaloga, V., Dyadyura, K., Rybalka, I., Pandova, I.: Implementation of integrated management system in order to enhance equipment efficiency. Manage. Syst. Prod. Eng. **27**(4), 221–226 (2019)
4. Zaloga, V., Dyadyura, K., Rybalka, I., Pandova, I., Zaborowski, T.: Enhancing efficiency by implementation of integrated management system in order to align organisational culture and daily practice. Manage. Syst. Prod. Eng. **28**(4), 304–311 (2020)
5. Dyadyura, K.O., Sukhodub, F.: Magnesium-based matrix composites reinforced with nanoparticles for biomedical applications. In: 2017 IEEE 7th International Conference Nanomaterials: Application & Properties (NAP), pp. 04NB14–1–04NB14–5 (2017). https://doi.org/10.1109/NAP.2017.8190327
6. Hrebenyk, L.I., Ivakhniuk, T.V., Sukhodub, L.F.: ZnS quantum dots encapculated with alginate: synthesis and antibacterial properties. In: 2017 IEEE 7th International Conference Nanomaterials: Application & Properties (NAP), pp. 04NB07–1–04NB07–7. IEEE, Piscataway, NJ (2017). https://doi.org/10.1109/NAP.2017.8190320
7. Pandová, I., Panda, A., Valíček, J., Harničárová, M., Kušnerová, M., Palková, Z.: Use of sorption of copper cations by clinoptilolite for wastewater treatment. Int. J. Environ. Res. Public Health **15**, 1364 (2018). https://doi.org/10.3390/ijerph15071364
8. Dyadyura, K., Hovorun, T.P., Pylypenko, O.V., Hovorun, M.V., Pererva, V.I. Influence of roughness of the substrate on the structure and mechanical properties of TiAlN nanocoating condensed by DCMS. In: 2017 IEEE 7th International Conference Nanomaterials: Application & Properties (NAP), pp. 01FNC10-1–01FNC10-4 (2017). https://doi.org/10.1109/NAP.2017.8190202
9. ISO/ASTM TR 52916:2022 Additive manufacturing for medical – Data – Optimized medical image data (2022)
10. Rafi, K., Zhonghong Liu, A., Di Prima, M., et al.: Regulatory and standards development in medical additive manufacturing. MRS Bull. **47**, 98–105 (2022)
11. ISO/ASTM 52950:2021 Additive manufacturing – General principles – Overview of data processing (2021)
12. ISO/ASTM 52915:2020 Specification for additive manufacturing file format (AMF). Version 1.2 (2020)

13. Vora, H.D., Sanyal, S.: A comprehensive review: metrology in additive manufacturing and 3D printing technology. Prog. Addit. Manuf. **5**, 319–353 (2020)
14. ISO/IEC FDIS 3532–1:2023 Information technology – Medical image-based modelling for 3D printing – Part 1: General requirements (2023)
15. ISO/IEC DIS 3532–2:2023 Information technology – 3D Printing and scanning – Medical image-Based modelling – Part 2: Segmentation (2023)
16. Schuh, J.C.L., Funk, K.A.: Compilation of international standards and regulatory guidance documents for evaluation of biomaterials, medical devices, and 3-D printed and regenerative medicine products. Toxicol. Pathol. **47**(3), 344–357 (2019)
17. Chen, H., Geng, L., Zhao, H., et al.: Image recognition algorithm based on artificial intelligence. Neural Comput. Appl. **34**, 6661–6672 (2022)
18. Wu, Y., Wang, D.H., Lu, X.T., et al.: Efficient visual recognition: a survey on recent advances and brain-inspired methodologies. Mach. Intell. Res. **19**, 366–411 (2022)
19. Pavan Kalyan, B., Kumar, L.: 3D printing: applications in tissue engineering, medical devices, and drug delivery. AAPS Pharm. Sci. Tech. **23**, 92 (2022)
20. Evans, L.M., Sözümert, E., Keenan, B.E., et al.: A review of image-based simulation applications in high-value manufacturing. Arch. Comput. Meth. Eng. **30**, 1495–1552 (2023)
21. Rybicki, F.J.: The impact of regulation, reimbursement, and research on the value of 3D printing and other 3D procedures in medicine. 3D Print Med. **8**, 6 (2022)
22. Yang, H., Chen, C.B., Kumara, S.: Heterogeneous recurrence analysis of spatial data. Chaos **30**(1), 013119 (2020)
23. Lin, G.S., Chai, S.K., Li, H.M., et al.: Vision-based patient identification recognition based on image content analysis and support vector machine for medical information system. EURASIP J. Adv. Signal Process. **2020**, 27 (2020)
24. Touzé, C., Vizzaccaro, A., Thomas, O.: Model order reduction methods for geometrically nonlinear structures: a review of nonlinear techniques. Nonlinear Dyn. **105**, 1141–1190 (2021)
25. Cartas-Fuentevilla, R., Juarez-Dominguez, A.J.C.: An internal Lorentz symmetry induces the background Lorentz symmetry in the dissipative dynamics. Eur. Phys. J. Plus. **138**, 217 (2023)

Experimental Studies on the Form Error Effect of the Part Mounting Surface on the Strength Quality Parameter of the Interference Fit Joints

Oleksandr Kupriyanov[1](✉), Roman Trishch[2], Dimitar Dichev[3], and Hanna Hrinchenko[1]

[1] Ukrainian Engineering Pedagogics Academy, 16, Universitetskaya St., Kharkiv 61003, Ukraine
a_kupriyanov@uipa.edu.ua

[2] National Aerospace University «Kharkiv Aviation Institute», 17, Chkalov St., Kharkiv 61070, Ukraine

[3] Technical University of Gabrovo, 4, H. Dimitar St., Gabrovo 5300, Bulgaria

Abstract. An experimental study was performed to investigate the effect of macrogeometric parameters of the mating surface forms on the strength of the cylindrical interference fit joint: taper, bow, barrel, and ovality. Fifteen couplings were made with a nominal diameter of 60 mm and a mating surface length of 70 mm. The bushings had a regular cylindrical form, and the CNC machining of shafts caused the geometry error. The relative geometric accuracy of the form was assumed to be the maximum allowable and was up to 60% of the tolerance zone. During modeling, the minimum allowable interference fit in the coupling Ø60 H8/u8 was assumed to be 0.04 mm to enhance the effect of form geometry errors. Ranked assembly was used to compensate for the interference fit in joints. The joints were assembled using thermal assembly. The strength was determined by axial pressing on the press with the press diagram recorded. The experiment results showed that form geometry error with maximum allowable values significantly affects the strength of the cylindrical interference fit. The taper reduces the strength to 0.87, barrel to 0.8, bow to 0.59, and ovality to 0.79 of coupling with perfect cylindrical surface geometry. Comparison of experimental results with those previously obtained by finite element modeling showed good agreement, within 10%. It is recommended that for critical couplings, a stricter limitation of the geometric error of the mating surfaces shall be provided.

Keywords: Sustainable Manufacturing · Interference Fit Joint · Form Deviation · Mating Surface · Strength · Pressing · Product Innovation

1 Introduction

Interference fit joints are widely used in industry. Their main quality parameters are direct tensile strength and rotational strength. The interference value is known to be the most significant effect on the strength of such a joint. Other parameters that have a significant effect include [1]:

V. Tonkonogyi et al. (Eds.): InterPartner 2023, LNME, pp. 369–378, 2024.
https://doi.org/10.1007/978-3-031-42778-7_34

- material and characteristics of the surface layer of mating components;
- microgeometry of mounting surfaces – roughness and its direction;
- available intermediate media – lubricants, adhesive layers.

Targeted experimental studies of the effect on the strength of macrogeometric parameters of mating surfaces, such as form geometry error, have not been found. It can be explained by the fact that the specialized execution of such studies requires precision machining of parts. Theoretical calculations in this direction are also insufficient.

The presence of a form geometry error can have a significant effect on the interference fit strength [2]. Experimental studies are required to verify the results obtained theoretically or by modeling.

These studies aim to determine the effect of form geometry errors on the strength of a cylindrical interference fit joint.

2 Literature Review

The strength calculation of interference fit joints is still often performed, assuming that mounting surfaces have precise dimensions and a perfect form. A single local measurement determines the interference fit value. In addition to the interference fit value, the main focus is on the geometric error at the micro level - the roughness.

Studies show that form defects have no negative effect on the joint strength when the joint material is in the elastic and the elastic-plastic zone [3]. The authors suggest that the possibility of permissible shape defects can be used to reduce the cost of manufacturing products.

Methods for calculating the strength of interference fit using mathematical equations do not consider the specific geometry of real components but are often used for ordinary cases. However, most engineers have moved to finite element calculations in commercial calculation programs. In [1], stresses in the shaft-hub assembly were studied for different diameters and lengths for transition fits. The occurrence and distribution of the plastic deformation zone were evaluated.

The strength of cylindrical interference fits is affected by various factors such as diameter clearance, assembly type, material type, operating temperature, load type, surface roughness, etc. In a study [4], the effect of clearances and different fits on strength was experimentally studied. The expected study result was that a significant reduction in strength values was observed when the clearance in the joint was increased. When an interference fit and an adhesive joint were used, the strength samples were approximately five times higher than when an interference fit alone was used.

When designing a product, the main objective of tolerance setting is to find a balance between manufacturing cost and product geometric quality. Stress calculations for interference fit designs are typically evaluated using traditional Lamé equations. The finite element method for stress analysis of interference fit joints provides more complete and accurate results than the traditional method. In addition to mounting surface characteristics, the joint strength is also affected by the geometric characteristics of joined components [5]. It is well known that for thin-walled rings, the strength is reduced.

Theoretical calculations provide a high, up to 18%, error in predicting the interference fit strength. Nevertheless, the authors have applied the finite element method results to produce an adequate theoretical model to consider the effect of non-contacting areas [6]. Prediction from experimental data rarely applies to products made of other materials and with other surface finishes, as the friction factor is difficult to predict. The paper stresses that only the press-fit diagram used since the beginning of the last century provides reliable strength results.

Earlier models implied perfect surface models and deviations, which did not consider form defects. In [7], new calculation models are considered: nominal dimensions, deviations from the perfect surface, and skin model shape. Considering increasing quality requirements, the surface defect model has advantages.

There is always some uncertainty in the friction factor, both in the theoretical prediction of the interference fit strength and in finite element modeling. An approximate value for the friction factor in contact areas is often assumed when modeling an interference fit for known materials. It isn't easy to accurately estimate this factor based on intuition, engineering judgment, or previous testing. The approach in [8] is to derive the friction factor from experimental data and then apply it to boundary conditions in the finite element method. In [9], the friction factor was determined by considering the surface roughness of the components and by introducing a non-classical hardening friction law that considers tangential micromovements of sliding.

Interference fit joints are used in pure form and reinforced with bolt clamping. In [10], numerical and experimental fatigue tests were performed on an aluminum alloy joint. The fatigue test results showed that the best improvement in fatigue life was achieved using a combination of a smaller interference fit and a larger bolt clamping force.

In [11], an elastic model with transverse isotropy was used for the calculation. In [12], the standard generic interference fit is discussed by analyzing stresses in the interference fit of the hole and axis, and the concept of allowable shear stress of the interference fit is analyzed. Interference fit joints are sometimes used with cylindrical keyed joints [13].

Increased joint strength can be achieved by increasing the interference and using adhesive interlayers such as anaerobic acrylics [14], epoxy resins, and polyurethanes [15]. The highest strength is achieved by combining several methods to increase the fit strength. High strength is achieved if necessary so that the product cannot be disassembled without mounting surface damage.

A multifactorial experiment is recommended for experimental investigations. The joint strength can be determined by axial shear or torsional strength. The fracture energy of a joint is composed of a frictional and a cohesion component [16]. The joint strength is linearly dependent on the contact pressure.

3 Research Methodology

3.1 Experimental Joints

This study deals with cylindrical interference fit joints. Regulatory documentation specifies a set of deviations of the whole surface (cylindricity tolerance), its individual sections, or individual geometric elements, regardless of the real surface form. To investigate the effect of the real surface form, we studied the effect of previously used particular types of form deviations on joint strength. These include cross-sectional profile deviations: taper, barrel, bow, and roundness deviations: ovality.

The relative geometric accuracy of the form can range from 60% of the tolerance range for normal (A) to less than 25% for exceptionally high geometric accuracy. If the form error in the interference fit assembly is not normalized, it can correspond to the normal geometric accuracy.

One of the objectives was to verify the modeling results obtained in [2]. Therefore, the dimensions, fit, and form geometry errors were chosen accordingly. The nominal fitting diameter of the joints was assumed as Ø60. The outside diameter of bushings in the finishing experiment was Ø90 mm, 70 mm long, excluding bushing No. 15, Ø120 mm, 80 mm long. The joint fitting diameter is because, with smaller values, the required heating temperature during thermal assembly increases considerably. High values increase the metal consumption of test pieces and, consequently, their cost.

The fit Ø60 H8/u8 was selected for the experiments. For the 8th quality grade, the normal (A) relative geometric accuracy corresponds to the 7th degree of accuracy. For Ø60, the form geometry error tolerance is 16 mm per side, 32 mm per diameter. During modeling, the minimum allowable interference fit in the joint was assumed to be 0.04 mm to enhance the effect of form geometry errors.

The coarsest of the fits used for interference fits was chosen because of the accuracy limitations of the available machining equipment. The form error on test specimens had to be produced within the tolerance zone of the fit; with higher accuracy, it was difficult to achieve the required form error specifically.

The CNC lathe machine artificially created the form error on test roll surfaces. The mounting surfaces of the bushings were polished. The parts were made of steel 45, and the roughness of the mounting surfaces was Ra0.8.

The choice of sample materials was based on the requirement that they correspond in their physical and mechanical properties to the materials used for interference fit joints. Steel 45 was selected for experimental samples as its magnetic properties and thermal conductivity correspond to the material properties of the real components, and it is well-machined.

Experimental samples are shown in Fig. 1. Form geometry errors are graphically enlarged for clarity. Bushings were assumed cylindrical (Fig. 1a), and the form geometry error was produced on the shafts. The cylindrical shaft (Fig. 1b) was compared with taper, bow, barrel, and oval shafts (Fig. 1c – Fig. 1f). Shanks were provided for the shafts to make it easier to extrude during the practical experimental test.

3.2 Experiments

During the experiment, 15 joints were manufactured, measured, assembled, heat-assembled, and then pressed, three of each type: no geometry error, taper, bow, barrel, and oval. The strength during axial pressing was studied.

An optimeter with a scale value of 0.001 mm was used to measure the outside diameter. The inner diameter was measured using an indicator-head gauge with a graduation value of 0.001 mm. The roughness value was measured with a double microscope. Ensuring the measurements' quality requires considering all involved factors [17]. The strength was determined by pressing experimental joints using a P50 press. The press diagram was recorded on millimeter paper using the system supplied with the press.

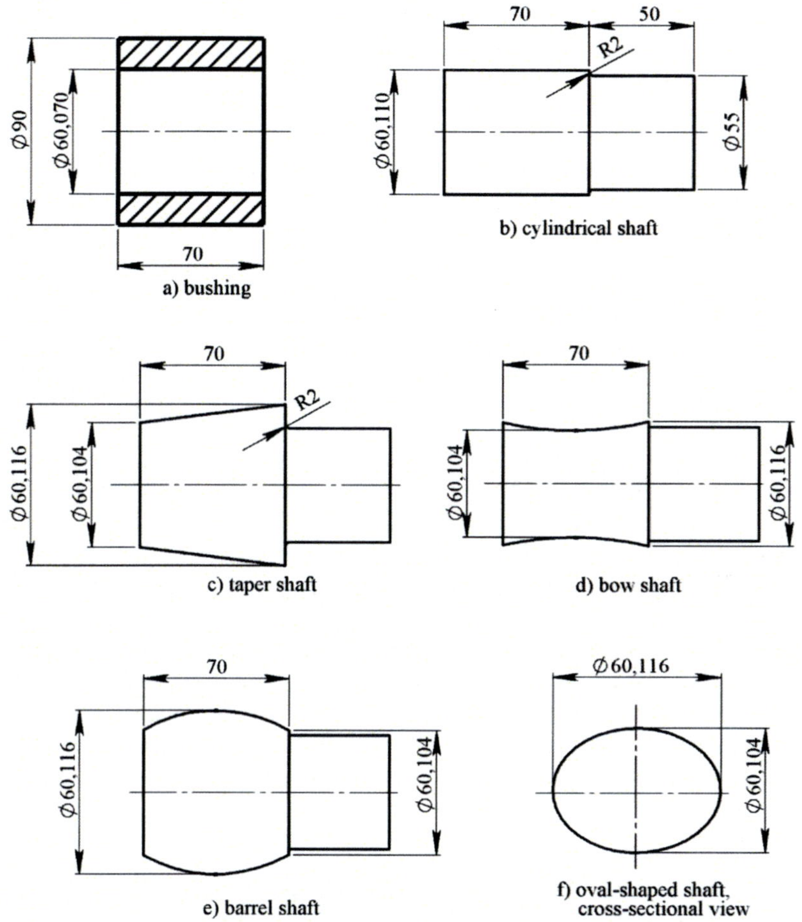

Fig. 1. Experimental samples.

Since the experiment aimed to determine the effect of form geometry errors on strength properties, all other parameters of test pieces, such as material, roughness, and finishing, were identical. However, during the fabrication of experimental specimens, it became clear that it was impossible to equalize the interference fit using available machining methods completely. At the same time, the interference fit value has a much more significant effect on joint strength than other factors.

For this reason, we used the ranking-based picking procedures suggested in [18, 19] for interference fit alignment. Two options were considered:

1. To pick all 15 sets, having a span of interference $N_{\min} = 0.025$; $N_{\max} = 0{,}041$.
2. To divide into two groups, pick the "best" 11 kits separately, having $N_{\min} = 0.03$; $N_{\max} = 0.43$, and separately the "rest" 4 sets.

It chose the second option, with the "best" samples having an average interference deviation of less than 0.01 mm from the nominal one. Those with a slightly lower accuracy of geometry error on the shafts were considered as the 'remainder'.

4 Results and Discussion

A photo of joints after thermal assembly is shown in Fig. 2a, after pressing in Fig. 2b. The characteristics and experimental strength of experimental joints are shown in Table 1, and relative figures as a diagram in Fig. 3. In Table 1, the number 1 represents shafts without geometrical errors, 2 represents shafts with the tapered mounting surface, 3 represents shafts with bow fit, 4 represents shafts with barrel fit, and 5 represents shafts with the oval fit. The second number is the number of the experimental shaft with the specified error. The bushings are numbered in order.

If we assume the joint strength without form geometry error as 1, the highest reduction in strength, to 0.59, was shown by joints with bow form. The taper showed a strength of 0.87. The barrel and ovality showed approximately equal strengths, 0.8 and 0.79, respectively.

Also, in Table 1, the last column shows the effect of form geometry error on the axial strength of interference fit joints obtained by finite element simulations in Ansys [2]. The results obtained by theoretical modeling and experimentally agree within 10%. The exception to this is the barrel form, where experimental results were 25% lower than theoretical results. It can be explained by the fact that the interference fit of experimental joints with barrel form was lower (0.0372 mm and 0.0303 mm) than the average interference fit of 0.04 mm. As we know, the interference fit value has the most significant effect on joint strength.

a)

b)

Fig. 2. Experimental joints before and after pressing.

The experiment has both strengths and weaknesses. The strengths include using a ranking-based picking procedure of parts in the kits, which made it possible to equalize the interference. As shown in Table 1, the deviations from the average interference were

Table 1. Picking results and pressing the strength of experimental samples.

Shaft	Bushing	Average shaft diameter, mm	Average bushing diameter, mm	Interference fit, mm	Deviation from 0.04	Strength, kN	Average strength, kN	Relative strength, experiment	Relative strength, modeling
1.1	9	60.112	60.070	0.0427	0.0027	157	166.5	1	1
1.3	7	60.113	60.074	0.0392	–0.0008	176			
2.1	14	60.110	60.067	0.0432	0.0032	137	144.5	0.87	0.8
2.2	3	60.111	60.069	0.0423	0.0023	152			
3.1	4	60.112	60.069	0.0432	0.0032	107	99	0.59	0.64
3.2	10	60.118	60.078	0.0393	–0.0007	103			
3.3	5	60.117	60.076	0.0408	0.0008	87			
4.1	12	60.104	60.067	0.0372	–0.0028	136	134	0.80	0.98
4.2	2	60.097	60.067	0.0303	–0.0097	132			
5.1	1	60.114	60.074	0.0393	–0.0007	115	131	0.79	0.82
5.2	6	60.111	60.067	0.0433	0.0033	147			
			Average	0.0401	7.6E-05				
Residual									
2.3	15	60.091	60.053	0.0375	–0.0025	244			
1.2	11	60.092	60.072	0.0198	–0.0202	129			
4.3	8	60.094	60.075	0.0193	–0.0207	106			
5.3	13	60.101	60.080	0.0210	–0.0190	95			
			Total	0.0359	–0.0041				

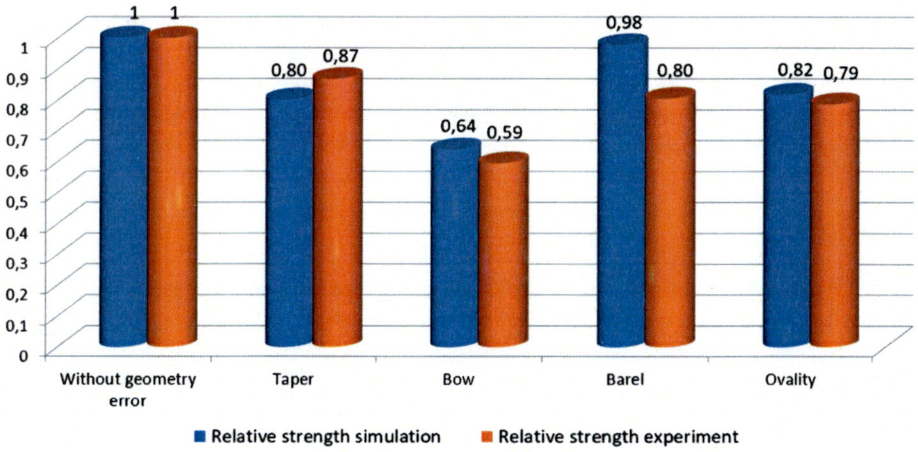

Fig. 3. Comparative results for axial shear strength of couplings with geometric error.

smaller than those in the "best" kits. It eliminated the effect of interference, one of the most significant strength factors, on the experiment results. Another positive point is the use of precision machining of the surfaces, which simulates the shape geometry errors as closely as possible.

The weakness of the experiment is the small number of experimental kits: Originally, there were three of each type of geometry error, and after selecting the "best" in the four groups remained two. Such a volume of experiments cannot be considered quite sufficient. The authors can explain this by the high cost of experimental samples, especially their precision machining.

In the experiments, in addition to artificially created geometry errors, the shape was also obtained naturally, due to inaccuracies in machining, so the processing of the experimental data and the total error for the kits were evaluated. So, the experiment had some scattering relative to the actual values.

In further research, although the amount of experimental data obtained is small, it is of practical use to establish the best functional dependencies for the analytical representation of the data [20].

5 Conclusions

In the article, a very precise experimental study of the influence of the form error of the part mounting surface on the strength quality parameter of the interference fit joints was carried out. The experiments have shown that geometric errors significantly affect the strength of the cylindrical interference fit. The most significant reduction in strength, up to 40%, is due to a bow. The taper reduces strength by 13%, and barrel shape and ovality have approximately the same effect, about a 20% reduction.

The results obtained by Ansys finite element modeling are in good agreement (within 10%) with the experimental data. It proves that using Ansys modeling for other sizes, materials, and form errors, finite element modeling with the proposed friction factor calculation can be used to estimate the interference fit strength.

Introducing a stricter standard for errors in the form geometry in manufacturing interference fit parts for critical couplings would be advisable. In this case, the relative geometric accuracy of the form should be less than 25% of the tolerance range.

References

1. Sen, S., Aksakal, B.: Stress analysis of interference fitted shaft–hub system under transient heat transfer conditions. Mater. Des. **25**(5), 407–417 (2004). https://doi.org/10.1016/j.matdes.2003.11.009
2. Kupriyanov, O.: Ansys simulation of the joint strength with the interference fit in the presence of the shape geometry error. In: Ivanov, V. et al. (eds.) Advances in Design, Simulation and Manufacturing. DSMIE 2018. Lecture Notes in Mechanical Engineering, pp. 72–78. Springer, Cham (2019). https://doi.org/10.1007/978-3-319-93587-4_8
3. Boutoutaou, H., Bouaziz, M., Fontaine, J.F.: Modeling of interference fits taking form defects of the surfaces in contact into account. Mater. Des. **32**(7), 3692–3701 (2011). https://doi.org/10.1016/j.matdes.2011.03.059
4. Sekercioglu, T., Gulsoz, A., Rende, H.: The effects of bonding clearance and interference fit on the strength of adhesively bonded cylindrical components. Mater. Des. **26**(4), 377–381 (2005). https://doi.org/10.1016/j.matdes.2004.05.016

5. Mucha, J.: Finite element modeling and simulating of thermomechanic stress in thermocompression bondings. Mater. Des. **30**(4), 1174–1182 (2009). https://doi.org/10.1016/j.matdes.2008.06.026
6. Wang, X., Lou, Z., Wang, X., Xu, C.: A new analytical method for press-fit curve prediction of interference fitting parts. J. Mater. Process. Technol. **250**, 16–24 (2017). https://doi.org/10.1016/j.jmatprotec.2017.06.022
7. Yan, X., Ballu, A: Tolerance analysis using skin model shapes and linear complementarity conditions. J. Manuf. Syst. **48**(A), 140−156 (2018). https://doi.org/10.1016/j.jmsy.2018.07.005
8. Seifi, R., Abbasi, K.: Friction coefficient estimation in shaft/bush interference using finite element model updating. Eng. Fail. Anal. **57**, 310–322 (2015). https://doi.org/10.1016/j.engfailanal.2015.08.006
9. Buczkowski, R., Kleiber, M.: A study of the surface roughness in elasto-plastic shrink fitted joint. Tribol. Int. **98**, 125–132 (2016). https://doi.org/10.1016/j.triboint.2016.02.021
10. Chakherlou, T.N., Abazadeh, B.: Experimental and numerical investigations about the combined effect of interference fit and bolt clamping on the fatigue behavior of Al 2024–T3 double shear lap joints. Mater. Des. **33**, 425–435 (2012). https://doi.org/10.1016/j.matdes.2011.04.045
11. Boutoutaou, H., Bouaziz, M., Fontaine, J.-F.: Modelling of interference fits with taking into account surfaces roughness with homogenization technique. Int. J. Mech. Sci. **69**, 21–31 (2013). https://doi.org/10.1016/j.ijmecsci.2013.01.012
12. Yao, H.X., Miao, E.M., Niu, P.C.: Selection of hole and axle interference fit tolerance. Appl. Mech. Mater. **80–81**, 475–479 (2011). https://doi.org/10.4028/www.scientific.net/AMM.80-81.475
13. Leonov, O., Shkaruba, N., Vergazova, Y., Khasyanova, D.: Justification of Keyed Joint Fits. J. Mach. Manuf. Reliab. **51**, 548–553 (2022). https://doi.org/10.3103/S1052618822060073
14. Castagnetti, D., Dragoni, E.: Experimental assessment of a micro-mechanical model for the static strength of hybrid friction-bonded interfaces. J. Adhes. **89**(8), 642–659 (2013). https://doi.org/10.1080/00218464.2012.747179
15. Gallio, G., Lombardi, M., Rovarino, D., Fino, P., Montanaro, L.: Influence of the mechanical behaviour of different adhesives on an interference-fit cylindrical joint. Int. J. Adhes. Adhes. **47**, 63–68 (2013). https://doi.org/10.1016/j.ijadhadh.2013.09.021
16. Ragni, M., Castagnetti, D., Dragoni, E.: Experimental validation of a simple shear strength model for hybrid friction-bonded interfaces. Int. J. Adhes. Adhes. **83**, 130–136 (2018). https://doi.org/10.1016/j.ijadhadh.2018.02.026
17. Dukendjiev, G., Diakov, D., Nikolova, H., Vassilev, V., Miteva, R.: Evaluation the performance of stationary coordinate-measuring systems with MSA methodology. In: 2022 XXXII International Scientific Symposium Metrology and Metrology Assurance (MMA), pp. 1−5. IEEE, Sozopol, Bulgaria (2022). https://doi.org/10.1109/MMA55579.2022.9992579
18. Kupriyanov, O., Lamnauer, N.: Improvement of the assembling technology for precision joints using the dimensional information. In: Tonkonogyi, V. et al. (eds.) Advanced Manufacturing Processes II. InterPartner 2020. Lecture Notes in Mechanical Engineering, pp. 52−60. Springer, Cham (2021). https://doi.org/10.1007/978-3-030-68014-5_6
19. Zhang, Y., Yang, M.S.: A coordinate SPC model for assuring designated fit quality via quality-oriented statistical tolerancing. Comput. Ind. Eng. **57**(1), 73–79 (2009). https://doi.org/10.1016/j.cie.2008.12.006

20. Lytvyn, O., Nechuiviter, O., Pershyna, I., Mezhuyev, V.: Input information in the approximate calculation of two-dimensional integral from highly oscillating functions (irregular case). In: Chertov, O., Mylovanov, T., Kondratenko, Y., Kacprzyk, J., Kreinovich, V., Stefanuk, V. (eds.) Recent Developments in Data Science and Intelligent Analysis of Information. ICDSIAI 2018. Advances in Intelligent Systems and Computing, vol 836, pp. 365–373. Springer, Cham (2019). https://doi.org/10.1007/978-3-319-97885-7_36

Improvement of Professional Competence of General Education Teachers for Engineering Curriculum

Olena Titova[1,2](✉), Petro Luzan[2], Tetiana Ishchenko[3], Maryna Kabysh[2], and Dmytro Homeniuk[2]

[1] Dmytro Motornyi Tavria State Agrotechnological University, 18, B. Khmelnytskyi Ave., Melitopol 72312, Ukraine
olena.titova@tsatu.edu.ua

[2] Institute of Vocational Education of the National Academy of Educational Sciences of Ukraine, 98a, Vito-Lithuanian Lane, Kyiv 03045, Ukraine

[3] State Institution "Scientific and Methodical Centre for Higher and Pre-Higher Education", 11, Smilianska St., Kyiv 02000, Ukraine

Abstract. The paper presents the results of the research on the problem of the purposeful development of professional skills of teachers in engineering education. The current research aimed to determine the learning conditions that could ensure the improvement of the pedagogical competence of general disciplines teachers at engineering and technical schools. The research was performed in two stages. The first stage was devoted to scientific publications, reports, and survey analysis to determine the relevant learning conditions. In the second stage, to determine the most significant learning conditions for the effective development of the general education teachers' professional skills, the initial list of learning conditions was examined and optimized by a group of 10 experts. To assess the unity of experts' opinions, the criterion of concordance was calculated. Thus, the experts determined the most significant circumstances predicted to develop general education teachers' pedagogical competence. The learning conditions were also ranked according to the degree of their influence on improving the general education teachers' professional skills. Creating such conditions could provide the teachers with continuous professional development for the efficient engineering students' training.

Keywords: Learning Conditions · Teaching Skills · General Disciplines · Engineering Students · Sustainability Education · Technical Schools · Education Policy

1 Introduction

Analyzing the role of general education for engineering students, scientists should consider that the goal of university education is not just training the students for specific purposes in their future job but equipping them with some universal skills they will need

V. Tonkonogyi et al. (Eds.): InterPartner 2023, LNME, pp. 379–388, 2024.
https://doi.org/10.1007/978-3-031-42778-7_35

in their lives [1]. To be successfully employed, engineering graduates should be able to analyze the situation, define a problem, communicate, propose a solution, and be a part of the team. Therefore, the teachers of general education disciplines have to organize an environment that would widen and deepen students' creative and professional potential for their engineering students. Since general course serves as the common starting point for engineering students, the teachers are responsible for such things as student learning enhancement, engagement in a professional atmosphere as well as interdisciplinary collaboration, treatment engineering beyond a technical solution, development of a creative learning environment which will allow them to think in a non-standard way for the decisions and consequences. Thus, general education teachers have to be well prepared for the function to organize students' effective individual work and work in a team [2], help them to identify and understand the knowledge and skills essential to succeed, show engineering students diverse paths in the field and engineers' larger role [3]. That requires implementing scientific approaches to the purposeful development of teaching skills for the teachers of general education disciplines and creating such circumstances that enable their continuous professional development for engineering students' high-quality and productive training.

2 Literature Review

The analysis of publications was focused on finding an answer to the question: what should be the process of the teacher's pedagogical skills development? The general answer could be retrieved from UNESCO teachers' training mission of creating and using knowledge for "inclusive, resilient and democratic societies" [4]. That requires teachers' treatment of the question of their professional competence as an integrated thing [5] that will provide sustainable development of society [6]. Therefore should be critically analyzed and supported with advanced theoretical concepts [7] and practical tools [8]. A great deal of research was devoted to the teacher's competence in engineering teacher training [9] as well as general education [10]. The findings referred to the approaches to the development of active learning context and competencies necessary for active learning [11], effective digital technologies application for engineering students [12], factors that influence online teaching competence [13], competence methods application in engineering education [14], innovation competencies as well as the ability to motivate students' innovativeness [15] and organize engineering students' outcomes assessment [14, 16]. Another vital aspect under study was the teachers' training to succeed in an inclusive classroom [17, 18].

For implementation in practice, the conclusions and provisions of the theory are usually formulated in the form of conditions (requirements) and rules. To solve practical problems, scientists are increasingly focusing on developing learning conditions [19] to influence individuals' acquirement of relevant skills [20]. The novelty of the current research was in analyzing the essence, meaning, and mechanisms of determining some learning conditions for an effective increase in the quality of the pedagogical activity of general education disciplines teachers at engineering schools. Here, the studied learning conditions were considered as the circumstances that stipulated specific directions of the development of the educational process and were created by a complex of systematic scientific and methodical measures and were predicted to provide improvement of professional competence of general education teachers at engineering schools.

It is worth agreeing with scientists [21, 22] to solve a scientific problem in education. One should consider a multitude of circumstances (dozens of learning conditions), but justify and experimentally verify only those with a specific scientific novelty. Thus, the current research aimed to determine the learning conditions that could ensure the growth of the pedagogical competence of general disciplines teachers at engineering and technical schools.

3 Research Methodology

Determination of the set of circumstances contributing to the development of teachers' competence for general education disciplines was carried out in two stages. In the first stage, a list of the most significant learning conditions was determined by analyzing scientific publications, reports, and surveys.

In the second stage, to determine the most significant conditions for the development of the teaching skills of general education teachers, the initial list of learning conditions was examined by experts and optimized. The group of experts included respondents who were chosen considering such requirements to an expert as a relatively independent administrative position, experience of practical work in the system of education, experience of scientific work, absence of irritability and pessimism, efficiency, confidence in judgments, natural curiosity, real-life position, natural sincerity, strict observance of morality, a sense of social responsibility. At the same time, the experience of expert evaluation was the primary criterion for selecting expert group members. The expert group of 10 people included specialists with experience in developing teachers' professional competence (scientists and educators involved in the training programs for future general education teachers, university and college managers, and engineering teachers).

For the expertise, an expert form was drawn up. It contained a list of proposed learning conditions and detailed instructions on how to work with them. The experts were asked to choose 5 of the most critical conditions among 19 determinants, which, in their opinion, have the most significant impact on the development of teaching skills of general education teachers.

The coefficient of importance for each of the learning conditions for the development of the teaching skills *the formula determined Mn*:

$$M_n = \frac{n_i}{m}, \tag{1}$$

where n_i is the number of experts who voted for the condition, m is the number of experts who participated in the study, and $m = 10$.

According to the results of the expert assessment, the most significant primary learning conditions were determined in the second stage. When the conditions were ensured, there would be a high probability of increasing the level of development of the teacher's professional competence. The list of such circumstances included conditions that had the weighting factor $M_n > 0.2$. The unity of experts' opinions was assessed using the criterion of concordance [23]. It is necessary to add that at this research stage. The authors considered the experts' comments regarding the formulation of the learning conditions. In order to determine the degree of influence (significance), the experts also ranked the learning conditions defined above.

4 Results and Discussion

Therefore, based on the results of the first stage of the learning conditions determination, a list of 19 conditions was compiled. The defined learning conditions contained circumstances that increased the probability of developing different teaching competence components effectively. At the same time, some conditions affected the same factors and dominated the circumstances of the development of individual components of teaching competence.

Examining the conditions at the second stage made it possible to identify 9 the most critical conditions that enabled the effective development of teachers' professional skills. Such many conditions were statistically significant for the research.

Thus, the experts attributed the following to the essential learning conditions for the development of teaching skills of general education teachers at engineering and technical schools:

1. Purposeful formation of positive motivation for teachers' self-development and self-improvement.
2. Priority of advanced learning technologies with equal access of all participants of the educational process to the technologies and resources.
3. Access to a flexible system of teachers' professional development.
4. Systematic mastery learning for teachers of general education disciplines.
5. Ensuring the integration of the content of general education subjects with the disciplines of general professional, as well as professional (engineering) theoretical and practical training.
6. Involvement of general education teachers in regular scientific and pedagogical activities.
7. Developing a creative educational environment is a factor for enhancing the creative abilities of all participants in the educational process.
8. Active interaction of the engineering education institutions with scientific institutions, educational and methodological centers, and other universities regarding implementing joint educational and scientific projects, organization of competitions, webinars, trainings, conferences, etc.
9. Expert and consultative support of general education teachers in self-educational development.

According to the instructions, each expert had to rank the learning conditions according to the degree of their importance for the development of the teaching skills of general education teachers (from 1 as the least important to 9 as the most important). It was indicated that if the conditions were equal regarding influence, the expert assigned them the same rank (Table 1).

As some conditions were assigned the same ranks, the "connected" (the same) ranks needed some clarification for the further calculation of the criterion of the unity of

Table 1 Expert survey data on the importance of pedagogical conditions

Condition No	Expert									
	I	II	III	IV	V	VI	VII	VIII	IX	X
1	7	8	4	3	5	7	5	6	7	3
2	3	2	3	1	1	2	2	3	3	1
3	1	1	2	1	2	3	2	1	1	1
4	2	1	1	1	2	1	1	2	1	1
5	4	3	3	2	1	4	2	3	2	2
6	8	7	7	4	4	6	4	8	8	3
7	3	4	5	5	3	5	3	4	4	2
8	5	6	8	6	6	8	6	5	5	4
9	6	5	6	7	7	9	7	7	6	4

experts' opinions (the criterion of concordance). It was because if there were no connected ranks, their sum would equal 45 ($1 + 2 + 3 + 4 + 5 + 6 + 7 + 8 + 9 = 45$). On the other hand, the sum of ranks obtained according to the first expert's data was smaller and equal to 44 ($7 + 3 + 1 + 2 + 4 + 8 + 3 + 5 + 6 = 44$). That happened because the expert considers conditions 2 and 7 equal from the perspective of their influence on the development of teachers' pedagogical skills, so they both received the 3rd rank. However, they occupy the 3rd and the 4th places. Therefore, those conditions were assigned a refined rank of 3.5 (7 / 2 = 3.5). The results of such refinement of ranks are presented as a matrix (Table 2).

The average sum of the ranks was calculated according to the formula:

$$\mathrm{di}^{-} = 1/2\ \mathrm{m} \cdot (\mathrm{n} + 1), \tag{2}$$

where m is the number of experts, $m = 10$, n is the number of conditions, $n = 9$. Then $di^{-} = 50$. To verify the correctness of the matrix, a checksum was determined for all rows, and the fulfillment of the equality was monitored.

$$\sum\nolimits_{j=1}^{n} x_{ij} = \frac{(1+n)n}{2}, \tag{3}$$

where n is the number of conditions, and x_{ij} is the range of the i condition from the j expert.

It was calculated that the control sum was equal to 45. The sums of numbers in each row turned out to be equal to each other and the control sum. Subsequently, the sum of the numbers in each column (rank sums) and the sum of the numbers in all columns were calculated.

Table 2 The matrix of the ranks of learning conditions for developing general education teachers' professional skills

Condition No	Expert										d_i	D_i	D_i^2
	I	II	III	IV	V	VI	VII	VIII	IX	X			
1	8	9	5	5	7	7	7	7	8	6.5	69.5	19.5	380.25
2	3.5	3	3.5	2	1.5	2	3	3.5	4	2	28	-22	484
3	1	1.5	2	2	3.5	3	3	1	1.5	2	20.5	-29.5	870.25
4	2	1.5	1	2	3.5	1	1	2	1.5	2	17.5	-32.5	1056.25
5	5	4	3.5	4	1.5	4	3	3.5	3	4.5	36	-14	196
6	9	8	8	6	6	6	6	9	9	6.5	73.5	23.5	552.25
7	3.5	5	6	7	5	5	5	5	5	4.5	51	1	1
8	6	7	9	8	8	8	8	6	6	8.5	74.5	24.5	600.25
9	7	6	7	9	9	9	9	8	7	8.5	79.5	29.5	870.25
$\sum$	45	45	45	45	45	45	45	45	45	45	450	0	5010.75

Notes: d_i is the sum of ranks according to the ranking of the i condition, D_i is the deviation from the average amount, and D_i^2 is the square of the deviation

The sum of the sums of the numbers in all rows was 450 and coincided with the sum of the sums of the numbers in all columns, so we could conclude that the equality was fulfilled:

$$\sum_{j=1}^{n}\sum_{i=1}^{m} xij = \sum_{i=1}^{m}\sum_{j=1}^{n} xij. \tag{4}$$

So, the matrix was built correctly. Next, we determined the unity of experts' opinions using the concordance coefficient K_{con}, which was calculated by the formula:

$$K_{con} = \frac{S}{\frac{1}{12}m^2\left(n^3 - n\right) - m\sum Ti}, \tag{5}$$

where S = 5010,75, n = 9; m = 10, Tj is the results of intermediate calculations:

$$\mathrm{T}j = \frac{1}{12}\sum\left(t^3 - t\right), \tag{6}$$

where t is the number of items repeated in the assessments of the j expert. After getting the Tj values ($T_1 = 0.5$, $T_2 = 0.5$, $T_3 = 0.5$, $T_4 = 2.0$, $T_5 = 1.0$, $T_6 = 0$, $T_7 = 2.0$, $T_8 = 0.5$, $T_9 = 0.5$, $T_{10} = 3.0$, $\sum \mathrm{T}j = 10.5$) the concordance coefficient was calculated:

$$K_{con} = \frac{5010,75}{\frac{1}{12}9^2\left(10^3 - 10\right) - 10 \cdot 10,5} = 0.76.$$

When the value of the concordance coefficient is in the range from 0 to 1, it shows the high unity of experts' opinions regarding the importance of the ranked learning conditions. To make sure that such a value of the concordance coefficient was not accidental, we used the $\chi 2$ Pearson consistency test. The chi-squared test was calculated by the formula:

$$\chi 2 = \frac{S}{\frac{1}{12} m * n(n+1) + \frac{\sum Ti}{n-1}}, \tag{7}$$

$$\chi 2 = \frac{5010, 75}{\frac{1}{12} 9 \cdot 10(10-1) + \frac{10,5}{10-1}} = 73.04.$$

Then the calculated value was compared with its critical value of 15.507 for the significance level $\alpha = 005$ for the degrees of freedom $df = n - 1 = 9 - 1 = 8$. Since the calculated value $\chi 2$ (73.04) significantly exceeds the critical value (15.507), with a probability of 95%, we could claim that the obtained value $Kcon = 0.76$ was a non-random value. Therefore the obtained results of the learning conditions ranking could be used in further research.

Let us analyze the status characteristics of the researched conditions for the development of pedagogical skills of general education teachers (Fig. 1).

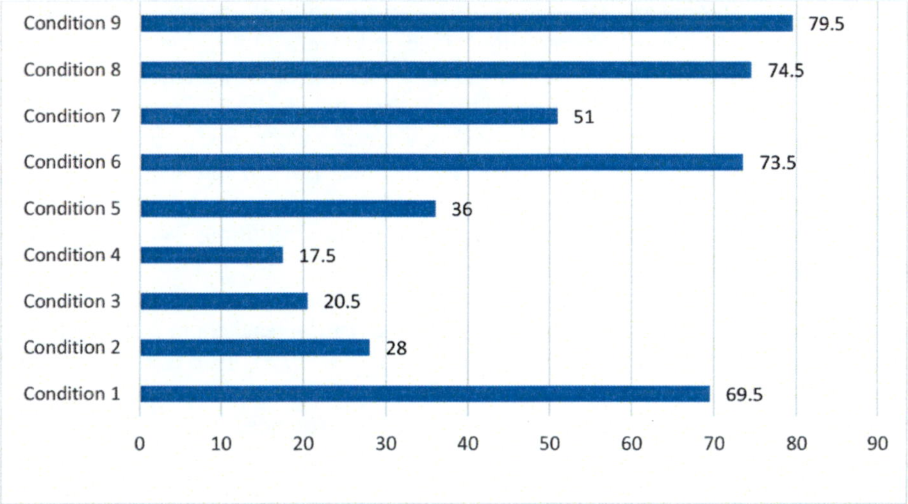

Fig. 1 The results of ranking the learning conditions for the development of pedagogical skills of general education teachers

The lowest number of ranks ($\sum = 17.5$) was scored by the condition "Systematic mastery learning for teachers of general education disciplines." Therefore, that condition took the first place as a circumstance that was predicted to increase the development of the teachers' professional skills.

The second place among the 9 learning conditions under consideration was taken by the condition “Access to a flexible system of teachers’ professional development” with the sum of ranks $\sum = 20.5$.

In third place ($\sum = 28.0$) in the ranked list of conditions for developing the pedagogical skill of general education teachers was the “Priority of advanced learning technologies with equal access of all participants of the educational process to the technologies and resources.” Indeed, to successfully apply advanced learning technologies (educational design technology, trainings, case study method, interactive technologies, etc.), an effective information and educational environment must function in an educational institution, which, in turn, requires teachers to improve their qualifications and organise their professional development constantly. Therefore, it is not surprising that such conditions took the following fourth and fifth places in the considered rating as “Ensuring the integration of the content of general education subjects with the disciplines of general professional, as well as professional (engineering) theoretical and practical training” ($\sum = 36.0$) and “Development of a creative educational environment as a factor for the enhancement of creative abilities of all participants of the educational process” ($\sum =$ 51.0).

The seventh and eighth place in the ranking was taken by the conditions “Involvement of general education teachers in regular scientific and pedagogical activities” and “Active interaction of the engineering education institutions with scientific institutions, educational and methodological centers, other universities regarding the implementation of joint educational and scientific projects, organization of competitions, webinars, trainings, conferences, etc.”, as their sums of the ranks were 73.5 and 74.5, respectively. In our opinion, such an experts’ view of the place and role of the specified factors in the considered set of conditions was caused by the fact that teachers of general education disciplines were extremely rarely involved in scientific work, performed pedagogical research, were the post-graduate students, etc. It is necessary to add that the practice of involving teaching staff in the implementation of joint projects and grants was also not common, and partnership with universities and scientific institutions was not typical for engineering education.

The last place among the considered learning conditions was occupied by the condition “Expert and consultative support of general education teachers in self-educational development,” with the highest number of ranks ($\sum = 79.5$). This fact was expected since self-education development was most often understood as a conscious, purposeful process of increasing the level of one’s professional competence based on the own development program. According to our observations, in Ukraine, the expert-consultative support of the specified processes in the engineering education system is only at the formation stage. It does not have an essential scientific and methodological basis.

5 Conclusions

The set of circumstances predicted to contribute to the development of the pedagogical competence of general education teachers was determined in two stages. In the first stage, a list of 19 learning conditions was formed. In the second stage, their examination was carried out, as well as the determination of the most significant circumstances predicted to improve general education teachers' professional skills. The experts ranked the learning conditions according to the degree of their significance. Thus, the final list included 9 ranged learning conditions: purposeful formation of positive motivation for teachers' self-development and self-improvement, priority of advanced learning technologies with equal access of all participants of the educational process to the technologies and resources, access to a flexible system of teachers' professional development, systematic mastery learning for teachers of general education disciplines, ensuring the integration of the content of general education subjects with the disciplines of general professional, as well as professional (engineering) theoretical and practical training, involvement of general education teachers in regular scientific and pedagogical activities, development of a creative educational environment as a factor for the enhancement of creative abilities of all participants of the educational process, active interaction of the engineering education institutions with scientific institutions, educational and methodological centres, other universities regarding the implementation of joint educational and scientific projects, organization of competitions, webinars, trainings, conferences, etc., expert and consultative support of general education teachers in self-educational development, etc.

So, the experts' conclusion confirmed the prediction that it was crucial for teachers of general education disciplines to regularly update their knowledge in a specific scientific field (knowledge of their subject), in psychology, pedagogy, and teaching methods, as well as master innovative learning technologies and constantly develop the ability to productively use advanced technologies and resources to be able to organize and keep a creative learning environment that could support the development of engineering students' potentials, engage them in a professional atmosphere, make them think in a non-standard way, practice working in a team as well as improve essential hard and soft skills.

Further research could be devoted to justifying and verifying learning conditions predicted to improve teachers' professional skills.

References

1. Walters, H.D., Bockorny, K.M.: Relevance of general education: an assessment of undergraduate business students. e-J. Bus. Educ. Sch. Teach. **12**(3), 34—43 (2018)
2. Skenes, K., Rabb, R.J., Washuta, N.J., Righter, J.: Teaching engineering in the general education curriculum. In: 2020 First-Year Engineering Experience, East Lansing, Michigan (2020). https://peer.asee.org/35779. Accessed on 21 Jan 2023
3. Fagette, P., Chen, S.-J., Baran, G., Samuel, S., Kiani, M.: Engineering a general education program: designing mechanical engineering general education courses. Innov. High. Educ. **38**, 117–128 (2012). https://doi.org/10.1007/s10755-012-9231-2
4. UNESCO: Social & Human Sciences. https://www.unesco.org/en/social-human-sciences. Accessed on 20 Jan 2023
5. Vare, P., et al.: Devising a competence-based training program for educators of sustainable development: lessons learned. Sustainability **11**(7), 1890 (2019). https://doi.org/10.3390/su11071890
6. Lozano, R., Barreiro-Gen, M., Lozano, F.J., Sammalisto, K.: Teaching sustainability in european higher education institutions: assessing the connections between competences and pedagogical approaches. Sustainability **11**(6), 1602 (2019). https://doi.org/10.3390/su11061602

7. Imara, K., Altinay, F.: Integrating education for sustainable development competencies in teacher education. Sustainability **13**(22), 12555 (2021). https://doi.org/10.3390/su132212555
8. Weng, S.-S., Liu, Y., Dai, J., Chuang, Y.-C.: A novel improvement strategy of competency for Education for Sustainable Development (ESD) of university teachers based on data mining. Sustainability **12**(7), 2679 (2020). https://doi.org/10.3390/su12072679
9. Harangus, K., Horváth, Z.-I., Erzsebet, S.: Existing and to be developed teacher competences in engineer teacher training. Contemp. Educ. Res. J. **10**(4) (2020). https://doi.org/10.18844/cerj.v10i4.5212
10. Ca-Nguyen, D., Phuong-Ngo, T., Thang-Ngoc, H., Hai-Luong, D., Lien-Nguyen, H., Thang-Nguyen, T.: A policy review of criteria and competencies for teaching staff of general education in Vietnam. Cogent Educ. **9**(1), 2133889 (2022). https://doi.org/10.1080/2331186X.2022.2133889
11. Neves, R., Lima, R., Mesquita, D.: Teacher competences for active learning in engineering education. Sustainability **13**(16), 9231 (2021). https://doi.org/10.3390/su13169231
12. Titova, O., Luzan, P., Sosnytska, N., Kulieshov, S., Suprun, O.: Information and communication technology tools for enhancing engineering students' creativity. In: Ivanov, V., Trojanowska, J., Pavlenko, I., Zajac, J., Peraković, D. (eds.) Advances in Design, Simulation and Manufacturing IV. DSMIE 2021. Lecture Notes in Mechanical Engineering, pp. 332–340. Springer, Cham (2021). https://doi.org/10.1007/978-3-030-77719-7_33
13. Aidoo, B., Macdonald, A., Gyampoh, A.O., Baah, K.A., Tsyawo, J.: Factors influencing teachers' online teaching competence in higher education. Soc. Educ. Res. **3**(1), 148–160 (2022). https://doi.org/10.37256/ser.3120221315
14. Cruz, M.L., Saunders-Smits, G.N., Groen, P.: Evaluation of competency methods in engineering education: a systematic review. Eur. J. Eng. Educ. **45**(5), 729–757 (2020). https://doi.org/10.1080/03043797.2019.1671810
15. Carvalho de, G.D.G., Corrêa, R.O., Carvalho de, H.G., Vieira, A.M.D.P., Stankowitz de, R.F., Kolotelo, J.L.G.: Competencies and performance of engineering professors: evidence from a Brazilian public university. Ingeniería E Investigación **38**(3), 33–41 (2018). https://doi.org/10.15446/ing.investig.v38n3.70998
16. Titova, O., Luzan, P., Davlatzoda, Q.Q., Mosia, I., Kabysh, M.: The taxonomy approach for engineering students' outcomes assessment. In: Tonkonogyi, V., Ivanov, V., Trojanowska, J., Oborskyi, G., Pavlenko, I. (eds.) Advanced Manufacturing Processes IV. InterPartner 2022. Lecture Notes in Mechanical Engineering, pp. 380–390. Springer, Cham (2023). https://doi.org/10.1007/978-3-031-16651-8_36
17. LaShorage, S., Karen, T.-B.: Enhancing teacher competency through co-teaching and embedded professional development. J. Educ. Train. Stud. **3**(3), 117–125 (2015)
18. Murphy, M., Destin, M.: Promoting inclusion and identity safety to support college success. The Century Foundation (2016). https://tcf.org/content/report/promoting-inclusion-identity-safety-support-college-success/. Accessed on 12 Jan 2023
19. Gagné, R.M.: The Conditions of Learning, 4th edn. Holt Rinehart & Winston, New York (1985)
20. Gagné, R.M., Medsker, K.L.: The Conditions of Learning: Training Applications. Harcourt Brace, Fort Worth (1996)
21. Koper, R.: Conditions for effective smart learning environments. Smart Learn. Environ. **1**(5) (2014). https://doi.org/10.1186/s40561-014-0005-4
22. Allensworth, E.M., et al.: Supporting social, emotional, academic development: Research implications for educators (2018). https://consortium.uchicago.edu/sites/default/files/2019-01/Supporting%20Social%20Emotional-Oct2018-Consortium.pdf. Accessed 14 Jan 2023
23. Glass, G.V., Hopkins, K.D.: Statistical Methods in Education & Psychology, 3rd edn. Allyn & Bacon, Boston (1996)

Fractal Dimension Using the Acoustic Infrared Thermal Method of Inspection of Non-metallic Heterogeneous Materials

Volodymyr Tonkonogyi[1], Maryna Holofieieva[1](✉), Oleksandr Levynskyi[1], Sergii Klimov[1], and Raul Turmanidze[2]

[1] Odessa Polytechnic National University, 1, Shevchenko Ave., Odessa 65044, Ukraine
mgolofeyeva@gmail.com

[2] Georgian Technical University, 77, Kostava St., 0160 Tbilisi, Georgia

Abstract. Various types of heterogeneous materials and structures are widely used in mechanical engineering. The widespread distribution of such materials cannot be imagined without reliable control methods. The paper uses the acoustic, infrared thermal imaging method to control non-metallic heterogeneous materials, which allows for determining structural features in solids that exhibit the effect of mechanical hysteresis and internal friction. The research was conducted on the features of the structure of a carbon-plastic plate with dimensions of 130 × 80 × 4 mm with defects made in advance: delamination and a defect obtained from shock loading. The method is based on the phenomenon of an increase in the temperature of the surface of a heterogeneous object in the defect zone under the influence of the energy of mechanical vibrations. At the same time, the carrier of information about structural defects is the thermogram, which is recorded using infrared devices. The difficulty of evaluating anomalies of the thermal field of the investigated surface of the product made of non-metallic heterogeneous material is that the result depends on the scale of measurements (on the scanning step). Therefore, the authors suggest using the so-called fractal approach invariant to the measurement scale. The box-counting method is used to determine the fractal dimension. It is shown that this method is sensitive to equiaffine transformations. For the unambiguousness of the measurement results, it is suggested to use known methods of mathematical statistics for their processing.

Keywords: Non-Metallic Heterogeneous Materials · Flaw Detection · Acoustic Infrared Thermal Imaging Control Method · Thermogram · Fractal Dimension · Process Innovation

1 Introduction

Nowadays, there are technical and economic prerequisites for the widespread use of non-metallic heterogeneous materials in the engineering industry [1]. The multi-level nature of such materials provides ample opportunities for synthesizing their physical and mechanical properties, which are manifested and, accordingly, can be evaluated through

V. Tonkonogyi et al. (Eds.): InterPartner 2023, LNME, pp. 389–398, 2024.
https://doi.org/10.1007/978-3-031-42778-7_36

interaction with other objects. Depending on the type of external influences (research methods used), the level of the specified properties may change. Therefore, it becomes clear that studying the behavior of non-metallic heterogeneous materials directly in structures is desirable. Moreover, the latter can be presented in the form of some system with its internal architecture with a hierarchy of structural levels and sublevels. In this case, it becomes possible to assess the contribution of each structural element to the operation of the specified structure. Simultaneously, primary attention should be paid to defects that occurred during the production of products or as a result of their operation. It is necessary to understand that the defects of the complex structure of non-metallic heterogeneous materials can be considered constructive. Moreover, their presence in products changes the behavior of materials during operation, including the mechanisms of defect development and the nature of destruction in general.

The effectiveness of using heterogeneous materials and constructions from them requires the measurement of reliable values of the parameters of the internal processes occurring in them at various stages of operation. At the same time, special attention should be paid to reliable, prompt, and cheap research. The research aims at developing the fundamentals of the method for processing thermograms using fractal dimension in flaw detection of products of non-metallic heterogeneous materials.

2 Literature Review

Heterogeneous materials [2] consist of at least two components and are characterized by a new set of properties that are different from the properties of structural elements. Moreover, more often than not, the property of additivity is not manifested when designing structurally complex materials. That is, the properties of individual components are transformed into the properties of their compositions through the organization of other structures [3]. Specialists in metallurgy, ceramics, polymer materials and their combinations with other natural materials, and structural building materials of various types and purposes deal with developing heterogeneous multicomponent materials [4].

It is known that replacing traditional materials with heterogeneous ones ensures a reduction in the material capacity of products several times while increasing their working life and reliability. In addition, the labor intensity is reduced due to improved manufacturability. Reduced plasticity should also be attributed to the features of non-metallic heterogeneous materials. This excludes the possibility of destruction of products due to deformation effects [5].

Common features of composite materials and structures are heterogeneity, heterophasicity, multicomponentity, an interface between phases (individual components), and the difference in physical and technical properties from the properties of their constituents [6, 7]. From this, it can be concluded that heterogeneous materials and structures can include systems consisting of at least two components that interact through or with the help of a separation boundary, the properties of which differ from the individual characteristics of the components [4].

The widespread distribution of such materials is unthinkable without reliable control methods. Low cost, speed, and the possibility of control outside the testing laboratories also play an essential role. The most interesting at the moment are nondestructive methods that make it possible to detect macroscopic technological defects. These include delamination, non-gluing, cracks, extraneous inclusions, and geometry violations during product assembly. Many features and physical phenomena determine the problems of nondestructive methods of control of non-metallic heterogeneous materials as a characteristic of this class of materials (non-magnetism, low electrical conductivity, reflection of ultrasonic waves from reinforcing elements). Surface cracks appear in such materials during exploitation, which cannot be detected by many nondestructive testing methods due to specific physical and mechanical properties. In addition, detecting many defects specific only to non-metallic heterogeneous materials, such as delamination of the matrix and reinforcement elements, is also compressed with difficulties in nondestructive testing. All this imposes known limitations on using classical research methods and flaw detection for the class of non-metallic heterogeneous materials [8].

It is promising to use the acoustic, infrared thermographic method of controlling the specified class of materials, which is suitable for determining structural features in solids that exhibit the effect of mechanical hysteresis and internal friction [9, 10]. These include shock and fatigue cracks, delamination, non-gluing and other defects without mutual penetration of materials in contact at the interface of phases [11].

The thermal energy distributed in the research object is used as an informative parameter in thermal control methods. The temperature field is a carrier of information regarding the features of heat transfer, which, in turn, depend on the presence of internal and external features [12]. Infrared devices are used to obtain such information (most often in the form of thermograms). Anomalies of thermal energy distribution on the thermogram are explained by defects in the structure of non-metallic heterogeneous materials [13].

The highest reliability of nondestructive testing is achieved with the maximum signal-to-noise ratio, which can be obtained at the optimal time by recording several consecutive thermograms and choosing the best one from the point of view of noise [14]. Such a thermogram has the status of an image to which all types of processing are applied. This concerns primary processing and segmentation, selection and description of contours of thermal anomalies, and image analysis [15]. The difficulty of evaluating anomalies of the thermal field of the investigated surface of a product made of non-metallic heterogeneous material is that the result depends on the scale of measurements (on the scanning step) [16]. Therefore, the authors suggest using the so-called fractal approach invariant to the measurement scale [17]. Fractal forms are widespread in nature. As for physical research, fractal features can be observed in the structure of received signals and fields. It should be noted that fractality is often manifested in functions that describe the distribution of physical quantities in time and space (for example, physical fields of various natures).

Recently, fractals began to be used related to heterogeneous materials and structures in the sense of some universal chemically structured unit, which contains information about the structure and properties of the studied material [18]. This approach opens up new opportunities in studying the properties and structure of the specified materials and designing constructions from them.

The concept of fractal (fractional) dimension, which is also known as the Hausdorff–Bezykovich dimension, was introduced by Mandelbrot for measuring fractal objects [19]. The latter include surfaces obtained by various methods of randomly distributed irregularities. Also, a mandatory requirement for fractals is a fractional, metric dimension, the value of which is strictly greater than topological.

There are various methods of determining the fractal dimension [20]. In the general case, they are reduced to the determination of the established characteristic parameter, based on the change of which judgments are made at different scale levels regarding the fractal dimension [21]. Moreover, to effectively use such an approach, it is necessary to consider the peculiarities of specific research objects [22].

3 Research Methodology

The research was conducted on the features of the structure of a carbon-plastic plate with dimensions of 130 × 80 × 4 mm with defects made in advance: delamination and a defect obtained from shock loading. Detection of defects was carried out by the acoustic, infrared thermometric method. For this purpose, the investigated plate was subjected to short-term action of the energy of mechanical vibrations (radiation electric power of 2 kW). As expected, substantial absorption of sound waves was observed in defective zones, which manifested in anomalies of the surface temperature field above them. That is, a defect of small thickness becomes a constantly operating source of heat during sound stimulation. This effect is observed for hidden defects and those that come to the surface. Corresponding thermograms were recorded with the help of infrared devices, namely the Fluke 9 thermal imager. Figure 1 shows the appearance of the investigated carbon plate.

Fig. 1. Appearance of the investigated carbon plate.

The resulting thermogram is shown in Fig. 2.

Temperature anomalies are visible on the thermogram. Layering is displayed in the upper left corner. In the center of the plate, there is a defect caused by impact load, around which an increased temperature is observed. The presence of microcracks explains this.

Considering the received thermogram as a fractal object, general ideas about them were used for analysis.

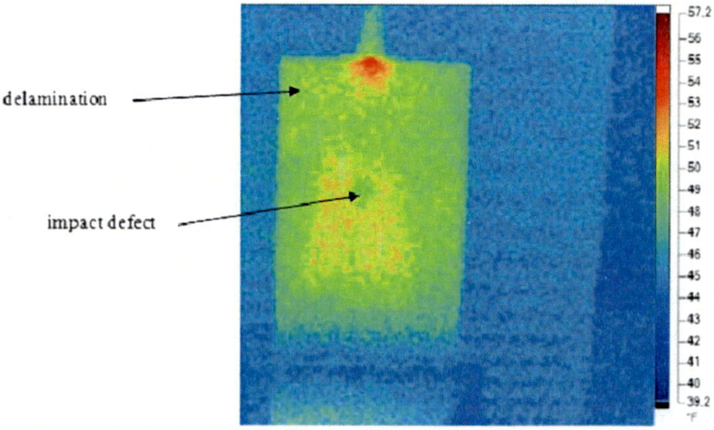

Fig. 2. Thermogram of the investigated carbon plate.

According to [14], a fractal object can be characterized by the measure of the set M_d of elementary segments h(δ) (elements, points) that fill (expose) the measuring object:

$$M_d = N(\delta) \cdot h(\delta) \tag{1}$$

where $N(\delta)$ is the number of elements needed to fill the object;

d – topological dimension;

$\delta < 1$ is a coefficient that depends on the scan step.

The elementary segments $h(\delta)$ are, in a way, a test function that acts as a measurement scale of the set [14]:

$$h(\delta) = A \cdot \delta^d \tag{2}$$

where A is the length of the straight line connecting the ends of the studied contour (one-dimensional case); the size of the projection area of the measuring surface (two-dimensional case); the size of the volume built on the projections of the side surfaces of the volume (three-dimensional case). For the case of thermogram processing, A is equal to the topological area of the selected area S_w.

On the other hand, the number of elements $N(\delta)$ can be represented by the Hausdorff-Bezykovich dimension D (fractal dimension:

$$N(\delta) = \delta^{-D} \tag{3}$$

Then, taking into account that M_d is actually a measure of the area of the studied fractal object S, (1) takes the form:

$$M_d = S = A \cdot \delta^{d-D} \tag{4}$$

From this, it can be concluded that with the equality of fractal and topological dimensions (at $d = D$), the value of M_d is finite, and $A = S_w = S$ for a two-dimensional surface.

A practical method of dimension calculation is box counting. During flaw detection, it was used to determine the fractal dimension of temperature field anomalies on the surface of non-metallic heterogeneous materials. This method is widely used to calculate the dimension of sets describing various objects because it is simple and intuitive.

The calculation procedure consists of the following stages:

- the research object is covered with a square grid with a known cell size;
- the number of cells covering a fragment of the studied object is counted. At the same time, the pair "cell side length" - "number of cells covering the object" is stored;
- the mesh is detailed; that is, the size of the cells covering the object decreases, and their number, accordingly, increases.
- the specified procedure is repeated many times.

The smaller δ is, the more cells can be used to cover the monitored area of the thermogram and the more accurate the value of the measured area. Then, for the fractal surface, taking into account (4), the set value of δ corresponds to the determined value of the area of the thermal field anomaly S_f:

$$S_f = S_w \cdot \delta^{2-D} \tag{5}$$

After plotting the dependence of ln S_f on lnδ, which has the form of a reversible S-shaped curve, the fractal dimension of the object under study is determined by the angle of inclination of its linear part.

4 Results and Discussion

Figure 3 shows the contours of the temperature field anomaly of the surface of the investigated carbon-plastic plate, which reflect the presence of the defect, as well as the dependence of ln S_f on lnδ to determine the fractal dimension. Excitation of thermal energy was carried out by applying the energy of mechanical vibrations to the studied material sample.

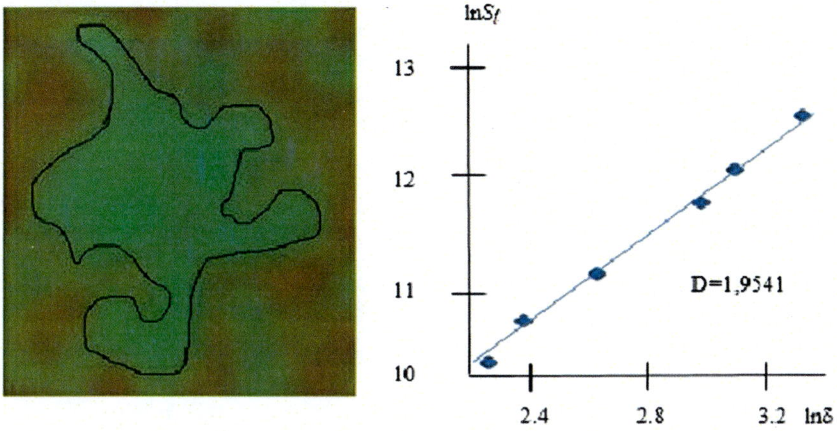

Fig. 3. Contours of the anomaly of the temperature field of the investigated carbon plate and the dependence of $\ln S_f$ on $\ln\delta$ for determining the fractal dimension.

When calculating the fractal dimension of temperature field anomalies, the question of the angle of their observation arises. The fact is that the surface itself, and, accordingly, the thermogram, can be turned at some angle. At the same time, the fractal dimension cannot change from such a change in the image, and the result of its calculation fully has the right to do so. Figure 4 shows the exact contours of the thermogram anomaly rotated by 90°, 180°, and 270° relative to its axis and the fractal dimension's calculated values (Fig. 4).

It can be seen that the result of using the box-counting method is sensitive to equiaffine transformations. This is observed in most cases. Exceptions are only objects whose fractal dimensions perfectly match the theoretical values, for example, an infinite straight line ($D = 1$) or a plane ($D = 2$).

It is also necessary to note the results' accuracy when rounded. The third digit after the comma already gives the difference in the calculated values. Thus, we have not one specific value of the fractal dimension but their set. The range of these values for the object under study is small, and it is possible to enter the statistical value of the fractal dimension of the temperature field anomaly of the surface of a non-metallic heterogeneous material, which is calculated using known methods of mathematical statistics. For this purpose, the fractal dimension of a series of images was evaluated for the studied contour, rotating it relative to its axis by a certain angle, going through a complete rotation cycle from 0° to 360°. After that, it is possible to obtain the value of the mathematical expectation and estimate the measurement uncertainty for the value of the fractal dimension, which will fully characterize the measurement result.

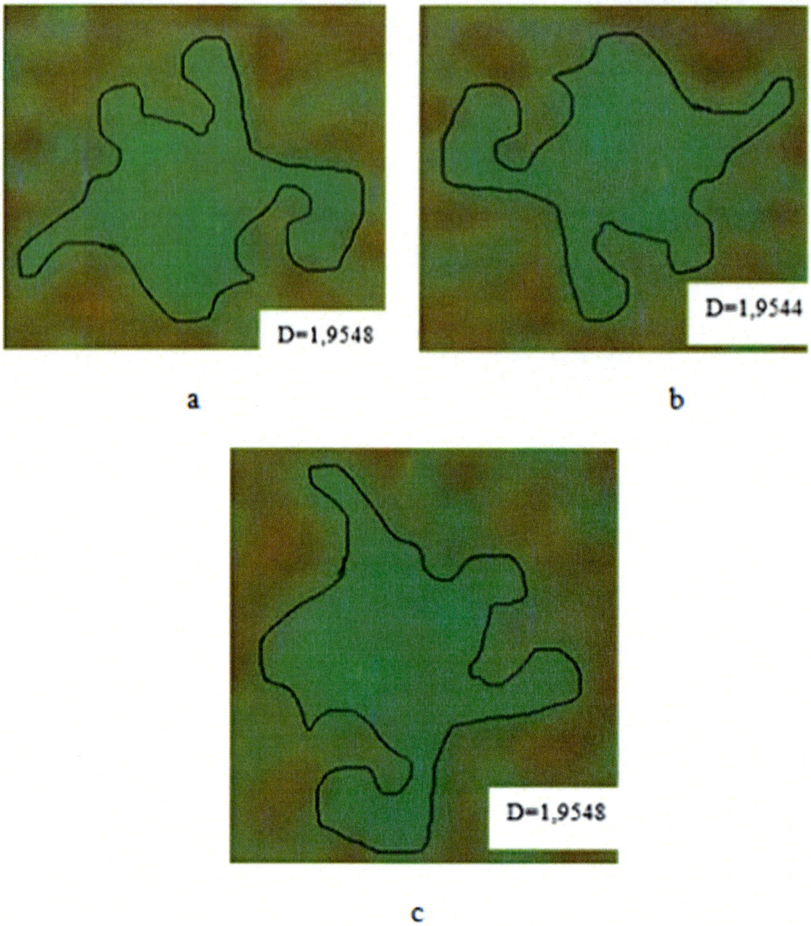

Fig. 4. Contours of the anomaly of the temperature field of the studied carbon plate, rotated by 90° (a), 180° (b), and 270° (c) relative to its axis and the value of their fractal dimensions.

5 Conclusions

The paper shows the perspective of using the acoustic infrared-thermal imaging method of controlling non-metallic heterogeneous materials. It is suitable for determining structural features in solids exhibiting mechanical hysteresis and internal friction. These include shock and fatigue cracks, delamination, non-gluing and other defects without mutual penetration of materials contacting at the phase interface. The method consists of establishing, using infrared equipment temperature anomalies of the object's surface made of heterogeneous material, which arise in defective zones under the influence of the energy of mechanical vibrations. The research was carried out on the features of the structure of a carbon-plastic plate with dimensions of 130 × 80 × 4 mm with defects made in advance: delamination and a defect obtained as a result of impact loading. Corresponding thermograms were recorded with the help of infrared devices, namely the Fluke 9 thermal imager.

To analyze surface thermal anomalies, the authors used the so-called fractal approach invariant to the measurement scale. The box-counting method is used to determine the fractal dimension, which is widely used to calculate the dimension of sets describing various objects, as it is simple and visual. It is shown that the result of using the box-counting method is sensitive to equiaffine transformations. For the unambiguousness of the measurement results, it is suggested to use known methods of mathematical statistics for their processing.

References

1. Holofieieva, M., Tonkonogyi, V., Kolesnik, M.: Control of the properties of materials with a multi-level structure. East Eur. Sci. J. **4**, 61–65 (2015)
2. Serna Moreno, M.D.C., Horta Muñoz, S., Ruiz Gracia, A.: Design of triaxial tests with polymer matrix composites. Polymers **14**(4), 837 (2022). https://doi.org/10.3390/polym14040837
3. Pendhari, S.S., Kant, T., Desai, Y.M.: Application of polymer composites in civil construction: a general review. Compos. Struct. **84**, 114–124 (2008). https://doi.org/10.1016/j.compstruct.2007.06.007
4. Dorofeyev, V., Myronenko, I., Pushkar, N.: The effect of technological damage on the properties and reliability of construction materials and structures. Appl. Mech. Mater. **908**, 149–156 (2022). https://doi.org/10.4028/p-lm1g4w
5. Fayzimatov, S.N., Xusanov, Y.Y., Valixonov, D.A.: Optimization conditions of drilling polymeric composite materials. Am. J. Eng. Technol. **3**(02), 22–30 (2021). https://doi.org/10.37547/tajet/Volume03Issue02-04
6. Aliha, M.R.M., Imani, D.M., Salehi, S.M., Shojaee, M., Abedi, M.: Mixture optimization of epoxy base concrete for achieving highest fracture toughness and fracture energy values using Taguchi method. Compos. Commun. **32**, 101150 (2022). https://doi.org/10.1016/j.coco.2022.101150
7. Lebedev, V.V., Miroshnichenko, D.V., Nyakuma, B.B., Moiseev, V.F., Shestopalov, O.V., Vyrovets, S.V.: Design of inorganic polymer composites for electromagnetic radiation absorption using potassium titanates. J. Eng. Sci. (Ukraine) **10**(1), C1–C8 (2023). https://doi.org/10.21272/jes.2023.10(1).c1
8. Sivakumar, S., Vignesh, V., Vijay Arasu, I., Venkatesan, G., Raja Mohamed Rabi, B., Adam Khan, M.: Experimental investigation on tensile and flexural properties of randomly oriented treated palmyra fibre reinforced polyester composites. Mater. Today: Proc. **46**, 7556–7560 (2021). https://doi.org/10.1016/j.matpr.2021.01.511
9. Bettencourt Da Silva, R., et al.: Analytical measurement: measurement uncertainty and statistics. Publications Office of the European Union, Luxembourg (Luxembourg) (2012). https://doi.org/10.2787/5825
10. Uysal, M.U.: Effects of cryogenically treated CFRP composite on the buckling behavior in the adhesively bonded beam. J. Eng. Sci. (Ukraine) **10**(1), D1–D7 (2023). https://doi.org/10.21272/jes.2023.10(1).d1
11. Mohapatra, D.K., Deo, C.R., Mishra, P.: Investigation of glass fiber influence on mechanical characteristics of natural fiber reinforced polyester composites: an experimental and numerical approach. Compos. Theory Pract. **3**, 123–129 (2022)

12. Vavilov, V.P., Shiryaev, V.V., Khorev, V.S.: Processing the results of active thermal control by wavelet analysis. Defectoscopy **4**, 70–79 (2011)
13. Maladague, X.: Theory and Practice of Infrared Technology for Nondestructive Testing. Wiley Series in Microwave and Optical Engineering, Wiley, New York (2001)
14. Gholizadeh, S.: A review of nondestructive testing methods of composite materials. Procedia Struct. Integr. **1**, 50–57 (2016). https://doi.org/10.1016/j.prostr.2016.02.008
15. Stanovska, I.: The compatibility of project management dynamic model as a measure of its efficiency. Bull. NTU KhPI. Series: New Solutions in Modern Technologies **10**(1335), 70−76 (2019). https://doi.org/10.20998/2413-4295.2019.10.09
16. Mandelbrot, B.B.: Fractal Geometry of Nature, 2nd edn. Times Books, New York (2002)
17. Family, F., Vicsek, T.: Dynamics of Fractal Surfaces. World Scientific, Singapore (1991)
18. Mishutin, A., Kroviakov, S., Pishev, O., Soldo, B.: Modified expanded clay light weight concretes for thin-walled reinforced concrete floating structures. Tehnicki glasnik – Tech. J. **11**(3), 121−124 (2017)
19. Torkhov, N.A., Bozhkov, V.G., Ivonin, I.V., Novikov, V.A.: Determination of the fractal dimension of the epitaxial n-GaAs surface in the local limit. Phys. Technol. Semicond. **43**(1), 38–47 (2009)
20. Xue, W., Chen, J.: Effect of fractal dimension of fine aggregates on the concrete chloride resistance. In: 5th International Conference on Durability of Concrete Structures, pp. 89−93. Shenzhen University, Shenzhen (2016)
21. Krasikova, I.E., Krasikov, I.V., Kartuzov, V.V., Muratov, V.B., Vasiliev, A.A.: Multifractal characteristics of hot-pressed composites from AlB12–AlN nanopowders. Nanosyst. Nanomater. Nanotechnol. **18**(1), 89–96 (2020)
22. Bisoi, A.K., Mishra, J.: On calculation of fractal dimension of images. Pattern Recogn. Lett. **22**, 631–637 (2001). https://doi.org/10.1016/S0167-8655(00)00132-X

Modeling of Thermal and Dynamic Conditions of Intermittent Grinding, Affecting the Quality Parameters of the Surface Layer of Machined Parts

Alexey Yakimov[1], Liubov Bovnegra[1], Kateryna Kirkopulo[1](✉), Yuliia Babych[1], and Viktor Strelbitskyi[2]

[1] Odessa Polytechnic National University, 1, Shevchenko Ave., Odessa 65044, Ukraine
kateryna.grygoryvna@gmail.com

[2] Odessa National Maritime University, 34, Mechnikova St., Odessa 65029, Ukraine

Abstract. This work is devoted to ensuring the required quality of the surface layer of the workpieces during the grinding operation, considering the thermal, physical, and dynamic phenomena that occur during intermittent grinding. The influence of the scale factor of the dimensions of the elements of macrotopography of the working surface of discontinuous circles on their cutting ability, heat release in the zone of contact of the abrasive tool with the workpiece, and parametric instability of the elastic system of the machine tool and, as a result, on the quality parameters of the surface layer of ground parts is established. The required quality parameters of the surface layer of parts in the operation of flat grinding with intermittent wheels can be achieved by reducing the scaling of the elements that make up their working macroprofile, i.e., the use of abrasive wheels on their working surfaces parallel to the axis of rotation. The design of a discrete wheel is proposed, which provides the required geometric and physical-mechanical parameters of the quality of the surface layer of workpieces during the grinding operation.

Keywords: Cutting Surface Topography · Interruptibility Factor · Discrete Circles · Manufacturing Innovation

1 Introduction

The grinding process is an essential final operation of the technological process of mechanical handling machine parts. The main drawback of the grinding process, which hinders labor productivity growth and worsens product quality, is the occurrence of high temperatures in the cutting zone.

Currently, the implementation of interrupted grinding is carried out using:

- textured circles (their dimensions of the depressions range from 0.01 to 1 mm);
- circles, on the working surface of which small radial cylindrical holes are applied;
- highly porous circles;
- composition circles.

V. Tonkonogyi et al. (Eds.): InterPartner 2023, LNME, pp. 399–408, 2024.
https://doi.org/10.1007/978-3-031-42778-7_37

The effect of interrupting the cutting process leads to a decrease in the temperature in the cutting zone. The widespread use of these circles is justified by the fact that vibrations do not accompany their work. Nevertheless, the absence of vibrations is a major drawback of these wheels since the presence of high-frequency vibrations in the cutting zone greatly facilitates the process of chip formation and contributes to the self-sharpening of grains as a result of their destruction and the formation of new cutting edges on them, as well as the timely removal of dull grains from the bond. This increases the durability of the wheels and contributes to an additional reduction in temperature in the cutting zone. These positive effects are inherent in traditional discontinuous circles with wide grooves (from 1 to 50 mm in size) through execution (from end to end). However, their use is constrained by the possibility of parametric resonance in the elastic system of the machine due to periodic changes in the rigidity of this system. Therefore, the study of the operation of traditional intermittent circles to minimize the possibility of high-amplitude oscillations in the elastic system of the machine is an urgent task.

2 Literature Review

Grinding is the most common final operation of the technological process for manufacturing machine-building parts. In the grinding process, the surface and near-surface layers of hardened workpieces being processed are subjected to thermal effects [1]. If the temperature in the contact zone of the grinding wheel with the workpiece reaches the tempering temperature in the surface layer of the workpiece, thermal damage is formed, known as a "grinding burn" [2].

Burns reduce fatigue life [3] and cause premature wear [4] and failure [5] of machined parts. The heat absorbed by the workpiece can be reduced using intermittent grinding wheels [6].

With periodic short-term interruptions of the cutting process, the temperature in the zone of contact of the abrasive wheel with the workpiece does not reach the maximum value since the duration of heat release is less than the time of thermal saturation [7]. Designs of intermittent circles can be different [8].

The intensity of the heat stress reduction depends largely on the size and location of cutting sections on the working cylindrical surface of the grinding wheel [9]. Periodic interruption of the cutting process leads to a change in the rigidity of the elastic system [10] of the surface grinding machine and the danger of the appearance of forced and parametric vibrations in it, which can lead to resonances. At resonance states of the machine's elastic system, it is impossible to provide the required geometric accuracy of machined surfaces.

At the stage of designing abrasive wheels with discontinuous working surfaces, it is necessary to consider not only the amount of heat emitted between the wheel and the working area, but also the possibility of undesirable vibrations in the elastic system of the grinding machine [11]. Therefore, predicting physical-mechanical and geometrical parameters of the quality of the surface layer of parts made of hardened steels at the stages of design of abrasive tools and the technological process [12] of their production is an urgent task.

3 Research Methodology

Theoretical studies were carried out based on the thermal physics of cutting and the theory of vibrations. Laboratory studies on the cutting ability of circles were carried out on a surface grinding machine 3G71M.

Grinding was carried out with a wheel 1A1 200 × 75 × 5 × 20 KP 160/125 Б8 100.

Flat specimens of P6M5 steel 8 mm wide and 150 mm long were processed.

Graphical dependencies were built using the Mapl 2015 computer mathematics system and the AutoCAD 2019 computer-aided design system.

Theoretical studies were carried out based on the thermal physics of cutting and the theory of vibrations. Laboratory studies on the cutting ability of circles were carried out on a surface grinding machine 3Г71M. Grinding was carried out with a wheel 1A1 200 × 75 × 5 × 20 KP 160/125 Б8 100. Flat specimens of P6M5 steel 8 mm wide and 150 mm long were processed.

In the process of grinding [13] with an abrasive wheel having the same radial grooves evenly distributed over its working cylindrical surface, the rigidity of the elastic system of the machine changes according to a periodic law, according to which, for the periods falling on the contact of the cutting segments with the workpiece being machined λ_L and the interruption of this interaction λ_l, the stiffness retains constant values, respectively equal to the sum $\left(\sigma + \frac{\sigma^*}{2}\right)$ and difference $\left(\sigma - \frac{\sigma^*}{2}\right)$, where σ is the reduced stiffness, σ^* is the pulsation depth (Fig. 1).

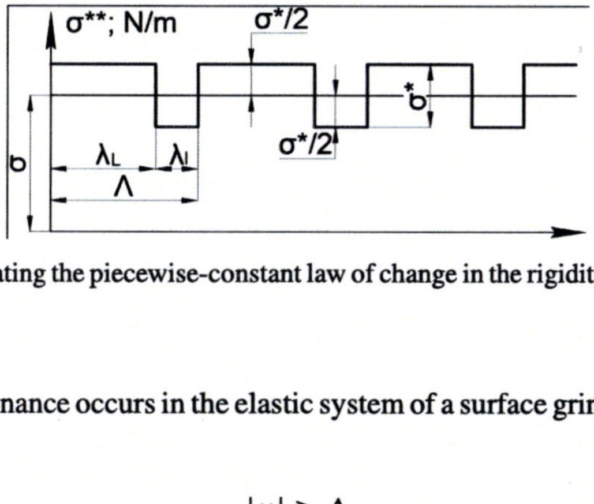

Fig. 1. Graph illustrating the piecewise-constant law of change in the rigidity of the elastic system of a surface grinder.

Parametric resonance occurs in the elastic system of a surface grinder if the following inequality is met:

$$|\mu| \geq \Delta \tag{1}$$

$$\mu = \frac{1}{\chi \cdot \varepsilon} \cdot \begin{bmatrix} sin\,\beta_L\,\lambda_L \cdot \left(\beta_L^2 \cdot sin\,\beta_l\,\lambda_l - \beta_L^2 \cdot sin\,\delta\right) - \\ -2\beta_L\beta_l \cdot cos\,\beta_L\,\lambda_l \cdot cos\,\beta_L\,\Lambda \end{bmatrix} \tag{2}$$

$$\Delta = \frac{1}{2} \cdot \left(1 + \frac{\beta_L\beta_l \cdot cos\,2 \cdot \beta_L \cdot \Lambda}{\chi^2 \cdot \varepsilon}\right) \tag{3}$$

$$\varepsilon = \zeta \cdot (\beta_L + \xi \cdot sin\,2 \cdot \beta_l\lambda_L) \tag{4}$$

$$\chi = e^{\xi \cdot \Lambda} \tag{5}$$

$$\Lambda = \frac{L}{\omega_{kr}} + \frac{l}{\omega_{kr}} = \lambda_L + \lambda_l, s \tag{6}$$

L, l – lengths of sections of the cylindrical surface of the grinding wheel [14], corresponding to the cutting protrusions and depressions, respectively, m (Fig. 1);

ω_{kr} – circumferential speed of the grinding wheel, m/s;

ξ is the coefficient characterizing the decrease in the depth of pulsation of parametric oscillations in time 1/s;

$$\beta_L = \sqrt{\frac{2 \cdot \sigma + \sigma^*}{2 \cdot M} - \xi^2} \tag{7}$$

$$\beta_l = \sqrt{\frac{2 \cdot \sigma - \sigma^*}{2 \cdot M} - \xi^2} \tag{8}$$

$$\sigma^* = \sigma \cdot \left(\frac{t - t_f}{t_f}\right) \tag{9}$$

M – reduced mass.

σ* – rigidity pulsation depth, equal to the difference between its maximum and minimum values, taken respectively during the periods of contact of the cutting segments with the material being processed and at time intervals in which there is no contact between the abrasive tool and the workpiece, kg/m (Fig. 1);

σ is the reduced rigidity of the elastic system of the machine, equal to the arithmetic mean between its minimum and maximum values, kg/m;

t, t_f are the depth of cut set on the machine and the actual depth of cut, taking into account the squeezing of the abrasive tool from the workpiece, respectively, m

β_L, β_l are the frequencies of oscillations occurring in time intervals characterized by the presence and absence of contact between the cylindrical surface of the grinding wheel and the flat surface of the workpiece, respectively, 1/s;

$$L = \frac{2 \cdot \pi \cdot R_{kr}}{n \cdot \left(1 + \frac{l}{L}\right)} \tag{10}$$

$$N = \frac{l}{L} \tag{11}$$

n – the number of cutting segments on the cylindrical working surface of the abrasive grinding wheel;

R_{kr} – grinding wheel radius, m;

λ_L, λ_l – time intervals corresponding to the presence and absence of contacts of fragments of the working cylindrical surface of the abrasive tool with a flat workpiece, respectively, s (Fig. 1).

4 Results and Discussion

By changing the dimensional relationships between the elements of the cylindrical working surface of the abrasive wheel and the number of these elements, it is possible to get rid of the risk of parametric resonance in the elastic system of the surface grinder.

Similarly, the risk of burns can be reduced by reducing the high local grinding temperatures to a level that does not exceed the tempering temperature in the surface layer of the workpiece [14].

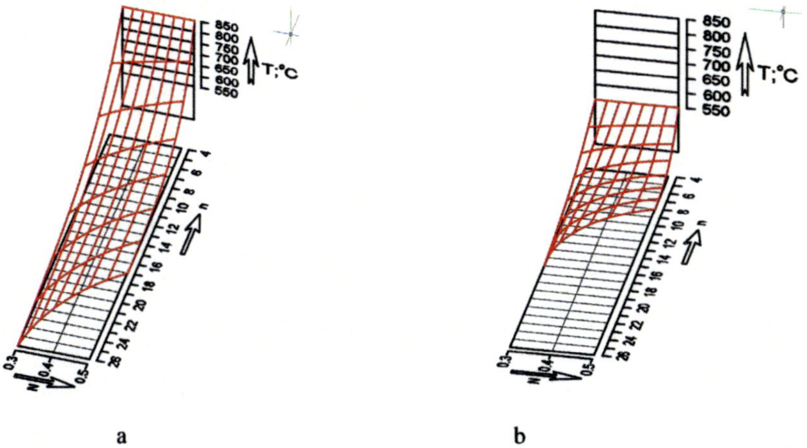

Fig. 2. Temperature drop 850 °C (a) and 550 °C (b) that occurred during grinding with a solid wheel, up to T = 400 °C when grinding with discrete wheels having different sets of lengths of active and passive segments, as well as their number on the working cylindrical surfaces of the wheels.

From Fig. 2, the temperature during grinding with a conventional wheel (without grooves) T = 850 °C (a) and T = 550 °C (b) can be reduced to a safe level T = 400 °C due to discretization of the working surfaces of the circle, varying the dimensional ratios between active and passive segments and their number.

It is possible to eliminate the risk of burns due to periodic short-term interruption of the cutting process, achieved by processing with discrete circles, in which: n = 17, N = 0.5; n = 20, N = 0.4; n = 26, N = 0.3 (Fig. 2a); n = 8, N = 0.5; n = 10, N = 0.4; n = 17, N = 0.3 (Fig. 2b).

From Fig. 3, the greater the required degree of reduction of local temperatures, the greater the discretization should be subjected to the working surface of the abrasive tool.

The graphs presented in Fig. 3 show that burn-free and resonance-free processing is ensured by using intermittent grinding wheels, the cutting topography of which is determined by certain predetermined dimensional ratios between its elements and the number of these elements.

The boundaries of non-resonant and non-burn zones follow similar patterns. This indicates that with the same parameters that determine the topography of the cutting surface [15] of a discontinuous circle, it is possible to provide such short-term force

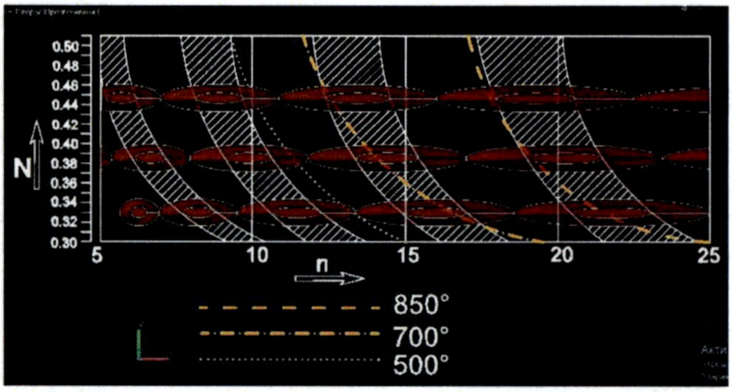

Fig. 3. Resonance-free zones located outside the hatched areas and burn-free zones located to the right of the dashed, dash-dotted, and dotted lines (at continuous grinding temperatures 850 °C, 700 °C and 500 °C respectively).

shock-cyclic and thermo-cyclic effects of the active segments of the circle on the workpiece, which will create conditions for reducing the temperature to a safe level (not exceeding the tempering temperature of the material being processed) and preventing parametric resonance [16].

Burn-free processing zones include various sets of geometric parameters N and n that define the topography of the working surfaces [17] of intermittent wheels during grinding with which the local temperatures that occur in the cutting zone do not exceed the temperatures at which structural and phase transformations begin to occur in the surface layer of the workpiece [18].

Each point on these lines corresponds to such a set of intermittent circle parameters n and N, at which the grinding temperature will be lowered to a safe limit compared to the temperature that occurs, ceteris paribus, during continuous grinding. Continuous grinding temperatures are respectively equal to 850 °C, 700 °C, and 500 °C.

Figures 4 and 5 show how the stiffness fluctuation depth σ* affects the nature of the surfaces described by the dependences $\mu = f\ (N, n)$ and $\Delta = f\ (N, n)$. A decrease in the parameter σ* leads to refinement and mutual approach of the zones of parametric instability, the boundaries of which are the lines of intersection of the surfaces μ and Δ.

In Fig. 4, the case is considered when the depth of stiffness pulsation σ* increases with an increase in the number of depressions on the working surface of the wheel, and in Fig. 6 – decreases. A unidirectional and multidirectional change in the parameters σ* and n leads to a significant change in the nature of the graphic dependences $\mu = f\ (n, N)$ and, consequently, to a change in the shapes and sizes of the parametric instability zones.

Figures 4 and 5 show graphs $\mu = f\ (n)$ calculated for various values of σ* using the Mapl2015 computer mathematics system. With an increase in the parameter σ* from 1.0 • 107 kg/m to 1.8 • 107 kg/m, the number of waves in the studied interval $6 \leq n \leq 16$ decreased from 4 to 3.

Figure 6 shows the dependence $(t - tf)/tf = f\ (n, N)$. The parameter $(t - tf)/tf$ characterizes the efficiency of grinding the material being processed by abrasive grains belonging to the working cylindrical surface of the grinding wheel. The smaller the

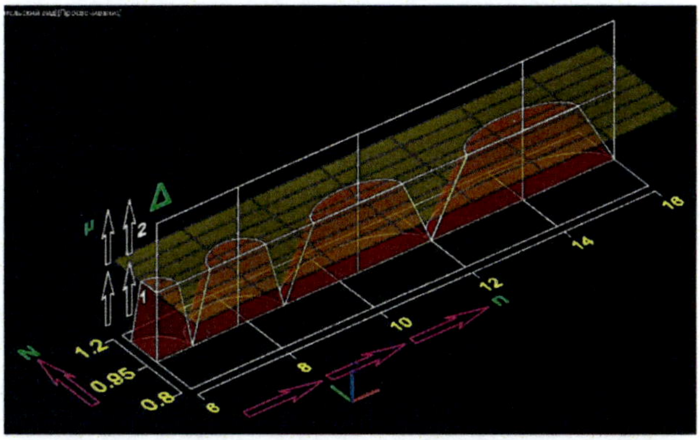

Fig. 4. Graphic dependencyΔ = f (n, N) before cutting off the tops of the ridges of the wavy surface described by the dependence μ = f (n, N) and after cutting off. In calculations, it was assumed that σ* increases with increasing n.

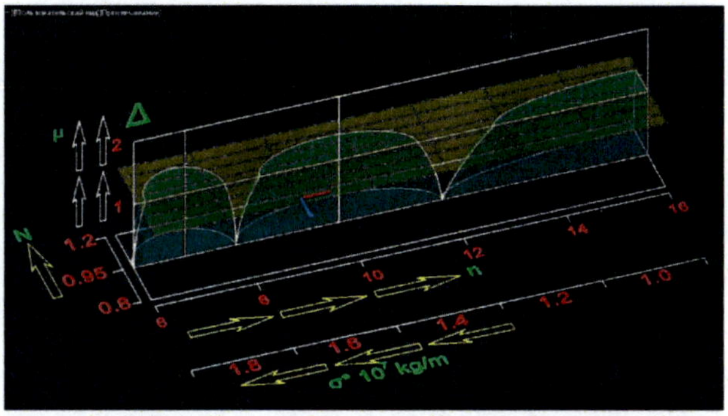

Fig. 5. Graphic dependency Δ = f (n, N) before cutting off the tops of the ridges of the wavy surface described by the dependence μ = f (n, N) and after cutting off. In calculations, it was assumed that σ* with increasing n.

numerical value of this parameter, the more efficiently the metal is removed by the cutting grains of the wheel. From formula (9), an increase in the ratio (t – tf)/tf (that is, a decrease in the cutting ability of the wheel) leads to an increase in the depth of stiffness pulsation σ* of the elastic system of the surface grinder.

In this case, the probability of the appearance of parametric resonance increases.

From the graph (t – tf)/tf = f (n, N), it can be seen that the dimensional ratios between the active and passive segments of the working cylindrical surface of the abrasive wheel have a significant impact on the efficiency of removing the material being processed. An increase in the numerical value of the ratio of the cavity length to the protrusion length leads to an increase in the cutting ability of the abrasive tool. In [15], it is not recommended to manufacture intermittent wheels with a numerical value of the parameter N > 1 since the operation of such wheels is accompanied by increased radial wear.

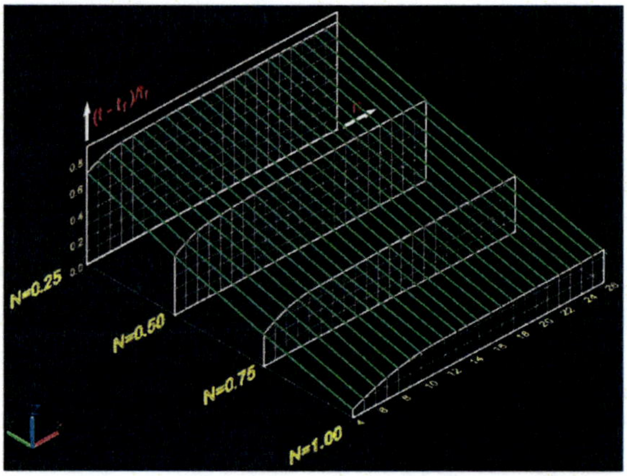

Fig. 6. The dependence of the parameter ((t/tf) − 1) on the parameters n and N, which determine the macrotopography of the cylindrical working surface of the intermittent grinding wheel.

Usually, when designing discrete abrasive wheels, such numerical values of the parameter N are assigned, at which the length of the cavity ι would fluctuate in the range $0.3 \bullet L \leq \iota \leq 0.5 \bullet L$.

From Fig. 6, it can be seen that with an increase in the discretization of the working surface of the circle, a slight increase in the parameter (t − tf)/tf is observed, i.e., decrease in the cutting ability of the abrasive tool.

The graph (t − tf)/tf = f (n) changes its character to the reverse: the discretization of the working surface of the abrasive wheel is accompanied by a slight increase in its cutting ability.

An increased number of depressions on the wheel leads to decreased elastic movements due to the delay in the spindle releases from the periodic short-term shock action of the abrasive tool on the workpiece. As a result, the cutting grains have more metal removal, and the load acting on them increases, contributing to their destruction and the formation of new sharp cutting edges.

5 Conclusions

The influence of the scale factor of the dimensions of the cutting elements of the macrotopography of the working surface of discontinuous circles on their cutting ability, heat release in the zone of contact of the abrasive tool with the workpiece, and the parametric instability of the elastic system of the machine tool and, as a result, on the quality parameters of the surface layer of ground parts has been established.

It has been established that the required parameters of the quality of the surface layer of parts for the operation of flat grinding with intermittent wheels are provided by decreasing scaling of the elements that make up their working relief, i.e., the use of abrasive wheels with multiple narrow holes on their working surfaces, parallel to the axis of rotation.

The design of a discrete wheel is proposed, which provides the required geometric and physical-mechanical parameters of the quality of the surface layer of workpieces during the grinding operation.

The possibility of a cumulative decrease in temperature in the cutting zone and undesirable high-amplitude oscillations in the elastic system of the machine is achieved by identifying similar patterns of their manifestation when grinding with traditional intermittent wheels.

References

1. Jiang, X., Liu, K., Li, M., Gong, P., He, H.: Grinding temperature and surface integrity of quenched automotive transmission gear during the form grinding process. Materials **15**, 7723 (2022). https://doi.org/10.3390/ma15217723
2. Uhlmann, E., Lypovka, P., Hochsild, L., Schröer, N.: Influence of rail grinding process parameters on rail surface roughness and surface layer hardness. Wear **366–367**, 287–293 (2016)
3. Lin, B., Zhou, K., Guo, J., Liu, Q.Y., Wang, W.J.: Influence of grinding parameters on surface temperature and burn behaviors of grinding rail. Tribol. Int. **122**, 151–162 (2018)
4. Del Re, F., Dix, M., Tagliaferri, F.: Grinding burn on hardened steel: characterization of onset mechanisms by design of experiments. Int. J. Adv. Manuf. Technol. **101**, 2889–2905 (2019)
5. Heinzel, C., Heinzel, V., Guba, N., Hüsemann, T.: Comprehensive analysis of the thermal impact and its depth effect in grinding. CIRP Ann. Manuf. Technol. **1**, 289–292 (2021)
6. Ryabenkov, I.A.: Theoretical substantiation of the technological possibilities of conventional and interrupted grinding. Cutting Tools Technol. Syst. **89**, 149–157 (2018)
7. Denkena, B., Grove, T., Göttsching, T.: Grinding with patterned grinding wheels. CIRP J. Manuf. Sci. Technol. **8**, 12–21 (2015)
8. Bogutsky, V.B., Shron, L.B.: On the expediency of using grinding wheels with intermittent profile in flat grinding operations. Progress. Technol. Syst. Mech. Eng. **2**(65), 10–15 (2019)
9. Azarhoushang, B., Daneshi, A., Lee, D.H.: Evaluation of thermal damages and residual stresses in dry grinding by structured wheels. J. Clean. Prod. **142**(4), 1922–1930 (2017). https://doi.org/10.1016/j.jclepro.2016.11.091
10. Aurich, I.C., Herzenstiel, P., Suderman, H., Magg, T.: High – performance dry grinding using a grinding wheel with a defined grain pattern. CIPP Ann. – Manuf. Technol. **57**(1), 357–362 (2018)
11. Tonkonogyi, V., Dašić, P., Rybak, O., Lysenko, T.: Application of the modified genetic algorithm for optimization of plasma coatings grinding process. In: Karabegović, I. (ed.) NT 2019. LNNS, vol. 76, pp. 199–211. Springer, Cham (2020). https://doi.org/10.1007/978-3-030-18072-0_23
12. Tonkonogyi, V., Sidelnykova, T., Dašić, P., Yakimov, A., Bovnegra, L.: Improving the performance properties of abrasive tools at the stage of their operation. In: Karabegović, I. (ed.) NT 2019. LNNS, vol. 76, pp. 136–145. Springer, Cham (2020). https://doi.org/10.1007/978-3-030-18072-0_15
13. Larshin, V., Lishchenko, N., Pitel, J.: Intermittent grinding temperature modeling for grinding system state monitoring. Appl. Asp. Inf. Technol. **3**(2), 48–73 (2020)
14. Oborskyi, G., Orgiyan, A., Tonkonogyi, V., Aymen, A., Balaniuk, A.: Study of dynamic impacts at combined operations of the thin turning and boring. In: Tonkonogyi, V., et al. (eds.) InterPartner 2019. LNME, pp. 226–235. Springer, Cham (2020). https://doi.org/10.1007/978-3-030-40724-7_23

15. Larshin, V., Lishchenko, N.: Gear grinding system adapting to higher CNC grinder throughput. In: MATEC Web of Conferences, vol. 226, p. 04033 (2018)
16. Naddachin, V.B.: Relationship between processing temperature and interruption in the grinding process. Cutting Tools Technol. Syst. **93**, 122–137 (2020)
17. Tonkonogyi, V., Yakimov, A., Bovnegra, L., Sidelnykova, T., Dašić, P.: The use of intermittent wheels, impregnated by the contact method to reduce the thermal stress of the grinding process. In: IOP Conference Series: Materials Science and Engineering, vol. 708, p. 012034(2019). https://doi.org/10.1088/1757-899X/708/1/012034
18. Usov, A., Tonkonogyi, V., Dašic, P., Rybak, O.: Modelling of temperature field and stress-strain state of the workpiece with plasma coatings during surface grinding. Machines **7**(1), 20 (2019). https://doi.org/10.3390/machines7010020

Study of the Scaling Effect of Cutting Elements of the Abrasive Wheels' Discretized Working Surface on Geometric and Physical-Mechanical Parameters of the Surface Layer Quality of Ground Parts

Oleksii Yermolenko[1], Fedir Novikov[1], Alexey Yakimov[2(✉)], Yuliia Babych[2], and Alla Toropenko[2]

[1] Simon Kuznets Kharkiv National University of Economics, 9-A, Nauky Ave., Kharkiv 61166, Ukraine
[2] Odessa Polytechnic National University, 1, Shevchenko Ave., Odesa 65044, Ukraine
yakimov_aa@ukr.net

Abstract. The paper proves that reducing the scale of the abrasive wheel's working surface discretization decreases the spindle unit's vibration level and creates processing conditions similar to ultrasonic vibrational grinding. It was established that the boundaries of the regions of parametric instability of the elastic system of a surface grinding machine and undesirable structural changes in the surface layer of the machined part have a similar character. This means that the geometric and physicomechanical quality indicators of the surface layer of the part during the grinding operation can be stabilized by scaling the macro-relief of the working surface of the discontinuous circles (proportional reduction of the lengths of the protrusions and recesses) and reducing the discretization step of this surface while simultaneously reducing the numerical value of the ratio of the width of the recess to the length of the protrusion within the specified limits $0{,}3 \leq N \leq 0{,}5$. It was established that interrupted cut-off wheels with many grooves parallel to the wheel's axis and having a through-hole design (from end to end) maintain their cutting ability over time much better than high-porosity wheels. Therefore, replacing these wheels with high-porosity ones is counterproductive. To ensure the required quality of the surface layer of the machined parts in flat grinding operations, interrupted abrasive wheels with a reduced scale of discretization of the working surface were developed.

Keywords: Stiffness Pulsation Depth · Discretization Degree · Zone Boundaries · Working Surface Macrotopography · Industrial Innovation

1 Introduction

Grinding is the final operation of machining, which forms the quality of the surface layer of parts and, as a consequence, their operational properties. Grinding is accompanied by thermal phenomena, which can cause structural heterogeneity, hardness reduction, and

V. Tonkonogyi et al. (Eds.): InterPartner 2023, LNME, pp. 409–418, 2024.
https://doi.org/10.1007/978-3-031-42778-7_38

formation of tensile residual stresses in thin surface layers of the workpiece. To prevent these grinding defects, abrasive wheels with a discrete working surface consisting of cutting and discontinuous sections are used. Interrupted contact of the abrasive wheel with the workpiece causes a change in the rigidity of the machine's elastic system and, consequently, the appearance of parametric resonance. This reduces the geometric accuracy of the parts and contributes to the appearance of cyclic burns in their surface layer. To manage the surface layer quality in the grinding operation, it is essential to establish the laws of occurrence of parametric resonance in the machine's elastic system and develop measures to prevent it.

It is also essential to analyze the thermal stress of the grinding process by abrasive wheels with discrete working surfaces and justify the conditions of its reduction. In this case, it is necessary to establish geometric parameters of the discrete working surface of the abrasive wheel, in compliance with which can significantly reduce the grinding temperature and the amplitude of parametric oscillations. This will ensure grinding by abrasive wheels with discrete working surfaces simultaneously stabilizing both physical and mechanical quality parameters of the workpiece surface layer and parameters of geometric accuracy of machining. The solutions obtained on this basis will have a critical scientific and practical importance for determining new technological capabilities of the grinding process, in terms of reducing its thermal and force stresses and the design of highly effective discrete abrasive wheels on ceramic bond for the operation of planar grinding.

2 Literature Review

In [1], an informative review of research related to the grinding of discontinuous (textured) wheels (TGW) is given [2]. In [3], a classification of discontinuous grinding wheels according to the width of the groove is made [1]. According to this classification, all discontinuous grinding wheels are divided into three groups: wheels with megarelief (groove width 0.5–100 mm), wheels with macrorelief (macrotextured wheels with groove width 0.1–0.5 mm), wheels with microrelief (microtextured wheels with groove width 0.01–0.1 mm). Interrupted circles with megarelief were first invented in the late twenties [4] and early thirties of the last century [5] (Fig. 1).

a b

Fig. 1. Abrasive wheel with slots parallel to its axis (a) and macrotopography of the working surface of the wheel in the form of a checkerboard.

In [6], all discontinuous circles are divided into segmented tools, assembled from individual segments with gaps (slots) between them, and textured tools, the working surface of which contains blank areas (without cutting grains) [7]. In [8], the temperature of discontinuous grinding decreases with increasing protrusions and troughs at a constant ratio of lengths [9]. The author concludes that it is advisable to use discontinuous wheels with many protrusions and troughs in deep grinding. Similar results were obtained by the authors of [10, 11]. In these works, the authors put forward and substantiate the conclusion about the expediency of replacing discontinuous wheels with highly porous abrasive wheels. Their conclusion is based on the fact that with a decrease in the size of protrusions and troughs, the length of the trough approaches the pore size of a highly porous wheel.

Therefore, it seems necessary and relevant to compare the technological capabilities of solid, discontinuous, and highly porous abrasive wheels based on theoretical and experimental research. In particular, this applies to the assessment of the stability of the cutting ability of the specified abrasive wheels in time, which determines the intensity of thermal and mechanical processes during grinding, and, accordingly, physical-mechanical and geometrical parameters of the surface layer quality of machined parts. It is essential to substantiate the conditions of occurrence and elimination of parametric resonance in the elastic system of the machine when grinding with discrete abrasive wheels. This will allow a scientifically justified approach to the management of thermal and mechanical processes in grinding and the design of high-performance discrete abrasive wheels on ceramic bonds for the operation of flat grinding. The work aims to identify the possibility of cumulative provision of physical-mechanical and geometric parameters of the quality of the surface layer of machined parts by scaling the working surface of discrete ceramic-bonded abrasive wheels in the operation of planar grinding.

3 Research Methodology

Theoretical studies were carried out based on the thermal physics of cutting and the theory of oscillations. Experimental studies were carried out on the surface grinding machine model 3G71M with abrasive wheels 250 × 25 × 75 24A F60 K10 V5 and 250 × 25 × 75 24A F60 K6 V5. Experiments to determine the influence of geometrical parameters of cutting macro-relief of the discontinuous wheel on the grinding temperature were carried out on samples of X12MF steel (Japanese analog SKD11, French analog Z160CDV12, English analog BD2, German analog X155CrVMo 12-1, US analog D2).

Experiments to compare the cutting ability of solid, highly porous, and discontinuous abrasive wheels were carried out on samples of 12X2N4A steel (Japanese analog of steel SNC B15, French analog 12NC15, English analog 655V13, German analog 14NiCr14, US analog 3415. Bulgarian analog 12ch2N4A, Polish analog 12H2N4A). Graphical relationships were constructed using the computer mathematics system Mapl 2015 and the computer-aided design system AutoCAD 2019.

10 × 20 × 30 mm flat specimens of X12M steel were used to measure the grinding temperature. The specimens were heat treated using the following regimes: quench heating at 1020–1040 °C and cooling in oil. The samples were preheated at 850 °C before heating to quench the temperature. After quenching, the samples were subjected to a low-temperature tempering in a thermostat at 170–180 °C for 2 h. After the above heat treatment, the samples had a hardness of 59–63 HRC.

4 Results and Discussion

Calculations of temperature fields formed during grinding by discontinuous grinding wheels with different numbers of cutting edges on their working surfaces were carried out according to the formula whose conclusion was based on the principle of superposition of temperature fields formed by the action of continuous but time-shifted positive and negative heat sources. In such a way, the temperature fields arising from the operation of single-cutting protrusions of a discontinuous grinding wheel were formed.

From Fig. 2 (a), it can be seen that the temperature of continuous grinding $T = 850$ °C can be reduced to $T = 400$ °C if 26 depressions ($n = 26$) are made on the working surface of the circle, and the numerical value of the ratio of their lengths to the lengths of the protrusions is kept equal to $N = 0.3$. If the value of this ratio is made equal to $N = 0.5$, then to achieve the same result ($T = 400$ °C), a smaller number of depressions ($n = 17$) on the abrasive tool will be required.

From Fig. 2 (b) it can be seen that if the temperature of continuous grinding is $T = 550$ °C, then the safe temperature $T = 400$ °C will be provided by intermittent wheels with the same values of the ratios of the dimensions of depressions and protrusions ($N = 0.3$; $N = 0.5$), but significantly fewer of them ($n = 17$; $n = 8$, respectively).

From Fig. 2, the same goal ($T = 400$ °C) can be achieved by increasing the number of protrusions n with a simultaneous decrease in the value of N. If both parameters (n and N) are simultaneously increased, then the temperature will be reduced to a lower level ($T < 400$ °C) [12].

The effect of geometric parameters (n and N), which determine the macro topography of the working surface of an intermittent abrasive wheel, on the grinding temperature T was studied experimentally [13].

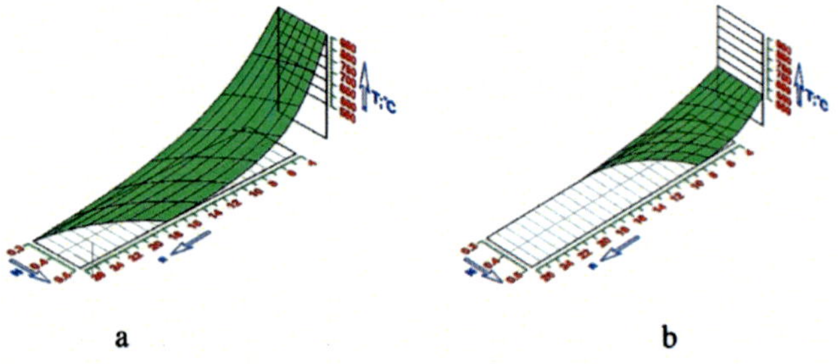

Fig. 2. Reducing the continuous grinding temperatures of 850 °C (a) and 550 °C (b) to the level $T = 400$ °C due to intermittent wheels with different designs of working surfaces.

Figure 3 shows two experimental dependencies:

- dependence of the temperature T of intermittent grinding on the number of depressions on the wheel with their constant length $L_1 = 22$ mm (b);
- dependence of temperature T on the length of troughs L_2 with their constant number $n = 6$ (a).

Samples of X12M steel (HRC 59–60) were ground on a surface grinder at $V_{cr} = 35$ m/s; $V_d = 10$ m/min; $t = 0.03$ m, without the use of cutting fluids around 24A F60

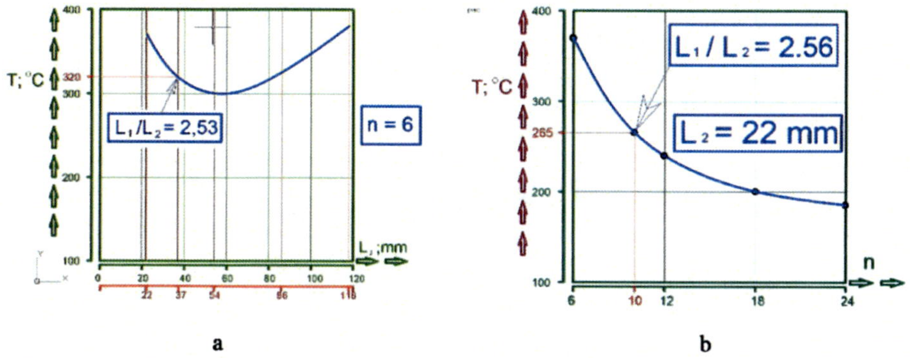

Fig. 3. Dependence of the grinding temperature on the length of the cavity L_2 (a) and the number of cutting protrusions *n* (b).

K 6 V5. From the analysis of the graphs $T = f(L_2)$ and $T = f(n)$, it follows that with an increase in the number of cavities at a constant value of the ratio of the length of the protrusion to the length of the cavity, equal to 2.56, the temperature in the cutting zone decreases.

Dependence $T = f(L_2)$ is extreme. This can be explained by the fact that an increase in the size L_2 of the passive area contributes to the cooling of the machined surface but also causes an increase in the cutting force due to the incoming allowance.

To prevent the deterioration of the geometric accuracy of ground surfaces caused by interrupted cutting, theoretical studies have been carried out to identify the occurrence of parametric resonance in the elastic system of a surface grinding machine caused by periodic changes in its rigidity. This condition has the following form: $|\mu| > (1 + \Pi)/2$, where μ and Π are dimensionless parameters, depending on the grinding conditions and geometrical parameters of macrotopography of the working surface of a discrete abrasive wheel. The boundaries of the areas of parametric instability of the machine tool elastic system are the lines of intersection of two surfaces describing the left and right parts of the condition of unstable operation of the surface grinding machine elastic system (Fig. 4). These lines outline the sets of such values of geometric parameters of the working macro-relief of the discontinuous abrasive wheel, at which in the elastic system of the machine tool a parametric resonance can arise.

It has been established that the value of the ratio N of the circumferential arc length of the air gap between two adjacent active segments of the abrasive wheel L_2 to the circumferential length of the cutting segment L_1, as well as their number n on the wheel, have a significant impact on the possibility of parametric resonance in the elastic system of a surface grinding machine (Fig. 4). The arcuate boundaries of the areas of parametric instability of the elastic system of the machine (Fig. 5) are similar to the boundaries of the zones, inside which there is a high probability of the appearance of thermal surface defects (Fig. 3, a,b).

The probability of exit from zones, including grinding defects caused by thermal or dynamic reasons, increases with the number of cutting protrusions on an intermittent abrasive wheel. It is possible to minimize the risk of parametric oscillations in the elastic

Fig. 4. Influence of the ratios of the lengths of the protrusions and depressions ($N = L_2/L_1$) and the number (n) on the working surface of the abrasive wheel on the parametric stability of the elastic system of the surface grinder μ.

system of the machine by using highly porous wheels or intermittent wheels, the cavities of which do not have an exit to the end. When grinding with such wheels, the constancy of contact of their cutting surfaces with the processed material is ensured. This is consistent with the work [14, 15].

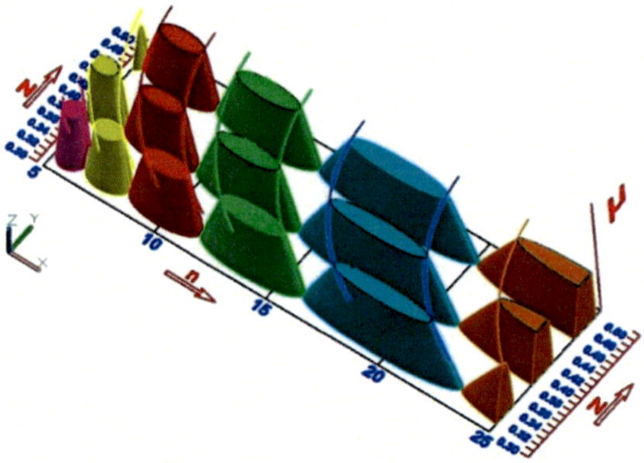

Fig. 5. Influence of the geometric parameters *N* and *n*, which determine the macrotopography of the working surface of intermittent abrasive wheels, on the geometric shape and dimensions of the zones of parametric instability of the elastic system of a surface grinder.

The physical and mechanical state of the surface layer of the workpieces largely depends on the stability of the cutting ability of abrasive wheels over time.

Experiments were conducted to compare the cutting ability of solid, highly porous and discontinuous abrasive wheels over time. The cutting ability of the wheels was estimated by the ratio of the tangential (P_z) component of the cutting force to its normal (P_y) component. Samples of steel 12X2H4A were processed on a cylindrical grinding machine for 30 min without transverse feed according to an elastic pattern (wheel pressing force $F_y = 1.2$ N/min) in the following mode: wheel rotation speed $V_{kr} = 35$ m/s, part rotation speed $V_d = 0.1$ m/s. We tested continuous and intermittent (trough length $L_2 = 24$ mm, protrusion length $L_1 = 36$ mm) wheels of the same characteristics 24A F60 K 6 V5 and a wheel with an open structure of 10 and medium soft hardness *K*, which can be attributed to a highly porous abrasive tool [16].

The ratio P_z/P_y characterizes the degree of blunting of the cutting grains of an abrasive tool. The graphs in Fig. 6 show that the cutting ability of the highly porous wheel after a 30-min grinding period is better than that of a solid wheel but worse than that of an interrupted one. The high cutting ability of a discontinuous circle can be explained by the fact that during its operation in the elastic system of the machine, forced high-frequency vibrations occur, which change the wear conditions of abrasive grains and reduce the likelihood of their adhesive setting with the material being processed. Reducing the scaling of the macrotopography of the working surface of abrasive wheels leads to a decrease in the likelihood of parametric resonance in the elastic system of the machine, a decrease in the heat stress of processing, and a decrease in the amplitude values of the high-frequency vibration of the tool, which has a positive effect on the quality indicators of the surface layer of the workpieces. Increasing the discretization degree of the cutting macrorelief wheel induces processing conditions close to ultrasonic vibratory grinding [17, 18]. Experimental graphs (Fig. 6) agree with the data given in [19].

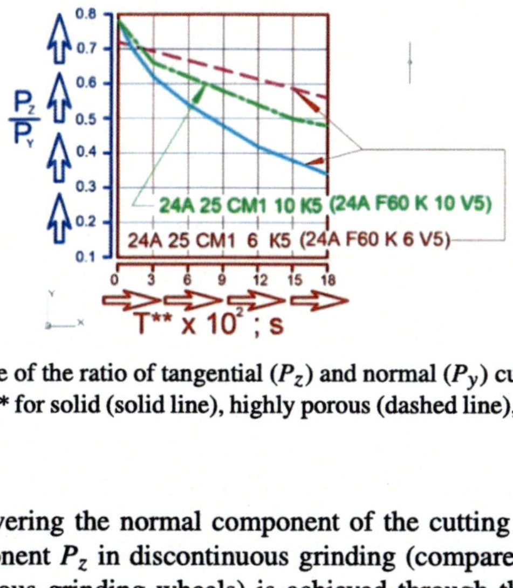

Fig. 6. The dependence of the ratio of tangential (P_z) and normal (P_y) cutting force components on the grinding time *T*** for solid (solid line), highly porous (dashed line), and interrupted (dotted line) abrasive wheels.

The effect of lowering the normal component of the cutting force P_y concerning the tangential component P_z in discontinuous grinding (compared with grinding with solid and highly porous grinding wheels) is achieved through the creation of forced high-frequency vibrations in the machine's elastic system, which significantly reduce resistance against chip flow and facilitate the chip-forming process.

Based on our research, we propose designs of interrupted wheels, the appearance of which is shown in Fig. 7 and 8.

They have a substantially increased number of working ledges and have created through-and-through hollows of varying lengths and configurations on the working surface of the wheel. This actually excludes the shocking character of the interaction of a wheel with processed material and provides parametric stability of the elastic system of

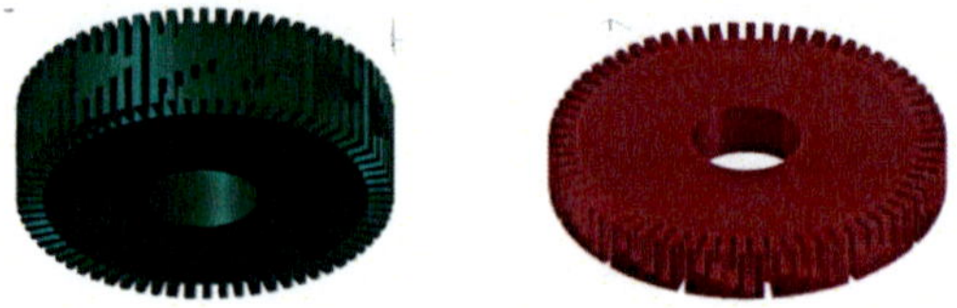

a b

Fig. 7. The working surface of the wheel with through slots and recesses of different lengths parallel to the axis of rotation of the abrasive tool (a), the working surface of the wheel consisting of cyclically repeating sections with monotonically changing lengths of recesses, transitioning to through slots (b).

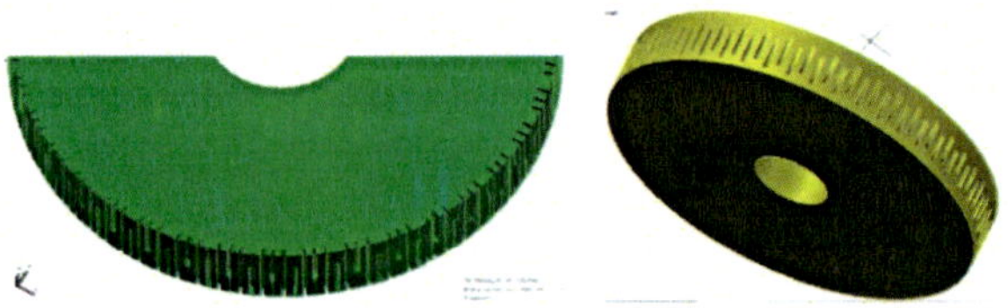

a b

Fig. 8. The working surface of the abrasive tool with a large number of recesses and slots parallel to the axis of the wheel (a), the working surface of the tool with a large number of recesses that do not extend to the ends of the abrasive tool (b).

surface grinding machine, and also simultaneously provides a decrease in the grinding temperature and improvement of quality of processing. The proposed designs of abrasive wheels can be effectively used in the design of technological processes of grinding. However, they are justified only based on theoretical studies and need experimental verification and clarification for specific processing conditions.

5 Conclusions

It has been established that reducing the scale of the abrasive wheel's working surface discretization decreases the spindle unit's vibration level and creates processing conditions similar to those of ultrasonic vibrational grinding. It has been established that the number and dimensions of the structural elements that make up the working surface of a discontinuous circle strongly influence the thermal stress in the grinding zone and the parametric stability of the grinding process. Studies have shown that the likelihood of undesirable structural changes and deterioration of the macro- and microgeometry of the surface layer of ground parts can be reduced by scaling the structural elements on the cylindrical surface of the abrasive wheel.

It has been established that the boundaries of the regions of parametric instability of the elastic system of a surface grinding machine and undesirable structural changes in the surface layer of the machined part have a similar character. This means that the geometric and physicomechanical quality indicators of the surface layer of the part during the grinding operation can be stabilized by scaling the macro-relief of the working surface of the discontinuous circles (proportional reduction of the lengths of the protrusions and recesses) and reducing the discretization step of this surface while simultaneously reducing the numerical value of the ratio of the width of the recess to the length of the protrusion within the specified limits $0.3 \leq N \leq 0.5$.

It has been established that interrupted cut-off wheels with many grooves parallel to the wheel's axis and having a through-hole design (from end to end) maintain their cutting ability over time much better than high-porosity wheels. When grinding with discrete wheels, the interrupted cutting process induces high-frequency oscillations in the elastic system of the machine, facilitating chip formation and promoting the timely destruction of dulled abrasive grains and the formation of new sharp cutting edges on them. Therefore, replacing these wheels with high-porosity ones is counterproductive.

To ensure the required quality of the surface layer of the machined parts in flat grinding operations, interrupted abrasive wheels with a reduced scale of discretization of the working surface have been developed. However, they are justified only based on theoretical studies and need experimental verification and clarification for specific processing conditions.

References

1. Li, H.N., Axinte, D.: Textured grinding wheels: a review. Int. J. Mach. Tools Manuf. **109**, 8–35 (2016). https://doi.org/10.1016/j.ijmachtools.2016.07.001
2. Morozov, A.V.: Substantiation of abrasive disc discretization with highly concentrated energy flows. Sci. Intensiv. Technol. Mech. Eng. **12**, 14–19 (2016). https://doi.org/10.12737/23482
3. Guo, B., Zhao, Q., Fang, X.: Precision grinding of optical glass with laser micro-structured coarse-grained diamond wheels. J. Mater. Process. Tech. **214**(5), 1045–1051 (2014). https://doi.org/10.1016/j.jmatprotec.2013.12.013
4. Sherk, H.E.: Slotted abrasive wheel. Patent US 2049874A. Miami Abrasive Products Inc., United States (1933)
5. Mosher, A.G.: Abrading wheel. Patent US 1736355A. United States, Syracuse, New York (1929)
6. Del Re, F., Dix, M., Tagliaferri, F.: Grinding burn on hardened steel: characterization of onset mechanisms by design of experiments. Int. J. Adv. Manuf. Technol. **101**, 2889–2905 (2019). https://doi.org/10.1007/s00170-018-3156-6
7. Gusev, V.G.: Aerodynamic streams at cylindrical internal grinding by the textured wheels. In: MATEC Web of Conference, vol. 298, p. 00018 (2019). https://doi.org/10.1051/matecconf/201929800018
8. Ryabenkov, I.A.: Theoretical substantiation of the technological possibilities of conventional and interrupted grinding. Cutting Tool Technol. Syst. **89**, 149–157 (2018)
9. Mao, C., et al.: Simulation and experiment of electroplated grinding wheel with orderly-micro-grooves. J. Manuf. Process. **79**, 284–295 (2022). https://doi.org/10.1016/j.jmapro.2022.04.063
10. Lishchenko, N.V., Larshin, V.P.: Geometric optimization of intermittent grinding wheels. Bull. Natl. Tech. Univ. Ukraine "Kyiv Polytech. Inst." **65**, 110–117 (2012). Series: Mechanical Engineering
11. Lishchenko, N.V., Larshin, V.P.: Temperature determination when grinding with intermittent and highly porous wheels. Sci. Notes: Interuniv. Collect. **40**, 150–158 (2013)
12. Yakimov, A.A.: Calculation of temperatures formed during gear grinding with disc wheels using the zero-degree method, taking into account multi-pass. High Technol. Mech. Eng. **1**(23), 220–230 (2013)

13. Novikov, F.V., Yakimov, A.V.: Physical and Mathematical Theory of Material Processing Processes and Engineering Technology. ONPU, Odessa (2003)
14. Rirsch, B., Aurich, J.C.: Influence of the macro-topography of grinding wheels on the cooling efficiency and the surface integrity. Procedia CIRP **13**, 8–12 (2014). https://doi.org/10.1016/j.procir.2014.04.002
15. Ding, N., Jiang, S., Duan, J., Liu, C., Cui, S.: Design of new slotted structured grinding wheel. In: Journal of Physics: Conference Series, vol. 1635, p. 012013 (2020). https://doi.org/10.1088/1742-6596/1635/1/012013
16. Starkov, V.K.: Highly porous CBN wheels for grinding without cooling. J. Superhard Mater. **35**(5), 56–62 (2013). https://doi.org/10.3103/S1063457613050055
17. Cao, Y., Ding, W., Zhao, B., Wen, X., Li, S., Wang, J.: Effect of Intermittent cutting behavior on the ultrasonic vibration-assisted grinding performance of Inconel 718 nickel-based superalloy. Precis. Eng. **78**, 248–260 (2022). https://doi.org/10.1016/j.precisioneng.2022.08.006
18. Ding, W., Huand, Q., Zhao, B., Chen, Q.: Wear characteristics of white corundum abrasive wheel in ultrasonic vibration-assisted grinding of AISI 9310 steel. Ceram. Int. **49**(8), 12832–12839 (2022). https://doi.org/10.1016/j.ceramint.2022.12.153
19. Marchuk, V.I., Hryniuk, S.V., Marchuk, I.V., Sachkovskaya, L.O.: Dynamic model of grinding circuit grinding process. Perspect. Technol. Devices **16**, 33–37 (2020). https://doi.org/10.36910/6775-2313-5352-2020-16-5

Mechanical Engineering

Analysis of Factors Affecting the Energy Efficiency of an Elevator Winch

Andrii Boiko, Elena Naidenko(✉), Oleksandr Besarab, and Oleksandr Bondar

Odessa Polytechnic National University, 1, Shevchenko Ave., Odessa 65044, Ukraine
a.o.boyko@opu.ua, alena2808@ukr.net

Abstract. Based on the energy conservation equation during the machine's movement, dependencies were obtained to determine the efficiency of the elevator winch for two modes of movement: at a steady speed and during start-up. The dependence for calculating the efficiency of an elevator winch in the start-up mode is determined based on the analysis of dynamic loads that occur when lifting the cabin. At the same time, it was determined that the efficiency of the winch for this mode depends not only on inertial loads but also on the rigidity of the rope. It has been established that with an increase in the rigidity of the hoisting rope, the maximum value of the force in the rope increases after the cabin is detached from the base, and, accordingly, the efficiency of the passenger elevator winch decreases in dynamic modes. It is shown that the efficiency of pulley blocks depends not only on the friction in the block support but also on the rigidity of the lifting rope: the greater the rigidity of the rope, the lower the efficiency of the mechanical block.

Keywords: Energy Efficiency · Elevator Winch · Pulley Block · Rigidity · Lifting Rope · Dynamic Loads · Dynamic Mode · Maximum Force Value · Industrial Growth

1 Introduction

The European Union supports lift owners and manufacturers in their efforts to reduce costs and save resources. Lift systems consume up to 10% of a building's electricity [1]. When designing lifts [2], the primary task is to develop machine mechanisms with the best quality indicators [3].

To assess the quality of the machine according to the principle of energy consumption when the lift performs the necessary work, the efficiency and the proportion of losses are used [4].

Consider the energy balance of the machine [5], based on which it is possible to find a relationship to determine the coefficient of performance (COP) and the share of losses for various periods of the machine's operation, as well as for the entire cycle of operation. The energy conservation equation can be written as the sum of the powers of the forces acting in the machine.

$$P_{\mathrm{d}} - P_{\mathrm{usful}} - P_{harm} - P_{\mathrm{rec}} \pm P_{\mathrm{el}} \pm P_G \pm P_{\mathrm{in}} = 0, \tag{1}$$

V. Tonkonogyi et al. (Eds.): InterPartner 2023, LNME, pp. 421–431, 2024.
https://doi.org/10.1007/978-3-031-42778-7_39

where P_d is the power developed by the engine of the machine, W; P_{usful} is the power of useful resistance forces, W; P_{harm} is the power of harmful resistance forces, W; P_{rec} is the instantaneous power of accumulating devices that allow storing the energy of braking forces, W; P_{el} is the power spent on the deformation of the links, W; P_G is the power of the gravity forces of the links, W; P_{in} is the power of inertia forces.

The signs in front of the numbers of the energy balance equation are chosen according to their physical meaning: the positive sign is in front of the power, the increase of which causes an increase in speed. The last members of the energy balance P_{el}, P_G, P_{in} can have positive and negative values at different times, i. e. can be combined with engine power or used for braking.

During the full cycle of the steady motion of the machine, the sum of the work of the last three terms of the energy balance (1) is equal to zero.

The criteria for evaluating the efficiency of energy consumption (engine operation) for a driving cycle (without recuperation) can be cycle efficiency

$$\eta_{cycly} = \left(\frac{A_{losses}}{A_d}\right)_{cycle}. \tag{2}$$

Dependencies are obtained to determine the efficiency of the elevator winch by expression (2) for the following modes of motion: with a steady speed

$$\eta_{d\ st} = \eta_{tr} \cdot \eta_r \cdot \eta_d \tag{3}$$

and during launch

$$\eta_{launch} = \left(\frac{A_{usful}}{A_d}\right)_{cycle} = \frac{W_c + A_c}{\frac{(W_c+A_c)+W_{ke}+W_{bl}+W_{ts}}{\eta_{te}\eta_r\eta_d} + 3(W_d + W_r)}, \tag{4}$$

where W_c is the kinetic energy of the accumulated nominal mass of the load moving at nominal speed, A_c is the work of the gravity of the load, taking into account the influence of the counterweight on the path of the cabin acceleration, W_{ke} is the total kinetic energy of the cabin and the counterweight, W_{bl} is the kinetic energy stored by the blocks, W_{ts} is kinetic energy stored by the traction sheave; W_d - kinetic energy determined by the rotor of the winch motor, W_r- the energy of the gear wheels of the reducer and the electromagnetic brake pulley, η_{te}- cable transmission efficiency, r.u; η_r– gearbox efficiency, r.u.; η_d- engine efficiency.

Analysis of expression (4) shows that to increase the efficiency of the elevator winch during the start-up period, it is necessary to reduce the mass of all moving parts of the system, the diameters of the blocks (as well as the number of blocks) and the traction sheave, motor rotor, gear wheels of the gearbox (or abandon the gearbox).

2 Literature Review

Authors [6] proposed a controller to dissipate the vibratory energy of an elevator. The paper explores synthesizing the sufficient expertise of a lift traffic analysis expert and implementing this expertise into software [7]. The Monte Carlo Simulation method

(MCS) [8] has been successfully used to find the value of the elevator round trip time under general traffic conditions [9]. To decrease the vibration of the lift cabin [10], an analytical model for calculating the time history response of the lift cabin will be constructed [11]. The reference sources [12, 13] are devoted to analyzing and calculating passenger elevators' dynamic loads due to the presence of elastic mechanical links in rope gears. Based on the research, several criteria for assessing the effect of elasticity on transient processes are proposed [14], allowing to evaluate this effect without cumbersome calculations [15], at the same time that various methods of vibration damping are suggested [16]. In many works, vertical and lateral vibrations are studied independently [17]. Lateral natural frequencies [18] were calculated using a lumped-parameter model where the rope was decomposed in a series of rigid links inter-connected by torsion springs. Authors [19] proposed a parameter optimization method to minimize car acceleration, and the articles [20, 21] proposed several control methods to suppress lateral vibrations of elevators represented by lumped-parameter models.

The drive unit of the crane has a significant impact on electricity consumption, as operating costs, the durability of the equipment, as well as driving comfort, i.e., smooth starts and accurate and smooth stops at stops, all depend on this. In electric cranes, these conditions are best met by gearless drives built on synchronous permanent-magnet motors [22].

3 Research Methodology

3.1 Dynamic Model of an Elevator Winch

In expression (4) for determining the efficiency of the elevator winch during the start-up, losses associated with the elastic deformation of the links of the kinetic chain of the winch, primarily ropes, are not considered. These losses can be determined by considering the proposed dynamic model of the elevator winch (Fig. 1) and determining the dynamic loads acting on the elevator hoisting ropes. Figure 1 shows: 1 – first design mass, 2 – second design weight, 3 – third design weight, 4 - traction sheave, 5 – cab with load, 6 – counterweight, C_p– rigidity of the rope suspension of the cab with load, $y_{c,}$ and y_p – the deformation of the cab with a load and the deformation of the counterweight, respectively, C_M is the rigidity of the metal structure at the point of lifting the load.

Let us consider the movement of the mechanical system "winch–load" (taking into account only the elasticity of the ropes) in the pre-separation (Fig. 2, a) and post-separation (Fig. 2, b) stages of movement.

The following designations are adopted in Fig. 2: m_M is the mass of the metal structure of the winch, reduced to the vertical deformation of the metal structure y_M, counted from the static position of the metal structure in the absence of a load; C_M is the rigidity of the metal structure at the point of lifting the load; m_1 is the mass of the rotating parts of the lifting mechanism, reduced to translational motion y_1, coinciding with the direction of movement of the load y_c; C_p – rigidity of the rope suspension of the load; m_c is the weight of the cargo; G_c is the weight of the cargo; P is the force of the engine of the lifting mechanism, reduced to the translational movement of the load; S_1 and S_2 are the total force in the hoisting ropes, respectively, before the load is lifted off and after the load is lifted off the base; F_1 and F_{12} – the force acting on the metal

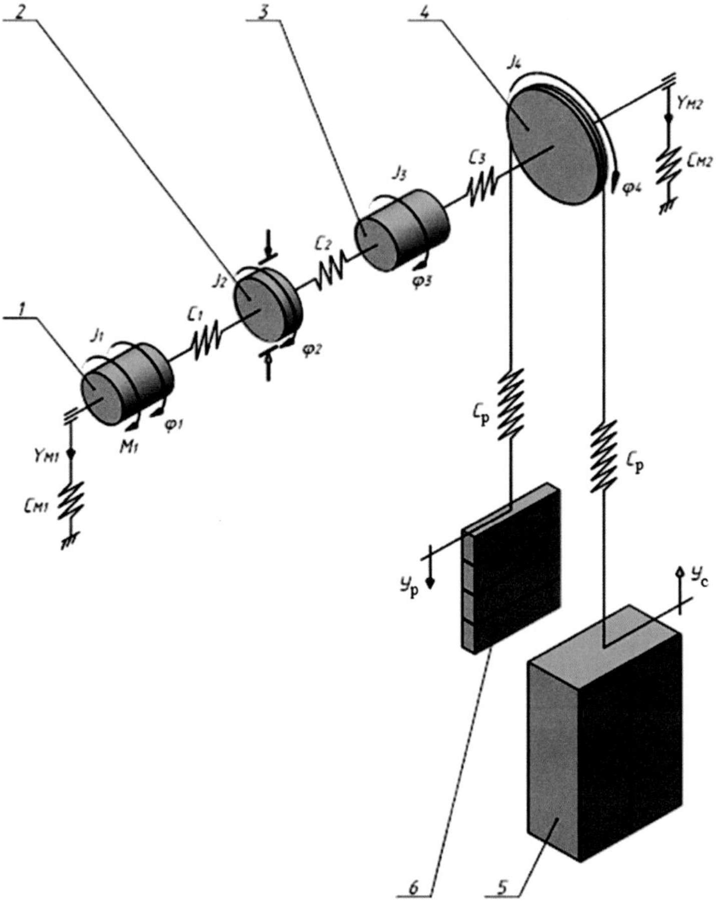

Fig. 1. Calculation dynamic scheme of the lifting mechanism of a passenger elevator.

structure of the winch, arising when lifting the load in the pre-separation and after the separation stages of movement, respectively.

3.2 Preseparation Stage of Movement

The equation of motion of the "winch-load" system in the pre-separation stage has the form

$$\left.\begin{aligned} m_M \ddot{y}_M &= S_1 - F_1 \\ m_1 \ddot{y}_1 &= P - S_1 \end{aligned}\right\}. \tag{5}$$

The reduced driving force of the lift winch motor

$$P = 2M \frac{i_p \eta_{\text{mech}} a}{D_M + d}, \tag{6}$$

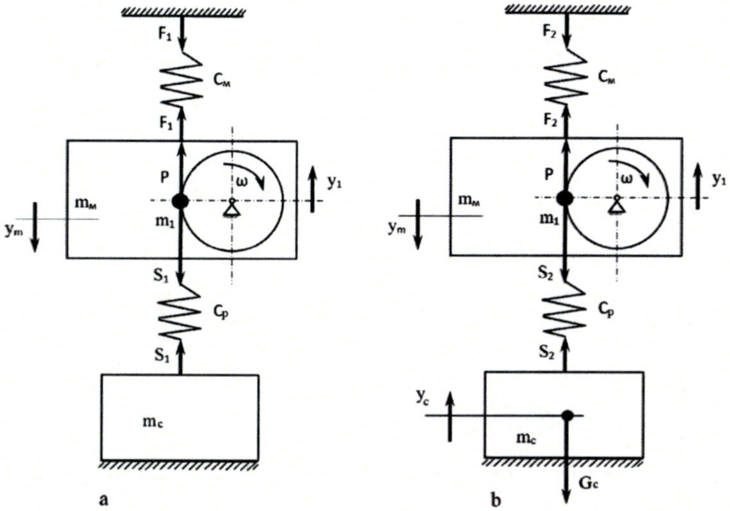

Fig. 2. Calculation scheme of the mechanical system "winch-load" in the pre-separation (a) and post-separation (b) stages of movement.

where M is the moment of the engine of the lifting mechanism, Nm; i_p is the gear ratio of the hoist gear, r.u.; η_{mech} is the efficiency of the lifting mechanism, r.u.; a is the multiplicity of the pulley blocks, r.u.; D_M – traction sheave diameter, m; d is the diameter of the rope, m.

Assuming the linearity of the working sections of the mechanical characteristics of asynchronous electric motors, the dependence of the change in the driving force at this stage can be determined.

$$P = P_0 - \beta \dot{y}_1, \tag{7}$$

where P_0 is the force at the moment of starting the winch engine, N; $\dot{y}_1$ is the speed of the reduced mass of the rotating parts of the lift hoist, m/s; β is the stiffness coefficient of the mechanical characteristic, Ns/m

$$\beta = \eta_{\text{mech}} \frac{2M_0}{\pi n_0} \left(\frac{i_p a}{D_M + d} \right)^2, \tag{8}$$

here M_0 is the starting torque on the motor shaft (determined under the condition of linearity of the mechanical characteristics of the AM), Nm; n_0 is the number of winch engine revolutions corresponding to its synchronous speed, rpm.

By substituting the values S_1, F_1, and P into (5), a system of differential equations is formed that describes the motion of a dynamic mechanical system at the preseparation stage

$$\left. \begin{aligned} m_M \ddot{y}_M + (c_p + c_M) y_M - c_p y_1 = 0 \\ m_1 \ddot{y}_1 + \beta \dot{y}_1 + c_\Pi (y_1 - y_M) = P_0 \end{aligned} \right\}. \tag{9}$$

The initial conditions for the considered stage of the system movement in the case of lifting a load with pickup have the form: $t_1 = 0$, $y_M = 0$, $\ddot{y}_M = 0$, $y_1 = 0$, $\dot{y}_1 = (\dot{y}_1)_0$.

In this case $(\dot{y}_1)_0$ – mass velocity m_1, corresponding to the idling speed of the winch engine, is equal to

$$(\dot{y}_1)_0 = \frac{\pi n_0 (D_M + d)}{i_p a}. \quad (10)$$

The condition for the transition to the postseparation stage of motion of a dynamic mechanical system is the equality

$$(S_1)_1 = G_c, \quad (11)$$

where $(S_1)_1$ is the force in the lifting ropes of the elevator at the end of the preseparation stage.

3.3 Post-Separation Stage of Movement

After the separation of the load from the base, the movement of the elevator system is described by the equations

$$\left. \begin{array}{r} m_M \ddot{y}_M = S_2 - F_2 \\ m_1 \ddot{y}_1 = P - S_2 \\ m_c \ddot{y}_c = S_2 - G_c \end{array} \right\}. \quad (12)$$

The force in the metal structure and the tension of the rope are determined by the expressions

$$F_2 = c_M y_M; \quad (13)$$

$$S_2 = c_p (y_1 - y_M - y_c). \quad (14)$$

Taking into account (13) and (14), the system of differentiated Eqs. (12) can be written in the following form

$$\left. \begin{array}{r} m_M \ddot{y}_M + (c_p + c_M) y_M - c_p y_1 + c_p y_c = 0 \\ m_1 \ddot{y}_1 + \beta \dot{y}_1 + c_p (y_1 - y_M - y_c) = P_0 \\ m_c \ddot{y}_c - c_p (y_1 - y_M - y_c) = -G_c \end{array} \right\}. \quad (15)$$

Initial conditions for a given stage of movement $t_2 = 0$, $y_M = (y_M)_1$, $\dot{y}_M = (\dot{y}_M)_1$, $y_1 = (y_1)_1$, $\dot{y}_1 = (\dot{y}_1)_1$, $y_c = 0$, $\dot{y}_c = 0$, where $(y_M)_1$, $(\dot{y}_M)_1$, $(y_1)_1$, $(\dot{y}_1)_1$ – displacements and velocities of the masses m_M and m_1 at the end of the pre-separation stage of the movement of the lifting mechanism of the elevator.

It is assumed that the dynamic deformation of the supporting metal structure differs little from the static deformation, i.e.

$$c_M y_M \approx c_p (y_1 - y_M - y_c). \quad (16)$$

This assumption introduces a minimal error in the result. The maximum value of the force in the hoisting ropes after the cab has been lifted off the base

$$S_{2max} = G_c + \frac{G_c m_c}{m_1 + m_c} \left\{ 1 - \cos\, arctg \left[-\frac{c (\dot{y}_1)_0}{p} \frac{(m_1 + m_c)}{G_c m_c} \right] \right\}_{2max}$$

$$+\frac{c(\dot{y}_1)_0}{p}\sin arctg\left[-\frac{c(\dot{y}_1)_0}{p}\frac{(m_1+m_c)}{G_c m_c}\right] \quad (17)$$

The efficiency of the elevator winch during the start-up, taking into account the reduced rigidity of the rope c_p and metal structures c_M, can be determined as

$$\eta_d=\frac{W_c+A_c}{S_{2\max}h_K}, \quad (18)$$

where h_K– is the path that the cabin passes during the start-up, m.

Analysis of expression (17) shows that with an increase in the rigidity of the rope c_p, the maximum value of the force in the hoisting ropes $S_{2\max}$ increases after the cabin is separated from the base, and, following (18), the efficiency of the elevator winch, decreases during the start-up.

4 Results and Discussion

To reduce rope bending and, accordingly, reduce bending stresses in the rope, thereby increasing its durability, it is recommended to use the minimum ratio between the diameter of the selected rope d and the diameter of the traction sheave D

$$\mathrm{D}\geq d\cdot e, \quad (19)$$

where e is the coefficient taken depending on the group of the operating mode of the mechanism and the type of lifting machine (for passenger elevators, $e = 40$).

When calculating steel ropes for strength, the diameter of the rope turns out to be large for a given lifting capacity of the elevator, and, accordingly, the diameter of the traction sheave according to expression (19) also turns out to be large, while the rope has greater rigidity.

Rope stiffness can be determined

$$c=\frac{E\cdot F}{l}, \quad (20)$$

where E is the tensile modulus, Pa; F is the cross-sectional area of the rope, m; l is the length of the rope, m.

Following expression (20), the stiffness of the rope increases with a decrease in the length of the rope, which occurs when the elevator car is raised. Therefore, the efficiency of the elevator winch during the start-up decreases as the car moves.

The rigidity of the rope also affects the efficiency of the fixed blocks used in the pulley blocks cab suspension. From consideration of the scheme of forces acting on a fixed block (Fig. 2, a), we can write the equality

$$S_2=S_1+\mathrm{W}+W_s, \quad (21)$$

where W is the resistance in the block bearings, reduced to the block rim, N.

As a result of the rigidity, the rope does not immediately enter its stream when it runs into the block. When it runs away, it does not immediately acquire a rectilinear position.

When determining the resultant tension forces and calculating the friction resistance in the block support, it can be taken with sufficient accuracy $S_1 = S_2$ (Fig. 3, a), since the resistance to the rigidity of the rope is not significant compared to the working tension.

The moment created by friction forces in a support with a diameter of d

$$M_s = 2S_1 \sin\frac{\alpha}{2} f \frac{d}{2}, \tag{22}$$

where f is the coefficient of friction in the block bearings, kg · s, and a is the angle of wrapping the block with a rope, rad.

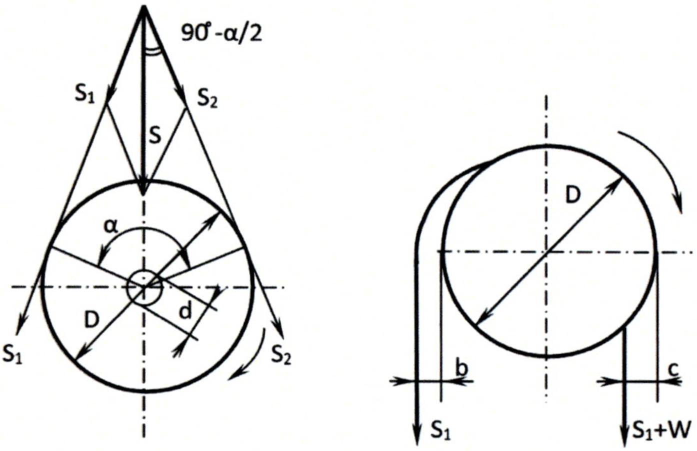

Fig. 3. Calculation scheme for determining the resistance on the cable transmission blocks.

If the gravity force of the load S_1 acts on the rope branch under tension G_c, then the efficiency of the fixed block will be the ratio of useful work $G_c h = S_1 h$ to work expended $S_2 h$ (here h - the height of the load)

$$\eta_{bl} = \frac{S_1 h}{S_2 h} = \frac{1}{1 + \phi + 2f \frac{d}{D} \sin\frac{\alpha}{2}}. \tag{23}$$

Analysis (23) shows that an increase in the rigidity of the rope ϕ and the angle of wrapping reduces the efficiency of the block.

Considering the influence of the rope stiffness and friction in the block support on the efficiency, it is impossible, for example, to agree with the decision of the KONE corporation, which went by increasing the multiplicity of the pulley blocks up to 6:1 and even 10:1. This solution makes it possible to reduce the load on the rope and reduce its diameter (and hence the rigidity of the rope). Hence, the diameter of the traction sheave and also significantly increase the nominal speed of the engine. Since the pulley blocks in this design are installed not only on the cab but also on the counterweight, this significantly increases the number of movable and non-movable blocks and significantly reduces the efficiency of the elevator. If we take the efficiency of one block as 0.98, then the total efficiency of the cable transmission is 0.98n, where n is the number of blocks. At n = 20, the efficiency of only cable transmission is 67%, and at n = 30, it is already 55%. Taking into account the efficiency of the engine, for example, 85%, the resulting efficiency of the winch in the mode of movement at a steady speed and at start-up is, respectively, 60% and 47%. In the case of cabin movement in dynamics, the resulting efficiency becomes catastrophically small.

Thus, the rope's rigidity significantly impacts the efficiency of the elevator winch, both in the steady state and in the start mode.

5 Conclusions

Based on the energy conservation equation during the machine's movement, dependencies were obtained to determine the efficiency of the elevator winch for two modes of movement: at a steady speed and during start-up.

The dependence for calculating the efficiency of an elevator winch in the start-up mode is determined based on the analysis of dynamic loads that occur when lifting the cabin. At the same time, it was determined that the efficiency of the winch for this mode depends not only on inertial loads but also on the rigidity of the rope.

It has been established that with an increase in the rigidity of the hoisting rope, the maximum value of the force in the rope increases after the cabin is detached from the base, and, accordingly, the efficiency of the passenger elevator winch decreases in dynamic modes.

It is shown that the efficiency of pulley blocks depends not only on the friction in the block support but also on the rigidity of the lifting rope: the greater the rigidity of the rope, the lower the efficiency of the mechanical block.

It is determined that when lifting the elevator car, the length of the rope between the point of contact of the rope on the traction sheave decreases. It is shown that the rigidity of the rope increases with a decrease in its length, so the efficiency of the elevator winch decreases as the cabin moves.

Further research can focus on developing mathematical models and computer simulations to simulate the behavior of pulley blocks with varying friction and rope rigidity.

References

1. Afonin, V., Badaljan, N., Maslakova, G., Mitrofanov, A., Chashchin, E.: Electricity consumption in lifts. In: 3rd International Conference on Control Systems, Mathematical Modeling, Automation and Energy Efficiency (SUMMA), pp. 1126–1128. Lipetsk, Russian Federation (2021). DOI: https://doi.org/10.1109/SUMMA53307.2021.9632037
2. Shuangchang, F., Jie, C., Xiaoqing, C.: Analysis of the hidden danger for old elevator safety. In: 2020 3rd International Conference on Electron Device and Mechanical Engineering (ICEDME), pp. 605–608. Suzhou, China, IEEE (2020). https://doi.org/10.1109/ICEDME50972.2020.00143
3. Boiko, A, Naidenko, E., Wang, Y.: Vibration damping of lifting mechanisms. In: Tonkonogyi, V., Ivanov, V., Trojanowska, J., Oborskyi, G., Pavlenko, I. (eds.) Advanced Manufacturing Processes IV. InterPartner 2022. LNME, pp. 403–413. Springer, Cham (2022). https://doi.org/10.1007/978-3-031-16651-8_38

4. Al-Sharif, L., Al-Ahdab, M.: A MATLAB/Simulink based journey-based lift energy consumption model. In: 11th Symposium on Lift & Escalator Technologies, pp. 19–29 (2020)
5. Kaczmarczyk, S., Blaszczyk, J., Lei, H., Smith, S.: Feasibility of an energy efficient fuel cell hybrid lift: the main concept and design. In: 11th Symposium on Lift & Escalator Technologies, pp. 59–68 (2020)
6. Mei, D., Du, X., Chen, Z.: Optimization of dynamic parameters for a traction type passenger elevator using a dynamic byte coding genetic algorithm. J. Mech. Eng. Sci. **223**(Part C), 595–605 (2009). https://doi.org/10.1243/09544062JMES1149
7. Furuya, O., Matsuzaki, R., Fujita, S.: Study on the proper performance of lift buffers in revised JIS A 4306 using non-linear damping characristic. In: 9th Symposium on Lift & Escalator Technologies, pp. 201–210 (2018)
8. Shima, Y., Furuya, O.: Study on a vibration reduction system for lift roller guides. In: 11th Symposium on Lift & Escalator Technologies, pp. 97–108 (2020)
9. Appleby, M., Peters, R.: The round trip time simulation: monte carlo implementation and consistency with other techniques. In: 11th Symposium on Lift & Escalator Technologies, pp. 151–160 (2020)
10. Bonopera, M., Chang, K., Lee, Z.-K.: State-of-the-art review on determining prestress losses in prestressed concrete girders. Appl. Sci. **10**, 72–57 (2020)
11. Nehe, R., Mahale, C., Patil, B.: Feasibility study of using a coupler skate encoder system for monitoring the real time status of lift door operation. In: 11th Symposium on Lift & Escalator Technologies, pp. 69–80 (2020)
12. Nguyen, T.X., Miura, N., Sone, A.: Analysis and control of compensation rope response in elevator system with time-varying length. In: 2017 11th Asian Control Conference (ASCC), pp. 905–910 (2017)
13. Pyatibratov, G., Danshina, A., Altunyan, L.: Optimal force compensating control of robotic lifting mechanisms. In: 2019 International Russian Automation Conference (RusAutoCon), pp. 1–5 (2019). https://doi.org/10.1109/RUSAUTOCON.2019.8867811
14. Wang, J., Tang, S.X., Krstic, M.: Lateral vibration suppression of a disturbed mining cable elevator with flexible guideways. In: 2020 59th IEEE Conference on Decision and Control (CDC), pp. 4436–4441 (2020)
15. Donner, P., Buss, M.: Cooperative swinging of complex pendulum-like objects: experimental evaluation. IEEE Trans. Rob. **32**(3), 744–753 (2016). https://doi.org/10.1109/TRO.2016.2560898
16. Zudilova, T.V., Ivanov, S.E., Ivanova, L.N.: The automation of electromechanical lift for disabled people with control from a mobile device. In: 2017 Computing Conference, pp. 668–674 (2017)
17. Anderle, M., Michiels, W., Čelikovský, S., Vyhlídal, T.: Damping a pendulum's swing by string length adjustment – design and comparison of various control methods. In: 2019 American Control Conference (ACC), pp. 4399–4405. Philadelphia, PA, USA (2019)
18. Watanabe, K., Yoshikawa, M., Ishikawa, J.: Damping control of suspended load for truck cranes in consideration of second bending mode oscillation. In: IECON 2018 - 44th Annual Conference of the IEEE Industrial Electronics Society, pp. 4561–4568. Washington, DC (2018). https://doi.org/10.1109/IECON.2018.8591232
19. Zhang, H., Zhang, R., On, K., Liu, L.: Variable universe fuzzy control of high-speed elevator horizontal vibration based on firefly algorithm and backpropagation fuzzy neural network. IEEE Access **9**, 57020–57032 (2021). https://doi.org/10.1109/ACCESS.2021.3072648
20. Kodani, N., Oushi, S., Takahashi, R., Hirata, H.: Transport control of a jib crane with a rotating cargo. Trans. Inst. Electr. Eng. Jpn. C **136**, 821–831 (2016). https://doi.org/10.1109/IECON.2017.8217456

21. Naidenko, E., Bondar, O., Boiko, A., Fomin, O., Turmanidze, R.: Control optimization of the swing mechanism. In: Tonkonogyi, V., Ivanov, V., Trojanowska, J., Oborskyi, G., Pavlenko, I. (eds.) Advanced Manufacturing Processes III. InterPartner 2021. LNME, pp. 13–21. Springer, Cham (2021). https://doi.org/10.1007/978-3-030-91327-4_2
22. BS EN ISO 25745-2:2015 Energy performance of lifts, escalators and moving walks Part 2: Energy calculation and classification for lifts (elevators) (2015)

Check for updates

Experimental Study of Longevity in the Metallic Structure of Boom for a Portal Crane of Seaport

Liubov Bovnegra[1], Andrey Pavlyshko[1], Oleksiy Nemchuk[2], Viktor Strelbitskyi[2](✉), and Isak Karabegovich[3]

[1] Odessa Polytechnic National University, 1, Shevchenko Avenue, Odesa 65044, Ukraine
[2] Odessa National Maritime University, 34, Mechnikova Street, Odessa 65029, Ukraine
vict141174@gmail.com
[3] Academy of Sciences and Arts of Bosnia and Herzegovina, 7, Bistrik Street, 71000 Sarajevo, Bosnia and Herzegovina

Abstract. The usage of cranes in the seaport is widespread and essential. However, long-term use unavoidably leads to failures of portal cranes' structural parts, severe damages, or total collapses, and is often followed by very high financial losses and serious injuries or crane-related fatalities. Sea portal cranes are operated under intensive cyclic loading, which leads to steel plasticity exhaustion. However, the latent and educated defects that are available during manufacturing and operation lead to a reduction in their useful life. They can serve as concentrators of fatigue cracks; their development is random. The work of quay crane operations for handling "shore-to-ship" was analyzed. For studies, flat beam specimens with one-side notch were used in the ferrite-pearlite class St38b2 steel and cut from the lower and rear shelves of the crane boom, as well as the right boom, used for 40 years in the river port. From the obtained experimental kinetic diagrams of the fatigue fracture of the material, it was found that an increase in the asymmetry of the cycle leads to an intensification of crack growth at values of 0.6 and 0.8. Also, there was no crack closure effect, a decrease in the fatigue threshold, and crack growth was intensified over the entire range. The metallographic evaluation of experimental samples showed the absence of intercrystalline corrosion.

Keywords: Portal Crane · Seaport · Defect · St38b2 Steel · Fatigue Crack · Cycle Asymmetry · Industrial Innovation

1 Introduction

Portal cranes are one of the most popular and effective types of modern lifting equipment in seaports [1]. Cranes are hazardous objects. Every year, the volumes of cargo transshipment in seaports increase [2], and portal cranes are loaded more [3]. So, in 2022 the ports of Reni and Izmail, the transshipment volume increased by 42 times compared to last year [4]. Because of this, the terminals must reload more cargo using the same transportation equipment [5]. One of the most essential aspects of cargo loading is operation safety which can decrease over time [6].

V. Tonkonogyi et al. (Eds.): InterPartner 2023, LNME, pp. 432–439, 2024.
https://doi.org/10.1007/978-3-031-42778-7_40

In Ukraine, there is an urgent matter of operating weight-lifting equipment, including portal cranes, whose standard operation life exceeds theirs, but those cranes are in working condition, despite being operated in the regime of intensive cyclic loading [7]. It should be noted that the cost of metal crane structures accounts for approximately 70% of the total cost of a crane. Special attention should be paid to crane metal structures that have worked out their normative life cycle. The situation with insufficient safety of operation of portal cranes, which has developed recently, is due not only to the wear of the crane fleet but also to the imperfection of the structural elements of portal cranes.

Thus, maintaining portal cranes' efficiency and reliability during long-term operation is an actual task of mechanical engineering.

2 Literature Review

When studying portal cranes from the point of view of service reliability [8], special attention needs to be paid to load-bearing elements of the metal structures [9] that are capable of gradual failures [10] and the occurrence of unacceptable damage [11] under the influence of cyclic loads [12]. Service reliability is the most significant parameter of lifting equipment quality [1–4]. Among the defects of the metal structure, the most dangerous is fatigue failure [13], concentrated in the zones of concentration of stresses of the elements [14], which arise and develop for several years [15] in the course of their operation [16]. In addition, authors' studies [17, 18] show that atmospheric corrosion is the factor of steel's hydrogenation that could lead to its hydrogen embrittlement. It is concerned with the development of dissipated damage at the nano- and microlevel under the combined action of operational stresses and hydrogen.

It should be noted that each case of crack propagation requires a separate study since the endurance of the structure significantly depends on its operating conditions [19], geometric and physical [20], and mechanical parameters [21]. In order to improve the safety and reliability of crane steel structures, enterprises generally adopt the method of regular maintenance and follow-up maintenance [22].

Therefore, as for portal cranes, it is essential to evaluate the influence of work and asymmetry of the load cycle in an aggressive environment on the fracture toughness characteristics of portal crane metal structures [23].

3 Research Methodology

The subject of the investigation was the 39-year-old portal crane "Sokol" manufactured by "Kranbau Eberswalde".

Depending on the loading process, the operation is divided into seven stages: lifting (hook), vertical lifting, horizontal transport, ramp down, vertical down, and boarding. The overall loading cycle during the shore-to-ship process was also assessed to differ slightly for the studied cranes.

For metal fragments to be tested, the stress range $\Delta\sigma e$ is measured (determined) on the surface of the rolled steel plate under a crane load close to the operating stress. It should be noted that the parameter $\Delta\sigma_e$ only reflects the change of the stress state caused by the loading of the crane and does not consider the stress caused by the specific unit weight. The results allowed us to compare the characteristics of different crane systems and their stresses during operation.

Specimens from selected parts of the portal crane were used for the study in the form of one-side notched beam specimens, made of rolled St38b2 (0.17C, 0.25Si, 0.50Mn, 0.035P, 0.035S) made rolled steel plates of thickness t = 10 mm, namely: the lower plate and the jib rear plate of the portal crane. Two different fragments are selected for each unit, and there will eventually be 10 metal samples.

Standard mechanical properties of the steel, such as yield stress σ_Y, ultimate stress σ_{UTS}, ultimate elongation δ, and reduction of area RA, were obtained by SSRT from cylindrical specimens with a working diameter of 5 mm cut along the rolling direction.

Consider the working voltage scheme in the crane element structure to select the sample interface; after cutting, finally, grind the sample (the size deviation does not exceed ± 0.02 mm). To better observe the crack movement, the sides of the working part were polished, allowing visual recording of the crack length with a microscope with an accuracy of 0.01 mm (Fig. 1).

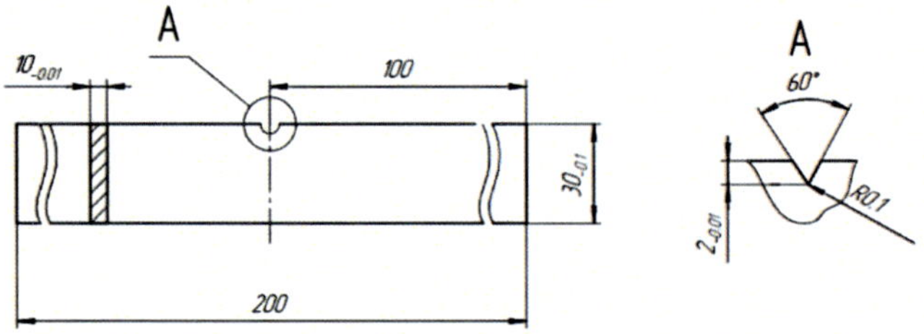

Fig. 1. Test samples.

This test method involves cyclic loading of acceptable notched specimens pre-cracked by fatigue. Load the specimen on the rigid testing machine at a frequency of 4 Hz by cantilever bending at an external temperature of 20 °C according to cyclic asymmetry R 0,1; 0,3; 0,6 and 0.8. The obtained values were averaged.

4 Results and Discussion

In the port cranes we tested, cracks were detected visually. In some cases, they have become the cause of abrupt fracture of an installation. The cracks initiated and propagated due to fatigue. The cracks initiated and propagated due to fatigue.

Based on the obtained experimental data crack propagation, Kinetic Fatigue Fracture Diagram (KFFD) ΔK – crack growth rate da/dN" were constructed (Fig. 2) according to the method [19].

Since the thicknesses of the test samples and metal structures are the same, it can be assumed with a certain probability that the crack growth rates for ΔK values should also be approximate. This makes it possible to use the obtained fatigue fracture kinetic diagrams for engineering calculations at the design stage.

The crack increment is a function of the stress intensity factor:

$$\frac{da}{dN} = C\Delta K^n \quad (1)$$

when coefficient C and the exponent n should be material constants.

The constant n in the Law of Paris model is related to the tilt angle of the experiment data lines (Fig. 2).

ΔK is related to the range of load changes $\Delta K = K_{max} - K_{min}$, corresponding successively to the range of external load changes.

The effect of the exercise scale is generally more pronounced at lower ranges of exercise intensity factors. The stress ratio effect tends to disappear for regions with higher stress intensity factors.

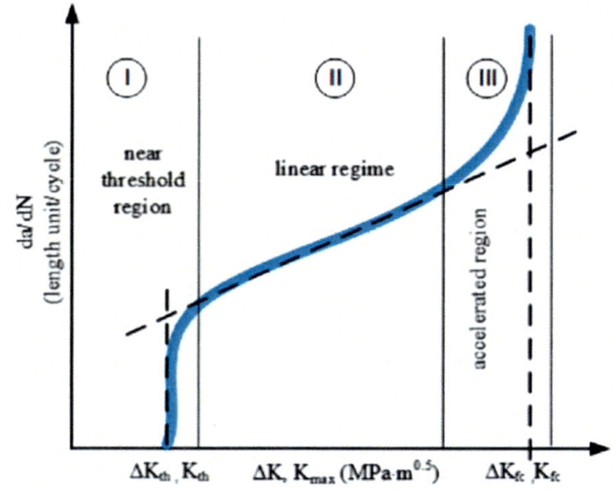

Fig. 2. A typical fatigue crack growth rate curve (kinetic fatigue failure diagram: KFFD).

Equation (1) is known as the Paris law, and it is widely used in many papers dealing with fatigue crack growth [20] and is valid for the following linear range of the crack growth curve (Fig. 2).

The kinetic fatigue failure diagram of samples is shown in Figs. 3 and 4.

Figures 3 and 4 show that the thresholds are limited on the left by the asymptote corresponding to the threshold $\Delta Kth = 6$ MPa·m$^{1/2}$.

The development of cracking ends when the stress factor reaches the critical value ΔKfc, above which cracking propagates unstable. Figures 2 and 3 show that an increase in the asymmetry of the loading cycle leads to an intensification of crack growth, particularly in the range of values $\Delta K = 7\ldots25$ MPa·m$^{1/2}$ speed increases by 1.5–2,3 fold.

The difference in fatigue fracture kinetics of steels with different load cycle asymmetries follows a general pattern [19], mainly explained by the decrease in the effectiveness of fatigue crack closure with increasing R values. Crack closure increases with the reduction in the stress ratio leading to lower fatigue crack growth rates. In the case of low loading cycle asymmetry, the operation does not significantly affect the fatigue crack growth rate, which is consistent with the research results [20], and the effect of fatigue crack closure can be explained by the increase of surface roughness. At higher stress ratios (such as R > 0.6), crack closure is likely to be negligible, and the fatigue cracks growth lines for each stress ratio are not perfectly parallel to each other, which means that crack closure also depends on the stress intensity factor domain level (Fig. 3, 4). At R ≥ 0.6, the effects of crack closure, fatigue threshold reduction, and crack growth intensification are no longer observed over the entire ΔK range (Fig. 3, 4).

Surface visual inspection and fracture metallographic analysis of the samples showed no oxide deposition in the critical crack growth rate region, which agrees with the findings [19].

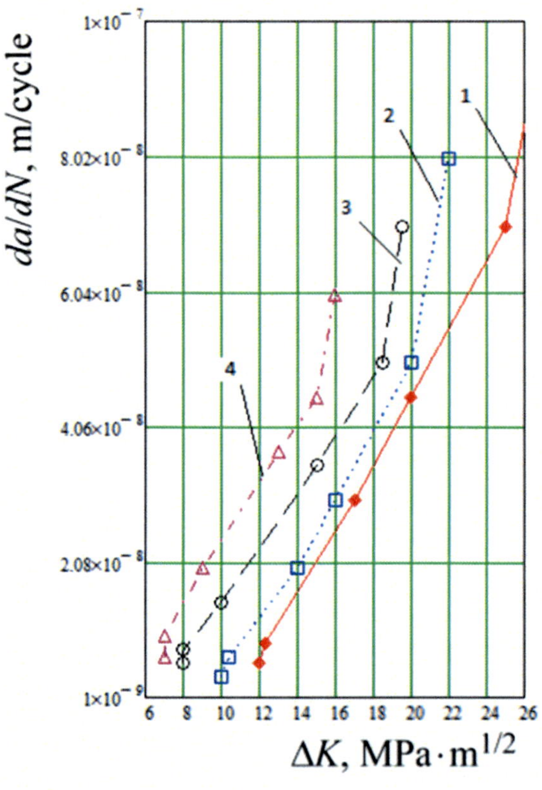

Fig. 3. Diagrams of fatigue crack growth on the lower shelf of a jib in St38b2 steel after service for R = 0.1 (1); 0,3 (2); 0,6 (3) and 0,8 (4).

Based on the results of experimental studies, it can be seen that:

1. The crack resistance characteristics of samples of metal structures made of st38b2 steel are influenced by the operating time and asymmetry of the metal loading cycle of portal cranes.

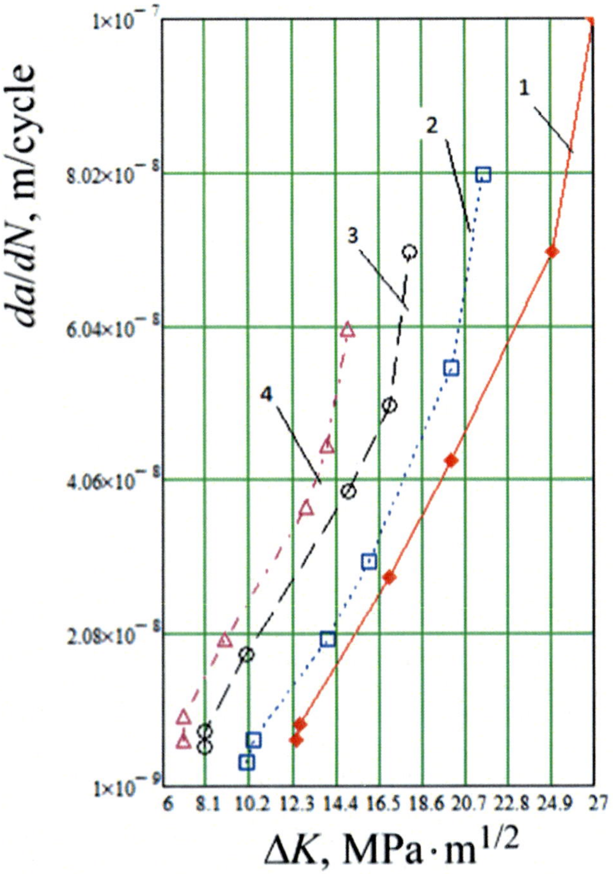

Fig. 4. Diagrams of fatigue crack growth on the back shelf of a boom in St38b2 steel after service for R = 0.1 (1); 0,3 (2); 0,6 (3) and 0,8 (4).

2. An increase in the asymmetry of the load cycle leads to an intensification of crack growth, particularly in the range of values K = 7…25 MPa·m$^{1/2}$ speed increases by 1.5–2.3 fold (Fig. 3, 4). At R = 0.8, there is already no crack closing effect, a decrease in the fatigue threshold, and an intensification of crack growth over the entire ΔK range.
3. Metallographic analysis of the samples showed the absence of intergranular corrosion.

5 Conclusions

In the works research of longevity in the metallic structure of boom of a portal crane who worked in the port for 40 years.

The main conclusions are as follows. The failure of a seaport portal crane boom under intensive cyclic loading is usually associated with material problems and latent defects. They can serve as concentrators of fatigue cracks; their development is random.

It was found that the long-term operation of the crane (40 years) in the seaport did not lead to intercrystalline corrosion.

An increase in the asymmetry of the loading cycle leads to an increase in the rate of development of cracks in the metal, resulting in premature destruction of the boom structures. We found that the speed increases by 1.5–2.3 times. In the region of large ΔK, the crack propagation rates reach a maximum. The resulting diagrams agree with the general patterns, mainly explained by a decreased efficiency of closing fatigue cracks with increased R values.

The resulting dependence can be used to control the state of cracks in the metal of portal cranes during their long-term operation.

References

1. Sorgenfrei, J.: Port business. 2nd edn. De/G Press, World Bank (2018). https://doi.org/10.1515/9781547400874
2. Kolisnichenko, V.: The sea ports of Ukraine processed 59 million tons of cargo in 2022. https://gmk.center/en/news/the-sea-ports-of-ukraine-processed-59-million-tons-of-cargo-in-2022. Accessed 10 May 2023
3. In 2022, the Ukrainian ports of the Danube increased the transshipment of grain by 42 times. https://graintrade.com.ua/en/novosti/u-2022-r-ukrainski-porti-dunayu-u-42-razi-zbilshili-perevalyuvannya-zerna.html. Accessed 10 May 2023
4. Liu, L., Guan, K., Wu, P.: Analysis and research on energy utilization efficiency of port portal crane. In: E3S Web of Conferences, vol. 341, p. 01024 (2022). https://doi.org/10.1051/e3sconf/202234101024
5. Pustovyi, V.M., Semenov, P.O., Nemchuk, O.O., Hredil, M.I., Nesterov, O.A., Strelbitskyi, V.V.: Degradation of steels of the reloading equipment operating beyond its designed service life. Mater. Sci. **57**(5), 640–648 (2022). https://doi.org/10.1007/s11003-022-00590-1
6. Lixin, R., Jun, Q., Li, Q., Weiping, O.: Research on safety performance evaluation method of portal crane based on reliability evaluation and risk assessment. In: E3S Web of Conferences, vol. 257, p. 01070 (2021). https://doi.org/10.1051/e3sconf/202125701070
7. Nemchuk, O., Hredil, M., Pustovoy, V., Nesterov, O.: Role of in-service conditions in operational degradation of mechanical properties of portal cranes steel. Procedia Struct. Integrity **16**, 245–251 (2019). https://doi.org/10.1016/j.prostr.2019.07.048
8. Faltinová, E., Mantič, M., Kuľka, J., Kopas, M.: Reliability analysis of crane lifting mechanism. Sci. J. Silesian Univ. Technol. Ser. Transp. **98**, 15−26 (2018). https://doi.org/10.20858/sjsutst.2018.98.2
9. Sagirov, Y., Artiukh, V., Mazur, V., Aleksandrovskiy, M.: Scientific and methodological bases of rational design of hoisting-and-transport machines metal structures. In: E3S Web of Conferences, vol. 164, p. 03005 (2020). https://doi.org/10.1051/e3sconf/202016403005
10. Okanminiwei, L., Oke, S.A.: Port equipment downtime prediction and lifetime data analysis: evidence from a case study. J. Ind. Eng. Manage. Syst. **14**(1), 8–18 (2021). https://doi.org/10.30813/jiems.v14i1.2362
11. Depale, B., Bennebach, M.: residual life of structures and equipment. Procedia Struct. Integrity **38**, 317–330 (2022). https://doi.org/10.1016/j.prostr.2022.03.033
12. Yang, R.: Interval non-probabilistic time-dependent reliability analysis of boom crane structures. J. Mech. Sci. Technol. **35**, 535–544 (2021). https://doi.org/10.1007/s12206-021-0112-4

13. Ren, L.-X., et al.: A review of fatigue life prediction method for portal crane. IOP Conference Series: Earth and Environmental Science, vol. 657, p. 012094 (2021). https://doi.org/10.1088/1755-1315/657/1/012094
14. Gubskyi, S., Yepifanov, V., Chukhlib, V., Basova, Y., Okun, A., Ivanova, M.: Integrated approach to determine operational integrity of crane metal structure. Periodica Polytech. Mech. Eng. **63**(4), 319–325 (2019). https://doi.org/10.3311/PPme.14064
15. Depale, B., Bennebach, M.: Residual life of steel structures and equipment: problems and application to cranes. Mech. Ind. **20**(8), 802 (2019). https://doi.org/10.1051/meca/2019047
16. Depale, B., Bennebach, M.: Residual life assessment (RLA) of cranes and steel structures: problems and strategy. In: Ha-Minh, C., Tang, A.M., Bui, T.Q., Vu, X.H., Huynh, D.V.K. (eds.) CIGOS 2021, Emerging Technologies and Applications for Green Infrastructure. Lecture Notes in Civil Engineering, vol. 203, pp. 863−872. Springer, Singapore (2022). https://doi.org/10.1007/978-981-16-7160-9_87
17. Nykyforchyn, H., et al.: Methodology of hydrogen embrittlement study of long-term operated natural gas distribution pipeline steels caused by hydrogen transport. Frat. Ed Integrita Strutt. **16**, 396–404 (2022). https://doi.org/10.3221/IGF-ESIS.59.26
18. Kartik, L., Vishwakarma, M.: A study of the effect of hydrogen on the fatigue behaviour of metals. In: IOP Conference Series: Materials Science and Engineering, vol. 1248, p. 012026 (2022). https://doi.org/10.1088/1757-899X/1248/1/012026
19. Christ, H.J.: Fatigue of Materials at Very High Numbers of Loading Cycles: Experimental Techniques, Mechanisms, Modeling and Fatigue Life Assessment. Springer, Wiesbaden (2018). https://doi.org/10.1007/978-3-658-24531-3
20. Ritchie, R.O., Liu, D.: Introduction to Fracture Mechanics. Elsevier, Amsterdam (2021)
21. Wu, W., et al.: Mechanostructures: rational mechanical design, fabrication, performance evaluation, and industrial application of advanced structures. Prog. Mater. Sci. **131**, 101021 (2023). https://doi.org/10.1016/j.pmatsci.2022.101021
22. Zhao, J., Gao, C., Tang, T.: A review of sustainable maintenance strategies for single component and multicomponent equipment. Sustainability **14**(5), 2992 (2022). https://doi.org/10.3390/su14052992
23. Zhang, B., Li, G.F., Huang, Z.Q., Tong, Y.T.: The research on mechanical performance of portal crane main steel structure based on sustainable development. In: E3S Web of Conferences vol. 257, p. 01069 (2021). https://doi.org/10.1051/e3sconf/202125701069

The Influence of Mass Absorption and Technological Damage of Concrete on the Contact Strength During the Restoration of Buildings and Structures

Vitaliy Dorofeev[1], Hanna Zinchenko[1(✉)], Maryna Holofieieva[1], Natalia Pushkar[2], and Stanislav Fic[3]

[1] Odessa Polytechnic National University, 1, Shevchenko Ave., Odessa 65044, Ukraine
a-golud6@ukr.net

[2] Odessa State Academy of Civil Engineering and Architecture, 4, Didrihson St., Odessa 65029, Ukraine

[3] Lublin University of Technology, 38 D, Nadbystrzycka St., 20-618 Lublin, Poland

Abstract. The materials of experimental studies of mass transfer in the contact zone of old and new concrete with the determination of the mass absorption of old concrete, depending on the composition of new concrete, are presented. In order to restore the serviceability and ensure the solidity of prefabricated monolithic structures, mass absorption in the contact zone of old concrete with new concrete was studied depending on the composition of the new concrete and the time of their contact. The value of mass absorption in the contact zone is experimentally determined, which varies with time and depends on the composition of the concrete mixture cast on the sample. The water-cement ratio maximizes the adhesion of old and new concrete in the contact zone. The amount of fine aggregate also significantly affects the mass absorption value. It has been confirmed that the value of mass absorption can be used to predict the values of the normal and tangential strengths of old and new concrete in prefabricated monolithic structures.

Keywords: Concrete · Experiment · Solidity · Concrete Composition · Contact Zone · Adhesion · Manufacturing Innovation

1 Introduction

The task of restoring buildings and structures partially destroyed due to terrorist acts that damaged concrete and reinforced concrete structures arose recently. To ensure the safety of buildings and structures, the contact between them must have a close margin of safety. Therefore, great attention in the design and production of works is given to the design of compressed joints [1], bendable [2], and their implementation in kind [3].

Also, the experimental properties of buildings and structures depend on the strength of the contacts, the solidity of the joints – reliability [4], durability [5], rigidity and stability [6] of the entire structure, the rigidity and crack resistance of composite structures, the degree of impermeability of the joints. The effect of bonding old concrete to

V. Tonkonogyi et al. (Eds.): InterPartner 2023, LNME, pp. 440–449, 2024.
https://doi.org/10.1007/978-3-031-42778-7_41

new depends on the structure of the old concrete [7] and its properties [8]. Only the cement paste, which has adhesive ability from the side of the new concrete, participates in the bonding process. Astringent mineral substances can be considered as a kind of stone adhesives and, like typical organic adhesives, they are colloidal systems, and their hardening is colloidal-chemical.

Coarse and fine aggregates in the surface layer of old concrete are also involved in the bonding process [9]. The chemical and mineralogical composition of aggregates and their physical properties determine one or another strength interaction with the binder of new concrete.

The adhesion process is complicated and depends on various factors that characterize the adhesive itself (chemical activity, plasticity) and the materials in contact with it.

2 Literature Review

In recent years, many works have appeared on a detailed study of ensuring economic efficiency in operation and saving resources. The issue of self-organization of the concrete structure, assessment of technological damage, development of technologies, and methods for imparting desired properties to concrete were also studied. The role of active elements of structures in their life cycle [10], the resistance of concrete and expanded clay concrete under periodic external influences [11], and the geometry of cracks were studied [12]. The work of prefabricated monolithic structures, the effect of carbonization of the concrete surface on adhesion [8], the bond of new concrete with the destroyed reinforced concrete structure [9], the effect of the composition of the cement composition on the geometry of cracks [13] were studied in detail. The working seam's influence on the contact zone's durability and the crack resistance of reinforced concrete columns were also studied [15]. The residual strength of damaged reinforced concrete beams was calculated [16], and the strength of interlayer bonds in the outer walls of prefabricated residential buildings was experimentally established [17]. The crack resistance of reinforced concrete structures, depending on the technological damage of concrete [18], and the strength of contacts of composite structures, depending on the technological damage using mathematical and statistical methods, were studied by the authors of this work [19]. However, the studies were conducted to identify individual factors' influence on the contact strength of prefabricated monolithic structures. Comprehensive studies of the influence of mass absorption and technological damage to concrete using the methods of planning an experiment have not been repeated.

The work aims to establish the effect of mass absorption and technological damage on the strength of the contact between old and new concrete, ensuring economic efficiency during operation and saving resources that are currently poorly understood.

This work aims to study the phenomenon of mass transfer in the contact zone of old and new concrete by determining the mass absorption of old and new concrete and studying its change depending on the contact time and composition of new concrete based on the planned experiment.

3 Research Methodology

To study the phenomenon of mass transfer in the zone of contact between old and new concrete, the value of mass absorption of old and new concrete and its change, depending on the contact time and the composition of the new concrete, were determined. Samples of old concrete measuring 150×150×75 mm, treated with metal brushes and water, were stored in a wet state. The samples were weighed on a balance with an accuracy of 0.01 g. After using a special frame, a mixture of new concrete of various compositions with a thickness of 75 mm was applied to the surface of the samples. The composition of the new concrete corresponded to the main plan [8]. A total of 108 samples were tested, 6 samples with each composition. The time of the beginning of concreting, exposure, followed by weighing, was recorded. In parallel, studies were carried out on the free absorption of water.

Technological damage to concrete was assessed by the damage coefficient - the ratio of the total length of surface cracks, L. to the sample area S, on which the damage was measured $K_s = L/S$ (cm^{-1}). The proposed method makes it possible to detect cracks with an opening width of 5 μm or more and a length of 2 mm or more. Cracks on the surface of the sample are the same defects and stress concentrators as in the volume, which makes it possible to judge their effect on the mechanical characteristics of the sample. The surface cracks are considered as a kind of imprint of volumetric processes.

For studying the effect of concrete composition on technological damage, the experiments were carried out according to the main plan [8] with the variation factors of the water-cement ratio W/C, cement consumption C, and structural coefficient r in the intervals of the main experiment.

4 Results and Discussion

4.1 The Mass Absorption of Old Concrete from New Concrete as a Function of Contact Time and Composition of the New Concrete

The amount of mass of the substance absorbed as a result of the mass transfer of the structure of the old concrete depends on the characteristics and rheology of the new concrete mixture, their structure, the state of the surface of the old concrete, the time, spent by the new concrete mixture on the surface of the sample of the old concrete.

It has been established that the value of mass absorption in the contact zone changes with time, reaching its highest value for each new concrete composition at different times. The mass absorption value changed most intensively, increasing exposure time from 20 to 40 min. This exposure was adopted in the main experiment to study the relationship between the mass absorption value and the contact strength. The results of the main experiment confirmed the assumption that the value of mass absorption in its change in time depends on the composition of the concrete mixture packed on the sample. The nature of this dependence may vary. Figure 1 shows the change in mass absorption of M_{ab} at exposures of 20 and 40 min.

The mass absorption value, as an integral characteristic that reflects the migration of mechanically bound water and the physical and mechanical processes occurring in the contact zone, also includes the amount of the solution part remaining in the pores, surface

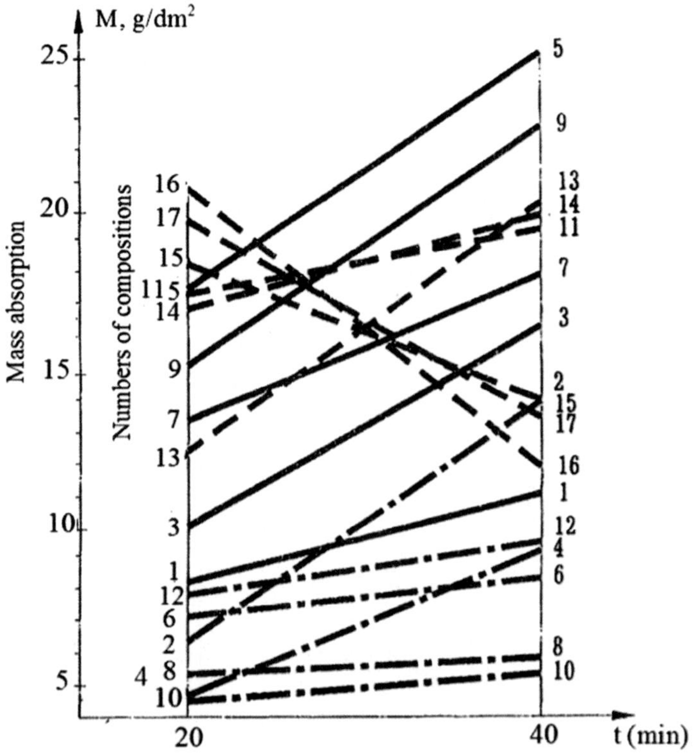

Fig. 1. Change in the mass absorption value in time 1…15 – numbers of compositions; _._ compositions with W/C – 0,48; compositions with W/C – 0,64; compositions with W/C – 0,80.

cracks, and microcapillaries. In studies of mass absorption at the same holding time of a concrete mixture of various compositions on samples of old concrete, the results were obtained that reflect the influence of these processes.

4.2 The Technological Damage of Old and New Concrete

The value of mass absorption in the zone of contact between old and new concrete, depending on the composition of the new concrete, varied from 5.7 to 25.7 g/dm^2. The coefficient of variation in the center of the plan was 13.4%. A polynomial model approximates the dependence of the mass absorption value:

$$M_{ab} = 15,714 + 3,760x_1 + 1,560x_2 - 1,880x_3 - 4,986x_1^2 - 1,886x_2^2 - 1,525x_1x_3 \quad (1)$$

The module of dependence of the mass absorption value on the variable factors of new concrete has the form of a polynomial of the second degree, contains linear effects from all variable factors, negative square effects from varying the water-cement ratio and cement consumption, as well as the negative effect from the interaction of W/C and r.

The geometric image of the response surfaces is an elliptical paraboloid. The non-parallelism of the surfaces during X_3 fixation, successively at three levels, reflects the influence of the synergism of the W/C and r factors on the mass absorption value (Fig. 2).

In general, the nature of the surfaces indicates a positive dependence of the mass absorption value on the water-cement deviation of the new concrete mixture.

In the transition from rigid mixtures to plastic ones, the value of mass absorption increases sharply: by 3… 3.5 times for low-cement concretes and more than 10 times for high-cement concretes. At the same time, the mass absorption value of the latter is generally 1.5…2.0 times less than for low-cement concretes.

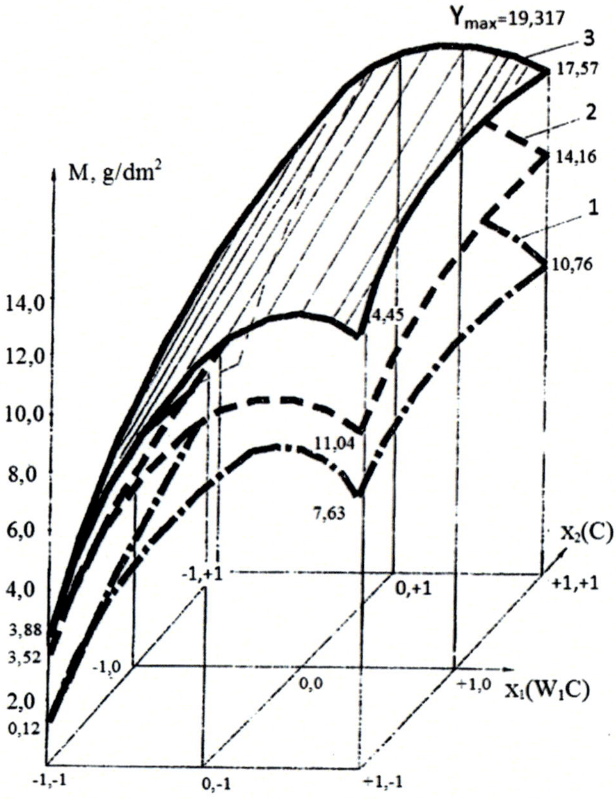

Fig. 2. Mass absorption in the contact zone: 1– at $x_3 = +1$; 2 – at $x_3 = 0$; 3 – at $x_3 = -1$.

With an increase in the proportion of sand in a mixture of aggregates, the mass absorption value decreases: for hard mixtures with low and medium cement consumption; it is 2–3 times; for high-cement concrete mixtures, 10%. For plastic mixtures, an increase in the proportion of sand in a mixture of aggregates reduces the mass absorption by 21… 26%, and for highly mobile and cast mixtures – by 38… 47%, depending on cement consumption. This reflects settling solid particles on the surface of old concrete with high mobility of new concrete mixtures.

In the region of maximum values, the change in factor X1 (water-cement ratios) has the most significant influence. The degree of its influence is 2.97 times higher than the influence of the cement consumption factor. The second rank regarding the degree of influence is the proportion of sand in the mixture of aggregates for the mass absorption value. Its increase reduces the amount of mass absorption. The degree of its influence is 1.43 times higher than the degree of influence of cement consumption.

The precast and cast-in-situ construction damage affects the strength and deformation characteristics. It is known that the damage caused by the action of low-cycle operational loads includes the initial damage to the precast and cast-in-situ construction. It includes damage to the material of the prefabricated and monolithic parts of the structure and damage to the seam and contact. Damage to the contact seam depends on the value of the contact strength of the concrete.

The K_{so} coefficient assessed technological damage to old concrete. At the time of strengthening, it was $43{,}78 \times 10^{-9} cv^{-1}$, and the variation coefficient was 40.9% in studies on 50 samples.

The coefficient of technological damage to new concrete varied depending on the composition within the range [from 7.4 to 134.9]$\times 10^{-2} cm^{-1}$, the coefficient of variation in the center of the plan for 9 samples was 10.2%.

The dependence of the value of the coefficient of technological damage on the variational factors is approximated by a polynomial of the 2nd degree (cm^{-1}).

$$
\begin{aligned}
K_s &= 58662 + 26,94x_2 - 19,10x_2 - 41,028x_1^2 + 15,882x_2^2 \\
&+5,672x_3^2 - 7,087x_1x_2 + 9688x_1x_3 - 24,087x_2x_3
\end{aligned}
\tag{2}
$$

Maximum technological damage $K_{s,max} = 151.909$ at $X_1 = 0.205$; $X_2 = 1.00$; $X_3 = -1.00$, which corresponds to the water-cement ratio $W/C = 0.607$, cement consumption $C = 500\ kg/m^3$, sand content in the mixture of aggregates $r = 0.30$.

The resulting polynomial contains linear effects from varying cement consumption and structural coefficient factors and quadratic effects from the interaction of three variable factors. The resulting model is approximated by the surfaces presented in Fig. 3, with X_3 stabilization at each of the three levels of the experiment plan. The intersection of the surfaces reflects the influence of the synergism of the variation factors.

The influence of the value of the water-cement ratio on the technological damage is different for different areas of study. For rigid concrete mixtures, under-consolidation of the concrete mix affected the surface quality of the new concrete. Few visible cracks in such a composite material; they pass through the mortar part of the concrete, bypassing the coarse aggregate. Such mixtures are characterized by an increase in the damage factor by 5…7 times with an increase in the water-cement ratio to a particular value.

In the transition to plastic mixtures, coarse aggregate grains are redistributed, their more rational and dense packing. As a result of the compaction of the concrete structure, the nature of technological damage changes. Cracks on the surface of such concrete are primarily short and chaotically located. They reflect the change in the influence of coarse filler due to the increase in its role in structure formation. A further increase in the mobility of low-cement concrete mixes from the accepted ones leads to a decrease in the damage coefficient by 30… 60% and changes the nature of the damage.

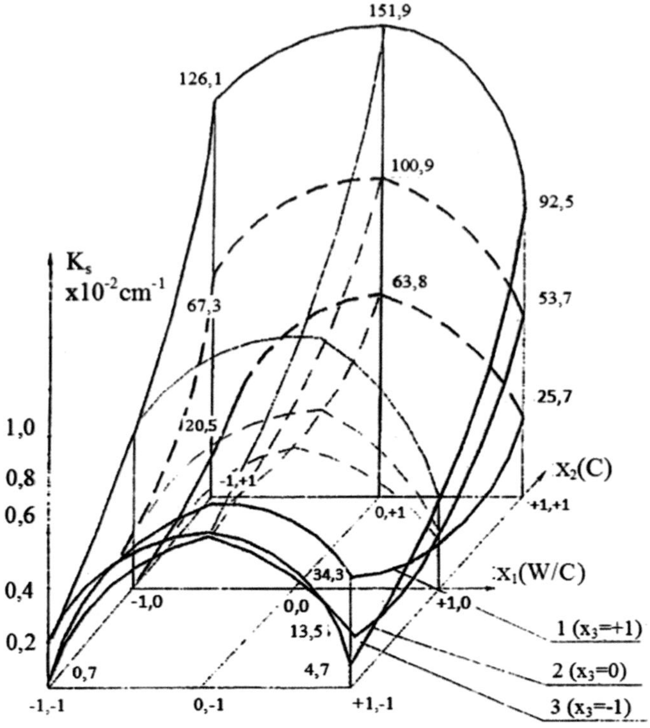

Fig. 3. Technological damage of the new concrete.

For high-quality concretes the nature of the change in technological damage is more pronounced. Suppose rigid mixtures are characterized by extended, mostly oriented in one, direction cracks with a common porous matrix material. In that case, plastic mixtures are characterized by the presence of short incomplete cracks, the location of which is chaotic, and the boundaries of structural blocks begin to appear. The final separation of the composite material into separate structure blocks (elements) is observed during the transition to cast mixtures. The nature of the distribution of cracks can be attributed to stochastic based on its repeatability with a change in scale.

An increase in the cement in the concrete mixture leads to a significant increase in technological damage (by 5… 10 or more times). The influence of cement consumption is more pronounced at the boundaries of the experiment in terms of water-cement ratio and is slightly smoothed out for plastic mixtures. For them, with an increase in cement consumption from the minimum to the maximum, the coefficient of technological damage increases by 1.1… 3 times. The influence of the proportion of sand in the mixture of aggregates on technological damage is different depending on the water-cement ratio of the composition of the new concrete. For rigid mixtures, with an increase in the proportion of sand, technological damage decreases by 5…6 times; for plastic concrete mixtures at high and medium consumption of cement, it decreases, and at low costs, it slightly increases with an increase in the proportion of sand in the aggregate mixture r.

For highly mobile and cast concrete mixtures, an increase in r has a different effect on the magnitude of technological damage. For low-cement concretes, the increase in K_s with an increase in the proportion of sand in the aggregate mixture manifests itself even more, and for other compositions with an average and high cement content, a decrease in K_s remains with an increase in the proportion of sand in the mixture of aggregates. The change in the sign of the influence of r for low-cement concrete compositions affects an increase in surface cracks due to the partial delamination of the concrete mixture.

In terms of the degree of influence of maximum values of K_s, as a result of the analysis of one-factor dependencies, cement consumption has the most significant influence (1.73 times higher than the water-cement ratio). The second, regarding the degree of influence, is the structural coefficient r. Its degree of influence is 1.65 times higher than the degree of influence of the water-cement ratio.

4.3 The Technological Damage of Old and New Concrete

A comparative analysis of the response surfaces R_{bt} (f_{ctd}), $R_{sh,j}$ ($f_{sh,j}$) and M_{ab} showed that j, the areas of maximum values for these dependences coincide, and the response maxima are in close areas, on isosurfaces, corresponding to the lower level r [8]. The isoclines of the responses of the dependences R_{bt}, R_{sh}, j, and M_{ab} visually represent the same influence of variation factors on the response values. All dependencies have a common direction of increasing levels of response surfaces from minimum to maximum. When combining the isosurfaces, it was revealed that there is a range of values of the variation factors for which the values of the parallel and tangential contact strength are close to the maximum. They correspond to the values of the mass absorption M_{ab}, which are also close to the maximum. This gives reason to believe a fair relationship exists between the contact strength of concrete of different ages and mass absorption in the contact zone. Mass absorption, as an integral characteristic, reflects both the properties of new, freshly laid concrete, the ratio of adhesive and cohesive forces that occur in the zone of contact with the surface of old concrete (the ratio of its surface and the ability of the structure of old concrete to absorb a certain amount of substance from freshly laid concrete mix). The value of mass absorption qualitatively equally reflects the normal and tangential contact strength change. Consequently, by the value of the mass absorption, it is possible to predict the values of the normal and tangential strength of the old and new concrete of prefabricated monolithic structures.

5 Conclusions

It has been established that the value of mass absorption in the contact zone changes with time, reaching its highest value for each new concrete composition at different times. The results of the main experiment confirmed the assumption that the value of mass absorption in its change in time depends on the composition of the concrete mixture packed on the sample.

In the region of maximum values, the water-cement ratio W/C has the most significant influence/ The degree of its influence is 2.97 times higher than the influence of the cement consumption factor. The second rank regarding the degree of influence is the proportion of sand in the mixture of aggregates for the mass absorption value. The degree of its influence is 1.43 times higher than the degree of influence of cement consumption.

The value of mass absorption qualitatively equally reflects the normal and tangential contact strength change. Therefore, by the value of the mass absorption, it is possible to predict the values of the normal and tangential strength of the old and new concrete of prefabricated monolithic structures.

References

1. Uzaeva, A.A., Uzaeva, S.A., Uzaev, M.A.: Monolithing the contact area and ensuring the adhesion of the old concrete to the new. In: Topical Issues of the Development of Modern Society, A Collection of Scientific Articles Based on the Materials of the 1st International Scientific and Practical Conference, pp. 100–103 (2016)
2. Le, T.Z., Glagolev, A.V.: Factors affecting the adhesion of old and new concrete. Exact Sci. **7**, 35–38 (2017)
3. Mirzazhonov, M.A., Otakulov, B.A.: Influence on the strength of the contact zone of the working joint of the exposure time of the new concrete. In: XLIII International Scientific and Practical Conference "International Scientific Review of the Problems and Prospects of Modern Science and Education", pp. 22–24. Problems of Science, Boston (2018)
4. Karpiuk, I., Danilenko, D., Karpiuk, V., Danilenko, A., Lyashenko, T.: Bearing capacity of damaged reinforced concrete beams strengthened with metal casing. Acta Polytech. **61**(6), 703–721 (2021). https://doi.org/10.14311/AP.2021.61.0703
5. Karpiuk, V., Syomina, Y., Antonova, D.: Bearing capacity of common and damaged CFRP-strengthened R. C. beams subject to high-level low-cycle loading. Mater. Sci. Forum **968**, 185–199 (2019). https://doi.org/10.4028/www.scientific.net/MSF.968.185
6. Blikharskyy, Y., Vashkevych, R., Kopiika, N., Bobalo, T., Blikharskyy, Z.: Calculation residual strength of reinforced concrete beams with damages, which occurred during loading. In: IOP Conference Series: Materials Science and Engineering, vol. 1021(1), p. 012012 (2021). https://doi.org/10.1088/1757-899X/1021/1/012012
7. Mirzazhonov, M.A., Otakulov, B.A.: Restoration of destroyed parts of concrete and reinforced concrete structures. Achievements Sci. Technol. **13**(35), 13–14 (2018)
8. Molodin, V.V., Anufrieva, A.E., Leonovich, S.N.: Influence of Carbonization of concrete surfaces on their adhesion with freshly-laid concrete. Sci. Tech. **20**(4), 320–328 (2021). https://doi.org/10.21122/2227-1031-2021-20-4-320-328
9. Molodin, V.V., Leonovich, S.N.: Bonding of recovery concrete with corrosion-destroyed reinforced concrete structure. Sci. Tech. **21**(1), 36–41 (2022). https://doi.org/10.21122/2227-1031-2022-21-1-36-41
10. Virovoj, V.M., Korobko, O.O., Zakorchemnij, Y.O., Urazmanova, N.F.: The role of active elements of the structure in the life cycle of building structures. In: Collection of Scientific Works of the Ukrainian State University of Railway Transport, vol. 186, pp. 46–54 (2019)
11. Vyrovoy, V., Korobko, O., Sukhanov, V., Zakorchemny, Y.: Resistance of concrete and expanded clay concrete under periodic external influences. In: MATEC Web of Conferences, vol. 23016, p. 03021 (2018). https://doi.org/10.1051/matecconf/201823003021
12. Vashpanov, Y., Con, I.-Y., Heo, G., Podousova, T., Kim, Y.S.: Determination of geometric of cracks in concrete by image processing. Adv. Civil Eng. **2019**, 159–165 (2019). https://doi.org/10.1155/2019/2398124
13. Szeląg, M.: The influence of cement composite composition on the geometry of their thermal cracks. Politechnika Lubelska, Lublin (2017)

14. Otakulov, B.A., Salimjonov, J.X.: Influence on durability of contact zone of working joint time of the endurance of a new concrete. EPRA ECEM **8**(5), 1–2 (2021). https://doi.org/10.36713/epra0414
15. Selejdak, J., Blikharskyy, Y., Khmil, R., Blikharskyy, Z.: Crack resistance RC columns strengthened by CFRP system. Key Eng. Mater. **878**, 127–133 (2021). https://doi.org/10.4028/www.scientific.net/KEM.878.127
16. Blikharskyy, Y., Vashkevych, R., Kopiika, N., Bobalo, T., Blikharskyy, Z.: Calculation residual strength of reinforced concrete beams with damages, which occurred during loading. In: IOP Conference Series: Materials Science and Engineering, vol. 1021(1), pp. 100–107 (2021). https://doi.org/10.1088/1757-899X/1021/1/012012
17. Krentowski, J.R., Knyziak, P., Mackiewicz, M.: Durability of interlayer connections in external walls in precast residential buildings. Eng. Fail. Anal. **121**, 113–121 (2021). https://doi.org/10.1016/j.engfailanal.2020.105059
18. Vitaliy, D., Natalia, P., Hanna, Z.: The influence of concrete structure on the destruction of reinforced concrete bended elements. In: B, Z. (ed.) EcoComfort 2020. LNCE, vol. 100, pp. 103–111. Springer, Cham (2021). https://doi.org/10.1007/978-3-030-57340-9_13
19. Dorofeev, V., Pushkar, N.: The bearing capacity of precast beams with vertical contact plain. In.: Blikharskyy, Z. (ed.) Proceedings of EcoComfort 2022. EcoComfort 2022. Lecture Notes in Civil Engineering, vol. 290, pp. 67−75. Springer, Cham (2022). https://doi.org/10.1007/978-3-031-14141-6_7

Influence of the Radius of Curvature of the Teeth on the Geometric and Functional Parameters of the Rotors of the Planetary Hydraulic Motor

Sergey Kiurchev[1] (✉), Volodymyr Kyurchev[1], Aleksandr Fatyeyev[2], Irina Tynyanova[2], and Krzysztof Mudryk[3]

[1] Dmytro Motornyi Tavria State Agrotechnological University, 18, B. Khmelnytskyi Ave., Melitopol 72310, Ukraine
tia_tgatu@ukr.net
[2] National Technical University Kharkiv Polytechnic Institute, 2, Kyrpycheva St., Kharkiv 61002, Ukraine
[3] University of Agriculture in Krakow, 21, Adama Mickiewicza Str., 31-120 Krakow, Poland

Abstract. Using planetary (orbital) hydraulic motors is advisable in drives with low rotational speed and high torque. The issue of stabilizing the output characteristics of these hydraulic motors by studying the effect of the radius of the curvature of teeth on a change in geometric and functional parameters is relevant. As a result of the research, a calculation scheme and a mathematical apparatus have been developed that describe the relationship between the geometric and functional parameters of the rotors, taking into account the influence of the radius of curvature of their teeth. The study found that for the nominal value of the tooth radius 4.5 mm, the gap in the critical pair is 0.02 mm, which borders on a shape error and leads to the jamming of the rotors. An analysis of the change in gaps between the corresponding pairs of teeth during wear shows that the range of their change in the critical pair of teeth (0.022–0.169 mm) is much less than in the pair located diametrically opposite (0.040–0.736 mm). Therefore, the possibility of jamming is not excluded under operating conditions due to the error in manufacturing gear contours. It has been established that the obtained patterns of influence of the radius of curvature of the planetary hydraulic motor rotor teeth on the change in its geometric and functional parameters will stabilize the planetary hydraulic motor output characteristics and ensure the efficient operation of these hydraulic machines.

Keywords: Energy Efficiency · Cycloidal Engagement · Diametrical Clearance · Center Distance · Wear · Pitch Circle Radius

1 Introduction

To date, planetary (orbital) hydraulic motors are used in various fields of mechanical engineering [1] in mechanisms with a rotary drive with low rotational speed and high torque [2, 3]. The principle of operation of a planetary (orbital) hydraulic motor is

V. Tonkonogyi et al. (Eds.): InterPartner 2023, LNME, pp. 450–461, 2024.
https://doi.org/10.1007/978-3-031-42778-7_42

based on the interaction of cycloidal engagement's internal and external rotors with teeth formed by circular arcs [4, 5]. Various errors in the approximation of the cycloidal contour of the tooth profiles of the inner and outer rotors, caused by design and manufacturing (technological difficulties), lead to a diametrical gap [6]. During the operation of a planetary hydraulic machine, due to the wear of the rotors, the diametrical clearance constantly increases, reaching its limit value [7]. An increase in the diametrical clearance due to a change in the radius of curvature of the teeth of the rotors caused by the wear of their working surfaces leads to additional movements of the inner rotor, which negatively affects the output characteristics of the planetary hydraulic motor [8].

In this regard, to stabilize the output characteristics of a planetary hydraulic motor, it is necessary to conduct in-depth studies of the relationship between the geometric and functional parameters of the cycloidal engagement formed by the teeth of the inner and outer rotors. Therefore, the question of studying the influence of the radius of curvature of the teeth on the change in the geometric and functional parameters of the rotors of a planetary hydraulic machine is very relevant.

2 Literature Review

The geometry of a hydrostatic gear pump [9] is described, and the force contact between the teeth of a hydrostatic gear pump is determined [10]. Modeling of the engagement of rotor profiles in parametric form was carried out [11], a rotor design program was developed [12], and a measuring system for diagnostics and operational control was proposed [13], identifying the most significant factors affecting the reliability and efficiency of hydraulic devices [14]. The influence of the parameters of the working fluid and the load on the output characteristics of the hydraulic motor was determined [15], the flow path of the labyrinth-screw pump was improved [16], models for modeling fluid aeration were described [17], the effect of the density of the medium on losses in the drainage channel was established [18]. However, using a technological drainage channel leads to losses of the pumped medium [19]. Issues related to the design of planetary (orbital) hydraulic motors were not considered.

It is known [20] that orbital and planetary hydraulic motors are effectively used in mechatronic systems of self-propelled vehicles. The disadvantages of these hydraulic machines include [21] uneven output characteristics due to the shape error of the elements of its rotor system [6] and the design features of the distribution system [22]. Mathematical [23] and dynamic [24] models of the processes occurring in the distribution system are proposed. The influence of the segmental [25], round [26], and oval [27] shape of distribution windows on the throughput [28] and output characteristics of the planetary hydraulic motor [29] is studied. The influence of the design features of the rotor system on the output characteristics of the planetary hydraulic motor has not been studied.

Mathematical [30] and dynamic [31] models of the processes occurring in the system of rotors during the operation of an orbital hydraulic motor [32] were developed, wear analysis was carried out using numerical simulation [33], the manufacturing error of gears was studied [34], the relationship between the error, contact force and friction [35], devices [36] and means [37] for controlling the error in the form of their manufacture have been developed, and experimental studies have been carried out [38]. The relationship between the geometric and functional parameters of the cycloidal gearing of the planetary hydraulic motor was not considered.

Thus, by analyzing publications devoted to the calculation, modeling, and design of planetary hydraulic motors, it is possible to state [39] that today there are practically no publications that reveal issues of research on the relationship between geometric and functional parameters of the cycloidal gearing of the planetary hydraulic motor. It should be noted that most of the publications are aimed at research on the design of gerotor hydraulic machines (mainly pumps). However, the operation of the rotors of a gerotor pump is fundamentally different from the operation of the rotors of a planetary (orbital) hydraulic motor [7, 8]. In connection with the foregoing, the presented article is devoted to research on the influence of the radius of curvature of the teeth on the change in the geometric and functional parameters of the rotors of a planetary hydraulic machine to stabilize the output characteristics of the planetary hydraulic motor.

3 Research Methodology

To study the influence of the radius of curvature of the teeth on the change in the geometric and functional parameters of the rotors of a planetary hydraulic machine, it is necessary to:

- to develop a calculation scheme and a mathematical apparatus for determining the relationship between the geometric and functional parameters of the rotors that characterize the cycloidal engagement of a planetary hydraulic motor when they are worn;
- to investigate the influence of the radius of curvature of the teeth on the change in the geometric and functional parameters of the rotors of a planetary hydraulic machine.

Let us clarify that the diametral gap is the gap G (Fig. 1) between the corresponding pairs of teeth of the inner and outer rotors, provided that the diametrically opposite tooth of the inner rotor is in contact with two teeth (rollers) of the outer rotor [7, 8]. The technical condition of the cycloidal engagement (and the hydraulic motor as a whole) during operation is determined by the change in the value of its diametrical clearance (due to wear). Therefore, to determine the technical condition of the planetary hydraulic motor, it is crucial to establish the relationship between the functional and geometric parameters of the cycloidal gear.

For further research, we will make the following assumptions:

- deviations of the profiles of the gear contours of the inner and outer rotors are equal to zero;
- change in the diametrical clearance (wear of the parts of the cycloidal gearing) is carried out by changing the radius of the pitch circle R_1 of the inner rotor;
- the radius of the pitch circle R_2 of the outer rotor and the radius of rounding of the teeth of the inner rotor r_1 and the outer rotor r_2 are unchanged.

Since the cycloidal linkage is symmetrical concerning the line of O_1O_2 centers, it is reasonable to consider only one, for example, its right half (Fig. 1). In this case, the numbers of the considered pairs of teeth of the inner and outer rotors are numbered from above, from left to right. Then the diametrical gap G will be located between the teeth of the "first" pair ($i = 1$), and the serial number of the diametrically opposite tooth will be determined by the expression $i = (Z_1 + 1)/2$, the angle δ between the line of centers O_1O_2 and the normal at the point of contact of the pair of teeth $i = (Z_1 + 1)/2$ constant ($\delta = \text{const}$).

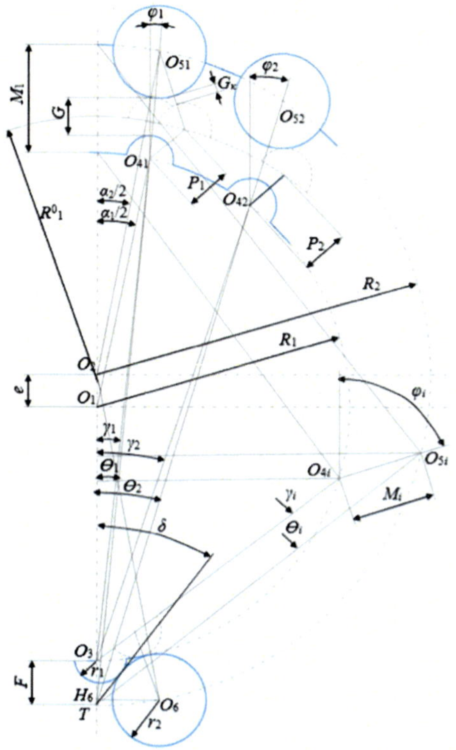

Fig. 1. Calculation scheme for determining the functional and geometric parameters of the cycloidal gear.

Considering the accepted assumptions, we determine the main geometric parameters characterizing the cycloidal engagement of the planetary hydraulic motor.

Gap G_i characterizes the relative position of the mating teeth of the inner and outer rotors:

$$G_i = \sqrt{A_i^2 + B_i^2} - (r_1 + r_2), \tag{1}$$

$$A_i = 2R_2 \cdot \cos^2 \frac{\pi \cdot (2i - 1)}{2Z_2} - 2R_1 \cdot \cos^2 \frac{\pi \cdot (2i - 1)}{2Z_1} - F,$$

$$B_i = R_2 \cdot \sin \frac{\pi \cdot (2i - 1)}{Z_2} - R_1 \cdot \sin \frac{\pi \cdot (2i - 1)}{Z_1},$$

$$F = \sqrt{(r_1 + r_2)^2 - R_2^2 \cdot \sin^2 \frac{\pi}{2Z_2}} + 2R_2 \cdot \sin^2 \frac{\pi}{2Z_2}, \quad (2)$$

where R_1 and R_2 are the radii of the pitch circles of the inner and outer rotors, respectively; r_1 and r_2 are the radii of curvature of the teeth of the inner and outer rotors, respectively; Z_1 and Z_2 are the numbers of teeth of the inner and outer rotors, respectively.

Parameter P_i characterizes the amount of parallel movement of the inner rotor until the i -th tooth of the outer rotor touches:

$$\mathrm{P}_i = 2 \cdot (r_1 + r_2) \cdot \sin \frac{\omega_i \cdot \lambda_i \cdot \delta}{2}, \quad (3)$$

where ω_i, λ_i, δ are auxiliary parameters describing the relationship of the geometric parameters of the rotors:

$$\omega_i = arctg \frac{(r_1 + r_2) \cdot \sin \delta + A_i}{(r_1 + r_2) \cdot \cos \delta - B_i}, \delta = \arccos \frac{R_2 \cdot \sin \frac{\pi}{2Z_2}}{r_1 + r_2},$$

$$\lambda_i = \arccos \frac{\sqrt{[(r_1 + r_2) \cdot \sin \delta + A_i] + [(r_1 + r_2) \cdot \cos \delta - B_i]^2}}{2 \cdot (r_1 + r_2)}.$$

The contact gap G_{cont} (i, j) between the teeth of the considered i-th pair at the contact of the j-th teeth of the inner and outer rotors is determined from the expression:

$$G_{cont}(i,j) = \sqrt{M_i^2 + P_j^2 - 2M_i - P_j \cdot \cos \frac{\omega_j - \lambda_j + \delta - 2\varphi_i}{2}} - (r_1 + r_2). \quad (4)$$

Since in a real (taking into account the manufacturing error) cycloidal engagement, the diametral clearance exists as a set of design parameters R_1, R_2, r_1, r_2, Z_1 and Z_2, then all considered geometric parameters (G_i, P_i and G_{cont}) studied as a function of design parameters and are defined by expressions (1), (3) and (4), respectively.

The relationship between the geometrical parameters was considered as a function of the center-to-center distance e of the inner and outer rotors. The distance between the rotors e_1 for theoretical cycloidal engagement, according to the recommendations of some researchers, was determined by the relationship of the radii of curvature of the teeth of its rotors $e_1 = (r_1 + r_2)/4$.

Since the parameter F changes during operation (during wear), then, taking into account (2), the center-to-center distance e_2 for theoretical cycloidal engagement is determined by the dependence:

$$\mathrm{e}_2 = R_2 - R_1 - \sqrt{(r_1 + r_2)^2 - R_2^2 \cdot \sin^2 \frac{\pi}{2Z_2}} - 2R_2 \cdot \sin^2 \frac{\pi}{2Z_2}. \quad (5)$$

The developed calculation schemes and mathematical apparatus make it possible to determine the relationship between the geometric and functional parameters of the rotors, taking into account the influence of the radius of curvature of the teeth, characterizing the change in the technical state of the cycloidal gearing of the planetary hydraulic motor during wear.

4 Results and Discussion

The study of the influence of the radius of curvature of the teeth on the change in the geometric and functional parameters of the rotors of the planetary hydraulic machine was carried out by simulating operating conditions using the Visual Simulator software package. Considering the symmetry of the gear profile arrangement, modeling was carried out only for one part of the cycloidal gearing (right).

When modeling, a pair of rotors (GPR-F planetary hydraulic motor) with the number of teeth equal to $Z_1 = 15$ and $Z_2 = 16$ for the inner and outer rotors was considered. Then the pair of teeth No. 1 is at the top right, and the pair of teeth No. 8 is diametrically opposite along the axis of symmetry (bottom). The nominal values of the teeth curvature radii are set to $r_1 = 4.5$ mm and $r_2 = 5.5$ mm, respectively, for the inner and outer rotors. When modeling, the wear of the working surfaces of the teeth of the rotors (an increase in the diametral clearance) was simulated by a decrease in the radius of the tooth r_1 of the inner rotor.

As a result of the simulation revealed the dependence of the change in gaps (Fig. 2, a) between the corresponding pairs of teeth of the cycloidal gearing of the planetary hydraulic motor for different values of the radius of the tooth r_1 of the inner rotor. It has been established that at the nominal value of the tooth radius $r_1 = 4.5$ mm, the gap in the pair of teeth No. 7 (critical in the pair under consideration) is 0.02 mm. With this value, the gap borders on the error in the gear profile's shape, which can lead to the rotors' jamming during operation.

With a decrease in the radius of curvature of tooth r1, the gap in the pair of teeth No. 7 increases, and at r1 = 4.0 mm, the gap is already 0.15 mm. It indicates an incorrect justification of the nominal value of the radius of curvature of the teeth of the GPR-F hydraulic motor. At the same time, even with a slight increase in the radius r1 = 4.65 mm, an interference of 0.003 mm occurs in the critical pair of teeth No. 7, which indicates that the hydraulic motor is inoperative even in the initial state.

The movement of the teeth of the inner rotor until contact with the corresponding teeth of the outer rotor shows (Fig. 2, b) that in the absence of an error in the shape of the tooth profiles of the rotors, contact in a pair of teeth No. 7 (when worn) is almost impossible. With a tooth rounding radius r1 = 4.5 mm, the gap in the critical pair of teeth No. 7 is 4.5 mm (curve 1), and with a decrease in radius to a value of r1 = 4.0 mm, the gap also decreases to a value of 2 mm (curve 2). It is explained by the fact that with a decrease in the radius of curvature of the tooth r1, the contour of the toothed surface of the inner rotor approaches the hypocyclidal one (the approximation error decreases).

By studying the relationship between the radii of curvature of the tooth of the inner rotor r_1 and its pitch circle R_1 (Fig. 3, curve 1), it was found that there are many inner rotors for each "conditional" outer rotor. When changing the radius of tooth curvature r_1 from 4.0…4.65 mm, the radius of the pitch circle R1 changes from 40.3 to 39.03 mm (Fig. 3, curve 1). When designing a planetary hydraulic motor, this makes it possible to justify the geometric parameters of its rotors with rational gaps between the corresponding pairs of teeth.

The existing relationship between the radii of curvature of the tooth r1 and the pitch circle R1 of the inner rotor at constant geometric parameters of the outer rotor (r2 and R2 - const) is characterized by significant changes in the center-to-center distance e between

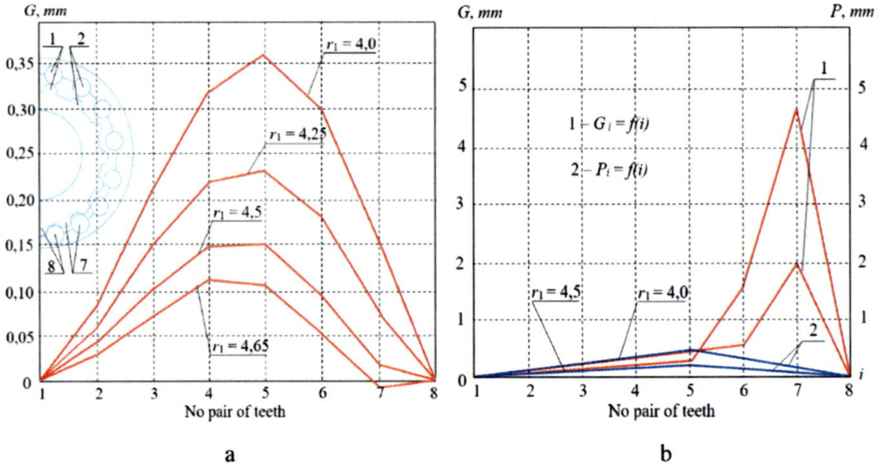

Fig. 2. Changing the gaps and the magnitude of the displacements of the corresponding pairs of teeth of the cycloidal gearing of the planetary hydraulic motor.

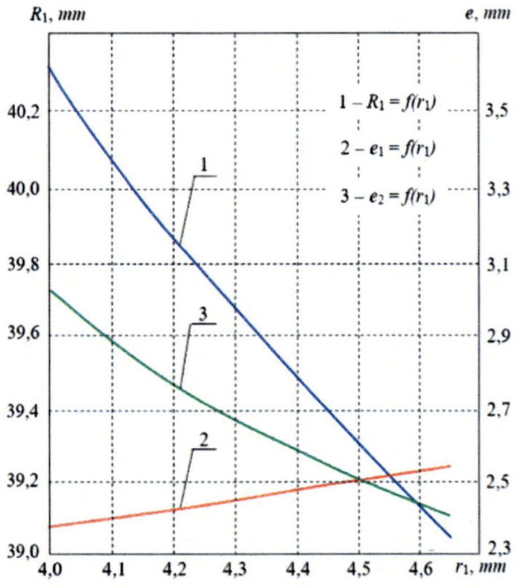

Fig. 3. Changing the geometric parameters of the cycloidal gearing from the radius r1 of the inner rotor tooth.

them. When the radius of tooth curvature r1 changes in the range from 4.0…4.65 mm, the center-to-center distance e changes in the aisles from 3.03 to 2.31 mm (Fig. 3, curve 2). Such a change in the center-to-center distance e, allows us to argue about the inconsistency of the recommendations of some researchers about its relationship with the radii of curvature of the teeth, described by the expression e = (r1 + r2)/4 (Fig. 3, curve 3).

An analysis of the change in gaps between the corresponding pairs of teeth of cycloidal engagement during wear shows (Fig. 4) that the range of gap changes in a pair of teeth No. 7 (0.022–0.169 mm) is significantly less than in a pair No. 2 (0.040–0.736 mm). Therefore, under operating conditions, the possibility of jamming the corresponding teeth of pair No. 7 is not excluded due to the error in manufacturing the shape of the gear contour of the outer and inner rotors [6].

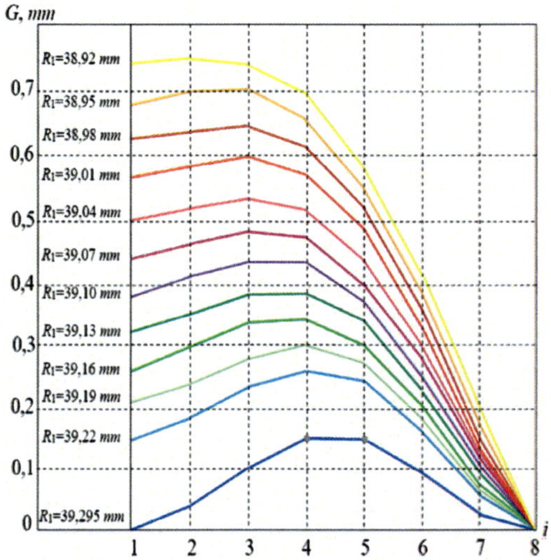

Fig. 4. Changing the gaps between the corresponding pairs of teeth of cycloidal gearing during wear.

The conducted studies have established the influence of the radius of curvature of the teeth of the planetary hydraulic motor rotors on the change in its geometric and functional parameters, which will ensure the stabilization of the output characteristics of the planetary hydraulic motor during its design. Knowledge of the patterns of change in the output characteristics of a planetary hydraulic motor from a change in diametrical clearance (from wear) will ensure more efficient operation of these hydraulic machines in real conditions.

5 Conclusions

As a result of the research, a calculation scheme and a mathematical apparatus have been developed that make it possible to describe the relationship between the geometric and functional parameters of the rotors, taking into account the influence of the radius of curvature of their teeth. The study shows that for a nominal tooth radius $r_1 = 4.5$ mm, the gap in the critical pair is 0.02 mm, which borders on the error in the shape of the gear profile and can lead to jamming of the rotors during operation.

An analysis of the change in gaps between the corresponding pairs of teeth of cycloidal engagement during wear shows that the range of change in gaps in a critical pair of teeth (0.022–0.169 mm) is much less than in a pair located diametrically opposite (0.040–0.736 mm). Therefore, under operating conditions, the possibility of jamming of the corresponding teeth due to the manufacturing error of the gear contour of the rotors is not ruled out.

It has been established that the regularities obtained will stabilize the output characteristics of the planetary hydraulic motor during its design, ensuring more efficient operation of these hydraulic machines in real conditions.

References

1. Panchenko, A., Voloshina, A., Milaeva, I., Luzan, P.: Operating conditions' influence on the change of functional characteristics for mechatronic systems with orbital hydraulic motors. In: Nadykto, V. (ed.) Modern Development Paths of Agricultural Production, pp. 169–176. Springer, Cham (2019). https://doi.org/10.1007/978-3-030-14918-5_18
2. Antoshchenkov, R., et al.: Determining the influence of geometric parameters of the traction-transportation vehicle's frame on its tractive capacity and energy indicators. Eastern-Eur. J. Enterp. Technol. **2**(7(116)), 60–61 (2022). https://doi.org/10.15587/1729-4061.2022.254688
3. Antoshchenkov, R., Bogdanovich, S., Halych, I., Cherevatenko, H.: Determination of dynamic and traction-energy indicators of all-wheel-drive traction-transport machine. Eastern-Eur. J. Enterp. Technol. **1**(7(121)), 7–14 (2023). https://doi.org/10.15587/1729-4061.2023.270988
4. Vecchiato, D., Demenego, A., Argyris, J., Litvin, F.L.: Geometry of a cycloidal pump. Comput. Methods Appl. Mech. Eng. **190**, 2309–2330 (2001). https://doi.org/10.1016/S0045-7825(00)00236-X
5. Bonandrini, G., Mimmi, G., Rottenbacher, C.: Theoretical analysis of internal epitrochoidal and hypotrochoidal machines. Proc. Inst. Mech. Eng. **223**, 1469–1480 (2009). https://doi.org/10.1243/09544062JMES1163
6. Panchenko, A., Voloshina, A., Milaeva, I., Panchenko, I., Titova, O.: The Influence of the form error after rotor manufacturing on the output characteristics of an orbital hydraulic motor. Int. J. Eng. Technol. **7**(4.3), 1–5 (2018). https://doi.org/10.14419/ijet.v7i4.3.19542
7. Panchenko, A., Voloshina, A., Luzan, P., Panchenko, I., Volkov, S.: Kinematics of motion of rotors of an orbital hydraulic machine. In: IOP Conference Series: Materials Science and Engineering, vol. 1021, p. 012045 (2021). https://doi.org/10.1088/1757-899X/1021/1/012045
8. Panchenko, A., Voloshina, A., Sadullozoda, S.S., Panchenko, I., Mitin, V.: The changes in the output parameters of planetary hydraulic machines with the increase in the gap between their rotors. In: Tonkonogyi, V., Ivanov, V., Trojanowska, J., Oborskyi, G., Pavlenko, I. (eds.) Advanced Manufacturing Processes IV. InterPartner 2022. LNME, pp. 540–551. Springer, Cham (2023). https://doi.org/10.1007/978-3-031-16651-8_51
9. Paffoni, B.: Pressure and film thickness in a trochoidal hydrostatic gear pump. Proc. Inst. Mech. Eng. **217**, 179–187 (2003). https://doi.org/10.1243/095441003769700744
10. Paffoni, B., Progri, R., Gras, R.: Teeth clearance effects upon pressure and film thickness in a trochoidal hydrostatic gear pump. Proc. Inst. Mech. Eng. **218**, 247–256 (2004). https://doi.org/10.1243/0954410041872799
11. Bonandrini, G., Mimmi, G., Rottenbacher, C.: Design and simulation of meshing of a particular internal rotary pump. Mech. Mach. Theory **49**, 104–116 (2012). https://doi.org/10.1016/j.mechmachtheory.2011.11.001
12. Choi, T.H., Kim, M.S., Lee, G.S., Jung, S.Y., Bae, J.H., Kim, C.: Design of rotor for internal gear pump using cycloid and circular-arc curves. J. Mech. Des. **134**, 011005 (2012). https://doi.org/10.1115/1.4004423
13. Artiomov, N., Antoshchenkov, R., Antoshchenkov, V., Ayubov, A.: Innovative approach to agricultural machinery testing. Eng. Rural Dev. **20**, 692–698 (2021)

14. Andrenko, P., Rogovyi, A., Hrechka, I., Khovanskyi, S., Svynarenko, M.: Characteristics improvement of labyrinth screw pump using design modification in screw. J. Phys: Conf. Ser. **1741**(1), 012024 (2021). https://doi.org/10.1088/1742-6596/1741/1/012024
15. Andrenko, P., Rogovyi, A., Hrechka, I., Khovanskyi, S., Svynarenko, M.: The influence of the gas content in the working fluid on parameters of the hydraulic motor's axial piston. In: Ivanov, V., Pavlenko, I., Liaposhchenko, O., Machado, J., Edl, M. (eds.) DSMIE 2021. LNME, pp. 97–106. Springer, Cham (2021). https://doi.org/10.1007/978-3-030-77823-1_10
16. Andrenko, P., Hrechka, I., Khovanskyi, S., Rogovyi, A., Svynarenko, M.: Improving the technical level of hydraulic machines, hydraulic units and hydraulic devices using a definitive assessment criterion at the design stage. J. Mech. Eng. (JMechE) **18**(3), 57–76 (2021). https://doi.org/10.24191/jmeche.v18i3.15414
17. Rundo, M.: Models for flow rate simulation in gear pumps: a review. Energies **10**, 1261 (2017). https://doi.org/10.3390/en10091261
18. Rogovyi, A., Korohodskyi, V., Neskorozhenyi, A., Hrechka, I., Khovanskyi, S.: Reduction of granular material losses in a vortex chamber supercharger drainage channel. In: Ivanov, V., Pavlenko, I., Liaposhchenko, O., Machado, J., Edl, M. (eds.) Advances in Design, Simulation and Manufacturing V. DSMIE 2022. Lecture Notes in Mechanical Engineering, pp. 218–226. Springer, Cham (2022). https://doi.org/10.1007/978-3-031-06044-1_21
19. Rogovyi, A., Neskorozhenyi, A., Krasnikov, S., Tynyanova, I., Khovanskyi, S.: Improvement of vortex chamber supercharger performances using slotted rectangular channel. In: Tonkonogyi, V., Ivanov, V., Trojanowska, J., Oborskyi, G., Pavlenko, I. (eds.) Advanced Manufacturing Processes IV. InterPartner 2022. Lecture Notes in Mechanical Engineering, pp. 552–561. Springer, Cham (2023). https://doi.org/10.1007/978-3-031-16651-8_52
20. Panchenko, A., Voloshina, A., Titova, O., Panchenko, I., Caldare, A.: Design of hydraulic mechatronic systems with specified output characteristics. In: Ivanov, V., Pavlenko, I., Liaposhchenko, O., Machado, J., Edl, M. (eds.) DSMIE 2020. LNME, pp. 42–51. Springer, Cham (2020). https://doi.org/10.1007/978-3-030-50491-5_5
21. Voloshina, A., Panchenko, A., Titova, O., Milaeva, I., Pastushenko, A.: Prediction of changes in the output characteristics of the planetary hydraulic motor. In: Tonkonogyi, V., Ivanov, V., Trojanowska, J., Oborskyi, G., Grabchenko, A., Pavlenko, I., Edl, M., Kuric, I., Dasic, P. (eds.) InterPartner 2020. LNME, pp. 744–754. Springer, Cham (2021). https://doi.org/10.1007/978-3-030-68014-5_72
22. Voloshina, A., Panchenko, A., Boltyansky, O., Panchenko, I., Titova, O.: Justification of the kinematic diagrams for the distribution system of a planetary hydraulic motor. Int. J. Eng. Technol. **7**(4.3), 6–11 (2018). https://doi.org/10.14419/ijet.v7i4.3.19544
23. Kiurchev, S., Luzan, P., Zasiadko, A., Radionov H., Boltianska, N.: Influence of the flow area of distribution systems on changing the operating parameters of planetary hydraulic motors. In: IOP Conference Series: Materials Science and Engineering, vol. 1021, p. 012037 (2021).https://doi.org/10.1088/1757-899X/1021/1/012037
24. Voloshina, A., Panchenko, A., Titova, O., Panchenko, I.: Changes in the dynamics of the output characteristics of mechatronic systems with planetary hydraulic motors. J. Phys: Conf. Ser. **1741**, 012045 (2021). https://doi.org/10.1088/1742-6596/1741/1/012045
25. Panchenko, A., Voloshina, A., Panchenko, I., Pashchenko, V., Zasiadko, A.: Influence of the shape of windows on the throughput of the planetary hydraulic motor's distribution system. In: Ivanov, V., Pavlenko, I., Liaposhchenko, O., Machado, J., Edl, M. (eds.) DSMIE 2021. LNME, pp. 146–155. Springer, Cham (2021). https://doi.org/10.1007/978-3-030-77823-1_15
26. Voloshina, A., Panchenko, A., Boltyansky, O., Titova, O.: Improvement of manufacture workability for distribution systems of planetary hydraulic machines. In: Ivanov, V., Trojanowska, J., Machado, J., Liaposhchenko, O., Zajac, J., Pavlenko, I., Edl, M., Perakovic, D. (eds.) DSMIE 2019. LNME, pp. 732–741. Springer, Cham (2020). https://doi.org/10.1007/978-3-030-22365-6_73

27. Voloshina, A., Panchenko, A., Panchenko, I., Zasiadko, A.: Geometrical parameters for distribution systems of hydraulic machines. In: Nadykto, V. (ed.) Modern Development Paths of Agricultural Production, pp. 323–336. Springer, Cham (2019). https://doi.org/10.1007/978-3-030-14918-5_34
28. Voloshina, A., Panchenko, A., Titova, O., Pashchenko, V., Zasiadko, A.: Experimental studies of a throughput of the distribution systems of planetary hydraulic motors. In: IOP Conference Series: Materials Science and Engineering, vol. 1021, p. 012054 (2021). https://doi.org/10.1088/1757-899X/1021/1/012054
29. Voloshina, A., Panchenko, A., Boltyansky, O., Zasiadko, A., Verkholantseva, V.: Improvement of the angular arrangement of distribution system windows when designing planetary hydraulic machines. In: Tonkonogyi, V., Ivanov, V., Trojanowska, J., Oborskyi, G., Pavlenko, I. (eds.) InterPartner 2021. LNME, pp. 53–63. Springer, Cham (2022). https://doi.org/10.1007/978-3-030-91327-4_6
30. Ding, H., Lu, J., Jiang, B.: A CFD model for orbital gerotor motor. In: IOP Conference Series: Earth and Environmental Science, vol. 15(6), p. 062006 (2012).https://doi.org/10.1088/1755-1315/15/6/062006
31. Panchenko, A., Voloshina, A., Titova, O., Panchenko, I., Zasiadko, A.: The study of dynamic processes of mechatronic systems with planetary hydraulic motors. In: Tonkonogyi, V., Ivanov, V., Trojanowska, J., Oborskyi, G., Grabchenko, A., Pavlenko, I., Edl, M., Kuric, I., Dasic, P. (eds.) InterPartner 2020. LNME, pp. 704–713. Springer, Cham (2021). https://doi.org/10.1007/978-3-030-68014-5_68
32. Panchenko, A., Voloshina, A., Sadullozoda, S.S., Boltyansky, O., Panina, V.: Influence of the design features of orbital hydraulic motors on the change in the dynamic characteristics of hydraulic drives. In: Ivanov, V., Pavlenko, I., Liaposhchenko, O., Machado, J., Edl, M. (eds.) Advances in Design, Simulation and Manufacturing V. DSMIE 2022. Lecture Notes in Mechanical Engineering, pp. 101—111. Springer, Cham (2022). https://doi.org/10.1007/978-3-031-06044-1_10
33. Furustig, J., Almqvist, A., Pelcastre, L., Bates, C.A., Ennemark, P., Larsson, R.: A strategy for wear analysis using numerical and experimental tools, applied to orbital type hydraulic motors. Proc. Inst. Mech. Eng. **230**, 2086–2097 (2016). https://doi.org/10.1177/0954406215590168
34. Bates, C.A., Broe-Richter, H.W., Bendlin, C.R., Ennemark, P.: The effect of an amorphous hydrogenated carbon-coated gear-wheel on a hydraulic orbital motor's efficiency over time. Proc. Inst. Mech. Eng., **234**, 1–14 (2018). https://doi.org/10.1177/1350650117752610
35. Strmčnik, E., Majdič, F.: The improvement of the total efficiency of the gerotor orbital hydraulic motor. In: Proceedings of the 11th International Fluid Power Conference, pp. 294—305. Aachen, Germany (2018). https://doi.org/10.18154/RWTH-2018-224639
36. Kiurchev, S., Abdullo, M.A., Vlasenko, T., Prasol, S., Verkholantseva, V.: Automated control of the gear profile for the gerotor hydraulic machine. In: Tonkonogyi, V., Ivanov, V., Trojanowska, J., Oborskyi, G., Pavlenko, I. (eds.) Advanced Manufacturing Processes IV. InterPartner 2022. Lecture Notes in Mechanical Engineering, pp. 32–43. Springer, Cham (2023). https://doi.org/10.1007/978-3-031-16651-8_4
37. Panchenko, A., Voloshina, A., Boltianska, N., Pashchenko, V., Volkov, S.: Manufacturing error of the toothed profile of rotors for an orbital hydraulic motor. In: Tonkonogyi, V., Ivanov, V., Trojanowska, J., Oborskyi, G., Pavlenko, I. (eds.) InterPartner 2021. LNME, pp. 22–32. Springer, Cham (2022). https://doi.org/10.1007/978-3-030-91327-4_3

38. Panchenko, A., Voloshina, A., Titova, O., Panchenko I.: The influence of the design parameters of the rotors of the planetary hydraulic motor on the change in the output characteristics of the mechatronic system. In: Journal of Physics: Conference Series, vol. 1741, p. 012027 (2021). https://doi.org/10.1088/1742-6596/1741/1/012027
39. Gamez-Montero, P.: A review of gerotor technology in hydraulic machines. Energies **12**(12), 2423 (2019). https://doi.org/10.3390/en121224232019

Check for updates

Assessment of the Life Cycle Cost and Improvement of the Parametric Series of Torque-Flow Pumps

Vladyslav Kondus[1,2](✉), Mykola Sotnyk[1], Andriy Sokhan[1], Serhii Antonenko[1], and Volodymyr Rybalchenko[1]

[1] Sumy State University, 2, Rymskogo-Korsakova St., Sumy 40007, Ukraine
v.kondus@pgm.sumdu.edu.ua
[2] Sumy Machine-Building Cluster of Energy Equipment, Sumy State University, 2, Rymskogo-Korsakova St., Sumy 40007, Ukraine

Abstract. The research developed and proposed criteria for a comprehensive assessment of the life cycle cost of a dynamic pump installation, which allows to practically and visually assess the value of energy efficiency (energy consumption indicator ε_{eff}) and mass-dimensional qualities (material capacity indicator εmat) of dynamic pumps. The practical application of the developed indicators made it possible to evaluate the value of energy efficiency and mass-dimensional qualities of the existing parametric series of TFP torque-flow pumps. It was determined that many torque-flow pumps significantly exceeded average energy consumption and material capacity indicators. An updated promising torque-flow pump TFP 25-28-2900 was developed. The energy efficiency of this pump at BEP mode is 0.4578. The design of the promising TFP 25-28-2900 pump made it possible to reduce the energy consumption indicator ε_{eff} up to 51.2% and the material capacity indicator ε_{mat} up to 74.6%. The research results correspond to the guidelines approved by the United Nations as Sustainable Development Goals, i.e., clean water and sanitation, affordable and clean energy, and industry, innovation and infrastructure.

Keywords: Sustainable Development Goals · Process Innovation · Energy Efficiency · Dynamic Pump · Torque-Flow Pump · TFP Pump · Power Consumption · Energy Consumption Indicator · Material Capacity Indicator · Parametric Series · Specific Speed

1 Introduction

The direction of human development in order to achieve the orientations approved by the United Nations as sustainable development goals (here and next – SDGs) [1] is the key to a fair and effective development of society in the context of the Industry 4.0 strategy [2]. Accordingly, special attention in the process of development of industrial equipment should be paid to the following directions [3]: “Clean water and sanitation” (goal 6), “Affordable and clean energy” (goal 7), “Industry, innovation and infrastructure” (goal 9).

V. Tonkonogyi et al. (Eds.): InterPartner 2023, LNME, pp. 462–476, 2024.
https://doi.org/10.1007/978-3-031-42778-7_43

The production processes of industrial enterprises are closely related to the need for fluid transportation. The level of electricity used by pumping equipment in the overall balance of energy consumption of enterprises proves that in most industrial enterprises' production processes, energy consumption by pumping equipment is a significant share and, in some cases, even dominant. In terms of industries, the proportion of pumping equipment energy consumption is [4]: in the oil and refining industry – 59%, in water supplying – 50%, in chemical – 31%, in the cellulose and paper industry – 26%, in metallurgy – 15%, in construction – 12%, in transport – 10%, and in the food industry – 6%, in the energy sector – 5% [5].

Given the above essential issues of pumps, pump units, and pumping equipment functioning is decreasing their energy consumption, sealing the pumped environment within the pumping installation, and preventing leaks to the external environment [6].

A significant issue in reducing energy consumption is increasing the energy efficiency of pumps, units based on them, and pumping equipment. Moreover, important issues should be understood regarding the direct pumping equipment design and their rational use in achieving the highest energy efficiency indicators.

Dynamic-type pumps have been widely used in the processes of different types of liquids transporting [7]. In turn, there are several varieties of dynamic pumps: centrifugal, diagonal, axial, and torque-flow pumps. Using one or another type of dynamic pump depends largely on the type of pumped environment and working parameters (head, flow rate [8]), which are required to ensure the inpatient operation of the pipeline.

It is important to note that most dynamic type pumps (centrifugal [9], diagonal, axial [10]) are not intended for pumping liquids with a high content of different types of inclusions. This is due to the use of front seals that form minimum gaps (0.2–0.4 mm) between the rotary and stator component of the flowing part. Thus, pumping liquids with inclusions, the value of which is greater than the width of the gap seal, is impossible because of the requirements of safety and durability of pumps and units on their basis [11, 12].

In turn, the torque-flow pumps (Fig. 1) design does not imply such front seals [13]. The operating body of the pump (impeller), which is driven by the rotor motion, is located in the boring or free chamber, which creates a wide passage channel in the pump flowing part. As a result, torque-flow pumps can transport liquids with a high content of up to 0.8 of the width of the free camera canal.

The main advantage of torque-flow pumps is their significantly increased resource. This is due to the much lower wear rate of the elements of their flowing parts while pumping liquids with inclusions. In addition, using such pumps, unlike the use of centrifugal or diagonal pumps for transporting liquids with fibrous inclusions, there is no clogging of the pump flowing part due to the winding of fibrous inclusions on rotary elements.

Due to their high reliability [14] and wear resistance [15], torque-flow pumps have found wide application in various industries, in particular: in chemical (crystalline suspensions, filtered sediments, latex, soda solutions, hot brines, aqueous mixtures, suspensions), petrochemical (catalytic tailings, benzene, drilling mud, unenriched ore, crushed coke, water-oil emulsions), mining (coal and other slurries, coke fuel, waste after ore enrichment, mine drainage), food (vegetable pomace, pig feed, fruit suspensions, bird trimmings with feathers, bone and food waste, crustaceans), pulp and paper industry (pulp, decoctions, wood shavings, grinding stone fragments, waste paper, straw, rags, kaolin), metallurgy (crushed slag, ash, soot, coke, sludge), operating processes of heat generating stations (ash deposits, filtered dirt), water treatment (untreated sewage, sludge, fertilizers, sand, gravel, sewage in sewage from slaughterhouses, municipal sewage), construction (aerated concrete slurries, sand, gravel, stone dust and marble chips with water, creosote mixtures), textile industry (paints with sludge, wastewater with fibers, glass wool, textile fiber, nitrocellulose), dredging and underwater extraction of soil (sand, gravel, rock, cleaning of lakes and harbors), etc.

a b

Fig. 1. The torque-flow pump operating process: a – simulating using the Solid Flow Simulation software; b – features of the operating process: red arrow – toroidal vortex, green arrow – flowing stream.

However, the main disadvantage of torque-flow pumps is some lower energy efficiency compared to centrifugal, diagonal, or axial pumps [16].

In addition, an essential problem in the existing parametric series of torque-flow pumps is their high material consumption compared to centrifugal, diagonal, and axial pumps [17]. This is explained by the lower frequency of rotation and the necessary larger overall dimensions of the operating body (impeller) to ensure the necessary head.

It should also be noted that the existing Ukrainian torque-flow pumps are equipped with unreliable packing seals. They pass part of the pumped product into the external environment, polluting the surrounding space [18].

Considering the above mentioned, the main task of the research is the general reduction of the life cycle cost [19], particularly for torque-flow pumps and pump units based on them, and improvement of their design in order to minimize the negative impact on the external environment [20, 21]. All of the above will make it possible to implement a set of measures to achieve several United Nations' Sustainable Development Goals, in particular, "Clean water and sanitation" (goal 6), "Affordable and clean energy" (goal 7), "Industry, innovation and infrastructure" (goal 9) [22, 23].

The following research tasks were set to reach this goal:

1. Development of criteria for a comprehensive assessment of the life cycle cost of pumps, pump units, and pumping equipment;
2. Comprehensive assessment of the cost of the life cycle of the existing parametric series of torque-flow pumps;
3. Determination of ways to reduce energy consumption by torque-flow pumps and reduce their material consumption;
4. Improve the existing parametric series of torque-flow pumps by developing an energy-efficient pump with the minimum necessary material capacity.

2 Literature Review

The so-called specific speed coefficient [24] significantly influences the energy efficiency of dynamic pumps. This indicator depends on the pump rotor's flow rate, head, and rotational frequency. It is determined by dependence:

$$n_s = \frac{3{,}65n\sqrt{Q}}{H^{3/4}}, \tag{1}$$

where n is the rotation frequency, rpm; Q – pump flow rate, m^3/h; H – pump head, m of water column.

It is known that this indicator's growth increases dynamic pumps' energy efficiency. Simultaneously, the maximum value of pump efficiency is achieved at $n_s \approx 120$, and in the range $n_s < 60$, the dynamic pump efficiency decreases to 50–60%.

A similar dependence is observed for torque-flow pumps [20]. The peculiarity of their operating process is that only a part of the fluid passes through the interblade channels of the impeller (Fig. 1). The other part receives energy by interacting with the toroidal vortex, which is formed by the force interaction of the impeller blades with the flow passing through it. In the context of a torque-flow pump, a toroidal vortex is called a "liquid blade" forming a force interaction with the flowing stream.

The design of torque-flow pumps with high energy efficiency is possible in specific speed values $n_s = 60$–170 [24].

Achieving a high technical level of torque-flow pumps, the main components of which are indicators of energy efficiency, reliability, and material consumption, can be achieved in different ways [25].

One of the main possibilities for influencing the energy efficiency indicators of a torque-flow pump is changes in its flowing part design, namely the operating body (impeller) and the housing. It is also essential that a pump housing include an inlet and outlet device.

A number of sources indicate the possibility of using so-called winglets, which partially cover the interblade channel of the impeller near its edge [25]. The peculiarity of this impeller design allows for reducing inclusions passed through the impeller interblade channel. On the other hand, this design reduces the proportion of liquid passing through the interblade channel, reducing the intensity of the toroidal vortex [26].

Increasing energy efficiency can be achieved by designing an impeller with a curvilinear blade profile [27]. In this case, the skeleton of the blade corresponds to the direction of the liquid inflow at each point of the flow. As a result, a reduction of hydraulic losses due to vortex formation is achieved [28].

In some cases, in order to reduce the head to the required level and, accordingly, to reduce the power consumption of the torque-flow pump, it is recommended to cut the impeller of the pump along with the outer diameter or the blade's edge [29]. The disadvantage of this method is the small range of cutting the impeller [30]. This is caused by a significant drop in efficiency (more than 5–10%) while cutting the impeller by a value greater than $0.1D_2$.

Considerable attention is also paid in the reviewed sources to the change in the design of the stator elements of the housing, in particular, the tap and the free chamber [31]. However, using such techniques, the influence of many factors on the integral parameters of the pump must be considered, particularly viscosity [32], and pulsating nature of the flow [33].

3 Research Methodology

At the empirical level of scientific research, numerical simulation (finite element analysis and methods of computational hydrodynamics) was used in the environment of the ANSYS software complex with further processing in the MathCAD algebra computer system.

The theoretical developments were verified using imitational simulation, the models based on the actual constructions of pumping equipment, and a comparison of the results of modeling integral characteristics with the passport data of torque-flow pumps and units based on them.

The university version of the ANSYS CFX software product was used to conduct numerical simulations. This software was developed based on the numerical solution method of the fundamental laws of hydromechanics.

This software has been repeatedly tested to resolve similar problems while simulation the working dynamic pumps. The simulation results with subsequent verification by performing a physical experiment showed sufficient accuracy in all cases. The difference in results did not exceed 5% in the operating range of the integral pump characteristics. Studies were performed for centrifugal pumps, including those with a low specific speed ns [34] and torque-flow pumps.

At the first stage, an energy-efficient torque-flow pump was developed according to the theoretical methods of torque-flow pump designing, which made it possible to improve the existing parametric series of torque-flow pumps (Fig. 2) for transporting liquids with inclusions [7, 13, 16].

In order to conduct research in the SolidWorks software environment, a solid-state model of the liquid environment of the designed pump flowing part was developed.

The parameters of the boundary conditions (wall roughness, inlet, outlet of the calculation area, turbulence model, size, and number of wall layers) were chosen considering the experience of designing pumping equipment by conducting numerical research [34].

The standard k-ε turbulence model was chosen as the turbulence model, which accurately described the operating process of torque-flow pumps in previous studies.

The flow rate was set at the pump inlet from 0.05 to 2.0 of Q_{bep} as a boundary condition. A pressure of 1 MPa was chosen as the boundary condition at the outlet from the calculation area.

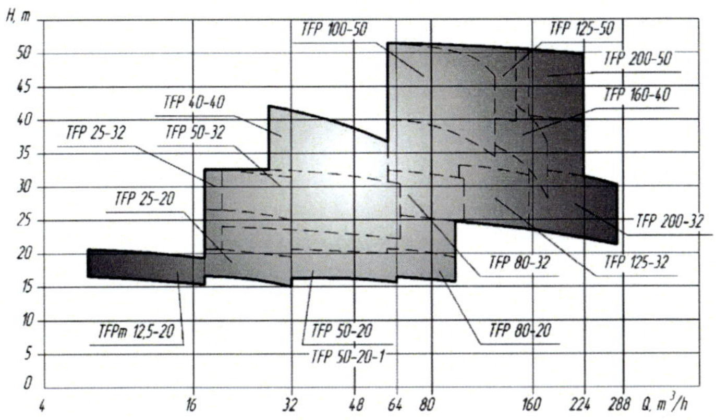

Fig. 2. A parametric series of TFP torque-flow pumps for transporting liquids with inclusions.

The total thickness of the wall layer, characterized by special prismatic layers of cells, is chosen to equal 0.05 mm, with a total number of 7 layers with a gradual increase in the thickness of each subsequent layer according to the logarithmic law.

The roughness of the solid surfaces of the pump flowing part elements is selected according to the pump's developed design features.

The convergence of integral characteristics in the ANSYS CFX Solver environment is achieved at the level of 1.0E-5 for RMS U-mom, RMS V-mom, and RMS W-mom, and at the level of 1.0E-8 for RMS P-Mass. In addition, monitoring the convergence of pump head, hydraulic and consumed power and efficiency were monitored. Variation of the parameters within no more than 3% was achieved, which is sufficient accuracy when conducting engineering calculations.

3.1 Methodology for Improving the Parametric Series of Torque-Flow Pumps

A comprehensive methodology for estimating the life cycle cost of pumping equipment has not been developed for today. The complete method of estimating the cost of the life cycle of pumping equipment is the determination of the total cost of funds at each stage of design, manufacture, installation, operation, repair, and disposal [20]. However, this indicator is indirectly affected by many external factors that do not depend on the operation of the pumping unit, such as inflation, the cost of energy resources, and the cost of materials and maintenance. As a result, it is possible to compare the life cycle cost of pumping installations using two or more pumps of different sizes or types only under the same external factors, which is not possible in most cases.

Comparing two pumps in terms of energy efficiency is also not entirely practical, as in some cases, a more energy-efficient pump may have significantly higher investment, maintenance, repair, and disposal costs than a less energy efficient one. As a result, in some cases, the total life cycle cost of a pump installation using a more energy-efficient pump will be greater than a less energy-efficient one.

Similarly, estimating a pump at its cost or market value is impossible because the electricity costs over a cheaper pump's life will often be significantly higher than a more expensive one.

Therefore, an essential factor in improving the existing pumping fleet is the development of clear and visible criteria for assessing the life cycle cost of pumping equipment, which would minimally depend on external factors.

As such criteria, it is proposed to include energy consumption indicators and pumping equipment's material capacity. Moreover, it should be noted that energy efficiency is understood as the ratio of hydraulic energy to the energy consumed by the pump at the so-called best efficiency point (here and next - BEP). As head and flow rate when calculating the hydraulic power, select the appropriate parameters of the pump at the BEP. However, in most cases, the pump parameters do not exactly match the parameters required for the pipeline network. In this case, part of the energy considered useful should be attributed to additional hydraulic losses.

Considering those mentioned above, the work proposes the introduction of a new parameter - the pump energy consumption indicator, which can be calculated according to the dependence:

$$\varepsilon_{eff} = \frac{N_c}{Q_n H_n}, \tag{2}$$

where Q_n is the flow rate of the pipeline network, m^3/s; N_n is the required head in the network, m of the water column; N_c – the pump consumed power at this mode.

The physical content of this indicator consists in determining the amount of power (W) needed to transport a unit of liquid volume (m^3/h) when creating a unit of head (m of water column).

In contrast to the usual efficiency indicator, this indicator takes into account not only hydraulic losses in the pump, but also the correct selection of the pump and the presence of the necessary pump in the existing parametric series.

In turn, the existing pumping equipment design methods do not consider material capacity indicators. It is customary to design pumps to achieve the highest efficiency at the BEP, while the material capacity is usually not evaluated at the designing stage.

To ensure the assessment of the material capacity of the pumping equipment, it is proposed to introduce a new criterion – the indicator of the material capacity of the pumping equipment, which is calculated according to the dependence:

$$\varepsilon_{mat} = \frac{M_p}{Q_n H_n}, \tag{3}$$

where M_p is the mass of the pump in kg.

The physical content of this indicator consists in determining the value of pump mass (kg), which is required to move a unit of liquid volume (m^3/h) when creating a unit of head (m of water column).

Ensuring the lowest possible values of the given energy consumption and material consumption indicators leads to minimizing resource costs for electricity and manufacturing the pump, respectively, which are key factors in reducing the total life cycle cost of the pump installation.

4 Results and Discussion

4.1 Improvement of the Parametric Series of Torque-Flow Pumps

A parametric series of TFP torque-flow pumps for pumping liquids with inclusions has been developed and is functioning (Fig. 2). The parameters and technical characteristics of these pumps are shown in Table 1.

Table 1. A parametric series of TFP torque-flow pumps for transporting liquids with inclusions: parameters and technical characteristics.

№	Pump size	Flow rate, m^3/h	Head, m	Consumption Power, W	Rotational frequency, rpm	Specific speed, n_s	Efficiency, %
1	TFP 25-20	25	20	3 586	1500	48	38
2	TFP 25-32	25	32	7 786	1500	34	28
3	TFP 40-40	40	40	10 900	3000	73	40
4	TFP 50-20	50	20	5 924	1500	68	46
5	TFP 50-32	50	32	11 474	1500	48	38
6	TFP 80-20	80	20	10 140	1500	86	43
7	TFP 80-32	80	32	18 358	1500	60	38
8	TFP 100-50	100	50	29 620	3000	97	46
9	TFP 125-32	125	32	30 278	1500	76	36
10	TFP 125-50	125	50	40 551	1500	54	42
11	TFP 160-40	160	40	40 558	3000	145	43
12	TFP 200-32	200	32	40 558	1500	96	43
13	TFP 200-50	200	50	64 881	1500	69	42

In Table 1, pumps for which the specific speed ns lies in the range from 90 to 150 are highlighted in green, for which, according to [13, 16], theoretically achievable maximum efficiency is 0.5–0.53; yellow – n_s in the range from 60 to 170 with the theoretically achievable maximum efficiency in the range 0.45–0.5, red – n_s in the range of ≤ 60 and ≥ 170, for which the theoretically achievable maximum efficiency does not exceed 0.45.

Indicators of energy consumption and material capacity calculated according to dependencies (2) and (3) are given in Table 2. For the convenience of estimating the pump life cycle cost, the average indicators of energy consumption and material capacity are calculated. The values of the pump parameters at the BEP are chosen as the necessary head and flow rate. In the future, while selecting a pump, the necessary values of the pipeline network parameters will be used. Values above the average are highlighted in red, close to the average – in yellow, and below the average – in green.

Table 2. Indicators of energy consumption and material capacity of torque-flow pumps of the existing parametric series.

№	Pump size	Flow rate, m^3/h	Head, m	Consumption Power, W	Mass, kg	Energy consumption indicator, ε_{eff}, W / (m $\cdot m^3/h$)	material capacity indicator, ε_{mat}, kg / (m $\cdot m^3/h$)
1	TFP 25-20	25	20	3 586	106	7,171	0,212
2	TFP 25-32	25	32	7 786	160	9,732	0,200
3	TFP 40-40	40	40	10 900	160	6,813	0,100
4	TFP 50-20	50	20	5 924	110	5,924	0,110
5	TFP 50-32	50	32	11 474	170	7,171	0,106
6	TFP 80-20	80	20	10 140	110	6,337	0,069
7	TFP 80-32	80	32	18 358	170	7,171	0,066
8	TFP 100-50	100	50	29 620	110	5,924	0,022
9	TFP 125-32	125	32	30 278	185	7,569	0,046
10	TFP 125-50	125	50	40 551	260	6,488	0,042
11	TFP 160-40	160	40	40 558	125	6,337	0,020
12	TFP 200-32	200	32	40 558	225	6,337	0,035
13	TFP 200-50	200	50	64 881	290	6,488	0,029
Average value:						6,882	0,081

Table 2 shows that the pumps TFP 25-20 and TFP 25-32 have indicators of energy consumption ε_{eff} and material capacity ε_{mat} much higher than the average value. In addition, the TFP 50-32 and TFP 80-32 pumps have energy consumption indicators ε_{eff} above average and material capacity ε_{mat} indicators close to average. Thus, we can conclude about the primary need for improving or developing updated pumps with these operating parameters.

With this in mind, it is worth noting that the high material capacity indicator of the TFP 25-20 and TFP 25-32 pumps indicate an urgent need to reduce the weight and size indicators, which can be most effectively achieved by developing pumps for these parameters with increased rotation frequency.

Torque-flow pumps use asynchronous electric motors with a short-circuited rotor as a drive. The rotation frequency of the magnetic field of the stator of such motors is equal to 750, 1000, 1500, and 3000 rpm. Since the existing TFP 25-20 and TFP 25-32 pumps use asynchronous electric motors with a short-circuited rotor with a stator magnetic field rotation frequency of 1500 rpm, it is planned to use electric motors with a stator magnetic field rotation frequency of 3000 rpm in improved pumps of this type.

A characteristic feature of the available parametric range of torque-flow pumps is that both pumps with a delivery of 25 m^3/h have higher than average energy consumption and material capacity indicators. Energy consumption will be even higher when using these pumps to transport liquids with the required pressure in the pipeline network N_n range of 23–30 m of the water column. That is why, while using a torque-flow pump TFP 25-32 to create a pressure of 28 m of water column, the energy consumption indicator will be $\varepsilon_{eff} = 11.097$ W/(m·m^3/h), and the material capacity will be $\varepsilon_{mat} = 0.229$ kg/(m·m^3/h).

Because of this, the primary design of an intermediate variant of a torque-flow pump with parameters: flow rate Q = 25 m^3/h, head H = 28 m of water column, and the rotation frequency of the motor stator magnetic field n = 3000 rpm looks promising.

The standard DIN EN ISO 2858:2010 defines dynamic pumps' main overall and connecting dimensions when they work with different operating parameters.

It should be noted that the main size that determines the overall dimensions of the pump is the impeller diameter. According to DIN EN ISO 2858:2010, increasing the engine frequency from n = 1450 rpm to n = 2900 rpm leads to a decrease of the impeller diameter by 1.97 times (from 315 mm to 160 mm). Thus, the weight and dimensions of the promising TFP 25-28 pump with the rotation frequency of the motor stator magnetic field n = 3000 rpm can be roughly considered to be approximately 3.95 times smaller than that of the existing TFP 25-32 pump. That is, the mass of such a pump will be approximately 40.6 kg.

The specific speed of the promising TFP 25-28 pump is $n_s = 75$. According to [13] maximally achievable efficiency of the torque-flow pump with $n_s = 75$ is 0.48. The power consumption of such a pump will be about 3973 W. The energy consumption indicator of such a pump will be equal to $\varepsilon_{eff} = 5.677$ W/(m·m^3/h), and the material capacity indicator will be equal to $\varepsilon_{mat} = 0.058$ kg/(m·m^3/h). Both indicators are significantly lower than the average of the corresponding indicators of the existing parametric series of torque-flow pumps.

4.2 Development of a Promising Torque-Flow Pump TFP 25-32-2900 for Transporting Liquids with Inclusions

The developed promising torque-flow pump (Fig. 3) is marked as follows – TFP 25-28-2900, where TFP is an abbreviation that means torque-flow pump, 25 – flow rate in m^3/h, 28 – head in m of water column, 2900 – rotation frequency of the electric motor shaft in rpm.

According to the numerical research results, the integral characteristics of the developed promising pump TFP 25-32-2900 were constructed (Table 3).

According to numerical research results, the efficiency of the developed pump at the BEP is 0.4578, which is 4.6% less than the maximum achievable. The pump head at the BEP is 27.48 m of water column, which is 1.85% less than the estimate. The obtained indicators are within the limits of permissible errors for engineering calculations, indicating the calculations' adequacy.

The values of energy consumption ε_{eff} and material capacity ε_{mat} of the promising torque flow pump TFP 25-28-2900 and their comparison with the corresponding indicators of existing torque-flow pumps are shown in Table 4. For a more precise comparison,

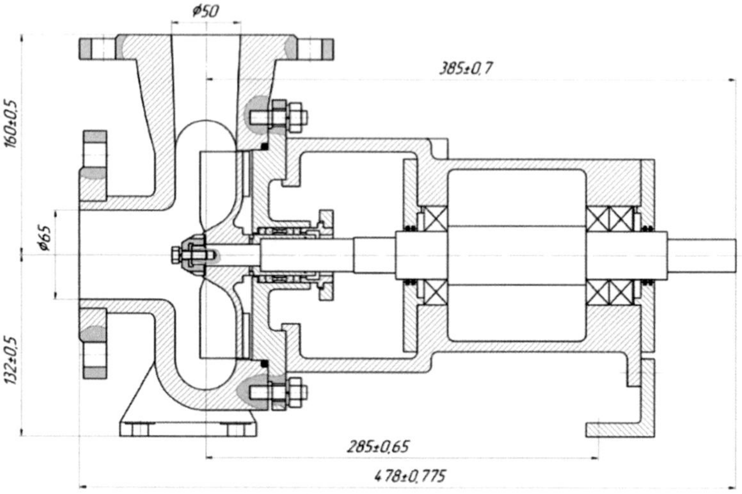

Fig. 3. Promising free-vortex pump TFP 25-32-2900.

Table 3. Integral characteristics of the promising TFP 25-32-2900 pump.

Duty point	Q	Mass flow	Head	Power	Efficiency	Power
	m^3/h	kg/s	m	W	%	kW
0.6	15	4.1542	29.05	3225	0.3682	3.2
0.8	20	5.5389	28.79	3593	0.4365	3.6
1	25	6.9236	27.48	4085	0.4578	4.1
1.2	30	8.3083	26.12	4611	0.4628	4.6
1.4	35	9.6931	24.76	5094	0.4635	5.1

row 2 of the table shows the indicators of energy consumption ε_{eff} and material capacity ε_{mat} when using the existing TFP 25-32 pump, whose shaft rotation frequency is 1450 rpm, within a flow rate of 25 m^3/h and required pressure in the pumpline network of 28 m of the water column.

Thus, the design of the promising torque-flow pump TFP 25-28 pump made it possible to reduce the energy consumption indicator from 11.097 W/(m·m^3/h) to 5.677 W/(m·m^3/h), i.e., 51.2%, and the material capacity index from 0.229 kg/(m·m^3/h) to 0.058 kg/(m·m^3/h), i.e., by 74.6%, when using torque-flow pumps of the existing parametric series to create a flow rate in the pipeline network $Q_n = 25$ m^3/h and a head $H_n =$ 28 m of water column. In general, it can be concluded that the design of the promising torque-flow pump TFP 25-28-2900 installation pump made it possible to reduce the total cost of the pump life cycle by 50–70%, depending on the conditions of the pump usage.

Table 4. Indicators of energy consumption εef and material capacity εmat of the promising torque flow pump TFP 25-28-2900 and their comparison with the corresponding indicators of existing torque-flow pumps.

Pump size	Flow rate, m^3/h	Head, m	Consumption Power, W	Mass, kg	Energy consumption indicator, ε_{eff}, W / (m $\cdot m^3/h$)	material capacity indicator, ε_{mat}, kg / (m $\cdot m^3/h$)
TFP 25-20	25	20	3 586	106	7,171	0,212
TFP 25-32	25	28	7 786	160	11,097	0,229
TFP 25-28-2900	25	28	4 167	40,6	5,677	0,058
Average value:					6,882	0,081

5 Conclusions

The main goal of the research was to reduce the overall cost of the life cycle of torque-flow pumps and pump units based on them. For this purpose, a number of tasks were set, which were successfully resolved in the course of the research.

According to the results of the research, the criteria for a comprehensive assessment of the pump installation's lifecycle cost were developed, allowing to practically and visually assess the value of energy efficiency (energy consumption indicator ε_{eef}) and mass-dimensional quality (material capacity indicator ε_{mat}) of dynamic pumps.

The practical application of the developed indicators made it possible to assess the value of energy efficiency and mass-dimensional qualities of the existing parametric series of torque-flow pumps. It was determined that the number of pumps significantly exceeds the indicators of energy consumption ε_{eff} and material capacity ε_{mat}. In particular, the existing TFP 25-32 pump, when operating at the BEP at the required head in the pipeline network, exceeds the average energy consumption ε_{eff} of the existing parametric series of torque-flow pumps by 1.41 times and the material capacity index ε_{mat} – by 2.47 times.

It was determined that due to the design of a torque-flow pump with an increased rotation frequency, it is possible to achieve an increase in the specific speed n_s of a torque-flow pump with a flow rate of $Q = 25\ m^3/h$ and a head of $H = 28$ m of water column from the value of $n_s = 34$ to $n_s = 75$. In turn, this made it possible to increase the maximum achievable efficiency of the pump from 0.35 to 0.48.

Based on the research results, an updated promising torque-flow pump TFP 25-28-2900 was developed. Its efficiency at the BEP is 0.4578, which is 4.6% less than the maximum achievable. The pump head at the BEP is 27.48 m of water column. The design of the promising TFP 25-28-2900 pump made it possible to reduce the energy consumption indicator comparable to torque-flow pumps using the existing parametric series up to 51.2% and the material capacity index up to 74.6%.

In the course of the research, it was established the need to improve the existing torque-flow pumps TFP 25-20 and TFP 25-32.

Acknowledgment. The authors of this paper acknowledge the National Research Foundation of Ukraine for the possibility of realizing the project "Development of design solutions and layout schemes of a parametric series of high-speed energy-efficient well pumps for the needs of enterprises in the field of critical infrastructure." (No. 2022.01/0096). The authors also appreciate the Public Union "Sumy Machine-Building Cluster of Energy Efficiency" (SMBCEEQ) and the International Association for Technological Development and Innovations (IATDI) for all the valuable support while conducting the research.

References

1. United Nations. General Assembly Resolution A/RES/70/1. Transforming Our World, the 2030 Agenda for Sustainable Development (2015). http://www.un.org/ga/search/view_doc.asp?symbol=A/RES/70/1&Lang=E. Accessed 11 May 2023
2. Luscinski, S., Ivanov, V.: A simulation study of Industry 4.0 factories based on the ontology on flexibility with using Flexsim® software. Manag. Prod. Eng. Rev. **11**(3), 74–83 (2020). https://doi.org/10.24425/mper.2020.134934
3. Zaleta, O.M., Povstyanoy, O.Y., Ribeiro, L.F., Redko, R.G., Bozhko, T.Y., Chetverzhuk, T.I.: Automation of optimization synthesis for modular technological equipment. J. Eng. Sci. (Ukraine) **10**(1), A6–A14 (2023). https://doi.org/10.21272/jes.2023.10(1).a2
4. Tverdokhleb, I., Semenov, A., Ivanyushin, A., Niemtsev, O., Rudenko, A.: Creating a standard size range as one of the factors reducing production time of modern pumping equipment. Appl. Mech. Mater. **630**, 137–142 (2014). https://doi.org/10.4028/www.scientific.net/AMM.630.137
5. Andrenko, P., Rogovyi, A., Grechka, I., Khovanskyy, S., Svynarenko, M.: Characteristics improvement of labyrinth screw pump using design modification in screw. In: Journal of Physics: Conference Series, vol. 1741, p. 012024 (2021). https://doi.org/10.1088/1742-6596/1741/1/012024
6. Syomin, D., Rogovyi, A.: Features of a working process and characteristics of irrotational centrifugal pumps. Procedia Eng. **39**, 231–237 (2020). https://doi.org/10.1016/j.proeng.2012.07.029
7. Antonenko, S., Sapozhnikov, S., Kondus, V., Chernobrova, A., Mandryka, A.: Creation a universal technique of predicting performance curves for small-sized centrifugal stages of well oil pump units. In: Journal of Physics: Conference Series, vol. 1741, p. 012011 (2021). https://doi.org/10.1088/1742-6596/1741/1/012011
8. Pavlenko, I., Ivanov, V., Gusak, O., Liaposhchenko, O., Sklabinskyi, V.: Parameter identification of technological equipment for ensuring the reliability of the vibration separation process. In: Knapcikova, L., Balog, M., Perakovic, D., Perisa, M. (eds.) MMS 2019. EICC, pp. 261–272. Springer, Cham (2020). https://doi.org/10.1007/978-3-030-34272-2_24
9. Dindorf, R., Wos, P.A.: Numerical solution of temperature distribution in the clearance and the sealing rings of the non-contact face seal. In: Journal of Physics: Conference Series, vol. 1741, p. 012032 (2021). https://doi.org/10.1088/1742-6596/1741/1/012032
10. Peczkis, G., Synowiec, T., Zahorulko, A.: New flowing system in cooling system axial-flow pump in a hard coal electric power plant. In: Journal of Physics: Conference Series, vol. 1741, p. 012006 (2021). https://doi.org/10.1088/1742-6596/1741/1/012006
11. Pavlenko, I.: Static and dynamic analysis of the closing rotor balancing device of the multistage centrifugal pump. Appl. Mech. Mater. **630**, 248–254 (2014). https://doi.org/10.4028/www.scientific.net/AMM.630.248

12. Andrusiak, V.O., Lugova, S.O., Medvid, S.A., Tkach, P.Yu., Rudenko, A.A.: Effect of front impeller seal leakages on centrifugal stage characteristics. In: Journal of Physics: Conference Series, vol. 1741, p. 012036 (2021). https://doi.org/10.1088/1742-6596/1741/1/012036
13. Kondus, V., Puzik, R., German, V., Panchenko, V., Yakhnenko, S.: Improving the efficiency of the operating process of high specific speed torque-flow pumps by upgrading the flowing part design. In: Journal of Physics: Conference Series, vol. 1741, p. 012023 (2021). https://doi.org/10.1088/1742-6596/1741/1/012023
14. Tkachuk, M.M., Grabovskiy, A., Tkachuk, M.A., Saverska, M., Hrechka, I.: A semi-analytical method for analysis of contact interaction between structural elements along aligned surfaces. Eastern-Eur. J. Enterp. Technol. **1**(7), 103 (2020). https://doi.org/10.15587/1729-4061.2020.193985
15. Rogovyi, A., Khovanskyy, S., Grechka, I., Pitel, J.: The wall erosion in a vortex chamber supercharger due to pumping abrasive mediums. In: Ivanov, V., et al. (eds.) DSMIE 2019. LNME, pp. 682–691. Springer, Cham (2020). https://doi.org/10.1007/978-3-030-22365-6_68
16. Gusak, O., Krishtop, I., German, V., Baga, V.: Increase of economy of torque flow pump with high specific speed. In: IOP Conference Series: Materials Science and Engineering, vol. 233, p. 012004 (2017). https://doi.org/10.1088/1757-899X/233/1/012004
17. Krishtop, I.: Creating the flowing part of the high energy-efficiency torque flow pump. Eastern-Eur. J. Enterp. Technol. **2**(74), 31–37 (2015). https://doi.org/10.15587/1729-4061.2015.39934
18. Kotenko, A., Herman, V., Kotenko, A.: Rationalisation of Ukrainian industrial enterprises in a context of using torque flow pumps on the basis of valuation of the life cycle of pumping equipment. Nauka i Studia **16**(126), 83–91 (2014)
19. Kiyko, S.G., Druzhinin, E.A., Koba, S.A., Haidabrus, B.V.: A mathematical background for information technology of project's processes integration taking into account risk factors. Metall. Min. Ind. **6**(3), 66–70 (2014)
20. Puzik, R.V., Kovaliov, I.O., Ratushnyi, O.V., Dzafarov, T.V., Petrenko, S.S.: The ways to increase the efficiency of the stage of low specific speed. In: Journal of Physics: Conference Series, vol. 1741, p. 012013 (2021). https://doi.org/10.1088/1742-6596/1741/1/012013
21. Prokopovych, I.V., Kokhanov, A.B., Khamitov, V.M., Tikhenko, V.M., Dašić, P.: Standardizing life cycle organization: a synergetic quality management approach. J. Eng. Sci. (Ukraine) **10**(1), B1–B7 (2023). https://doi.org/10.21272/jes.2023.10(1).b1
22. Plotkin, J., Levchenko, N., Shyshkanova, G., Levchenko, S.: Development of energy enterprises in the context of green transformation. J. Eng. Sci. (Ukraine) **10**(1), G22–G33 (2023). https://doi.org/10.21272/jes.2023.10(1).g3
23. Kotliar, A., et al.: Ensuring the economic efficiency of enterprises by multi-criteria selection of the optimal manufacturing process. Manag. Prod. Eng. Rev. **11**(1), 52–61 (2020). https://doi.org/10.24425/mper.2020.132943
24. Panchenko, V., Ivchenko, A., Dynnyk, O., Drach, O.: Increasing the technical level of a torque flow pump by changing the geometry of a flowing part. Technol. Audit Prod. Reserves **3/1**(41), 10–21 (2018)
25. Machalski, A., Skrypacz, J., Szulc, P., Blonski, D.: Experimental and numerical research on influence of winglets arrangement on vortex pump performance. In: Journal of Physics: Conference Series, vol. 1741, p. 12019 (2021). https://doi.org/10.1088/1742-6596/1741/1/012019
26. Quan, H., Fu, B., Li, R.: Research stage and development tendency of vortex pump. Fluid Mach. **9**(44), 36–40 (2016)
27. Kostornoi, A.S., Tkach, P.Yu., Bondariev, O.O., Podopryhora, N.N.: Design of impeller blades in the intermediate stage of centrifugal pump to a preset shape of meridional flow pattern. In: Journal of Physics: Conference Series, vol. 1741, p. 012005 (2021). https://doi.org/10.1088/1742-6596/1741/1/012005

28. Pavlenko, I., Kulikov, O., Ratushnyi, O., Ivanov, V., Pitel̆, J., Kondus, V.: Effect of impeller trimming on the energy efficiency of the counter-rotating pumping stage. Appl. Sci. **13**(2), 761 (2023). https://doi.org/10.3390/app13020761
29. Kulikov, A., Ratushnyi, O., Kovaliov, I., Mandryka, A., Ignatiev, O.: Numerical study of the centrifugal contra rotating blade system. In: Journal of Physics: Conference Series, vol. 1741, p. 012008 (2021). https://doi.org/10.1088/1742-6596/1741/1/012008
30. Rogovyi, A., Korohodsky, V., Khovanskyy, S., Hrechka, I., Medvediev, Y.: Optimal design of vortex chamber. In: Journal of Physics: Conference Series, vol. 1741, p. 012018 (2021). https://doi.org/10.1088/1742-6596/1741/1/012018
31. Gerlach, A., Thamsen, P., Wulff, S., Jacobsen, C.: Design parameters of vortex pumps: a meta-analysis of experimental studies. Energies **10**(1), 58 (2017). https://doi.org/10.3390/en10010058
32. Voropay, A.V., Menshykov, O.V., Povaliaiev, S.I., Sharapata, A.S., Yehorov, P.A.: Modeling a viscoelastic support considering its mass-inertial characteristics during non-stationary vibrations of the beam. J. Eng. Sci. (Ukraine) **10**(1), D8–D14 (2023). https://doi.org/10.21272/jes.2023.10(1).d2
33. Rogovyi, A., Korohodsky, V., Medviedev, Y.: Influence of Bingham fluid viscosity on energy performances of a vortex chamber pump. Energy **218**, 119432 (2021). https://doi.org/10.1016/j.energy.2020.119432
34. Krishtop, I., German, V., Gusak, A., Lugova, S., Kochevsky, A.: Numerical approach for simulation of fluid flow in torque flow pumps. Appl. Mech. Mater. **630**, 43–51 (2014). https://doi.org/10.4028/www.scientific.net/AMM.630.43

Method of Assessing the Optimality of the Mechanical Characteristics of Foams

Olena Mikulich(✉)

Lutsk National Technical University, 75, Lvivska St., Lutsk 43018, Ukraine
shyprao@gmail.com

Abstract. The article is devoted to developing a method for assessing the optimality of the mechanical characteristics of foam materials based on using the Cobb-Douglas model within the framework of Cosserat mechanics. The results of experimental studies for foamed polymer materials with closed cells were used for modeling. The Cobb-Douglas model used for modeling allows accounting for the nonlinear dependence of the Shear modulus on the density and cell sizes of the material. The developed approach allows the evaluation of the influence of material density and pore sizes on changes in shear-rotational waves in the medium. Based on the approaches proposed in the work, the analytical-numerical modeling method is convenient and practical for foam, porous, and other structurally heterogeneous materials. The advantage of this approach is the ability to evaluate the mechanical characteristics of foam-structure materials widely used in production without special laboratory tests. This approach significantly expands the scope of the application of foamed polymers.

Keywords: Polyurethane Foam · Cosserat Elasticity · Shear-Rotation Wave · Cobb–Douglas Model · R&D Investment

1 Introduction

Currently, there is a significant increase in the use of foam, porous, and other structurally heterogeneous materials in various spheres of life, production, and medicine. Such materials' low density and high vibration-absorbing and heat-retaining characteristics explain this.

The results presented by the scientific researchers, most of which are experimental, are related to the evaluation of mechanical behavior or the determination of mechanical or strength characteristics of specific foam materials. This does not allow generalizing the obtained results to other types of foams.

However, the possibility of selecting optimal characteristics of structurally heterogeneous materials to optimize vibration-absorbing, heat-insulating, and other characteristics of foamed polymers is also of scientific interest.

For developing a methodology for assessing the optimality of the microstructure of foam materials, it is necessary to use analytical approaches within the framework of refined models of continuous mechanics based on applying the Cosserat continuum

V. Tonkonogyi et al. (Eds.): InterPartner 2023, LNME, pp. 477–484, 2024.
https://doi.org/10.1007/978-3-031-42778-7_44

model. In addition, developing such methods requires constructing models that would account for the nonlinear dependence of input parameters.

Therefore, the paper will propose a method for evaluating the optimality of the mechanical characteristics of foam materials based on the use of the Cobb-Douglas model and Cosserat mechanics apparatus.

2 Literature Review

Using foam materials with different structures, densities, and porosity types has several advantages compared to similar non-foamed plastic parts. Among the advantages, the reduction of material consumption and better mechanical, thermal and physical properties should be noted.

Therefore, many scientific studies are devoted to experimental studies of changes in the structure of foam materials in order to optimize them. For example, the results of a study of the effect of adding an extruder to microporous foam on the change in the material's mechanical properties are presented [1]. In [2] it is proposed a new additive method for producing polymer foams based on direct bubble recording. This approach allows obtaining polymer foams with the same cell sizes, volume fraction, and cohesion. The study [3] deals with optimizing a polypropylene-wollastonite composite's thermal and mechanical properties based on experiments selected using the Box-Behnken approach. The advantages of using isocyanate, which has unsaturated bonds and high activity, in polyurethane foam production are investigated in [4].

In the works of R.S. Lakes, the mechanical characteristics of foam materials within the Cosserat elasticity are determined. The effect of rotational-shear deformations of microparticles of the medium was accounted for in determining the scale effects of polyurethane foams. In [5], the microstructure elastic constants of the material were determined based on experimental studies of foams within the framework of Cosserat elasticity under wave loads. In static experiments, the effect of size during compression of the studied materials is not observed. Using the Cosserat elasticity allows us to establish experimentally that the deformation of the foam is significantly reduced compared to the predictions of classical elasticity both in torsion [6] and in bending [7].

In [8] it is developed a technique for analytically determining the speed of propagation of shear-rotation waves within the Cosserat elasticity, which allows an account of the influence of the microstructure of the material. Therefore, we will use the approach proposed in [8] to develop methods for optimizing the effective characteristics of foam materials.

3 Research Methodology

In [8], dependencies for determining the propagation speeds of shear-rotation waves in a microporous medium were obtained. Numerical calculations of the change in the propagation speed of rotation waves depending on the frequency of the applied load were obtained for polyurethane foams WF300, WF110, and WF51.

However, the development of the method for optimizing the effective characteristics of structurally heterogeneous materials, particularly foams, which could be investigated on the basis of analytical dependencies, is also of great interest. This is also justified by the fact that the developed technologies for obtaining polyurethane foams will allow them to significantly expand and optimize their application with the possibility of modeling changes in the main mechanical characteristics when changing the density and porosity of the material.

In the work [8], it is shown that with low-frequency loads, the main contribution to wave propagation is made by a rotational-shear wave with a speed V_I:

$$V_I^2 = \frac{c_2^2 + c_3^2}{\frac{c_2^2}{2\ell_C^2\omega^2} - \aleph - 1 + \sqrt{\left(\frac{c_2^2}{2\ell_C^2\omega^2} - \aleph - 1\right)^2 + \left(\frac{1}{\omega^2} - \frac{1}{\omega_*^2}\right)\frac{\aleph}{\ell_C^2}}}. \quad (1)$$

Since

$$c_2^2 = \frac{G}{\rho};\ c_3^2 = \frac{\kappa}{2\rho};\ \ell_C^2 = \frac{\gamma}{2\kappa};\ \aleph = \frac{c_2^2 + c_3^2}{c_5^2};\ \omega_*^2 = \frac{2\kappa}{J}. \quad (2)$$

where c_2 is the speed of shear (transverse) wave in the framework of classical elasticity [8], c_3 is the speed of rotation wave in the framework of Cosserat elasticity [9], ℓ_C is the scale factor in the framework of Cosserat elasticity [9], $\aleph$ is couple constant [8] determined the relationship between shear and rotation speeds in the Cosserat elasticity, ω_* is the characteristic frequency, G is shear modulus, ρ is the material density, γ, κ are elastic constants, which described an isotropic constrained Cosserat elastic solid, J is unit volume rotational inertia [9].

Analysis of the results of experimental studies conducted by R.S. Lakes and his colleagues [10] shows that using Cosserat elasticity allows accounting for the influence of the microstructure and heterogeneity of the material within the framework of analytical studies. Using an analytical approach is very important for this type of problem, as it allows not only to avoid the accumulation of errors when performing numerical calculations but also to assess the accuracy of calculations at each stage.

Since the use of the Cosserat continuum model allows us to account for the influence of rotational-shear deformations of microparticles of the environment [8], we will use this approach to model the dependence of the mechanical characteristics of the foam material by the density and size of the foam cells. Such an approach in the selection of dependent parameters is justified by the fact that it allows an account of the influent of both the approaches of the classical theory of elasticity (material density ρ) and Cosserat elasticity (cell size h). This approach is also characterized by universality for various types of two-component polyurethane foams formed by mixing polyurethane and foaming agent, which are widely used in production.

Let us investigate the change in shear modulus G in polyurethane foams WF51, WF110, and WF300 using the results of experimental studies [11, 12]. Here, the values of such elastic characteristics within the Cosserat continuum as dimensional characteristics in bending ℓ_b and coupling number N were determined. The corresponding values of elastic characteristics are given in Table 1.

The nonlinearity of the changing of shear modulus G for different types of foams is shown in Fig. 1.

Table 1. Values of polyurethane foam characteristics [10].

	WF300	WF110	WF51
Shear modulus G *(MPa)*	285	104	65
Characteristic length, bending ℓ_b *(mm)*	0.77	0.33	0.55
Coupling number N^2	0.04	0.04	0.01
Cell size *(mm)*	0.65	0.5	0.4
Density ρ *(kg/m^3)*	380	340	60

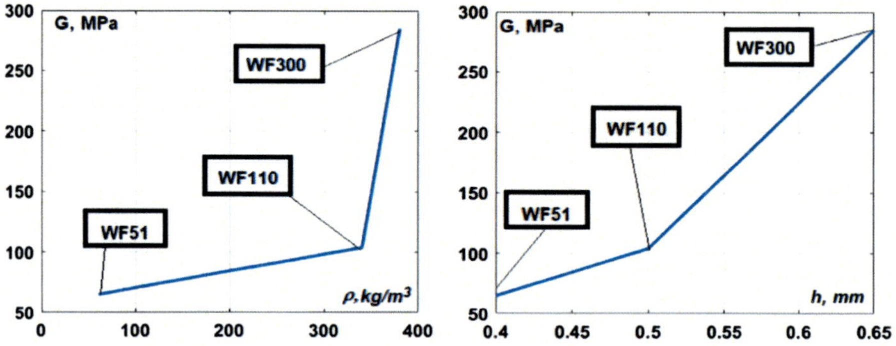

Fig. 1. Dependence of shear modulus on density (left) and cell size (right).

Figure 1 shows that the dependence of the shear modulus G on the density ρ and cell size h is nonlinear. Therefore, we use the Cobb-Douglas [12] model to build a model of the dependences of the shear modulus of the foam material by the density and cell sizes of the material. According to the Cobb-Douglas model, we present the dependence of the change in the shear modulus in the form:

$$G = a_G \cdot h^{b_G} \cdot \rho^{c_G}. \tag{3}$$

The values of constants $a_G,\ b_G,\ c_G$ we find by using of regression analysis method by the logarithmic dependence obtained by formula (3):

$$\ln G = \ln a_G + b_G \cdot \ln h + c_G \cdot \ln \rho. \tag{4}$$

Solving the problem comes down to solving the system of equations, which is written in matrix form as

$$C = W^{-1} \cdot Q, \tag{5}$$

where $Q = \begin{pmatrix} \sum lnG_i \\ \sum lnG_i \cdot \ln h_i \\ \sum lnG_i \cdot \ln \rho_i \end{pmatrix}$; $W = \begin{pmatrix} N & \sum \ln h_i & \sum \ln \rho_i \\ \sum \ln h_i & \sum \ln^2 h_i & \sum \ln \rho_i \ln h_i \\ \sum \ln \rho_i & \sum \ln \rho_i \ln h_i & \sum \ln^2 \rho_i \end{pmatrix}$; $C = \begin{pmatrix} a_G \\ b_G \\ c_G \end{pmatrix}$, W^{-1}.

4 Results and Discussion

The proposed approach allows to obtain values of constants for determining the shear modulus based on the results of experimental studies [11] given in Table 1. Having solved the matrix Eq. (5), we determine the coefficients in formula (3):

$$a_G = 4.259 \cdot 10^{21}, \; b_G = 3.9425, \; c_G = -0.236. \tag{6}$$

The values of the shear modulus are obtained using formula (3) based on the value of the constants (6) practically math with the values given in Table 1:

$$G_{\exp} = 285.358 \; MPa \; (WF300), G_{\exp} = 104.128 \; MPa \; (WF110), \\ G_{\exp} = 65.056 \; MPa \; (WF51)$$

In addition, this approach allows not only determining the value of the shear modulus at different density and cell size values both within and outside the studied ranges. The results of the corresponding calculations are shown in Fig. 2 (left) for the case when the density of the material varies from 50 kg/m^3 to 400 kg/m^3, and the cell sizes vary from 0.3 mm to 1 mm. Also interesting are the results of modeling the influence of the change in the shear modulus due to the material's cell sizes at fixed density values. The corresponding numerical calculations are shown in Fig. 2 (right).

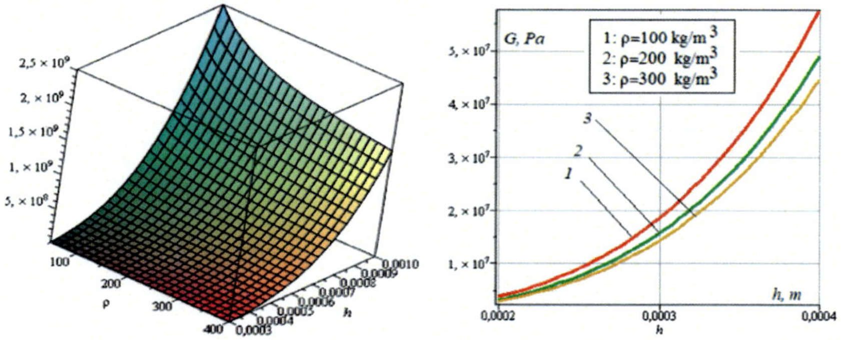

Fig. 2. Modelling of Shear modulus changing.

Analysis of the numerical results shown in Fig. 2 not only confirms the effectiveness of using the Cobb-Douglas function for modeling the change in the Shear modulus of a foam material and is consistent with the data of experimental studies (Fig. 1), but also allows optimization of the effective characteristics of foam materials. Thus, doubling the cell sizes of the material for the case of constant density increases the value of the Shear modulus by 4–5 times.

In addition, the proposed approach based on modeling the mechanical characteristics of the material can be used to optimize the vibration-absorbing characteristics of the foam. For this purpose, we will construct a formula for determining the speed of shear-rotation wave propagation, defined by formula (1), as a function accounted for the influence of the microstructure of the material:

$$V_I = V_I(\rho, \ h, \ \omega). \tag{7}$$

For the case of honey-comb cell models of foam, the unit volume rotational inertia is written based on the fundamental principles of Cosserat elasticity [9]:

$$J = \frac{2}{3}h^2\rho. \tag{8}$$

Characteristics of length in bending ℓ_b and couple number N are evaluated by formulas [11]:

$$\ell_b^2 = \frac{\gamma}{4G}, \ N^2 = \frac{\kappa}{2G + \kappa}. \tag{9}$$

Modeling of characteristics within the Cosserat elasticity was based on the Cobb-Douglas model according to the algorithm described above. These characteristics were chosen as functions of density and cell size. The modeling results in the shear-rotation waves' propagation speed as a function of frequency are shown in Fig. 3.

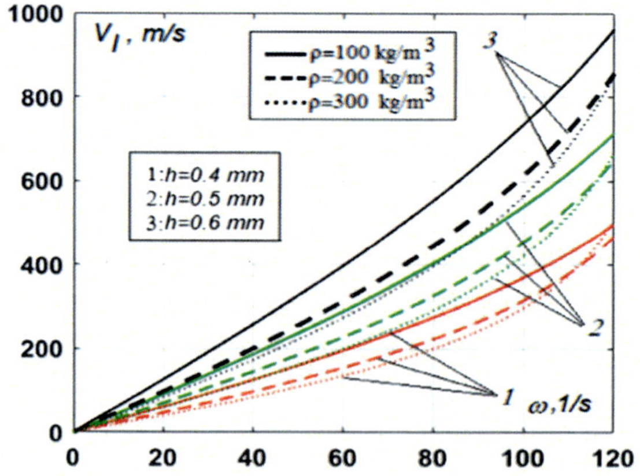

Fig. 3. Modelling of changing shear-rotation speeds on frequency.

Analysis of the numerical results of Fig. 3 shows that an increase in the Shear modulus of the material at the same cell sizes leads to an increase in the speed of propagation of shear-rotation waves in the Cosserat elasticity but an increase in the size of the cells in the case of unchanged material density contributes to a decrease in the speed of propagation of shear-rotation waves.

Such a simulation analysis allows for optimizing both the operational characteristics of the material and determining the optimal values of the density and porosity of the foam to obtain the most desired results.

5 Conclusions

Assessment of the impact of changes in the propagation speed of shear-rotational waves in structurally heterogeneous materials plays an essential role in optimizing the operating parameters of the corresponding elements of constructions made of such materials and in predicting the optimal foam microstructure. Using the apparatus moment theory of elasticity based on Cosserat model continuum allows modelling using analytical approaches and considering the influence of the material's microstructure. The analytical-numerical modeling method based on the approaches proposed in the work is convenient and practical for foam, porous, and other structurally heterogeneous materials. The advantage of this approach is the ability to evaluate the mechanical characteristics of foam materials widely used in production without specific laboratory tests. This approach significantly expands the areas of application of foamed polymers.

Acknowledgment. The work performs within the state grants of applied research № 0122U001064 "Methodology of predicting mechanical behavior and optimizing the effective characteristics of foam and porous materials". The team of authors expresses their sincere gratitude to the Ministry of Education and Science of Ukraine.

References

1. Azimi, H., Jahani, D., Mohebifar, A., Yazdan, M.: Optimization of mechanical properties of PP-polymer foam fabricated via extruder. J. Appl. Res. Chem. Polymer Eng. **5**(1), 105–120 (2021)
2. Visser, C., Amato, D., Mueller, J., Lewis, J.: Architected polymer foams via direct bubble writing. Adv. Mater. **31**(46), 1904668 (2019)
3. Leontiadis, K., et al.: Optimization of thermal and mechanical properties of polypropylene-wollastonite composite drawn fibers based on surface response analysis. Polymers **14**(5), 924 (2022)
4. Chen, S., Lei, S., Zhu, J., Zhang, T.: The influence of microstructure on sound absorption of polyurethane foams through numerical simulation. Macromolecural Theory Simul. **30**(5), 2000075 (2021)
5. Lakes, R.: Experimental evaluation of micromorphic elastic constants in foams and lattices. Z. Angew. Math. Phys. **74**, 31 (2023)
6. Lakes, R.: Softening of cosserat sensitivity in a foam: warp effects. Int. J. Mech. Sci. **192**, 106125 (2021)
7. Lakes, R.: Cosserat shape effects in the bending of foams. Mech. Adv. Mater. Struct. 2086328 (2022). https://doi.org/10.1080/15376494.2022.2086328

8. Mikulich, O.: Wave propagation speed analysis in polyurethane foams. In: Tonkonogyi, V., Ivanov, V., Trojanowska, J., Oborskyi, G., Pavlenko, I. (eds.) Advanced Manufacturing Processes IV, pp. 465–472. Springer, Cham (2023). https://doi.org/10.1007/978-3-031-16651-8_44
9. Erofeev, V.I.: Wave Processes in Solids with Microstructure. World Scientific, Singapore (2003)
10. Rueger, Z., Lakes, R.: Experimental study of elastic constants of a dense foam with weak cosserat coupling. J. Elast. **137**, 101–115 (2019)
11. Lakes, R., Huey, B., Goyal, K.: Extended poisson's ratio range in chiral isotropic elastic materials. Physica Status Solidi (b) **259**(12), 2200336 (2022)
12. Mahaboob, B., Ajmath, K., Venkateswarlu, B., Narayana, C., Praveen J.: On Cobb-Douglas production function model. In: AIP Conference Proceedings, vol. 2177, p. 020040 (2019)

Numerical Simulation of the Natural Frequencies Dependence of Turbine Blade Vibrations on Single-Crystal Anisotropy

Yevhen Nemanezhyn[1,2](✉), Gennadiy Lvov[2], and Yuriy Torba[1]

[1] State Enterprise "Ivchenko-Progress", 2, Ivanova Str., Zaporizhzhia 69068, Ukraine
Yevhen.Nemanezhyn@infiz.khpi.edu.ua

[2] NTU "Kharkiv Polytechnic Institute", 2, Kyrpychova Str., Kharkiv 61002, Ukraine

Abstract. The subject of study of this article is one of the key tasks that appear in the development and operation of advanced aircraft gas turbine engines, namely the problem of ensuring the dynamic strength of their parts. Dynamic strength directly impacts the reliability and service life of the engine since most defects are caused by dynamic stresses from jump-like loads that increase significantly under resonance conditions. The turbine blades are one of the most highly loaded engine parts. At the design stage, it is necessary to evaluate and prevent the possibility of resonant vibrations of these blades throughout the entire range of operation of the aircraft gas turbine engine. To regulate the frequency characteristics of the blades to prevent dangerous resonant mode shapes of vibrations that arise from different harmonics of the exciting force, it is necessary to carry out a complex of various technological or structural changes. One of the most progressive and used methods for manufacturing turbine blades is single-crystal casting, which produces monocrystals with anisotropic properties. In this study, the authors developed a method for determining the elastic characteristics of a single crystal when the crystallographic system of directions is rotated. With the help of finite-element analysis, the dependence of its natural frequencies and mode shapes on elastic constants of a single crystal was investigated on the example of a typical blade model. Calculations were carried out using the ANSYS software package and the Maple computing complex.

Keywords: Elastic Constants · Monocrystalline Alloy · Crystallographic Orientation · Cubic Symmetry · Aircraft Engine · Process Innovation

1 Introduction

One of the biggest problems that arise in developing and operating promising aviation gas turbine engines is ensuring the dynamic strength of its parts. Dynamic strength directly impacts engine life and reliability since most defects are caused by dynamic stresses from jump-like loads, which increase slightly when hitting resonance. Since the turbine blade is one of the most responsible and loaded parts of an engine, at the design stage, it is necessary to assess and prevent the possibility of resonant vibrations in the

V. Tonkonogyi et al. (Eds.): InterPartner 2023, LNME, pp. 485–497, 2024.
https://doi.org/10.1007/978-3-031-42778-7_45

entire range of operation of the aircraft gas turbine engine [1]. Forced vibrations of the turbine blade arise due to time-varying gas-dynamic forces from the gas flow and are periodic in nature, as the rotor rotation frequency determines them. As a spatial elastic system, a turbine blade has an infinite discrete number of natural frequencies and mode shapes. Each form of vibration corresponds to its frequency, and their combination forms a spectrum. As a rule, vibrations with the highest amplitude in a certain frequency range are of interest in practice. As it is known, each mode of vibrations and its associated natural frequency are determined by the displacement scheme with the corresponding position of the nodal lines and depend on the blade's material, shape and size, and the conditions for its fastening. Analysis of natural frequencies and mode shapes at the design stage allows predicting dangerous resonance modes of an aircraft gas turbine engine to take measures in forecasting blade resonances and determine their points of biggest stresses [2].

When the engine is in operating condition, forces acting on the blades of gas turbines periodically change over time. If the excitation force's vibration frequency coincides with the blade's vibration frequency, then a phenomenon called resonance occurs. As a result of the appearance of resonant vibrations in any mode of engine operation, the stresses on the blades increase significantly. Consequently, a significant characteristic of the blades is the spectrum of their vibration frequencies.

Because of the anisotropy of the mechanical properties of the single crystal, the crystallographic orientation significantly affects the spectrum of natural frequencies of the blades. The presence of the engine on the impeller of blades with a different orientation increases the dispersion of natural frequencies [3].

Based on the material mentioned above, we can conclude that an urgent task is to study the effect of the anisotropy of the elastic properties of monocrystalline turbine blades on their natural frequencies and vibration modes.

2 Literature Review

The problem of studying the influence of the anisotropy of monocrystalline elastic characteristics on eigenfrequencies and vibration modes of single-crystal turbine blades is raised in many literature sources. Article [4] provides detailed information on the study of the properties of the foreign Inconel 625 alloy, and in particular, great attention is paid to the consideration of the diffraction and elastic constants of single crystals obtained from the Inconel 625 alloy, which were determined experimentally at room and elevated temperatures using the phenomenon of neutron diffraction. Authors of the publication [5] discovered that the value of the modulus of elasticity of a single crystal tends to decrease, and the anisotropy of the elastic properties of a single crystal increases along with rising temperature, which indicates the importance of taking into account the features of the texture of the material on the distribution of stresses in it at high temperatures. Researchers of the source [6] additionally investigated the influence of the dynamically recrystallized structure of hot-deformed superalloy Inconel 625 on its stress-strain high-temperature properties. They found the appropriate grain boundary nucleation mechanism corresponding to the requirements of optimal strength characteristics.

The article [7] provides meaningful information on the experimental data on the temperature dependence of the components of the elastic yield matrix S_{11}, S_{12}, S_{44} of cubic single-crystals of the foreign heat-resistant alloy on a nickel base MAR-M200. The authors obtained relations and dependencies for single crystals with arbitrary crystallographic orientation and presented the results of calculations of the elasticity characteristics of single crystal samples. In the publication [8], the authors give generalized information about the anisotropy of elastic characteristics of single-crystal alloys. Research [9] emphasizes the importance of determining the elastic parameters of monocrystalline turbine blades to understand better the mechanical behavior of a single-crystal alloy under the action of cyclic fatigue loads. In [10], the authors describe a method for evaluating all elastic constants of cubic crystals considering the temperature dependence of nickel elastic characteristics to check the reliability of the embedded-atom method potential. In [11], researchers study the effects of hydrogen on the elastic parameters of Ni-based cubic crystals. Article [12] provides information on the characteristics of another well-known foreign nickel monocrystalline alloy, CMSX-4, which is actively used to manufacture gas turbine blades. The elastic properties of this alloy were studied using the sound resonance method in the temperature range from normal to 1300 °C. The monograph [13], compared to [8], comprehensively describes ability of monocrystals to withstand loading at an operating temperature close to its melting point and their substantial resistance to mechanical degradation over extended periods of time under vast mechanical loading. In [14], the authors raise the problem of determining the extremum values of Poisson's ratio for monocrystals with cubic symmetry. Researchers use derived analytical expressions to calculate Poisson's ratio for many known modern cubic crystals. Publication [15], in addition to the information of [12], describes the temperature dependence of elastic parameters of isolated γ' and matrix phases of superalloy CMSX-4, which were measured and compared to each other in room temperature conditions.

Open-source literature contains information on the mistuning of mode and frequency vibrations calculations, such as [16]. Those researches, in common, describe dealing with random structural perturbations, including using intentional mistuning patterns to mitigate the harmful effects of random mistuning. Although those publications are pretty relevant, they generally don't suggest how to avoid the resonance problem. In theory, there are several options for avoiding large amplitude resonance-associated vibrations. First, changing an operating speed may contribute to avoiding resonance conditions, but this method tends to be very restrictive. The alternative route is to use a damper that can be attached to the turbine blade to reduce overall stress level and amplitude [17]. Various types of dampers are widely used in industrial manufacturing. However, their efficiency at high frequencies dramatically reduces. Another alternative way of avoiding resonance is redesigning the construction of the blade itself, as suggested in [18]. This method is quite effective and often used in manufacturing. The opposite side of its effectiveness is that the specifics of blade redesign processes are not available in open sources because these technologies are a trade secret of commercial companies. The guided tuning of turbine blades methodology described in [19] uses general results from the perturbation solution to the eigenvalue problem to provide a strategy for a redesign. Although this method is quite comprehensive, its use is limited for compressor blades with isotropic properties, so it cannot be applied to complex high-pressure cooled heat-resistant turbine blades with anisotropic mechanical characteristics.

An analysis of publications has shown a significant gap in the application of modern finite element analysis methods to determine the natural frequencies of turbine blades made of single-crystal alloys. In particular, there are no publications on studies of the influence of the orientation of crystallographic axes on the natural frequency spectrum of cooled blades. Given that the use of finite element software systems is becoming the main trend in the practice of leading manufacturers of gas turbines, the urgent need arises to develop a methodology for researching the dependence of changing elastic properties of single-crystal turbine blades on its eigenfrequencies and vibration modes.

3 Research Methodology

3.1 Elastic Properties of Single-Crystal Alloys

In the macro mechanical analysis of products made of single-crystal alloys, their material is considered a homogeneous anisotropic continuum. The relationship between stresses and strains in an anisotropic body is described by Hooke's law [20] in the tensor form:

$$\varepsilon_{ij} = a_{ijkl} \cdot \sigma_{kl}, \ (i, j, k, l = 1, 2, 3), \tag{1}$$

where ε_{ij}, σ_{kl} are components of strain and stress tensors in the adopted coordinate system, a_{ijkl} is a tensor of the 4th rank, which characterizes the material's elastic properties. For an ideal elastic body, the tensor of elastic constants has the following symmetry conditions:

$$a_{ijkl} = a_{jikl} = a_{klij}. \tag{2}$$

As a consequence of this condition, the number of different elastic constants is equal to 21. The number of elastic constants decreases depending on the type of symmetry of the material. For single crystals with cubic symmetry, three independent elastic constants remain.

The natural frequencies of blades made of single-crystal alloys depend on the crystal axes' orientation relative to the blades' design. To obtain new scientific results on establishing such dependencies, it is necessary to obtain relations between the components of the elasticity tensor of a single crystal in various coordinate systems. The same relationships are also needed to identify the elastic constants of single-crystal alloys based on the results of physical experiments. The necessary formulas for alloys with cubic symmetry are derived below.

The number of nonzero components of the elasticity tensor depends on the orientation of the selected coordinate system in relation to the planes of symmetry of the material. Let's choose the original (crystallographic) coordinate system in such a way that its unit vectors i_1, i_2, i_3, were directed along the crystallographic directions [100], [010], [001] of the elementary cubic lattice.

To identify all components of the elasticity tensor, it is necessary to use the formulas to transform the components of the 4th rank tensor when the coordinate axes are rotated.

We will introduce a new, rotated relative to the crystallographic coordinate system with unit vectors i'_1, i'_2, i'_3. The matrix of direction cosines gives the position of the new coordinate system relative to the original one:

$$\alpha_{ij} = i'_i \cdot i_j. \tag{3}$$

The components of this matrix are the cosines of the angles between the vectors of the new system i'_i, and the old one i_k. Then, the components of the tensor of elastic constants in the new coordinate system are related to the components in the crystallographic system by the following relations [20]:

$$a'_{ijkl} = \alpha_{im} \cdot \alpha_{jn} \cdot \alpha_{ks} \cdot \alpha_{lt} \cdot a_{mnst}. \tag{4}$$

If it is necessary to make a transition from an arbitrary coordinate system to a crystallographic one, the inverse relations are used:

$$a_{mnst} = \alpha_{im} \cdot \alpha_{jn} \cdot \alpha_{ks} \cdot \alpha_{lt} \cdot a'_{ijkl}. \tag{5}$$

A more revealing representation of the structure of physical relationships of anisotropic bodies is provided using the so-called Voigt's notation when the matrix form of symmetric 4th rank tensor is applied. When using the matrix form of Hooke's law, coordinate stresses and strains are written as vector columns, and a symmetric square 6x6 matrix gives the relationship between them.

$$\begin{pmatrix} \varepsilon_x \\ \varepsilon_y \\ \varepsilon_z \\ \gamma_{xy} \\ \gamma_{yz} \\ \gamma_{xz} \end{pmatrix} = \begin{pmatrix} d_{11} & d_{12} & d_{13} & d_{14} & d_{15} & d_{16} \\ & d_{22} & d_{23} & d_{24} & d_{25} & d_{26} \\ & & d_{33} & d_{34} & d_{35} & d_{36} \\ & & & d_{44} & d_{45} & d_{46} \\ & Sym. & & & d_{55} & d_{56} \\ & & & & & d_{66} \end{pmatrix} \times \begin{pmatrix} \sigma_x \\ \sigma_y \\ \sigma_z \\ \tau_{xy} \\ \tau_{yz} \\ \tau_{xz} \end{pmatrix}. \tag{6}$$

The matrix form of elastic constants is often used when presenting the results of physical experiments with samples of different crystallographic orientations. In addition, stiffness or compliance matrices are used in modern software complexes for structural analysis of products from anisotropic materials, where 21 elastic constants need to be specified.

In the technical literature, special notations are often used: modulus of elasticity E, Poisson's ratio ν, and shear modulus G, which are related to the components of the compliance matrix by the following relations:

$$d_{11} = \frac{1}{E_x}; d_{12} = -\frac{\nu_{yx}}{E_y} = -\frac{\nu_{xy}}{E_x}; d_{44} = \frac{1}{G_{xy}}. \tag{7}$$

The methodology for modal analysis of turbine blades includes the determination of the elastic properties of single-crystal alloys and the determination of the components of the tensor of elastic constants for various orientations of the crystallographic axes, the creation of finite element models of cooled blades, and the study of the effect of crystallographic orientation on the natural frequencies of the blades. The flowchart of the research is shown in Fig. 1.

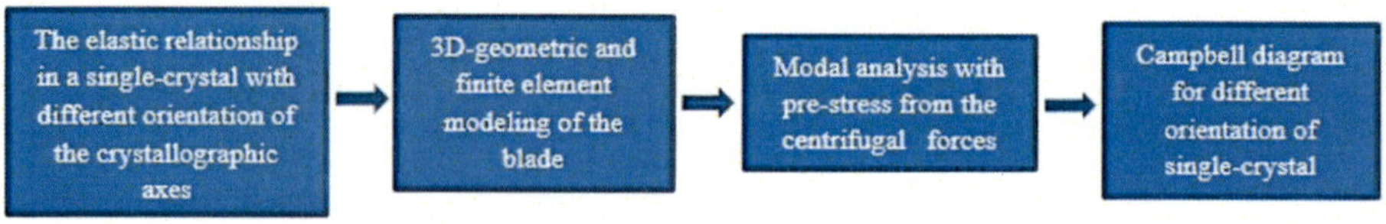

Fig. 1. The flowchart of the research methodology.

3.2 Identification of Elastic Constants of Single Crystals with Cubic Symmetry

Turbine blades from monocrystalline alloys are usually manufactured so that their longitudinal axis coincides with one of the crystallographic directions, for example, the [001] direction. Changing the orientation of the other two directions relative to the blade's design can influence its mechanical properties. In the coordinate system that coincides with the crystallographic directions, the matrix of elastic compliances contains three independent parameters $d_{12} = d_{13} = d_{23}$ and $d_{44} = d_{55} = d_{66}$. Other components of the matrix are, at the same time, zero. In this study, we will consider a partial case of changing the matrix parameters of elastic compliances when the selected system of single crystal directions is rotated by 45°. The components of the matrix that change when the system of directions is rotated by 45° are determined by the following ratios:

$$
\begin{aligned}
d'_{11} = d'_{22} &= \frac{1}{2}d_{11} + \frac{1}{2}d_{12} + \frac{1}{4}d_{44};\\
d'_{12} &= \frac{1}{2}d_{11} + \frac{1}{2}d_{12} - \frac{1}{4}d_{44};\\
d'_{44} &= 2d_{11} - 2d_{12}.
\end{aligned}
\tag{8}
$$

Ratios (8) are used to form the initial data in the finite-element analysis of natural frequencies of the blades from single-crystal alloys with different orientations.

3.3 Elastic Properties of a Typical Heat-Resistant Monocrystalline Alloy

For turbine blades made of monocrystalline heat-resistant nickel alloys, one factor affecting the range of natural vibration frequencies is the temperature dependence of their elastic characteristics. It is known that nickel alloys are designed to work in very high temperatures (1000 °C and more) and their characteristics change with increased temperature in different engine operating modes.

We will give an example of the identification of elastic characteristics based on experimental data for a typical heat-resistant nickel alloy [7]. Using ratios (7) from experimental graphs, elastic moduli, Poisson's ratios, and shear moduli for two temperatures were calculated with the help of Maple Release 2021.0. The calculated elastic characteristics for a typical nickel alloy for different temperatures are shown in Table 1.

Table 1. Elastic properties of a typical heat-resistant alloy.

T, °C	$(E_x = E_y = E_z) \cdot 10^5$, MPa	$\nu_{xy} = \nu_{yz} = \nu_{xz}$	$(G_{xy} = G_{yz} = G_{xz}) \cdot 10^5$, MPa
20	1,388	0,388	1,256
1100	0,892	0,396	0,877

4 Results and Discussion

4.1 Determination of Natural Frequencies of Turbine Blade Vibrations

As an example of forecasting resonance forms and vibration frequencies of turbine blades, we present the Modal analysis of a cooled high-pressure turbine blade model. The solid-state 3D model of the blade is built using the capabilities of the Siemens NX 20.0 graphics complex, as shown in Fig. 2a.

After building the geometric model of the object, its finite-element model was created using ANSYS Workbench 19.2. The type of element used in this case is the volumetric spatial element Tetrahedron (SOLID 187). The finite-element mesh was constructed with the Patch Conforming method. The size of the element is 0,5 mm. Generated finite-element mesh on a blade model is shown in Fig. 2b.

In this calculation, the applied properties of the material were previously calculated based on the available experimental data, as mentioned in Table 1. The calculated density was taken as 8.75 kg/m^3. The constructed finite-element mesh consists of 217555 elements and 333831 nodes.

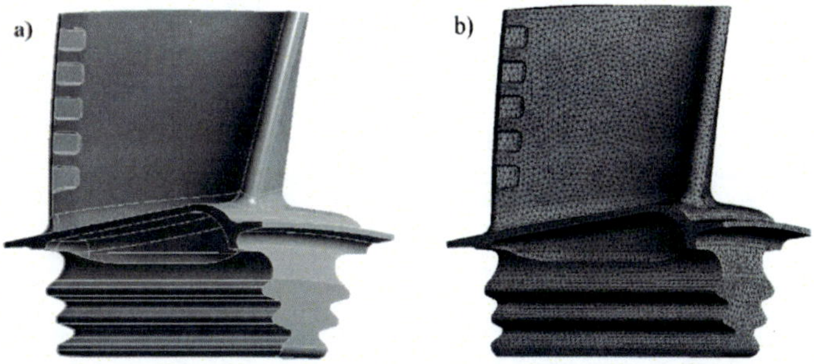

Fig. 2. 3D models of a cooled high-pressure turbine blade: a – solid-state; b – finite-element.

Stresses caused by centrifugal forces affect the natural frequencies and mode shapes of the blades. To consider this effect, the ANSYS software package uses a special procedure for correcting the structural stiffness matrix.

The centrifugal force is accepted as a pre-stress effect for modal analysis, and the absence of displacements on the planes in contact with the turbine disc is used. Figure 3 shows the calculation results for a temperature of 20 °C of the first four natural frequencies and mode shapes of the blade in the form of total displacements.

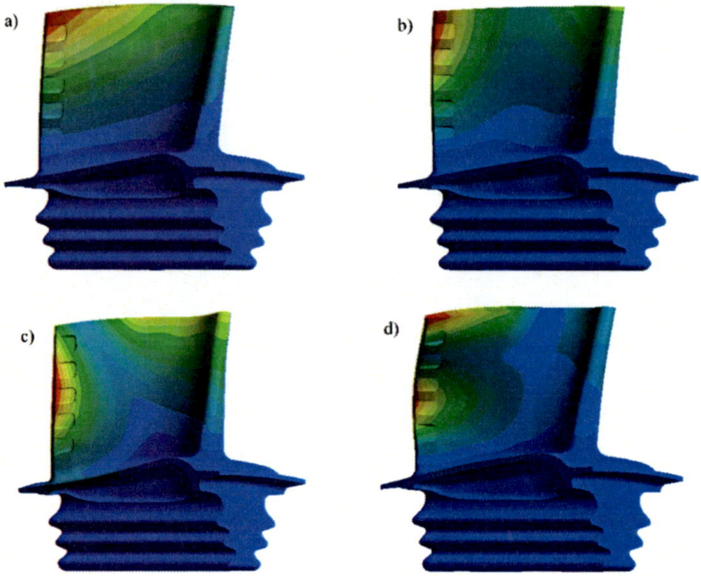

Fig. 3. Results of the modal analysis of the blade: a – the first mode shapes, $f_1 = 11252$ Hz; b – the second mode shapes, $f_2 = 18946$ Hz; c – the third mode shapes, $f_3 = 24130$ Hz; d – the fourth mode shapes, $f_4 = 32011$ Hz.

Calculations of the natural frequencies and mode shapes of the turbine blade at high temperature (1100 °C) were also carried out. The results of these calculations are shown in Table 2.

Table 2. Natural frequencies of blade vibrations for different temperatures.

Calculated natural frequencies of vibrations		
Number in order of the form of vibrations	$T = 20$ °C	$T = 1100$ °C
	$f_{20\,°C}$, Hz	$f_{1100\,°C}$, Hz
1	11252	9177,9
2	18946	15571
3	24130	19706
4	32011	26247

According to the obtained calculation data, it can be seen that with a relatively equal Poisson ratio $\nu_{xy} \approx \nu_{yz} \approx \nu_{xz} \approx const$, the decrease of the shear modulus G and the modulus of elasticity E, when the temperature on the turbine blade increases, significantly affects the value of the natural frequencies of the blade vibrations.

4.2 Construction of Blade Resonance Diagram and Prediction of Dangerous Resonance Modes

The conducted numerical experiments were used for building a blade resonance diagram (Campbell diagram). When constructing this diagram, the influence of centrifugal force was considered (in the Static Structural module, the rotor's rotation frequency was specified for the X-component). Centrifugal forces on the diagram are displayed as a percentage of the maximum rotor revolutions: 0, 25%, 50%, 75%, 100%, and 105% (centrifugal force at $N = 50000$ rpm).

Dependences of resonance frequency values and mode shapes of the blade on the harmonics of the exciting force at different rotor rotation frequencies are shown in Fig. 4. From the calculation. It can be seen that for blades with such type of design, the natural vibration frequencies slightly increase under the influence of centrifugal force.

The calculations were done for a cooled high-pressure turbine blade of a typical turboshaft aircraft gas turbine engine. Usually, the excitation of resonant vibrations of turbine blades occurs from several main factors. The excitation source of such vibrations can be irregularities and impulses that occur due to the specifics of the operation of certain parts of the gas turbine engine. Examples of such triggers can be burners of the combustion chamber, nozzle devices located in front of the blades, and nozzle devices of the following stages of the turbines, which receive the oncoming flow of the gas-air mixture. The harmonics of the excitation force indicated on the Campbell diagram correspond to various structural elements of a typical engine. The specified factors lead to K-harmonics multiples of the fundamental frequency $K = 1$ appearing in the spectrum of exciting forces. Harmonics K = 14, 16, 19, 23 refer to such types of structural elements as combustion chamber burners, nozzles of the current turbine stage, and subsequent stages. Harmonics of the exciting force K = 28, 32, 40 correspond to the total exciting forces from the action of typical various factors in the process of engine operation. Corresponding black straight lines $K = 14,\ldots,40$ are shown in the Campbell diagram. The blue lines represent the dependence of the first four resonance frequencies of vibrations on the rotor rotation frequency. The harmonics used for the calculation in the article were taken from the strain measurement results of the existing prototype engine.

4.3 Application of Proposed Methodology for Avoiding Dangerous Resonance Modes

Taking the calculated elastic characteristics of the material as a basis, it is possible to investigate how the rotation of the selected crystallographic system of directions by 45° affects the natural frequencies and mode shapes of the blades. Using relations (8) and it is possible to determine the elastic compliances when the selected single-crystal direction system is rotated by 45°.

With the use of calculated data in ANSYS Workbench, we can estimate the influence of the determined elastic characteristics, due to the rotation of the angle of the selected crystallographic system of single-crystal directions by 45°, on the natural frequencies of blade vibrations, as shown in Table 3. The influence of centrifugal forces was also taken into account.

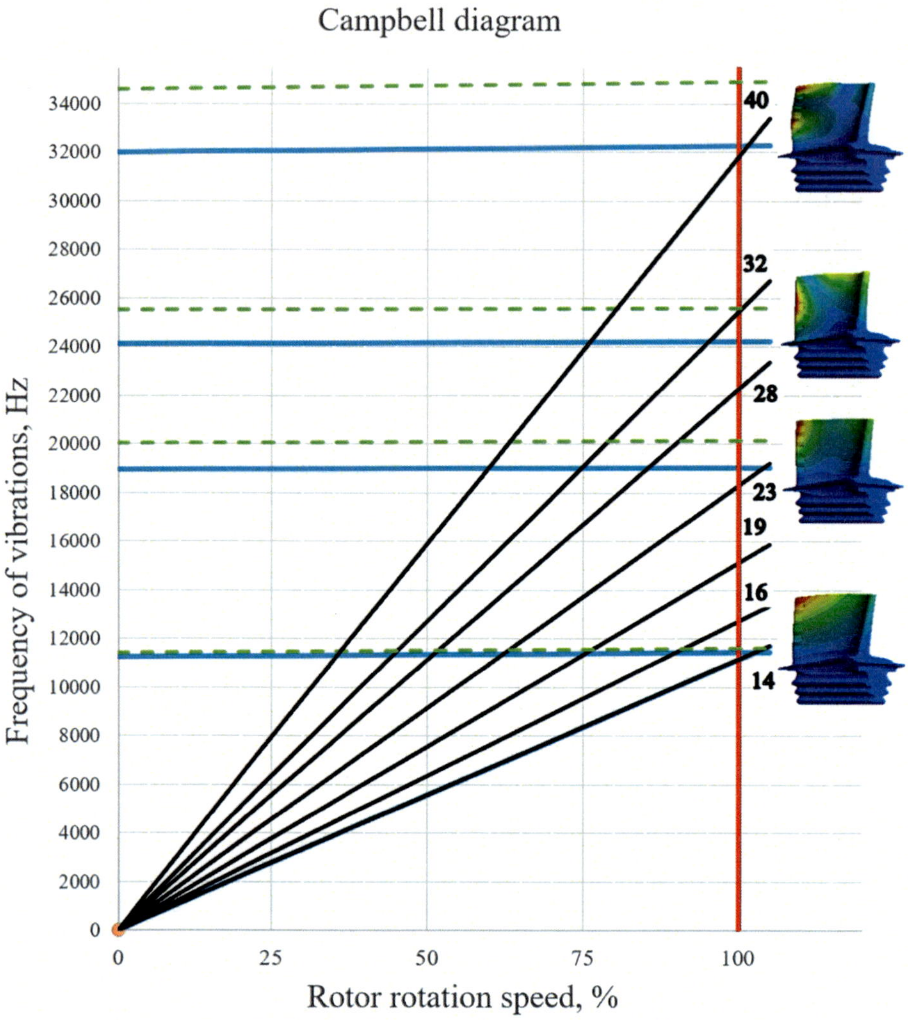

Fig. 4. Resonance diagram for a cooled blade of a high-pressure turbine.

The obtained calculation results were taken into account when constructing the Campbell diagram (green dashed lines in Fig. 4). It can be seen from the diagram that the natural frequencies of vibrations of the blade have shifted, and this helps to avoid resonance modes from several harmonics of the excitation forces. It has been discovered that the manufacture of blades with different crystallographic orientations can significantly affect the values of the spectrum of resonant frequencies of the blades. In the considered example, the rotation of the crystallographic directions of the monocrystal by 45° creates the possibility of shifting the second natural frequency at the K = 23 harmonic by 800 Hz and the fourth natural frequency from the resonance at the K = 40 harmonic by 2400 Hz. The change in the first natural frequency was only 173 Hz, which is insignificant compared to the excitation frequency at the K = 14 harmonic, being

Table 3. The results of the calculation of the natural frequencies of vibrations of the blade when the angle of rotation of the selected crystallographic system of directions is changed.

Calculated values of natural frequencies of vibrations		
Number in order of the form of vibrations	Initial crystallographic system f_{init}, Hz	The crystallographic system of directions is rotated by 45° $f_{45°}$, Hz
	$T = 20$ °C	
1	11252	11427
2	18946	20054
3	24130	25515
4	32011	34607
	$T = 20$ °C + centrifugal force	
1	11439	11612
2	19057	20157
3	24238	25603
4	32301	34923

equal to 11400 Hz. But at the same time, the rotation of the crystallographic directions by 45° couldn't help in avoiding the danger of resonance at the K = 32 harmonic.

Thus, the analysis of received results allows us to establish that they are very promising and outline the task of further research on the optimal choice of the angle of rotation of crystallographic directions in developing technology for turbine blade manufacturing. Also, using the finite-element method creates further opportunities for analyzing the fatigue strength of cooled turbine blades with a detailed account of the stress concentration appearing from the complexity of their cross-sectional shapes.

5 Conclusions

The method of determining natural frequencies and mode shapes of gas turbine monocrystalline blades using finite element analysis software complexes has been developed. For cooled blades with a complex cross-sectional profile, finite element analysis has an advantage over methods that use continuous functions to approximate displacement shapes (for example, the Rayleigh-Ritz method [21] and the Galerkin method [22]). Modern software systems have powerful tools for building finite element models that allow taking into account the smallest features of complex geometric shapes of cooled blades. The developed technique includes the procedure for determining the components of the tensor of elastic constants and their presentation in matrix form, which is used in modern software complexes of finite element analysis. To make it possible to rationally choose the orientation of crystallographic directions during the manufacture of blades, convenient ratios of the transformation of elastic constants were obtained.

An obligatory stage in the design of turbine blades is to consider the effect of centrifugal forces on natural frequencies. For uncooled blades with a solid cross-section, such an effect is manifested only for relatively long blades. In the case of cooled blades with thin-walled regions of the cross-section, the effect of centrifugal forces can be noticeable for short blades, necessitating a detailed analysis of this effect.

On a typical example of a cooling high-pressure turbine blade, a modal analysis was performed considering the stress state due to the action of centrifugal forces. It was found that centrifugal forces cause a slight change in natural frequencies for such blades, compared to a decrease in the modulus of elasticity in the crystallographic directions at operating temperatures.

Campbell diagram was created to analyze possible resonance states. The possibility of the emergence of resonance modes from various multiple harmonics of the exciting forces has been discovered.

It has been established that manufacturing blades with different crystallographic orientations make it possible to change the spectrum of resonant frequencies of the blades significantly. In the considered example, the rotation of the crystallographic directions by 45° makes it possible to shift the fourth natural frequency from the resonance at the $K = 40$ harmonic by 2400 Hz, the second natural frequency at the $K = 23$ harmonic by 800 Hz. The change in the first natural frequency was only 173 Hz, which is small compared to the excitation frequency at the $K = 14$ harmonic equal to 11400 Hz. At the same time, the rotation of the crystallographic directions by 45° led to the danger of resonance at the $K = 32$ harmonic.

This analysis allows us to outline the task of further research on the optimal choice of the rotation angle of crystallographic directions in the development of technology for manufacturing turbine blades. Also, using the finite element method opens up opportunities for analyzing the fatigue strength of cooled blades with a detailed account of the stress concentration arising from the complexity of their cross-sectional shapes.

References

1. Ibrahim, T.K., Mohammed, M.K., Al-Door, W.H.A., et al.: Study of the performance of the gas turbine power plants from the simple to complex cycle: a technical review. J. Adv. Res. Fluid Mech. Therm. Sci. **54**, 228–250 (2019)
2. Ibrahim, T.K., Rahman, M., Mohammed, M., Basrawi, F.: Statistical analysis and optimum performance of the gas turbine powerplant. Int. J. Autom. Mech. Eng. **13**(1), 3215–3225 (2016). https://doi.org/10.15282/ijame.13.1.2016.8.0268
3. Luan, X., Qin, H., Liu, F., et al.: The mechanical properties and elastic anisotropies of cubic Ni_3Al from first principles calculations. Crystals **8**(8), 307–318 (2018). https://doi.org/10.3390/cryst8080307
4. Wang, Z., Stoica, A.D., Ma, D., Beese, A.M.: Diffraction and single-crystal elastic constants of Inconel 625 at room and elevated temperatures determined by neutron diffraction. Mater. Sci. Eng. **674**, 406–412 (2016). https://doi.org/10.1016/j.msea.2016.08.010
5. Aba-Perea, P.E., Pirling, T., Withers, P.J., Kelleher, J., Kabra, S., Preuss, M.: Determination of the high temperature elastic properties and diffraction elastic constants of Ni-base superalloys. Mater. Des. **89**, 856–863 (2016). https://doi.org/10.1016/j.matdes.2015.09.152

6. Li, D., Guo, Q., Guo, S., Peng, H., Wu, Z.: The microstructure evolution and nucleation mechanisms of dynamic recrystallization in hot-deformed Inconel 625 superalloy. Mater. Des. **32**, 696–705 (2011). https://doi.org/10.1016/j.matdes.2010.07.040
7. Yang, S.W.: Elastic constants of a monocrystalline nickel-base superalloy. Ibid **16A**(4), 661–665 (1985). https://doi.org/10.1007/BF02814240
8. Jinlai, L., Lihua, Y., Yizhou, Z., Jinguo, L., Xiaofeng, S.: Anisotropy of elasticity of a Ni-base single crystal superalloy. Acta Metall. Sin. **56**(6), 855–862 (2020). https://doi.org/10.11900/0412.1961.2019.00355
9. Hou, N.X., Gou, W.X., Wen, Z.X., et al.: The influence of crystal orientations on fatigue life of single crystal cooled turbine blade. Mater. Sci. Eng. A **492**(1,2), 413–418 (2008). https://doi.org/10.1016/j.msea.2008.03.043
10. Wen, M., Barnoush, A., Yokogawa, K.: Calculation of all cubic single crystal elastic constants from single atomistic simulation: hydrogen effect and elastic constants of nickel. Comput. Phys. Commun. **182**(8), 1621–1625 (2011). https://doi.org/10.1016/j.cpc.2011.04.009
11. Jamal, M., Asadabadi, S.J., Ahmad, I., et al.: Elastic constants of cubic crystals. Comput. Mater. Sci. **95**, 592–599 (2014). https://doi.org/10.1016/j.commatsci.2014.08.027
12. Epishin, A., Fedelich, B., Finn, M., et al.: Investigation of elastic properties of the single-crystal nickel-base superalloy CMSX-4 in the temperature interval between room temperature and 1300 °C. Crystals **11**(2), 152–160 (2021). https://doi.org/10.3390/cryst11020152
13. Reed, R.C.: The Superalloys: Fundamentals and Applications. Cambridge University Press, UK (2006). https://doi.org/10.1017/CBO9780511541285
14. Epishin, A.I., Lisovenko, D.S.: Extreme values of the Poisson's ratio of cubic crystals. Tech. Phys. **61**, 1516–1524 (2016). https://doi.org/10.1134/S1063784216100121
15. Siebörger, D., Knake, H., Glatzel, U.: Temperature dependence of the elastic moduli of the nickel-base superalloy CMSX-4 and its isolated phases. Mater. Sci. Eng. A **298**, 26–33 (2001). https://doi.org/10.1016/S0921-5093(00)01318-6
16. Hemberger, D., Filsinger, D., Bauer, H.-J.: Investigations on maximum amplitude amplification factor of real mistuned bladed structures. In: Proceedings of the ASME Turbo Expo 2012, ASME Turbo Expo 2012: Turbine Technical Conference and Exposition, pp. 1—12. Copenhagen (2012). https://doi.org/10.1115/GT2012-68084
17. Sanliturk, K.Y., Ewins, D.J., Stanbridge, A.B.: Underplatform dampers for turbine blades: theoretical modeling, analysis, and comparison with experimental data. J. Eng. Gas Turbines Power **123**(4), 919–929 (2003). https://doi.org/10.1115/1.1385830
18. Ewins, D.J.: Control of vibration and resonance in aero engines and rotating machinery –an overview. Int. J. Press. Vessels Pip. **87**(9), 504–510 (2010). https://doi.org/10.1016/j.ijpvp.2010.07.001
19. Duong, L., Murphy, K.D., Kazerounian, K.: Guided tuning of turbine blades: a practical method to avoid operating at resonance. J. Vib. Acoust. **135**(5), 1–5 (2013). https://doi.org/10.1115/1.4024761
20. Lurie, A.I.: Theory of Elasticity. Springer, Berlin (2005). https://doi.org/10.1007/978-3-540-26455-2
21. Jeong, S., Yoo, H.H.: Nonlinear structural analysis of a flexible multibody system using the classical Rayleigh-Ritz method. Int. J. Non-Linear Mech. **110**, 69–80 (2019). https://doi.org/10.1016/j.ijnonlinmec.2019.01.011
22. Zhang, J., et al.: Topology optimization of periodic mechanical structures with orthotropic materials based on the element-free Galerkin method. Eng. Anal. Bound. Elements **143**, 383–396 (2022). https://doi.org/10.1016/j.enganabound.2022.06.01

Experimental Studies of the Wear on the Rotors' Working Surfaces of a Planetary Hydraulic Motor

Anatolii Panchenko[1](✉), Angela Voloshina[1], Roman Antoshchenkov[2], Ivan Halych[2], and Szymon Głowacki[3]

[1] Dmytro Motornyi Tavria State Agrotechnological University, 18, B. Khmelnytskyi Ave., Melitopol 72310, Ukraine
anatolii.panchenko@tsatu.edu.ua

[2] State Biotechnological University, 44, Alchevskikh Str., Kharkiv 61001, Ukraine

[3] Warsaw University of Life Sciences, 166, Nowoursynowska Str., 02-787 Warsaw, Poland

Abstract. During the operation of planetary hydraulic motors, wear of the working surfaces of their rotors occurs. Therefore, the experimental study of the effect of wear on the working surfaces of the planetary hydraulic motor rotors on the dynamics of changes in its output characteristics is relevant. Studies show that with an increase in the diametral clearance, the theoretical torque decreases by 3.5%, changing from 1450 N·m to 1400 N·m. The theoretical rotation frequency has significant deviations up to 47%, varying from 150 min^{-1} to 80 min^{-1}. The theoretical flow rate of overflows reaches 10 l/min at a critical value of a diametral gap of 0.4 mm. When changing the diametral clearance, the real torque sharply decreases by 42% to 720 N·m, and the real flow rate of the overflows sharply increases to 7.5 l/min at a maximum value of the diametral gap of 0.25 mm. The decrease in the real rotation frequency by 65.5% occurs quite dynamically in the range from 145 min^{-1} to 50 min^{-1}. Such a change in the real rotation frequency occurs due to "additional" movements of the inner rotor and an increase in the actual flow rate of overflows. Experimental torque, leakage flow, and rotation frequency change similarly to real values, respectively, having the exact numerical values. It has been established that the critical value of the diametral gap is 2.5 times less than the same value for a planetary hydraulic motor with the theoretical rotor.

Keywords: Critical Gap · Block Diagram · Torque · Flow Rate · Rotation Frequency · Energy Efficiency

1 Introduction

Planetary (orbital) hydraulic motors are increasingly used in various national economy sectors. For example, in the drive of rotary cleaners of the road surface of communal harvesting equipment [1], in the drives of conveyors and rotary devices of grain harvesters [2], in the drives of steering and running systems of self-propelled equipment, as well

V. Tonkonogyi et al. (Eds.): InterPartner 2023, LNME, pp. 498–508, 2024.
https://doi.org/10.1007/978-3-031-42778-7_46

as in production lines of various types of enterprises [3]. Planetary hydraulic motors, in terms of their technical characteristics [4], occupy an intermediate niche between axial piston [5] and radial piston hydraulic motors [6]. High torque in operating modes and low rotation frequency of the output shaft make it possible to use orbital hydraulic motors without a gearbox for the direct drive of various actuators of self-propelled machines [7]. During the operation of planetary hydraulic motors, wear of the working surfaces of their rotors occurs (an increase in the diametrical gap between the rotors). It is known [4] that an increase in the diametrical gap due to wear of the working surfaces of the rotors is accompanied by two phases of their interaction. The first phase of interaction between the inner and outer rotors is characterized by their self-sealing, which ensures the separation of the high-pressure zone from the low-pressure zone. When the value of the diametrical gap exceeds the critical value, the second phase of the interaction of the rotors begins with their functional depressurization [4]. Theoretical studies have established [4] that when the gap is less than the critical value, the dependences of the functional parameters are parallel to the abscissa axis. In the case when the gap is more significant than the critical one, there is a sharp deterioration in the functional parameters of planetary hydraulic motors [4].

In this regard, the issue of experimental study of the effect of wear on the working surfaces of the rotors of a planetary hydraulic motor on the dynamics of changes in its output characteristics is an actual scientific direction.

Therefore, the presented article is aimed at solving an actual scientific direction related to the issue of experimental research on the effect of wear on the working surfaces of the rotors of a planetary hydraulic motor on the dynamics of changes in its output characteristics.

2 Literature Review

For the design and optimization of gear machines, simulation models have been proposed [8], a kinematic analysis has been carried out [9], and the gaps between the teeth and their deformation upon contact have been taken into account [10]. The theory of rotor engagement is presented [11], the kinematic [12] and geometric [13] parameters of gear rotor profiles are substantiated, the effect of the span angle of the tooth profile on the distribution of the contact force on the inner and outer rotors [14] is studied, the contact stresses in the rotor teeth are calculated [15] and on rotational elements [16], approaches have been proposed to reduce them [17]. The displacement analysis [18] and the instantaneous flow rate [19] of gerotor pumps were determined. A rotor profile has been proposed that makes it possible to increase the fluid flow at low speed [20]. However, in this case, pressure pulsations are observed, which lead to unstable operation of the entire system [21]. The issues of studying the kinematics of moving the rotors of planetary hydraulic motors and their self-sealing (sealing of high and low-pressure zones) were not considered experimentally.

The influence of the viscosity properties of the liquid on the efficiency of hydraulic motors at low speeds and high torque [22] was determined, and the flow characteristics of pumps were calculated [23]. The effect of abrasive wear on the reliability and durability of pumping equipment has been studied [24]. An increase in reliability and durability due to using vortex-chamber jet blowers is considered [25]. Estimates of the technical level of volumetric hydraulic machines have been established [26]. However, the effect of wear of the rotor system on the technical level of planetary hydraulic motors has not been considered.

A model of wear in hydraulic motors of the gerotor type was proposed [27], the wear of the contact surfaces of the rotors was analyzed [28], and special attention was paid to modeling gaps and determining leakage in the gearing of an orbital hydraulic motor [29] and hydraulic machines with involute and cycloidal gears [30]. The gap between the teeth of the rotors associated with the presence of leaks was studied [31]. An approach to leak analysis for various types of hydraulic pumps is presented [32]. A new gerotor pump design with a variable gap is proposed [33]. A predictive model of internal leakage was developed, taking into account the physical gap between the rotors [34], a mathematical model for predicting the efficiency of flow and volume [35], and the parameters of the rotor system were substantiated to predict changes in the output characteristics of a planetary hydraulic motor [36]. Multicriteria optimization of the gear profile of rotors [37] and flow rate and wear rate non-uniformity [38] are presented. The influence of the wear of the working surfaces of the rotors (the gap between the rotors) of the planetary hydraulic motor on the change in its output characteristics has not been studied.

3 Research Methodology

To experimentally determine the effect of wear on the working surfaces of the rotors of a planetary hydraulic motor on the dynamics of changes in its output characteristics with a change in the diametrical gap between its rotors, it is necessary to:

- to develop a mathematical apparatus and a block diagram for modeling the dynamics of a change in the output parameters of a planetary hydraulic motor from wear (diametric gap) of the working surfaces of its rotors;
- to carry out theoretical (by modeling) and experimental (bench) studies of the dynamics of changes in the output characteristics of a planetary hydraulic motor with a change in the diametral gap.

When analyzing the change in parameters that determine the functional characteristics of a planetary hydraulic motor [4], attention is drawn to changes in three parameters - torque M_{tor}, rotation frequency n of the output shaft of the hydraulic motor, and leakage flow rate Q_{leak}. Changes of these parameters (M_{tor}, n, Q_{leak}) of a planetary hydraulic motor, depending on the wear of the working surfaces of its rotors (from the diametral gap), occur entirely differently than in a conventional positive displacement hydraulic machine [4].

Therefore, this paper presents studies of changes in a planetary hydraulic motor's functional parameters (Mtor, n, Qleak) depending on the diametral gap (from wear) carried out at three levels. The first level involves research for a planetary hydraulic motor whose rotors have a theoretical cycloidal mesh profile. Under the theoretical profile of the cycloidal engagement, we mean such a gear profile that does not have an error in the shape of its contour (geometrically correct). The second level involves research for a hydraulic motor with a rotor and an accurate cycloidal gearing profile. The gear profile's real cycloidal contour is characterized by an error in its manufacture, which causes unforeseen changes in the diametral clearance. The third level involves carrying out experimental studies of a planetary hydraulic motor, the rotor of which has an error in the form of manufacture.

For theoretical studies of the first and second levels, we have developed a mathematical apparatus that allows us to describe the effect of diametral clearance (wear) on the dynamics of changes in the output characteristics of a planetary hydraulic motor. Depending on the change in the diametral gap, the theoretical value of the torque M_{tor} is determined by the product of the resulting force from the action of the pressure of the working fluid F on the shoulder of its application h [4]:

$$M_{torT} = F \cdot h, \tag{1}$$

where $F = 2 \cdot (e \pm G/2) \cdot \Delta p \cdot b \cdot (z1 + 1)$; $h = e + R1$; e is the center-to-center distance between the rotors; G is the diametrical gap between the rotors; Δp is liquid pressure difference; b is the width of the rotors; z1, z2 are the number of teeth of the inner and outer rotors, respectively; R1 is the radius of location of the centers of the teeth of the inner rotor [4].

The real value of the torque M_{tor}, taking into account the error in the shape of the manufacture of the rotors, is determined by the change in the value of the shoulder $h = h_i$ [4]:

$$h_i = e + R_1 - \left[\left(e + R_{1i} \cdot sin\frac{3\pi}{z_1} \right) \cdot \frac{R_2 \cdot sin\frac{\pi}{z_2}}{r_1 + r_2} \pm \frac{G}{2} \right] \cdot C_{er}, \tag{2}$$

where R_2 is the radius of the location of the centers of the teeth of the outer rotor; r_1 and r_2 are the radii of curvature of the teeth of the inner and outer rotors, respectively; C_{er} is the coefficient taking into account the shape error.

During operation, at values of the diametrical gap G less than the critical G_{lim} ($G < G_{lim}$), there are no leaks in the hydraulic motor due to the self-sealing of the discharge zone from the drain zone [4]. When the value of the diametral gap G is greater than the critical value G_{lim} ($G > G_{lim}$), leakage occurs between the rotors [4]. Theoretical flow rates Q_{leak}, along the diametral gap, are characterized by the flow rate of fluid through a slot of height G', formed by the cylindrical surfaces of the teeth of the inner and outer rotors in the zone of formation of the diametrical gap G. The real values of the flow rate Q_{leak} are characterized by an additional flow rate of liquid through a slot with a height G", in the zone opposite to the diametrical gap G, due to a change in the value of the shoulder $h = h_i$ and are determined by the dependence:

$$Q_{leak} = \mu \cdot b \cdot \sqrt{\frac{2 \cdot \Delta p}{\rho}} \cdot (G' + G'') \tag{3}$$

where μ is the coefficient characterizing the geometry of the flow path; ρ is the density of the working fluid.

The theoretical rotation frequency n_{vol} of the shaft of a planetary hydraulic motor in the presence of a diametrical clearance [4] is determined by the expression:

$$n_{val} = \frac{n_{theor}(\pi - 2 \cdot t_1 \cdot n_{theor} \cdot z_1 \cdot z_2)}{\pi - t_1 \cdot n_{theor} \cdot z_1 \cdot z_2}, \tag{4}$$

$$t_1 = \frac{G \cdot R_1 \cdot b_1}{2 \cdot n_{theor} \cdot V_0 \cdot cos\,\delta},$$

where t_1 is the "ascent" time of the inner rotor; n_{theor} is the rotation frequency in the absence of a diametral clearance; V_0 is the working volume of the hydraulic motor; δ is the angle of inclination of the "floating" normal of the inner rotor.

When changing the value of the shoulder $h = h_i$, the real rotation frequency n'_{vol} of the shaft of the planetary hydraulic motor is determined by the expression:

$$n'_{val} = \frac{Q_{val} - Q_{leak}}{V_0}, \tag{5}$$

where Q_{val} is the working fluid's actual (given) flow rate.

Experimental studies of changes in the functional parameters (M_{tor}, n, Q_{leak}) of a planetary hydraulic motor, depending on the diametral gap (wear), were carried out under GOST 20719-83 (Hydraulic motors. Acceptance and methods of tests). The studies were carried out for the planetary hydraulic motor PRG-22, with a working volume V_0 = 630 cm^3.

4 Results and Discussion

Research of change of functional parameters (Mtor, n, Qleak) of a planetary hydromotor was carried out at three levels. At the first and second levels, theoretical studies were carried out by modeling the operation of planetary hydraulic machines using the block modeling system VisSim. The input parameters for modeling were taken as follows: the working volume of the hydraulic motor V0 = 630 cm^3, pressure drop $\Delta p = 16$ MPa.

To simulate operating conditions, a block diagram has been developed that allows determining the change in the output parameters of a planetary hydraulic motor when the working surfaces of its rotors are worn out (Fig. 1). The output parameters of the planetary hydraulic motor are the torque determined by block 1, the leakage flow rate determined by block 2, and the rotation frequency determined by block 3, where the above developed mathematical apparatus is implemented.

Changes in the functional parameters of the planetary hydraulic motor (M_{tor}, Q_{leak}, n), for the theoretical cycloidal engagement, depending on the change in the diametrical gap G (Fig. 2), are characterized by two sections $G < G_{lim}$ and $G > G_{lim}$ (at the value of $G_{lim} = 0.4$ mm [4].

The theoretical torque M_{tor} over the entire range of diametral gap G from 0 to 0.6mm is characterized by a straight line. It has slight deviations (up to 3.5%) from its nominal value, varying in the range from 1450 N·m to 1400 N·m (Fig. 2, curve 1).

The theoretical flow rate Q_{leak} (Fig. 2, curve 2) in the first section, when the diametral gap G changes in the range from 0 to 0.4 mm ($G < G_{lim}$), is absent (equal to zero). In the second section, with an increase in the diametral gap G from 0.4 mm to 0.6 mm, the rotors "depressurize" [4], and the theoretical flow rate of the leakages Q_{leak} is characterized by a parabolic dependence, sharply increasing from 0 to 10 l/min (Fig. 2, curve 2).

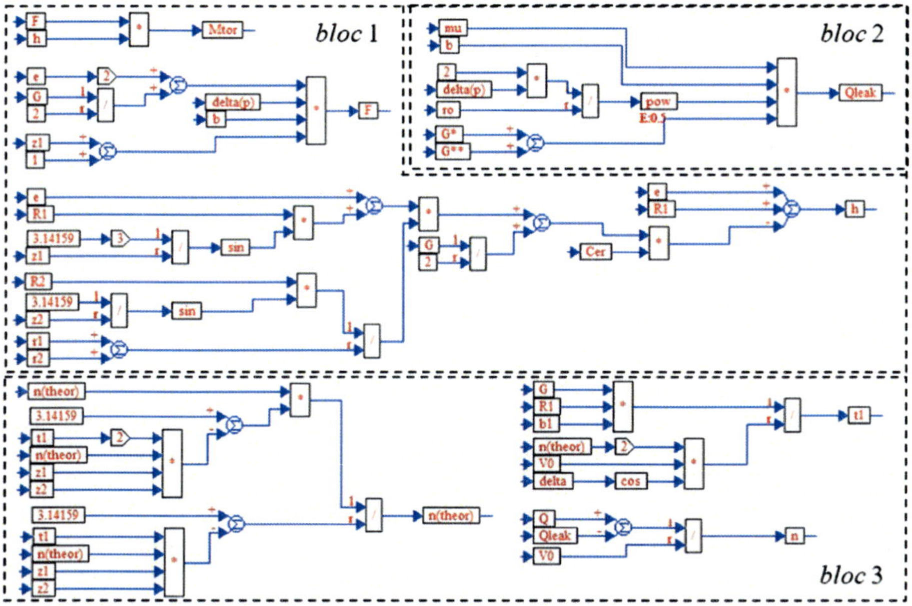

Fig. 1. Block diagram of modeling the change in the output parameters of a planetary hydraulic motor in case of wear of the working surfaces of its rotors.

Changes in the theoretical rotation frequency n in the first section (Fig. 2, curve 3), with a change in the diametral gap G in the range from 0 to 0.4 mm ($G < G_{lim}$), are linear while changing significantly (up to 29%) in the range from 150 min^{-1} to 107 min^{-1}. The reason for this change in the theoretical rotation frequency n is the additional displacement of the inner rotor [4]. In the second section, with an increase in the diametral gap from $G_{lim} = 0.4$ mm to $G_{max} = 0.6$ mm and the theoretical flow rate of leakages Q_{leak}, as a result of the "depressurization" of the rotors, the change in the theoretical rotation frequency n occurs already according to a parabolic law and only in this section it decreases (by 25.3%) from 107 min^{-1} to 80 min^{-1} (Fig. 2, curve 3).

Changes in functional parameters for real cycloidal engagement are also characterized by two sections (Fig. 2, curves 4, 5, and 6). In a real cycloidal engagement, the division of plots occurs at a value of $G_{lim} = 0.16$ mm. It is explained by the fact that when modeling the processes occurring in a real cycloidal engagement, the error in the shape of the gear contour of the rotors was taken into account, which causes a discrete decrease in the shoulder h_i of the application of the resulting pressure force of the working fluid F, described by expression (2). This process, in the second section (at $G > G_{lim}$), is characterized by sharp changes in real functional parameters (M_{tor}, n, Q_{leak}).

The real torque $M_{tor,}$ in the range of diametrical gap G from 0 to 0.16 mm, is characterized by a straight line and has almost no deviation from its original value. In the studied (first) section, M_{tor} = 1320 N·m (Fig. 2, curve 4). In the second section (at $G > G_{lim}$), the real torque M_{tor} sharply decreases (by 42%) to M_{tor} = 720 N·m at the maximum value of the diametral gap G_{max} = 0.25 mm (Fig. 2, curve 4).

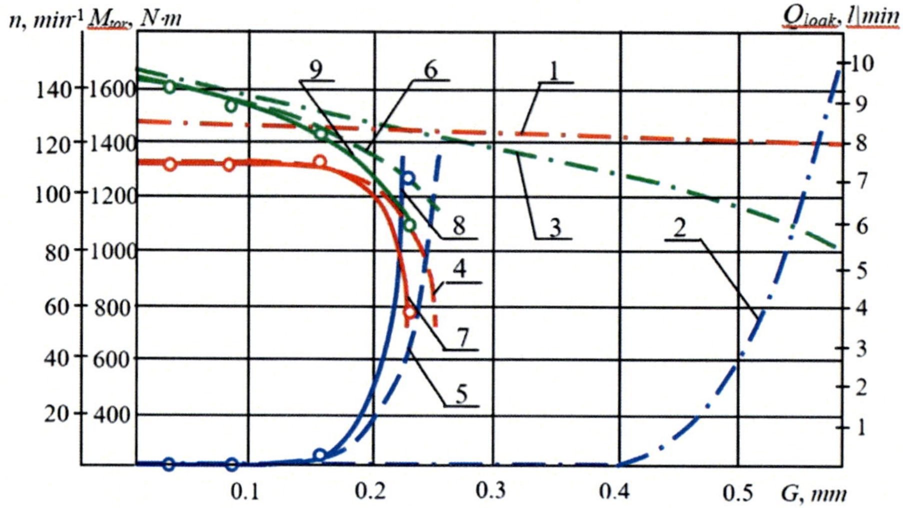

Fig. 2. Dependence of the change in the functional parameters of planetary hydraulic motors on the diametral clearance (wear): 1, 2, and 3 are curves of change in the theoretical torque, theoretical leakage flow, and theoretical rotation frequency, respectively [4]; 4, 5 and 6 are curves of change of real torque, the real flow rate of leaks and real rotation frequency, respectively; 7, 8 and 9 are experimental curves of torque, leakage flow, and rotation frequency, respectively.

The real Q_{leak} flow rate (as well as the theoretical one) is absent when the diametral gap G varies from 0 to 0.16 mm (in the first section) (Fig. 2, curve 5). In the second section, with an increase in the diametral gap from G_{lim} = 0.16 mm to G_{max} = 0.25 mm, as a result, the shoulder hi of the application of the resulting force decreases, and the actual flow rate of the leakages Q_{leak} increases sharply. At the maximum value of the diametral gap G_{max} = 0.25 mm, as a result of the "depressurization" of the rotors [4], the value of the actual flow rate Q_{leak} reaches 7.5 l/min (Fig. 2, curve 5).

The decrease in the actual rotation frequency n in the first and second sections occurs according to a parabolic law without sharp deviations but rather dynamically (Fig. 2, curve 5). When changing the diametral gap in the range from 0 to 0.25 mm, as a result of "additional" movements of the inner rotor [4] and an increase in the actual flow rate of Q_{leak}, the value of the actual rotation frequency n decreased by 65.5%, changing in the range from 145 min^{-1} to 50 min^{-1} (Fig. 2, curve 5).

It should be noted that the results of modeling the first and second levels are quite close regarding the nature of the regularities but differ significantly in quantitative indicators. The critical value of the diametrical gap G_{lim} for the first level is 2.5 times higher than the same value for the second level and, accordingly, is equal to G_{lim} = 0.4 mm and G_{lim} = 0.16 mm. It is because the theoretical cycloidal gearing, according to the modeling condition, does not have an error in the handicap of the gear contour. When simulating real gearing, it is set due to a manufacturing error.

The third level involves experimental studies of a planetary hydraulic motor whose rotor has errors in the shape of the gear profile due to manufacturing. Experimental studies were carried out using a PRG-22 planetary hydraulic motor with a power of 22 kW and a working volume of $V_0 = 630$ cm^3. The nominal differential pressure Δp for this hydraulic motor $\Delta p = 16$ MPa provides a torque $M_{tor} = 1300$ N·m, and the actual flow rate of the working fluid $Q_{val} = 96$ l/min allows you to get the nominal rotation frequency n = 145 min^{-1}.

The experimental torque M_{tor}, leakage flow rate $Q_{leak,}$ and rotation frequency n in the first and second sections change similarly to the real values of M_{tor}, $Q_{leak,}$ and n, respectively, having the exact numerical values (Fig. 2, curve 7 and 4). However, the limiting value of the diametrical gap G_{max} for the experimental values (n) is 8% less than the same values in real studies and, respectively, are equal to $G_{max} = 0.23$ mm and $G_{max} = 0.25$ mm (Fig. 2, curve 7 and 4).

The coincidence of the results of experimental data and the results of modeling the operation of a planetary hydraulic motor with real rotors indicates the reliability of the theoretical prerequisites for the influence of wear on the working surfaces of rotors on the dynamics of changes in the output characteristics of a planetary hydraulic motor. The critical value of the diametrical gap G_{lim} for the first level is 2.5 times higher than the same value for the second and third levels, which is respectively equal to $G_{lim} = 0.4$ mm and $G_{lim} = 0.16$ mm. These differences show that the technology of manufacturing the gear surfaces of the planetary hydraulic motor rotors has a significant impact on the dynamics of its output characteristics. Therefore, the improvement of manufacturing technology and the modernization of the design of the toothed surface of the rotors of the planetary hydraulic motor will more than double its service life.

5 Conclusions

As a result of the research, a mathematical apparatus and a block diagram have been developed that allow simulation of the dynamics of a change in the output parameters of a planetary hydraulic motor from the wear of the working surfaces of its rotors.

Studies of the dynamics of changes in the output characteristics of a planetary hydraulic motor depending on wear (diametrical gap) show that the theoretical torque and rotation frequency, when the diametrical clearance changes from 0 to 0.6 mm, is characterized by a straight line. The theoretical torque has deviations of up to 3.5%, varying from 1450 N·m to 1400 N·m, and the theoretical rotation frequency has significant deviations of up to 47%, varying from 150 min^{-1} to 80 min^{-1}. Theoretical leakage flow reaches 10 l/min at a critical diametral gap of 0.4 mm.

The actual torque and leakage flow rate in the diametral clearance range from 0 to 0.16 mm is characterized by a straight line. In the second section, the real torque sharply decreases (by 42%) to $M_{tor} = 720$ N·m, and the real flow rate of the overflows sharply increases to 7.5 l/min with a maximum diametral gap of 0.25 mm. The decrease in the actual rotation frequency by 65.5% occurs dynamically according to the parabolic law, changing from 145 min^{-1} to 50 min^{-1}.

Experimental studies of the planetary hydraulic motor have established that the experimental torque, leakage flow rate, and rotation frequency change similarly to real values, respectively, having the exact numerical values. The limiting value of the diametrical gap, for experimental values, is 8% less than the same values in real studies and, respectively, are equal to 0.23 mm and 0.25 mm.

It has been established that the critical value of the diametral gap for the results of experimental studies is 2.5 times less than the same value for modeling the operation of a planetary hydraulic motor with theoretical rotors, respectively equal to 0.16 mm and 0.4 mm. These differences show that the technology of manufacturing the gear surfaces of the planetary hydraulic motor rotors has a significant impact on the dynamics of changes in its output characteristics.

References

1. Nadykto, V., Kyurchev, V., Chaplinskyi, A., Ayubov, A.: Ways to increase the traction efficiency of modular draft device. In: IOP Conference Series: Materials Science and Engineering, vol. 1021, no. 1, p. 012043 (2021). https://doi.org/10.1088/1757-899X/1021/1/012043
2. Bulgakov, V., et al.: Experimental investigation of plow-chopping unit. Agric. (Switz.) **11**(1), 1–14 (2021). https://doi.org/10.3390/agriculture11010030
3. Nadykto, V., Kyurchev, V., Bulgakov, V., Findura, P., Karaiev, O.: Influence of the plough with tekrone mouldboards and landsides on ploughing parameters. Acta Technol. Agric. **23**(1), 40–45 (2020). https://doi.org/10.2478/ata-2020-0007
4. Panchenko, A., Voloshina, A., Sadullozoda, S.S., Panchenko, I., Mitin, V.: The changes in the output parameters of planetary hydraulic machines with the increase in the gap between their rotors. In: Tonkonogyi, V., Ivanov, V., Trojanowska, J., Oborskyi, G., Pavlenko, I. (eds.) InterPartner 2022. LNME, pp. 540–551. Springer, Cham (2023). https://doi.org/10.1007/978-3-031-16651-8_51
5. Andrenko, P., Rogovyi, A., Hrechka, I., Khovanskyi, S., Svynarenko, M.: The influence of the gas content in the working fluid on parameters of the the hydraulic motor's axial piston. In: Ivanov, V., Pavlenko, I., Liaposhchenko, O., Machado, J., Edl, M. (eds.) DSMIE 2021. LNME, pp. 97–106. Springer, Cham (2021). https://doi.org/10.1007/978-3-030-77823-1_10
6. Zhang, X., Zhang, J., Xu, B., Yang, Z., Zhao, Q., Zhang, H.: The effect of slotted hole on minimum oil film thickness of piston in radial piston hydraulic motor. In: Proceedings of the ASME/BATH 2021 Symposium on Fluid Power and Motion Control, FPMC2021-69937, V001T01A054 (2021). https://doi.org/10.1115/FPMC2021-69937
7. Roquet, P., Raush, G., Berne, L.J., Gamez-Montero, P.-J., Codina, E.: Energy key performance indicators for mobile machinery. Energies **15**(4), 1364 (2022). https://doi.org/10.3390/en15041364
8. Rundo, M.: Models for flow rate simulation in gear pumps: a review. Energies **10**, 1261 (2017). https://doi.org/10.3390/en10091261
9. Paffoni, B.: Pressure and film thickness in a trochoidal hydrostatic gear pump. Proc. Inst. Mech. Eng. G **217**, 179–187 (2003). https://doi.org/10.1243/095441003769700744
10. Paffoni, B., Progri, R., Gras, R.: Teeth clearance effects upon pressure and film thickness in a trochoidal hydrostatic gear pump. Proc. Inst. Mech. Eng. G **218**, 247–256 (2004). https://doi.org/10.1243/0954410041872799
11. Bonandrini, G., Mimmi, G., Rottenbacher, C.: Theoretical analysis of internal epitrochoidal and hypotrochoidal machines. Proc. Inst. Mech. Eng. C **223**, 1469–1480 (2009). https://doi.org/10.1243/09544062JMES1163

12. Bonandrini, G., Mimmi, G., Rottenbacher, C.: Theoretical analysis of an original rotary machine. J. Mech. Des. **132**(2), 024501 (2010). https://doi.org/10.1115/1.4000698
13. Bonandrini, G., Mimmi, G., Rottenbacher, C.: Design and simulation of meshing of a particular internal rotary pump. Mech. Mach. Theory **49**, 104–116 (2012). https://doi.org/10.1016/j.mechmachtheory.2011.11.001
14. Hsieh, C.F.: Fluid and dynamics analyses of a gerotor pump using various span angle Designs. J. Mech. Des. **134**(12), 121003 (2012). https://doi.org/10.1115/1.4007703
15. Kwon, S.-M., Kim, C.-H., Shin, J.-H.: Analytical wear model of a gerotor pump without hydrodynamic effect. J. Adv. Mech. Des. Syst. Manuf. **2**, 230–237 (2008). https://doi.org/10.1299/jamdsm.2.230
16. Ivanović, L., Rakić, B., Stojanović, B., Matejić, M.: Comparative analysis of analytical and numerical calculations of contact stresses at rotational elements of gerotor pumps. Appl. Eng. Lett. **1**, 1–7 (2016)
17. Ivanović, L.: Reduction of the maximum contact stresses by changing geometric parameters of the trochoidal gearing teeth profile. Meccanica **51**, 2243–2257 (2016)
18. Yang, D.C.H., Yan, J., Tong, S.-H.: Flowrate formulation and displacement analyses for deviation function-based gerotor pumps. Proc. Inst. Mech. Eng. C **225**, 480–487 (2011). https://doi.org/10.1243/09544062JMES2203
19. Liu, H., Lee, J.-C., Yoon, A., Kim, S.-T.: Profile design and numerical calculation of instantaneous flow rate of a gerotor pump. J. Appl. Math. Phys. **3**, 92–97 (2015). https://doi.org/10.4236/jamp.2015.31013
20. Hao, C., Wenming, Y., Guangming, L.: Design of gerotor oil pump with new rotor profile for improving performance. Proc. IMechE Part C **230**, 592–601 (2016). https://doi.org/10.1177/0954406215618228
21. O'Shea, C.: Hydraulic flow ripple cancellation using the primary flow source. In: Proceedings of the BATH/ASME 2016 Symposium on Fluid Power and Motion Control, FPMC2016-1783, V001T01A044. Bath, UK (2016). https://doi.org/10.1115/FPMC2016-1783
22. Michael, P., Burgess, K., Kimball, A., Wanke, T.: Hydraulic fluid efficiency studies in low-speed high-torque motors. In: SAE 2009 Commercial Vehicle Engineering Congress & Exhibition, SAE Technical Papers 2009-01-2848. SAE International, USA (2009). https://doi.org/10.4271/2009-01-2848
23. Andrenko, P., Rogovyi, A., Hrechka, I., Khovanskyi, S., Svynarenko, M.: Characteristics improvement of labyrinth screw pump using design modification in screw. J. Phys.: Conf. Ser. **1741**(1), 012024 (2021). https://doi.org/10.1088/1742-6596/1741/1/012024
24. Rogovyi, A., Neskorozhenyi, A., Krasnikov, S., Tynyanova, I., Khovanskyi, S.: Improvement of vortex chamber supercharger performances using slotted rectangular channel. In: Tonkonogyi, V., Ivanov, V., Trojanowska, J., Oborskyi, G., Pavlenko, I. (eds.) InterPartner 2022. LNME, pp. 552–561. Springer, Cham (2023). https://doi.org/10.1007/978-3-031-16651-8_52
25. Rogovyi, A., Korohodskyi, V., Neskorozhenyi, A., Hrechka, I., Khovanskyi, S.: Reduction of granular material losses in a vortex chamber supercharger drainage channel. In: Ivanov, V., Pavlenko, I., Liaposhchenko, O., Machado, J., Edl, M. (eds.) DSMIE 2022. LNME, pp. 218–226. Springer, Cham (2022). https://doi.org/10.1007/978-3-031-06044-1_21
26. Andrenko, P., Hrechka, I., Khovanskyi, S., Rogovyi, A., Svynarenko, M.: Improving the technical level of hydraulic machines, hydraulic units and hydraulic devices using a definitive assessment criterion at the design stage. J. Mech. Eng. (JMechE) **18**(3), 57–76 (2021). https://doi.org/10.24191/jmeche.v18i3.15414
27. Furustig, J., Larsson, R., Almqvist, A., Bates, C.A., Ennemark, P.: A wear model for EHL contacts in gerotor type hydraulic motors. Proc. Inst. Mech. Eng. C **229**, 254–264 (2015). https://doi.org/10.1177/0954406214534204
28. Ivanović, L., Mackić, T., Stojanović, B.: Analysis of the instantaneous friction coefficient of the trochoidal gear pair. J. Balk. Tribol. Assoc. **22**, 281–293 (2016)

29. Bigliardi, E., Francia, M., Milani, M., Montorsi, L., Paltrinieri, F., Stefani, M.: A combined methodology for studying the axial balancing mechanism of orbit annular hydraulic machines. IFAC-PapersOnLine **28**, 427–432 (2016). https://doi.org/10.1016/j.ifacol.2015.05.110
30. Maiti, R., et al.: Leakage past active contacts in involute and cycloidal gear hydrostatic units. In: Proceedings of the 14th Scandinavian International Conference on Fluid Power, Tampere, Finland (2015)
31. Gamez-Montero, P.J., Garcia-Vilchez, M., Raush, G., Freire, J., Codina, E.: Teeth clearance and relief grooves effects in a trochoidal-gear pump using new modules of GeroLAB. J. Mech. Des. **134**, 054502 (2012). https://doi.org/10.1115/1.4006440
32. Inaguma, Y.: A practical approach for analysis of leakage flow characteristics in hydraulic pumps. Proc. Inst. Mech. Eng. C **227**, 980–991 (2013). https://doi.org/10.1177/0954406212456933
33. Hsieh, C.F.: Flow characteristics of gerotor pumps with novel variable clearance designs. J. Fluid. Eng. **137**(4), 041107 (2015). https://doi.org/10.1115/1.4029274
34. Harrison, J., Aihara, R., Eisele, F.: Modeling gerotor oil pumps in 1D to predict performance with known operating clearances. SAE Int. J. Engines **9**, 1839–1846 (2016). https://doi.org/10.4271/2016-01-1081
35. Ivanović, L.T., Veličković, S.N., Stojanović, B.Ž, Kandeva, M., Jakimovska, K.: The selection of optimal gerotor pump parameters by applying factorial experimental design. FME Trans. **45**, 159–164 (2017)
36. Voloshina, A., Panchenko, A., Titova, O., Milaeva, I., Pastushenko, A.: Prediction of changes in the output characteristics of the planetary hydraulic motor. In: Tonkonogyi, V., et al. (eds.) InterPartner 2020. LNME, pp. 744–754. Springer, Cham (2021). https://doi.org/10.1007/978-3-030-68014-5_72
37. Robison, A., Vacca, A.: Multi-objective optimization of circular-toothed gerotors for kinematics and wear by genetic algorithm. Mech. Mach. Theory **128**, 150–168 (2018). https://doi.org/10.1016/j.mechmachtheory.2018.05.011
38. De Martin, A., Jacazio, G., Sorli, M.: Optimization of gerotor pumps with asymmetric profiles through evolutionary strategy algorithm. Machines **7**, 17 (2019). https://doi.org/10.3390/machines7010017

The Form of a Spiral Spring in a Free State

Serhii Pylypaka[1], Vyacheslav Hropost[1], Tetiana Kresan[2], Tatiana Volina[1,3](✉), and Volodymyr Vasyliuk[2]

[1] National University of Life and Environmental Sciences of Ukraine, 15, Heroyiv Oborony Str., Kyiv 03041, Ukraine
t.n.zaharova@ukr.net

[2] Separate Subdivision of the National University of Life and Environmental Sciences of Ukraine "Nizhyn Agrotechnical Institute", 10, Shevchenko Str., Nizhyn 16600, Ukraine

[3] Sumy National Agrarian University, Kondratieva Str., Sumy 16040021, Ukraine

Abstract. This article deals with a spiral spring made of tape material with a rectilinear cross-section. A classic example of the use of this type of spring is mechanical type clockwork and starter mechanisms of internal combustion engines with manual start. If one end of a metal ruler is pinched, and a specific force is applied to the other end, it will bend under the action of the generated moment. The magnitude of the moment depends on the applied force and the ruler's length. Therefore, the amount of deflection will increase with the increase of these parameters, and the shape of the ruler may take the form of a spiral. After the external load influence is terminated, the ruler will take its initial shape. That is, it will become straight. However, such spiral springs do not exist in practice. In its free state, the spring also has a spiral shape for compact sizes. This article uses the theory of elastic bending of rods for large deflections to determine this form. In addition, the initial curvature of its elastic axis in the free state must be considered for the spring. Simplifying linear bending formulas cannot be used to calculate the shape of the spring. The calculation of the shape of a spiral spring in a free state based on a given final shape with an applied moment is conducted in the article. The nonlinear bending theory is applied using the corresponding differential equations.

Keywords: Elastic Axis · Arc Length · Force Moment · Curvature · Differential Equations · Product Innovation

1 Introduction

One of the widespread elements of machines is a spring. The operation of complex mechanisms and structures is impossible without this simple element. The spring accumulates and absorbs energy due to elastic deformation under the influence of the external load. After removing the load, the spring releases the stored energy and returns to its original shape. Spring designs differ in their shape depending on the purpose. They work on stretching-compression, torsion, and bending. Their task also includes preventing mechanisms from the influence of excessive loads during shocks and vibrations.

V. Tonkonogyi et al. (Eds.): InterPartner 2023, LNME, pp. 509–517, 2024.
https://doi.org/10.1007/978-3-031-42778-7_47

Despite the apparent simplicity of the spring, its manufacturing technology must meet specific requirements for its functionality. First, it is necessary to find the shape of its elastic axis in a free state so that it is compact. But the main task is to avoid or reduce friction between adjacent turns when winding it to the final state. If such a shape of the elastic axis in the free state is found, then the metal strip is forcibly deformed under the found axis, for example, using a curvilinear slot in the mandrel. The heat treatment followed by tempering removes internal stresses so that after removal from the mandrel, the strip remains the same, i.e., with a specific initial curvature. Thus, the shape of the elastic axis of the spiral spring in the free state has great importance, which determines the urgency of the task of finding it.

2 Literature Review

Various methods can perform geometric modeling of various objects and processes. So, the paper [1] describes an approach to the solid modeling geometric objects in the form of an organized three-parameter set of points in three-dimensional space. The article [2] describes more detailed principles of solid modeling in point calculus, including defining geometric bodies as an organized set of points in space. Paper [3] generalizes a method for determining the curves – edges of the return of the torse surface, using the geometric properties of the point definition of the curve in the plane and spatial simplices. Examples of constructing geometric models of torse surfaces, for which algebraic and transcendental spatial curves were used as the return edge, are given. Besides, geometric modeling of bistable composite tape springs for aerospace applications is shown in [4]. Moreover, analytical studies of coupling between stress-strain and moment-curvature of thin curved shells have been used to estimate the total stored strain energy in different tape spring configurations in the article.

Increasing the durability and reliability of parts and connections has always been relevant because improvements and repairs are cheaper than manufacturing. Thus, in [5], a method of improving the quality of press joints is proposed, which consists of introducing specific intermediate layers between the contacting surfaces of parts. The results of a practical study of coatings applied using classical technology and an aluminum electrode are given in [6], showing higher thermal resistance than the last. The method of studying the defective structure of coatings is given in [7]. The research [8] presents a formalized methodology for solving the problem of creating fundamentally new materials with increased surface wear resistance and relatively high strength and viscosity. The prospect of applying plastic processing technology using a UV laser to produce typical elements is outlined in [9]. In addition, numerical modeling is successfully applied to the processes of heat exchange and hydrodynamics [10]. Mathematical and experimental modeling methods, heat, and mass transfer processes can be applied in heat accumulators with phase transformations, where heat sources are made in the form of staggered tube bundles, which have been studied [11]. The research carried out by the authors has enormous practical value, despite the high cost of their implementation. One should remember that in some cases, such scientific problems can be solved much easier and cheaper utilizing geometric methods. It should be noted that the modeling of curvilinear contours is widely used in many fields: aerodynamics [12], hydraulics [13,

14] and pneumatics [15], agriculture [16, 17], and chemical process technology [18], fixture design [19], etc.

Usually, in buildings, a simplified theory of linear bending [20] is used to determine the deflection of beams since the deflection of such building structures is minimal compared to their length.

A great deal is being written and said about springs and their construction features. The design and manufacturing of based composite helical springs for increasing automobile fuel efficiency and emissions gas regulation are proposed in [21]. The authors of [22] proposed multiscale shear failure mechanisms, which will benefit the optimization of the structural design to maintain the structural integrity of composite tape-spring in aerospace applications. According to it, such spring is stable in extended and coiled configurations.

Based on the foregoing, the article aims to solve the problem of finding the elastic axis of a tape spiral spring in a free state, which takes a given shape under the action of a concentrated force.

3 Research Methodology

Firstly, let's assume that the strip is completely elastic. That is, it completely recovers its shape after the force is stopped. In this regard, it is necessary to consider the bending of a cantilevered fixed straight strip under the action of a concentrated force P applied at its end. A metal ruler can serve as such a strip. The force of the weight of the strip is neglected since it is much smaller than the force P and its effect on the deformation of the strip will be insignificant. It will be even more true if the strip deflection, not in the vertical but in the horizontal plane, is considered.

It is a well-known fact of the theory of resistance of materials [1] that during elastic bending of the rod, the curvature k of its axis is directly proportional to the applied moment M and inversely proportional to the stiffness of the rod EI. In turn, stiffness is the product of Young's modulus E, which characterizes the property of the material, at the moment of inertia I of the cross-section of the rod, which characterizes its ability to resist bending:

$$k(s) = \frac{M(s)}{EI}, \tag{1}$$

where s is the length of the rod from its end to the current point.

Before applying the force, the strip is straight, and the curvature of its axis is equal to zero. After applying force, the strip bends, and its axis becomes curved.

It is common knowledge that curvature is the limit of the angle ratio between tangents at the ends of an arc element to the length of this element. So, formula (1) will be rewritten as follows:

$$\frac{d\alpha}{ds} = k(s) = \frac{M(s)}{EI}. \tag{2}$$

Equation (1) is the natural equation of the elastic axis of the strip. There is a transition from the natural equation to the coordinate form of the record:

$$x = \int cos\,\alpha\,(s)ds;\; y = \int sin\,\alpha\,(s)ds, \tag{3}$$

where the angle α can be found through the integration of expression (2).

4 Results and Discussion

It is advisable to consider a cantilevered fixed strip with a concentrated force P applied to its end. Figure 1a shows the diagram of the application of force P before the start of its action. The cross-section of the strip has a rectangular shape with dimensions a and b (see Fig. 1, b). The applied force P causes a moment M = Ps, where the current value s of the length of the strip is counted from the end of the strip in the direction of its pinching.

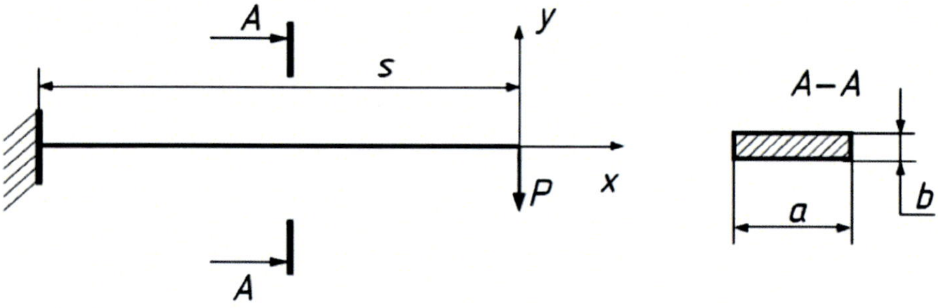

a b

Fig. 1. Graphic illustrations of the bending of a cantilevered fixed strip by a concentrated force *P*: a) scheme of application of force *P* at the end of the strip; b) designation of the dimensions of the cross-section of the strip.

Under the action of the force P, the moment M = Ps is forced, the strip is deformed, and its elastic axis acquires a curvilinear shape. The angle α between the tangents can estimate the curvature of the axis to it at the beginning and the current point M (Fig. 2, a), taking into account the current value s of its arc length. Taking into account that M = Ps, the value of the angle α can be determined by integrating the expression (2):

$$\alpha = \int \frac{Ps}{EI} ds = \frac{Ps^2}{2EI}. \tag{4}$$

Let's take a metal strip with a cross-section of a = 3 sm, b = 0.3 sm, and a length of 1 m. The material of the strip was taken with Young's modulus $E = 2{,}1 \cdot 10^{11}$ Pa. The moment of inertia of a rectangular cross-section was found by the well-known formula $I = ab^3/12$ in m^4. The stiffness of the strip is a constant value found after the transition of the cross-sectional dimensions from sm to m: $EI = 14{,}175\ N \cdot m^2$. The constant tracking force P remains perpendicular to the elastic axis of the strip during its bending. The value of the angle α at the current point M of the elastic axis for any value of the arc s can be found by the formula (4). For example, when P = 40 N and s = 0.5 m, the angle is α = 20.2, and when s = 1 m, the angle is α = 80.8 (see Fig. 2a). At constant values of force P and stiffness EI, the angle α according to (4) grows directly proportional to the square of the length of the elastic axis. The elastic axis of the strip after applying a force P = 800 N to it is shown in Fig. 2, b. At the same time, the angle is α = 1616.2, i.e., the end of the strip has made 4.5 revolutions.

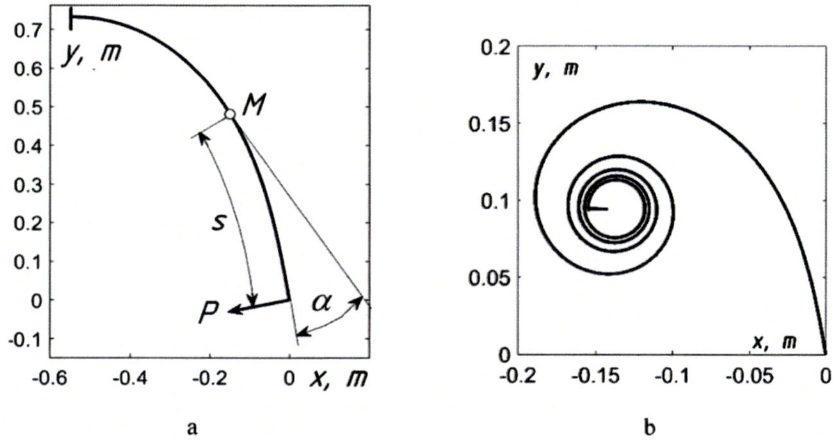

Fig. 2. The shape of the elastic axis of the cantilevered fixed strip due to the action of the concentrated force *P*: a) *P* = 40 N; б) *P* = 800 N.

From Fig. 2, b, it can be seen that a straight strip 1 m long has acquired the shape of a spiral, which fits into a square with a side 20 sm. However, the density of turns of such a spiral is very uneven: the distance between them decreases rapidly as it approaches the axis around which the twisting occurs.

Let's take a strip in the form of a spiral with an equal density of turns. Archimedes' spiral meets this condition but does not have the natural equation k = k(s). There is another curve that can satisfy this condition – the involute of a circle of radius a. Its natural equation has the form $k = \frac{1}{\sqrt{2as}}$, where the density of adjacency of turns to each other depends on the constant a. However, according to this equation, as the arc length s increases, the point moves in a spiral away from the twisting axis. Still, it is necessary to receive the opposite situation. For this, the following equation should be written:

$$k = \frac{1}{\sqrt{2a(s_0 - s)}}, \tag{5}$$

where s_o is the entire length of the spiral.

By integrating the expression (5), the dependence α = α(s) can be found. After substituting the found dependence in (3), the parametrical equations of the spiral will be obtained. It is built in Fig. 3a, at a = 0.0001 m, s_o = 1 m.

With the accepted value a = 0.0001 m, i.e., a = 0.1 mm, the distance between adjacent turns is 0.6 mm. Let's take a metal strip for making a spring 1 cm wide and 0.3 mm thick (a = 0.01 m, b = 0.003 m) if it takes the form of an involute under the action of the applied force P (that is, in the established state) (Fig. 3, a). Then there will be a gap equal to 0.3 mm between the tape turns.

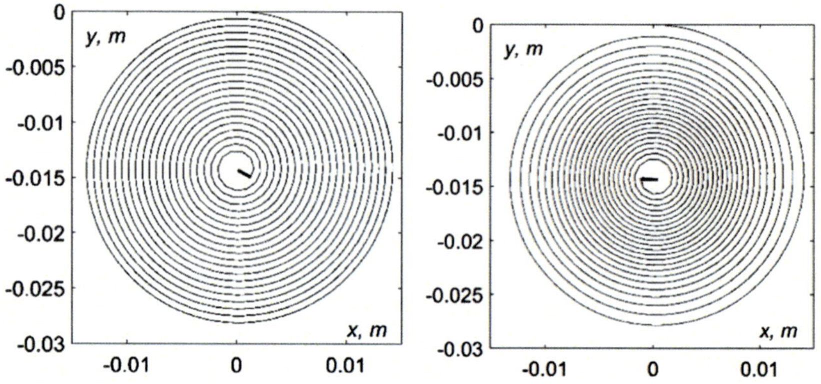

Fig. 3. The shape of the elastic axis of the spiral spring: a) in the free state, the involute of the circle is the elastic axis; b) the elastic axis of the spring after applying a force $P = 0.15$ N.

First, let's consider what shape the elastic axis of the spring will take if its curvature in the free state has the shape of the involute of a circle (Fig. 3, a). Due to the application of the moment $M = Ps$, its curvature will increase. According to (2):

$$\frac{d\alpha}{ds} = \frac{1}{\sqrt{2a(s_0 - s)}} + \frac{Ps}{EI}. \tag{6}$$

Integration of expression (6) gives the dependence $\alpha = \alpha(s)$:

$$\alpha = \int \left[\frac{1}{\sqrt{2a(s_0 - s)}} + \frac{Ps}{EI} \right] ds = -\sqrt{\frac{2(s_0 - s)}{a}} + \frac{Ps^2}{2EI}. \tag{7}$$

Substitution of the dependence $\alpha = \alpha(s)$ (7) in (3) gives the parametrical equations of the elastic axis of the spring after applying moment $M = Ps$ to the spring, the shape of which in the free state is the involute of a circle (see Fig. 3, a). Numerical integration of the equations allows obtaining the shape of the elastic axis of the spring after the applied force $P = 0.15$ N (see Fig. 3, b). It is clear from Fig. 3 that the density of the winding's contact with each other increases as the length s of the spring increases, that is, as it approaches the center of twisting.

From dependence (7), the value of the angle $\alpha = 141$ rad in the free state (at $P = 0$) and $\alpha = 156$ rad after the applied force $P = 0.15$ N can be derived. The difference is 15 rad, which corresponds to 2.4 full turns of the spring winding.

We then review the inverse problem: to obtain the shape of the elastic axis of the spring in the free state, which after the applied force P will take the form of an involute (see Fig. 3, a). For this particular case, it is advisable to apply a moment with the opposite sign to the elastic axis in the form of an involute of a circle, that is, to "untwist" the spring, which in its free state has the shape of an involute of a circle. It means taking the sign "–" before the second term in expressions (6) and (7). Having repeated all the actions, as in the first case, one can find the shape of the curve (Fig. 4, a). It shows that the density of the turn contact increases in the opposite direction.

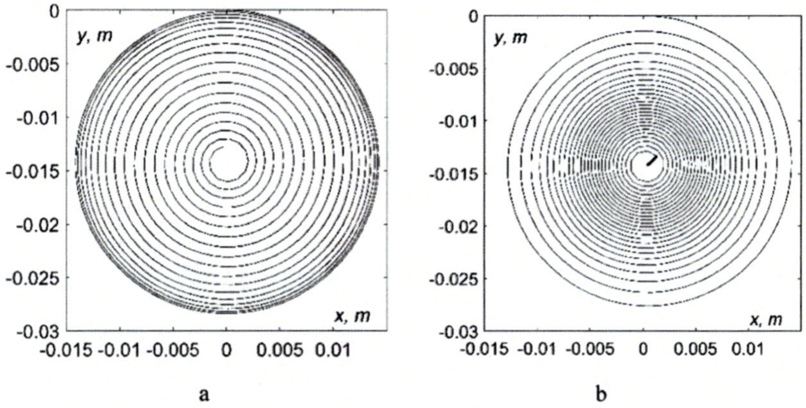

Fig. 4. The shape of the elastic axis of the spiral spring: a) in a free state; b) after applying force $P = 0.3$ N.

The angle α for the spring in the free state (Fig. 4, a) according to (7), taking into account the "–" sign is $\alpha = 126$ rad, and after applied force $P = 0.15$ N, the angle value will increase to the value $\alpha = 141$ rad (this concerns the involute of a circle and is found earlier). Thus, the difference is also 15 rad. If the elastic axis of the spring in the free state has the shape shown in Fig. 4a, then after the applied force $P = 0.15$ N, it will take the form of an involute (Fig. 3, a). If a force $P = 0.15$ N is applied to it, it will take the form shown in Fig. 3, b. Generally, applying a force of $P = 0.3$ N to the spring in its free state (Fig. 4, a), we will get the "twisted spring" (Fig. 3, b). If the force $P = 0.3$ N is applied to the spring in its free state (Fig. 3, a), then the "twisted spring" shown in Fig. 4, b will be received. In such a case, the density of turns near the spring twist center may be such that there will be contact between adjacent turns, which will cause friction during the spring operation. For both cases, the twist angle will be the same and equal to 30 rad.

5 Conclusions

For manufacturing spiral tape springs, it is necessary to have the shape of an elastic axis in a free state. It should provide the absence of friction between adjacent turns while twisting the spring. If the elastic axis of the spring in the free state is constructed in the form of a spiral with the same density of turns, then when it is twisted, the same density is disturbed. The distance between adjacent turns decreases in the direction of the spring twist center, which can lead to friction between adjacent turns. To avoid this, it is necessary to form the elastic axis of the spring in a free state in the form of a spiral, in which the distance between adjacent turns in the direction of the spring twist center should increase. The calculation of the shape of the elastic axis of the spiral tape spring depends on the dimensions of the cross-section of the tape, Young's modulus of the material, and the applied force that creates the moment. The calculation is based on the theory of non-linear bending resistance of materials.

References

1. Konopatskiy, E., Bezditnyi, A.: Solid modeling of geometric objects in point calculus. In: CEUR Workshop Proceedings, vol. 3027, pp. 666–672 (2021). https://doi.org/10.20948/graphicon-2021-3027-666-672
2. Konopatskiy, E.V., Bezditnyi, A.A., Lagunova, M.V., Naidysh, A.V.: Principles of solid modelling in point calculus. J. Phys.: Conf. Ser. **1901**, 012063 (2021). https://doi.org/10.1088/1742-6596/1901/1/012063
3. Konopatskiy, E., Bezditnyi, A., Litvinov, A.: Geometric modeling of torse surfaces in BN-calculus. J. Phys.: Conf. Ser. **1791**, 012050 (2021). https://doi.org/10.1088/1742-6596/1791/1/012050
4. Shahryarifard, M., Golzar, M., Tibert, G.: Viscoelastic bistable tape spring behavior modeled by a two-dimensional system with rigid links, springs and dashpots. Compos. Struct. **258**, 113181 (2021). https://doi.org/10.1016/j.compstruct.2020.113181
5. Martsynkovskyy, V., Tarelnyk, V., Konoplianchenko, I., Gaponova, O., Dumanchuk, M.: Technology support for protecting contacting surfaces of half-coupling – shaft press joints against fretting wear. In: Ivanov, V., et al. (eds.) DSMIE 2019. LNME, pp. 216–225. Springer, Cham (2020). https://doi.org/10.1007/978-3-030-22365-6_22
6. Gaponova, O., et al.: Estimating qualitative parameters of aluminized coating obtained by electric spark alloying method. In: Pogrebnjak, A.D., Novosad, V. (eds.) Advances in Thin Films, Nanostructured Materials, and Coatings. LNME, pp. 249–266. Springer, Singapore (2019). https://doi.org/10.1007/978-981-13-6133-3_25
7. Pogrebnjak, A.D., et al.: Positron annihilation studies of defect structure of (TiZrHfNbV)N nitride coatings under Xe14+ 200 MeV ion irradiation. Mater. Lett. **303**, 130548 (2021). https://doi.org/10.1016/j.matlet.2021.130548
8. Tarelnyk, V., Konoplianchenko, I., Tarelnyk, N., Kozachenko, A.: Modeling technological parameters for producing combined electrospark deposition coatings. Mater. Sci. Forum **968**, 131–142 (2019). https://doi.org/10.4028/www.scientific.net/MSF.968.131
9. Antoszewski, B., Tofil, S., Scendo, M., Tarelnik, W.: Utilization of the UV laser with picosecond pulses for the formation of surface microstructures on elastomeric plastics. In: IOP Conference Series: Materials Science and Engineering, vol. 233, p. 012036 (2017). https://doi.org/10.1088/1757-899X/233/1/012036
10. Gorobets, V., Trokhaniak, V., Bohdan, Yu., Antypov, Ie.: Numerical modeling of heat transfer and hydrodynamics in compact shifted arrangement small diameter tube bundles. J. Appl. Comput. Mech. **7**(1), 292–301 (2021). https://doi.org/10.22055/JACM.2020.31007.1855
11. Antypov, Ie., Gorobets, V., Trokhaniak, V.: Experimental and numerical investigation of heat and mass transfer processes for determining the optimal design of an accumulator with phase transformations. J. Appl. Comput. Mech. **7**(2), 611–620 (2021). https://doi.org/10.22055/JACM.2020.34893.2524
12. Gorobets, V., Trokhaniak, V., Masiuk, M., Spodyniuk, N., Blesnyuk, O., Marchishina, Ye.: CFD modeling of aerodynamic flow in a wind turbine with vertical rotational axis and wind flow concentrator. INMATEH – Agric. Eng. **64**(2), 159–166 (2021). https://doi.org/10.35633/inmateh-64-15
13. Fesenko, A., Yevsiukova, F., Basova, Y., Ivanova, M., Ivanov, V.: Prospects of using hydrodynamic cavitation for enhancement of efficiency of fluid working medium preparation technologies. Periodica Polytech. Mech. Eng. **62**(4), 269–276 (2018). https://doi.org/10.3311/PPme.11877
14. Gorobets, V., Trokhaniak, V., Antypov, I., Serdiuk, A.: Investigation of preparation processes of liquid feed mixtures in rotary pulsating apparatus. In: Blikharskyy, Z. (ed.) EcoComfort 2020. LNCE, vol. 100, pp. 118–126. Springer, Cham (2021). https://doi.org/10.1007/978-3-030-57340-9_15

15. Merzliakov, I., Pavlenko, I., Chekh, O., Sharapov, S., Ivanov, V.: Mathematical modeling of operating process and technological features for designing the vortex type liquid-vapor jet apparatus. In: Ivanov, V., et al. (eds.) DSMIE 2019. LNCE, pp. 613–622. Springer, Cham (2020). https://doi.org/10.1007/978-3-030-22365-6_61
16. Spirin, A., Borysiuk, D., Tsurkan, O., Tverdokhlib, I., Veselovska, N., Edl, M.: Ways of intensification of grass seed production. J. Eng. Sci. (Ukr.) **10**(1), F11–F19 (2023). https://doi.org/10.21272/jes.2023.10(1).f3
17. Rogovskii, I.L., Titova, L.L., Trokhaniak, V.I., Borak, K.V., Lavrinenko, O.T., Bannyi, O.O.: Research on a grain cultiseeder for subsoil-broadcast sowing. INMATEH – Agric. Eng. **63**(1), 385–396 (2021)
18. Ochowiak, M., Wlodarczak, S., Pavlenko, I., Janecki, D., Krupinska, A., Markowska, M.: Study on interfacial surface in modified spray tower. Processes **7**(8), 532 (2019). https://doi.org/10.3390/pr7080532
19. Karpus, V.E., Ivanov, V.A.: Choice of optimal configuration of modular reusable fixtures. Eng. Res. **32**(3), 213–219 (2012). https://doi.org/10.3103/S1068798X12030124
20. Voropay, A.V., Menshykov, O.V., Povaliaiev, S.I., Sharapata, A.S., Yehorov, P.A.: Modeling a viscoelastic support considering its mass-inertial characteristics during non-stationary vibrations of the beam. J. Eng. Sci. (Ukr.) **10**(1), D8–D14 (2023). https://doi.org/10.21272/jes.2023.10(1).d2
21. Apsar, G., Musthak, M., Ahmed, J.: Study of factors affecting tape-wound composite helical spring prepared by E-glass/epoxy by using taguchi method and statistical distributions. Int. J. Compos. Mater. **10**(1), 10–17 (2020). https://doi.org/10.5923/j.cmaterials.20201001.02
22. Wang, B., et al.: In-situ multiscale shear failure of a bistable composite tape-spring. Compos. Sci. Technol. **200**, 108348 (2020). https://doi.org/10.1016/j.compscitech.2020.108348

Improved Methods for Diagnosing an Autotransformer with a Defect in a High-Voltage Bushing

Sergey Zaitsev, Victor Kishnevsky, Gennadii Oborskyi, Valentin Tikhenko, and Aleksandr Volkov(✉)

Odessa National Polytechnic University, 1, Shevchenko Ave., Odessa 65044, Ukraine
sdjavdet@ukr.net, aleksandr.tuff@gmail.com

Abstract. The article is devoted to the improvement of methods for diagnosing an oil-filled power autotransformer of the ATDCTN-200000/330/110/10 type with an internal defect in an oil-filled high-voltage bushing of the GMTPA-45-330/1000U1 type. Purpose: improving the reliability of the results of diagnosing an oil-filled power autotransformer with an internal defect in its oil-filled high-voltage bushing by improving the methods of diagnosing, taking into account the results of electrical tests of this equipment and analyzing samples of mineral oil from them. Studies of the electrical characteristics of an oil-filled power autotransformer and an oil-filled high-voltage bushing and mineral oils' physicochemical, thermophysical, and electrical properties have been conducted. Numerical values of indicators of electrical characteristics for an oil-filled power autotransformer and its high-voltage input are determined, and for mineral transformer oils, numerical values of indicators are determined - breakdown voltage; flash point of oil vapor in an open crucible; acid number; water-soluble acids; dielectric loss tangent; density; moisture contents; the content of dissolved diagnostic gases, additives "Ionol", furan compounds. The analysis of the studies makes it possible to increase the reliability of the results of technical diagnostics of a power electric oil autotransformer with an internal defect in a high-voltage oil-filled bushing.

Keywords: Mineral Transformer Oil · Electrical Characteristics · Gas Chromatography · Dissolved Gases · Water Content · Acid Number · Breakdown Voltage · Process Innovation

1 Introduction

The reliability of modern systems for producing, distributing, and consuming electricity largely depends on the reliability of high-voltage oil-filled electrical equipment (OFEE). Failures of OFEE with oil-paper electrical insulation (OPEI) lead to economic damage in power systems. Accidents in OFEE associated with electrical breakdown of capacitor electrical insulation of high-voltage bushings are the most frequent and economically costly. Considering its current state, the requirements for the maintenance of OFEE determine the need to improve the technical control systems and diagnostics of such

V. Tonkonogyi et al. (Eds.): InterPartner 2023, LNME, pp. 518–527, 2024.
https://doi.org/10.1007/978-3-031-42778-7_48

equipment [1]. Determining the technical state of OFEE with OPEI is necessary for making decisions related to the operation, maintenance, and repair of OFEE; engineering and economic modeling in determining the most cost-effective alternative to restore or maintain OFEE [2]. The main diagnostic parameters for monitoring the technical condition of an oil-filled power autotransformer (OFPA) or an oil-filled high-voltage bushing (OFHVB) with OPEI are [3]: electrical resistance and dielectric loss tangent of electrical insulation, electrical capacitance of electrical windings and electrical insulation, absorption coefficient [4]; physicochemical, thermophysical, electrophysical (PCTE) properties of mineral transformer oil (MTO) [5]; the content of dissolved gases [6] H_2, CH_4, C_2H_2, C_2H_4, C_2H_6, CO_2, CO, O_2, N_2, air in MTO [7]. Thus, improving methods for monitoring the technical condition and diagnostics of OFEE with OPEI is relevant.

The work aims to increase the reliability of the results of diagnosing an oil-filled power autotransformer with an internal defect in its oil-filled high-voltage bushing by improving the methods of diagnosing, taking into account the results of electrical tests of these types of equipment, also it analyzes samples of mineral oil from them.

2 Literature Review

The main problems and tasks of ensuring the uniformity of measurements when using modern methods and means of monitoring the PCTE properties of mineral power oils and their degradation products for diagnosing OFEE to ensure its reliability are considered in [8, 9]. When studying the electrophysical properties of MTO, the indicator "electrical breakdown voltage (breakdown voltage)" is determined [10, 11]. Chromatography methods, including gas chromatography (GCh), determine the content of such dissolved components in MTO as gases [12, 13], "Ionol" additive [14, 15], furan compounds (FC) of types 2-acetylfuran (2ACF), 2-furfural (2FAL), 2-furyl alcohol (2FOL), 5-methyl-2-furfural (5MEF) [16, 17] and [18]. The diagnosis of OFPA or OFHVB with OPEI based on the results of the analysis of MTO samples is performed using the appropriate diagnostic models when determining the contents of dissolved gases in MTO [19, 20], the "Ionol" additive [5, 21], H_2O [15], FC [22]. Diagnostic models based on the results of the determination of the content of dissolved gases in MTO samples make it possible to determine in OFPA or OFHVB with OPEI: defects of a thermal and/or electrical nature (for example, partial electrical discharges (PED); presence of electrically conductive X-wax in MTO, leading to progressive MTO destruction [4, 21]); oxidative aging, pyrolysis and carbonization of MTO; formation of carbon particles in MTO. In this case, the following are used: the results of determining the values of the current concentrations of dissolved gases in the MTO and the rates of their accumulation – the methods of "boundary levels of concentrations", "ratios of concentrations of characteristic gases" (including the "tabular method", "the method of graphic images of defects"), "the method determining the rate of change in the concentrations of dissolved gases in MTO" [20, 22]. The mechanism of OPEI degradation in OFPA under the influence of various factors, including O_2 and H_2O impurities during MTO aging, was considered in [23, 24]. When PED impulses occur in the local volumes of the MTO: a) shock waves are formed with a propagation velocity of up to $2 \cdot 10^5$ m/s and a pressure at the shock wave front of up to $2 \cdot 10^4$ MPa; electric current pulses with a duration of 1–4000 ns in the frequency range of 0.1–40 MHz; light emission; gas bubbles; acoustic effects; b) the temperature rises; plasma-chemical reactions and degradation of MTO and impurities in it – organic acids, alcohols, aldehydes, ketones, FCs occur [4, 8]. It is known that: a) the excess of CO and CO_2 concentrations above the limit values indicate oxidative processes in the MTO volume when OPEI and OFHVB are overheated; b) in case of unsatisfactory OFHVB electrical test results, it is recommended to determine the FC content in the MTO: if their total content is more than 1 mg/kg, then a defect is assumed – overheating of the OPEI in OFHVB [19, 22]; c) under the influence of electric and thermal fields on OPEI, FCs accumulate in the MTO, and the values of their concentrations are used to diagnose OPEI. In this case, the concentrations of FCs of the 2FAL type are mainly considered [22]. In [25], the results of laboratory studies of the effect of electric voltage on the degradation of MTO and OPEI with the formation of dissolved gases and FC in the volume of MTO are presented. The effect of electric voltage on the degradation of these FCs in the MTO volume has not been studied. Thus, the above information indicates the incompleteness of information about the impact of defects in OFHVB with OPEI, which has internal defects of an electrical and/or thermal nature and is in contact with the MTO with its outer surface, on the reliability of diagnosing defects associated with internal structural elements of the OFPA itself with OPEI; electric discharges on the degradation of FC, formed during the destruction of OPEI under the influence of electric and thermal fields in the MTO environment in OFPA or OFHVB. At the same time, there is no information on the effect of a high-voltage pulsed electric current on the degradation of FC in MTO in the presence of dissolved H_2, which is formed during the degradation of MTO. All this requires the implementation of relevant research.

3 Research Methodology

Objects of research: methods for diagnosing OFPA and its OFHVB based on the results of physicochemical, thermophysical, electrophysical, and GCh studies of MTO samples from this OFEE. Research subjects: OFPA type ATDCTN-200000/330/110/10; OFHVB types GMTPA-45-330/1000 U1, GMTA-90-110/2000 U1; operational and fresh MTO grades GK; components dissolved in MTO: gases (H_2, CH_4, C_2H_2, C_2H_4, C_2H_6, CO_2, CO, O_2, N_2, TAC (total air content), TGC (total gas content)); "Ionol" additive; furan compounds 2FOL, 2FAL, 2ACF, 5MEF and the sum of their concentrations ΣC_F; H_2O in MTO. Investigated electrical characteristics: a) for OFPA: R_{60} and R_{15} – electrical resistance of electrical insulation; absorption coefficient K_a; dielectric loss tangent of electrical windings tgδ; electric capacitance C of electric windings [3]; b) for OFHVB: dielectric loss tangent of the main electrical insulation $\text{tg}\delta_1$; electric capacitance C_1 of the main electrical insulation; electrical resistance R_i of a special output; dielectric loss tangent of the electrical insulation of the measuring capacitor $\text{tg}\delta_2$; electrical capacitance C_2 of the electrical insulation of the measuring capacitor; electrical resistance R_2 of the measuring output [3]. The gas chromatograph was used to determine the following contents in the MTO: i-th gases C_g, TAC, TGC [7, 26]; additives "Ionol" C_i [5]; furan compounds C_F and the sum of their concentrations ΣC_F. The studied characteristics of MTO samples from OFPA and OFHVB: electrical breakdown voltage (breakdown voltage) U, a flash point of MTO vapors in a closed crucible t_f, acid number AN, water-soluble acids WSA, dielectric loss tangent $\text{tg}\delta_o$, density ρ, water content W [5, 21];

the content of C_i additives "Ionol" [21]; content of C_g gases, TGC, TAC [7, 26]; the content of furan compounds C_f and the sum of their concentrations ΣC_f. The effect of a high-voltage pulsed electric current on the degradation of the furan compound 2FAL in MTO in the presence of dissolved H_2 was studied: initial content $C_F = 30$ mg/kg for 2FAL; saturation of MTO with gaseous hydrogen up to a concentration of 5% volume was performed at a temperature of 20 °C; the measurements were performed according to the procedure with the number of electrical breakdowns $N = 144$ in sealed test cell, the duration of the each electrical breakdown was no more than 0.02 s, and the time interval between each electrical breakdown was 5 min, sampling of MTOs to determine the content of the furan compound 2FAL was performed sequentially after every $N = 12$ electrical breakdowns. OFPA and OFHVB were diagnosed based on the results: studies of electrical characteristics for OFPA and OFHVB [1, 3]; PCTE studies of the properties of MTO samples from OFPA and OFHVB [5, 21], GCh studies of MTO samples from OFPA and OFHVB [7, 26]. To apply the GCh methods to diagnose OFPA and OFHVB, the following values were used: concentrations of C_g gases, TGC, TAC (method of "boundary concentration levels"); CH_4, C_2H_4, C_2H_2 ("Duval's triangle method"); ratios of gas concentrations C_2H_2/C_2H_4, CH_4/H_2, C_2H_4/C_2H_6 ("tabular method"); H_2/H_2, C_2H_6/H_2, C_2H_4/H_2, C_2H_2/H_2 ("method of graphic images of defects" – with the main dissolved in MTO gas - H_2); C_2H_2/C_2H_6, C_2H_4/C_2H_6 ("ETRA method"); CO_2/CO, O_2/N_2; water concentrations W; concentrations of C_i additive "Ionol"; concentrations of C_F and ΣC_F for FC [5, 19].

4 Results and Discussion

The research OFPA (MTO weight 67000 kg) was operated in 1992. After 214 months of operation, OFPA electrical tests were performed to determine the values of the main indicators R_{60}, R_{15}, K_a, tgδ, C, following the requirements of regulatory documents [3]. The results of electrical tests showed that OFPA complies with the established standards [3]. At the same time, the results of GCh analyses of MTO samples from OFPA indicate the presence of an internal defect in it. The PCTE studies of the properties of MTO samples from OFPA showed that: PCTE properties of MTO samples correspond to the standards [21]; the value of H_2O concentration in MTO increased from $W = 12.3$ g/t to $W = 15.4$ g/t (corresponds to the standards); the sum of ΣC_F for FC in MTO does not exceed the detection thresholds of 0.2 mg/kg [28], which indicates the absence of OPEI degradation in OFPA [22]; the content of the "Ionol" additive decreased from the initial value $C_i = 0.35$ wt % to $C_i = 0.31$ wt % and corresponds to the norm [21]. At the same time, electrical tests were carried out on OFHVB types GMTPA-45-330/1000U1 (phase "A"), GMTPA-45-330/1000U1 (phase "C"), GMTA-90-110/2000U1 (phase "A"), GMTA-90-110/2000U1 (phase "B"), GMTA-90-110/2000U1 (phase "C") to determine the values of indicators $tg\delta_1$, $tg\delta_2$, R_i, R_2, C_I, C_2 [3]. Tests have shown that these OFHVB comply with the regulations [3]. Electrical tests OFHVB type GMTPA-45-330/1000U1 (phase "B") when determining the values of indicators $tg\delta_1$, $tg\delta_2$, R_i, R_2, C_I, C_2, showed that: indicators $tg\delta_1 = 1.546\%$ (norm $tg\delta_1 = 0.15$–1.0%) and $tg\delta_2 = 5.008\%$ (norm $tg\delta_2 < 1\%$) do not meet the standards [3]; OFHVB is faulty and damaged [3]. Table 1 presents the results of PCTE studies of the properties of MTO samples from damaged OFHVB.

Table 1. Results of physicochemical, thermophysical, and electrophysical studies of the properties of MTO samples from damaged OFHVB type GMTPA-45-330/1000U1 (phase "B").

$\Delta\tau_1$,	ρ, g/cm^3	AN, mg KOH on 1g MTO	t_f, °C	W, g/t	U, kV	$tg\delta_o$, %, at 70 °C/ 90 °C	C_i, % mass
Norm	–	< 0.1	>135	<20	>55	<2.5/5.0	>0.1
120*	0.853	0.005	–	12.3	–	2.0/–	0.34
95	0.852	abs	138	15.4	52	2.96/5.1**	0.22

In Table 1: Norm – normalized value; * – running time after OFHVB commissioning in OFPA, month; $\Delta\tau1$ is the operating time after the previous analysis of the MTO sample, month; abs. – the value is less than its determination threshold; ** – the value does not meet the standards.

The results of PCTE studies of the properties of MTO samples from damaged OFHVB showed (Table 1) that: the index $tg\delta_o$ (70 °C/90 °C) = 2.96/5.1 does not meet the standards; the content of H_2O increased from $W = 11.2$ g/t to $W = 15.4$ g/t (corresponds to the standards); indicator $U = 52$ kV approaches its boundary value; the content of the "Ionol" additive decreased from $C_i = 0.34$ wt % to $C_i = 0.22$ wt % and corresponds to the norm [21]. Based on the results of GCh analyses of MTO samples from OFPA, it was found that: a) the concentrations of gases dissolved in MTO are H_2 = 588 ppm (above the norm 100 ppm), CH_4 = 26.4 ppm, C_2H_2 = 0.3 ppm, C_2H_4 = 27.5 ppm (above the norm 15 ppm), C_2H_6 = 7.1 ppm, CO_2 = 1711 ppm (above the norm 1500 ppm), CO = 107 ppm, TGC = 31800 ppm (above the norm 20000 ppm), CO_2/CO = 16 (out of the norm 3–10). At the same time: a) depending on the duration of operation of the OFPA, the concentrations of H_2, CO_2, and CO gases dissolved in the MTO are constantly increasing, and the concentration of the dissolved gas C_2H_4 in the MTO remains practically constant; for OFPA, technical condition level 2 [19]; b) after replacing the damaged OFHVB with a functional OFHVB of a similar type, and after degassing the MTO in the OFPA, putting the OFHVB into operation, the concentrations of dissolved gases in the MTO in the OFPA after 3 months of its operation have the values H_2 = 28 ppm, CH_4 = 3.6 ppm, C_2H_2 = 0 ppm, C_2H_4 = 8.9 ppm, C_2H_6 = 2.3 ppm, CO_2 = 660 ppm, CO = 27 ppm, TGC = 14200 ppm, and correspond to the standards [19]. This points to the absence of internal defects in OFPA and to the influence of the damaged OFHVB on the concentrations of H_2, C_2H_4, and CO_2 gases dissolved in the MTO before its replacement with a functional OFHVB. Table 2 presents the results of GCh analyses of MTO samples from damaged OFHVB.

It follows from Table 2 that, depending on the duration of OFHVB operation, the concentrations of H_2, C_2H_2, C_2H_4, and C_2H_6 gases dissolved in the MTO practically do not change until defects are detected in this OFHVB based on electrical tests. Before removing the damaged OFHVB from an operation, the concentrations of dissolved gases H_2, CH_4, C_2H_6, and TGC in the MTO, and the CO_2/CO ratio, increased, exceeding the limit values.

Table 2. Results of GCh analyses of MTO samples from OFPA.

$\Delta\tau_2$,	Concentration, C_f, ppm								CO_2/CO
	H_2	CH_4	C_2H_2	C_2H_4	C_2H_6	CO_2	CO	TGC	
BL	600	140	1	30	60	3000	800	60000	3–10
120*	160	35	0	1.2	14	1840	250	17100	7.4
34	147	48.4	0	5.9	11	1107	127	25900	8.7
55	43887**	14560**	0	5.9	1536**	982	8.0	85200**	122.8**

In Table 2: BL – boundary level; * – running time after OFHVB commissioning in OFPA, month; $\Delta\tau_2$ is the operating time after the previous analysis of the MTO sample, month; ** – values do not meet the standards [19].

At the same time, the concentrations of dissolved gases CO_2, CO in the MTO constantly decreased and were within the limit values [19] until defects were detected in this bushing based on electrical tests.

Due to the high content of dissolved H_2 in MTO, it is assumed that in the process of the occurrence of a "PED" type defect in OFHVB, it is possible: the occurrence of chemical reactions of hydrogenation of CO and CO_2 with a decrease in their concentrations and the formation of CH_4 and H_2O in MTO [4]; the addition of H_2 at double and triple carbon bonds in unsaturated hydrocarbon molecules or the dissociation of MTO hydrocarbons at C-C bonds [4].

Table 3 presents the results of GCh determination of TAC and FC content in MTO samples from damaged OFHVB.

Table 3. Results of GCh determination of TAC and FC contents in MTO samples from damaged OFHVB.

$\Delta\tau_3$,	Component							
	O_2, ppm	N_2, ppm	TAC, ppm	2FOL, mg/kg	2ACF, mg/kg	5MEF, mg/kg	2FAL, mg/kg	ΣC_F, mg/kg
BL	<10000	–	<60000	1	1	1	1	1
120*	300	32000	32300	abs.	abs.	abs.	abs.	abs.
89	250	24000	24250	0.3	0.3	0.4	1.8**	2.8**

In Table 3: BL – boundary level; * – running time after OFHVB commissioning in OFPA, month; $\Delta\tau_3$ is the operating time after the previous analysis of the MTO sample, month; abs. – values are less than the threshold for determining the measured indicator; ** – values do not correspond to the norms [22].

The results of determining the contents of O_2, N_2, TAC, and FC in MTO samples from OFHVB (Table 3) show that: OFHVB was leak-tight during operation; the concentration of 2FAL and the sum of concentrations ΣC_F for FC exceeded the norm. These points to the destruction of OPEI in OFHVB [22].

When a suspected defect is detected in OFPA following [19, 20]: a) "Tabular method" – a defect of the "Low energy electric discharges" type is assumed, and the temperature in the defect zone is less than 150 °C; b) by the "Graphic image" method – a defect of the "PED" type is assumed; c) "ETRA method" – defects of the type "Heating more than 700 °C" or "Heating and PED" are assumed; d) "Duval's triangle method" – a defect of the type "Heating more than 700 °C" is assumed. Since, according to the results of studies of the electrical characteristics of the OFPA, it was recognized as serviceable and operable, the presence of electrical and thermal defects inside this OFPA may be associated with internal defects in its OFHVB. When using the diagnostic methods "ETRA" and "Duval's Triangle", OFPA established the alleged type of defect "Heating over 700 °C".

This type of defect should be accompanied by the formation of heterogeneous carbon particles in the MTO and dissolved C_2H_2 in the MTO. However, no heterogeneous carbon particles were found in MTO, and the content of dissolved C_2H_2 in MTO in OFPA (Table 2) ranges from 0.3 ppm to 0.7 ppm, which corresponds to technical condition level 2 (defect is assumed) [19]. At the same time, the concentrations of individual FCs and the sum ΣC_F of their concentrations in MTO does not exceed the determination thresholds of 0.2 mg/kg in MTO. It indicates that: OPEI in OFPA has not undergone degradation [19, 22]; the use of diagnostic methods "ETRA" and "Duval's Triangle" requires clarification of the boundary conditions of its application.

When a suspected defect is detected in OFHVB (data from Table 2, Table 3) under [19, 20]: a) "Tabular method" – a defect of the "Low-energy electric discharges" type is assumed, the temperature in the defect zone is less than 150 °C, at $O_2/N_2 = 0.01$, high-intensity oxidative processes occur in the MTO; b) by the "Graphic image" method – a defect of the "PED" type is assumed; c) "ETRA method" – defects of the type "Heating up to 300 °C" or "Heating and PED" are assumed; D) "Duval's triangle method" – a defect of the "PED" type is assumed. Based on the research results, it is recommended to perform the determination of FC contents in MTO samples from OFHVB with OPEI in cases where the results do not meet the standards for: OFHVB electrical tests [3]; PCTE studies of the properties of MTO samples from OFHVB [21]; GCh of analyzes of MTO samples for at least one dissolved gas [19].

When examining the outer surface and internal structural elements of the damaged OFHVB after opening the OFPA, defects were found: electrical and thermal on the outer surface of the OFHVB in contact with the MTO; electrical and thermal character with traces of the formation of X-wax and carbon particles in OPEI.

Figure 1 shows the dependence of the content of C_F for FC type 2FAL in MTO on the number of electrical breakdowns N in the volume of MTO saturated with hydrogen at an average breakdown voltage $U = 65$ kV. On Fig. 1 above, the curve shows the numerical values of the content of CF with the dimension in mg/kg. It can be assumed that if 2FAL is formed in the MTO as a result of the degradation of the OFEE solid cellulose electrical insulation, then the impact of a high-voltage pulsed electric current on the MTO leads to the degradation of 2FAL and a decrease in its concentration in the MTO.

From Fig. 1, it follows that an increase in the values of N leads to a decrease in the values of the concentrations of C_F in MTO. It will reduce the reliability of OFEE

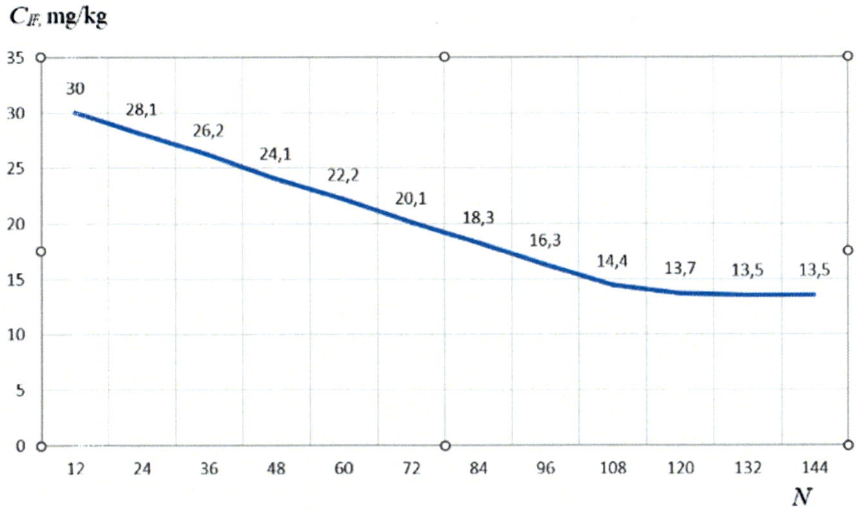

Fig. 1. Dependence of the CF content of the furan compound 2FAL in MTO on the number of electrical breakdowns N in the volume of MTO.

diagnosis based on determining the concentration of 2FAL in MTO without considering the results of other OFEE studies (physical and chemical analysis of MTO.

5 Conclusions

In the work, a complex of experimental studies of the technical condition of OFPA and its damaged OFHVB with OPEI was carried out. The electrical characteristics of OFPA and its OFHVB were obtained, and the PCTE method was used to determine the properties of MTO samples and the contents of dissolved gases TAC and FC in them. To identify the nature of the alleged defects of a thermal and/or electrical nature, the use of the "Tabular method", the "ETCPA method", and the "Duval Triangle method" is proposed. When examining the damaged OFHVB, defects were found: electrical and thermal on the outer surface of the OFHVB in contact with the MTO; an electrical and thermal character with traces of the formation of X-wax and carbon particles in OPEI. It makes it possible to assess the degree of influence of the damaged OFHVB on the presence of internal defects in the OFPA. Relationships between the concentrations of dissolved components in MTO in OFPA and in its damaged OFHVB, depending on the duration of their operation, have been studied. The effect of a high-voltage pulsed electric current on the degradation of FC in MTO in the presence of dissolved H_2, which is formed during the degradation of MTO, is shown. The proposed improvement of methods for diagnosing OFA consists in the fact that it is recommended to determine the content of FC in MTO samples from OFHVB with OPEI in cases where the results do not meet the standards for at least one controlled parameter when performing: OFHVB electrical tests; PCTE studies of the properties of MTO samples; GCh analyzes of MTO samples. It is recommended to simultaneously determine the content of FC, "Ionol" additive, and dissolved gases in the MTO for diagnosing OFPA. It may be promising to study the effect of high-voltage pulsed electric current on the MTO and the degradation of 2FOL, 2FAL, 2ACF, and 5MEF co-existing in its volume.

References

1. Abramov, V., Brzhezytskyi, V., Protsenko, O.: Acceptance and operational tests of electrical equipment. NTUU "KPI", Kyiv (2015)
2. Gobrei, R., Rubanenko, O., Grimud, G.: Technical diagnostics, testing and measurement of electrical equipment in the conditions of installation, adjustment and operation. Part 1. NTUKTS AEE, Kyiv (2008)
3. Electrical equipment testing standards: SOU-N EE 20.302:2007. OEP "GRIFRE", Kyiv (2007)
4. Humenyuk, O., Rubanenko, O., Ostapchuk, O., Talover'ya, V., Shapovalov, Yu.: Technology of repair and operation of high-voltage inputs and their design features. SE "NTUKC", Kyiv (2012)
5. Fluids for electrotechnical applications. Mineral insulating oils for electrical equipment. IEC publication 60296:2020. Edition 5.0. International Standard. Geneva, Switzerland (2020)
6. Acceptance, application and operation of transformer oils. Norms of quality assessment: SOU-N EE 43-101:200. NTUKC AEE, Kyiv (2018)
7. IEEE C57.12.90-2015/Cor 1-2017. IEEE Standard Test Code for Liquid-Immersed Distribution, Power, and Regulating Transformers Corrigendum 1: Editorial and Technical Corrections. Institute of Electrical and Electronics Engineers, USA (2017)
8. Zaitsev, S., Kishnevskyi, V., Oborsky, G., Prokopovich I.: Modern methods of control of energy oils and their degradation products to ensure the reliability of operation of oil-filled electrical equipment of power stations and networks. Ecology, Odessa (2019)
9. Zaitsev, S., Kishnevskyi, V., Savich, S.: Development of gas chromatography methods for determining the content of dissolved components in energy oils. Eastern Eur. J. Adv. Technol. **6/6**(72), 34–42 (2014). https://doi.org/10.15587/1729-4061.2014.29389
10. Ishaka, M., Jusoha, M., Adnana, J., Rahmanb, R.: Lightning impulse breakdown voltage RBDPO with the presence of metallic particles under a uniform electric field. J. Kejuruteraan SI **4**(2), 127–134 (2021). https://doi.org/10.17576/jkukm-2021-si4(2)-19
11. Bahrim, F., Rahman, N F., Haris, H., Salim, N.A.: Effects of electrical insulation breakdown voltage and partial discharge. In: IOP Conference Series: Materials Science and Engineering, vol. 341, p. 012020 (2018). https://doi.org/10.1088/1757-899X/341/1/012020
12. Xiang, C., Zhou, Q., Li, J., Huang, Q., Song, H., Zhang, Z.: Comparison of dissolved gases in mineral and vegetable insulating oils under typical electrical and thermal faults. Energies **9**, 312 (2016). https://doi.org/10.3390/en9050312
13. Tang, X., Wang, W., Zhang, X., Wang, E., Li, X.: On-line analysis of oil-dissolved gas in power transformers using fourier transform infrared spectrometry. Energies **11**, 3192 (2018). https://doi.org/10.3390/en11113192
14. Korobeynikov, S., Lyutikova, M.: Investigation of technological operations affecting the determination of the concentration of the Ionol additive in insulating oil of high-voltage equipment. Probl. Energeticii Regionale **1**(36), 96–105 (2018). https://doi.org/10.5281/zenodo.1217282
15. Zaitsev, S., Kishnevskyi, V., Shulyak, I.: Development of a gas chromatographic method for determining ionol and water in energy oils by the method of additives. Eastern Eur. J. Adv. Technol. **2/6**(74), 21–28 (2015). https://doi.org/10.15587/1729-4061.2015.40896
16. Thango, B., Jordaan, J., Nnachi, A.: Assessment of transformer cellulose insulation life expectancy based on oil furan analysis (case study: South African transformers). Adv. Sci. Technol. Eng. Syst. J. **6**(6), 29–33 (2021). https://doi.org/10.25046/aj060605

17. El-Aal, R., Helal, K., Hassan, A.M., Dessouky, S.: Prediction of transformers conditions and lifetime using furan compounds analysis. IEEE ACCESS **7**, 102264–102273 (2019). https://doi.org/10.1109/ACCESS.2019.2931422
18. Abu-Siada, A., Lai, S.P., Islam, S.: A novel fuzzy-logic. Approach for furan estimation in transformer oil. IEEE Trans. Power Deliv. **27**(2), 469–474 (2012). https://doi.org/10.1109/TPWRD.2012.2186986
19. Methodological guidelines: SOU-N EE 46.501:2006. Diagnostics of oil-filled transformer equipment based on the results of chromatographic analysis of free gases taken from the gas relay and gases dissolved in insulating oil. OEP "GRIFRE", Kyiv (2007)
20. IEEE Std C57.104 ™-2019. IEEE Guide for the Interpretation of Gases Generated in Mineral Oil-Immersed Transformers. Transformers Committee of the IEEE Power and Energy Society. The Institute of Electrical and Electronics Engineers, USA (2019)
21. Acceptance, application and operation of transformer oils. Norms of quality assessment: SOU-N EE 43-101:2009. NTUKC AEE, Kyiv (2018)
22. Methodological guidelines: SOU-N MEV 40.1-21677681-64:2012. Examination of the technical condition and determination of the residual resource of solid insulation of oil-filled equipment. OEP "GRIFRE", Kyiv (2012)
23. Emsley, A., Xiao, X., Heywood, R.: Degradation of cellulose insulation in power transformers. Pt. 3. Effects of oxygen and water on ageing in oil. IEEE. Proc. Sci. Meas. Technol. **147**(3), 115–119 (2001). https://doi.org/10.1049/ip-smt:20000021
24. Abdi, S., Harid, N., Safiddine, L., Boubakeur, A., Haddad, A.: The correlation of transformer oil electrical properties with water content using a regression approach sifeddine. Energies **14**, 2089 (2021). https://doi.org/10.3390/en14082089
25. Cennamo, N., De Maria, L., D'Agostino, G., Zeni, L., Pesavento, M.: Monitoring of low levels of furfural in power transformer oil with a sensor system based on a POF-MIP platform. Sensors **15**, 8499–8511 (2015). https://doi.org/10.3390/s150408499
26. Methodical instructions: SOU-N EE 46.302:2006. Preparation and carrying out of chromatographic analysis of free gases taken from a gas relay and gases dissolved in insulating oil of oil-filled electrical equipment. OEP "GRIFRE" (2007)

Process Engineering

Energy Efficiency of Combined Heating Systems Based on Heat Pumps for Private Residential Buildings Under the Climatic Conditions of Ukraine

Dmytro Konovalov[1,2], Halina Kobalava[1(✉)], Mykola Radchenko[3], Maxim Karpoff[4], and Yuriy Shapovalov[3]

[1] Admiral Makarov National University of Shipbuilding, 44, Ushakov Ave., Kherson 73003, Ukraine
halyna.kobalava@nuos.edu.ua

[2] Norwegian University of Science and Technology, 1, Høgskoleringen, 7034 Trondheim, Norway

[3] Admiral Makarov National University of Shipbuilding, 9, Heroes of Ukraine Ave., Mykolayiv 54025, Ukraine

[4] Heat Energy Group LTD, 48, Hretska St., Kherson 73003, Ukraine

Abstract. Many technical solutions for residential heating systems were analyzed to determine the efficiency of electric heating systems compared to traditional solutions. These included: using only a gas boiler in a weather-dependent heat control mode, using only an electric boiler in a weather-dependent heat control mode, combining a gas boiler and heat pump in weather-dependent mode with the gas boiler included, and combining an electric boiler and heat pump in weather-dependent mode. Empirical relationships for the heat transfer coefficient and heat load were obtained to calculate the heating system based on the initial inlet heat-carrying fluid and outside air temperatures. Characteristics for circuit solutions of the heating system were determined for combinations of an electric boiler, a heat pump, and a gas-condensing boiler at different heat-carrying fluid temperatures. It should be noted that using an electric boiler with a heat pump is rational, as this reduces electricity consumption per electric boiler by 14,000 to 19,000 kW · h (35–50%) during one heating period.

Keywords: Energy Efficiency · Heat Pump · Heating System · Climatic Conditions

1 Introduction

Increasing demand for energy resources drives the current state of the world economy. The growth of world industrial production and active energy consumption has led to the uncontrolled use of non-renewable energy sources. Consequently, humanity has come close to a global crisis in the world's fuel reserves [1] and the environmental situation [2]. The rise in world gas prices has created the conditions for a shift towards

V. Tonkonogyi et al. (Eds.): InterPartner 2023, LNME, pp. 531–540, 2024.
https://doi.org/10.1007/978-3-031-42778-7_49

active electricity consumption. Electricity generation offers inexhaustible possibilities compared to other types of energy, as alternative sources can be used, and electric heating is the most environmentally friendly. Reducing the combustion of natural gas and replacing it with electricity produced by nuclear power plants can reduce carbon dioxide emissions [3], which will help reduce environmental pollution [4]. Electric heating is becoming increasingly popular in Western Europe; for example, in Spain, this type of heating is used in 30% of buildings and structures; in Finland and France – more than 40%, and in Norway – more than 80%. Electric heating has recently been developed in the USA and Canada [5]. However, the electric heating market in Ukraine is developing at a slower pace. There is a steady increase in gas and electricity prices in Ukraine, and this trend is expected to continue in the coming years.

When analyzing the dynamics of the increase in energy costs, it is essential to note that the growth rate of the cost of electricity is much lower than that of natural gas and oil. Moreover, since electricity is provided to consumers at night at a significantly lower rate than during the day, this creates a favorable environment for using electrical energy to provide high-quality heat to residential facilities. National energy systems often experience uneven distribution of loads during the day, with peak loads during the day and excess electricity at night. To address this issue, preferential tariffs are offered during nighttime hours, allowing electricity use at lower prices. By combining electric heating with traditional heat engineering equipment and heat pumps, efficient consumption of fuel and energy resources can be ensured in the municipal (housing and communal) energy sector.

2 Literature Review

Humanity has been using various forms of energy for centuries, with an increasing amount needed to support modern life. However, we also know that not all energy sources are sustainable, with valuable resources decreasing, gas prices rising [1], and environmental pollution worsening [2]. Ukraine actively seeks ways to improve energy efficiency [6] and conservation [7]. For example, the European Union has committed to reducing its annual primary energy consumption by 20% by 2020. In 2014, the EU Energy Strategy for the period up to 2030 was adopted, which set the following goals for energy saving and efficiency [8]: reducing greenhouse gas emissions by at least 40% from 1990 levels, or up to 45% under favorable conditions; increasing the share of renewable energy in final energy consumption up to 27%; achieving a 27% improvement in energy efficiency.

Several European countries are already undergoing an energy transition, which aims to increase the use of renewable energy sources and move away from fossil fuels (oil and gas) and nuclear energy while promoting responsible energy use. Ukraine is currently in the process of implementing reforms to improve its energy sector.

The measures mentioned in the text aim to reduce heat losses and improve the efficiency of heat supply systems. Here is a summary of the measures mentioned:

1. Reduce energy and heat consumption for own needs [9].
2. Use modern equipment with higher heat generation efficiency, such as condensing boilers and heat pumps.

3. Use heat energy metering units.
4. Use co- and trigeneration heating networks.
5. Insulate networks to reduce heat loss in the environment.
6. Shorten the path of the heat-carrying fluid from the producer to the consumer of thermal energy (mini-boiler house near the house).
7. Optimize hydraulic modes of heating networks [10].
8. Reduce leaks in networks.
9. Properly insulate heated premises.
10. Use systems for local adjustment of heating devices [11].
11. Transfer houses to the mode of zero heat consumption for heating (the temperature inside is maintained by internal heat release (Recovery) and good insulation (windows with airgel)) [12].
12. Use heat energy metering units.

The heat carrier for the heating system can be any medium that can accumulate thermal energy and change thermal properties, is mobile, cheap, does not worsen sanitary conditions in the premises, and allows for the regulation of the release of heat.

The most widely used heating systems include water, water vapor, and air, all of which meet the above requirements. Today, ensuring the energy efficiency of buildings is a relevant and extremely important direction for reducing fuel and energy consumption and ensuring the state's energy security [11].

Energy costs in buildings, such as educational institutions, healthcare facilities, children's institutions, preschools, residential buildings, etc., can be significantly reduced by implementing various energy efficiency measures [13]. Some examples include additional thermal insulation of the building envelope, automatic control, hydraulic balancing of the heating system, installing thermostatic valves on radiators [14], and implementing energy management procedures [15].

The energy efficiency of a building depends on many factors, including the building's envelope (such as the characteristics of building materials, insulation, and windows), technical installations (heating, ventilation, hot water supply, lighting, etc.) [16], and operation and maintenance procedures [17].

For each factor, relevant measures should be evaluated, and energy savings, investments, and profitability should be calculated to determine the energy efficiency potential of the building [18]. It is also essential to consider the mutual influence of factors on each other. The energy efficiency improvement potential assessment should include ranking various measures according to their cost-effectiveness [19].

The study aims to determine the heating system's efficiency [20] compared to traditional solutions [21]. To achieve this goal, the following tasks should be accomplished: analyze the meteorological data of the region where the building is located and determine the technical solutions for the electric heating system; evaluate the effectiveness of solutions for electric heating systems of a residential building in comparison with traditional methods and identify the most rational ones.

3 Research Methodology

In this work, a comparative computational research method was used to determine the thermodynamic and energy efficiency of the system using a computational model developed by the author.

Outside air parameters, including temperature (t_{aa}), were obtained from the electronic resource "meteomanz.com" [22]. The website contains a database of hydrometeorological conditions from registered meteorological stations dating back to 2000. The date and time are in UTC format, and the station's location must be entered to obtain the necessary data. Approximately 50 stations in Ukraine provide data to the database. It is appropriate to establish the outdoor temperature's dependence on the boiler room's weather-dependent operation to analyze the influence of weather conditions. Using "meteomanz.com," data were obtained for the heating periods of three years, from 2018 to 2020, with an interval of 3 h. Meteorological parameters such as air temperature, humidity, pressure, and wind were obtained in tabular form.

The object of study is a large single-story house outside the garden that includes the implementation of architectural techniques for capturing and accumulating a large amount of heat from the environment. The house has warm floors and is sheathed with insulation and various heat carriers to maintain comfort, covering an area of over 200 sq.m.

Heat control equipment: Vitocal 200-S heat pump with a heat output of 15.5 kW; Vitodens 100-W parapet-type gas condensing boiler with a heat output of 26 kW; two flat solar collectors Vitosol 100-FM (area of each is 2.3 m^2).

The main source of heat energy is the Vitocal 200-S heat pump, the Vitodens 100-W parapet gas condensing boiler is used to cover peak loads. To prepare hot water from solar collectors, a capacity of Vitocell water heaters – 100W with a capacity of 300 L is installed in the furnace. In addition, it is possible to heat the water with the heat-carrying fluid of a gas boiler or heat pump. The heat-carrying fluid is supplied to the radiator heating system, underfloor heating, and ventilation by three Pico 25/1–6 circulation pumps installed in the Meibes pump groups (Fig. 1).

Several heating system schemes were proposed to analyze the effectiveness of the use of electric heating. The existing scheme with a Vitodens 100-W type gas boiler with a thermal output of 26 kW and two-stage power control (30% and 100%) in weather-compensated mode was taken as the base one. It is proposed to use a double-circuit electric boiler, VITOTRON 100, for the electric heating system, with a thermal power of 30 kW. It can operate weather-dependently with a smooth electric power adjustment from 10% to 100%.

The use of the boiler is suggested together with the use of the Vitocal 200-S heat pump. In this case, the heat pump must operate in a weather-dependent mode (depending on the outdoor temperature), and boilers (gas or electric) are switched on at peak loads on the heating system at outdoor temperatures below 0 °C. It is primarily because the coefficient of performance (COP) at low outdoor temperatures decreases significantly, and using a heat pump becomes unprofitable.

An analysis was carried out for four circuit solutions of the heating system:

1. The use of only a gas boiler in a weather-dependent heat regulation mode.

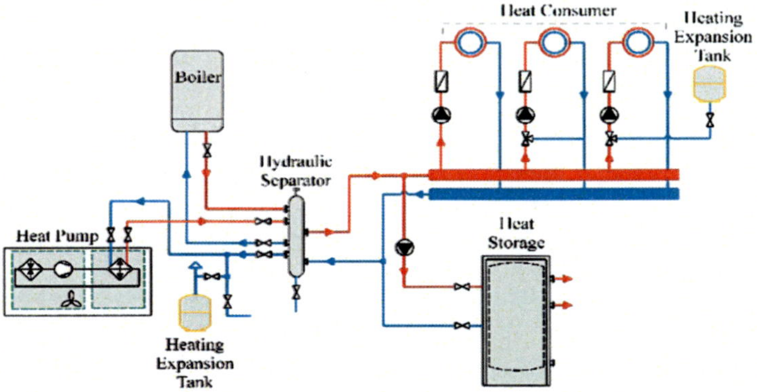

Fig. 1. Heating system with the combined operation of a heat pump and a boiler.

2. The use of only an electric boiler in a weather-dependent heat regulation mode.
3. Combined use of a gas boiler and a heat pump in a weather-dependent mode, where the gas boiler turns on at outdoor temperatures below $t_{aa} = 0$ °C and compensates for the heat deficit.
4. The combined use of an electric boiler and a heat pump in a weather-dependent mode, where the electric boiler turns on at outdoor temperatures below $t_{aa} = 0$ °C and compensates for the heat deficit.

Additionally, solar panels can be used for hot water supply during the summer. To calculate the characteristics of the Vitocal 200-S heat pump, the obtained experimental data provided in the public domain were analyzed [23, 24]. An analysis of the experimental data showed that the COP and the corresponding amount of useful heat Q decrease with a decrease in external temperature and increase with a decrease in the temperature of the heat-carrying fluid (heating system water). At $t_{aa} = 0$ °C, the amount of useful heat is $Q_{hp} = 8$–14 kW and COP = 2.2–3.5, and at $t_{aa} = 10$ °C, the amount of useful heat is $Q_{hp} = 11$–17 kW and COP = 2.8–4.4 (Fig. 2).

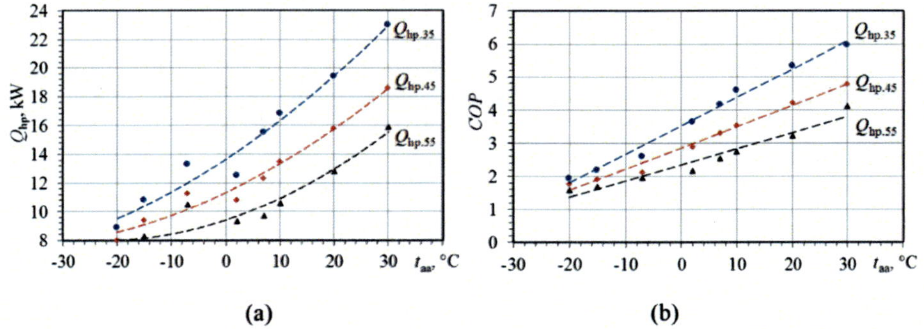

Fig. 2. Experimental values of heat load Q_{hp} (a) and the heat transfer coefficient COP (b) depending on the outdoor temperature t_{aa} at different water temperatures t_{w1}.

To simulate the characteristics of the heating system (heat load, natural gas consumption, electrical energy consumption, etc.) depending on the changes in outdoor temperature, the following input conditions were adopted:

1. The maximum heat load of an electric boiler or a gas boiler corresponds to the outdoor air temperature $t_{aa} = -20$ °C.
2. For a heating system that combines a boiler and a heat pump, the outdoor air temperature at which the boiler is turned on (10% of the rated load) is $t_{aa} = 0$ °C.
3. The calculations are made for the heating period from November to March, every three hours.
4. The base scheme for comparing other options for the heating system is a scheme with a gas boiler.
5. The initial temperature of the heat-carrying fluid at the inlet to the heating system is $t_{w1} = 35$ °C, 45 °C, or 55 °C.

4 Results and Discussion

The following conclusions can be made after analyzing four circuit solutions for the heating system during the 2020–2021 heating period. When operating with a gas boiler alone, natural gas consumption during the heating period averages $V_{gb} = 0.5–1.9$ m³/h (Fig. 3). However, with combined use with a heat pump, natural gas consumption is reduced to up to $V_{gb.hp} = 1.4$ m³/h. The total consumption of natural gas for the entire period is $\Sigma V_{gb} = 3.4 \cdot 10^3$ m³ for gas heating and $\Sigma V_{gb.hp} = 0.3 \cdot 10^3$ m³ for combined use (with a heat pump), resulting in a reduction of $3.1 \cdot 10^3$ m³.

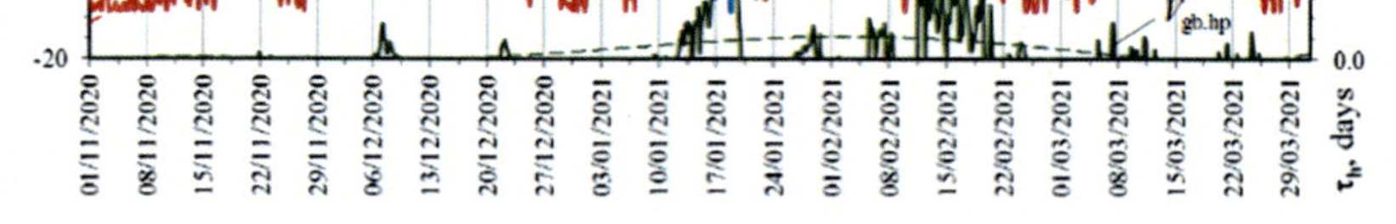

Fig. 3. Dependences of the outdoor air temperature t_{aa}, natural gas consumption for a scheme with a gas boiler V_{gb}, and for a combined scheme with a heat pump $V_{gb.hp}$.

When considering the use of an electric boiler and the combined use of a boiler and a heat pump, it can be observed that the load on the electric boiler is $N_{eb} = 5–23$ kW. In this case, for the combined version, the electrical load consists of the load on the boiler and the heat pump compressor. The load on the heat pump is $N_{hp} = 4.0–5.0$ kW, which practically remains constant, and the load on the electric boiler is episodic, amounting to $N_{eb.hp} = 12.0$ kW. Consequently, the total electrical load is $N_{eb.hp+hp} = 4.0–17.0$ kW.

The total consumption of electrical energy for a heating system with an electric boiler is $\Sigma N_{eb} = 35{,}000$ kW · h, the load on the heat pump for the combined scheme is $\Sigma N_{hp} = 14{,}500$ kW · h, and the load on the electric boiler for the combined scheme is $\Sigma N_{eb.hp} = 300$ kW · h. In this case, the total electrical energy consumption for the combined scheme will be $\Sigma N_{eb.hp+hp} = 14{,}800$ kW · h (Fig. 4).

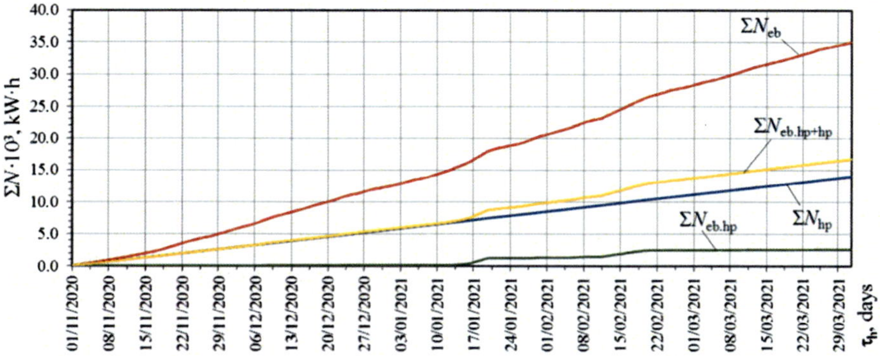

Fig. 4. Dependences of the total consumption of electrical energy ΣN_{eb} for a scheme with an electric boiler, for an electric boiler $\Sigma N_{eb.hp}$ and for a heat pump ΣN_{hp} in the combined scheme with a heat pump, and the total consumption of an electric boiler and a heat pump $\Sigma N_{eb.hp+hp}$ in the combined scheme.

Modeling regarding the change in the initial temperature of the heat-carrying fluid shows that the COP coefficient values are the smallest at $t_{w1} = 55$ °C and amount to COP = 2.0–3.0. (Fig. 5, a).

At the same time, the heat load of the heat pump is $Q_{hp} = 9.0$–11.0 kW, with the electrical load on the heat pump compressor and electric boiler $N_{eb.hp+hp} = 4.0$–20.0 kW. When the initial temperature drops to $t_{w1} = 45$ °C (Fig. 5, b), the COP coefficient value is COP = 2.5–4.0, the heat load of the heat pump is $Q_{hp} = 10.0$–14.0 kW with the electrical load on the heat pump compressor and electric boiler $N_{eb.hp+hp} = 4.0$–17.0 kW. The initial temperature $t_{w1} = 35$ °C corresponds to large values of the COP coefficient COP = 3.0–4.8 and the smallest values of the electrical load. In this case, the heat load of the heat pump is $Q_{hp} = 12.0$–18.0 kW with the electrical load on the heat pump compressor and electric boiler $N_{eb.hp+hp} = 4.0$–17.0 kW (Fig. 5, c).

The monthly electricity consumption reduction reaches 2,015–4,945 kW · h, which corresponds to 42–57% compared to the basic scheme. The lowest indicators of reduction of electrical energy correspond to November, which is 2,015 kW · h (48%). Higher average air temperatures during the day explain this distribution by months. Thus, it can be concluded that using an electric boiler is only rational when combined with a heat pump. The consumption of electrical energy for the combined scheme is $\Sigma N_{eb.hp+hp} =$ 15,000–17,000 kW · h, which is 35–50% less than when using only an electric boiler during one heating period.

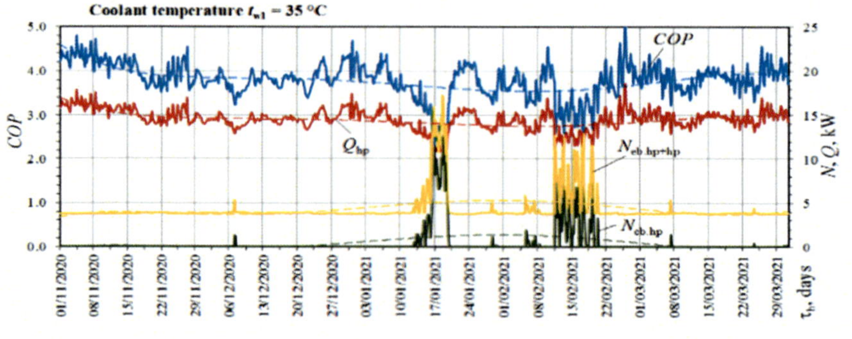

(a)

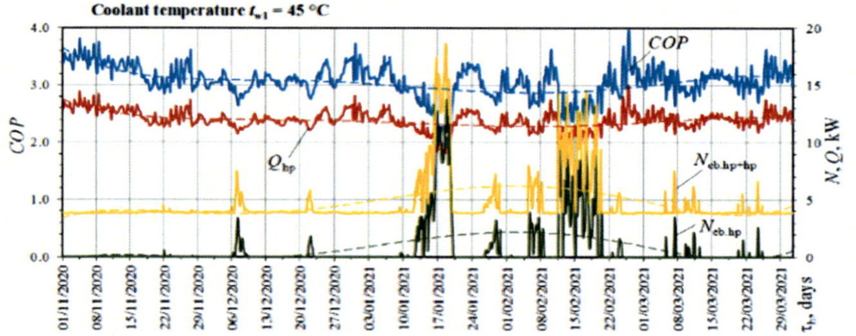

(b)

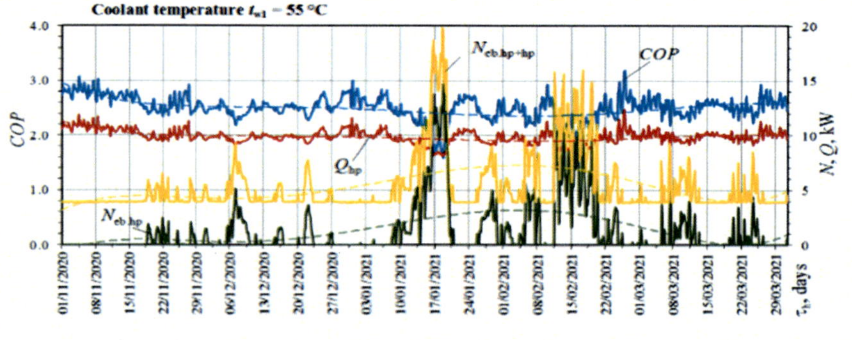

(c)

Fig. 5. Dependences of the COP, the heat load of a heat pump Q_{hp}, the load on the electric boiler for a combined scheme with a heat pump $N_{eb.hp}$, the total load of the electric boiler and heat pump $N_{eb.hp+hp}$ at the water temperature in the heating system t_{w1}, °C: 35 (a); 45 (b); 55 (c).

5 Conclusions

The article provided the main dependencies of the heat engineering calculation and proposed technical solutions for the electric heating system. Four circuit solutions of a heating system in combination with an electric boiler, a heat pump, and a condensing gas boiler were analyzed, including the use of only a gas boiler in a weather-dependent heat control mode, the use of only an electric boiler in a weather-dependent mode of heat regulation, combined use of a gas boiler and a heat pump in a weather-dependent mode, and combined use of an electric boiler and a heat pump in a weather-dependent mode.

Characteristics were obtained for these circuit solutions at different heat-carrying fluid temperatures. The analysis showed that combining an electric boiler with a heat pump is the most rational solution, reducing the consumption of electric energy for the electric boiler. The monthly electricity consumption reduction reaches 2,015–4,945 kW · h compared to the basic scheme. Electrical energy consumption for the combined scheme is 15,000–17,000 kW · h, 35–50% less than using only an electric boiler during one heating period.

References

1. Holechek, J.L., Geli, H.M.E., Sawalhah, M.N., Valdez, R.: A global assessment: can renewable energy replace fossil fuels by 2050. Sustainability **14**, 4792 (2022). https://doi.org/10.3390/su14084792
2. Mahmood, N., Wang, Z., Hassan, S.T.: Renewable energy, economic growth, human capital, and CO2 emission: an empirical analysis. Environ. Sci. Pollut. Res. Int. **26**(20), 20619–20630 (2019). https://doi.org/10.1007/s11356-019-05387-5
3. Alam, F., Sarkar, R., Chowdhury, H.: Nuclear power plants in emerging economies and human resource development: a review. Energy Procedia **160**, 3–10 (2019). https://doi.org/10.1016/j.egypro.2019.02.111
4. Coyle, E.D., Simmons, R.A.: Understanding the Global Energy Crisis. Purdue University Press, West Lafayette (2014)
5. Barbosa,M., Peteves, E., Vázquez Hernández, C., Gonzalez Aparicio, I.: EU coal regions – Opportunities and challenges ahead. Publications Office, European Commission, Joint Research Centre (2018)
6. Yang, Z., et al.: Analysis of efficiency of thermopressor application for internal combustion engine. Energies **15**(6), 2250 (2022). https://doi.org/10.3390/en15062250
7. Yu, Z., et al.: Investigation of thermopressor with incomplete evaporation for gas turbine intercooling systems. Energies **16**(1), 20 (2023). https://doi.org/10.3390/en16010020
8. Pakare, I., Gravelsins, A., Lauka, D., Blumberga, D.: Estimating energy efficiency increase in national district heating network. Energy Rep. **7**, 401–409 (2021). https://doi.org/10.1016/j.egyr.2021.08.088
9. Capehart, B.L., Turner, W.C., Kennedy, W.J.: Guide to Energy Management. Seventh Edition. The Fairmont Press (2012)
10. Zhang, Y., Johansson, P., Kalagasidis, A.S.: Applicability of thermal energy storage in future low-temperature district heating systems – case study using multi-scenario analysis. Energy Convers. Manag. **244**, 114518 (2021). https://doi.org/10.1016/j.enconman.2021.114518

11. D'Agostino, D., Parker, D., Melià, P.: Environmental and economic implications of energy efficiency in new residential buildings: a multi-criteria selection approach. Energ. Strat. Rev. **26**, 100412 (2019). https://doi.org/10.1016/j.esr.2019.100412
12. Yüksek, I., Karadayi, T.T.:Energy-Efficient Building Design in the Context of Building Life Cycle. InTech (2017). https://doi.org/10.5772/66670
13. Muresan, A., Attia, S.: Energy efficiency in the Romanian residential building stock: a literature review. Renew. Sustain. Energy Rev. **74**, 349–363 (2017). https://doi.org/10.1016/j.rser.2017.02.022
14. Ziemele, J., Pakere, I., Blumberga, D.: The future competitiveness of the non-emissions trading scheme district heating systems in the baltic states. Appl. Energy **162**, 1579–1585 (2016). https://doi.org/10.1016/j.apenergy.2015.05.043
15. Jouhara, H., Nannou, T.K., Anguilano, L., Ghazal, H., Spencer, N.: Heat pipe based municipal waste treatment unit for home energy recovery. Energy **139**, 1210–1230 (2017). https://doi.org/10.1016/j.energy.2017.02.044
16. Hamid, A.A., Farsäter, K., Wahlström, Å., Wallentén, P.: Literature review on renovation of multifamily buildings in temperate climate conditions. Energy Build. **172**, 414–431 (2018). https://doi.org/10.1016/j.enbuild.2018.04.032
17. Azizi, S., Nair, G., Olofsson, T.: Adoption of energy efficiency measures in renovation of single-family houses: a comparative approach. Energies **13**, 6042 (2020). https://doi.org/10.3390/en13226042
18. Mahmoud, M., Ramadan, M., Naher, S., Pullen, K.R., Olabi, A.-G.: The impacts of different heating systems on the environment: a review. Sci. Total Environ. **766**, 142625 (2021). https://doi.org/10.1016/j.scitotenv.2020.142625
19. Ingersoll, J.G., Arasteh, D.K.: Energy efficiencies of heat pumps in residential buildings. Energy Build. **5**(4), 253–262 (1983). https://doi.org/10.1016/0378-7788(83)90013-0
20. Gradziuk, P., Siudek, A., Klepacka, A.M., Florkowski, W.J., Trocewicz, A., Skorokhod, I.: Heat pump installation in public buildings: savings and environmental benefits in underserved rural areas. Energies **15**, 7903 (2022). https://doi.org/10.3390/en15217903
21. Gaur, A.S., Fitiwi, D.Z., Curtis, J.: Heat pumps and our low-carbon future: a comprehensive review. Energy Res. Soc. Sci. **71**, 101764 (2021). https://doi.org/10.1016/j.erss.2020.101764
22. Meteomanz.com. http://meteomanz.com/index?l=1. Accessed 25 June 2021
23. Viessmann VITOCAL 200-S. https://www.manualslib.com/manual/723545/Viessmann-Vitocal-200-S.html. Accessed 16 June 2021
24. Mugnini, A., Coccia, G., Polonara, F., Arteconi, A.: Variable-load heat pumps: impact of the design and control parameters on the actual operation conditions. TECNICA ITALIANA-Italian J. Eng. Sci. **65**(2–4), 371–377 (2021). https://doi.org/10.18280/ti-ijes.652-434

Evaluation of Dust Concentration Using Computer Measurement Technologies

Gennadii Oborskyi, Vladimir Gugnin, Liudmyla Perperi(✉), Ganna Goloborodko, and Volodymyr Goloborodko

Odessa Polytechnic National University, 1, Shevchenko Ave., Odessa 65044, Ukraine
plm.mvms@gmail.com

Abstract. This work presents the creation of a program for analyzing measurement results in the Scilab software environment, which simulates the process of determining dust concentration using computer measurement technologies. The article considers methods and approaches to computer modeling using software tools. A virtual device was created to implement the task. Requirements for drawing up an air sampling plan and conditions for sampling were determined. The developed virtual device allows for processing the received measurement information in three ranges. Data arrays processed by the device program are presented in graphic form to visualize dust concentration assessment using a color scale of dustiness with the output of topographical graphs of dust distribution in the room. The use of modern computer technologies in the educational process allows students of technical specialties to gain practical skills and develop the ability to analyze and interpret the received information.

Keywords: Air Quality · Concentration Topographical Graphs · Virtual Instruments · Scilab · Process Innovation

1 Introduction

It is known that city dwellers usually spend most of their time indoors [1], which is approximately 80–90% of the total time [2]. Many studies indicate that indoor air can be more polluted than outside [2, 3]. This is explained by imperfect air exchange and specific sources of pollution (heating systems, specific uses of premises, such as sports halls, swimming pools, schools, universities, and workshops of industrial enterprises) [2, 4]. Many researchers pay attention in their studies to the issue of the concentration [5] of solid particles in the air [6]. One of the essential parameters of the indoor microclimate is settled and suspended dust [7].

This is due to the proven negative impact of dust on human health. As stated in the source [8]: "Particulate matter (PM) is an air pollutant identified by the International Cancer Institute of the World Health Organization (WHO) as a group 1 carcinogen that has been shown to cause cancer in humans. Fine dust (can be divided into fine dust with a diameter of less than 10 μm (PM 10) and ultrafine dust (PM 2.5) with a diameter of less than 2.5 μm according to its size, and the accumulation of ultrafine dust can

V. Tonkonogyi et al. (Eds.): InterPartner 2023, LNME, pp. 541–549, 2024.
https://doi.org/10.1007/978-3-031-42778-7_50

quickly deteriorate indoor air quality". WHO's global recommendations on air quality note the equal importance of air quality monitoring and assessment for implementing the guidelines [8].

Numerous devices for measuring various factors are utilized to monitor air quality. Given the recent trends related to digital transformation in the international scientific infrastructure, digital transformation in metrology will bring many benefits [9], for example, speeding up the time to market of measurement products and services [10], reducing delays related to approval processes, which in turn will promote innovation, flexibility and stability of products [11]. When solving the tasks of digital transformation not only in industry but also in the field of higher education in the conditions of pandemics and war in Ukraine, it is relevant to consider the issue of using modern information technologies to create virtual devices using computer environments for modeling processes and conducting virtual experiments [12].

This thematic study aims to develop a program for automating the processing of measurement results and their visualization through topographic graphs of room contamination levels. This will enable researchers and engineers to effectively conduct modeling and measurement of air quality control processes by utilizing mathematical, numerical, and analytical methods with sequential graphical visualization in a safe environment.

2 Literature Review

Air pollution is a growing problem in many countries, with PM being one of the main contributors to poor air quality. PM refers to tiny particles in the air that can negatively impact human health, especially for people with respiratory problems. This is also stated in the paper [6]. To solve the problem of dust concentration measuring, the authors analyzed the reliability of inexpensive devices for measuring fine dust based on sensors for monitoring the concentration of solid particles in the atmosphere. The authors proposed formulas for correcting errors in the operation of inexpensive sensors.

This research [1] briefly overviews the analytical methods used to assess indoor air pollutants detected in household dust samples. The main objective was to detect and analyze hazardous substances' presence, quantity, and distribution. Monitoring PM concentration is critical to maintaining satisfactory indoor air quality, and tools for rapidly detecting dust particle concentrations are a major trend [7], so computer-based measurement technologies play an increasingly important role in this process.

The authors of the article [11] note that today more and more attention is paid to the design of devices with the help of virtual tools and made a thorough review of open-source virtual tools, including Scilab. The Scilab package is a computer mathematics system with an open-source license. The article [13] also notes that virtual instruments occupy a special place in the measurement world, providing simple integration and flexible adjustment. They are also proposed to be used as educational tools for engineering students, which, according to the authors, will stimulate innovation in developing virtual devices and consolidating theoretical knowledge.

In papers [12, 13], the authors conduct a comprehensive analysis to evaluate the efficacy of using virtual tools in contemporary educational settings. They thoroughly examine the impact and effectiveness of incorporating virtual tools in learning. Along with many advantages, the main disadvantage is the lack of practicing communicative competence due to the lack of live communication. With the popularization of inexpensive and high-performance computing resources, digital tools have gradually become the mainstay of education. Using virtual instrument equipment in education saves costs and improves the quality and effect of training and research [14].

Computer modeling become more popular as a methodological approach to research organization. One of the key advantages of computer measurement technologies is the ability to simulate the process of determining dust concentration. It is a valuable educational tool that allows students to gain hands-on experience with monitoring equipment without needing expensive or hazardous materials. In this article, we will discuss modeling the process of determining the concentration of dust using computer measurement technologies and its importance in air quality monitoring.

3 Research Methodology

The input parameter for modeling the process are files with the results of measuring the number of dust particles of different sizes in a cubic meter of air. The concentration of dust particles is measured using the "Walcom HT-9600" air dustiness tester [15] (Fig. 1, left).

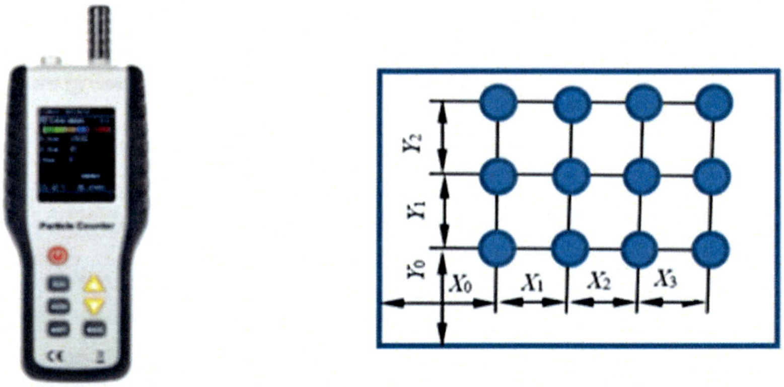

Fig. 1. Air dust tester Walcom HT-9600 [15] (left) and the image diagram of quantitative air sampling points (right).

This tester has three channels for measuring the concentration of dust particles in the ranges from 0 to 0.3 μm, more than 0.3 to 2.5 μm, and more than 2.5 to 10.0 μm, and it is also equipped with a Micro USB data transfer port on the computer. A program was created to implement the program for measurement results processing. It simulates determining the dust concentration in residential or industrial premises. Simulation of a virtual device for measuring the concentration of aerosol particles in industrial premises was carried out in the Scilab software environment [11, 16]. An air sampling plan is drawn up for each room (Fig. 1, right).

The air sampling plan should include:

- specifying sampling points (control points);
- specifying the frequency (periodicity) of control for each point;
- specifying the number of samples at each point;
- indication of the time (moments) of the working day (shift) when sampling is carried out;
- indication of the nature of the technological process at the time of sampling;
- designation of critical control points.

Sampling must be organized to minimize its influence on the production process. The number and location of control points are set depending on the area of the production premises and the nature of the technological operations performed in it. The control of the content of microparticles in the air is mainly carried out in those working areas of industrial premises where based the most likely sources of atmospheric air pollution are located (places with a large number of personnel and places with an increased risk of dust formation), or to those working areas where there are various substances, auxiliary reagents, and finished product, which are in direct contact with the environment.

When determining the content of particles in the air, the minimum number of air sampling points (N) is determined by rounding up to a more significant number:

$$N = \sqrt{S},$$

where S is the room area (m^2), in which air sampling is carried out.

The number of sampling points should be at least two. When the content of particles in rooms with vertical unidirectional and non-unidirectional air flow is determined, the sampling points should be evenly distributed in a plane parallel to the plane of the floor or vertically, if necessary. In rooms with non-vertical unidirectional airflow, the plane of sampling points should be perpendicular to the airflow direction. Suppose some sampling points fall on places occupied by equipment when the content of particles in the air is determined. In that case, they are moved to places most critical to the requirements for protecting products (technological process) from microparticles.

Determination of the content of particles in the air of industrial premises was carried out by using the Walcom HT-9600 air dust tester. The measuring instrument allows taking into account the content of particles in the air should be chosen based on the size of the microparticles to be measured, as well as the required accuracy of the measurement. The device used to monitor the particulate content must have a valid calibration certificate.

Control of the content of particles in the air should be carried out in strict accordance with the sampling plan. When taking samples should be observed:

- sampling isokineticity – equality of air velocities in the sampler and the flow under investigation;
- sampling isoaxiality – the sampler must be oriented so that the airflow direction entering the sampler coincides with the direction of movement of the unidirectional air flow from which the sample is taken. If the flow direction is unclear, the sampler is oriented vertically upwards.

The volume V (m^3) of the air sample at each point of sampling when determining the content of particles in the air must be at least the value necessary for the registration of 20 particles:

$$V = \frac{20}{N},$$

where N is the maximum permissible number of particles within the specified size for a certain cleanliness class of the room or zone.

At each point for each channel of the"Walcom HT-9600″ device, at least 3 measurements are performed, which are recorded in the device's memory.

For each room and the i-th sampling point, the average arithmetic value of A and the calculated concentration of particles (particles/m^3) is calculated according to the formula:

$$A_i = \frac{1}{m} \cdot \sum\nolimits_{i=1}^{m} (a_i),$$

where m is the number of measurements at each sampling point;

a_i is the measurement result (i).

After carrying out measurements according to the scheme (Fig. 1), the results are recorded in a file for each channel of the device, as shown, for example, for the first channel of measuring the concentration of dust particles in the ranges from 0 to 0.3 μm:

```
5.1198e+07  4.8023e+07  4.94641e+07  4.89963e+07
5.1656e+07  4.8588e+07  5.16401e+07  5.02986e+07
5.2629e+07  5.2459e+07  5.34371e+07  5.11813e+07
```

Similar arrays are created for channels measuring the concentration of dust particles in the ranges of more than 0.3 to 2.5 μm and more than 2.5 to 10.0 μm.

From the aggregate A_i, the average arithmetic value A of the calculated concentration of particles in 1 m^3 of air in the entire room is calculated according to the formula:

$$A = \frac{1}{n} \cdot \sum\nolimits_{i=1}^{n} (A_i),$$

where n is the sampling points number.

The point estimate of the mean square deviation S of the result of the measured concentration is calculated according to the formula:

$$S = \sqrt{\frac{\sum_{i=1}^{n} (A_i - A)^2}{n(n-1)}}.$$

The virtual function finds the value of the t factor in the table for a confidence probability is 0.95.

The calculated concentration of C particles in one m^3 of the air of the industrial premises, taking into account the confidence interval, is calculated by the formula:

$$C = A \pm t \cdot S.$$

The calculated values of the calculated concentration of particles C in one m^3 of the air of the industrial premises and A_i of the average arithmetic value of the calculated concentration of particles at each sampling point should not exceed the maximum permissible number of particles, which is provided for premises of a certain cleanliness class.

4 Results and Discussion

When the program starts, a window will appear with an offer to enter the value of the confidence probability (Fig. 2). The window can be fielded with the required probability value, and then the button [OK] should be pressed. The value of the confidence probability is entered to obtain the result of the dust concentration measurement for each channel of the device.

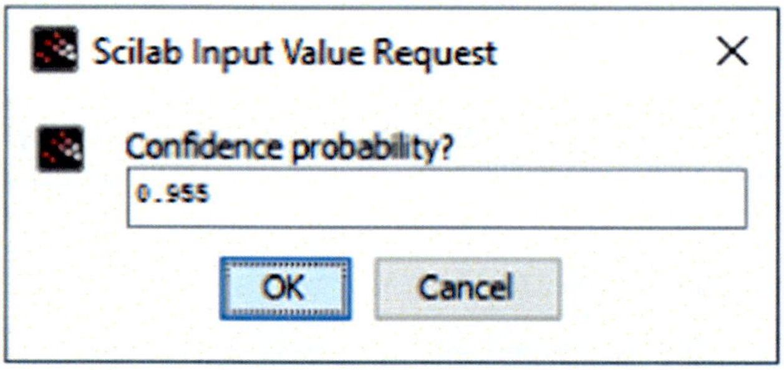

Fig. 2. Entering the confidence probability value.

After the calculation is finished, the calculation results and the topographical graph of the distribution of dust in the room according to each channel of the measuring device will be presented in the graphic window (Fig. 3).

After rounding the calculation results, the measurement result for each channel can be recorded separately. For example, the dust measurement result for the third channel will look like this: $C = 12000 \pm 160$ dust particles per cubic meter. In the right part of the panel (see Fig. 3) of each channel, there is a pollination concentration palette. A color scale of dustiness is shown to the right of the palette for a visual assessment of dust concentration. The program processes the array of data for each channel separately and displays the processing results in a graphic window, but the program code for all three channels is almost the same, only the comments and references to the corresponding file, where the measurement results in each channel of the device are recorded, are different.

The algorithm for executing the program for measuring the concentration of dust in the air for each channel requires four actions to be performed in sequence. The first recognition of the original data matrix and transformation of the string-type matrix into the numeric-type one. Then the array of data is displayed in the graphic window of the program. Next, the characteristics of the sample of measurement results for each channel are calculated. Finally, the topographical graphs of dust distribution in the room are displayed.

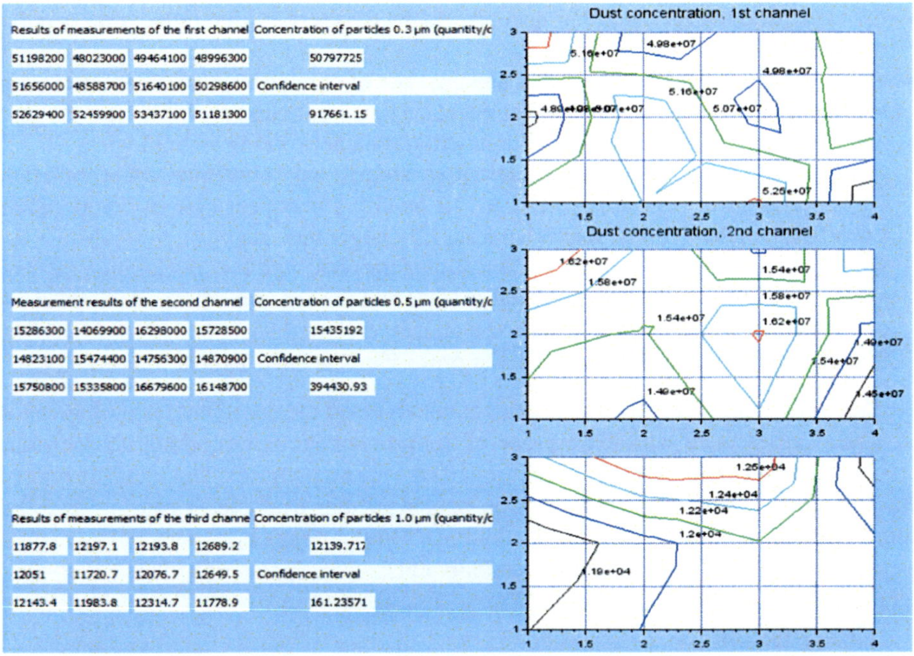

Fig. 3. Graphical window with the results of calculating the dustiness of the room.

The following algorithm is performed to find a confidence interval for the particle concentration. First, the matrix of measurement results is divided into three-row vectors. Next, their values are displayed in the graphic window. Then, using the function "mean, "the average concentration of dust particles is calculated for each device channel for the entire room. Using the function "stdev", a point estimate of the mean square deviation of the dustiness value is calculated. The next step is, by using the function "cdft", the value of the Student coefficient is determined. This probability is entered in the input box (for example, 0.955, as shown in Fig. 2) and is determined using the functions "evstr" and "x_dialog". Finally, data processing visualization is performed using the "contour" function to construct topographical graphs depicting dust distribution within each channel's premises (Fig. 3). Since each line of the graph corresponds to one concentration value of the dust, the topographical graphs provide a visual representation of the dust distribution and facilitate the identification of major sources of varying dust magnitudes.

The authors may, if necessary, present the code to all interested parties who apply these tools in their work, especially for use to meet educational needs.

Considering the aforementioned, it can be stated that incorporating innovations in the monitoring procedures for particulate matter concentration and assessing air pollution by employing computer-based measurement technologies is becoming increasingly significant. This observation has also been highlighted in works [11, 14].

If we compare the application of virtual devices with traditional ones across various sectors of economic activity, we can highlight the following advantages, which have also been noted in sources [12, 14, 16]:

- the task (design) of functions necessary directly for the user rather than rigidly determined by the manufacturer;
- ability to apply quick data retrieval and analysis of obtained measurement data with sequential application of a graphical interface for visualization;
- ability to apply virtual devices in conjunction with networking technologies;
- the use of the freely distributed Scilab programming environment allows significant cost savings in the development of an air quality assessment system compared to LabVIEW and MATLAB, which have slightly larger toolsets;
- the potential for application in the educational process, both for developing practical skills in students for designing virtual devices and for conducting experimental evaluations based on data obtained from the already created device. It also contributes to understanding the ecological aspects of preserving the environment's air quality.

Designing virtual instruments using computer-based measurement technologies for air quality assessment will address one of the problematic issues related to decision-making based on evaluating topographic charts in various premises. This can help reduce the dust level or minimize its impact on personnel. For example, it can improve the ventilation system in buildings, install additional filters, or implement personal protective measures against dust contamination, among others.

5 Conclusions

Modeling the process of determining the concentration of dust using computer-aided measurement technologies is a valuable educational tool for students majoring in "information and measurement technologies", which provides students with hands-on experience with the equipment and methods used in air quality monitoring and allows them to experiment with various variables in a safe and controlled environment. The proposed program for modeling the process of determining air dustiness is used during laboratory work and practical classes, including distance learning. Students can better understand the importance of maintaining good air quality using simulations. In addition to the educational process, this technology is essential to create dust control reports on industrial premises.

According to the topographic graph, utilizing the Scilab package in the industry to solve indoor air quality control problems has limited possibilities of tracking changes in dynamics. To solve this problem, applying another graphical programming environment (e.g., LabVIEW with a license) with more variable tools is possible.

Future research could focus on integrating computer-based measurement technologies with hardware and networking technologies to enable real-time monitoring and assessment processes. The incorporation of hardware and networking technologies can enhance the efficiency of data collection, analysis, and communication, leading to improved overall system performance and accuracy.

References

1. Orecchio, S., Amorello, D., Indelicato, R., Barreca, S., Orecchio, S.: A short review of simple analytical methods for the evaluation of PAHs and PAEs as indoor pollutants in house dust samples. Atmosphere **13**(11), 1799 (2022). https://doi.org/10.3390/atmos13111799

2. Gad, A., Saleh, A., Farhat, H.I., Dawood, Y.H., Abd El Bakey, S.M.: Spatial distribution, contamination levels, and health risk assessment of potentially toxic elements in household dust in Cairo city, Egypt. Toxics **10**, 466 (2022). https://doi.org/10.3390/toxics10080466
3. Alotaibi, M.O., et al.: Pollution indexing and health risk assessment of heavy-metals-laden indoor and outdoor dust in elementary school environments in Riyadh, Saudi Arabia. Atmosphere **13**(3), 464 (2022). https://doi.org/10.3390/atmos13030464
4. Szulc, J., Cichowicz, R., Gutarowski, M., Okrasa, M., Gutarowska, B.: Assessment of dust, chemical, microbiological pollutions and microclimatic parameters of indoor air in sports facilities. Int. J. Environ. Res. Public Health **20**, 1551 (2023). https://doi.org/10.3390/ijerph20021551
5. Šarkan, B., Gnap, J., Loman, M., Harantová, V.: Examining the amount of particulate matter (PM) emissions in urban areas. Appl. Sci. **13**, 1845 (2023). https://doi.org/10.3390/app13031845
6. Cho, E.-M., et al.: Reliability of low-cost, sensor-based fine dust measurement devices for monitoring atmospheric particulate matter concentrations. Int. J. Environ. Res. Public Health **16**, 1430 (2019). https://doi.org/10.3390/ijerph16081430
7. Yu, X., Shi, Y., Wang, T., Sun, X.: Dust-concentration measurement based on Mie scattering of a laser beam. PLoS ONE **12**(8), e0181575 (2017). https://doi.org/10.1371/journal.pone.0181575
8. WHO global air quality guidelines: particulate matter (PM2.5 and PM10), ozone, nitrogen dioxide, sulfur dioxide and carbon monoxide. World Health Organization (2021). https://apps.who.int/iris/handle/10665/345329. Accessed 23 June 2023
9. Gasiorowski-Denis, E. ISO – Why metrology matters in the digital era. https://www.iso.org/contents/news/2022/05/metrology-in-digital-era-1.html. Accessed 23 June 2023
10. Grasso Toro, F., Lehmann, H.: Brief overview of the future of metrology. Meas. Sens. **18**, 100306 (2021). https://doi.org/10.1016/j.measen.2021.100306
11. Gupta, N., Jaiswal, D.: Virtual instruments in the open source world? A review. Int. J. Sci. Res. **5**(2), 617–621 (2016)
12. Oborskyi, H., Gugnin, V., Perperi, L., Goloborodko, G., Goloborodko, V.: Simulation modelling in the tasks of digital engineering in the creation of information-measuring systems. Proc. Odessa Polytech. Univ. **1**(65), 129–136 (2022). https://doi.org/10.15276/opu.1.65.2022.15
13. Bilous, O., Hovorun, T., Berladir, K., Dunaeva, M.: Ensuring the quality of training engineers in a virtual environment. In: Tonkonogyi, V., et al. (eds.) Advanced Manufacturing Processes II. InterPartner 2020. LNME, pp. 765–774. Springer, Cham (2021). https://doi.org/10.1007/978-3-030-68014-5_74
14. Zheng, X., Dong, F.: Application of virtual instrument technology in the teaching of embedded system course. Int. Trans. Electr. Energy Syst. **2022**, 6721450 (2022). https://doi.org/10.1155/2022/6721450
15. Air dust tester Walcom HT-9600. https://hti-instrument.com/products/ht-9600-pm2-5detector. Accessed 23 June 2023
16. Campbell, S.L., Chancelier, J.P., Nikoukhah, R.: Modeling and Simulation in Scilab/Scicos with Scicoslab 4.4, 2nd edn. Springer, New York (2011). https://doi.org/10.1007/978-1-4419-5527-2

Check for updates

Using the Similarity Theory in Designing Vibroconveyor Dryer for Grain

Igor Palamarchuk[1], Vladislav Palamarchuk[2], Mikhailo Mushtruk[1](✉), Evgenii Shtefan[3], and Ievgenii Petrychenko[4]

[1] National University Life and Environmental Sciences of Ukraine, 15, Heroes of Defense Str., Kyiv 03041, Ukraine
mixej.1984@ukr.net

[2] Vinnytsia Institute of Trade and Economics of State University of Trade and Economics, 87, Soborna Str., Vinnitsia 21000, Ukraine

[3] National Technical University of Ukraine, 37, Peremohy Ave., Kyiv 03056, Ukraine

[4] Uman National University of Horticulture, 1, Institutska Str., Uman 20300, Ukraine

Abstract. The article describes the developed vibration conveyor dryer. This device takes advantage of the flow form while functioning, provides automatic production, realizes intensive infrared action in the fluidized-product layer, and creates energy-saving and constant flow for the technological load. The decomposition of the product mass under the influence of alternating loads leads to a reduction of internal friction and viscosity in the technological environment, which allows for increasing the heat transfer coefficient as much as possible. The process of mixing friable particles of products during their transportation in the working zone by a wave propulsion engine provides even heat treatment, which eliminates overheating of the surface layer under the action of energy-intensive infrared irradiation. This complex technological process is determined by certain factors: the coefficient of mass return, the mass of products, the speed of the conveyor belt, the size of the raw material particles, the moisture content, the coefficient of thermal conductivity, the volume performance of the process for the removed moisture, the specific temperature of the vapor, the amplitude of the oscillation of the wave propeller. It allows recommending the regime parameters and the design series of projected thermo-radiation dryers with vibration-wave transport of products when varying the main influence factors.

Keywords: Infrared Irradiation · Rape Seed · Dimensional Theory · Vibro-waving Transport · Oscillatory System · Product Innovation · R&D Investment

1 Introduction

Infrared drying is carried out by transferring the product's heat from the radioactive energy source, which either uses gas burners or special incandescent lamps (80–90% of which energy is converted into infrared radiation energy) or special radiation surface -

V. Tonkonogyi et al. (Eds.): InterPartner 2023, LNME, pp. 550–560, 2024.
https://doi.org/10.1007/978-3-031-42778-7_51

slabs. In the heat exchange process, the radiant flow penetrates partially into the capillary-porous bodies to a depth of 0.1–2 mm and is almost absorbed due to a series of reflections from the walls [1]. In this case, the heat transfer coefficient has a significant length of the drying process compared to the convective or conductive method, which decreases 30–100 times [2]. High-frequency jet drying is based on the fact that the molecules located between two plate-like electrodes are driven by oscillatory motion under the influence of an alternating electric field. This motion of the molecules causes the material to heat over its entire thickness [3]. The inner layers of the product are heated more than the external due to the contact of the latter with the environment. As a result, moisture movement from the center to the periphery increases [4].

In the same direction when drying humidity changes, which means that the gradients of temperature and humidity coincide on the sign, contributing to the migration of moisture from the center to the surface. Therefore, the speed of such drying significantly exceeds the speed of convective drying, although its implementation requires 3 ÷ 4 times more power consumption [5]. In addition, the high thermal stresses acting on the surface layer of products when moving in the working area are also problematic. Constructive improvement of thermo-radiation dryers in recent years has allowed for equalizing the energy consumption during their operation with convection machines [6]. Using vibration and wave effects has significantly reduced the heat load on the product layers. Consequently, it caused the urgency of developing promising schemes of processes and equipment for drying in the conditions of infrared irradiation.

2 Literature Review

There are undoubtedly conveyor machines among the machines that accumulate all the given signs of efficient technological equipment. The word "conveyor" is derived from the English "convey" - transport and means a conveyor or a continuous machine for moving loose lumped and piece cargo [7].

With the increase in the mechanization of production processes and the introduction of more efficient technological equipment into production, conveyor machines perform not only a pure transport function but also are used as transfer devices in technologically automated lines for the manufacture, processing of parts, and assembly of assemblies [8]. Due to high productivity, the directivity of action, and automation of control, conveyors become an integral part of the modern technological process; they establish and regulate the pace of production, ensure its rhythmicity, and increase labor productivity and output [9].

For such a machine, the following essential properties are characteristic: continuity of operation, the presence of a load-carrying body, and others. With the development of the current method of organizing production, the following properties of conveyors take on weight: the proportionality, dynamism, and direction of the process are realized with their help of them [10]. In modern conveyor production, some new defining faces arise associated with the spread of the application areas of this equipment. This is the synchronism and proportionality between the primary and auxiliary operations, the absence of a contradiction between transport and technological movements, the universality of the machine, and the possibility of implementing a complex technological action in the composition of automated production lines [11].

An essential sign of a conveyor vibrating technological machine determines the possibility of its existence, as such, is the presence of a vibration technological action with the current method of organizing the machine operation.

So the movement acts as a single, basic, and auxiliary technological movement. Vibration action is the main dynamic factor of the machine's operation. It aims to implement a pure transport movement in vibrating delivery conveyors, technological movement, or a combination of recent operations in vibrating technological machines [12]. The way of realization of transport and technological movements, that is the main components of the production process, basically determines the variety of the design of conveyor technological machines.

The proportionality between transport and technological movement serves as a classification feature and determines the process equipment's perfection and design features. On this basis, cars with separate, concomitant, and compatible performance of transport and technological action are distinguished [13].

Existing vibrating conveyors are used mainly for moving loads in different directions. Transport-technological machines are also being developed. They are efficiently carried out in transporting and processing the moving bulk product: drying, extraction, crystallization, and other heat-mass-exchange processes. Such technological systems, as a rule, are noted for the complexity of the design of their elements, and the need for special transporting devices often negates the advantages of the current processing regime [14]. In addition, significant problems are the provision of mutual functioning of transport and technological links and the possibility of increasing the level of mechanization and automation of such systems. Therefore, the search for efficient vibrating conveyor schemes for transport-technological processing of loose agricultural and food raw materials, the analysis of which is presented in Fig. 1, becomes urgent.

The efficiency of implementation, high potential, and prospects for improving these processes are observed precisely from the creative unification of these areas, as evidenced by the work of both known vibration technologists and designers A. Babichev, P. Bernick, I. Blekhman, I. Goncharevich, V. Nadutyi, V. Poturaeva, A. Spivakovsky, A. Chervonenko and recognized founders of technologies and equipment for food and processing silk-pig production, O. Burdo, G. Limonova, V. Popov, P. Rebinder, Z. Rogov, V. Stabnikov, A. Sokolenko and other outstanding scientists [15, 16].

The aim of the scientific work is the development of energy-efficient and reliable technological schemes for the implementation of vibration conveyor heat exchange processing of bulk products based on analysis of the main features, structural components, and development trends of vibrating transport-technological machines; determination of the kinematic and power characteristics of the non-vibrating nature of the vibrating systems; substantiation of rational regimes of raw material advancement in the processing zone.

3 Research Methodology

To study the influence of technological, energy, and structural parameters during infrared irradiation on the efficiency of moisture removal from the product, an experimental setup was created, the scheme of which is presented in Fig. 1.

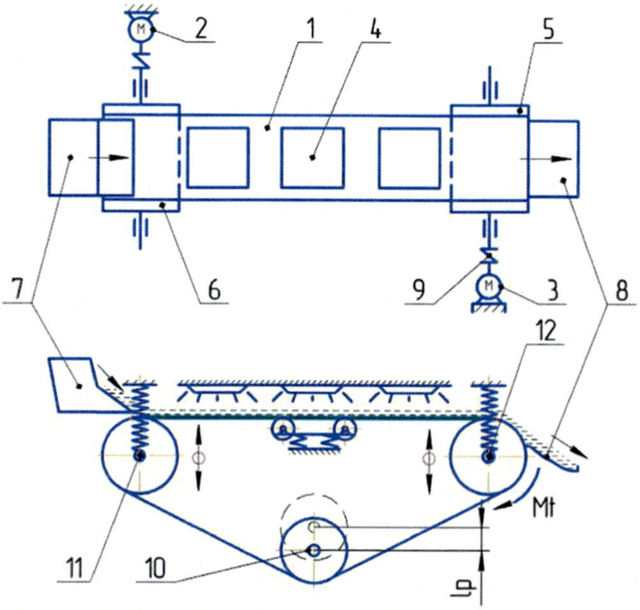

Fig. 1. Scheme of the conveyor vibrowave infrared dryer: 1 – belt; 2, 3 – exciter engines; 4 – infrared emitters; 5, 6 – rollers; 7 – feeder; 8 – receiving hopper; 9 – flexible coupling; 10 – tensioning roller; 11, 12 – unbalance vibration exciter.

When conducting research, a certain amount of whole rapeseed or soybean grain is fed to the belt, obtaining specific loading values equal to 2.5; 3.5; 5 kg/m^2, determining the mass of products per 1 m^2 of conveyor belt area. After that, we set the tape in motion at a speed of 0.13, 0.33, or 0.54 cm/s. The installation starts with a switch; the infrared illuminator's power regulator is set to the required power. The next step is to turn on the fan and the infrared illuminator using the appropriate switches. In the future, indicators of the mass and temperature of air and product will be recorded every minute. To measure air and product temperatures, we use a chromel-alumel thermocouple connected to a potentiometer, which displays temperature data on the pointer using a switch.

During the processing of food raw materials in the field of infrared radiation under the conditions of open working space, the fact is that electromagnetic waves from the radiation source penetrate the product to a depth of 2 mm and are partially or entirely absorbed in it. At the same time, electromagnetic energy is transformed into thermal energy, which causes product heating. In addition, the product is heated convectively, i.e., by the hot air of the working space with a temperature of 175–350 °C. This temperature is characterized by a short-term effect on products in conditions of constantly renewing their surface layer during transportation along the processing zone. During the movement of the tape under the infrared emitters, the product receives exposure to radiation of a specific power of 100, 200, or 300 W. At the same time. We turn on one, two, or three emitters along the tape. The power of the emitters is regulated by changing the current with the help of the laboratory autotransformer AOSN-20-220-75. The temperature of the product layer is determined with the help of a pyrometer, the indicator of which is observed through the windows in the casing of the installation, which are located after each of the emitters in the direction of the tape. Grain moisture is determined with the help of an AQUA-15 moisture meter, as well as from the difference in the weight of the product sample and the weight of the same sample after drying it in an electric drying cabinet. The vibration frequency of the vibration wave driver of the infrared dryer under study is measured using a UNI-T UT372 frequency meter.

The area of the conveyor belt determined the area of the updated surface for one layer of products. The number of product layers was calculated considering the thickness of the processed material and the average grain size.

Amplitude frequency, power, and energy characteristics can be singled out among the main parameters of the studied vibrating machines that optimize processing modes. Amplitude-frequency characteristics make it possible to identify the areas of resonance peaks and the established mode of operation of executive bodies of machines. The power and energy characteristics make it possible to estimate the energy consumption of the drive and build the power balance of vibration exciters. A comprehensive analysis of the specified parameters allows for finding modes corresponding to the maximum system dynamics with minimum energy consumption per drive [17]. When performing experimental studies, German Robotron equipment was used to evaluate the kinematic parameters of vibration, electromechanical equipment to ensure a slow change in the frequency of rotation of the drive shaft of the vibration exciter, and a mechanism for stepwise changes in the eccentricity of the drive shaft was developed. With the help of this equipment, the values of the main parameters that characterize the imbalance of the oscillating system, namely the equilibrium of the unbalanced inertial forces and the equilibrium moment of these forces, were determined. Variation of these parameters by magnitude and direction allows us to note such main imbalance factors as directional force and moment, which can vary in magnitude, but not in direction; circular force and moment, which can change only in direction; elliptic force and moment that can vary both in magnitude and direction at a constant value.

When processing the results of experimental studies, statistical analysis was used, which is based on the theory of random errors and proposed by the English mathematician V.S. Gosset (Student).

4 Results and Discussion

The equipment for drying products in high-frequency jets includes a carrier conveyor, electrodes with a power supply system, and devices for transforming the jet of industrial frequency into high-frequency and voltage current. The developed installation contains a flexible load-carrying body on which a running or standing wave is created at the work of mechanical vibroscopes 2, 3 (Fig. 1).

Such a wave promotes the transportation of products from feeder 7 and its intense mixing. It reduces the thermal intensity of the surface layer while maintaining a sufficiently high flow rate. At the same time, it is enough to provide oscillations only for rollers 5 and 6 to maintain the high kinetics of the investigated process. It significantly reduces the drive's power consumption compared to the vibration conveyor installation. Based on the presented scheme, the experimental-industrial model of infrared Vibro-wave drying was [18].

The main factors influencing the process, their experimental values, and the dependences of their changes, which were obtained with the help of the developed experimental model, were analyzed to create a mathematical model.

The mass transfer coefficient β is determined from the formula:

$$\beta = \frac{\Pi_V}{\Delta X \cdot S_g}, \text{ m/s} \tag{1}$$

where S_g is the surface of the irradiated grain: $S_g = 0{,}1413\ m^2$;

$$\Delta X = \frac{m_a}{m_h} = X_p - X_i \tag{2}$$

where a – a mass of air coolant; $m_в$ – a mass of lost humid.

Then we calculate the mass exchange criteria for the process.

The Pecle criterion is defined as:

$$Pe = \frac{\nu_C \cdot d}{a} \tag{3}$$

where a = 12,6 • 10^{-8} m^2/s (a coefficient of temperature conductivity of the grain); d – diameter of the grain.

The Stanton number could be found by using the following formula:

$$St = \frac{\beta}{\nu_C} \tag{4}$$

The Burdo [1] criterion is calculated by using the following formula:

$$Bu = \frac{E_I}{Q_B} \tag{5}$$

where $E_I = N_I \cdot \tau$ is the energy released by infrared irradiation; $N_I = N_{IR} \cdot N_P$ – total energy consumption;

$N_P = 300$ W – the power of the drive mechanism, N_{IR} – power when the unit of radiators is operated;

$Q_B = m_B \cdot r$ – the amount of heat needed to evaporate the moisture mass mB; **r** - the specific heat of steam generation: r = 2.3 · 10^6 J/kg (m^2/s^2).

So, the Burdo criterion looks as follows: $Bu = \frac{N_I \tau}{m_B r}$.

For further calculations, we use the value of the volume performance of the dryer for the removed moisture, which is defined as $\Pi_V = V_m \cdot \Pi_B$,

where Π_B mass productivity of the dryer for the removed moisture:

$\Pi_B = \frac{m}{\tau}$; $V_m = 1/\rho = 0,849$ m^3/kg – the volume of a unit mass of removed moisture.

Since

$$\Pi_V = \frac{m_B}{\rho_B \tau} \cdot \frac{N_I}{N_I r} = \frac{N_I}{\rho_B \tau} \cdot \frac{1}{Bu} \tag{6}$$

Then

$$Bu = \frac{N_I}{\Pi_V \rho_B \tau} \tag{7}$$

The basis of this method lies in the second theorem of the similarity of Federmann-Buckingham, which enables the processing of experimental data in the form of a general criterion equation [19].

The intensity of the studied drying process is determined by the coefficient of mass transfer β, the magnitude of which is significantly influenced by such parameters as the specific mass of products of P_S, the speed of the conveyor belt ν_c, the size of the raw material particles d_g, the moisture density ρ_B, the coefficient of temperature conductivity a, the volume performance of the process for moisture P_V, specific vaporization temperature r. In addition, the course of the study process under the action of microwave irradiation is determined by the power of irradiation $N_{оп}$ concerning the energy Q_B for vapor formation and, accordingly, the value of r.

Thus, the desired functional dependence can be represented in the form

$$\beta = f\left(\Pi_V, N_I, P_S, \nu_C, a, \rho_B, r, d_g\right). \tag{8}$$

For a given factor space, the number of variables 9 equals the quantity of dimensions 3. Therefore, by the π-theorem of the number of dimensionless complexes, 9 − 3 = 6.

Following the principle of dimensional analysis, dependence (8) is represented as a power series.

$$\beta = \Pi_V^k, N_I^l, P_S^m, \nu_C^n, a^\delta, \rho_B^P, r^\varepsilon, d_g^\eta \tag{9}$$

To implement the mathematical analysis of Eq. (8), we construct a matrix based on which the following system of algebraic equations is formed.

$$\begin{cases} l + m + p = 0 \\ 3k + 2l - 2m + n + 2\delta - 3p + 2\varepsilon + \eta = 1 \\ -k - 3l - n - \delta - 2\varepsilon = -1 \end{cases}$$

$$\begin{gathered} \text{So} \Rightarrow l = -m - p;\ 2k - l - 2m + \delta - 3p + \eta = 0;\ \delta = 1 - k - 3l - n - 2\varepsilon \\ m = 2k - 2p + \delta + \eta;\ \eta = m - 2k + 2p - \delta \end{gathered} \tag{10}$$

Taking into account expressions (9), the dependence (8) was presented as following

$$\begin{gathered} \beta = \Pi_V^k, N_I^l, P_S^m, \nu_C^n, a^{(1-k-3l-n-2\varepsilon)} \\ \rho_B^{(-l-m)}, r^\varepsilon, d_g^{(m-2k+2p-\delta)} \\ or \\ \beta = \left(\frac{\Pi_V}{ad_g^2}\right)^k \cdot \left(\frac{N_I}{a^3\rho}\right)^l \cdot \left(\frac{P_S \cdot d_g}{\rho}\right)^m \cdot \left(\frac{\nu_C}{a}\right)^n \cdot \left(\frac{r}{a^2}\right)^\varepsilon \cdot a \cdot d_g^{(2p-\delta)} \end{gathered} \tag{11}$$

We rewrite (10) as follows

$$\frac{\beta}{\nu} = \frac{a}{\nu_C}\left(\frac{\nu_C d_g}{a}\right)^k \cdot \left(\frac{N_I}{\Pi_V \rho r}\right)^l \cdot \left(\frac{\Pi_V \cdot r}{a^3}\right)^l \cdot \left(\frac{P_S \cdot d_g}{\rho}\right)^m \cdot \left(\frac{r}{a^2}\right)^\varepsilon \cdot d_g^{(2p-\delta-)}. \tag{12}$$

Taking into account the dependencies mentioned above for the similarity criteria, we transform (11) into the following

$$St = A \cdot Pe^{n} \cdot Bu^{l} \cdot k.$$

$k = \frac{1}{\nu C}\left(\frac{\Pi_V \cdot r}{a^3}\right)^{l} \cdot \left(\frac{P_S \cdot d_g}{\rho_B}\right)^{m}$ is a function determined by the technological parameters that determine the productivity of the investigated process;

A – constant value determined by the parameters of the object of processing:

$$A = a\left(\frac{r}{a^2}\right)^{\varepsilon} \cdot d_g^{(2p-\delta-m)}$$

As a result

$$St = \left(\frac{rcd_g^2}{a^2}\right)^{\varepsilon} \cdot Pe^{n} \cdot Bu^{l}\left(\frac{a}{\nu c d_g}\right) \cdot \left(\frac{\Pi_V \cdot r \cdot d_g}{a^3}\right)^{l} \cdot \left(\frac{P_S \cdot d_g}{\rho \cdot d_g}\right)^{m} \quad (13)$$

The indices n, l, and A derived from the results of the graph-analytical analysis using the following method [20].

According to experimental data, a graph of the function $St = f(Pe)$ was constructed, from which we find the power index $n = tg(\alpha)$. Similarly, the second indicator of power $l = tg(\gamma)$ was determined from the function graph $\frac{St}{Pe^n} = f(Bu)$.

From the graph of the function $\frac{St}{Pe^n Bu^l} = f(k)$, we find the value of the parameter A.

Using the obtained database [9, 10], we find the investigated mass transfer process equation in the criterion form.

$$St = A \cdot Pe^{n} \cdot Bu^{l} \cdot \frac{1}{\nu} \cdot \left(\frac{\Pi_V \cdot r}{a^3}\right)^{l} \cdot \left(\frac{P_S \cdot d}{\rho}\right)^{m} \quad (14)$$

As a result of the graphic-analytical analysis, it was determined that: $n = 1{,}08$; $l = 1{,}2$; $m = 1{,}56$.

Using the compound Eq. (14) and the developed program, we find the recommended range of parameters of the working regime for infrared drying of rape in the moving layer of products (Fig. 2).

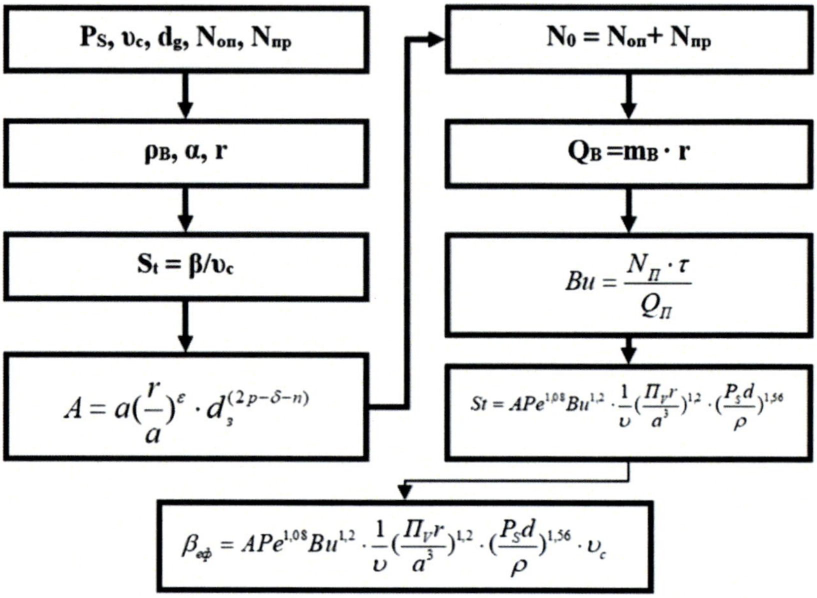

Fig. 2. Block diagram of the calculation of the intensity of mass transfer in the infrared drying of rape crops.

5 Conclusions

Intensification of processes of drying of friable lubricants of raw materials in modern technological processes is realized with the application of radiation and inductive methods of heat transfer, when combined with convection of energy carrier, when creating pseudo-liquefied layers of products, in particular, under the action of vibrational impulses. According to the experimental data, using the method of "dimensional analysis" and the Fedman-Buckingham theorem, a criterion equation for the process of mass transfer under infrared drying in a moving product layer, which is determined by the criteria of Pekle, Stanton, Burdo, is a function that characterizes the productivity of the investigated process; and allows you to form a recommended set of working mode.

References

1. Şevik, S., Aktaş, M., Dolgun, E.C., Arslan, E., Tuncer, A.D.: Performance analysis of solar and solar-infrared dryer of mint and apple slices using energy-exergy methodology. Sol. Energy **180**, 537–549 (2019). https://doi.org/10.1016/j.solener.2019.01.049
2. Capece, M., Larson, J.: Improving the effectiveness of the conical screen mill as a dry-coating process at lab and manufacturing scale. Pharm. Res. **39**, 3175–3184 (2022). https://doi.org/10.1007/s11095-022-03196-z
3. Delfiya, D.A., Prashob, K., Murali, S., et al.: Drying kinetics of food materials in infrared radiation drying: a review. J. Food Process. Eng. **45**(6), e13810 (2022). https://doi.org/10.1111/jfpe.13810

4. Sun, Q., Zhang, M., Mujumdar, A.S.: Recent developments of artificial intelligence in drying of fresh food: a review. Crit. Rev. Food Sci. Nutr. **59**(14), 2258–2275 (2019). https://doi.org/10.1080/10408398.2018.1446900
5. Selimefendigil, F., Özcan Çoban, S., Öztop, H.F.: Convective drying of a moist porous object under the effects of a rotating cylinder in a channel. J. Therm. Anal. Calorim. **141**(5), 1569–1590 (2020). https://doi.org/10.1007/s10973-019-09140-5
6. Mukhitdinov, J.P., Safarov, E.X.: Reviewing technologies and devices for drying grain and oilseeds. Oilseeds Focus **8**(1), 32–35 (2022). https://doi.org/10.51346/tstu-02.21.3-77-0014f
7. Czako, P., Zajác, P., Čapla, J., et al.: The effect of UV-C irradiation on grape juice turbidity, sensory properties, and microbial count. Potravinarstvo Slovak J. Food Sci. **12**(1), 1 (2018). https://doi.org/10.5219/856
8. Ha, S.-Y., Liu, J.-G.: A simple proof of the Cucker-Smale flocking dynamics and mean-field limit. Comm. Math. Sci. **7**, 297–325 (2019). https://doi.org/10.4310/cms.2009.v7.n2.a2
9. Zhao, C., He, B., Liu, J., Han, Y., Wen, B.: Design method of dynamic parameters of a self-synchronization vibrating system with dual mass. J. Multi-body Dyn. **232**(1), 3–20 (2018). https://doi.org/10.1177/1464419316689643
10. Sukhenko, Y., Mushtruk, M., Vasyliv, V., Sukhenko, V., Dudchenko, V.: Production of pumpkin pectin paste. In: Ivanov, V., et al. (Eds.) Advances in Design, Simulation, and Manufacturing II (DSMIE-2019). LNME, pp. 805–812. Springer, Cham (2019). https://doi.org/10.1007/978-3-030-22365-6_80
11. Casali, B., Brenna, E., Parmeggiani, F., Tessaro, D., Tentori, F.: Enzymatic methods for the manipulation and valorization of soapstock from vegetable oil refining processes. Sustain. Chem. **2**(1), 74–91 (2021). https://doi.org/10.3390/suschem2010006
12. Yang, H., Wang, Z., Zhang, T., Du, F.: A review on vibration analysis and control of machine tool feed drive systems. Int. J. Adv. Manuf. Technol. **107**(1–2), 503–525 (2020). https://doi.org/10.1007/s00170-020-05041-2
13. Karapuzha, A.S., Fraser, D., Zhu, Y., et al.: Effect of solution heat treatment and hot isostatic pressing on the microstructure and mechanical properties of Hastelloy X manufactured by electron beam powder bed fusion. J. Mater. Sci. Technol. **98**, 99–117 (2022). https://doi.org/10.1016/j.jmst.2021.04.059
14. Kim, H., Wu, Z., Eedugurala, N., Azoulay, J.D., Ng, T.N.: Solution-processed phototransistors combining organic absorber and charge transporting oxide for visible to infrared light detection. ACS Appl. Mater. Interfaces **11**(40), 36880–36885 (2019). https://doi.org/10.1021/acsami.9b08622
15. Gómez-Estaca, J., et al.: Assessment of a healthy oil combination structured in ethyl cellulose and beeswax oleogels as animal fat replacers in low-fat, PUFA-enriched pork burgers. Food Bioprocess Technol. **12**(6), 1068–1081 (2019). https://doi.org/10.1007/s11947-019-02281-3
16. Singh, R., Kaushik, R., Gosewade, S.: Bananas as underutilized fruit having huge potential as raw materials for food and non-food processing industries: a brief review. Pharma Innov. J. **7**(6), 574–580 (2018). https://doi.org/10.1007/s11947-019-02281-3
17. Thiel, E., Ziegler, M., Studemund, T.: Localization of subsurface defects in uncoated aluminum with structured heating using high-power VCSEL laser arrays. Int. J. Thermophys. **40**(2), 17 (2019). https://doi.org/10.1007/s10765-018-2478-9
18. Sukhenko, Y., Sukhenko, V., Mushtruk, M., Litvinenko, A.: Mathematical model of corrosive-mechanic wear materials in technological medium of food industry. In: Ivanov, V., et al. (Eds.) Advances in Design, Simulation, and Manufacturing II (DSMIE-2018). LNME, pp. 498–507. Springer, Cham (2018). https://doi.org/10.1007/978-3-319-93587-4_53

19. Dehghani, S., Haghighi, M.: Influence of various irradiation time on sono-functionalization of zirconia-doped mesoporous-silica by sulfuric acid for biofuel production from waste cooking oil. Waste Biomass Valorization **22**, 1–14 (2019). https://doi.org/10.1007/s12649-019-00715-9
20. Dhinesh, B., Raj, Y., Kalaiselvan, C., Krishna Moorthy, R.: A numerical and experimental assessment of a coated diesel engine powered by high-performance nano biofuel. Energy Convers. Manage. **171**, 815–824 (2018). https://doi.org/10.1016/j.enconman.2018.06.03

Energy Characteristics of the Oil Vortex Chamber Supercharger

Andrii Rogovyi[1](✉), Serhiy Lukianets[1], Sergey Krasnikov[2], Iryna Hrechka[1], and Oleksandr Shudryk[1]

[1] National Technical University Kharkiv Polytechnic Institute, 2, Kyrpychova St., Kharkiv 61000, Ukraine
asrogovoy@ukr.net

[2] Kharkiv National Automobile and Highway University, 25, Yaroslava Mudrogo St., Kharkiv 61002, Ukraine

Abstract. Very often, deposits with a great depth of occurrence are used irrationally due to pumping equipment's absence or low efficiency. As the efficiency of classic pumps decreases, it becomes expedient to search for and use new pumping technologies based on jet technology. Research of oil vortex chamber superchargers can lead to advanced designs with high efficiency. The purpose of the paper is to determine the integral characteristics of oil pumping by vortex chamber pump and to determine the effect on the characteristics of dissolved gas and various overpressures in the suction channel. The study consisted of three stages: experimental adequacy confirmation of the proposed fluid flow mathematical models in the vortex chamber pump, simulation of pump characteristics when it pumps oil with dissolved gas, and obtaining pump characteristics. It was found that the oil vortex chamber pump efficiency exceeds the efficiency of the serial industrial jet pump of the direct-flow type by 2.5 times, and the pump's overall size is reduced by more than 3 times. As the suction pressure increases, there is a linear decrease in the relative useful pressure at the outlet of the pump in the absence of static pressure at the outlet.

Keywords: Vortex Chamber Supercharger · Experiment · Numerical Simulation · Oil · Gas Content · Energy Efficiency

1 Introduction

The problem of insufficient energy resources in European countries requires the development of high-viscosity oil deposits located at a great depth [1]. Very often, deposits with a great depth of occurrence are used irrationally due to the pumping equipment's absence or low efficiency [2]. Therefore, the problem of adjusting wells producing high-viscosity oils is quite relevant [3]. Improving the oil pumping efficiency is difficult due to the significant variety of composition, physical properties, and impurities that oil can contain [4]. Therefore, the efficiency of the used pump may not meet the designed parameters [5] and be significantly less than the same pumps when pumping water or oil from the hydro drives [6]. As the efficiency of classic pumps decreases, it becomes expedient to search for and use new pumping technologies based on jet technology [7].

V. Tonkonogyi et al. (Eds.): InterPartner 2023, LNME, pp. 561–570, 2024.
https://doi.org/10.1007/978-3-031-42778-7_52

Therefore, jet technology allows us to avoid many disadvantages of pumping with decreased efficiency. Therefore, it becomes expedient to improve jet pumps and search for new types, which will significantly increase the efficiency of oil pumping.

The work aims to determine the integral characteristics of oil pumping by the VCP and the effect on the characteristics of dissolved gas under different overpressures in the suction channel.

2 Literature Review

Jet pumps can be divided into two main types: direct-flow jet pumps and pumps with vortex chamber-vortex chamber pumps (VCP) [8]. The first is already used in the oil and gas complexes [9], and the second was developed for pumping bulk mediums. Therefore, their research and characteristics for oil pumping have not been conducted to date. In general, jet technology has indisputable advantages: durability, reliability, and simplicity of construction [10]. However, jet pumps have very low efficiency [11]. Their efficiency for pumping oil is often in the range of 7–15%. Modern devices based on renewable energy sources can make it possible to build constructions based on jet pumps autonomously and efficiently [12, 13]. On the other hand, with a significant decrease in the efficiency of classical pumps [14], the efficiency of jet pumps practically does not depend on the medium they pump. Therefore, further development of vortex chamber superchargers and research into their behavior during oil pumping can lead to perfect designs with high efficiency [15].

There are several reasons for using jet technology in oil wells: installing the jet ejector above the dynamic level of the well allows to stabilize of the supply of the electric centrifugal pump; extraction of oil gas from the annular space of the well, which facilitates the rise of products to the surface [16].

VCPs generally have significantly higher efficiency pumping bulk mediums than direct-flow jet pumps. Two work processes can be implemented in VCP [8]. These work processes differ in energy parameters at the pump outlet and are implemented by changing the geometric parameters of two main channels: the tangential channel of the pump outlet and the tangential inlet channel (the pump scheme is shown in Fig. 2). The VCP consists of a vortex chamber and four inlet and outlet channels: two axial suction channels and two tangential inlet and outlet channels. The positive influence of the centrifugal force allows counting on an increase in the efficiency of the jet pump with an increase in the hydraulic energy of the output flow. Theoretical studies and experimental characteristics confirm it. The dynamics of the mixing process [17] practically do not affect the pump characteristics [18]. The wear of the pump flow part elements practically does not affect the reliability and durability since the wear only leads to a deterioration in the hydromechanical properties [19].

Previous studies of devices with the vortex chamber showed that the gas contained in the liquid could be released from it in the low-pressure zone and worsen the operating conditions of the vortex chamber devices. This gas forms vortex cords and interferes with the implementation of the work process [20]. For pumping oil, the presence of gas also leads to the need to consider this factor. Therefore, it becomes relevant to study oil VCP, considering the possible gas dissolved in oil. In addition, the specifics of oil pumping and the location of the jet pump in the well require investigating the effect of suction channel excess pressure on the pump characteristics.

3 Research Methodology

Based on the specified aim, the following research tasks should be solved:

- mathematical modeling of the fluid flow in the VCP
- computation model
- model verification
- experimental study of the flow in the VCP
- the experimental and calculated data were compared with further modeling based on the CFD approach and the determining the integral characteristics of oil pumping by the VCP and the effect on the characteristics of dissolved gas under different overpressures in the suction channel.

The study consisted of three stages (Fig. 1): experimental confirmation of the proposed fluid flow mathematical models' adequacy in the vortex chamber pump; simulation of pump characteristics during pumping of oil containing dissolved gas, and obtaining pump characteristics.

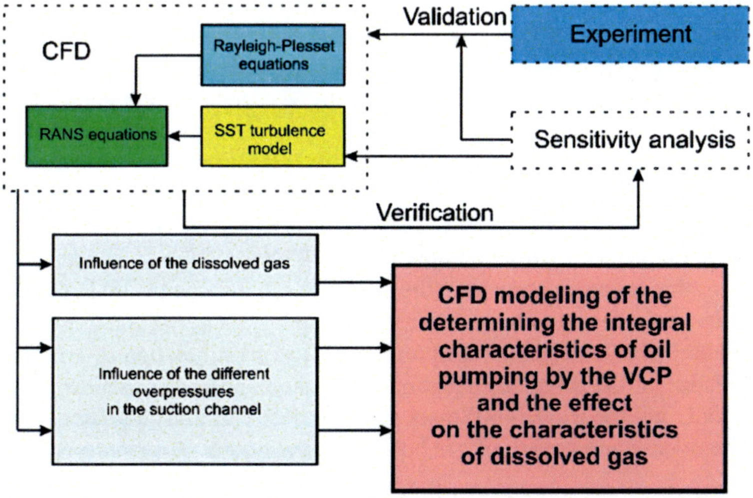

Fig. 1. A flowchart of the methodology applied in the present analysis.

Experimental investigations were carried out on a specially created experimental installation with a transparent VCP (Fig. 2), which made it possible to check adequacy at the qualitative and quantitative levels. The devices used for experimental studies passed metrological checks and had an error not exceeding 1%. The manometers were used to determine pressures and flow rates using Venturi tubes indirectly. Flow rates and pressures were determined in each of the four VCP channels. The main VCP geometric dimensions: the diameter of the vortex chamber is 50 mm relative to the diameter of the vortex chamber throat, the diameters of the tangential inlet channels are 0.86, and the outlet diameters are 1.14. Thus, this pump implements the operating mode with oil suction through both axial channels.

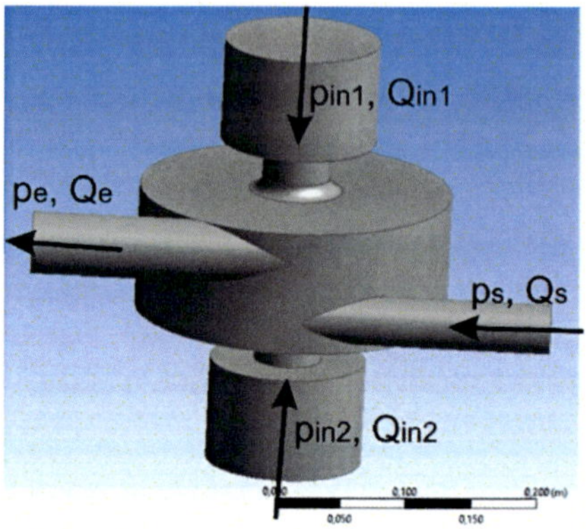

Fig. 2. Calculation scheme of the vortex chamber pump.

Mathematical modeling was carried out in the OpenFoam application programs. A three-dimensional model [21] and its meshing [22] were created [23]. OpenFOAM is free open-source CFD software. The model adequacy based on comparing calculation results with experimental data was confirmed by comparing pressures, flow rates, and efficiency. The mathematical model equations were the Reynolds-averaged Navier-Stokes equations, the Rayleigh-Plesset equations for gas motion studying, the continuity equation, and the SST turbulence model equations. During the research, an analysis of the solution sensitivity of the mesh was carried out. Solving the equations was done in a stationary setting, and the solution process ended when the residuals reached values of 10–5. In addition, achieving a stable flow rate in all channels was considered a mandatory condition for the calculation end (the flow rate did not change by iteration for 100 iterations). In total, the solution process required approximately 3000 iterations.

The research was conducted based on the numerical simulation of a mathematical model consisting of the Reynolds equations [24], the continuity equation, the SST turbulence model, and the Rayleigh-Plesset equation [25]. The last equation was used to calculate the release and growth of gas bubbles in the liquid:

$$R_B \frac{d^2 R_B}{dt^2} + \frac{3}{2}\left(\frac{dR_B}{dt}\right)^2 + \frac{2\sigma}{\rho_f R_B} = \frac{p_\nu - p}{\rho_f} \tag{1}$$

where R_B is the bubble radius; p_ν is the vapor pressure; p is the ambient pressure; ρ_f is fluid density; σ is the surface tension at the gas-liquid interface.

To compare the VCP performance characteristics, two types of calculations were performed: the first calculation is the influence of pressure at the suction channel - a mixture of oil and gas with a gas content of 5% without considering the Rayleigh-Plesset equation, i.e., without a cavitation model. The second simulation was performed according to the above mathematical model of cavitation.

The boundary conditions were as follows: dynamic pressure was set at the input tangential channel, at the input axial channels – an open boundary with a pressure of 96000 Pa, allowing the liquid to enter the calculation domain and leave it without complicating the calculation. To simulate the pressure influence at the suction channels, the pressure value at this boundary varied in a wide range from 0 to 96 kPa. Zero pressure is set at the outlet tangential channel. The range of changes for pressure at outlet channel: 0–500 kPa. Dynamic pressure is set at the inlet channel. Physical properties of the liquid for calculation: oil with a density of $\rho = 850\ \text{kg/m}^3$ and dynamic viscosity of $\mu = 0,004 Pa \cdot s$. Flow rates at all the channels were calculated during the calculations.

4 Results and Discussion

4.1 The Adequacy of the Mathematical Model

To check the adequacy of the proposed mathematical model, a comparison of the VCP characteristics when it pumps air and water with the experimental characteristics was carried out (Fig. 3).

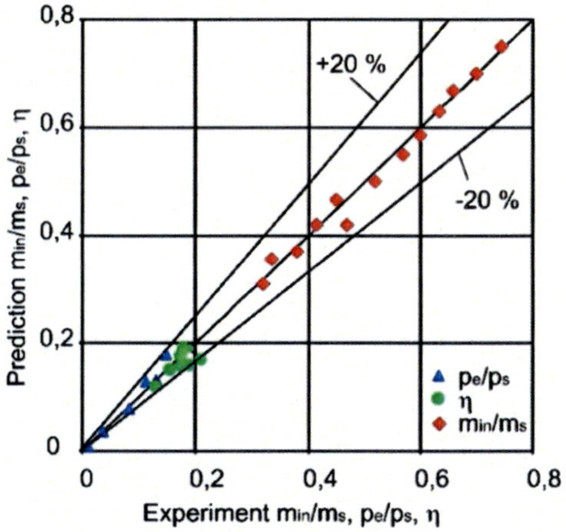

Fig. 3. The errors in determining the pump integral parameters when operating on water.

The results analysis of the calculation errors showed that the determination of the flow rates has a maximum error of 9%, efficiency – 17.5%, and pressure ratio – 16.6%. However, the maximum errors correspond to small values of these ratios, which are not points of efficient pump operation. The errors of the calculated values at the operating points did not exceed 8% for all parameters. The maximum measurement error is not exceeded 1% for flow rates and 0.5% for pressures.

4.2 Simulation Results

To determine the pressure influence at the suction channel, the study was carried out for five different pressures: 96 kPa, 80 kPa, 60 kPa, 30 kPa, and 0 concerning the pressure at the inlet tangential channel, the characteristics with $p_{in} = 96$ kPa (at the inlet suction channel) and $p_e = 0$ (at the outlet tangential channel): 0.27; 0.23; 0.18; 0.094 and 0, respectively.

An analysis of the maximum flow rate (at the inlet axial channels) shows that as the pressure at the suction channel increases, the flow rate also increases. It can be seen in Fig. 4, which presents the VCP characteristics. Increasing the gauge pressure at the suction channel to 0.27 increases the maximum suction flow rate by 44%.

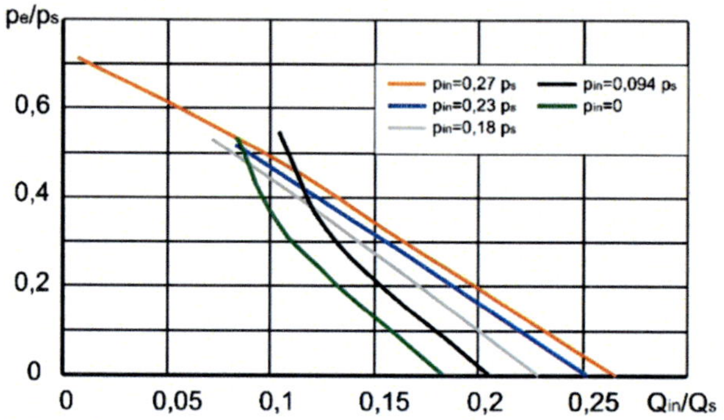

Fig. 4. VCP performances for different suction pressure.

An increase in the relative suction flow rate does not allow for increased pump efficiency due to decreased outlet pressure. The equation determined the pump efficiency:

$$\eta = \frac{p_e - p_{in} + \frac{\rho}{2}\left(V_e^2 - V_{in}^2\right)}{p_s - p_e + \frac{\rho}{2}\left(V_s^2 - V_e^2\right)} \cdot \frac{Q_{in}}{Q_s} \tag{2}$$

Increasing the suction pressure reduces the relative net pressure at the pump outlet by 30%. The relative net pressure is calculated by the ratio of the net pressure to the working one according to the general Eqs. (3) and (4) for jet pumps.

The net pressure:

$$p_e^* = p_e - p_{in} + \frac{\rho}{2}\left(V_e^2 - V_{in}^2\right) \tag{3}$$

The working pressure:

$$p_s^* = p_s - p_e + \frac{\rho}{2}\left(V_s^2 - V_e^2\right) \tag{4}$$

A decrease in the relative net outlet pressure reduces the pump efficiency. In the area of relatively small suction flow rates, the efficiency and relative pressure increase for pumping options without using gauge pressure at the suction axial channel.

Summarized information on the effect of the gauge pressure increase at the suction channel is shown in Fig. 5. There is a linear decrease in the relative net pressure at the pump outlet in the absence of static pressure at the outlet. Also, there is an almost linear

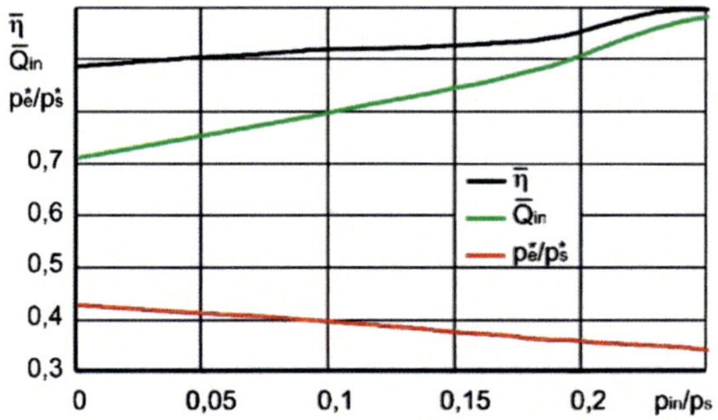

Fig. 5. The influence of suction channel gauge pressure on the integral VCP characteristics.

increase in suction flow rate, but with a greater inclination angle of the performance, which leads to an increase in relative efficiency.

The serial industrial direct-flow jet type pump NSFB 39/45 efficiency is approximately 3%. The obtained efficiency of the vortex chamber pump is more than 2.5 times higher than the efficiency of NSFB 39/45. In addition, due to the use of the vortex chamber, the overall pump size is reduced by more than 3 times.

Figure 6a, b show the distribution of oil axial velocities in the pump flow part. The suction pattern is generally symmetrical relative to the horizontal plane in the case of the calculation without considering cavitation (Fig. 6a).

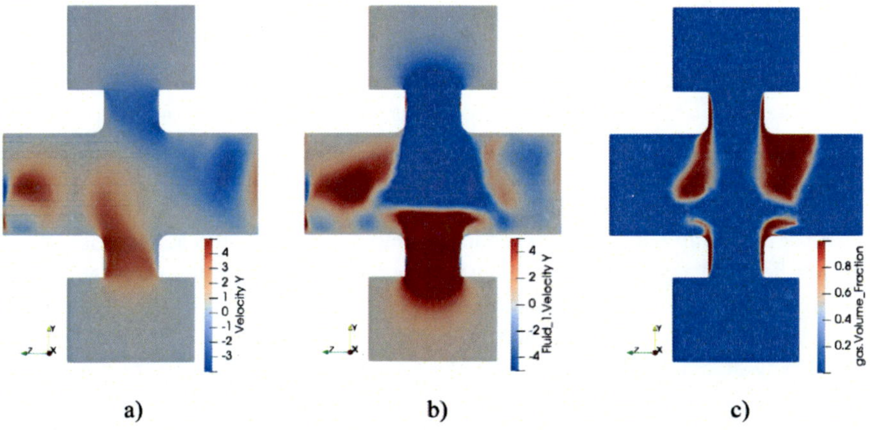

Fig. 6. The oil axial velocity distribution in the VCP (a, b) and gas volume concentration distribution (c).

It is possible to note the symmetry of the flow pattern relative to the vertical plane as well in the case of the calculation with the cavitation model (Fig. 6b).

The simulation analysis shows a slight change in gas concentration for the case of the model without cavitation for gas release and a significant release of gas from oil in the case of calculation according to the model with cavitation (Fig. 6c).

Features of two-phase fluid modeling allow to study of the kinematic parameters of each phase movement, and based on the results of the velocities calculation, it can be stated that neglecting the cavitation model leads to the fact that the gas concentration in the flow part of the VCP is very small and does not exceed 1%. It can be considered an inadequate result because the physical experiments prove that a decrease in pressure near the rotation axis of the vortex chamber nevertheless leads to a significant release of gas from the liquid. For further calculations, it is necessary to use the Rayleigh-Plesset equation.

The integral parameters calculation confirmed that the problem of obtaining flow patterns is the mismatch of the calculation results according to different models. It should be noted that the value of the pump efficiency based on the cavitation model turned out to be significantly lower than without considering the cavitation model. It can be explained by the release of gas from the oil, which creates problems with oil suction into the pump. Another reason for the decrease in efficiency is that in Eq. 2, density is used to determine efficiency. In the case of a large gas concentration, the density of the mixture is significantly different from the density of oil, which can affect the accuracy of the efficiency calculation.

5 Conclusions

Based on mathematical modeling, by jointly solving the Reynolds, Rayleigh-Plesset equations, the SST turbulence model, and the continuity equation, the integral characteristics of oil transportation by a vortex chamber pump and the influence of the suction pressure.

The effect of the Rayleigh-Plesset equation on the kinematic parameters of gas movement in the vortex chamber of the VCP was revealed. The absence of the Rayleigh-Plesset equations in the mathematical model leads to the absence of a vortex cord in the center of the vortex chamber, which contradicts the visual observation of the phenomenon in the experiment and the experiments of other authors for vortex devices.

With the suction pressure increase, there is a linear decrease in the relative net pressure at the pump outlet in the absence of static pressure at the outlet. Also, there is an almost linear increase in suction flow rate, but with a greater inclination angle of the performance, which leads to an increase in relative efficiency.

The vortex chamber pump efficiency is found to be more than 2.5 times higher than the efficiency of serial industrial direct-jet pumps. The overall pump size is reduced by more than 3 times.

References

1. Bazaluk, O., Dubei, O., Ropyak, L., Shovkoplias, M., Pryhorovska, T., Lozynskyi, V.: Strategy of compatible use of jet and plunger pump with chrome parts in oil well. Energies **15**(1), 83 (2021). https://doi.org/10.3390/en15010083
2. Voloshina, A., Panchenko, A., Titova, O., Milaeva, I., Pastushenko, A.: Prediction of changes in the output characteristics for the planetary hydraulic motor. In: Tonkonogyi, V., et al. (Eds.) Advanced Manufacturing Processes II (InterPartner 2020). LNME, pp. 744–754. Springer, Cham (2021). https://doi.org/10.1007/978-3-030-68014-5_72

3. Panevnyk, D.: Study on characteristics of a downhole vortex jet pump. Rom. J. Petrol. Gas Technol. **2** (2021). https://doi.org/10.51865/JPGT.2021.02.03
4. Toteff, J., Asuaje, M., Noguera, R.: New design and optimization of a jet pump to boost heavy oil production. Computation **10**(1), 11 (2022). https://doi.org/10.3390/computation10010011
5. Panchenko, A., Voloshina, A., Luzan, P., Panchenko, I., Volkov, S.: Kinematics of motion of rotors of an orbital hydraulic machine. IOP Conf. Ser.: Mater. Sci. Eng. **1021**, 012045 (2021). https://doi.org/10.1088/1757-899X/1021/1/012045
6. Voloshina, A., Panchenko, A., Titova, O., Pashchenko, V., Zasiadko, A.: Experimental studies of a throughput of the distribution systems of planetary hydraulic motors. IOP Conf. Ser.: Mater. Sci. Eng. **1021**, 012054 (2021). https://doi.org/10.1088/1757-899X/1021/1/012054
7. Merzliakov, I., Pavlenko, I., Chekh, O., Sharapov, S., Ivanov, V.: Mathematical modeling of operating process and technological features for designing the vortex type liquid-vapor jet apparatus. In: Ivanov, V., et al. (Eds.) Advances in Design, Simulation and Manufacturing II (DSMIE 2019). LNME, pp. 613–622. Springer, Cham (2020). https://doi.org/10.1007/978-3-030-22365-6_61
8. Rogovyi, A., Korohodskyi, V., Neskorozhenyi, A., Hrechka, I., Khovanskyi, S.: Reduction of granular material losses in a vortex chamber supercharger drainage channel. In: Ivanov, V., Pavlenko, I., Liaposhchenko, O., Machado, J., Edl, M. (Eds.) Advances in Design, Simulation and Manufacturing V (DSMIE 2022). LNME, pp. 218–226. Springer, Cham (2022). https://doi.org/10.1007/978-3-031-06044-1_21
9. Asuaje, M., Toteff, J., Noguera, R.: Evaluation of a jet-pump for the improvement of oil-water flow in pipeline loops using CFD tools. In: E3S Web of Conferences, vol. 321, p. 02012 (2021). https://doi.org/10.1051/e3sconf/202132102012
10. Mochalin, I., Cai, J., Berezhnyi, O., Brazhenko, V.: Computational approach to the turbulent boundary layer of a rotating porous cylinder with strong suction based on an algebraic turbulence model. Eng. Appl. Comput. Fluid Mech. **14**(1), 1472–1488 (2020). https://doi.org/10.1080/19942060.2020.1835737
11. Merzliakov, I., Pavlenko, I., Ochowiak, M., Ivanov, V., Agarwal, P.: Flow modeling in a vortex chamber of a liquid–steam jet apparatus. Processes **10**(5), 984 (2022). https://doi.org/10.3390/pr10050984
12. Hnatov, A., et al.: Development of an unified energy-efficient system for urban transport. In: 2020 6th IEEE International Energy Conference (ENERGYCon), pp. 248–253. IEEE (2020). https://doi.org/10.1109/ENERGYCon48941.2020.9236606
13. Minakova, K., Zaitsev, R.: Photovoltaic thermal PV/T systems: increasing efficiency method. In: 2021 IEEE 2nd KhPI Week on Advanced Technology (KhPIWeek), pp. 303–306. IEEE (2021). https://doi.org/10.1109/KhPIWeek53812.2021.9570090
14. Voloshina, A., Panchenko, A., Titova, O., Panchenko, I.: Changes in the dynamics of the output characteristics of mechatronic systems with planetary hydraulic motors. J. Phys.: Conf. Ser. **1741**, 012045 (2021). https://doi.org/10.1088/1742-6596/1741/1/012045
15. Rogovyi, A., Neskorozhenyi, A., Krasnikov, S., Tynyanova, I., Khovanskyi, S.: Improvement of vortex chamber supercharger performances using slotted rectangular channel. In: Tonkonogyi, V., Ivanov, V., Trojanowska, J., Oborskyi, G., Pavlenko, I. (Eds.) Advanced Manufacturing Processes IV (InterPartner 2022). LNME, pp. 552–561. Springer, Cham (2023). https://doi.org/10.1007/978-3-031-16651-8_52
16. Panevnik, D.A., Panevnik, A.V., Krehel̆, R., et al.: Determination of jet pump performance when eliminating borehole hydrates. Chem. Technol. Fuels Oils **58**, 45–49 (2022). https://doi.org/10.1007/s10553-022-01349-9
17. Sokolov, V., Porkuian, O., Krol, O., Stepanova, O.: Design calculation of automatic rotary motion electrohydraulic drive for technological equipment. In: Ivanov, V., Trojanowska, J.,

Pavlenko, I., Zajac, J., Peraković, D. (Eds.) Advances in Design, Simulation and Manufacturing IV (DSMIE 2021). LNME, pp. 133–142. Springer, Cham (2021). https://doi.org/10.1007/978-3-030-77719-7_14

18. Panchenko, A., Voloshina, A., Titova, O., Panchenko, I., Zasiadko, A.: The study of dynamic processes of mechatronic systems with planetary hydraulic motors. In: Tonkonogyi, V., et al. (Eds.) Advanced Manufacturing Processes II (InterPartner 2020). LNME, pp. 704–713. Springer, Cham (2021). https://doi.org/10.1007/978-3-030-68014-5_68
19. Marchenko, A., Tkachuk, M.A., Kravchenko, S., Tkachuk, M.M., Parsadanov, I.: Experimental tests of discrete strengthened elements of machine-building structures. In: Tonkonogyi, V., et al. (Eds.) Advanced Manufacturing Processes (InterPartner 2019). LNME, pp. 559–569. Springer, Cham (2020). https://doi.org/10.1007/978-3-030-40724-7_57
20. Simpson, A., Ranade, V.V.: Flow characteristics of vortex based cavitation devices: computational investigation on influence of operating parameters and scale. AIChE J. **65**(9), e16675 (2019). https://doi.org/10.1002/aic.16675
21. Krol, O., Sokolov, V.: Modeling of spindle node dynamics using the spectral analysis method. In: Ivanov, V., Trojanowska, J., Pavlenko, I., Zajac, J., Peraković, D. (Eds.) Advances in Design, Simulation and Manufacturing III (DSMIE 2020). LNME, pp. 35–44. Springer, Cham (2020). https://doi.org/10.1007/978-3-030-50794-7_4
22. Tkachuk, M.M., Grabovskiy, A., Tkachuk, M.A., Zarubina, A., Lipeyko, A.: Analysis of elastic supports and rotor flexibility for dynamics of a cantilever impeller. J. Phys.: Conf. Ser. **1741**, 01204 (2021). https://doi.org/10.1088/1742-6596/1741/1/012043
23. Krol, O., Sokolov, V.: Research of toothed belt transmission with arched teeth. Diagnostyka **21**(4), 15–22 (2020). https://doi.org/10.29354/diag/127193
24. Smirnov, P.E., Menter, F.R.: Sensitization of the SST turbulence model to rotation and curvature by applying the Spalart-Shur correction term. J. Turbomach. **131**(4), 041010 (2009). https://doi.org/10.1115/1.3070573
25. Tiwari, G., Kumar, J., Prasad, V., Patel, V.K.: CFD investigation for surface roughness effects on the hydrodynamics of cavitating turbulent flow through a low head prototype Francis turbine. J. Appl. Fluid Mech. **15**(5), 1593–1607 (2022). https://doi.org/10.47176/jafm.15.05.33301

Numerical Simulation of a Modified Nozzle for Cold Spraying

Oleksandr Shorinov[1](✉), Andrii Volkov[1], Anatolii Dolmatov[1], and Kostyantyn Balushok[2]

[1] National Aerospace University "Kharkiv Aviation Institute", 17, Chkalova Str., Kharkiv 61070, Ukraine
o.shorinov@khai.edu
[2] Motor Sich JSC, 15, Motorobudivnykiv Ave., Zaporizhzhia 69068, Ukraine

Abstract. The purpose of the paper is to study the velocity of aluminum and nickel powder particles in the channel of a modified nozzle for low-pressure cold spraying and compare it with the values obtained for a standard nozzle using the method of computational gas dynamics. The authors propose a hypothesis about the possibility of improving the properties of coatings during low-pressure cold spraying by providing higher values of the velocity of the powder particle at the moment of contact with the substrate. Based on the standard cold spray nozzle, a new geometry of the flow part of the nozzle has been developed. Simulations were performed using the computational gas dynamics method in the ANSYS FLUENT software package. Based on the simulation results, the flow velocity distribution patterns in the standard and modified nozzles were obtained, and graphs of the velocity of aluminum and nickel particles in these flows were constructed. It was found that the velocity of aluminum and nickel particles at the moment of contact with the substrate for the modified nozzle is 14.6% and 12%, respectively, higher than the velocity values for the standard nozzle under the same initial conditions. The novelty of the obtained results is that the influence of the geometry of the cold spray nozzle channel on the particle impact velocity is demonstrated. The practical value is that the obtained results can be used in developing technological recommendations for cold spraying of coatings for a modified nozzle and its further optimization.

Keywords: Product Innovation · Particle Velocity · Gas Flow · Computational Simulation · Coating · Optimization

1 Introduction

The study of the interaction of a particle with a high-speed gas flow in a nozzle is an essential stage in the development of technological processes for coating spraying. The search for rational spraying parameters to achieve the required values of the particle velocity and temperature at the moment of contact with the substrate is necessary in terms of both the possibility of particle adhesion to the substrate and ensuring a given set of physical and mechanical properties of the coating.

V. Tonkonogyi et al. (Eds.): InterPartner 2023, LNME, pp. 571–579, 2024.
https://doi.org/10.1007/978-3-031-42778-7_53

It is obvious that many factors influence the course of processes inside the nozzle, and one-factor experiments, accompanied by extensive time and financial costs, do not always make it possible to reveal the patterns of processes in the nozzle. In this case, numerical simulation is of considerable interest not only from the point of view of determining the parameters of the two-phase flow in the nozzle and when flowing into the substrate but also with the possibility of optimizing the geometry of the flow channel of the nozzle. The modern level of computer technology allows wide usage of direct optimization methods for profiling the flow part of nozzles [1], studying two-phase flows [2], processes of high-speed collision of particles with the substrate [3], stress-strain state of the coating-substrate system [4], etc. It provides an opportunity to consider more factors compared to analytical models of specific processes and apply numerical optimization algorithms that require many calculations.

One of the most critical efficiency criteria that characterize any gas spraying based on the contact of a high-speed particle with a substrate is the particle's kinetic energy. In this regard, the particle's kinetic energy at the nozzle outlet should be selected as the target function for optimizing the process, and the characteristic particle size and linear gas flow rate from the nozzle should be selected as optimization parameters. The powder particle velocity is the most convenient parameter for optimizing cold spraying.

2 Literature Review

For cold spraying, as for other gas spraying methods, the particle velocity [5] and temperature [6] are the most critical parameters that affect the coating formation process and their physical and mechanical properties. To successfully fix the powder particle during cold spraying, it is necessary to transfer sufficient energy (thermal and kinetic) for the deformation and formation of adhesive-cohesive bonds [7].

The use of thermal energy is limited since negative consequences are possible with increased gas temperature to specific values due to the increased sensitivity of some powder materials to high operating temperatures [8]. It is especially critical for low-pressure cold spraying, where coatings are formed using compressed air as a working gas [9]. This disadvantage can be eliminated by using inert gases (nitrogen, helium) for spraying. However, their use is most widespread in high-pressure cold spraying systems [10]. Overheating of the powder material leads to the porosity of the coatings, which results in their low-performance characteristics.

The kinetic energy has two parameters that affect its value: the mass of the sprayed powder particles and the velocity of these particles. The mass of a powder particle depends on the density of the material from which it is obtained and its size. The study of the effect of material on particle velocity is presented in [11]. The authors show that as the material density increases, the particle velocity decreases under the same initial conditions of the working gas. The impact of powder particle size on their temperature and velocity characteristics is shown in [12]. It is demonstrated that as the particle size decreases, its velocity at the outlet in and out of the nozzle increases. However, simultaneously, for such particles, the presence of a reverse shock wave from the substrate surface significantly affects its braking during the passage of this area and deviation from the trajectory. In addition, as the particle size decreases, the velocity required for its adhesion to the substrate (critical velocity) increases [13].

The particle velocity of the sprayed powder is one of the main parameters for optimizing the spraying process. The velocity of powder particles upon collision directly affects the microstructure [14], physical and mechanical characteristics of coatings [7], powder utilization rate [15], etc.

Supersonic nozzles are the most essential element of equipment for cold spraying. Laval nozzles are most widely used for cold spraying [16]. When designing nozzles, the purpose is to achieve the optimal geometry for spraying a wide range of materials, i.e., to obtain a universal nozzle. However, it is often necessary to develop nozzles for a particular powder material of a certain fraction [17].

Analyzing the existing nozzle geometries for cold spraying and other process features, two important factors that had long been ignored were noted. First, one of the features of the cold spraying process is the relatively small distances between the nozzle outlet section and the substrate. Secondly, the geometry of existing nozzles for cold spraying is calculated on the principle of achieving the maximum flow rate at the nozzle section, which creates a powerful flow that, reflecting off the substrate, slows down the particles of the sprayed powder, and prevents the particles from effectively taking over the flow velocity.

3 Research Methodology

3.1 Development of 3D-Models of Nozzles

Schematic images of standard and modified nozzles for cold spraying are shown in Fig. 1 and Fig. 2.

The difference between the modified and the standard nozzle is described below. Compressed gas is supplied to the nozzle inlet, which accelerates to supersonic in the closed part. After the critical Sect. 1 and the expanding Sect. 2, the nozzle contains a section with an increased cross-sectional area 3 with a constant diameter, which transfers the front of maximum flow acceleration from the nozzle section to the channel. This results in the powder particles fed into the nozzle accelerating more efficiently in the flow, faster than in a standard nozzle under the same conditions. In addition, the modified nozzle's proposed geometry reduces the flow's shock wave reflected from the spray surface in which particle inhibition is observed.

The section with an increased cross-sectional area 4 can be made of ceramic to increase the nozzle life and reduce energy costs by reducing the heat exchange of the gas flow with the environment.

3.2 Governing Equations

Using numerical simulations for two-phase gas-solid flow, the airflow field is obtained by solving the Navier–Stokes differential equations using the standard k-ε turbulence model and the non-equilibrium wall function. The second phase (sprayed powder) is assumed to consist of spherical particles dispersed in a continuous phase (gas). In addition to solving the air transport equation, particle trajectories are calculated using the Lagrange multiplier technique to solve the particle motion equation.

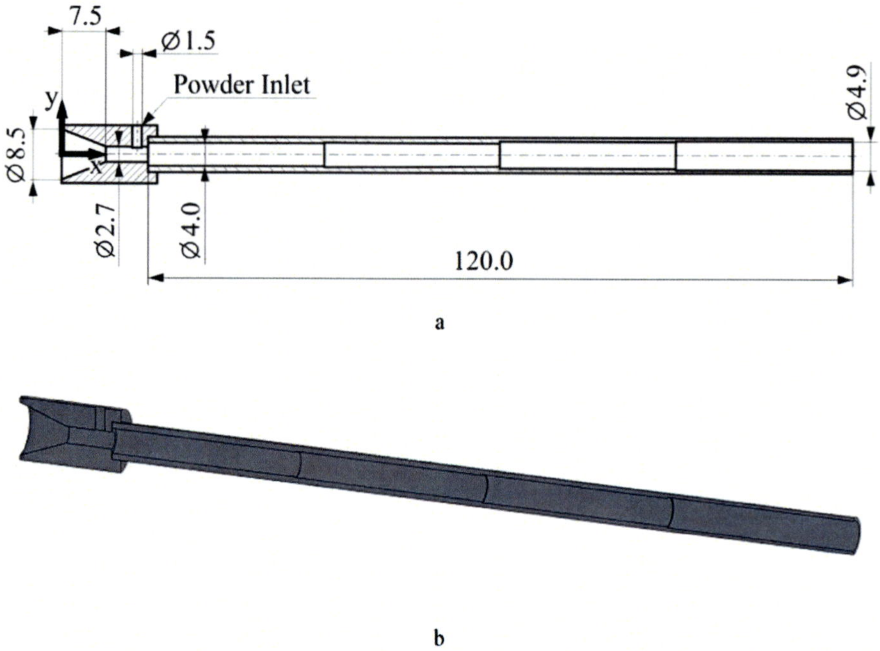

Fig. 1. Schematic image (*a*) and a 3D model (*b*) of a standard nozzle SK-20 for cold spraying of low-pressure machine DYMET-405.

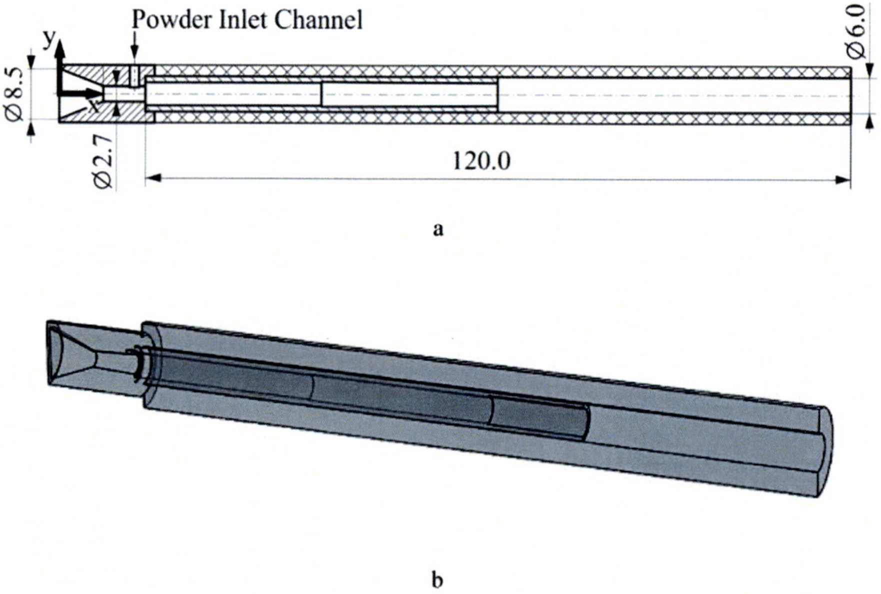

Fig. 2. Schematic image (*a*) and a 3D model (*b*) of a modified nozzle for cold spraying.

The basic equations describing turbulent flow are provided in [18]. The system of equations includes conservation of mass, momentum, and stress equations. Calculation of a powder particle velocity in a flow requires writing down the equation of the balance of forces acting on the particle [19]:

$$\frac{dV_{p,i}}{dr} = f_D\left(V_{\infty,i} - V_{p,i}\right) + \frac{g_i\left(\rho_p - \rho_\infty\right)}{\rho_p} + F_i, \tag{1}$$

where $f_D\left(V_{\infty,i} - V_{p,i}\right)$ is the resistance force per unit mass of the particle.

$$f_D = \frac{18\mu}{\rho_p d_p^2}\frac{C_D Re}{24}. \tag{2}$$

Here, Re is the relative Reynolds number:

$$Re = \frac{\rho_\infty d_p\left|V_{p,i} - V_{\infty,i}\right|}{\mu}. \tag{3}$$

Resistance coefficient C_D is a function of the relative Reynolds number:

$$C_D = \alpha_1 + \frac{\alpha_2}{Re} + \frac{\alpha_3}{Re^2}, \tag{4}$$

where $\alpha_1, \alpha_2, \alpha_3$ are constants that are used in some ranges of *Re* numbers.

Combining Eq. (1) and (4) provides the particle velocity at each point along the trajectory, and the trajectory itself is described as follows:

$$\frac{dx_i}{dr} = V_{p,i}. \tag{5}$$

3.3 Initial and Boundary Conditions

After the development of the nozzle models, simulations were performed at an initial temperature of T_0 equal to 723 K for aluminum powder particles with an initial pressure of $P_0 = 1.0$ MPa. The distance from the nozzle outlet section to the substrate was 15 mm to simulate a shock wave when the jet flow collided with the substrate. The calculation model of the modified nozzle is shown in Fig. 3.

The simulation was performed in a two-dimensional nozzle plane, the geometry of which was considered completely symmetrical. Calculations were performed with an accepted ambient pressure of 0.1 MPa and a temperature of 293 K. The turbulence intensity was 5%, and the boundary layer was considered turbulent.

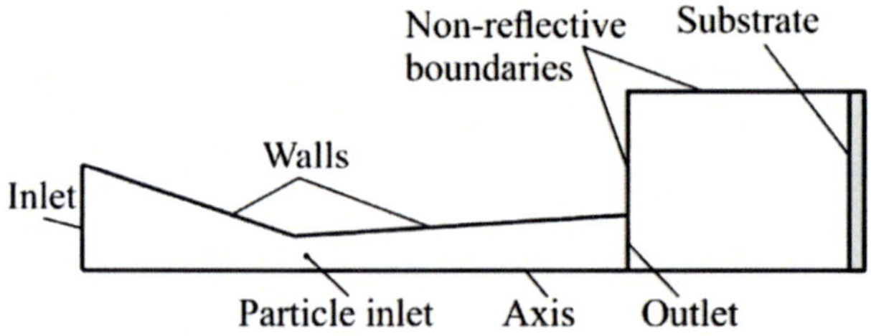

Fig. 3. Design model of the modified nozzle.

4 Results and Discussion

Figure 4 and Fig. 5 show the numerical simulation results with a contour image of the flow rate change at T0 = 723 K and P0 = 1.0 MPa in the channel of standard and modified nozzles.

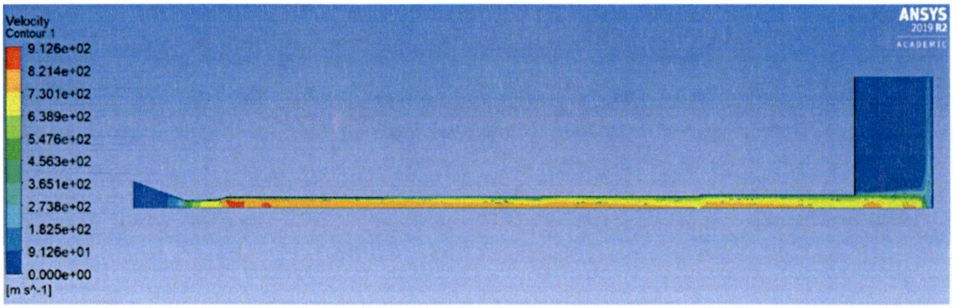

Fig. 4. Results of numerical simulation for the SK-20 nozzle with a contour image of the flow rate change at T0 = 723 K and P0 = 1.0 MPa.

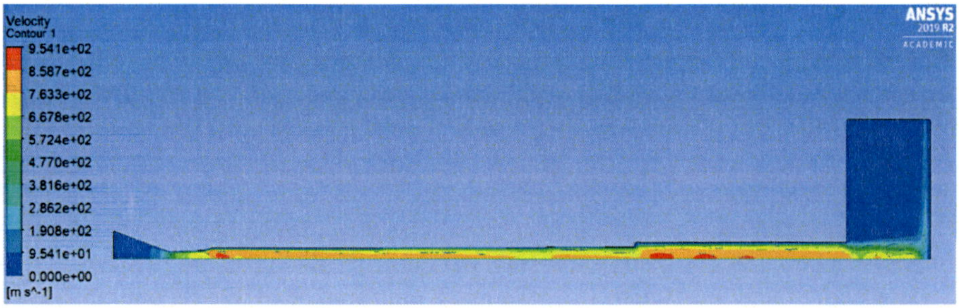

Fig. 5. Numerical simulation results for a modified nozzle with a contour image of the flow rate change at T0 = 723 K and P0 = 1.0 MPa.

When comparing the results obtained, it can be seen that the acceleration of the gas flow for a standard SK-20 nozzle occurs along the critical cross-section in the expansion part of the nozzle (see Fig. 4). For the modified nozzle, the initial flow acceleration is also observed along the critical nozzle cross-section (see Fig. 5). The reduced length of the expansion part of this nozzle and an additional section with a constant diameter allows for further acceleration of the flow closer to the outlet section of the nozzle.

Figure 6 and Fig. 7 show the graphs of the velocity change for nickel and aluminum particles with a size of 25 μm in the channel of SK-20 and the modified nozzle at $T_0 = 723$ K and $P_0 = 1.0$ MPa.

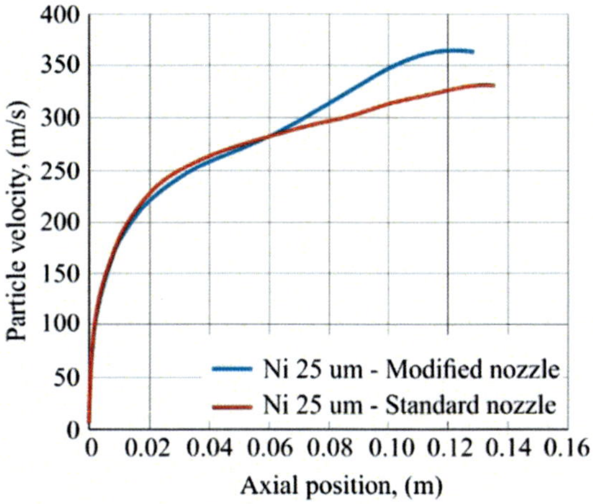

Fig. 6. Graph of the velocity change for nickel particles with a size of 25 μm for the SK-20 nozzle and the modified nozzle at $T_0 = 723$ K and $P_0 = 1.0$ MPa.

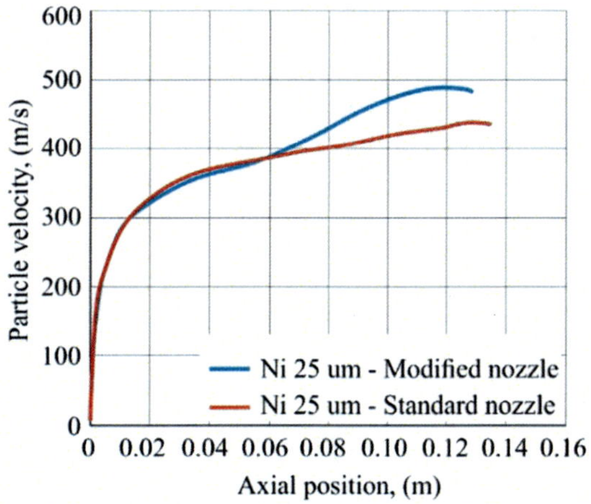

Fig. 7. Graph of the velocity change for aluminum particles with a size of 25 μm for the nozzle SK-20 and the modified nozzle at $T_0 = 723$ K and $P_0 = 1.0$ MPa.

As seen in Fig. 6, the velocity of a 25 μm nickel particle at the exit from the modified nozzle was 364 m/s. In contrast, the particle velocity was 325 m/s for the standard SK-20 nozzle at the same initial temperature and gas pressure at the nozzle inlet. For an aluminum particle with a diameter of 25 μm, an increase in velocity for the modified nozzle compared to the standard nozzle is also observed: 492 m/s and 429 m/s, respectively (Fig. 7).

As noted earlier, the velocity is significant for achieving the required deformation energy of particles of the sprayed powder and, as a result, achieving the necessary physical and mechanical properties of the coating. More intense deformation of particles in contact with the substrate ensures better contact with the surface and higher adhesive strength values. If we consider further coating growth, the higher particle velocity additionally tampers the pre-formed coating layers, reducing the number and size of pores between the powder particles in the coating [20]. In addition, an increase in the particle velocity in contact with the substrate positively affects the powder utilization rate [15].

5 Conclusions

A numerical simulation of the acceleration of the gas flow and powder particles in this flow in the channel of a modified supersonic nozzle for low-pressure cold spraying is performed. To evaluate the results obtained, simulations were also performed for the standard nozzle of the DYMET-405 cold spray machine under the same initial conditions. The simulation showed that the proposed nozzle provides better particle acceleration than the standard one. The velocity of aluminum particles increased from 429 m/s to 492 m/s (14.6% increase) when replacing the standard nozzle with the modified one; for nickel particles with a diameter of 25 μm – from 325 m/s to 364 m/s (12.0% increase) at the same initial temperature and gas pressure at the nozzle inlet of 723 K and 1 MPa, respectively. The results confirmed the hypothesis that increasing the particle velocity characteristics during cold spraying is possible by modifying the supersonic nozzle channel. An increase in the particle velocity at the moment of contact with the substrate positively affects both the coating properties and the powder utilization rate. Further research steps include the manufacture and experimental study of the physical and mechanical properties of the coatings and comparison with the results obtained for the standard nozzle.

Acknowledgment. O. S. and A. V. acknowledge the support from the project funded by the National Research Foundation of Ukraine under grant agreement No. 2020.02/0119.

References

1. Hu, W., Tan, K., Markovych, S., Cao, T.: Research on structure and technological parameters of multi-channel cold spraying nozzle. Eastern-Eur. J. Enterp. Technol. **5**(1(113)), 6–14 (2021). https://doi.org/10.15587/1729-4061.2021.242707
2. Shorinov, O.V., Polyviany, S.A.: Simulation of gas flow with nanocomposite carbon-containing powders in supersonic nozzle. Metallofiz. Noveishie Tekhnol. **44**(5), 601–611 (2022). https://doi.org/10.15407/mfint.44.05.0601
3. Hu, W.J., Tan, K., Markovych, S., Cao, T.T., Liu, X.L.: Research on the adhesive mechanism of Al + Ti mixed powders deposited on Ti6Al4V substrate by CS using Abaqus/Explicit. Metallofiz. Noveishie Tekhnol. **44**(5), 613–621 (2022). https://doi.org/10.15407/mfint.44.05.0613

4. Shorinov, O.: Finite element analysis of thermal stress in Cu2O coating synthesized on Cu substrate. Arch. Mater. Sci. Eng. **115**(2), 58–65 (2022). https://doi.org/10.5604/01.3001.0016.0753
5. Silvello, A., Cavaliere, P.D., Albaladejo, V., Martos, A., Dosta, S., Cano, I.G.: Powder properties and processing conditions affecting cold spray deposition. Coatings **10**(2), 91 (2020). https://doi.org/10.3390/coatings10020091
6. Nastic, A., Jodoin, B., Poirier, D., Legoux, J.-G.: Particle temperature effect in cold spray: a study of soft particle deposition on hard substrate. Surf. Coat. Technol. **406**, 126735 (2021). https://doi.org/10.1016/j.surfcoat.2020.126735
7. Singh, S., Singh Raman, R.K., Berndt, C.C., Singh, H.: Influence of cold spray parameters on bonding mechanisms: a review. Metals **11**(12), 2016 (2021). https://doi.org/10.3390/met11122016
8. Chen, C., Xie, Y., Huang, R., Deng, S., Ren, Z., Liao, H.: On the role of oxide film's cleaning effect into the metallurgical bonding during cold spray. Mater. Lett. **210**, 199–202 (2018). https://doi.org/10.1016/j.matlet.2017.09.024
9. Zhang, L., Yang, S., Lv, X., Jie, X.: Wear and corrosion resistance of cold-sprayed Cu-based composite coatings on magnesium substrate. J. Therm. Spray Technol. **28**, 1212–1224 (2019). https://doi.org/10.1007/s11666-019-00887-9
10. Singh, H., Kumar, M., Singh, R.: An overview of various applications of cold spray coating process. Mater. Today Proc. **56**(5), 2826–2830 (2022). https://doi.org/10.1016/j.matpr.2021.10.160
11. Assadi, H., Kreye, H., Gärtner, F., Klassen, T.: Cold spraying – a materials perspective. Acta Mater. **116**, 382–407 (2016). https://doi.org/10.1016/j.actamat.2016.06.034
12. King, P.C., Jahedi, M.: Relationship between particle size and deformation in the cold spray process. Appl. Surf. Sci. **256**(6), 1735–1738 (2010). https://doi.org/10.1016/j.apsusc.2009.09.104
13. Assadi, H., et al.: On parameter selection in cold spraying. J. Therm. Spray Technol. **20**, 1161–1176 (2011). https://doi.org/10.1007/s11666-011-9662-9
14. Khamsepour, P., Moreau, C., Dolatabadi, A.: Numerical simulation of the effect of particle and substrate preheating on porosity level and residual stress of as-sprayed Ti6Al4V components. J. Therm. Spray Technol. **31**, 70–83 (2022). https://doi.org/10.1007/s11666-021-01286-9
15. Canales, H., Cano, I.G., Dosta, S.: Window of deposition description and prediction of deposition efficiency via machine learning techniques in cold spraying. Surf. Coat. Technol. **401**, 126143 (2020). https://doi.org/10.1016/j.surfcoat.2020.126143
16. Buhl, S., Breuninger, P., Antonyuk, S.: Optimization of a Laval nozzle for energy-efficient cold spraying of microparticles. Mater. Manuf. Processes **33**(2), 115–122 (2018). https://doi.org/10.1080/10426914.2017.1279322
17. Cao, C., Li, W., Zhang, Z., Yang, X., Xu, Y.: Cold spray additive manufacturing of Ti6Al4V: special nozzle design using numerical simulation and experimental validation. Coatings **12**(2), 210 (2022). https://doi.org/10.3390/coatings12020210
18. Aziz, T.N., Raiford, J.P., Khan, A.A.: Numerical simulation of turbulent jets. Eng. Appl. Comput. Fluid Mech. **2**(2), 234–243 (2008). https://doi.org/10.1080/19942060.2008.11015224
19. Yin, S., Meyer, M., Li, W., Liao, H., Lupoi, R.: Gas flow, particle acceleration, and heat transfer in cold spray: a review. J. Therm. Spray Technol. **25**, 874–896 (2016). https://doi.org/10.1007/s11666-016-0406-8
20. Rokni, M.R., Nutt, S.R., Widener, C.A., Champagne, V.K., Hrabe, R.H.: Review of relationship between particle deformation, coating microstructure, and properties in high-pressure cold spray. J. Therm. Spray Technol. **26**, 1308–1355 (2017). https://doi.org/10.1007/s11666-017-0575-0

Author Index

V. Tonkonogyi et al. (Eds.): InterPartner 2023, LNME, pp. 581–583, 2024.
https://doi.org/10.1007/978-3-031-42778-7